D1293377

Metamorphosis

Postembryonic Reprogramming of Gene Expression in Amphibian and Insect Cells

CELL BIOLOGY: A Series of Monographs

EDITORS

Volumes published since 1984

Gary S. Stein and Janet L. Stein (editors). RECOMBINANT DNA AND CELL PROLIFERATION, 1984

Prasad S. Sunkara (editor). NOVEL APPROACHES TO CANCER CHEMOTHERAPY, 1984

B. G. Atkinson and D. B. Walden (editors). CHANGES IN EUKARYOTIC GENE EXPRESSION IN RESPONSE TO ENVIRONMENTAL STRESS, 1985

Reginald M. Gorczynski (editor). RECEPTORS IN CELLULAR RECOGNITION AND DEVELOPMENTAL PROCESSES, 1986

Govindjee, Jan Amesz, and David Charles Fork (editors). LIGHT EMISSION BY PLANTS AND BACTERIA, 1986

Peter B. Moens (editor). MEIOSIS, 1986

Robert A. Schlegel, Margaret S. Halleck, and Potu N. Rao (editors). MOLECULAR REGULATION OF NUCLEAR EVENTS IN MITOSIS AND MEIOSIS, 1987

Monique C. Braude and Arthur M. Zimmerman (editors). GENETIC AND PERINATAL EFFECTS OF ABUSED SUBSTANCES, 1987

E. J. Rauckman and George M. Padilla (editors). THE ISOLATED HEPATOCYTE: USE IN TOXICOLOGY AND XENOBIOTIC BIOTRANSFORMATIONS, 1987

Heide Schatten and Gerald Schatten (editors). THE MOLECULAR BIOLOGY OF FERTILIZATION, 1989

Heide Schatten and Gerald Schatten (editors). THE CELL BIOLOGY OF FERTILIZATION, 1989

Anwar Nasim, Paul Young, and Byron F. Johnson (editors). MOLECULAR BIOLOGY OF THE FISSION YEAST, 1989

Mary P. Moyer and George Poste (editors). COLON CANCER CELLS, 1990

Gary S. Stein and Jane B. Lian (editors). MOLECULAR AND CELLULAR APPROACHES TO THE CONTROL OF PROLIFERATION AND DIFFERENTIATION, 1991

Victauts I. Kalnins (editor). THE CENTROSOME, 1992

Carl M. Feldherr (editor). NUCLEAR TRAFFICKING, 1992

Christer Sundqvist and Margareta Ryberg (editors). PIGMENT-PROTEIN COMPLEXES IN PLASTIDS: SYNTHESIS AND ASSEMBLY, 1993

David H. Rohrbach and Rupert Timpl (editors). MOLECULAR AND CELLULAR ASPECTS OF BASEMENT MEMBRANES, 1993

Danton H. O'Day (editor). SIGNAL TRANSDUCTION DURING BIOMEMBRANE FUSION, 1993

John K. Stevens, Linda R. Mills, and Judy E. Trogadis (editors). THREE-DIMENSIONAL CONFOCAL MICROSCOPY: VOLUME INVESTIGATION OF BIOLOGICAL SPECIMENS, 1994

Lawrence I. Gilbert, Jamshed R. Tata, and Burr G. Atkinson (editors). METAMORPHOSIS: POST-EMBRYONIC REPROGRAMMING OF GENE EXPRESSION IN AMPHIBIAN AND INSECT CELLS, 1996

R. John Ellis (editor). THE CHAPERONINS, 1996 (in preparation)

Metamorphosis

Postembryonic Reprogramming of Gene Expression in Amphibian and Insect Cells

Edited by

Lawrence I. Gilbert
Department of Biology
College of Arts and Sciences
University of North Carolina at Chapel Hill
Chapel Hill, North Carolina

Jamshed R. Tata
National Institute for Medical Research
London, United Kingdom

Burr G. Atkinson
Department of Zoology
University of Western Ontario
London, Ontario, Canada

ACADEMIC PRESS
San Diego New York Boston London Sydney Tokyo Toronto

Front cover illustration depicts the unified concept of insect and amphibian postembryonic development as illustrated by the major morphological changes that take place during metamorphosis, which are obligatorily regulated by the hormonal signals of ecdysteroids and thyroid hormones, respectively. Illustration courtesy of J. R. Tata and J. W. Brock, MRC National Institute for Medical Research, Mill Hill, London, United Kingdom.

This book is printed on acid-free paper. 

Academic Press, Inc.
A Division of Harcourt Brace & Company
525 B Street, Suite 1900, San Diego, California 92101-4495

United Kingdom Edition published by
Academic Press Limited
24-28 Oval Road, London NW1 7DX

Library of Congress Cataloging-in-Publication Data

Metamorphosis : postembryonic reprogramming of gene expression in amphibian and insect cells / edited by Lawrence I. Gilbert, Jamshed R. Tata, Burr G. Atkinson.
p. cm. -- (Cell biology series)
Includes index.
ISBN 0-12-283245-0 (alk. paper)
1. Insects--Metamorphosis. 2. Amphibians--Metamorphosis. 3. Insects--Molecular genetics. 4. Amphibians--Molecular genetics. I. Gilbert, Lawrence I. (Lawrence Irwin), date. II. Tata, Jamshed R. (Jamshed Rustom) III. Atkinson, Burr G. IV. Series.
QL494.5.M48 1996
595.7'0334--dc20 95-20630
CIP

PRINTED IN THE UNITED STATES OF AMERICA
96 97 98 99 00 01 BB 9 8 7 6 5 4 3 2 1

Contents

3 Ecdysone-Regulated Chromosome Puffing in *Drosophila melanogaster*

Steven Russell and Michael Ashburner

4 Chromosome Puffing: Supramolecular Aspects of Ecdysone Action

Markus Lezzi

5 Molecular Aspects of Ecdysteroid Hormone Action

Peter Cherbas and Lucy Cherbas

6 Molecular Aspects of Juvenile Hormone Action in Insect Metamorphosis

Lynn M. Riddiford

7 Metamorphosis of the Cuticle, Its Proteins, and Their Genes

Judith H. Willis

8 Metamorphosis of the Insect Nervous System

James W. Truman

9 Gene Regulation in Imaginal Disc and Salivary Gland Development during D*rosophila* Metamorphosis

Cynthia Bayer, Laurence von Kalm, and James W. Fristrom

10 Genes Involved in Postembryonic Cell Proliferation in *Drosophila*

Elizabeth L. Wilder and Norbert Perrimon

II Amphibians

11 Endocrinology of Amphibian Metamorphosis

Jane C. Kaltenbach

12 Neuroendocrine Control of Amphibian Metamorphosis

Robert J. Denver

13 Hormonal Interplay and Thyroid Hormone Receptor Expression during Amphibian Metamorphosis

Jamshed R. Tata

14 Thyroid Hormone-Regulated Early and Late Genes during Amphibian Metamorphosis

Yun-Bo Shi

15 Reprogramming of Genes Expressed in Amphibian Liver during Metamorphosis

Burr G. Atkinson, Caren Helbing, and Yuqing Chen

16 Switching of Globin Genes during Anuran Metamorphosis

Rudolph Weber

17 Hormone-Induced Changes in Keratin Gene Expression during Amphibian Skin Metamorphosis

Leo Miller

Contributors

Numbers in parentheses indicate the pages on which the authors' contributions begin.

Michael Ashburner (109), Department of Genetics, University of Cambridge, Cambridge CB2 3EH, England

Burr G. Atkinson (539), Molecular Genetics Unit, Department of Zoology, University of Western Ontario, London, Ontario, Canada N6A 5B7

Cynthia Bayer (321), Division of Genetics, Department of Molecular and Cell Biology, University of California, Life Science Addition 589, Berkeley, California 94720

Yuqing Chen (539), Molecular Genetics Unit, Department of Zoology, University of Western Ontario, London, Ontario, Canada N6A 5B7

Lucy Cherbas (175), Department of Biology, Indiana University, Bloomington, Indiana 47405

Peter Cherbas (175), Department of Biology, Indiana University, Bloomington, Indiana 47405

Nicholas Cohen (625), Department of Microbiology and Immunology, University of Rochester Medical Center, Rochester, New York 14642

Robert J. Denver (433), Department of Biology, Natural Science Building, University of Michigan, Ann Arbor, Michigan 48109

James W. Fristrom (321), Division of Genetics, Department of Molecular and Cell Biology, University of California, Life Science Addition 589, Berkeley, California 94720

Lawrence I. Gilbert (59), Department of Biology, College of Arts and Sciences, University of North Carolina at Chapel Hill, Chapel Hill, North Carolina 27599

Caren Helbing (539), Molecular Genetics Unit, Department of Zoology, University of Western Ontario, London, Ontario, Canada N6A 5B7

Jane C. Kaltenbach (403), Department of Biological Sciences, Clapp Lab, Mount Holyoke College, South Hadley, Massachusetts 01075

Markus Lezzi (145), Institute for Cell Biology, Swiss Federal Institute of Technology, Zürich, Switzerland

Leo Miller (599), Department of Biological Sciences, University of Illinois at Chicago, Chicago, Illinois 60690

Norbert Perrimon (363), Department of Genetics, Howard Hughes Medical Institute, Harvard Medical School, Boston, Massachusetts 02115

Lynn M. Riddiford (223), Department of Zoology, University of Washington, Seattle, Washington 98195

Louise A. Rollins-Smith (625), Department of Microbiology and Immunology and of Pediatrics, Vanderbilt University School of Medicine, Nashville, Tennessee 37232

Steven Russell (109), Department of Genetics, University of Cambridge, Cambridge CB2 3EH, England

Robert Rybczynski (59), Department of Biology, University of North Carolina at Chapel Hill, Chapel Hill, North Carolina 27599

František Sehnal (3), Institute of Entomology, Czechoslovak Academy of Sciences, 370 05 České Budějovice, Czech Republic

Yun-Bo Shi (505), Laboratory of Molecular Embryology, National Institute of Child Health and Human Development, National Institutes of Health, Bethesda, Maryland 20892

Jamshed R. Tata (465), Division of Developmental Biochemistry, National Institute for Medical Research, Mill Hill, London NW7 1AA, England

Stephen S. Tobe (59), Department of Zoology, Ramsay Wright Zoological Laboratories, University of Toronto, Toronto, Canada M5S 1A1

James W. Truman (283), Department of Zoology, University of Washington, Seattle, Washington 98195

Petr Švácha (3), Institute of Entomology, Czechoslovak Academy of Sciences, 370 05 České Budějovice, Czech Republic

Laurence von Kalm (321), Division of Genetics, Department of Molecular and Cell Biology, University of California, Life Science Addition 589, Berkeley, California 94720

Rudolph Weber (567), Oberwohlenstrasse 43, CH-3033 Wohlen, Switzerland

Elizabeth L. Wilder (363), Department of Genetics, Howard Hughes Medical Institute, Harvard Medical School, Boston, Massachusetts 02115

Judith H. Willis (253), Department Cellular Biology, University of Georgia, Athens, Athens, Georgia 30602

Katsutoshi Yoshizato (647), Yoshizato Morphomatrix Project of ERATO, JRDC, Saijo Misonou 242-37, Higashi-Hiroshima 739,

Japan, and Department of Biological Sciences, Hiroshima University, Higashi-Hiroshima 739, Japan

Jan Zrzavý (3), Institute of Entomology, Czechoslovak Academy of Sciences, 370 05 České Budějovice, Czech Republic

Preface

"Metamorphosis," edited by William Etkin and Lawrence I. Gilbert in 1968, concentrated on the morphological and physiological aspects of the phenomenon in arthropods, lower chordates, and amphibians, although there was one chapter on the hormonal control of salivary gland puffing (Kroeger) and one each on the biochemical aspects of insect (Wyatt) and amphibian (Frieden) metamorphosis. "Metamorphosis: A Problem in Developmental Biology," edited by Lawrence I. Gilbert and Earl Frieden, appeared in 1981. The insect portion emphasized biochemical and endocrinological events with a chapter on macromolecular alterations (Sridhara) and two chapters dealing with the current status of hormone action on *Drosophila* imaginal discs (Fristrom) and cell lines (O'Connor and Chang). A similar emphasis appeared in the amphibian section, with macromolecular changes (Smith-Gill and Carver) and hormone action (Frieden) being major contributions. This volume on metamorphosis is primarily molecular, with several introductory chapters to provide the basis for contributions concerned with events at the level of the gene. Thus, each volume reflects the state of the life sciences at a particular time, and, remarkably, the phenomenon of metamorphosis continues to be an excellent model for those interested in the control of developmental and endocrinological processes, whether they utilize the paradigms of morphology or molecular biology.

Although chapters on insect and amphibian metamorphosis have been grouped separately, as in the earlier volumes, the reader will discern the evolutionary conservation of several mechanisms of hormonal regulation of development. For example, the receptors for ecdysteroids and thyroid hormones are remarkably similar, such that their artificially produced chimeric forms can equally effectively activate target genes in cells of vertebrate and invertebrate organisms with reciprocal ligands. The reader will also find postembryonic developmental processes that are different in insect and amphibia, as, for example, the sequential and multiple stages of insect metamorphosis in contrast to a single step of metamorphic progression in amphibia.

This volume is designed for research scientists and biology graduate students and should be a useful adjunct to biology courses dealing with development, physiology, biochemistry, and molecular biology. It was not planned as a classic textbook, but we certainly recommend it for scholars interested in growth and development, from undergraduate juniors to senior researchers.

L. I. Gilbert
J. R. Tata
B. G. Atkinson

I

Insects

Note: As pointed out by Peter Karlson [*Eur. J. Entomol.* **92,** 7–8 (1995)], who was most responsible for the purification and characterization of ecdysone, *ecdysone* is the trivial name for a specific compound while *ecdysteroid* is the generic term for this class of compounds. Unfortunately, there is a growing tendency to use the term *ecdysone* in a generic sense rather than *ecdysteroid* and since this usage has been accepted by a number of prominent journals, the editors have allowed authors to use either *ecdysteroid* or *ecdysone* to define the class of compounds that is closely related structurally to ecdysone. Each of the authors has defined the use of *ecdysone* as a generic term where appropriate and we hope that this is not confusing to the reader.

1

Evolution of Insect Metamorphosis

FRANTIŠEK SEHNAL,*,† PETR ŠVÁCHA,*
AND JAN ZRZAVÝ†

* Institute of Entomology
Czech Academy of Sciences
370 05 České Budějovice, Czech Republic

† Faculty of Biological Sciences
University of South Bohemia
370 05 České Budějovice, Czech Republic

I. INTRODUCTION

The terms *larva* and *metamorphosis* are used in different animal groups for unrelated developmental phenomena of various ontogenetic timing. The only common denominator seems to be the occurrence in the life cycle of morphological changes which, according to the observer, are significant, and/or conspicuous, and/or sudden enough to deserve to be named metamorphic. The word *metamorphosis* originates from the Greek, and primarily implies morphological change in a broad sense. *Webster's Dictionary* defines metamorphosis in animals as "a marked and more or less abrupt change in the form or structure of an animal during postembryonic development." However, *postembryonic* may have diverse definitions. In the study of invertebrates it is commonly used for ontogeny accomplished after hatching, as will be used in this chapter.

In many aquatic animals with planktonic dispersive early stages, the embryonic period is very short, and the animal leaves the protective coverings very early (often at the organization level comparable with the gastrula or even the blastula stages). Such early free-living stages are sometimes called "primary larvae," and, according to some, may belong to metazoan ancestral features (for a review see Willmer, 1990). Well-known examples of such "primary" larvae include the trochophora-type larvae in annelids, molluscs, and some other phyla, or the dipleurula-type larvae in echinoderms and hemichordates; the crustacean nauplius stage serves a similar purpose and is sometimes compared with trochophora, but its degree of organization is much higher. A more or less abrupt morphogenesis toward the final form (often connected with a transition from the pelagic to sessile or benthic life, sometimes through somewhat distinct transitional forms which may have special names), accompanied by degeneration of the specialized larval structures, ensues when the larva reaches a certain size threshold and/or on action of certain environmental cues. This type of metamorphosis is widespread and is found in most metazoan phyla. It is usually finished long before reaching the reproductive stage: metamorphosis and sexual maturation are primarily two independent phenomena.

In other animals the stages corresponding to the simple "primary" larvae and the following transitional forms are more or less embryonic (up to a disappearance of metamorphic changes in the so-called "direct development"). At least in some groups this is the result of a secondary "embryonization." A series of gradual embryonizations can be found, e.g., in crustaceans—some retain the primitive type of development beginning with a typical nauplius or metanauplius while others exhibit direct development (having embryonized all "larval" stadia).

Secondarily, however, the postembryogenesis of such "embryonized" animals could become metamorphic through divergent adaptations of postembryonic stages to different life-styles. The immature premetamorphic stages of such animals may then be called "secondary larvae." Insect metamorphosis undoubtedly evolved as a consequence of postembryonic adaptive changes. Because of their primarily terrestrial life, the insects have eggs rich in yolk and hatch as highly organized arthropods. Without doubt the development of primitive insects did not show any pronounced metamorphic changes, as documented by the extant primarily wingless groups and by some fossil Paleozoic pterygotes.

Proper use of metamorphic insects as models for research on general developmental problems requires appreciation of insect diversity and of the ecological and endocrinological mechanisms that apparently led to the evolution of metamorphosis and its various modifications. This chapter should promote and facilitate evaluation of experimental data on such a comparative basis.

We have evaluated data obtained on diverse insect taxa but since only a limited number of papers can be cited, we often refer to a single review rather than to several original articles. Examples are provided, usually those we know best, and no attempt is made to list all reported cases.

II. METAMORPHOSIS IN INSECTS

A. Hexapod Development

There are broad gaps regarding concept and terminology between "classical" entomologists and experimental biologists working on insect models. General textbooks often bring together data from various sources without critical evaluation. As a result, the evolution of insect development often emerges as a very complicated and almost discontinuous process (especially concerning the hemimetaboly–holometaboly transition) burdened with unnecessary specific terms. We will summarize a typical "average textbook" approach to insect postembryonic development; some controversial points will be discussed later.

Postembryogenesis of the Hexapoda (= Insecta s.lat.) is epimorphic (the juvenile hatches from the egg with a complete number of body segments) except for the Protura which are anamorphic (postembryonic increase in the number of abdominal segments). Collembola (springtails), Protura, and Diplura (all often grouped together as Entognatha), as well as the primarily wingless "true" insects, the Archaeognatha (bristletails) and Zygentoma (silverfish), develop by a gradual increase in size without any sudden allometric changes and are considered ametamorphic (*ametabolous*). With the probable exception of Protura (small soil-dwelling hexapods whose development is poorly known), all primarily wingless hexapods continue to molt after attaining sexual maturity.

The winged insects (Pterygota) terminate molting on reaching the winged adult stage, except for the Ephemeroptera (mayflies) which have two fully winged flying instars: the nonreproducing subimago and the imago. With the exception of the Holometabola, pterygotes have juveniles (called larvae, nymphs, or rarely naiads in aquatic forms) in which, beginning from a certain (usually early) point in postembryonic development, the "adult" organs (mainly the wings and genitalia) gradually start to develop. Nevertheless, the change between the last larval instar and the adult (imago; or the subimago in mayflies) is subjectively considered significantly greater than the changes between the larval instars. These insects are thus acknowledged as having an "incomplete" metamorphosis and are named *hemimetabolous* (usually including Ephemeroptera). The aquatic larvae of Ephemeroptera and Odonata (dragonflies), but not Plecoptera (stoneflies), are considerably different from the adults (and the mayfly subimago); this is sometimes interpreted as larval specializations for the aquatic environment which is thus considered secondary (see Pritchard *et al*, 1993; this opinion, however, is not universally accepted).

The juveniles of Holometabola (always called larvae, even if they are aquatic) conspicuously differ from the adult stage and usually have a different habitat and food niche. They never have external rudiments of wings and genitalia, and these and some other imaginal organs develop (or are replaced) from the so-called imaginal discs (described as invaginated groups of undifferentiated "embryonic" cells). A nonfeeding, usually hidden, or somehow protected instar of limited (or entirely suppressed) locomotory activity, which allows tissue degradation and rebuilding, occurs between the last larval instar and the adult. This instar, the pupa, has external rudiments of wings and genitalia. Such metamorphosis is called "complete," and this type of development is named *holometabolous.* In addition to the monophyletic Holometabola, a superficially similar type of metamorphosis (with several wingless larvae and one to three entirely or partly quiescent preadult instars with, or in whiteflies even without, external wing rudiments, followed by the adult) has developed independently in some Paraneoptera [namely, Thysanoptera (thrips), Aleyrodomorpha (whiteflies), and males of Coccomorpha (scale insects)]. Such groups are often called *neometabolous* (or various special names have been applied to individual types; see Section II,C), but sometimes also holometabolous (Chapman, 1969; Gillott, 1980), although this is not meant to imply the presence of imaginal discs; the groups do not show the difference between larval and adult habits characteristic of the Holometabola.

B. Metamorphosis, Sexual Maturation, and "Final Instar"

Some authors use the word *metamorphosis* for the sum of all morphological changes that occur during insect postembryogenesis, whereas others refer to it only for the change from larva to adult (cf. Rockstein, 1956; Chapman, 1969; Kukalová-Peck, 1991). This chapter will not apply the term to the sum of postembryonic changes in an insect and will require "a marked and abrupt change in the form or structure during postembryonic development." Consequently, *structural* changes comparable to those occurring between the larva and the adult may occur at various points during postembryonic development. The admittedly subjective formulation "marked and abrupt change" concerns individual organs or tissues (wings, genitalia, legs, digestive tract, etc.) instead of the whole animal. An insect is regarded as metamorphic if at least some of its parts show changes which can be so named (see Davies, 1966). Thus, in the cockroach, for example, the "metamorphosis concerns wings but not legs" (Sehnal, 1985), and the cockroach is a metamorphic insect.

Primarily wingless insects continue molting after attaining sexual maturity, and there is no predetermined "final" instar. This must be considered a primitive character. In modern metamorphic pterygotes, with the exception of the mayflies, there is always only one flying instar, namely the adult (if the wings have not been secondarily reduced or lost), which never molts again. Thus, the greatest morphological change typically occurs between the last larval instar and the adult in the hemimetabolous insects or, secondarily, in two steps (larva/pupa and pupa/adult) in the holometabolous insects; in any case at the very end of development. This arrangement has one important yet seldom realized consequence. Three major, and primarily probably independent, events occur in approximately the same period: (i) "structural" metamorphosis, (ii) obligate termination of molting cycles (establishment of a predetermined final instar), and (iii) acquisition of competence for reproduction (i.e., reaching the adult stage).

We are aware of the general usage of the word "metamorphosis" for the combined complex of the just named events, but when dealing with the *evolution* of insect metamorphosis it is essential to distinguish between them, since originally they may not have coincided (see Section V,A); in modern Ephemeroptera the "structural" metamorphosis (last larva → subimago) is still separate from the other two events that occur during the subimago → imago molt. At the present state of knowledge, distinguishing the three events is difficult or impossible in the endocrinological sections, and various regulational phenomena observed in this period that are routinely associated with "structural" metamorphosis may have been primarily connected with sexual maturation or the termination of molting, and only secondarily employed for the control of morphological changes. For example, both the rise, during the prepupal peak, and the decline, at the end of adult development, of hemolymph ecdysteroids in *Manduca sexta* (see Gilbert *et al.,* Chapter 2, this volume), are selectively used as a trigger for the cell death of various larval motoneurons (Weeks and Levine, 1990), and yet none of these ecdysteroid changes can be viewed as having evolved primarily for that purpose. To avoid ambiguity in using the term *metamorphosis,* the adjective *structural* will be added when emphasizing the exclusion of the two latter events.

C. Definitions and Terminology

Insects: The ideas about phylogeny and the higher classification of hexapods, and indeed the exact content of the taxon "Insecta" itself, have been the subject of frequent and significant changes (see Štys and Zrzavý, 1994). In accordance with the prevailing usage, in this chapter the term "(true) insects" does not include the entognathous hexapod groups (Protura, Collembola, and Diplura). Thus, here "Insecta" is synonymous with Ectognatha and includes the primarily wingless Archaeognatha and Zygentoma, and the Pterygota. The entognathous groups will not be discussed; they do not have a metamorphosis, and very little is known about the hormonal regulation of their development.

Types of postembryonic development: The great diversity of terms describing various types of insect development is reduced in this chapter (see Fig. 1 for the main types). To avoid ambiguity, *holometaboly* is restricted to Holometabola, and the term *neometaboly* is applied to *all* independently evolved similar types of postembryogenesis with a resting instar(s) in Paraneoptera (including "parametaboly," "remetaboly," and "allometaboly"). *Prometaboly* identifies the unique development of mayflies and hypothetical developments of fossil pterygotes with several "postmetamorphic" instars. Under *hemimetaboly,* postembryogenesis is a more or less gradual enlargement of wing pads, no resting stages, and a single winged adult instar (this simplest type is called "paurometaboly" by some authors), but this term also includes some extreme cases, such as certain aphids in which the wing pads occur as late as in the last larval instar but which have no resting instar ("homometaboly" of some authors). For a different classification see Nüesch (1987).

Juveniles: All pterygote immature stages except for the mayfly subimago, the holometabolan pupa, and the inactive instars of the neometabolous Paraneoptera are called *larvae* (i.e., the terms "nymph" and "naiad" are avoided).

Adult: The instar in which reproduction takes place; only very exceptionally can some insects reproduce paedogenetically as *larvae,* e.g., *Micromalthus* (Coleoptera) or some Cecidomyiidae (Diptera), but even those species do have normal adult forms. There may be long periods in the life of an adult during which it is unable to reproduce (e.g., during diapause). In most sexually reproducing insects the adult also undertakes copulation, but in some cases females may copulate as larvae [e.g., Heteroptera: Polyctenidae (Ferris and Usinger, 1939)]. The definition of the adult instar is thus functional; e.g., as soon as reproduction is taken over by the subimago in some mayflies (see Section III,B,1), that instar is called an adult. Difficulties with this definition might occur in some taxa outside Insecta as presently defined (e.g., in some Entognatha) which retain adult molting and alternate reproducing and nonreproducing instars. Unfortunately, nothing is known about regulation of this phenomenon.

Molting: The whole molting period begins with *apolysis* (separation of the epidermis from the old cuticle), followed by a more or less pro-

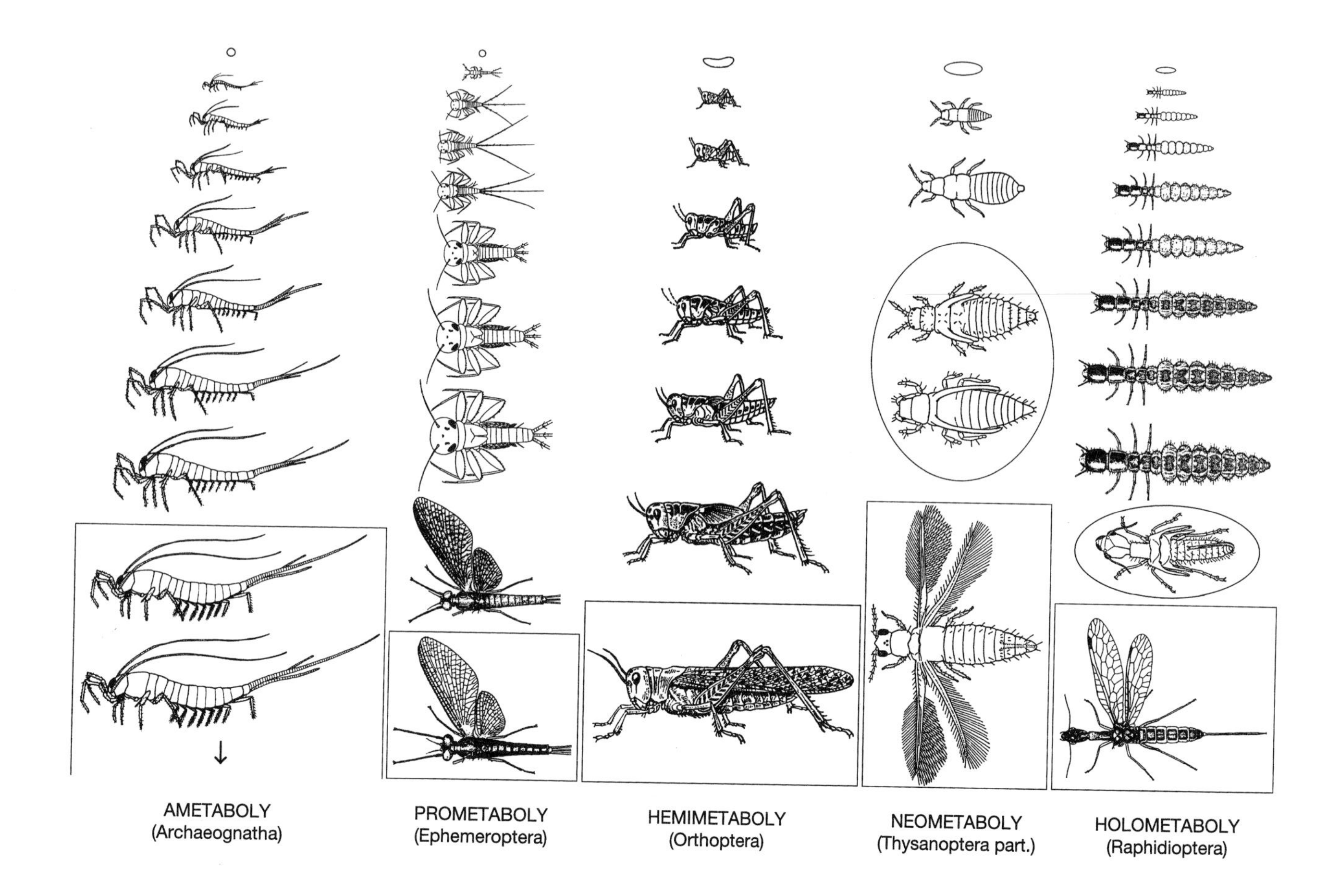
AMETABOLY
(Archaeognatha)
PROMETABOLY
(Ephemeroptera)
HEMIMETABOLY
(Orthoptera)
NEOMETABOLY
(Thysanoptera part.)
HOLOMETABOLY
(Raphidioptera)

nounced morphogenesis and deposition of the new cuticle, and ending with *ecdysis* (shedding of the old cuticle). Exceptionally, the ecdysis step can be omitted in one or more instars which thus remain permanently pharate (e.g., in the cyclorrhaphous flies the pupa remains enclosed in the puparium which is the sclerotized last larval cuticle).

D. Methodology of Evolutionary Studies

It is necessary to stress here some simple but often neglected methodological principles concerning evolutionary interpretation of *any* data, including morphological, physiological, behavioral, ecological, and developmental. For a detailed discussion of these principles see, for example, the work of Forey *et al.* (1992). (i) A hypothesis must be made about cladogenesis (splitting of phylogenetic lines) before attempting to establish a scientific hypothesis about anagenesis (evolutionary deviation from the ancestral state); characters cannot be compared without knowledge of the phylogenetic relationships of the species which carry them (Fig. 2). (ii) Cladogenetic position of a taxon and its anagenetic "advancement" may not be, and usually are not, in accord. The earliest evolutionary branch within a taxon may not be "primitive," i.e., possessing most of the primitive character states. (iii) Each organism is a mosaic of both primitive and derived characters; no taxon is entirely "primitive" or "advanced." (iv) No character per se can be labeled as primitive or derived: statements about evolutionary polarity of the character states can be made only after rigorous phylogenetic analysis; the same character state may be primitive in one group but derived in the other. (v) Evolution does not produce consistently more and more complex bodies, and therefore, simple may not be primitive. (vi) Evolution is often but not always conservative so the most common character state is not necessarily primitive, e.g., viviparity within the mammals. (vii) Any developmental stage can become a subject of evolutionary change so that no universal

---

Fig. 1. Five basic types of development of modern insects. Resting stages are encircled. Adults are boxed. The primarily wingless ametabolous insects continue molting as adults and do not have a predetermined final instar.

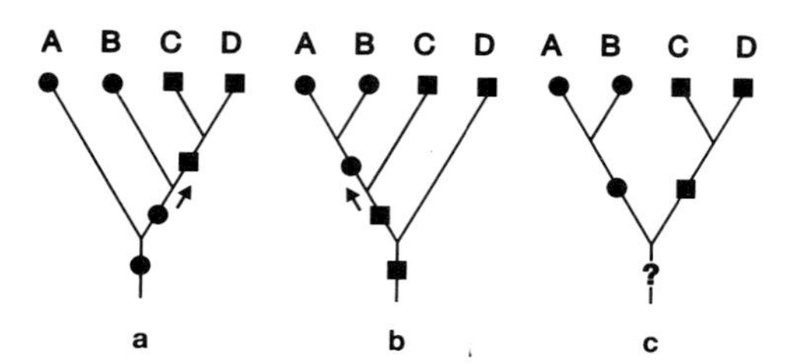

Fig. 2. The same distribution of two states of a character (●, ■) among four taxa (**A, B, C, D**) may be evolutionarily interpreted in several ways, depending on the phylogenetic (cladistic) relationships of the taxa. Note that in (a) and (b) the simplest hypotheses with only one evolutionary change (arrow) require ● and ■, respectively, as the ancestral states, whereas situation (c) does not allow reconstruction of the ancestral condition without additional information.

"biogenetic law," e.g., the Haeckelian recapitulation, is valid; consequently, larvae or embryos can be and often actually are much more derived than adults of the same species. The larva of species A may be called primitive or derived after comparison with larvae B and C, but not with adult A.

III. INSECT PHYLOGENY AND TYPES OF POSTEMBRYONIC DEVELOPMENT

There is no doubt that true insects (Ectognatha = Insecta s.str.; Fig. 3) are a monophyletic group and that they split primarily into Archaeognatha and Dicondylia; the latter further branched into Zygentoma and Pterygota. The phylogenetic position of the Paleozoic Monura is uncertain; according to Kukalová-Peck (1991), they belong to the Dicondylia (but see Bitsch, 1994).

The following comparison of developmental patterns in both fossil (marked with a dagger) and recent orders (for a summary see Fig. 4) indicates the possible evolution of metamorphosis. Fossil groups are classified according to Kukalová-Peck (1991); for different opinions see the work of Rohdendorf and Rasnitsyn (1980) and Carpenter (1992).

The following symbols will be used in the schemes of postembryonic development: J, juvenile; A, adult; 0, wings absent; w, wings represented by external winglets or "wing pads" which do not allow flight; d, invagi-

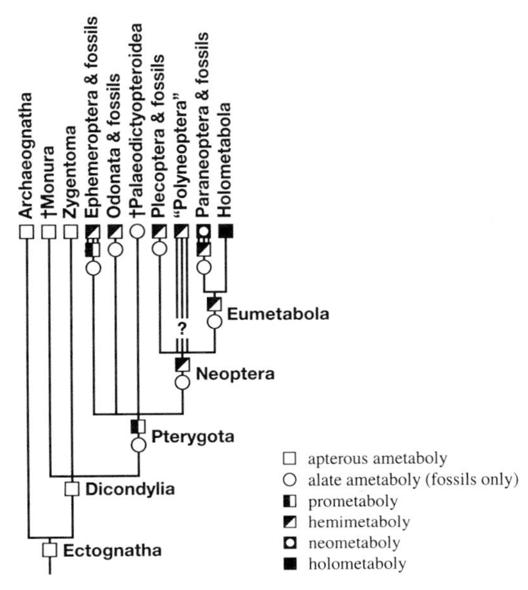

Fig. 3. Phylogenetic relationships of major groups of Ectognatha (= Insecta s. str.) with developmental types. Triple lines indicate multiple parallel evolution *of a developmental type* ("Polyneoptera" are also polyphyletic). Note that while the distribution of developmental types in modern insects permits a relatively simple evolutionary scenario, metamorphosis may have originated only once, and at most three times—in Ephemeroptera, Odonata, and Neoptera—inclusion of the "alate ametaboly" based on fossils as interpreted and classified by Kukalová-Peck (1992) makes the situation much more complicated. Metamorphosis must have originated at least seven times, at least twice in "Polyneoptera".

nated wing discs; W, flying instars; Q, nonfeeding resting ("quiescent") instars; triple or double bars mark morphological changes which can be considered more (≡) or less (=) metamorphic. Numbers of instars in the schemes are arbitrary except for the Paraneoptera.

A. Archaeognatha, Fossil Monura, and Zygentoma

All primarily wingless insects share the same mode of postembryogenesis, the *ametaboly* (J0-J0-J0-J0-J0-J0-A0-A0-A0- . . .). The number of

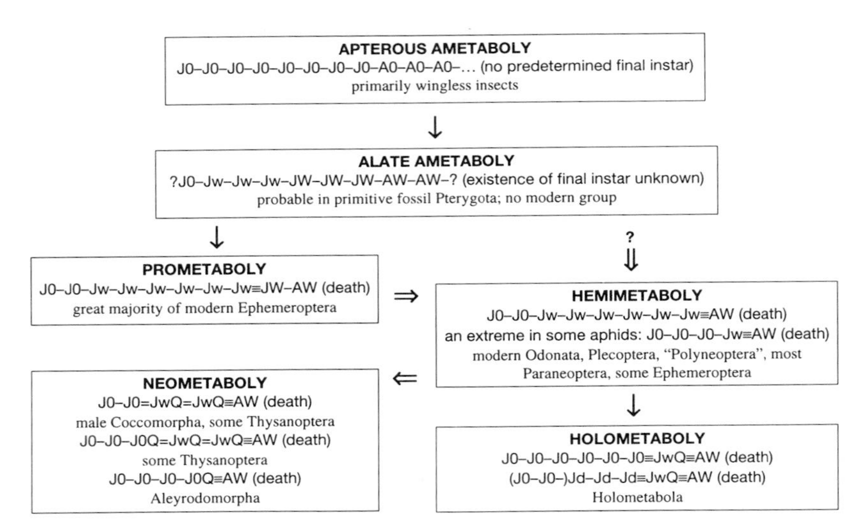

Fig. 4. An overview of insect developmental types and their presumed evolutionary relationships. Simple arrow, single origin; double arrow, multiple origin. For details and explanation of symbols, see Section III.

immature ("larval") instars is rather high (about 8–14) followed by numerous adult instars (there may be up to about 50 instars altogether); no predetermined final instar.

B. Pterygota

The winged insects are undoubtedly a monophyletic group with rather uncertain relationships between the higher (supraordinal) taxa. Traditional splitting into Palaeoptera (mayflies, dragonflies, and some fossil orders) and Neoptera (all other winged insects) is often contested, especially by those who regard dragonflies as a sister group of the Neoptera, a concept supported by molecular phylogenetics (see Kristensen, 1991). However, the "Palaeoptera hypothesis" has obtained considerable support from an analysis of wing venation and articulation by Kukalová-Peck (1991). For the present purpose, it is reasonable to treat all major pterygote clades as independent groups, three paleopteran (ephemeroids, odonatoids, and the Paleozoic plant-sucking paleodictyopteroids) and one neopteran.

1. Ephemeroptera

Prometaboly (J0-J0-J0-Jw-Jw-Jw-Jw-Jw-Jw ≡ JW-AW): two short-living and nonfeeding flying instars, the subimago (JW) and imago (AW). Larvae aquatic, with well-developed wing pads, number of instars is high and unstable (up to > 40), both testes and ovaries are fully developed in older larvae, and the larvae contain mature eggs and spermatozoa before metamorphosis (Soldán, 1979a,b). In some cases the subimago–imago molt is incomplete (some body parts, e.g., the wings, do not shed the subimaginal cuticle). *Hemimetaboly* (J0-J0-J0-Jw-Jw-Jw-Jw-Jw-Jw ≡ AW) occurs in a few genera (Polymitarcyidae partim, Behningiidae, Prosopistomatidae, and Palingeniidae partim) in which the final molt was suppressed (even apolysis does not occur) and the neotenic subimago reproduces, effectively becoming the adult (see Edmunds and McCafferty, 1988). Such species are usually prometabolous in males and hemimetabolous in females, but in a few species the original adult instar was lost in both sexes. In mayflies this is a rare and derived type of development which originated in parallel several times. *Ametaboly* [?J0-Jw-Jw-Jw-Jw-JW-JW-AW(?-AW-AW- . . .)] seems to have existed in primitive fossil mayflies: articulated wings developed gradually during a number of instars; older juveniles were probably terrestrial and were able to fly (Kukalová-Peck, 1991).

2. Odonata and Fossil Relatives

Hemimetaboly with aquatic larvae. Number of instars high and unstable (about 7–15), larvae rather different from adults, wing pads change their orientation during larval development. *Ametaboly* may have occurred in the Paleozoic †Protodonata.

3. Fossil Palaeodictyopteroidea

Ametaboly: terrestrial juveniles shared the habitat and feeding strategy with adults; development of articulated wings was gradual (see Kukalová-Peck, 1991) (Fig. 5A).

4. Neoptera

Monophyly of this group is accepted by all modern authors. Traditionally, they are divided into Polyneoptera, Paraneoptera, and Holometabola. Most modern phylogenetists agree that Paraneoptera

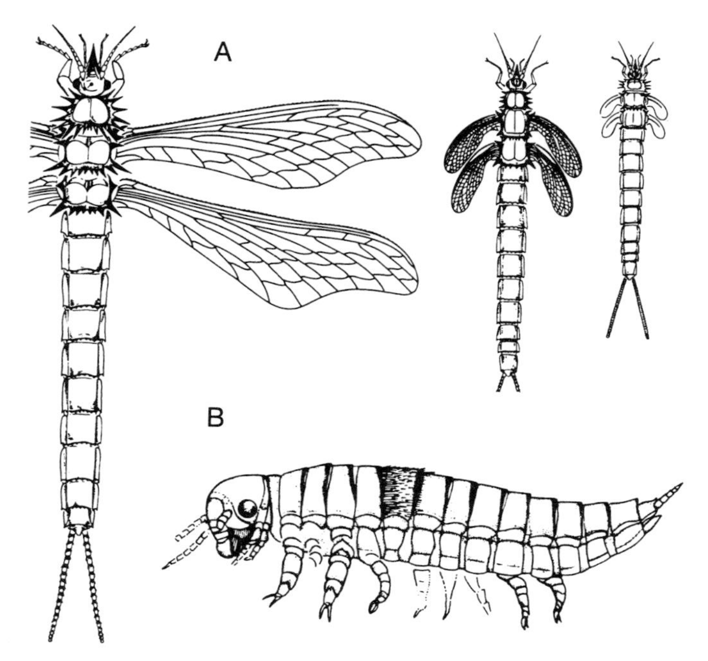

Fig. 5. (A) Adult and two larval stages of *Mischoptera* sp. (Palaeodictyopteroidea; upper carboniferous) showing gradual wing development (alate ametaboly). (B) Presumed oldest known holometabolous larva (upper carboniferous); note abdominal leglets, relatively long multisegmented antennae, and presence of structures which may be compound eyes. [Reprinted from Kukalová-Peck (1991) with permission from CSIRO Australia.]

(± Zoraptera) and Holometabola are each monophyletic and that they together also form a monophyletic group (Eumetabola), whereas the polyneopterans represent an unnatural assemblage of primitive neopterans whose phylogenetic relationships are uncertain. Comparing views published by Boudreaux (1979), Kristensen (1991), and Kukalová-Peck (1991, 1992), we conclude that the "Polyneoptera" consists of *at least* three independent clades (nomenclature of Kukalová-Peck, 1992), the "Pleconeoptera" (including Plecoptera), "Orthoneoptera" (including Orthoptera s. str. = Caelifera, Grylloptera = Ensifera, and Phasmoptera), and "Blattoneoptera" (including Blattaria, Mantodea, Isoptera); the positions of Notoptera (= Grylloblattodea), Dermaptera, and especially of Embioptera and Zoraptera are uncertain. Whatever the correct phylogenetic position of Plecoptera is, they represent the most primitive neopteran stock in most respects, probably also in regard to their development. The remaining orders are developmentally very similar and will be treated together as (probably polyphyletic) "Polyneoptera," including also the Embioptera, Notoptera, Dermaptera, and Zoraptera.

a. Plecoptera and Fossil Relatives. *Hemimetaboly* with a high and variable number of larval instars (>30 in some cases), larvae rather similar to adults, although they are usually aquatic while adults are terrestrial with a single exception. Wing pads of some stonefly larvae have "vague remnants" of wing articulation sclerites (Kukalová-Peck, 1991). Some adults do not feed. *Ametaboly* possibly occurred in Paleozoic forms; the older juveniles were probably terrestrial.

b. "Polyneopteran" Groups. *Hemimetaboly* with an unstable number of larval instars (about 5–10); larvae and adults usually share the same habitat. In larvae of the Orthoptera s. str. and Grylloptera the wing pads change their orientation during larval development.

c. Paraneoptera (= Acercaria; Including Fossil "Protorthoptera" and Some Other Paleozoic Groups, but Excluding Zoraptera). This group, at least its modern representatives, is undoubtedly monophyletic. Relationships among the three major lines, Psocoptera + Phthiraptera, Thysanoptera, and Hemiptera (including Psyllomorpha, Aleyrodomorpha, Aphidomorpha, Coccomorpha, Auchenorrhyncha, Coleorrhyncha, Heteroptera) remain uncertain.

Hemimetaboly with a rather stable and low number of larval instars [4–6, rarely also 3 or 2 (for a review see Štys and Davidová-Vilímová, 1989)]. Development of wing pads usually begins from the second or third instar, but sometimes they occur as late as in the ultimate larval instar (some Aphidomorpha: J0-J0-J0-Jw ≡ AW). Similar to the Holometabola, the larvae lack ocelli, with the exception of some Heteroptera, even if they are present in the adults. Habitats and feeding niches are not altered during postembrygenesis. *Neometaboly* with 1–3 immobile and nonfeeding pupa-like stages ("propupae" and/or "pupae") in which considerable histolysis and histogenesis takes place. This occurs in Thysanoptera (J0-J0 = JwQ = JwQ ≡ AW or J0-J0-J0Q = JwQ = JwQ ≡ AW), male Coccomorpha (J0-J0 = JwQ = JwQ ≡ AW; adults do not feed), and Aleyrodomorpha (J0-J0-J0-J0Q ≡ AW; the second to fourth instars are sessile, the initial phase of the fourth "pupal" instar feeds). *Ametaboly* occurred in some Paleozoic "protorthopterans," whose paraneopteran affinities were suggested by Kukalová-Peck (1991). Remnants of such development may be retained in the Coleorrhyncha whose larvae have wing pads separated from the dorsal thoracic sclerites by sutures.

d. Holometabola (= Oligoneoptera, = Endopterygota). Holometabola are certainly monophyletic; they comprise four major clades whose relationships are not clear: Coleopterida (including †Protocoleoptera, Coleoptera), Neuropterida (including Megaloptera, Raphidioptera, Neuroptera), Hymenoptera, and Mecopterida (including Lepidoptera, Trichoptera, Mecoptera, Siphonaptera, Diptera, and some fossil orders). The parasitic order Strepsiptera is usually regarded as a sister group of the Coleoptera, or even a coleopteran in-group (Crowson, 1981). However, results of 18S rDNA sequencing led Whiting and Wheeler (1994) to propose that Strepsiptera is a sister group of the Diptera (but see Carmean and Crespi, 1995). While their position within Holometabola will require further study, the holometabolan nature of the Strepsiptera [sometimes doubted (see Kristensen, 1991)] can now be considered proven.

Holometaboly (J0-J0-J0-J0-J0-J0 ≡ JwQ ≡ AW). Larvae significantly differ morphologically and usually also ecologically from adults and lack external wing pads and rudiments of genitalia. Some groups (Coleoptera partim, Hymenoptera partim, Trichoptera, Lepidoptera, Diptera partim) have developed invaginated "imaginal discs" for wings and often also for other organs: (J0-J0-)Jd-Jd-Jd ≡ JwQ ≡ AW. Larval ocelli are absent except for a median ocellus in some Mecoptera (Byers, 1991) and maybe also in some Coleoptera (Gnaspini, 1993). Larvae possess special photoreceptive organs, the lateral stemmata (for a review see Gilbert, 1994), which might probably be derived from a part of the compound eyes (identical innervation, sometimes similar structure). True compound eyes develop only at metamorphosis when the stemmata are withdrawn and come to lie near the eye stalks where they remain functional as photoreceptors. Compound eyes occur only in larval mosquitoes *along* with larval stemmata (Melzer and Paulus, 1991); the so-called larval compound eyes in some Mecoptera are multiple stemmata. Larval antennae are usually reduced and rarely have more than four segments. The oldest known fossil which is considered to be a holometabolous larva (Fig. 5B) is more primitive, having numerous abdominal leglets, relatively long multisegmented antennae, and perhaps ocelli and compound eyes. The primitive number of larval instars is probably high and unstable [e.g., 7–12 instars occur in various Megaloptera and Raphidioptera (Neunzig and Baker, 1991; Tauber, 1991; O. Pultar, personal communica-

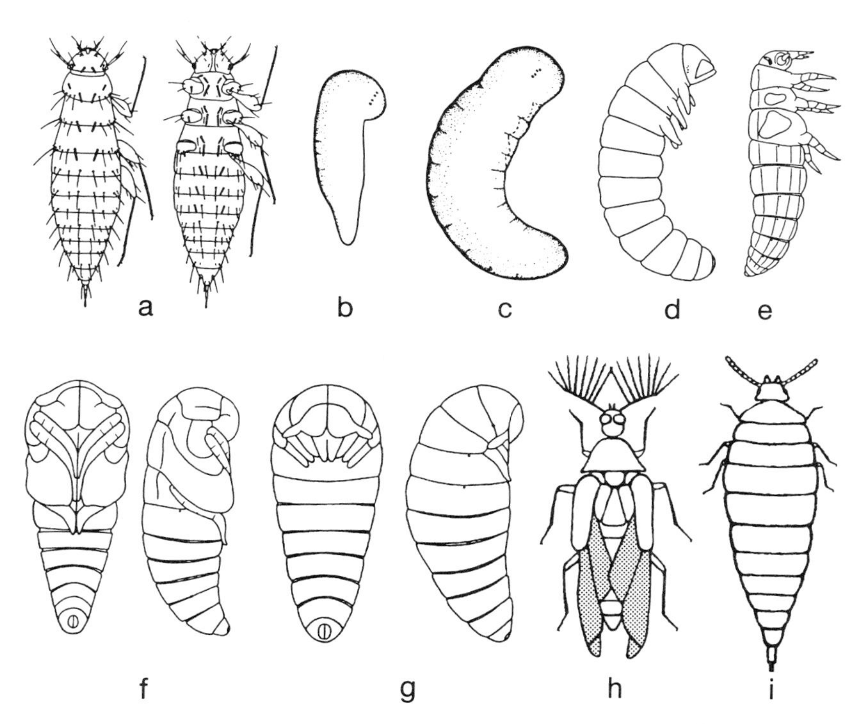

Fig. 6. Development of an endoparasitic beetle *Rhipidius quadriceps* (Rhipiphoridae) parasitizing cockroaches. The first instar (a; dorsal and ventral, body length 0.45–0.5 mm) is a structurally complex oligopod larva which penetrates the cuticle of the host and molts into an extremely simple legless larva (b; 0.4 mm). This larva, after some engorgement without molting (c), molts into a third larval type with short stump-like legs (d; this type may comprise more than one instar and reaches 4–6.5 mm). Another molt produces a fourth larval type with long segmented clawed legs (e) which leaves the host and pupates freely on or under stones, in bark crevices, etc. The obtect male (f; ventral and lateral) and female (g) pupae produce winged males (h; 4.5–6 mm) and wingless larviform females (i; 4–6.5 mm). Note the extreme differences between larval instars, particularly the extreme morphological simplification (including a loss of distinct segmentation) following the first molt. [Reprinted with permission from Besuchet (1956).]

tion)]. Metamorphosis occurs in two steps (last larva/pupa, pupa/adult); the last preadult instar, the pupa, does not feed, is more or less immobile, and possesses inarticulated wing pads and external rudiments of genitalia. Several exopterous preadult instars have been suggested to occur in males of some Strepsiptera (see Section V,C,2,d). Larvae of some groups change their morphology substantially one or more times before reaching the pupa. This phenomenon is here collectively called "hypermetamorphosis," although some authors distinguish "polymetaboly" and "hypermetaboly" based on the absence or presence of one or even more resting inactive *larval* instars. Pronounced cases of hypermetamorphosis are usually associated with specialized modified predatory habits or parasitism: Neuroptera: Mantispidae, Coleoptera: Meloidae [most species with inactive larval instars (see Selander, 1991)], Rhipiphoridae and a few other groups, all Strepsiptera, many Hymenoptera, some Diptera, and seldom also members of other holometabolous orders (see Clausen, 1940) (Fig. 6). Adults of some Trichoptera, Lepidoptera, Diptera, Coleoptera, and Megaloptera do not feed.

IV. ROLE OF HORMONES IN DEVELOPMENT OF MODERN INSECTS

Available information on insect neurohormones is not sufficient to allow analysis of their possible role in the origin of metamorphosis. This chapter thus concerns ecdysteroids and juvenile hormones (JHs), which have been the most studied and were found to govern the postembryonic life of recent insects (see Gilbert *et al.,* Chapter 2, this volume). We take a comparative approach, emphasizing hormone titer changes and hormonal effects at the organismic level. Differences in the action of different structural types of JH and ecdysteroids, and the conversion of their precursors to active hormones in the peripheral tissues are not dealt with, although we realize that these phenomena may have also been of importance in the evolution of metamorphosis.

Ecdysteroids occur in all Arthropoda and regular surges in their secretion induce the periodic molts (see Gilbert *et al.,* Chapter 2, this volume; Cherbas and Cherbas, Chapter 5, this volume); they are also important for regeneration, reproduction, and probably also for various morphogenetic processes occurring both during embryonic and postembryonic development (Spindler, 1989). In contrast, the JHs seem to be restricted to insects, in which they typically prevent metamorphosis in the immatures and stimulate reproduction in the adults (see Riddiford, Chapter 6, this volume). Substances closely related to the most common JH III, i.e., methyl 10,11-epoxy-*trans, trans, cis*-farnesoate, were identified in Crustacea; farnesylacetone and hydroxyfarnesylacetone from the androgenic gland inhibit vitellogenesis and cause male-specific accumulation of astaxanthin (Ferezou *et al.,* 1977), and methyl *trans, trans, cis*-farnesoate from the mandibular organ (Laufer *et al.,* 1987) affects duration of the molt cycles and delays metamorphosis (Chang, 1993). It is likely that the sesquiterpene hormones appeared in arthropod phylogeny before Crustacea and Hexapoda diverged (Cusson *et al.,* 1991).

A. Insect Endocrine Glands

The retrocerebral endocrine system of insects consists of (i) several groups of neuroendocrine cells in the nervous system; (ii) corpora cardiaca (CC) where some of the brain neurosecretion is released and additional hormones are produced; (iii) corpora allata (CA) secreting JH and serving also as a neurohemal organ; and (iv) glands providing ecdysteroids for late embryogenesis, larval development, and varying sections of metamorphosis. Watson (1963) demonstrated that the endocrine system of Zygentoma and apparently also of Archaeognatha is homologous to that of most pterygote insects. Ephemeroptera are unique by having the glands regarded as CA in an atypical position and not connected to the CC by nerves (Arvy and Gabe, 1953a).

Several types of possibly homologous ecdysteroid-producing glands are distinguished in insect larvae (for details see Novák, 1966): (i) compact ventral glands of Archaeognatha, Zygentoma, Ephemeroptera, Odonata, and various "polyneopteran" groups are located ventroposteriorly within or just behind the head; (ii) diffuse prothoracic glands of Blattaria, Isoptera, Paraneoptera, and most Holometabola are often close to the first spiracles (these spiracles are in the mesothorax, but in some insects the spiracular area is pushed into prothorax, and the spiracles are then commonly but incorrectly called "prothoracic"); and (iii) the peritracheal glands of Diptera are adjacent to the dorsal blood vessel just behind the head (in cyclorrhaphous Diptera these glands blend dorsally with the fused CA and ventrally with the fused CC, forming together a compact and multifunctional ring gland). Here all these glands will be called "prothoracic" glands.

B. Hormones in Postembryonic Development of Primarily Wingless Insects

The firebrat, *Thermobia domestica* (Zygentoma), produces JH III (Baker *et al.,* 1984) and ecdysteroids apparently identical to those of pterygote insects (Rojo de la Paz *et al.,* 1983). Watson (1967) examined changes in JH production by the CA of the juvenile firebrat with the aid of a bioassay. He found that CA activity declined in the third instar and then rose again. This profile of JH secretion was consistent with changes in the volume of CA cells that was also minimal in the third instar. Using the CA volume/body weight ratio as an approximation, Watson concluded that the JH titer was highest in the first and second juvenile instars, and gradually declined due to the negatively allometric growth of the CA. Watson suggested that the presumed JH decline in the third instar allowed the appearance of the integumental scales that are lacking in young firebrats and first appear in the fourth instar. The control of this change in epidermal function was also examined by transplanting fragments of integument from newly hatched juveniles into adults: the implants secreted new cuticle simultaneously with their hosts and the cuticle bore scales. This was interpreted as being due to an effect of JH deficiency at the time of molt induction in the adults.

These data may indicate that the occurrence of scales during early preadult development is a morphogenetic process controlled by JH analogous to the larval–adult metamorphosis of pterygotes (see later). However, other interpretations are possible. For example, the appearance of scales may depend on the supply of dietary nutrients that become available in the third instar, i.e., younger larvae do not feed and live from their yolk reserves (Watson, 1967). There is no doubt that firebrat juveniles contain JH, but the hormonal control of postembryonic morphogenesis by JH still remains to be proven. The hemolymph ecdysteroid titer probably undergoes cyclic changes similar to those in the adults (Fig. 7A).

Fig. 7. Hemolymph JH and ecdysteroid titers (A) and ovarian ecdysteroid content (B) during an adult instar of *Thermobia domestica.* [After Bitsch *et al.* (1979) and Hagedorn (1989).]

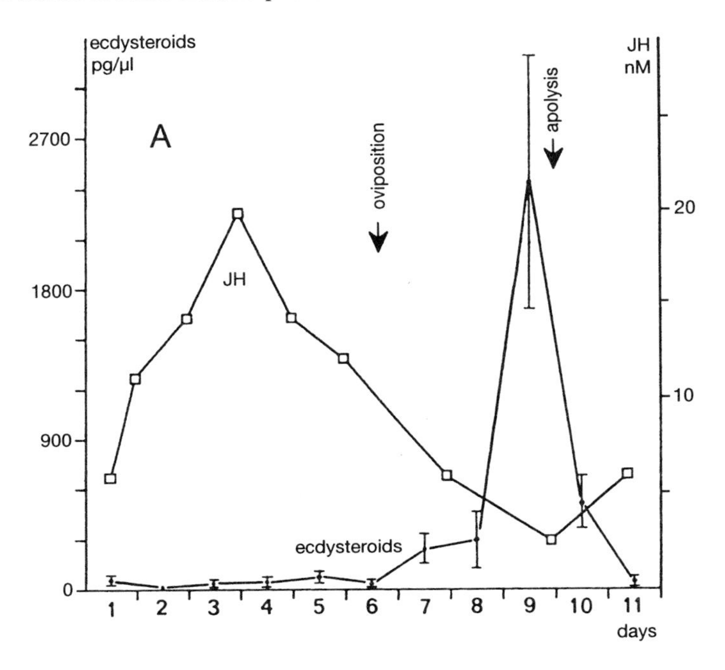
ecdysteroids
pg/µl
JH
nM
A
2700
1800
900
0
20
10
oviposition
apolysis
JH
ecdysteroids
1
2
3
4
5
6
7
8
9
10
11
days

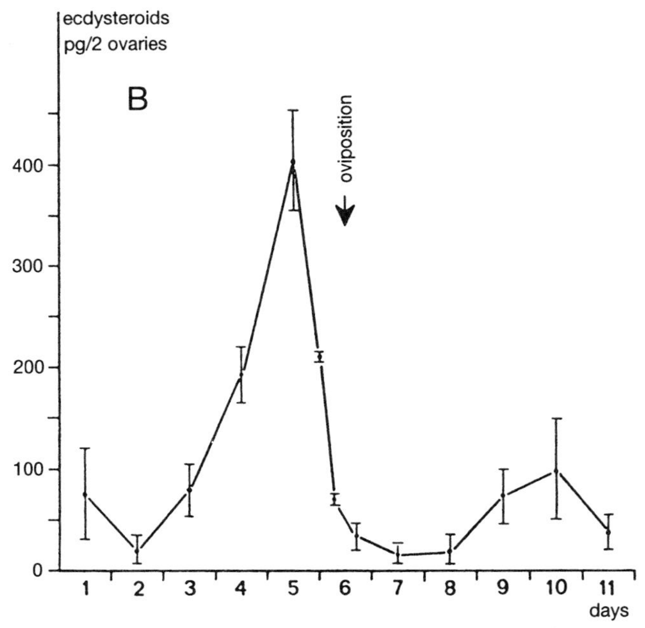
ecdysteroids
pg/2 ovaries
B
400
300
200
100
0
oviposition
1
2
3
4
5
6
7
8
9
10
11
days

C. Hormones in Pterygote Insects

1. Hormone Titers during Larval Development

Developing larvae always contain a low, and in some cases fluctuating level of ecdysteroids but only a distinct rise in the second half of each intermolt period elicits the molting process. Such a pattern of ecdysteroid titer changes was demonstrated in all major pterygote clades (mayflies have not been examined). Species differences in the concentration and composition of ecdysteroids and the short-term titer fluctuations are not well understood.

Because of methodological difficulties, JH titers were established only in few insect species. JH production from the explanted CA is often taken as a measure of their secretory capacity and consequently, as an indication of JH titer changes. A good correlation usually exists between these data and actual JH determination in the hemolymph. In most insects the titer of JHs seems to remain high from hatching until the last larval molt. Broad JH titer peaks, overlapping with the molt-inducing surges of ecdysteroids, were detected in some species. The high JH titer until the penultimate larval instar and the virtual absence of JH during most of the last larval instar (Akai and Rembold, 1989) may reflect a primitive situation in metamorphic pterygotes; indirect evidence indicates that similar titer changes occur in Odonata, in various "Polyneoptera," and in some Paraneoptera and Holometabola. Derived situations are found in advanced insect groups, represented by *Apis* (Hymenoptera) in which the JH titer is maintained at a nearly steady level from late embryos until the early last larval instar in worker larvae, whereas in queen larvae it rises to a level five times higher; in both cases the titer declines only temporarily when the larvae terminate feeding (Rembold *et al.,* 1992).

2. Hormones in Control of Larval Allometry

Larval development is characterized by differences in the growth rates of various body parts. Such allometry in hemimetabolous insects is manifested by the relative enlargement of wing pads, rudiments of genitalia, compound eyes, etc. The allometric growth seems to continue in the supernumerary larval instars induced with juvenoid application in "Polyneoptera," Paraneoptera, and possibly also in Odonata (see Sehnal,

1983). Continuous allometry probably restricts the number of extra molts because at a certain body size the disproportionality between different body parts becomes incompatible with survival. For example, the continuous development of ovaries in aphid supernumerary larvae results in the formation of embryos that cannot be accommodated in the body. The aphid dies, often following rupture of the the body wall. In contrast to such observations on various hemimetabolous insects, Wigglesworth (1952) reported for *Rhodnius* (Heteroptera) that exogenous JH erases anisometry between the growth of genital rudiments and wing pads compared to that of the body, but in our studies with other true bugs we could not confirm this singular observation.

Larval development of many Holometabola is associated with little change in the relative proportions of different body parts. In *Galleria* (Lepidoptera), the imaginal discs and the gonads enlarge in larvae more than some other organs and this allometry is maintained when larval development is extended by supplying the larvae with CA implants (Sehnal, 1968). However, it is obviously suppressed when several extra larval molts are induced with high doses of juvenoids. Numerous supernumerary larval molts with body enlargement but without a change in body form may occur in some other Lepidoptera and also in Megaloptera and certain Coleoptera. In most Holometabola, however, this effect is impossible: larval allometry and/or some metamorphic changes are not inhibited by juvenoids and only defective supernumerary larvae are produced (see Sehnal, 1983).

3. Larval Polymorphism

In addition to general allometric differences, two adjacent larval instars may become truly diversified. A profound morphological difference is very common between the first instar larvae that may not feed and whose major role is dispersal and the second instar larvae in which the larval growth is initiated. Striking changes in coloration and morphology that occur at certain molts of some caterpillars are also well known. Extreme morphological changes result in hypermetamorphosis (see Section III,B,4,d).

Thorough examinations of JH action on this *sequential polymorphism* are lacking. In several lepidopteran species we noticed, however, that morphological and sometimes also coloration changes between the first and second instars are insensitive to exogenous JH. Similarly, the change in head capsule color between the penultimate and ultimate larval instars of *Samia* (Lepidoptera) is not prevented with JH treatments but it does not occur in starved larvae which molt at a body size corresponding not to the last, but to the penultimate instar (F. Sehnal, unpublished observations).

Hormonal control of changes in larval appearance and development was shown only in cases of *parallel polymorphism,* when morphological and other differences occur in alternative developmental pathways. Larvae of many insects can occur in two or even more forms depending on the environmental conditions that act by altering hormone concentrations. The implication of hormones in the phase, seasonal, and caste polymorphism was clearly demonstrated (Hardie and Lees, 1985). In the honey bee, the worker and queen larvae are distinguished by their JH levels (Rembold *et al.,* 1992).

4. Larval–Adult Transformation in Hemimetabolous Groups

Larval development is terminated after a predetermined number of molts or when the insect reaches a constitutive size for metamorphosis, and a new pattern of hormone titers sets in (Sehnal, 1985). Similar changes in hormone titers during the larval–adult transformation were established in various "Polyneoptera" and Paraneoptera. Some data are also available on Odonata but nothing is known about the hormonal control of metamorphosis in Ephemeroptera and Plecoptera. Results obtained in the species examined allow the following generalizations. The beginning of the last larval instar is characterized by a decline in JH production which is usually fast, but the reduced JH titer may persist for some time (Szibbo *et al.,* 1982). A low JH level is the crucial requirement for metamorphosis. When a high titer is maintained by supplying the insects with CA implants, exogenous JH or juvenoids, metamorphosis is invariably prevented, even though the supernumerary larvae may be "abnormal" due to continuous allometry. In a reverse experiment, when the JH titer is reduced prematurely by removing the CA from the penultimate or earlier instar larvae, these larvae undergo precocious metamorphosis into miniature adults. Only in the very young larvae does allatectomy not elicit metamorphosis immediately, but after one or two additional larval instars, presumably because some organs need at least one more larval molt to acquire the competence for metamorphosis (Pflugfelder, 1952).

A prolonged period, when compared to previous instars, of a low ecdysteroid titer at the beginning of the last larval instar appears to be another requirement for successful metamorphosis (Steel and Vafopoulou, 1989). The titer eventually rises to a single molt-inducing peak that is usually higher than those noted before larval molts. In a few insects such as *Rhodnius,* it is preceded by a gradual rise in the ecdysteroid titer (Beaulaton *et al.,* 1984) which may appear as a small, separate peak (Steel *et al.,* 1982). Ecdysteroids seem to be mostly supplied by the prothoracic glands as in larval development, but investigations on *Gryllus* (Grylloptera) revealed that an important contribution is provided by the integument (Gerstenlauer and Hoffmann, 1955; see Gilbert *et al.,* Chapter 2, this volume).

5. Larval–Pupal–Adult Transformation in Holometabola

The metamorphosis of most Holometabola also depends on a drop in JH titer in the last larval instar. Metamorphosis is suppressed when the missing JH is replaced by exogenous JH or juvenoids. Treated insects molt into supernumerary larvae but these often possess "pupal" features (Sehnal, 1983). It is difficult to decide whether small deviations such as slightly everted wing discs or enlarged antennae are a consequence of continuing allometric growth or whether the initial metamorphic changes are programmed to occur at a certain body size irrespective of JH. Both possibilities exist; the second one is exemplified in chironomids and mosquitoes (Diptera: Culicomorpha), in which the differentiation of some imaginal organs such as wings and appendages in the last larval instar cannot be prevented by JH application.

The significance of JH for the proper progress of larval–pupal transformation probably varies even among related taxa, as exemplified by the Lepidoptera. For example, in *Bombyx* (Bombycoidea) the JH titer remains low until the pharate adult stage, except for a minor peak around the termination of cocoon spinning. This transient JH occurrence may play a role in the timing of pupal ecdysis but does not affect structural changes since animals deprived of their CA develop into normal pupae and adults. However, in some other lepidopterans, such as *Pieris* (Papilionoidea), the prepupal JH peak is high (Mauchamp *et al.,* 1981) and apparently controls the progress of morphogenesis in the last larval instar. JH elimination by depriving larvae of their CA leads to the appearance of scales and other imaginal features in pupae (Nayar, 1954). There are indications that both *Bombyx* and *Pieris* "types" of the regulation of metamorphosis by JH exist in various orders of Holometabola, and it is difficult to know which type is evolutionarily more advanced.

Larval–pupal transformation of highly derived insects, such as Coleoptera: Curculionoidea, Diptera: Brachycera, and Hymenoptera: Apocrita, is never prevented with juvenoids, indicating that it is *not* controlled by a drop of JH titer. Indeed, in the honey bee the JH titer remains high (in workers) or even increases (in queens) at the start of the last larval instar, then slowly decreases to a very brief period of near JH absence at the end of feeding, but rises again during cocoon spinning and apolysis (Rembold *et al.,* 1992). The presence of JH in the last instar is obviously important for the proper progress of metamorphosis because ligation of the last instar honey bee larvae, which effectively removes their CA, causes the appearance of adult features at the next, normally pupal, molt (Schaller, 1952). In *Drosophila,* the CA exhibits a high capacity for JH synthesis in the last larval instar (Dai and Gilbert, 1991), and the JH body titer culminates in the postfeeding larvae (Sliter *et al.,* 1987).

The pattern of ecdysteroid titer changes in the last larval instar of Holometabola is rather uniform. It begins with a plateau of a low ecdysteroid level, whose importance for the change from larval development to metamorphosis was shown in the silkworm (Gu and Chow, 1993). The change from the larval to the metamorphic pattern of ecdysteroid secretion in *Galleria* larvae is influenced by JH. When development is elicited in decapitated larvae by brain implants in the presence of JH, ecdysteroids rise rapidly to a peak and a larval molt follows. When development is elicited in the absence of JH, ecdysteroids remain low for several days before rising to a high peak that causes the pupal molt (Sehnal *et al.,* 1986). In various species the ecdysteroid titer rises to a small peak around the termination of feeding and only then to the high, molt-inducing surge toward the end of the instar. The small peak was implicated in tissue reprogramming (Riddiford, 1985), but some tissues are certainly reprogrammed before or after this peak, which is absent in some Holometabola (Smith, 1985). The molt-inducing ecdysteroid surge is usually, but not always, higher than in the previous instars. Numerous experiments with larval ligations showed that prothoracic glands are an indispensable source of ecdysteroids for pupation in most Holometabola, but this does not exclude the possibility that other organs participate in ecdysteroid production. In some beetles (Delbecque and Sláma, 1980) the prothoracic glands of last instar larvae become nonfunctional and all ecdysteroids are derived from alternative sources. In extreme cases, such as the mosquito, the body epidermis and not the prothoracic glands seems to provide ecdysteroids for the entire period of postembryonic development (Jenkins *et al.,* 1992).

The pupal–adult transformation of Holometabola seems to be characterized by JH absence until the pharate adult stage and by a large ecdysteroid peak in the first third, rarely in the middle, of the pupal instar (Smith, 1985). The peak typically exceeds maximal larval concentrations of ecdysteroids. Another difference from larval instars is a long period of low ecdysteroid concentration between the peak and adult ecdysis. Since prothoracic glands usually degenerate shortly after the pupal ecdysis, most ecdysteroids found in the pupae of many insects are derived from secondary sources (Delbecque *et al.,* 1990; see Gilbert *et al.,* Chapter 2, this volume).

6. Linkage of Structural Metamorphosis to Maturation and Molt Termination

Structural metamorphosis is hormonally synchronized in modern pterygotes with the attainment of sexual maturity and the cessation of molting (see Section II,B). The decline of JH with a subsequent rise in ecdysteroids, as they occur during metamorphosis, commonly triggers in males the completion of spermiogenesis and in females the separation of ovarioles and the formation of egg chambers that is often immediately followed by previtellogenesis (Raabe, 1986). Induction of these processes with ecdysteroids in the absence of JH has been demonstrated *in vitro.* The same hormonal changes stimulate development of the extragonadal parts of the reproductive system such as oviducts.

The dependence of sexual maturation on a decrease in JH titer and a rise in ecdysteroids is probably an ancient phenomenon and resembles the control of previtellogenesis in the reproductive cycles of adult *Thermobia* (Bitsch and Bitsch, 1988). This control pattern was modified under selective pressures either for rapid reproduction in newly emerged adults or, to the contrary, for a delay of reproduction until sufficient nutrition is obtained by adult feeding. In the first case, the sperm as well as eggs are formed in the larvae, and both previtellogenesis and vitellogenesis probably occur in response to ecdysteroids, but irrespective of the presence or absence of JH. This independence of JH was proven experimentally in Phasmoptera [*Carausius* (Pflugfelder, 1937)], Embioptera [*Hoembia* (F. Sehnal, unpublished observations)], and Diptera: Culicomorpha [*Chironomus* (Malá *et al.,* 1983)]. In many moths, such as *Bombyx,* egg formation occurs during the pupal instar and is also independent of JH but requires ecdysteroids (Tsuchida *et al.,* 1987). The other extreme, i.e., the shift of previtellogenesis until after metamorphosis, has been well demonstrated in mosquitoes (Diptera: Culicidae), in which previtellogenesis is completed only in response to JH release after the adult takes a blood meal (Hagedorn, 1989).

Exposure to a high ecdysteroid titer in the absence of JH triggers degeneration of the prothoracic glands that deliver ecdysteroids in the larvae (Wigglesworth, 1955; Gilbert, 1962; see Gilbert *et al.,* Chapter 2, this volume). The glands become nonfunctional in the last larval instar of hemimetabolous insects and in the pupal instar of Holometabola, but there are exceptions to this "generalization." In certain Holometabola the glands cease to function already in larvae, usually in the last larval instar when the decision for metamorphosis is made, and further development is driven by ecdysteroids produced elsewhere, e.g., imaginal discs and possibly all epidermal tissues (Delbecque *et al.,* 1990). On the other hand, prothoracic gland degeneration is delayed in some cockroaches (Blattaria), in locusts (Orthoptera), and in the nonreproductive castes of termites (Isoptera). Morphologically coherent prothoracic glands are also retained in the mayfly subimago (Arvy and Gabe, 1953b) and it is reasonable to assume that they deliver ecdysteroids for the final molt, after which they regress.

The absence of molting in adult pterygotes is in some cases due to a lack of a sufficient ecdysteroid concentration rather than to the insensitivity of the epidermis. Incomplete molting (without ecdysis) was induced with prothoracic gland implants or exogenous ecdysteroids in adult cockroaches, bugs, and even silkmoths (see Sehnal, 1985, and references therein). In some adults, however, considerable amounts of ecdysteroids occur in the hemolymph but molting does not take place. We cannot exclude differential epidermal sensitivity to different kinds of ecdysteroids but it seems more likely that some time after imaginal ecdysis the epidermis cannot any longer undergo apolysis and produce a new and complete cuticle.

D. Hormone Action at Tissue Level

Two aspects of hormone action at the tissue level which are important for considerations of the origin of insect metamorphosis are briefly discussed below (see also Sehnal, 1980).

1. Tissue Reprogramming

Cells may change their function during postembryonic development and these changes may or may not be related to cell divisions. The amount of such changes is usually much greater in the last larval instar, and also in the holometabolan pupa, than in the preceding instars. Metamorphic development can be regarded as "the manifestation of sequential polymorphism" (Highnam, 1981) produced by the same genome. For example, an epidermal cell produces successively the larval, pupal, and imaginal cuticles (see Willis, Chapter 7, this volume). Some authors regard the transition from larval to pupal and then to the adult functional state as a developmental process during which the epidermis acquires the state of terminal differentiation at which time it produces the imaginal cuticle. Sláma (1975) maintained that each step (larval, pupal, adult) is strictly determined and can be neither omitted nor mixed with the other steps; i.e., hormones merely control whether a cell remains at the present step or advances to the next one. A contrasting view first expressed by Piepho (1951) is that epidermal cells are totally controlled by hormones that determine the type of cuticle currently produced: at high JH the larval cuticle, at moderate JH the pupal cuticle, and in the absence of JH, the imaginal cuticle (see Gilbert *et al.,* Chapter 2, this volume).

Both of these views are regarded as extremes, each accentuating only one aspect of the phenomenon under study. The following observations demonstrate that both the preceding ontogenic history driving the endogenous developmental program and the hormone titers determine the type of secreted cuticle. (i) Hormonally induced reversal from a late to an earlier cuticle type is possible in some cases (Wigglesworth, 1954), but occurs much less readily than the normal succession and requires extensive cell divisions. (ii) The change from one to the next functional state is gradual and may be blocked with JH at various phases, which results in the production of cuticle bearing features of two successive stages [e.g., larval–pupal composite cuticle (see Willis *et al.,* 1982)]. (iii) The change in epidermal cell capacity from one cuticle type to the next requires a certain time period and precocious induction of cuticle secretion results in an imperfect cuticle (Sláma, 1975). (iv) Under experimental conditions, the epidermis can proceed from the larval to adult secretory state with the omission of pupal cuticle secretion (Sehnal, 1972). These data can be interpreted as showing that the progress from embryonic to adult cuticle secretion reflects the ontogeny of the epidermis during which the cuticle is produced repeatedly in response to hormones which can, under experimental conditions and to various degrees, revert the epidermis to an earlier ontogenetic stage.

2. *Larval and Imaginal Organs*

The change in the type of cuticle secretion is usually associated with epidermal cell division but its significance is still questionable. Functional changes in some organs clearly depend on extensive cell proliferation by which some parts of the organ, which performed specialized larval functions, are replaced. On the other hand, the silk glands of *Antheraea* (Lepidoptera) degenerate except for their outlets whose cells, without divisions, form a new gland whose secretion dissolves the maxillary enzyme cocoonase (Selman and Kafatos, 1974). We stress that both the larval silk gland and the new pupal–adult gland are sequential modifications of the labial glands whose original function was the secretion of saliva.

The highly specialized "larval" organs often contain histoblasts whose participation in larval function is negligible but which generate the organ functioning in the adult. The larval midgut of some Holometabola even undergoes two major reconstructions to form successively the pupal and adult midguts, all originating from the histoblasts of the regenerative nidi. Histoblasts often form multicellular imaginal discs, but even in this case they must be viewed as part of an organ functioning in the larvae and not as undifferentiated embryonic primordia persisting until metamorphosis (see Section V,C,2,c).

In summary, a working concept can be accepted of calling organs "larval" or "imaginal" (this simple division is imprecise—some "larval" organs occur only in certain instars; there are special "pupal" organs in Holometabola, etc.) when we intend to emphasize their particular larval or imaginal function, but a strict distinction in fact does not exist. Both are derived from tissues established before hatching, except that some cells reach "terminal differentiation" and are destroyed before the adult stage. The progress of differentiation is hormonally regulated. Typically, "larval" organs require for their functioning the presence of JH that simultaneously restrains differentiation of the "adult" organs. The surge of ecdysteroids and also often the JH decline in the last larval instar cause "larval" organs to degenerate and "adult" structures to attain the functional state.

V. EVOLUTION OF INSECT METAMORPHOSIS

A. Origin of Metamorphosis: Ecomorphological Aspects

The origin of *insect* metamorphosis was certainly connected to the progressive evolution of wings and flight. No other significant differences exist between ametabolous primarily wingless Ectognatha (i.e., Archaeognatha and Zygentoma) and "primitive" metamorphic Pterygota; the gradual development of their genitalia, for example, is fairly similar. The origin of flight undoubtedly provided for the evolutionary success of the Pterygota, but the morphological source and functional sequence of wing origin remain unresolved and are not discussed in this chapter (see Kukalová-Peck, 1991; Wootton, 1992; Kingsolver and Koehl, 1994; Brodsky, 1994). According to Kukalová-Peck (1991), almost all Paleozoic forms exhibited gradual wing development, i.e., no particular metamorphic instar, and possessed several flying instars, some of them sexually immature ("older nymphs," "subimagos"). If so, the functional wing cannot be considered a *primarily adult* character—a conclusion supported also by the development of modern mayflies (see later).

The wing precursors must have had a different function to be selected for before flight ability evolved. The first short-distance, probably gliding, flights could be employed for escaping from predators, easier moving from plant to plant, etc. Later, the evolution of the capability for active long-distance flights was presumably of importance mainly in adults for finding a mate or for establishing a new population in a remote environment. The original (unknown) function of the wing precursors may have become incompatible with the perfection of the more important flight function and/or may have been lost due to a change of habits, e.g., a transition of juveniles to the aquatic environment. Interestingly, many ancient taxa like Ephemeroptera, Odonata, Plecoptera, or Megaloptera have aquatic larvae. The winglets in the nonflying, early juveniles thus became a subject of negative selection as their diminution allowed easier locomotion and molting. The reduction of wing rudiments in the juveniles rendered them less vulnerable and capable of invading new niches. The resulting diversification between early and late instars led to the origin of metamorphosis.

Additional changes that led to the situation found in modern pterygotes included establishing a predetermined final instar and fixing reproductive activities into this instar. The succession of the occurrence of all these changes is uncertain and they may have been originally isolated and mutually independent (see Fig. 4). All *modern* pterygotes have only one adult instar, and it would be very difficult to detect the existence of multiple, probably very similar, adult instars in the fossil material. However, some fossils and the development of modern Ephemeroptera show that structural metamorphosis to a flying form originally might precede sexual maturation by one or more molts. The mayfly subimago primitively does not reproduce and there are no indications that this should be a secondary condition. We therefore consider the subimago a flying immature instar, not a remnant of adult molts occurring in primarily wingless insects. The evolution "prometaboly → hemimetaboly" has occurred in some Ephemeroptera due to the omission of the original adult instar and to a neotenic transfer of reproduction to the subimago (see Section III,B,1). It is quite possible that intermediate states in insect evolution from ametaboly to hemimetaboly (from which neometaboly and holometaboly can be derived) were always prometabolous or similar (Fig. 4). If the structural metamorphoses of various pterygotes were homologous (which is uncertain), the mayfly subimago could best be compared with the adult of other pterygotes, being the first postmetamorphic instar (see discussion in Edmunds and McCafferty, 1988). The holometabolan pupa is considered to be an originally *premetamorphic* instar (see Section V,C,2,b). Thus, the adult instar of pterygotes other than Ephemeroptera may have originated as a neotenically reproducing first postmetamorphic subimago. Difficult molts may have favored molt restrictions in the flying forms (Maiorana, 1979), although the existence of a predetermined final instar is not limited to flying insects but occurs also in other arthropods. Once further molting was blocked, the final adult instar was prone to undergo morphological modifications (e.g., degree of sclerotization, morphological complexity, etc.) incompatible with further molting. Many of these modifications evolved to further perfect the ability to fly.

In various pterygote adults the wings became secondarily reduced or completely absent. The secondary loss of wings in various insects has never led to the restoration of adult molting. In the hemimetabolous groups, wing loss frequently resulted in the disappearance of striking differences between the larva and adult (e.g., wingless Embioptera, Notoptera, Dermaptera: Hemimerina, termite workers, Phthiraptera), allowing some authors to use the term "secondary ametaboly" (Imms, 1938). A similar "secondary ametaboly" also exists in some forms retaining only the larva-like wing pads in adults [some Heteroptera (Štys and Davidová-Vilímová, 1989)]. In most of these forms various other structures (mainly genitalia, but, for example, also the scent gland system in most Heteroptera) retain changes which can be considered metamorphic, but if the external genitalia are reduced as in female Embioptera which are considered neotenic (Ross, 1970), no metamorphic changes may remain, at least in the external morphology. Nevertheless, the term "secondary ametaboly" is not accepted here.

Kukalová-Peck (1991) interpreted the fossil material as evidence that metamorphosis evolved several times independently from ametabolous ancestors with gradual wing development (almost all Paleozoic groups; see Section III and Fig. 3), but adaptive reasons why such evolutionary changes should have occurred in parallel only *after* the long Paleozoic era are unclear. While the independent origin of mayfly prometaboly is plausible, the proposal of the multiple parallel origin of metamorphosis of a similar type (with a single flying final adult instar) in several pterygote evolutionary lines is based exclusively on the fossil record and may change with its reclassification or reinterpretation.

B. Evolution of Number of Instars and Instar Homology

Insects, as well as other arthropods, have an external cuticle which must be shed periodically. Regular molts bring some cyclical events into insect development. This does not mean that insects develop discontinuously, i.e., the cuticle is merely a periodically removed secretion and the development of the insect body within the cuticle is continuous, although the types and rates of changes differ at various developmental stages. Molts are primarily necessary for, and determined by, growth, but they also allow external morphological changes to take place.

The periodic molts are associated with both advantages (*combined* possibility of growth and cuticular protection, possibility of regeneration including renewal of physicochemical properties of the cuticle) and disadvantages (time loss, difficult and risky ecdysis particularly in structurally complex insects, reduced locomotory capabilities and sensory input before ecdysis, high vulnerability due to soft cuticle after ecdysis). Insects tend to reduce the number of molts if they can, e.g., if they have acquired other protective measures (hidden way of life, defensive secretions, etc.). Such insects, typically some holometabolous larvae, can afford soft extensible cuticle and limited regeneration capabilities and undergo few molts. In adults the size increase and prolonged life span should result in enhanced reproductive output, but in modern pterygotes the advantages have apparently been outweighed by the difficulty of molting of the flying stage. Few insects have acquired a very long adult life without molting, e.g., several years in some beetles to several decades in protected termite reproductives.

The growth and morphogenetic changes seem to be largely independent of the number and the timing of the molting cycles. The amount of morphological change between two successive instars ranges from minimal in most larvae to the extreme (e.g., in the adult molt of whiteflies), and it is possible to regard the postembryonic morphogenesis of an insect as one unit in which a molt could be inserted at practically any stage. The actual number of instars and the timing of molts in a given species evolved under a combination of intrinsic and environmental pressures determining the relevance of advantages and disadvantages of molting. This view is consistent with great differences in the number of instars in related taxa, variation in the instar number in some species, and the experimentally possible induction of a molt at virtually any time.

A high and often unstable number of immature instars, possessing sclerotized cuticle of limited extensibility, apparently characterized ancestral Ectognatha and Pterygota (see Section III). Reduction of the instar number is evident in various phylogenetic lines, rendering impossible homologies among individual instars in phylogenetically remote groups. In some species with a variable number of instars one cannot distinguish *which* instar has been added or deleted. Tanaka (1981) demonstrated that the cockroach *Blattella* reared under optimal conditions may reach the adult stage, of similar average size, through either five or six instars with no homology between instars (the increases in cranium size differ in both groups). This may be due to setting the prothoracicotropic hormone (see Gilbert *et al.,* Chapter 2, this volume) release to different size increments, which may be a very simple and widespread mechanism underlying changes in the number of instars during development. Only exceptionally, in some taxa with a low and highly stabilized number of instars which can be reliably distinguished throughout the group, may it be possible to consider "instar homology" for deviated instar numbers. Štys and Davidová-Vilímová (1989) used the degree of wing pad development as a marker of the morphogenetic stage reached in a given instar, but use of such a single marker of course assumes that there were no heterochronic shifts of the development of the structure in question.

C. Retarded Wing Development and Resting Instars

Selective pressures against the presence of larval winglets, which caused evolution of the hemimetaboly from the supposed primary alate ametaboly, continued, leading to further suppression of larval wing development and/or to establishment of specialized inactive nonfeeding instars in some groups. There are two modes of such development, primarily driven by different evolutionary forces: neometaboly (in some Paraneoptera) and holometaboly (in Holometabola).

1. Neometaboly

The primitive development of modern Paraneoptera is six-instar hemimetaboly as seen in most Psocoptera. The general evolutionary trend within Paraneoptera is reduction of the number of instars and compression of wing development to later instars. External wing occurrence is—together with further shortening of development—sometimes delayed up to the last preadult instar as in some aphids or even up to the adult stage as in whiteflies. The delayed wing development may (as in whiteflies) or may not (as in aphids) be coupled with the origin of resting stages. On the contrary, thrips and male scale insects have a resting instar(s) but their wing development begins from the third instar similarly as in the typically hemimetabolous Heteroptera.

These modifications in life cycle and wing development are probably linked to a reduction in the number of larval instars (development of wings in a four-instar ontogeny, with at least the first instar wingless, can hardly be "gradual") and with remarkable shortening of development. Under such conditions, extreme allometry of wing growth certainly favored the evolution of resting nonfeeding instars in the mostly sedentary plant-sucking insects whose life style was a suitable preadaptation.

2. Holometaboly

a. Origin of Holometaboly. Since the great majority of known insect species belong to Holometabola, the appearance of holometabolous development must have provided some evolutionary advantages. The ancestral pterygotes are likely to have lived in the open, at least in their later ontogenetic stages. The origin of holometabolous development is usually envisioned as being connected to the adaptation of juveniles to life inside a substrate that provided protection as well as access to new food sources. Such habits, usually together with a high and unstable number of instars and relatively slow developmental rate, occur in some recent "primitive" holometabolan orders: Megaloptera (larvae aquatic, occur in mud, sand, or under stones), Raphidioptera (bark crevices, decaying wood, soil litter), Mecoptera (soil litter, moss, aquatic in mud; known groups, however, have a lower number of instars), and many subtaxa (often considered primitive) of other holometabolan orders.

Such larval life exerted a strong selection pressure for the streamlined "worm-like" flexible body shape. External wing rudiments, however modified, obtruded movement, and their appearance was postponed to the very end of development. As the difference between the larva and adult increased, the last larva gradually evolved into a nonfeeding exopterous instar with progressively reduced locomotory capabilities—the pupa (see later). A direct transformation of the larva lacking external wing rudiments into a flying adult would have probably required a prolonged period of inactivity anyway, as in the case of the whitefly "puparium", and the protected larval habitat may have permitted evolution of such a vulnerable instar. Freely exposed pupae are very rare even among recent Holometabola. The pupae of many primitive holometabolans (Neuropterida, Trichoptera, known Mecoptera) are relatively active (and, in particular, allow various activities of the pharate imago like crawling, biting, swimming, etc.) and are much more similar to the larva and adult than are most pupae of advanced orders such as Lepidoptera or Diptera. The pupal stage cannot be regarded as primarily "designed to ensure survival of the insect over periods of adverse seasonal changes" (Granger and Bollenbacher, 1981). Pupae of the primitive type with free legs and wing pads and usually poor sclerotization usually do not serve such purpose, and the protected pupae of Lepidoptera or many Diptera represent an advanced feature within Holometabola.

As a rule, typical holometabolous larvae are assumed to grow isometrically (larval instars similar to each other). First, it is not so since on closer examination many significant proportional differences exist between young and later instars. Second, exceptions are so numerous that the "rule" in fact does not exist. First instars, the function of which is often to disperse and/or find suitable food, may differ markedly from the following instars. In many sawflies (Smith and Middlekauff, 1987) or in some beetles (P. Švácha, unpublished observations) the last *larval* instar which is specialized for a winter diapause and/or finding a suitable pupation site does not feed, may be relatively inactive, and strongly differs morphologically from the preceding instars. But extreme morphological changes may occur at any stage of the development, and in some extremely hypermetamorphic parasitoids, virtually all larval instars may differ completely from each other (see Section III,B,4,d and Fig. 6). Such larval polymorphism is unequaled by any hemimetabolous species. Holometaboly is thus characterized by a large ecomorphological difference between the adult and *any* of the larval forms rather than by a morphological constancy of the larva.

The Paraneoptera and Holometabola reach similarly advanced stages during postembryogenesis [delayed and then accelerated wing development, extensive histolysis and histogenesis in metamorphosis, delayed development of gonads (see Heming, 1970)], and some paraneopterans are obviously more advanced than primitive holometabolans—compare, e.g., whiteflies (Weber, 1934) and alderflies (Geigy, 1937; Ochsé, 1944). However, their evolutionary histories may have been different: adaptations of larvae and adults to different life strategies in the primitive Holometabola with long and many-instar development versus shortening of development (including reduction of the number of instars) and subsequent "sessilization" connected with plant sucking in the derived groups of the Paraneoptera.

b. Evaluation of Theories on Origin of Pupa. The homology and evolutionary origin of the holometabolan pupal stage has been the subject of considerable dispute (Fig. 8). According to a theory first proposed by Berlese, holometaboly is a product of desembryonization; the larvae are supposedly equivalent to various embryonic stages of the hemimetabolous groups, and the pupa is viewed as a cumulation of the hemimetabolous larval ("nymphal"—this theory necessitates using the term "nymph" for the exopterous hemimetabolous immatures) instars. A more pronounced modification, but also based on "embryonization" and "cumulation of nymphal instars", is the theory of Heslop-Harrison (1958), who assumes existence of both externally wingless "larvae" and exopterous "nymphs" in the ancestral pterygotes; hemimetaboly is explained by embryonization of the larvae, holometaboly by the cumulation of the "nymphal" instars into the pupa.

Late development of a generalized primitive insect embryo proceeds through the following stages: *protopod* (only head and thoracic appendages present, abdomen slightly differentiated), *polypod* (abdominal segmentation and appendages well differentiated), and *oligopod* (abdominal

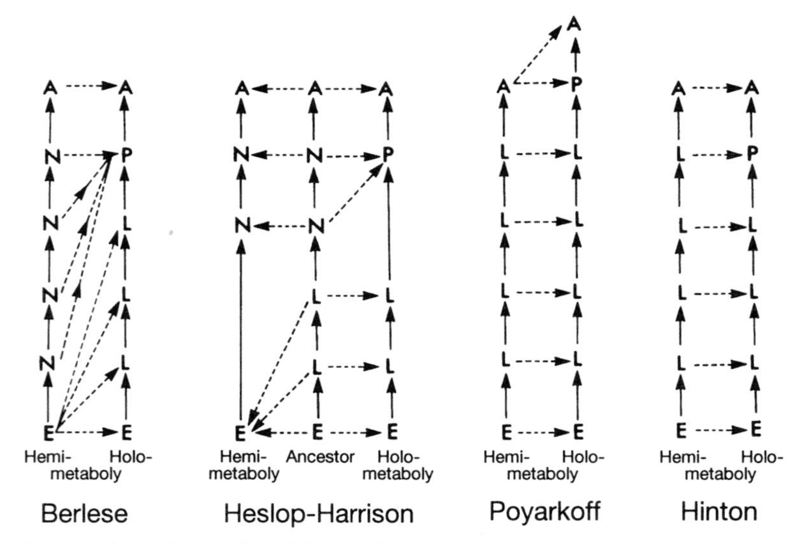

Fig. 8. The main theories of the relationships between hemi- and holometaboly and on the origin of the holometabolan pupa. E, egg; L, larva; N, nymph (an immature instar with external wing rudiments in the theories of Berlese and Heslop-Harrison); P, pupa; A, adult. Solid arrows, ontogeny; dashed arrows, changes leading to evolution of holometaboly from hemimetaboly (or to evolution of hemi- and holometaboly from the presumed ancestral situation in Heslop-Harrison's theory). For details see Section V,C,2,b. [Modified after Gillott (1980).]

appendages secondarily reduced). Larvae ("nymphs") of the hemimetabolous groups which have rudiments of "adult" organs (wings, genitalia) are considered to be in a *postoligopod* stage. This simplified scheme is the cornerstone of the Berlese theory which assumes that (i) the eggs of Holometabola are relatively smaller and contain less yolk reserves than those of the hemimetabolous groups; (ii) as a consequence, Holometabola leave the egg precociously, at one of the just named embryonic stages in the hemimetabolous groups; and (iii) the pupal instar comprises all the changes which in hemimetabolous groups occur during larval ("nymphal") development.

It is true that in comparison with some Crustacea possessing the nauplius larva, insect development is much more "embryonized," a feature connected with the terrestrial life. A reversal of this process would therefore be conceivable. However, there is no evidence that the first assumption of the Berlese theory is valid (Hinton, 1963). A great variability of size and yolk supply exists in the eggs of both holo- and hemimetabolous insects depending on their life strategy, and the degree of organization reached at hatching is not strictly and exclusively dependent on either egg size or yolk supply.

Concerning the second assumption, *oligopod* larvae are common (and perhaps primitive) in the Holometabola, and since one or several initial instars of the hemimetabolous groups also lack external wings and genitalia (and are thus effectively oligopod, not postoligopod, a fact obviously ignored by Berlese's and related theories), there is no basic difference between the holo- and hemimetabolous larvae except for the strongly retarded occurrence of the wings and genitalia in the former. In those Holometabola whose larvae possess imaginal discs in prefinal larval instars, even the larval growth of (invaginated) wings and genitalia has been more or less restored. There are also no differences in embryonic molts which exist both in holo- and hemimetabolous insects (Polivanova, 1979).

The presence of abdominal appendages in some holometabolous larvae was used in the Berlese theory as evidence of hatching at the *polypod* stage. This interpretation is incompatible with the following facts. (a) Fossil evidence indicates that abdominal leglets were present in many Paleozoic pterygotes. Polypod larvae thus might be primitive and not derived ("desembryonized"). Note that the earliest fossil interpreted as a holometabolous larva is depicted as having numerous abdominal leglets (Kukalová-Peck, 1991) (Fig. 5B). (b) Abdominal appendages in most, if not all, modern holometabolous larvae are, nevertheless, newly acquired structures. In contrast to Berlese's assumption, they undoubtedly evolved many times in parallel (for Mecopterida see Hinton, 1955) and many of them do not occur in the regions where there were legs in the polypod ancestors. This concerns Mecoptera as suggested by Snodgrass (1954), many Coleoptera [in the cerambycid beetle *Agapanthia* the first and second instar larvae even have their *dorsal* abdominal locomotory protuberances modified into distinct muscular prolegs (Duffy, 1951)], and the homology of many other larval abdominal appendages (including those of Megaloptera) to legs is very doubtful. Abdominal appendages which are not derived from the original legs cannot of course be used to support the Berlese theory. (c) Even in cases when the larval abdominal appendages are direct descendants of the embryonic abdominal appendages as in the larvae of sawflies (Ivanova-Kasas, 1959), their occurrence in larvae should be interpreted as a heterochronic extension of their existence. The overall degree of structural differentiation of sawfly larvae or caterpillars is incomparably higher than that of hemimetabolous polypod-stage embryos and is similar to what occurs in hemimetabolous larvae.

The *protopod* stage was suggested to occur in the first instar larvae of some hymenopterous parasites, e.g., in some Platygasteridae, but this suggestion is questionable because all larvae of Hymenoptera: Apocrita are legless and the first instar larvae of parasitic Hymenoptera are notorious for their various extreme adaptations, of which some are connected with reduced or indistinct external segmentation (see, e.g., Clausen, 1940). In addition, the larvae are not similar to any particular embryonic stage of the hemimetabolous insects, as discussed by Snodgrass (1954). The proposition of Novák (1991) that the first instar larvae of hypermetamorphic beetles are protopod is wrong because the first instar larvae in question are typical fully segmented oligopod larvae and are in many respects more "differentiated" structurally than the following instars (Fig. 6). Such cases demonstrate that the relative simplification of many holometabolous larvae is not the result of an overall "desembryonization" as required by the Berlese theory, but of organ-specific adaptation. Larval specializations present from the first instar naturally enforced late embryonic modifications (Snodgrass, 1954).

To sum up, both the first and second assumptions of the Berlese theory have no foundation, and this renders the third assumption (the origin of the holometabolan pupa by cumulation of hemimetabolous "nymphal" instars) also invalid. No support can be found for the Berlese theory, and this theory and its derivatives such as the gradient factor theory (Novák, 1991) should be abandoned.

The other two main theories, those of Poyarkoff (1914; originally supported also by Hinton, 1948) and Hinton (1963), represent a different concept by acknowledging equivalency between holo- and hemimetabolous larvae, but differ in the interpretation of the pupa.

Extensive reconstruction within a single instar is a complicated process. This led Poyarkoff to view the pupa as a necessary interstitial "mold" which allows proper development of adult musculature (particularly flight). This "mold" supposedly originated by an adult "duplication," i.e., by an insertion of an additional *imaginal* instar (Poyarkoff, 1914). We cannot agree. First, as demonstrated by the whiteflies, a metamorphosis from a wingless juvenile instar directly into a flying adult is possible, and no pupal "mold" is necessary. Hinton (1963) brought further evidence against the concept of the necessity of a "pupal mold." Second, we have argued earlier that the holometaboly evolved from the hemimetaboly via an extreme delay in wing development. It is hardly possible to assume a loss of external wings in *all* larval instars and a simultaneous "insertion" of a new adult-like exopterous instar. It is much simpler and equally effective to interpret the pupa as the end of *larval* development (see Section V,C,2,a), i.e., as a modified resting *larval* instar with retained external wing pads, whose primary role is to provide an intermediate step between the increasingly different larva and adult. This is consistent with relatively small differences between larva and pupa in primitive holometabolous orders, particularly in some Neuropterida. We therefore support the Hinton theory over the Poyarkoff theory, with the proviso that it is not possible to exactly homologize individual instars. There is no reason to view the pupa as a derivation of an imaginal instar.

Once established, the pupal instar made feasible the complete reconstruction of *any* organ, a possibility widely exploited by more advanced holometabolous groups where the degree of tissue rebuilding is often extreme. However, although *extensive* histolysis and histogenesis were facilitated by the origin of the resting instars in the Holometabola and neometabolous Paraneoptera, these phenomena are not necessarily associated either with the resting stages (for a description of muscle histolysis and histogenesis in metamorphosing dragonfly labium, see Munscheid, 1993), or with metamorphosis [cell death regularly and abundantly occurs in the insect epidermis during nonmetamorphic molting cycles (Wigglesworth, 1954)], or with molting [e.g., the cycles of degeneration and regeneration of flight muscles in some adult insects (Johnson, 1969)].

c. Imaginal Discs. In Holometabola, some organs are reduced or entirely suppressed in the larval stage. Imaginal discs (IDs; see Bayer *et al.*, Chapter 9, this volume) are described as invaginated regions of undifferentiated epidermis (with some other associated tissues such as tracheoles) which at metamorphosis evaginate and give rise to "adult" structures such as wings, genitalia, or legs of legless larvae or replace certain "larval" organs. Some aspects of the IDs have been reviewed (Švácha, 1992). We can summarize these as follows. (i) The epidermal portions of IDs are monolayers continuous with the externally exposed "larval" epidermis. (ii) ID cells produce cuticle in synchrony with the larval molt/intermolt cycle. Cuticle secretion by IDs was denied by most modern authors (see, e.g., Oberlander, 1985), who either ignored the cuticular lamella or called it an "extracellular matrix." (iii) No larval organs are "replaced" at metamorphosis, but corresponding larval and imaginal organs are sequentially homologous. The situation in larvae of Diptera: Cyclorrhapha (and to a lesser extent in larvae of some other derived Holometabola), where most of the "larval" epidermal cells become highly polyploid and degenerate at metamorphosis, is very specialized and exceptional, apparently enabling an extremely rapid larval development, but even here it is imprecise to speak about the replacement of larval organs by adult structures. For example, the "wing" disc in *Drosophila* contains not only the wing itself, but also certain other parts of the appropriate thoracic segment (Nöthiger, 1972; Oberlander, 1985; Williams *et al.,* 1993). Thus, for example, the dorsal part of the larval mesothorax does not represent the *whole* tergum, but its remaining part is hidden in the "wing" disc and at metamorphosis it supplies cells for the regions undergoing extensive cell death. (iv) Insect epidermis is normally composed of diploid, dividing cells. It is the polyploid "larval" epidermal cells (*not* the disc cells) that have deviated from the standard and represent an evolutionary novelty in the cyclorrhaphous Diptera or some Hymenoptera. (v) IDs must therefore be regarded as a normal part of the *larval* epidermis, albeit usually invaginated, and their cells cannot be called "undifferentiated" or "embryonic." IDs are equally "useless" to a holometabolan larva as are, for example, the wing pads or rudiments of genitalia to a hemimetabolous larva. The cells of these rudiments would then have to be also called undifferentiated, but this is equivalent to saying that the "right" function of a cell is the last one performed. Although functional cell reversals are rare due to the generally "unidirectional" nature of insect development, some exceptional cases demonstrate that the cellular changes may not be unidirectional and reversals are possible if evolutionarily advantageous. In a lepidopteran larva, Franzl *et al.* (1984) described cells which produce a lenticle, probably a secretory structure, at every *other* larval molt, and alternation occurs also between various cells, i.e., some produce a lenticle in the odd instars, others in the even ones. Finally, even if a cell becomes altered during ontogeny to a point of "no return," the situation is certainly *evolutionarily* reversible, e.g., by a heterochronic shift.

Although IDs are considered characteristic (if not indicative) of the larvae of Holometabola, hence the name "Endopterygota," wing IDs were not found in the penultimate larval instars of primitive holometabolous taxa [Megaloptera, Raphidioptera, Neuroptera, Mecoptera, Hymenoptera: Symphyta (Švácha, 1992; P. Švácha, unpublished observations)] and they are absent in many Coleoptera and even in some Diptera. Wing IDs thus do not belong to the Holometabola ground plan and must have evolved independently in various holometabolous lines. In many Holometabola, even the pupal wings develop as external outpocketings under the last larval cuticle, without undergoing any invaginated phase (see, e.g., Tower, 1903). In other words, holometaboly evolved from hemimetaboly through an extreme delay of external wing occurrence as suggested by Snodgrass (1954), not by an "inpocketing" of external wing rudiments as suggested, e.g., by Kukalová-Peck (1991), and the holometabolan juveniles were primarily apterous, not "endopterous." It was primarily not the diversification between the larval and adult habits and the associated morphological differentiation, i.e., the origin of holometaboly, which led to the development of typical IDs, but probably a selection for quick development. The origin of imaginal discs restored the possibility of wing growth during larval life *without* impairing the adaptive features of holometabolous larvae.

Summarizing the results of this and the preceding section, the primitive holometabolous development appears much less different from the hemimetaboly than is usually assumed from the highly derived holometabolan models. This means that some development traits, e.g., endocrinological, of the Holometabola may be rather primitive, and we may safely compare developmental regulation of holometabolous and hemimetabolous insects.

d. Stepsiptera: Holometabolans with Several Exopterous Juveniles? At least some males of the Stylopidia (the more advanced of the two strepsipteran suborders; females are neotenic and permanently endoparasitic, male puparia remain within the host and only adult males emerge and live freely) were described as having more than one preadult *exopterous* instar (Kinzelbach, 1967, and references therein) (Fig. 9). This raised doubts about the holometabolan nature of the Strepsiptera (see Section III,B,4,d) since there is always only one exopterous immature instar in the development of Holometabola, namely the pupa. These data require confirmation, and some of the supposed apolysed cuticles may represent what is usually called "ecdysial membranes." All the supposed exopterous "instars" remain pharate, the apolysed cuticles are not shed until the adult male emergence. No mention of more than one male pupal instar can be found in a special study of ecdyses of *Elenchus* (Kathirithamby *et al.,* 1984), and also in the work of Parker and Smith (1933, 1934), who studied a representative of the more primitive strepsipteran suborder Mengenillidia, where last instar larvae leave their hosts

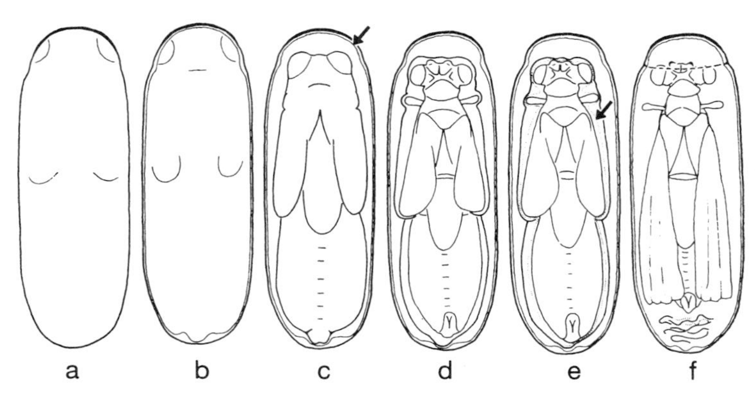

Fig. 9. Metamorphosis of male *Xenos vesparum* (Strepsiptera: Stylopidia) interpreted by Kinzelbach as an occurrence of several exopterous preadult instars: (a) last larva; (b) prepupa; (c) pupa; (d) subimago; (e) adult within the previous cuticles; and (f) adult after shedding the pupal and subimaginal cuticles (but still enclosed in the puparium). However, the events may be interpreted differently: (a) last instar larva; (b) pupal apolysis; (c) formation of the larval ecdysial membrane (arrow); (d) adult apolysis; (e) formation of the pupal ecdysial membrane (arrow); and (f) shedding of the pupal cuticle and its ecdysial membrane. Such an interpretation assumes only one exopterous preadult instar, the pupa (c), and is consistent with the supposed ecdysial membranes being apparently very fine, and also with the fact that the larval ecdysial membrane (prepupal cuticle of Kinzelbach) is not shed when the adult hatches from the pupa. [Reproduced with permission from Kinzelbach (1967).]

and both winged males and wingless females emerge from the puparia and live freely.

We therefore do not believe that male Stylopidia have several exopterous juveniles, but a possibility of such secondary reversal, which would be extremely interesting from the developmental point of view, cannot be excluded. It is well known that in some holometabolous larvae, various subnormal external conditions produce exopterous larval instars, the so-called prothetelic larvae which usually also have some other organs slightly changed in the pupal direction. Some beetle larvae show a great tendency to produce exopterous prothetelic larvae under just slightly suboptimal conditions, and very slightly prothetelic specimens can sometimes pupate and produce apparently normal adults (P. Švácha, unpublished observations). The possible supernumerary exopterous instars of the stylopidian males would thus represent a fixed evolutionary reversal, unique in the Holometabola and possibly connected with the entirely endoparasitic life and the host's hormonal milieu, since stylopidian males pupate within a *living* host.

D. Endocrinological Aspects

To understand the evolution of insect metamorphosis, one must (i) recognize developmental mechanisms restraining wing development during early postembryogenesis and promoting it later; (ii) explain coupling of "structural" metamorphosis to sexual maturation and the termination of molting; and (iii) account for the origin of enormous divergence in holometabolous larvae. Since at least the first two phenomena are controlled by hormones, the origin of metamorphosis can be sought in changes of hormone secretion or sensitivity (see Gilbert *et al.,* Chapter 2, this volume).

The first hormonal theory of the evolution of metamorphosis was by Novák (1956), who proposed that tissue and cellular morphogenesis is due to differential loss of a "gradient factor." During insect larval development this "factor" is replaced by JH. JH secretion allegedly ensues in hemimetabolans at the "postoligopod" developmental stage, while in Holometabola it starts at the "oligopod," "polypod," or even the "protopod" stages. The respective stage is then maintained throughout the larval period. Morphogenesis resumes as metamorphosis only when the rate of JH production does not suffice to replace the loss of the "gradient factor" in fully grown larvae. The theory is still being entertained (Novák, 1991) but its assumptions are not supported by recent experimental data. The claim that the appearance of JH in embryos causes hatching at different stages of morphogenesis is incompatible with the results of JH measurements in the eggs of diverse insects; other serious objections are presented in our treatment of the Berlese theory (see Section V,C,2,b). Similarly, the speculations of JH titer fluctuations in the last larval instar were not supported by JH quantification in several species.

We are unable to provide a comprehensive explanation for the origin of insect metamorphosis but are offering an idea that might be useful in future research. We envisage that insect metamorphosis could evolve in the following way. In the ametabolous Palaeozoic pterygotes, the gradual change from a nonflying hatchling, presumably with either minute or no wing rudiments, into a flying form certainly required that the wing primordia grew faster than most of the insect body. Only this differential growth rate presumably determined the body size at which the ability to fly was achieved. The growth was punctuated by regular molts controlled by ecdysteroids but JH was also present at this stage of evolution. It has been suggested that its original role may have been the regulation of reproduction (Novák, 1956; Sehnal, 1985), but the occurrence of JHs in late embryos of modern Zygentoma indicates that the postembryonic life of the ametabolous insect ancestors could have proceeded in the presence of JHs. JH may have provided additional control of the progress of the molt cycle by modulating the timing and level of ecdysteroid secretion. A similar interplay of JHs and ecdysteroids is seen in the adults (juveniles have not been examined) of modern Zygentoma (Fig. 7). Through this interplay, i.e., by influencing the length of time available for morphological changes during each molting cycle, JHs could exert certain effects on the rate of morphogenesis.

If we accept that ancestral insects contained fluctuating amounts of JH throughout their life, it is not hard to imagine that JHs became employed in slowing down the development of wings when environmental pressures favored their suppression in the immatures, in contrast to their full functionality in the adults. Acquisition of a direct morphogenetic function by JHs was probably a pivotal moment in the evolution of insect metamorphosis. Our assumption that this function was newly acquired in ancient pterygotes rests only on the lack of evidence that it is present in recent Zygentoma. By no means can we exclude the possibility that JHs and related compounds possessed some morphogenetic activity already in the apterygote insect ancestors, as indicated by the presence and action of epoxyfarnesoate in recent Crustacea. This possibility is also supported by the information on juvenoids, which are structurally extremely diversified and cause some effects that are unrelated to the known roles of JHs (Sláma *et al.,* 1974). For example, juvenoids (and exogenous JHs) inhibit insect embryogenesis and exert morphogenetic effects in animals in which JH has not been detected.

Once the development of wings in the immatures was suppressed by JH, primeval metamorphosis was born. The morphogenetic action of JHs was probably easily extended to other tissues and the retardation of imaginal development allowed the evolution of specific larval features required for adaptations to diverse niches. These features included the capacity to perform slightly different functions in larvae and adults [e.g., production of different cuticles by the epidermis (see Willis, Chapter 7, this volume)], as well as regional specializations for larval functions (e.g., gills in the larvae of Ephemeroptera, Odonata, and Plecoptera). In most insects, the presence of JH in larvae also delayed the differentiation of gametes. On the other hand, processes which did not interfere with larval adaptations, such as the gradual growth of appendages or compound eyes in hemimetabolans, were not subjected to JH control. Increasing diversity between larva and adult was probably first bridged over several instars in which JH could be absent only temporarily. The origin of a single metamorphic instar might have been linked to the evolution of a new mechanism for the suppression of JH production or rather for a new pattern of titer interactions between JH and ecdysteroids.

With a few exceptions (see Section IV,A), the glands producing ecdysteroids in recent pterygotes degenerate in the last larval instar in response to their own product, the ecdysteroids acting in absence of JH. Subjecting these glands to the control of JH was apparently an important event that linked metamorphosis to the cessation of molting (see Section V,A). The timing of gland degeneration to the last larval instar in most hemimetabolous and many holometabolous insects indicates that the just-mentioned linkage is an ancient phenomenon and was probably crucial for the transition from prometaboly to hemimetaboly. The loss of epidermal sensitivity to ecdysteroids in the flying forms seems to have been of minor importance.

Holometaboly is characterized by the absence of wing pads in larvae and by two metamorphic instars. The appearance of external wing rudiments was shifted to the pupa. Ecdysteroid secretion from source(s) other than the prothoracic glands was demonstrated in the pupae of several recent species (see Gilbert *et al.,* Chapter 2, this volume), but the significance of this fact for the evolution of metamorphosis can be assessed only when more data are obtained. The origin of the pupa, discussed in Section V,C,2,b, paved the way for further specialization of the larva. The secretory capacity of epidermis in some cases diversified enormously and the formation of imaginal discs (see Section V,C,2,c) allowed the specialization of some cells for functioning only in larvae. These "larval" cells often became polyploid, presumably to increase their functional output, but lost the capacity to divide and are thus usually destined to die at metamorphosis, while the "adult" cells retain the capacity for rapid proliferation and diversification.

The degeneration of prothoracic glands at metamorphosis required changes in the hormonal regulation of reproduction. The original pattern may be preserved in recent Zygentoma and is exemplified in the female firebrat (Fig. 7). The female mates (i.e., accepts the spermatophore;

Zygentoma do not physically copulate) shortly after each ecdysis when the spermathecal content is discarded with the old cuticle, and the presence of the spermatophore stimulates JH secretion via the neuroendocrine system (Rohdendorf and Watson, 1969). Circulating JH provokes vitellogenesis (Bitsch and Bitsch, 1984) that is linked to ecdysteroid production by the ovaries. Ovarian ecdysteroids are not released into the hemolymph but are deposited in the developing eggs (Rojo de la Paz *et al.,* 1983). Oviposition in the middle of the instar is associated with a decrease in JH titer coupled to stimulation of ecdysteroid production from the prothoracic (ventral) glands. The edysteroid titer in the hemolymph rises to a peak that causes molting but also the formation and previtellogenic development of new follicles in the ovaries (Bitsch and Bitsch, 1988). The new follicles are ready to respond to JH immediately after the next ecdysis.

In the evolution of pterygotes, hormonal conditions causing metamorphosis, i.e., high ecdysteroid and low JH titers, were also favorable for the stimulation of previtellogenesis that was probably completed before adult emergence. Metamorphic insects in which the adult stage was reduced to a single instar probably reproduced only once originally, like recent Ephemeroptera. Similarly as in most recent insects of diverse clades, the adults contained JH which stimulated vitellogenesis and egg production, and their ovaries generated ecdysteroids which were deposited into eggs. However, circulating ecdysteroids which would promote previtellogenesis were lacking. New sources of hemolymph ecdysteroids which were needed for this function in the adults may have evolved only after metamorphosis was established and prolonged reproduction within a single imaginal instar proved advantageous. Some insects developed a mechanism for ecdysteroid liberation from the ovaries while others began to use tissues employed in ecdysteroid production during pupal development. The inventory of ecdysteroid resources in the adults of recent pterygotes, which is still incomplete, will probably reveal considerable diversity.

VI. CONCLUSIONS

The preceding text combined morphological and endocrinological approaches in an attempt to formulate the simplest possible scenario of the origin of insect metamorphosis. We are aware that this scenario is more conservative than the usual approach. We hope to provoke arguments and stimulate research on phylogenetically important insect groups such as bristletails, silverfish, or mayflies. Crucial outcomes of our analysis of available data are the following.

(i) All insects, with some negligible and highly derived exceptions, hatch at a similar stage of ontogenic development; therefore, the origin of holometaboly cannot be sought in "desembryonization."

(ii) Insect metamorphosis is an evolutionary novelty associated with the evolution of wings and results from divergent adaptations of juveniles and adult stages. Metamorphosis may have evolved several times within the Pterygota.

(iii) Holometabola are characterized by the delay in the appearance of external wing rudiments until the last preadult instar, the pupa, which is viewed as a resting larva, comparable to the resting instars of thrips and some hemipterans. Wing rudiments usually develop as epidermal evaginations under the cuticle of the preceding, last larval, instar. The invaginated imaginal discs present in the larvae of *some* Holometabola are a secondary adaptation, allowing growth of wing primordia during larval development.

(iv) Originally, there may have been no tight link between structural metamorphosis, by which the insects acquire the ability to fly, and sexual maturation or obligate termination of molting.

(v) Suppression of adult molting in evolution may have been due to the specialized thoracic morphology of good fliers which is incompatible with ecdysis. The mayfly subimago represents a relic flying juvenile; however, the omissioin of the imago and the neotenic transfer of reproduction to the subimago in some mayflies suggests that such may have been the origin of hemimetaboly.

(vi) JHs and ecdysteroids coordinate structural metamorphosis with the attainment of sexual maturity and the cessation of molting. It is assumed that their roles in ancestral Pterygota were similar as in recent Zygentoma, and major changes allowing the evolution of metamorphosis included (a) acquisition of morphogenetic function by JHs in the juveniles; (b) degeneration of prothoracic glands in response to ecdysteroids in the absence of JH; and (c) diversification in the hormonal control of gametogenesis and vitellogenesis. The presumably original condition for metamorphosis, i.e., the drop in JH titer, has been secondarily lost in some advanced holometabolan groups.

(vii) Application of phylogenetic methodology to endocrinological research on recent insects can reveal relationships relevant to an understanding of the evolution of metamorphosis. Cladistic analysis will be possible only if presently available information is complemented by investigations on the hormonal control of development and reproduction in the following taxa: (a) Archaeognatha and/or Zygentoma that are the closest living ametabolous relatives of the metamorphic pterygote insects; (b) Ephemeroptera, whose development is unique in that structural metamorphosis occurs one molt before the final molt that produces the adult; (c) pterygotes closest to the supposed ground plans of the two distinct metamorphic schemes, hemimetaboly (e.g., Plecoptera) and holometaboly (e.g., Megaloptera); and (d) hypermetamorphic insects whose study should verify if the inhibition of postembryonic morphogenesis by JHs is limited to the differentiation of imaginal characters and does not affect sequential larval polymorphism. If so, the regulatory mechanisms of such polymorphism are of great interest.

REFERENCES

Akai, H., and Rembold, H. (1989). Juvenile hormone levels in *Bombyx* larvae and their impairment after treatment with an imidazole derivative KK-42. *Zool. Sci.* **6,** 615–618.

Arvy, L., and Gabe, M. (1953a). Données histo-physiologiques sur la neuro-secretion chez les Paleoptères (Ephemeroptères et Odonatès). *Z. Zellforsch.* **38,** 591–610.

Arvy, L., and Gabe, M. (1953b). Données histophysiologiques sur la neuro-secretion chez quelques Ephemeroptères. *Cellule* **55,** 203–222.

Baker, F. C., Lanzrein, B., Miller, C. A., Tsai, L. W., Jamieson, G. C., and Schooley, D. A. (1984). Detection of only JH III in several life-stages of *Nauphoeta cinerea* and *Thermobia domestica. Life Sci.* **35,** 1553–1560.

Beaulaton, J., Porcheron, P., Gras, R., and Cassier, P. (1984). Cytophysiological correlations between prothoracic gland activity and hemolymph ecdysteroid concentrations in *Rhodnius prolixus* during the fifth larval instar: Further studies in normal and decapitated larvae. *Gen. Comp. Endocrinol.* **53,** 1–16.

Besuchet, C. (1956). Biologie, morphologie et systématique des *Rhipidius* (Col. Rhipiphoridae). *Mitt. Schweiz. Entomol. Ges.* **29,** 73–144.

Bitsch, J. (1994). The morphological groundplan of Hexapoda: Critical review of recent concepts. *Ann. Soc. Entomol. Fr.* **30,** 103–129.

Bitsch, C., and Bitsch, J. (1984). Antigonadotropic and antiallatotropic effects of precocene II in the firebrat, *Thermobia domestica. J. Insect Physiol.* **30,** 462–470.

Bitsch, C., and Bitsch, J. (1988). 20-Hydroxyecdysone and ovarian maturation in the firebrat *Thermobia domestica* (Thysanura: Lepismatidae). *Arch. Insect Biochem. Physiol.* **7,** 281–294.

Bitsch, J., Rojo de la Paz, A., Mathelin, J., Delbecque, J. P., and Delachambre, J. (1979). Recherches sur les ecdysteroides hemolymphatiques et ovariens de *Thermobia domestica* (Insecta Thysanura). *C. R. Hebd. Séances Acad. Sci., Ser. D* **289,** 865–868.

Boudreaux, H. B. (1979). "Arthropod Phylogeny, with Special Reference to Insects." Wiley, New York.

Brodsky, A. K. (1994). "The Evolution of Insect Flight." Oxford University Press, Oxford.

Byers, G. W. (1991). Mecoptera. *In* "The Insects of Australia" (Commonwealth Scientific and Industrial Research Organization, ed.), Vol. 2, pp. 696–704. Melbourne University Press, Carlton, Australia.

Carmean, D., and Crespi, B. J. (1995). Do long branches attract flies? *Nature* (*London*) **373,** 666.

Carpenter, F. M. (1992). Hexapoda. *In* "Treatise on Invertebrate Paleontology" (R. L. Kaesler, ed.), Vols. 3–4, pp. 1–677. Geological Society of America, Boulder, Colorado/ University of Kansas, Lawrence.

Chang, E. S. (1993). Comparative endocrinology of molting and reproduction: Insects and crustaceans. *Annu. Rev. Entomol.* **38,** 161–180.

Chapman, R. F. (1969). "The Insects: Structure and Function." English Universities Press, London.

Clausen, C. P. (1940). "Entomophagous Insects." McGraw-Hill, New York.

Crowson, R. A. (1981). "The Biology of the Coleoptera." Academic Press, New York.

Cusson, M., Yagi, K. J., Ding, Q., Duve, H., Thorpe, A., McNeil, J. N., and Tobe, S. S. (1991). Biosynthesis and release of juvenile hormone and its precursors in insects and crustaceans: The search for a unifying arthropod endocrinology. *Insect Biochem.* **21,** 1–6.

Dai, J., and Gilbert, L. I. (1991). Metamorphosis of the corpus allatum and degeneration of the prothoracic glands during the larval–pupal–adult transformation of *Drosophila melanogaster:* A cytophysiological analysis of the ring gland. *Dev. Biol.* **144,** 309–326.

Davies, R. G. (1966). The postembryonic development of *Hemimerus vicinus* Rehn & Rehn (Dermaptera: Hemimeridae). *Proc. R. Entomol. Soc. London A* **41,** 67–77.

Delbecque, J. P., and Sláma, K. (1980). Ecdysteroid titres during autonomous metamorphosis in a dermestid beetle. *Z. Naturforsch., C* **35C,** 1066–1080.

Delbecque, J.-P., Weidner, K., and Hoffmann, K. H. (1990). Alternative sites for ecdysteroid production in insects. *Invertebr. Reprod. Dev.* **18,** 29–42.

Duffy, E. A. J. (1951). Dorsal prolegs and extreme cephalic modification in the first-instar larva of *Agapanthia villosoviridescens* Deg. (Col., Cerambycidae). *Entomol. Monthly Mag.* **87,** 313–318.

Edmunds, G. F., Jr., and McCafferty, W. P. (1988). The mayfly subimago. *Annu. Rev. Entomol.* **33,** 509–529.

Ferezou, J. P., Berreur-Bonnenfant, J., Tekitek, A., Rojas, M., Barbier, M., Suchy, M., Wipf, H. K., and Meusy, J. J. (1977). Biologically active lipids from the androgenic gland of the crab *Carcinus maenas. In* "Marine Natural Products Chemistry" (D. J. Faulkner and W. H. Fanical, eds.), pp. 361–366. Plenum, New York.

Ferris, G. F., and Usinger, R. L. (1939). The family Polyctenidae (Hemiptera; Heteroptera). *Microentomology* **4,** 1–50.

Forey, P. L., Humphries, C. J., Kitching, I. J., Scotland, R. W., Siebert, D. J., and Williams, D. M. (1992). "Cladistics: A Practical Course in Systematics." Clarendon, Oxford.

Franzl, S., Locke, M., and Huie, P. (1984). Lenticles: Innervated secretory structures that are expressed at every other larval moult. *Tissue Cell* **16,** 251–268.

Geigy, R. (1937). Beobachtungen über die Metamorphose von *Sialis lutaria* L. *Mitt. Schweiz. Entomol. Ges.* **17,** 144–157.

Gerstenlauer, B., and Hoffmann, K. H. (1995). Ecdysteroid biosynthesis and ecdysteroid titer during larval–adult development of the Mediterranean field cricket, *Gryllus bimaculatus* (Ensifera: Gryllidae) *Eur. J. Entomol.* **92,** 81–92.

Gilbert, C. (1994). Form and function of stemmata in larvae of holometabolous insects. *Annu. Rev. Entomol.* **39,** 323–349.

Gilbert, L. I. (1962). Maintenance of the prothoracic gland by the juvenile hormone in insects. *Nature (London)* **193,** 1205–1207.

Gillott, C. (1980). "Entomology." Plenum, New York.

Gnaspini, P. (1993). Brazilian Cholevidae (Coleoptera), with emphasis on cavernicolous species. III. *Dissochaetus* larvae, with description of a new feature. *Rev. Bras. Entomol.* **37,** 545–553.

Granger, N. A., and Bollenbacher, W. E. (1981). Hormonal control of insect metamorphosis. *In* "Metamorphosis. A Problem in Developmental Biology" (L. I. Gilbert and E. Frieden, eds.), pp. 105–137. Plenum, New York.

Gu, S.-H., and Chow, Y.-S. (1993). Role of low ecdysteroid levels in the early last instar of *Bombyx mori. Experientia* **49,** 806–809.

Hagedorn, H. H. (1989). Physiological roles of hemolymph ecdysteroids in the adult insects. *In* "Ecdysone: From Chemistry to Mode of Action" (J. Koolman, ed.), pp. 279–289. Thieme, Stuttgart.

Hardie, J., and Lees, A. D. (1985). Endocrine control of polymorphism and polyphenism. *In* "Comprehensive Insect Physiology, Biochemistry and Pharmacology" (G. A. Kerkut and L. I. Gilbert, eds.), Vol. 8, pp. 441–490. Pergamon, Oxford.

Heming, B. S. (1970). Postembryonic development of the female reproductive system in *Frankliniella fusca* (Thripidae) and *Haplothrips verbasci* (Phlaeothripidae) (Thysanoptera). *Misc. Publ. Entomol. Soc. Am.* **7,** 199–234.

Heslop-Harrison, G. (1958). On the origin and function of the pupal stadia in holometabolous Insecta. *Proc. Univ. Durham Philos. Soc., Ser. A* **13,** 59–79.

Highnam, K. C. (1981). A survey of invertebrate metamorphosis. *In* "Metamorphosis. A Problem in Developmental Biology" (L. I. Gilbert and E. Frieden, eds.), pp. 43–73. Plenum, New York.

Hinton, H. E. (1948). On the origin and function of the pupal stage. *Trans. R. Entomol. Soc. London* **99,** 395–409.

Hinton, H. E. (1955). On the structure, function and distribution of the prolegs of the Panorpoidea, with a criticism of the Berlese–Imms theory. *Trans. R. Entomol. Soc. London* **106,** 455–545.

Hinton, H. E. (1963). The origin and function of the pupal stage. *Proc. R. Entomol. Soc. London A* **38,** 77–85.

Imms, A. D. (1938). "A General Textbook of Entomology." 4th ed. Methuen, London.

Ivanova-Kasas, O. M. (1959). Die embryonale Entwicklung der Blattwespe *Pontonia* [sic] *capreae* L. (Hymenoptera, Tenthredinidae). *Zool. Jahrb., Abt. Anat. Ontog. Tiere* **77,** 193–228.

Jenkins, S. P., Brown, M. R., and Lea, A. O. (1992). Inactive prothoracic glands in larvae and pupae of *Aedes aegypti:* Ecdysteroid release by tissues in the thorax and abdomen. *Insect Biochem. Mol. Biol.* **22,** 553–559.

Johnson, C. G. (1969). "Migration and Dispersal of Insects by Flight." Methuen, London.

Kathirithamby, J., Spencer Smith, D., Lomas, M. B., and Luke, B. M. (1984). Apolysis without ecdysis in larval development of a strepsipteran, *Elenchus tenuicornis* (Kirby). *Zool. J. Linn. Soc.* **82,** 335–343.

Kingsolver, J. G., and Koehl, M. A. R. (1994). Selective factors in the evolution of insect wings. *Annu. Rev. Entomol.* **39,** 425–451.

Kinzelbach, R. (1967). Zur Kopfmorphologie der Fächerflügler (Strepsiptera, Insecta). *Zool. Jahrb., Abt. Anat. Ontog. Tiere* **84,** 559–684.

Kristensen, N. P. (1991). Phylogeny of extant hexapods. *In* "The Insects of Australia" (Commonwealth Scientific and Industrial Research Organization, ed.), Vol. 1, pp. 125–140. Melbourne University Press, Carlton, Australia.

Kukalová-Peck, J. (1991). Fossil history and evolution of hexapod structures. *In* "The Insects of Australia" (Commonwealth Scientific and Industrial Research Organization, ed.), Vol. 1, pp. 141–179. Melbourne University Press, Carlton, Australia.

Kukalová-Peck, J. (1992). The "Uniramia" do not exist: The ground plan of the Pterygota as revealed by Permian Diaphanopterodea from Russia (Insecta: Palaeodictyopteroidea). *Can. J. Zool.* **70,** 236–255.

Laufer, H., Borst, D., Baker, F. C., Carrasco, C., Sinkus, M., Reuter, C. C., Tsai, L. W., and Schooley, D. A. (1987). Identification of a juvenile hormone-like compounds in a crustacean. *Science* **235,** 202–205.

Maiorana, V. C. (1979). Why do adult insects not moult? *Biol. J. Linn. Soc.* **11,** 253–258.

Malá, J., Sehnal, F., and Laufer, H. (1983). Development of *Chironomus thummi* (Diptera) treated with highly active juvenoids. *Acta Entomol. Bohemoslov.* **80,** 1–12.

Mauchamp, B., Lafont, R., Pennetier, J.-L., and Doumas, J. (1981). Detection and quantification of the juvenile hormone I during the post-embryonic development of *Pieris brassicae. In* "Regulation of Insect Development and Behavior" (F. Sehnal, A. Zabża, J. J. Menn, and B. Cymborowski, eds.), pp. 199–206. Technical University Press, Wrocław, Poland.

Melzer, R. R., and Paulus, H. F. (1991). Morphology of the visual system of *Chaoborus crystallinus* (Diptera, Chaoboridae). 1. Larval compound eyes and stemmata. *Zoomorphology* **110,** 227–238.

Nayar, K. K. (1954). Metamorphosis in the integument of caterpillars with omission of the pupal stage. *Proc. R. Entomol. Soc. London A* **29,** 129–134.

Neunzig, H. H., and Baker, J. R. (1991). Order Megaloptera. *In* "Immature Insects" (F. W. Stehr, ed.), Vol. 2, pp. 112–122. Kendall/Hunt, Dubuque, Iowa.

Nöthiger, R. (1972). The larval development of imaginal disks. *In* "The Biology of Imaginal Disks" (H. Ursprung and R. Nöthiger, eds.), pp. 1–34. Springer-Verlag, Berlin.

Novák, V. J. A. (1956). Versuch einer zusammenfassender Darstellung der postembryonalen Entwicklung der Insekten. *Beitr. Entomol.* **6,** 205–493.

Novák, V. J. A. (1966). "Insect Hormones." Methuen, London.

Novák, V. J. A. (1991). Role of the gradient factor in arthropod morphogenesis. *In* "Morphogenetic Hormones of Arthropods: Roles in Histogenesis, Organogenesis and Morphogenesis" (A. P. Gupta, ed.), pp. 3–43. Rutgers University Press, New Brunswick, New Jersey.

Nüesch, H. (1987). Metamorphose bei Insekten: Direkte und indirekte Entwicklung bei Apterygoten und Exopterygoten. *Zool. Jahrb., Abt. Anat. Ontog. Tiere* **115,** 453–487.

Oberlander, H. (1985). The imaginal discs. *In* "Comprehensive Insect Physiology, Biochemistry and Pharmacology" (G. A. Kerkut and L. I. Gilbert, eds.), Vol. 2, pp. 151–182. Pergamon, New York.

Ochsé, W. (1944). Experimentelle und histologische Beiträge zur innern Metamorphose von *Sialis lutaria* L. *Rev. Suisse Zool.* **51,** 1–82.

Parker, H. L., and Smith, H. D. (1933). Additional notes on the strepsipteron *Eoxenos laboulbenei* Peyerimhoff. *Ann. Entomol. Soc. Am.* **26,** 217–233.

Parker, H. L., and Smith, H. D. (1934). Further notes on *Eoxenos laboulbenei* Peyerimhoff with a description of the male. *Ann. Entomol. Soc. Am.* **27,** 468–479.

Pflugfelder, O. (1937). Bau, Entwicklung und Funktion der Corpora allata und cardiaca von *Dixippus morosus. Z. Wiss. Zool.* **149,** 477–512.

Pflugfelder, O. (1952). "Entwicklungsphysiologie der Insekten." Geest & Portig, Leipzig, Germany.

Piepho, H. (1951). Über die Lenkung der Insektenmetamorphose durch Hormone. *Verh. Dtsch. Zool. Ges.* **7b,** 62–75.

Polivanova, E. N. (1979). Embryonization of ontogenesis, origin of embryonic moults and types of development in insects. *Zool. Zh.* **58,** 1269–1280 (in Russian, English abstract).

Poyarkoff, E. (1914). Essai d'une théorie de la nymphe des Insectes Holométaboles. *Arch. Zool. Exp. Gén.* **54,** 221–265.

Pritchard, G., McKee, M. H., Pike, E. M., Scrimgeour, G. J., and Zloty, J. (1993). Did the first insects live in water or in air? *Biol. J. Linn. Soc.* **49,** 31–44.

Raabe, M. (1986). Insect reproduction: Regulation of successive steps. *Adv. Insect Physiol.* **19,** 30–151.

Rembold, H., Czoppelt, C., Grüne, M., Lackner, B., Pfeffer, J., and Woker, E. (1992). Juvenile hormone titers during honey bee embryogenesis and metamorphosis. *In* "Insect Juvenile Hormone Research" (B. Mauchamp, F. Couillard, and J. C. Baehr, eds.), pp. 37–43. Institut National de la Recherche Agronomique, Paris.

Riddiford, L. M. (1985). Hormone action at the cellular level. *In* "Comprehensive Insect Physiology, Biochemistry and Pharmacology" (G. A. Kerkut and L. I. Gilbert, eds.), Vol. 8, pp. 37–84. Pergamon, Oxford.

Rockstein, M. (1956). Metamorphosis: A physiological interpretation. *Science* **123,** 534–536.

Rohdendorf, B. B., and Rasnitsyn, A. P. (1980). "Istoricheskoe Razvitie Klassa Nasekomykh" ("Evolution of the Class Insecta"). Nauka, Moscow. (In Russian.)

Rohdendorf, E. B., and Watson, J. A. L. (1969). The control of reproductive cycles in the female firebrat, *Lepismodes inquilinus. J. Insect Physiol.* **15,** 2085–2101.

Rojo de la Paz, A., Delbecque, J. P., Bitsch, J., and Delachambre, J. (1983). Ecdysteroids in the haemolymph and the ovaries of the firebrat, *Thermobia domestica:* Correlations with integumental and ovarian cycles. *J. Insect Physiol.* **29,** 323–329.

Ross, E. S. (1970). Embioptera. *In* "Taxonomist's Glossary of Genitalia in Insects" (S. L. Tuxen, ed.), pp. 72–75. Munksgaard, Copenhagen.

Schaller, F. (1952). Effets d'une ligature postcephalique sur le developpement de larves agees d'*Apis mellifica. Bull. Soc. Zool. Fr.* **77,** 195–204.

Sehnal, F. (1968). Influence of the corpus allatum on the development of internal organs in *Galleria mellonella* L. *J. Insect Physiol.* **14,** 73–85.

Sehnal, F. (1972). Action of ecdysone on ligated larvae of *Galleria mellonella* L. (Lepidoptera): Induction of development. *Acta Entomol. Bohemoslov.* **69,** 143–155.

Sehnal, F. (1980). Action of juvenile hormone on tissue and cell differentiation. *In* "Regulation of Insect Development and Behaviour" (F. Sehnal, A. Zabża, J. J. Menn, and B. Cymborowski, eds.), pp. 463–482. Technical University Press, Wrocław, Poland.

Sehnal, F. (1983). Juvenile hormone analogues. *In* "Endocrinology of Insects" (R. G. H. Downer and H. Laufer, eds.), pp. 657–672. Liss, New York.

Sehnal, F. (1985). Growth and life cycles. *In* "Comprehensive Insect Physiology, Biochemistry and Pharmacology" (G. A. Kerkut and L. I. Gilbert, eds.), Vol. 2, pp. 1–86. Pergamon, Oxford.

Sehnal, F., Delbecque, J.-P., Maróy, P., and Malá, J. (1986). Ecdysteroid titres during larval life and metamorphosis of *Galleria mellonella. Insect Biochem.* **16,** 157–162.
Selander, R. B. (1991). Order Coleoptera: Family Meloidae. *In* "Immature Insects" (F. W. Stehr, ed.), Vol. 2, pp. 530–534. Kendall/Hunt, Dubuque, Iowa.
Selman, K., and Kafatos, F. C. (1974). Transdifferentiation in the labial gland of silk moths: Is DNA required for cellular metamorphosis? *Cell Differ.* **3,** 81–94.
Sláma, K. (1975). Some old concepts and new findings on hormonal control of insect morphogenesis. *J. Insect Physiol.* **21,** 921–955.
Sláma, K., Romaňuk, M., and Šorm, F. (1974). "Insect Hormones and Bioanalogues." Springer-Verlag, Vienna.
Sliter, T. J., Sedlak, B. J., Baker, F. C., and Schooley, D. A. (1987). Juvenile hormone in *Drosophila melanogaster:* Identification and titer determination during development. *Insect Biochem.* **17,** 161–165.
Smith, S. L. (1985). Regulation of ecdysteroid titer: Synthesis. *In* "Comprehensive Insect Physiology, Biochemistry and Pharmacology" (G. A. Kerkut and L. I. Gilbert, eds.), Vol. 7, pp. 295–341. Pergamon, Oxford.
Smith, D. R., and Middlekauf, W. W. (1987). Order Hymenoptera: Suborder Symphyta. *In* "Immature Insects" (F. W. Stehr, ed.), Vol. 1, pp. 618–649. Kendall/Hunt, Dubuque, Iowa.
Snodgrass, R. E. (1954). Insect metamorphosis. *Smithson. Misc. Collect.* **122,** 1–124.
Soldán, T. (1979a). The structure and development of the male internal reproductive organs in six European species of Ephemeroptera. *Acta Entomol. Bohemoslov.* **76,** 22–23.
Soldán, T. (1979b). The structure and development of the female internal reproductive system in six European species of Ephemeroptera. *Acta Entomol. Bohemoslov.* **76,** 353–365.
Spindler, K.-D. (1989). Hormonal role of ecdysteroids in Crustacea, Chelicerata and other arthropods. *In* "Ecdysone: From Chemistry to Mode of Action" (J. Koolman, ed.), pp. 290–295. Thieme, Stuttgart.
Steel, C. G. H., and Vafopoulou, X. (1989). Ecdysteroid titer profile during growth and development of arthropods. *In* "Ecdysone: From Chemistry to Mode of Action" (J. Koolman, ed.), pp. 221–231. Thieme, Stuttgart.
Steel, C. G. H., Bollenbacher, W. E., Smith, S. L., and Gilbert, L. I. (1982). Haemolymph ecdysteroid titres during larval–adult development in *Rhodnius prolixus:* Correlations with moulting hormone action and brain neurosecretory cell activity. *J. Insect Physiol.* **28,** 519–525.
Štys, P., and Davidová-Vilímová, J. (1989). Unusual numbers of instars in Heteroptera: A review. *Acta Entomol. Bohemoslov.* **86,** 1–32.
Štys, P., and Zrzavý, J. (1994). Phylogeny and classification of extant Arthropoda: Review of hypotheses and nomenclature. *Eur. J. Entomol.* **91,** 257–275.
Švácha, P. (1992). What are and what are not imaginal discs: Reevaluation of some basic concepts (Insecta, Holometabola). *Dev. Biol.* **154,** 101–117.
Szibbo, C. M., Rotin, D., Feyereisen, R., and Tobe, S. S. (1982). Synthesis and degradation of C16 juvenile hormone (JH III) during the final two stadia of the cockroach, *Diploptera punctata. Gen. Comp. Endocrinol.* **48,** 25–32.
Tanaka, A. (1981). Regulation of body size during larval development in the German cockroach, *Blattella germanica. J. Insect Physiol.* **27,** 587–592.
Tauber, C. A. (1991). Order Raphidioptera. *In* "Immature Insects" (F. W. Stehr, ed.), Vol. 2, pp. 123–125. Kendall/Hunt, Dubuque, Iowa.

Tower, W. L. (1903). The origin and development of the wings of Coleoptera. *Zool. Jahrb., Abt. Anat. Ontog. Tiere* **17,** 517–572.

Tsuchida, K., Nagata, M., and Suzuki, A. (1987). Hormonal control of ovarian development in the silkworm, *Bombyx mori. Arch. Insect Biochem. Physiol.* **5,** 167–177.

Watson, J. A. L. (1963). The cephalic endocrine system in the Thysanura. *J. Morphol.* **113,** 359–373.

Watson, J. A. L. (1967). The growth and activity of the corpora allata in the larval firebrat, *Thermobia domestica* (Packard) (Thysanura, Lepismatidae). *Biol. Bull.* (Woods Hole, Mass.) **132,** 277–291.

Weber, H. (1934). Die postembryonale Entwicklung der Aleurodinen (Hemiptera–Homoptera). Ein Beitrag zur Kenntnis der Metamorphosen der Insekten. *Z. Morphol. Oekol. Tiere* **29,** 268–305.

Weeks, J. C., and Levine, R. B. (1990). Postembryonic neuronal plasticity and its hormonal control during insect metamorphosis. *Annu. Rev. Neurosci.* **13,** 183–194.

Whiting, M. F., and Wheeler, W. C. (1994). Insect homeotic transformation. *Nature* (London) **368,** 6.

Wigglesworth, V. B. (1952). Hormone balance and the control of metamorphosis in *Rhodnius. J. Exp. Biol.* **29,** 620–631.

Wigglesworth, V. B. (1954). "The Physiology of Insect Metamorphosis." Cambridge University Press, Cambridge, England.

Wigglesworth, V. B. (1955). The breakdown of the thoracic gland in the adult insect *Rhodnius prolixus. J. Exp. Biol.* **32,** 485–491.

Williams, J. A., Paddock, S. W., and Carroll, S. B. (1993). Pattern formation in a secondary field: A hierarchy of regulatory genes subdivides the developing *Drosophila* wing disc into discrete subregions. *Development* **117,** 571–584.

Willis, J. H., Rezaur, R., and Sehnal, F. (1982). Juvenoids cause some insects to form composite cuticles. *J. Embryol. Exp. Morphol.* **71,** 25–40.

Willmer, P. G. (1990). "Invertebrate Relationships: Patterns in Animal Evolution." Cambridge University Press, Cambridge, England.

Wootton, R. J. (1992). Functional morphology of insect wings. *Annu. Rev. Entomol.* **37,** 113–140.

2

Endocrine Cascade in Insect Metamorphosis

LAWRENCE I. GILBERT,* ROBERT RYBCZYNSKI* AND STEPHEN S. TOBE†

* Department of Biology
University of North Carolina at Chapel Hill
Chapel Hill, North Carolina

† Department of Zoology
Ramsay Wright Zoological Laboratories
University of Toronto, Toronto, Ontario, Canada M5S 1A1

I. INTRODUCTION

The apparently abrupt morphological transitions typifying insect metamorphosis have bemused, bewildered, and stimulated the intellectual curiosity of laymen, authors, philosophers, and scientists dating from the early written records of civilization (e.g., Aristotle in the golden days of Greek civilization, the domestication of the commercial silkworm, *Bombyx mori,* in China thousands of years B.C.). It is now known that these seemingly abrupt morphological transitions as viewed externally are in reality smooth continua of precisely regulated events (see Sehnal *et al.,* Chapter 1, this volume) controlled by titers of several hormones, the endocrine sources of which have been investigated in great detail since the 1930s.

The original concept of an endocrine axis consisting of the brain (source of prothoracicotropic hormone, PTTH), prothoracic glands (source of ecdysteroids), and the corpus allatum (source of juvenile hormone, JH) remains intact (Fig. 1), although studies have revealed that there are extraprothoracic gland sources of ecdysteroids; that ecdysone is not the major product of lepidopteran prothoracic glands; that there are several JHs; and that the corpus allatum at times secretes a JH precursor rather than JH *per se.*

Specific neurosecretory cells in the insect brain synthesize a neuropeptide, PTTH, that is transported to the corpora allata in the Lepidoptera, the latter glands also acting as neurohemal sites (Agui *et al.,* 1980). Once released into the hemolymph as a result of environmental cues (e.g., photoperiod, temperature), PTTH acts on the prothoracic glands to stimulate ecdysteroid synthesis (see Gilbert, 1989), and 3-dehydroecdysone is released into the hemolymph where it is reduced by a ketoreductase to ecdysone (Warren *et al.,* 1988a,b; Sakurai *et al.,* 1989), which itself has the dual role of hormone and prohormone. The ecdysone is converted to the principal molting hormone, 20-hydroxyecdysone, by target tissue ecdysone 20-monooxygenases (Smith *et al.,* 1983; Smith, 1985). How 20-hydroxyecdysone (and ecdysone) exerts its effects at the molecular level will be discussed in detail in the other chapters in this section on insects.

In addition to sequestering PTTH, the corpora allata synthesize and secrete JH (Fig. 1), and the timing and quantity of JH at target cells lead to the modulation of ecdysteroid action in the sense that JH can influence the quality of the molt, i.e., larval–larval, larval–pupal, and

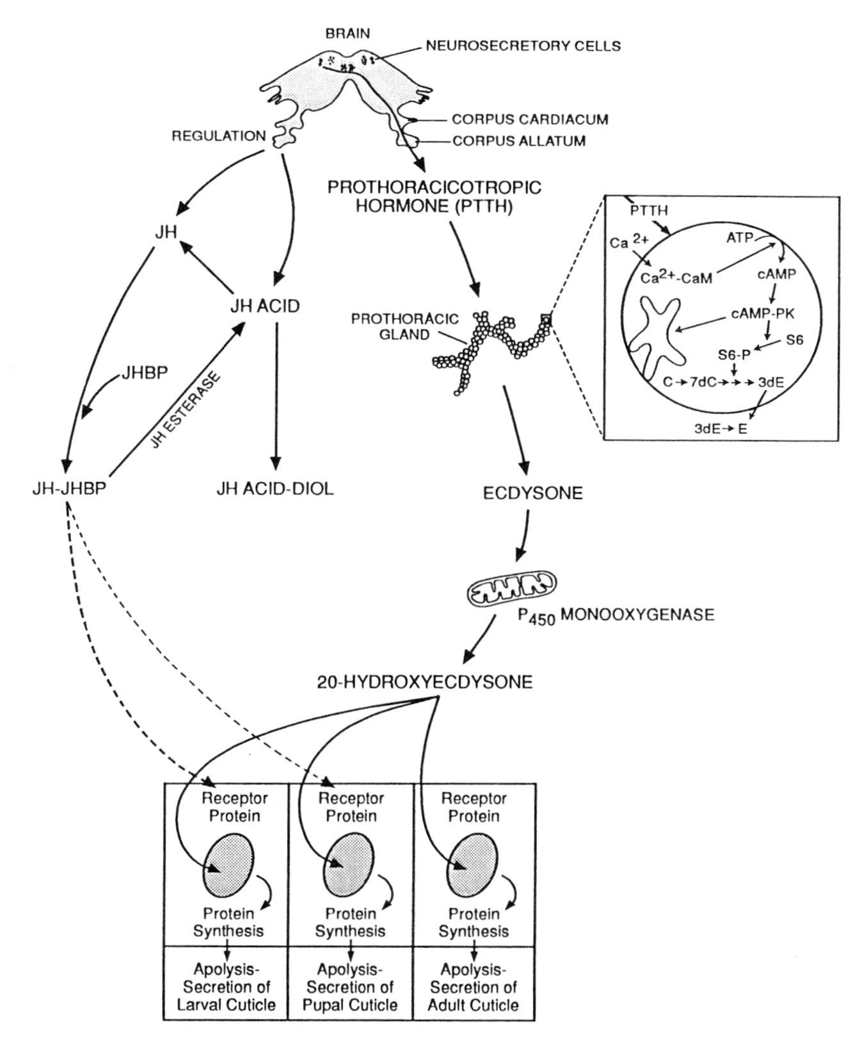

Fig. 1. Endocrine control of metamorphosis. Most of the data contributing to this scheme were derived from studies on *Manduca sexta,* although the scheme applies to all insects in a general sense. Note that in the case of *Manduca,* JH acid rather than JH is released from the corpus allatum toward the end of the last larval stage. C, cholesterol; 7dC, 7-dehydrocholesterol; 3dE, 3-dehydroecdysone; E, ecdysone; BP, binding protein; CaM, calmodulin.

pupal–adult (see Riddiford, Chapter 6, this volume; Willis, Chapter 7, this volume). Thus, metamorphosis is regulated by ecdysteroids initiating the molting process (the producer) and the titer of JH determining the result of the molt (the director).

Although so many of the accepted concepts in insect endocrinology are derived from the exquisite studies of the late V. B. Wigglesworth more than half a century ago on the blood-sucking bug *Rhodnius prolixius,* that model insect must share the glory with silkworms (*Bombyx mori, Hyalophora cecropia*) and more recently the sphingid, *Manduca sexta.* Studies on the Lepidoptera by Bounhiol, Fukuda, Joly, Piepho, Williams, and others using microsurgical techniques in the mid-1940s have set the stage for present-day investigations at the cellular, biochemical, and molecular levels. Indeed, *M. sexta* may be the best-studied insect endocrinological model since the 1970s, but with all its attributes of large size, relatively short life cycle, ability to be timed precisely, etc., little is known of its genetics. Therefore, in recent years an increasing number of investigators have turned to *Drosophila melanogaster* about whose genetics we know more than perhaps any other higher organism. As will be seen in succeeding chapters we have emphasized *Manduca* and the flies *Drosophila* and *Chironomus* (both species having polytene chromosomes). Therefore, this chapter deals mainly with studies on these insects or their relatives, e.g., *Bombyx mori.* This choice should not in any way detract from the classic studies conducted by giants in this field on orthopterans, hemipterans, coleopterans, etc.

It is our purpose to review the concepts of how hormones control insect metamorphosis and supply the basic information required for the succeeding chapters, most of which have a molecular bent. It is exciting to ponder, and indeed investigate, how a hormone–receptor complex interacts with the genome to produce specific gene products, the composite of which interplay and modulate growth and development. However, it is also important to know something of the chemistry of the hormones involved and their physiological effects, as well as how they are degraded, transported, enter target cells, etc. For a more detailed analysis of these latter aspects of insect endocrinology the reader is referred to the work of Gilbert and Frieden (1981).

One of the most exciting advances in insect endocrinology has been the characterization of many peptides, more than 50 at this writing, the roles for several having been elucidated. For the most part, they appear to be involved in homeostatic mechanisms, e.g., lipid and carbohydrate metabolism. These peptides likely play sensitive regulatory roles in modulating metabolic processes, and therefore indirectly do impinge on growth, development, and metamorphosis. However, because of lack of space they will not be considered in this volume which is basically centered on PTTH, ecdysteroids, and JH. Finally, one of the lessons learned over the years is how little hormones and transductory mechanisms have changed during evolution (see Gilbert *et al.,* 1988; Lafont, 1992). "In the course of evolution not very many new small molecules have appeared since the Cambrian era, and hormonal mechanisms have evolved by particular groups of animals adapting available and often ubiquitous molecules to special tasks. Under this view, the evolution of hormonal coordination involves primarily the evolution of target mechanisms with especial sensitivity to specific small molecules which may have been present from earliest times" (Schneiderman and Gilbert, 1959).

II. PROTHORACICOTROPIC HORMONE

A. Introduction

In the early 1920s, Kopeć's experiments on the gypsy moth revealed that the brain plays a central role in coordinating insect growth, molting, and metamorphosis, thus demonstrating that the nervous system is an endocrine organ, and laying the foundations for the development of the field of neuroendocrinology. Using simple techniques such as ligation of larvae into anterior and posterior sections and brain extirpation, as well as excellent deductive reasoning, he showed that the larval–pupal molt of *Lymantria dispar* was controlled by the brain during a critical period in the last instar larval stage and that this control was not dependent on intact nervous connections to regions posterior to the brain (Kopeć, 1922). This work was subsequently repeated and expanded on and culminated in the identification of specific brain neurosecretory cells as a source of a molt-eliciting neuropeptide, prothoracicotropic hormone (Agui *et al.,* 1979). PTTH coordinates molting indirectly by regulating the synthesis of ecdysteroids by the prothoracic gland (Fig. 1) or its homologs, e.g., the *Drosophila* ring gland (Fig. 6).

B. Characterization of Prothoracicotropic Hormone

Kopeć's studies, demonstrating a brain-derived, circulating factor that controlled insect development, preceded the isolation of PTTH by about

70 years. Following labor-intensive efforts by Professors Ichikawa and Ishizaki that spanned several decades and utilized millions of silkworm (*Bombyx*) brains, the *Bombyx* PTTH was purified and characterized. Oligonucleotide probes based on the nearly complete amino acid sequence were used to obtain a cDNA clone (Kawakami *et al.,* 1990). The deduced amino acid sequence of the cDNA (109 residues total) matched exactly the 104 amino acids obtained from the purified, sequenced PTTH and provided the identity of the last 5 amino acids, which were not obtained from the peptide sequencing (Fig. 2). A genomic clone for *Bombyx* PTTH was also obtained (Adachi- Yamada *et al.,* 1994); the gene spans ≈ 3 kb, contains five exons separated by four introns, and is present at one copy per haploid genome. Both the cDNA and genomic clone sequences provide the basis for the prediction that PTTH is synthesized as a precursor protein of 224 amino acids (Fig. 2A), starting with a putative signal sequence of 28 amino acids and terminating with a 109 amino acid sequence that is the active PTTH subunit. The putative signal sequence and the carboxy-terminal, active PTTH subunit are separated by two peptide fragments which are delimited by basic amino acid proteolytic cleavage sites (Kawakami *et al.,* 1990). These two peptides (≈ 2 and 6 kDa) are without demonstrated function, although Ishizaki and Suzuki (1992) suggested that they could modulate JH synthesis in the cells of the corpus allatum. The mature, secreted PTTH appears to be a homodimer (monomer ≈ 16.5 kDa) containing inter- and intramonomer disulfide bonds (Fig. 2A), the latter being necessary for retention of prothoracicotropic activity (Ishibashi *et al.,* 1994). The monomer also contains a potential glycosylation site but studies with PTTH expressed in *Escherichia coli* indicate that glycosylation may not be necessary for biological activity (Kawakami *et al.,* 1990).

Preparations containing PTTH activity have been obtained from a number of other insects, including *Drosophila* (Henrich *et al.,* 1987), and especially *M. sexta* (see Bollenbacher and Granger, 1985). In *Manduca* as in *Bombyx,* partially purified PTTH activity exhibits an apparent molecular mass of 25–30 kDa (Bollenbacher *et al.,* 1984) which may be dissociable into two 16.5-kDa subunits under reducing condition (Muehleisen *et al.,* 1994). Partial amino acid sequence from this reduced, putative subunit indicated no similarity with *Bombyx* PTTH but instead showed a rather surprising possible relationship to vertebrate retinol-binding proteins (Muehleisen *et al.,* 1993). At present it is unclear whether PTTHs will ultimately prove to be a group of dissimilar proteins with the same function or a family of related but nonidentical polypep-

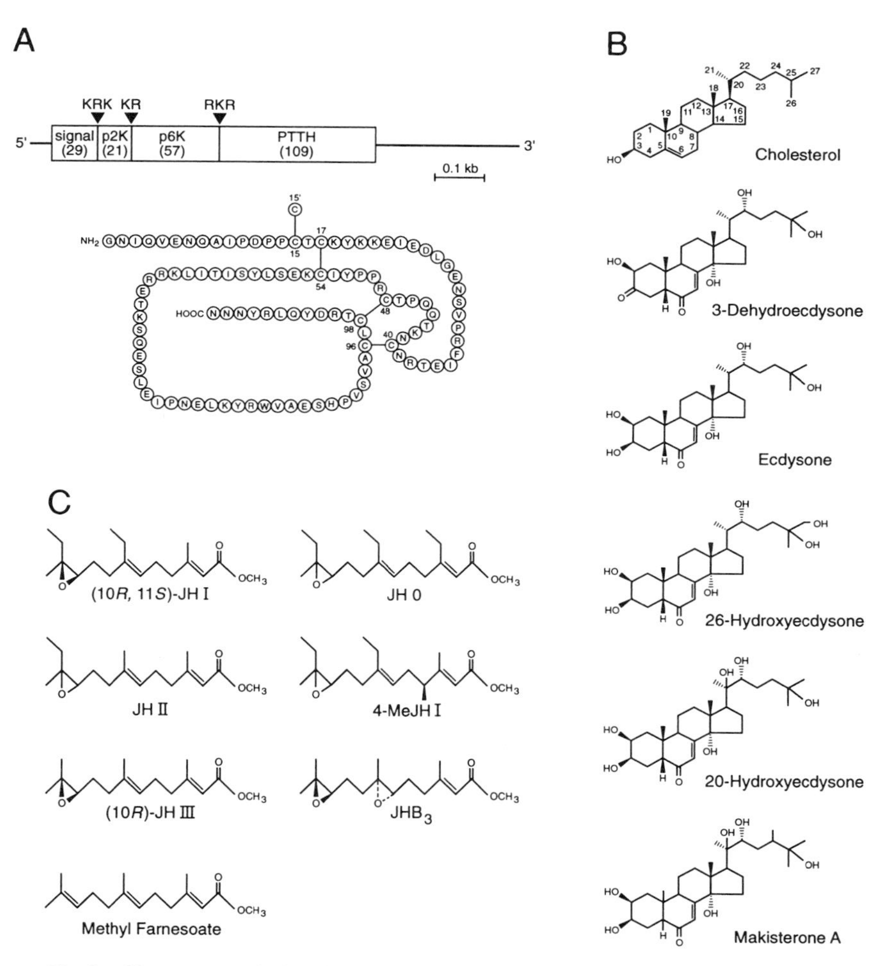

Fig. 2. Hormones and related molecules that play critical roles in the control of molting and metamorphosis. (A) The structure of *Bombyx* big PTTH. The upper diagram indicates the predicted organization of the initial translation product. [Reprinted with permission from Kawakami *et al.* Copyright (1990) American Association for the Advancement of Science.] The lower diagram shows the location of inter- and intracellular disulfide bonds. [Reprinted with permission from Ishibashi *et al.* Copyright (1994) American Chemical Society.] (B) The structure of cholesterol and some major ecdysteroids. (C) The structure of the various juvenile hormones and methyl farnesoate. JH I and JH II are almost entirely restricted to the Lepidoptera, JHB_3 to the cyclorraphan Diptera while JH III is ubiquitous in insects.

tides, some of which might have very different functions, e.g., *Bombyx* PTTH can be extracted from adult brains, a stage devoid of prothoracic glands (see Gilbert *et al.,* 1981).

C. Synthesis and Release of Prothoracicotropic Hormone

The sites of PTTH synthesis and release differ in both *Manduca* and *Bombyx.* PTTH is synthesized primarily by four large neurosecretory cells (the prothoracicotropes), two per hemisphere, whose soma reside in the dorsolateral region of the brain (Agui *et al.,* 1979; Mizoguchi *et al.,* 1990). PTTH is then transported down the axons of these neurons, through the corpus cardiacum, to their termination in the corpus allatum; the corpus allatum is also the source of JH (see Section IV). As expected, PTTH appears to be transported in electron-dense neurosecretory granules to the axonal endings in the corpus allatum for storage and release (Dai *et al.,* 1994b; unpublished observations). It is not known where the processing of the prohormone to the mature secreted form is accomplished but it is likely that this occurs in the axon as found in analogous vertebrate systems.

PTTH mRNA levels per brain are nearly constant during the fifth larval instar of *Bombyx* (Adachi-Yamada *et al.,* 1994) but the titer of PTTH varies during development. In the *Bombyx* brain, the PTTH concentration is highest at the beginning of the fifth (final) larval instar and declines during later larval and pupal stages (Ishizaki *et al.,* 1983). In contrast, PTTH activity in *Manduca* brains increases ≈ 60% between day 3 of the fifth larval instar and day 1 of the pupal stage (O'Brien *et al.,* 1986). Changes in circulating PTTH have been determined by bioassay in *Bombyx* fifth instar larvae and in *Manduca* fourth and fifth instar larvae. In the *Manduca* fourth (penultimate) larval instar, a single peak of circulating PTTH occurs, but in the final instar two peaks are observed (Bollenbacher and Gilbert, 1980). The first elicits a small peak of ecdysteroid production (Bollenbacher *et al.,* 1975), the so-called commitment (or reprogramming) peak, that is responsible for determining that the next ecdysteroid surge will elicit a pupal rather than a larval molt (see Section III). The second peak of circulating PTTH stimulates the larger ecdysteroid surge that actually initiates the complex developmental cascade that ends in ecdysis. In contrast, *Bombyx* fifth instar larvae exhibit five peaks of PTTH release based on *in vitro* assays (Shirai *et al.,* 1993) and four peaks using a direct immunoassay for PTTH (Dai *et al.,* 1995).

The first of these four or five peaks appears to have no ecdysteroidogenic effect (Shirai *et al.,* 1993) and might instead play a role in regulating the growth status of the prothoracic gland, including the concentration of ecdysteroidogenic enzymes (see later).

In *Bombyx,* PTTH mRNA is found only in the brain as judged by conventional Northern blot analysis but use of the reverse transcriptase–polymerase chain reaction to detect rare messages indicated a slight expression in *Bombyx* gut and perhaps silk gland and epidermis (Adachi-Yamada *et al.,* 1994). Westbrook *et al.* (1993), using antisera to a putative *Manduca* PTTH, have also reported its presence outside the lateral neurosecretory cells and the corpus allatum. However, prothoracicotropic activity has not been reported from these other locations, raising the possibility that an immunologically cross-reactive molecule other than PTTH may be expressed in these other sites. In this regard, it is notable that *Bombyx* PTTH immunoreactivity has been observed in *Manduca* (Gray *et al.,* 1994; Dai *et al.,* 1994b) and *Drosophila* (Žitňan *et al.,* 1993), although *Bombyx* PTTH is without PTTH activity in *Manduca* (Muehleisen *et al.,* 1993; R. Rybczynski, D. O'Reilly, A. Mizoguchi, and L. I. Gilbert, unpublished observations) and *Drosophila* (V. C. Henrich and L. I. Gilbert, unpublished observations).

The prothoracicotropes that produce PTTH apparently receive information from the external (e.g., photoperiod, temperature, etc.) and internal environment of the insect, and when the appropriate conditions are met, release PTTH from their termini in the corpus allatum. Factors that influence PTTH release include the time from the last molt or meal, implying that insects have an internal clock that can be reset or entrained by the physiological state of the animal at critical time windows, as well as by photoperiod (see Bollenbacher and Granger, 1985). How and where these influences are sensed and then "transmitted" to the neurons that synthesize PTTH is not known. Shirai *et al.* (1994) have provided evidence that the (muscarinic) acetylcholinergic pathway is involved in PTTH release in *Bombyx,* supporting earlier studies in *Manduca* that also implicated the cholinergic system (see Gilbert, 1989).

D. Prothoracicotropic Hormone Action

1. Prothoracicotropic Hormone Receptor

The only confirmed target of PTTH is the prothoracic gland, although as noted earlier, the presence of PTTH in the adult brain suggests that other tissues may be regulated in some way by PTTH, e.g., ovarian maturation in *Manduca* seems to require a brain-derived factor (Sroka and Gilbert, 1971). The prothoracic gland has been well studied in *Manduca* (see Section III), where it occurs as a pair of glands with about 220 monotypic cells, surrounded by a basal lamina. Although no candidate PTTH receptor(s) has been isolated or cloned from the prothoracic glands of any insect, the PTTH–prothoracic gland axis has many similarities to vertebrate steroid hormone-producing pathways, such as the adrenocorticotropic hormone (ACTH)–adrenal gland system (see Gilbert *et al.*, 1988). By analogy it is probable that PTTH binds to a receptor that spans the plasma membrane multiple times, contains an extracellular ligand-binding domain, and has an intracellular domain that binds G-protein heterotrimers. It is likely also that the PTTH receptor is expressed primarily by prothoracic gland cells, but expression in other cell types is likely, especially in adult insects. Our lack of knowledge about PTTH receptors represents a major lacuna in understanding the control of prothoracic gland function.

2. *PTTH and Second Messengers*

PTTH stimulates increased ecdysteroid production in the prothoracic glands via a cascade of events that has yet to be elucidated completely. Early studies on *Manduca* revealed a correlation between circulating ecdysteroid titers and adenylate cyclase activity in the prothoracic gland (Vedeckis *et al.*, 1976), suggesting a role for cAMP, and also, that at some developmental periods, a cAMP-independent pathway might be involved. Further studies *in vitro* revealed that PTTH elicited rapid increases in gland cAMP levels, readily measured < 5 min after initiating stimulation (Fig. 3A), but that phosphodiesterase inhibitors were necessary to detect the increased cAMP at some stages (Smith *et al.*, 1984). Later studies confirmed that endogenous-soluble phosphodiesterase activity was high in pupal and early and late fifth instar larval glands, and reached a nadir in mid-fifth instar (Smith and Pasquarello, 1989).

Calcium serves as an important second messenger in many cellular systems and plays a role in peptide stimulation of some vertebrate steroid hormone-producing tissues (see, e.g., Veldhuis and Klase, 1982). In the *Manduca* prothoracic gland, calcium is clearly pivotal in the response to PTTH. Glands incubated in Ca^{2+}-free medium, or medium with a calcium chelator or a calcium channel blocker, exhibit a greatly attenuated production of cAMP and ecdysteroids in response to PTTH (Smith

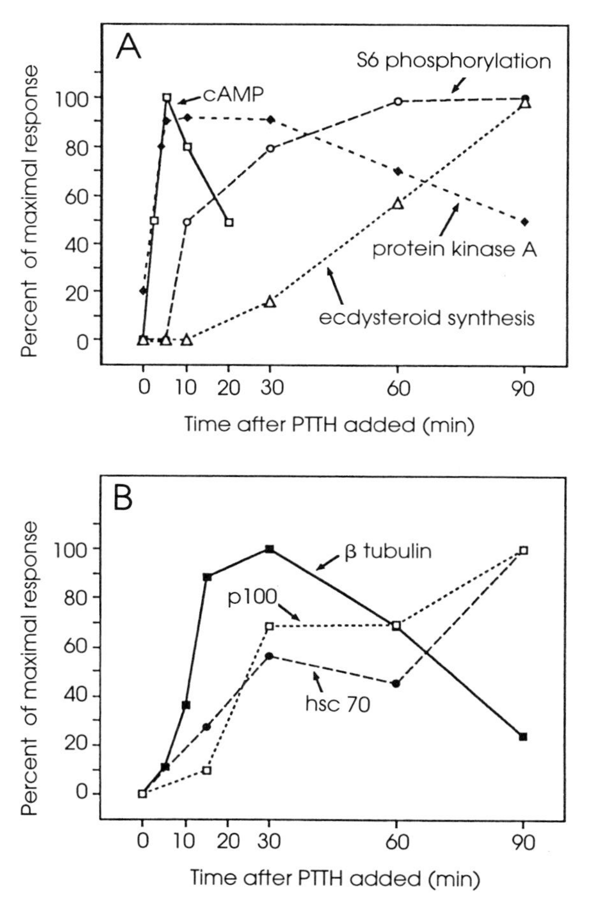

Fig. 3. Temporal sequence of the acute biochemical response of *Manduca* prothoracic glands to PTTH *in vitro.* (A) Transductory events (cAMP generation, protein kinase A activation, S6 phosphorylation) and ecdysteroid synthesis. [Reprinted from *Insect Biochem. Mol. Biol.* **22,** Rountree *et al.,* Prothoracicotropic hormone regulates the phosphorylation of a specific protein in the prothoracic glands of the tobacco hornworm, *Manduca sexta,* 353–362. Copyright (1992), with kind permission from Elsevier Science Ltd., The Boulevard, Langford Lane, Kidlington OX5 1GB, UK.] (B) Proteins whose syntheses are specifically stimulated by PTTH: ß-tubulin, hsc 70, and p100. [Based on Rybczynski and Gilbert (1995a,b).]

et al., 1985). Later studies have implicated the mobilization of internal as well as external Ca^{2+} stores in the PTTH response (Smith and Gilbert, 1989). Smith *et al.* (1985) showed also that Ca^{2+}-deprived glands failed to generate cAMP in response to PTTH, indicating that cAMP production was downstream of Ca^{2+} mobilization in the PTTH transductory cascade. However, glands incubated in Ca^{2+}-free medium and challenged with a cAMP analog responded with a normal increase in ecdysteroid synthesis, intimating that in contrast to ovarian granulosa cells (Veldhuis and Klase, 1982), Ca^{2+} was not important in the transductory pathway distal to cAMP production.

The composite observations suggested that the PTTH-dependent cAMP production by prothoracic glands was generated by a Ca^{2+}–calmodulin-sensitive adenylate cyclase. Meller *et al.* (1988, 1990) did indeed demonstrate the presence of a membrane-associated, calmodulin-sensitive adenylate cyclase in the prothoracic gland of *Manduca* and, furthermore, found evidence for guanine nucleotide-binding protein (G protein) involvement in adenylate cyclase activation (Meller *et al.*, 1988, 1990; Meller and Gilbert, 1988). The interaction between calmodulin and G protein (presumably $G_s\alpha$) is complicated and varies during the final instar. In the first half of this period, calmodulin can activate prothoracic gland adenylate cyclase and can facilitate G-protein activation of adenylate cyclase in *in vitro* experiments. Subsequently, prothoracic gland G-protein activation of adenylate cyclase is refractory to the presence of calmodulin in such assays (Meller *et al.*, 1990). Calcium still apparently plays a role in the PTTH transductory cascade after the first half of the fifth instar since incubation of pupal glands in Ca^{2+}-free medium inhibits PTTH-stimulated ecdysteroidogenesis and higher levels of Ca^{2+}–calmodulin can still activate adenylate cyclase in prothoracic gland membrane preparations (Smith *et al.*, 1985; Meller *et al.*, 1988).

Regardless of the complicated, developmentally dynamic relationships among calcium, calmodulin, G proteins, and adenylate cyclase, it is clear that PTTH elicits increased cAMP formation in prothoracic glands. Increases in intracellular cAMP levels can lead to the activation of cAMP-dependent protein kinases (PKAs) and subsequent protein phosphorylation (Fig. 3A). As expected, the *Manduca* prothoracic gland expresses PKAs, primarily of the type II form (Smith, 1993). Administration of a JH analog on the first day of the fifth instar in *Manduca* results in a $\approx$ 2-day delay of the wandering phase that precedes pupation and a 45% decrease in prothoracic gland PKA levels (Smith *et al.*, 1993); these effects may reflect JH modulation of prothoracic gland protein synthesis as controlled by the phosphorylation state of the ribosomal protein S6 (Rountree *et al.*, 1987; see later).

PTTH-stimulated PKA activity appears to be necessary for PTTH-stimulated ecdysteroidogenesis since such ecdysteroid synthesis by prothoracic glands challenged with a PKA-inhibiting cAMP analog is substantially inhibited (see Smith, 1993). Activation of PKA in PTTH-stimulated prothoracic glands is rapid ($\approx$ 90% of maximum within 5 min) (Smith *et al.,* 1986) and correlates with the rapid generation of cAMP described earlier (Fig. 3A). Several PTTH-dependent protein phosphorylations have been described for *Manduca* prothoracic glands (see, e.g., Rountree *et al.,* 1987, 1992; Combest and Gilbert, 1992; Song and Gilbert, 1994). Gilbert *et al.* (1988) speculated that the most striking and consistent of these phosphoproteins, p34 (molecular mass $\approx$ 35 kDa), might be the ribosomal protein S6 and this identification has been confirmed using the S6 kinase inhibitor rapamycin and two-dimensional PAGE analysis of ribosomal proteins (Fig. 4A) (Song and Gilbert, 1994, 1995). The phosphorylation of S6 has been correlated with an increased translation of specific mRNAs in several mammalian cell types (see, e.g., Duncan and McConkey, 1982). In *Manduca,* rapamycin inhibits both PTTH-stimulated S6 phosphorylation and ecdysteroidogenesis, suggesting that S6 is an integral player in the PTTH transductory cascade. Consistent with this view is the observation that PTTH-stimulated S6 phosphorylation can be readily detected before the PTTH-stimulated increase in ecdysteroid synthesis occurs (see Fig. 3A).

Other PTTH-stimulated phosphorylation substrates have not been studied as extensively. However, evidence has suggested that another peptide (molecular mass $\approx$ 52 kDa) is also transiently phosphorylated in prothoracic glands in response to PTTH, at least at some developmental stages (Song and Gilbert, 1995). It is likely that a full understanding of PTTH-stimulated phosphorylations will only result from careful developmental and time course studies since some of these events appear to be stage specific and/or have very short half-lives.

3. PTTH and Protein Synthesis

A characteristic of vertebrate peptide-stimulated steroid hormone production is a requirement for protein synthesis, i.e., blocking protein synthesis inhibits the increase in steroidogenesis normally resulting from peptide treatment. A similar requirement for protein synthesis, and possibly RNA synthesis, has been described for PTTH-stimulated ecdysteroidogenesis in *Manduca* (Smith *et al.,* 1987; Keightley *et al.,* 1990; Rybczynski and Gilbert, 1995a). In vertebrates, such observations have led to the identification of three small labile proteins believed to facilitate the delivery and binding of cholesterol to a mitochondrial P450 enzyme,

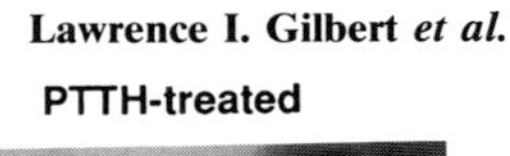

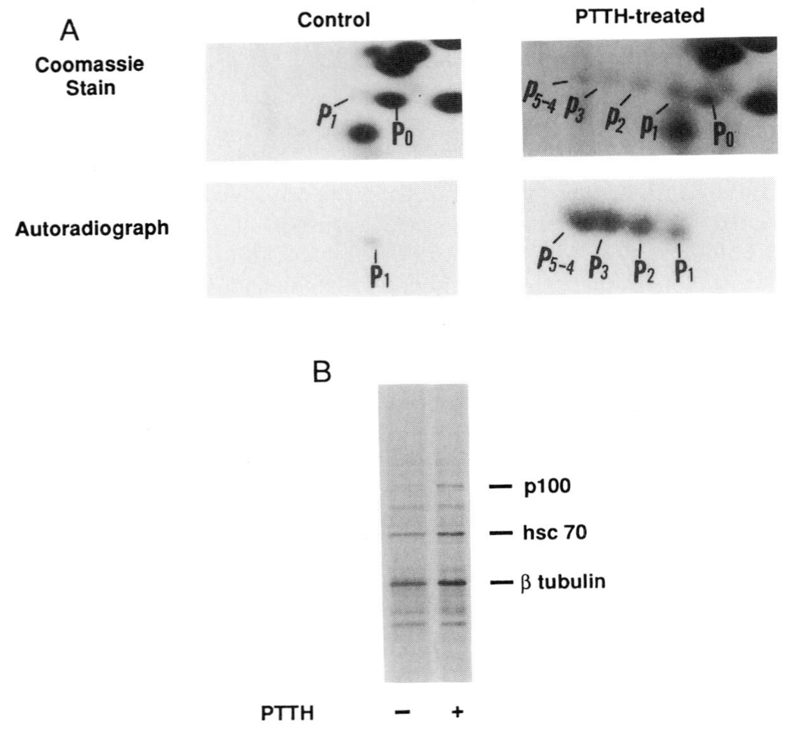

Fig. 4. The effects of PTTH on S6 phosphorylation and specific protein synthesis in *Manduca* prothoracic glands *in vitro.* (A) Patterns of PTTH-stimulated S6 phosphorylation as visualized by two-dimensional PAGE of ribosomal proteins followed by Coomassie staining and autoradiography. [Reprinted from *Insect Biochem Mol. Biol.* **25,** Song and Gilbert, Multiple Phosphorylation of ribosomal protein S6 and specific protein synthesis are required for prothoracicotropic hormone-stimulated ecdysteroid biosynthesis in the prothoracic glands of *Manduca sexta,* 591–602. Copyright (1995), with kind permission from Elsevier Science Ltd., The Boulevard, Langford Lane, Kidlington OX5 1GB, UK.] (B) The specific stimulation of p100, hsc 70, and ß-tubulin synthesis as visualized by SDS-PAGE followed by autoradiography (R. Rybczynski and L. I. Gilbert, unpublished observations; see also Rybczynski and Gilbert, 1994, 1995a,b).

the first and rate-limiting step in vertebrate steroid hormone synthesis (see Orme-Johnson, 1990). A cDNA clone for a *Manduca* homolog of one of these three vertebrate steroid-activating proteins, the peripheral (mitochondrial) benzodiazepine receptor ligand (diazepam binding inhibitor), has been obtained from a larval midgut library (Snyder and Feyereisen, 1993). The RNA for this small protein (predicted molecular mass ≈ 9,700 Da) is apparently expressed in a number of *Manduca* tissues, including the prothoracic gland where it is rare, but its function(s) awaits elucidation. In insects the rate-limiting step in ecdysteroid synthesis does not involve cholesterol but rather another sterol (see Section III), so it would not be surprising if proteins regulating ecdysteroid synthesis differ from those in verterbrates.

In *Manduca,* Keightley *et al.* (1990) and Rybczynski and Gilbert (1994) have described one and three proteins, respectively, whose translation is specifically stimulated relative to other proteins in PTTH-challenged prothoracic glands (Figs. 3B and 4B). Two of these proteins have now been identified. One, molecular mass ≈ 70 kDa, is a constitutively expressed (cognate) member of the 70-kDa heat shock protein family (hsc 70: Rybczynski and Gilbert, 1995b). A second, molecular mass ≈ 52 kDa, is a ß-tubulin of unknown isoform (Rybczynski and Gilbert, 1995); this is probably the ≈ 60-kDa peptide described by Keightley *et al.* (1990) and may be the 52-kDa phosphoprotein mentioned earlier. A third protein (molecular mass ≈ 100 kDa; Rybczynski and Gilbert, 1994) remains unidentified but it appears to be neither a component of the cytoskeleton nor an integral membrane protein (R. Rybczynski and L. I. Gilbert, unpublished observations).

Heat shock proteins play important roles in protein folding and in the import of nascent proteins into intracellular organelles (see Becker and Craig, 1994). Since PTTH can stimulate general protein synthesis (see Fig. 3B and later) in addition to stimulating the specific synthesis of the three peptides just discussed, increased translation of one or more such protein "chaperones" would be expected. The steady increase in PTTH-stimulated hsc 70 synthesis is also consistent with a role for this protein in long-term effects (e.g., growth) of PTTH on the prothoracic glands (Fig. 3B). A role in ecdysteroid synthesis for a cytoskeletal element like ß-tubulin might also be expected given the many observations that cytoskeleton-altering drugs change (stimulate or inhibit) basal and/or peptide-stimulated hormone production by steroid hormone-producing vertebrate cells, presumably by altering the efficiency of intracellular trafficking of precursors or export of product (see, e.g., Rainey *et al.,* 1985). However, such effects might result from broad alterations in cellular processes beyond the steroid synthesis pathway. For instance, in *Manduca* prothoracic glands, preliminary data indicate that doses of taxol sufficient to alter ecdysteroid synthesis also partially inhibit protein synthesis (R. Rybczynski and L. I. Gilbert, unpublished observations). What is striking about PTTH-stimulated ß-tubulin synthesis in prothoracic glands is the rapidity of the response, detectable *in vitro* in 10 min or less in contrast to hsc 70 or p100 synthesis (Fig. 3B), and the very short half-life of the polypeptide when PTTH is removed ($t_{1/2} \approx 35$ min) (Rybczynski and Gilbert, 1995a). These features suggest an important, perhaps novel, function for ß-tubulin in the prothoracic glands.

4. PTTH as a Trophic Factor

A number of studies have revealed that PTTH preparations or cAMP analogs stimulate general protein synthesis in the *Manduca* prothoracic gland at most stages studied (Keightley *et al.,* 1990; Rybczynski and Gilbert, 1994, 1995a). Three observations indicate that general PTTH-stimulated protein synthesis results from a branch of the transductory cascade that is distinct from the pathway leading to the activation of ecdysteroidogenesis and the synthesis of the three specific PTTH-stimulated proteins. First, PTTH stimulation of day 1 fifth instar prothoracic glands results in increased ecdysteroid synthesis and increased hsc 70, p100, and ß-tubulin synthesis, but no detectable increase in general protein synthesis (Rybczynski and Gilbert, 1994). Second, at later stages, challenging prothoracic glands with the calcium ionophore A23187, alone or with dibutyryl-cAMP, elicits steroidogenesis and a relative increase in synthesis of hsc 70, p100, and ß-tubulin, but a decrease in overall protein synthesis (Rybczynski and Gilbert, 1994). Third, PTTH-stimulated general protein synthesis lags behind ecdysteroid and ß-tubulin synthesis. Changes in the latter two variables can be detected within 15–20 min of PTTH stimulation, but similarly sized changes in overall protein synthesis require 60 min or more (Fig. 3B) (Keightley *et al.,* 1990; Rybczynski and Gilbert, 1994, 1995a). PTTH may, therefore, modulate or control the growth of the prothoracic gland, perhaps independently of its ability to elicit ecdysteroidogenesis and could play a role in regulating the levels of ecdysteroidogenic enzymes, analogous to peptide hormone regulation of enzymes responsible for vertebrate steroid hormone synthesis (see Simpson and Waterman, 1988). Additional factors such as JH could determine whether PTTH stimulates or inhibits gland growth, ecdysteroid synthesis, or both.

E. Small Prothoracicotropic Hormone and Bombyxin

The size heterogeneity of active PTTH has caused some confusion. Brain extracts from a number of species have yielded PTTH activity with molecular masses of 4–7 kDa (small PTTH), in addition to the ≈ 30-kDa PTTHs (big PTTH) (see Gilbert *et al.,* 1981; Bollenbacher and Granger, 1985). One such small molecule isolated from the brain of *Bombyx* and termed bombyxin was purified and cDNA and genomic clones obtained. Bombyxin is actually a family of related peptides coded for by separate genes and expressed in four pairs of medial neurosecretory brain cells whose axons terminate in the corpus allatum, the same neurohemal site for big PTTH (see Ishizaki and Suzuki, 1992). Bombyxin is a dimer joined by disulfide bonds and, based on amino acid sequence, is related to the insulin family of polypeptides. Paradoxically, bombyxin has no prothoracicotropic effects on *Bombyx* (or *Manduca;* see later) but exhibits such activity in *Samia cynthia ricini* (see Ishizaki and Suzuki, 1992). Bombyxin-like proteins are found in a number of species based on immunological analysis. In *Manduca,* a bombyxin-like molecule has been identified using immunoblotting techniques, and partial purification of this molecule has shown that it is distinguishable from *Manduca* small PTTH (A. Mizoguchi and L. I. Gilbert, unpublished observations).

In *Manduca,* small PTTH is found in similar amounts in larval and pupal brains (O'Brien *et al.,* 1986). However, pupal prothoracic glands are relatively refractory to stimulation by small PTTH, in contrast to early fifth instar (larval) glands (Bollenbacher *et al.,* 1984). Watson *et al.* (1993) have shown that partially purified small PTTH, like big PTTH, stimulates cAMP accumulation in larval glands. Costimulation of larval prothoracic glands with big and small PTTHs did not result in an additive synthesis of ecdysteroids, suggesting a convergence of transductory paths.

Since *Manduca* prothoracic glands are most sensitive to small PTTH at the same stage at which the glands are most susceptible to artifactual stimulation (Bollenbacher *et al.,* 1983), the physiological significance of small PTTHs must be considered carefully. Small PTTH could bind to a prothoracic gland receptor that is developmentally regulated, and its major effects could be on tissues other than the prothoracic gland, as suggested by the expression of small PTTH at times when the prothoracic gland is refractory to it. That is, its primary role may not be modulation of ecdysteroidogenesis by the prothoracic gland.

Further complicating an understanding of the regulation of ecdysteroidogenesis and molting by PTTHs are two sets of observations. First, Gray *et al.* (1994) found that the *Manduca* prothoracicotropes also express a peptide that is very similar in deduced amino acid composition to *Bombyx* PTTH (see also Dai *et al.,* 1994b). This *Bombyx*- like PTTH apparently does not activate *Manduca* prothoracic glands, which, as noted previously, are also refractory to authentic brain-derived or recombinant *Bombyx* PTTH. Second, molecules with prothoracicotropic activity apparently also reside in the gut (proctodeum) of two moth species (Gelman *et al.,* 1991) and can stimulate ecdysteroid synthesis by prothoracic glands *in vitro.* However, the physiological significance of these "ecdysiotropins" is presently unknown.

The largest gaps in our knowledge of PTTH action concern the nature and distribution of the PTTH receptor(s) and the identity and function of the proteins (S6?) that seemingly control and/or enhance the rate of ecdysteroid synthesis in PTTH-stimulated prothoracic glands. The first question awaits studies on *Bombyx* since only the PTTH of that insect is available in pure form, or the purification of another PTTH (*Manduca*?), while the second requires knowledge of the biosynthetic route of ecdysteroids. It is to the latter problem that we now turn.

III. ECDYSTEROIDS

A. Introduction

That ecdysteroids, particularly 20-hydroxyecdysone, elicit the molt is no longer in question and has been established as a central dogma of the field. What may not be so obvious is that in contrast to vertebrate systems, almost the entire insect is the target of ecdysteroids, e.g., regulation of the growth of motor neurons (Prugh *et al.,* 1992), control of choriogenesis (Bellés *et al.,* 1993), stimulation of the growth and development of imaginal discs (see Bayer *et al.,* Chapter 9, this volume), initiation of the breakdown of larval structures during metamorphosis (see Truman, Chapter 8, this volume), elicitation of the deposition of cuticle by the epidermis (see Willis, Chapter 7, this volume), etc. Beyond these "intramural" effects, ecdysteroids can regulate polydnavirus DNA replication which is integrated into wasp chromosomal DNA (Webb and Summers, 1993) and appear to act as a defensive secretion in at least one marine arthropod (Tomaschko, 1994).

It is fitting that recent breakthroughs on the mechanism of action of ecdysteroids were accomplished using *Drosophila melanogaster* (see Russell and Ashburner, Chapter 3, this volume; Cherbas and Cherbas, Chapter 5, this volume) since it was a bioassay developed with another cyclorrhaphan dipteran, *Calliphora erythrocephala,* based on larval ligation and ecdysteroid-elicited pupariation, that was so well utilized for the initial crystallization of ecdysone and then 20-hydroxyecdysone in the mid-1950s (Butenandt and Karlson, 1954). Since that time, and with the advent of radioimmunoassays and high-performance liquid chromatography coupled with mass spectrometry (Gilbert *et al.,* 1991), it has been possible to identify and characterize a host of ecdysteroids (Fig. 2B), their precursors, and metabolites (see Rees, 1985; Grieneisen *et al.,* 1993). Since many of these ecdysteroids have been discussed in chemical terms (Grieneisen, 1994), it is not practical to dwell on their structures and distribution here. However, a few points must be considered since this discussion provides the base for several succeeding chapters. Through a variety of bioassays some structure–activity relationships of ecdysteroids are now known (Gilbert and King, 1973) (Fig. 2B). The *cis* A–B ring junction is essential regardless of whether a hydrogen atom or a hydroxyl group is the 5ß substituent, as is the 6-oxo-7-ene system in the B ring. The 3ß- and 14α-hydroxyl groups are required for high activity *in vivo* while the presence or absence of hydroxyls at C-2, C-5, or C-11 does not appear to affect biological activity. The only essential feature of the side chain appears to be the $22\beta_F$-hydroxyl. Generally, it is difficult to quantify the relative biological activity of ecdysteroids in bioassays since their activity is enhanced if they are degraded more slowly than other ecdysteroids just as the potency of JH is enhanced when administered in a vehicle that protects it from JH esterases, e.g., peanut oil (Gilbert and Schneiderman, 1960; see Section IV,C). Further, some ecdysteroids are rapidly converted to more active ecdysteroids, e.g., ecdysone and deoxyecdysones to 20-hydroxyecdysone, etc. (see Gilbert and King, 1973).

B. Molting Hormones

1. *Hormone–Prohormone Relationships*

Although ecdysone was the first of the ecdysteroids to be crystallized and characterized and thought to be the insect molting hormone in the mid-1950s, it is actually converted to the principal molting hormone, 20-hydroxyecdysone, by tissues peripheral to the prothoracic glands (Fig. 1), a reaction mediated by an ecdysone 20-monooxygenase (Smith *et al.,*

1983). We now know that in some insects, particularly the Lepidoptera, as exemplified by *Manduca,* the major if not sole ecdysteroid synthesized and secreted by the prothoracic glands is 3-dehydroecdysone (Warren *et al.,* 1988a,b; Kiriishi *et al.,* 1990) which is converted to ecdysone by a ketoreductase in the hemolymph (Sakurai *et al.,* 1989); the resulting ecdysone is then hydroxylated to 20-hydroxyecdysone in target tissues. Depending on dietary conditions, the ring gland of *D. melanogaster* synthesizes ecdysone and another ecdysteroid, 20-deoxymakisterone A (Redfern, 1983, 1984; Pak and Gilbert, 1987); the latter compound is converted to makisterone A peripherally, analogous to the 3-dehydroecdysone → ecdysone → 20-hydroxyecdysone sequence of reactions in the Lepidoptera. The complexity of the issue is enhanced when one considers that insects like *Drosophila* also have the ability to convert ecdysone to "metabolites" such as ecdysone-22-fatty acid acylesters, 3-dehydroecdysone, 26-hydroxyecdysone, and ecdysonic acid (see, e.g., Grau and Lafont, 1994). Is it possible that all are not degradation products in the classic sense, but may play unknown physiological roles? When ecdysone or 20-hydroxyecdysone is injected into a test insect and these ecdysteroids are even used in an *in vitro* system where the tissue or organ is capable of metabolizing the primary ecdysteroid, it is not possible to state with assurance that the resulting effect is due directly to the injected ecdysteroid. Could target cells contain a multiplicity of ecdysteroid receptors (see Cherbas and Cherbas, Chapter 5, this volume)? Could discrete ecdysteroids play different roles in the molting process?

3-Dehydroecdysone may play a physiological role in *Drosophila* as well as in *Manduca* since it appears to be the major metabolite of injected ecdysone in adult males, whereas 3-dehydro-20-hydroxyecdysone is a major "degradation" product of 20-hydroxyecdysone (Grau and Lafont, 1994). Although these 3-dehydroecdysteroids are also present in larvae, can elicit specific puffing patterns in the polytene chromosomes of the salivary gland (see Russell and Ashburner, Chapter 3, this volume), and can bring about the expression of a specific fat body gene so as to stimulate the synthesis of a specific protein (Richards, 1978), they have not yet been proven to be active hormones.

Examples of other ecdysteroids possibly acting as hormones come from the work on *Manduca* where 26-hydroxyecdysone has been proposed to be influential in modulating or eliciting specific events during embryogenesis (Warren *et al.,* 1986; Dorn *et al.,* 1987) and from the qualitative and quantitative analyses of ecdysteroids during the pupal–adult metamorphosis of *Manduca* (Warren and Gilbert, 1986). The latter study showed the presence in the hemolymph and gut of progressively oxidized metabolites of ecdysone which itself was the result of the reduction of the primary product of the prothoracic glands, 3-dehydroecdysone. Data indicate that different steroid hydroxylases are induced by progressively oxidized substrates and that ecdysteroids other than 20-hydroxyecdysone may modulate processes in the developing (pharate) adult moth. The kinetics of ecdysteroid metabolism reveal three major ecdysteroid peaks in the hemolymph during the 19-day pupal–adult transformation; the temporal order is ecdysone, 20-hydroxyecdysone, and 20,26-dihydroxyecdysone (Fig. 5). Ecdysone is the only free ecdysteroid present during the first 4 days of development and since some important developmental events occur during this time, e.g., apolysis, the data suggest most strongly that ecdysone is a hormone in its own right as well as a precursor of 20-hydroxyecdysone (see also Tanaka *et al.,* 1994). The titer of the latter begins to rise on day 5 and peaks about day 9. 20,26-Dihydroxyecdysone reaches its peak titer on about day 12. The 19-day period of pupal–adult metamorphosis includes the development of adult nervous, digestive, and muscle systems as well as the development of genitalia, wings, compound eyes, etc., all displaying their own developmental kinetics. The composite data support the proposition that ecdysone and 20-hydroxyecdysone are hormones with specific and critical roles to play in orchestrating insect developmental processes, and to these we may add 20,26-dihydroxyecdysone, 26-hydroxyecdysone in the *Manduca* embryo, 3-dehydro-20-hydroxyecdysone and 3-dehydroecdysone in *Drosophila* as well as makisterone A. Whether separate ecdysteroid receptors are present for these putative active ecdysteroids is an important question for future study. Even the decrease in ecdysteroid titer prior to adult eclosion is an important message to the organism since a specific cascade of behavioral events leading to eclosion is elicited by a peptide, eclosion hormone, which in turn is synthesized and released in response to the falling ecdysteroid titer (Truman, 1985, and Chapter 8, this volume).

2. *Effects Due to Decreasing Titers*

Although most of the work on ecdysteroid action focuses on events accompanying increases in the ecdysteroid titer, it should not be overlooked that important physiological and molecular events accompany the down side of the ecdysteroid titer peak or occur on the withdrawal of ecdysteroid from *in vitro* paradigms (see Truman, Chapter 8, this volume; Bayer *et al.,*

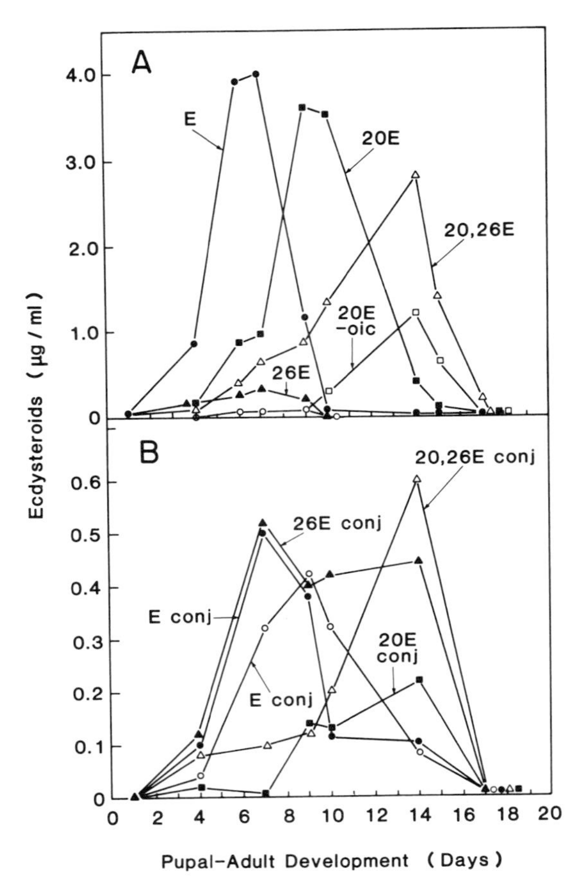

Fig. 5. (A) Concentrations of free ecdysteroids in female hemolymph during the pupal–adult development of *Manduca sexta*. E, ecdysone; 20E, 20-hydroxyecdysone; 26E, 26-hydroxyecdysone; 20, 26E, 20,26-dihydroxyecdysone; 20E-oic, 20-hydroxyecdysonic acid. (B) Concentrations of high polarity-conjugated ecdysteroids in female hemolymph during the pupal–adult development of *M. sexta*. E conj, ecdysone conjugate (each detected by a different antiserum, one specific for the ring structure and the other for the side chain). [Reprinted from *Insect Biochem.* **16,** Warren and Gilbert, Ecdysone metabolism and distribution during the pupal–adult development of *Manduca sexta,* 65–82. Copyright (1986), with kind permission from Elsevier Science Ltd., The Boulevard, Langford Lane, Kidlington OX5 1GB, UK.]

Chapter 9, this volume). Research on *Drosophila* has uncovered a number of genes that respond to decreasing titers of ecdysteroids (see Russell and Ashburner, Chapter 3, this volume; Cherbas and Cherbas, Chapter 5, this volume; Bayer *et al.,* Chapter 9, this volume) and is not described here. In the case of *Manduca* the DOPA decarboxylase gene that is so important in insect cuticle tanning and sclerotization (see Willis, Chapter 7, this volume) is an example of regulation by a decreasing ecdysteroid titer (Hiruma and Riddiford, 1993). Using subtractive hybridization, Mészáros and Morton (1994) have cloned a gene from *Manduca* that is expressed just prior to pupation, a time when the ecdysteroid level is relatively low after peaking about 2 days previously. It is of interest that this gene is expressed only in tracheal epithelial cells in a variety of tissues and the mRNA appears just before larval, larval–pupal, and pupal–adult molts, in all cases after the ecdysteroid peak had declined. Further, exogenous application of 20-hydroxyecdysone inhibited the expression of the gene that is somewhat homologous to the *knirps*-related gene of *Drosophila* which appears to be a secondary response gene (see Russell and Ashburner, Chapter 3, this volume; Cherbas and Cherbas, Chapter 5, this volume). Since the cells expressing this gene synthesize and secrete cuticular proteins, the authors suggest that it may be involved in cuticle formation. Similarly, Vogt *et al.* (1993) found that the expression of two proteins specific to the adult *Manduca* olfactory system also appears to be controlled by falling levels of ecdysteroid. Incubation of antennae at earlier stages in the absence of 20-hydroxyecdysone resulted in premature expression of these two proteins, odorant binding protein, and pheromone-metabolizing aldehyde oxidase. Inclusion of hormone in these incubations prevented this early expression. Low ecdysteroid levels may also play a role in decreasing the titer of JH so that metamorphosis can ensue (Gu and Chow, 1993). This interplay between the two hormones seems to occur throughout the life cycle of the insect. For example, the small pulse of hemolymph ecdysteroids one-third through the last larval instar of *Manduca* (Bollenbacher *et al.,* 1975) in the absence of significant quantities of JH reprograms the cells of the insect so that they undergo metamorphosis to the pupal state when they encounter the large surge several days later (Riddiford, 1976a,b). Thus, this small peak is a key metamorphic event since cells and tissues exposed to it in the absence of significant JH are transformed into pupal cells and tissues, i.e., they synthesize pupal gene products when challenged by the next large ecdysteroid surge. Metamorphosis is usually thought of as the obvious change of the organisms at ecdysis, but at the cellular level it is controlled by this subtle ecdysteroid peak. Indeed, the small reprogramming peak can only occur when the JH titer falls below a certain level since JH may prevent PTTH secretion by acting directly on the brain (Nijhout and Williams, 1974; Rountree and Bollenbacher, 1986).

As described in Section II,D, one reason that the control of ecdysteroidogenesis by PTTH and other possible factors is not yet well understood is the paucity of data on the reactions comprising the ecdysteroid biosynthetic pathway. S6 phosphorylation is surely involved (Fig. 4A), but at the level of which specific reactions? What is known is summarized in the following section.

C. Biosynthesis of Ecdysteroids

The reader is referred to excellent reviews by Rees (1985) and Grieneisen (1994) on ecdysteroid biosynthesis, and only a few highlights will be summarized here. In most organisms every carbon atom in cholesterol (Fig. 2A) is derived from either the methyl or carboxyl carbon of acetate, but insects (and other arthropods) are incapable of this synthesis due to one or more metabolic blocks between acetate and cholesterol. Thus, sterols are required in the diet, either cholesterol per se for omnivorous and carnivorous insects or ß-sitosterol in the case of phytophagous insects, with the latter being dealkylated (removal of C-24 ethyl group) to cholesterol (see Gilbert, 1967; Grieneisen, 1994).

The first step in the conversion of cholesterol to ecdysone via 3-dehydroecdysone is the stereospecific removal of the 7ß hydrogen to form 7-dehydrocholesterol, a sterol relegated to the prothoracic glands of *Manduca* and other Lepidoptera. This cholesterol 7,8-desaturating activity in the prothoracic glands of *Manduca* is cytochrome P450 dependent, perhaps via 7ß-hydrocholesterol (Grieneisen *et al.,* 1993). When ^{3}H-labeled 7-dehydrocholesterol is incubated with prothoracic glands *in vitro,* there is excellent conversion to both 3-dehydroecdysone and ecdysone (Warren *et al.,* 1988a,b; Grieneisen *et al.,* 1991); the kinetics of conversion are highly dependent on developmental stage and experimental paradigm. The desaturation to 7-dehydrocholesterol is probably not PTTH dependent, but the neuropeptide may moderate enzyme activity responsible for the transformation of 7-dehydrocholesterol to the next, yet unidentified, sterol in the ecdysone biosynthetic pathway (see Grieneisen, 1994).

A number of postulated intermediates exist between 7-dehydrocholesterol and 7-dehydroecdysone, e.g., 5α-sterol intermediates, 3-oxo-Δ^4 intermediates, and Δ^7-$5\alpha,6\alpha$-epoxide intermediates (see Grieneisen, 1994), but their intermediacy remains conjectural. In contrast, more is known about the terminal hydroxylations necessary for the synthesis of the polyhydroxylated ecdysteroids. The enzymes responsible for mediating the hydroxylations at C-2, C-22, and C-25 appear to be classic cytochrome P450 enzymes, with the former two being mitochondrial and the latter microsomal. The sequence of hydroxylation is C-25, C-22, and C-2 (Hetru *et al.,* 1978; Kappler *et al.,* 1989).

Once formed, 3-dehydroecdysone is converted to ecdysone through the mediation of a hemolymph ketoreductase and the ecdysone is then transformed to the major molting hormone 20-hydroxyecdysone at peripheral (target) tissues under the control of a mitochondrial and/or microsomal ecdysone 20-monooxygenase, depending on the species (see Smith, 1985). The complete biosynthetic scheme has not been elucidated due to the difficulty of identifying the extremely short-lived intermediates from minute quantities of tissues and the less than handful of laboratories actively engaged in such investigations. Needless to say, without the entire sequence of reactions in hand, it is not possible to identify those rate-limiting reactions that may be controlled by hormones, neuromodulators, or the nervous system.

D. Control of Prothoracic Glands

Most of the research conducted on the control of prothoracic gland activity, i.e., ecdysteroidogenesis, has been conducted on the Lepidoptera, particularly *Manduca;* the most striking example is PTTH stimulation as noted in Section II,D. However, other hormones such as JH and ecdysteroids impinge on the prothoracic glands, and some glands are innervated, although the exact role of the nervous system remains conjectural.

Research done in the 1960s showed that the implantation of active corpora allata into brainless pupae elicited development and indeed resulted in pupal–adult intermediates, indicating the presence of a substance secreted by the corpora allata with PTTH activity as well as JH activity (Ichikawa and Nishiitsutsuji-Uwo, 1959; Williams, 1959). The idea of PTTH in these glands appeared bizarre at that time, although we now know that these glands are indeed the neurohemal sites for PTTH in lepidopterans (Agui *et al.,* 1980). However, the same results were obtained with JH extracts (Gilbert and Schneiderman, 1959), demonstrating that it was indeed JH that could stimulate the prothoracic glands since there would be no PTTH in these lipophilic extracts, and neither corpora allata implantation nor JH injection stimulated development in isolated pupal abdomens devoid of prothoracic glands. Since that time, these results have been confirmed repeatedly and expanded (see Gilbert and King, 1973) so that it is now generally accepted that JH can both stimulate and inhibit ecdysteroidogenesis by the prothoracic glands, depending on the developmental stage examined (see Sakurai and Gilbert, 1990).

As important as gland stimulation paradigms are, it is equally important for the evolutionary life of the insect that the ecdysteroid titer decrease, either by enhanced catabolic processes or more likely by feedback inhibition. It has been known for many years that 20-hydroxyecdysone can stimulate prothoracic gland activity (Williams, 1952; Siew and Gilbert, 1971; Sakurai and Williams, 1989), analogous to the situation in premetamorphic amphibians where thyroxine triggers the release of TSH from the adenohypophysis. The exact meaning of these observations is obscure but perhaps it is one means of ensuring that all 220 cells of the prothoracic gland, and both glands of the pair, secrete simultaneously to enhance the precision of secretion and action of this critical hormone. On the other hand, the glands are structured for excellent intracellular communication, e.g., large and developmentally regulated gap junctions (Dai *et al.,* 1994a).

That a negative feedback loop exists has been demonstrated exquisitely by Sakurai and Williams (1989), who showed that developmentally appropriate high rates of ecdysteroidogenesis by *Manduca* larval and pupal prothoracic glands are inhibited both *in vivo* and *in vitro* by either ecdysone or 20-hydroxyecdysone. This mechanism is in effect prior to pupation, just after the ecdysteroid titer had reached its zenith a few days previously. [An analogous situation exists in the scheme of PTTH activation of the prothoracic glands where the glands become refractory to PTTH during diapause (Bowen *et al.,* 1984).] This mechanism is considered to be a short loop (direct) suppression, although the possibility of long loop negative feedback mechanisms (indirect, e.g., via the brain) has also been suggested (see Sakurai and Gilbert, 1990).

E. Prothoracic Gland Degeneration: Alternate Sources of Ecdysteroids

One reason that most adult insects do not molt is that the prothoracic glands degenerate early in adult development, although a few cells are recoverable in some species after adult ecdysis (Herman and Gilbert, 1966). The absence of JH during the metamorphic molt to the adult appears to be the stimulus for gland degeneration (Wigglesworth, 1955; Gilbert, 1964). Since the presence of JH at the molt prevents the degeneration of many larval and pupal organs and other structures, it is likely that the prothoracic gland is simply another larval or pupal structure whose maintenance is JH dependent.

In the higher flies, the cyclorrhaphan Diptera, the ring gland is prominent in the larva and is a composite endocrine gland composed of discrete regions representing the analogs of the lepidopteran prothoracic glands, corpus allatum, and corpus cardiacum. As in most insects, the prothoracic gland portion of the ring gland degenerates once adult development has gained momentum. This process of gland degeneration was studied at both the functional and the ultrastructural levels by Dai and Gilbert (1991) as summarized in Fig. 6. Data from studies *in vitro* of ring gland ecdysteroidogenic capacity as well as ultrastructural analysis showed that glands from early wandering larvae had the highest basal rate of ecdysteroid biosynthesis, but by puparium formation there is a dramatic decrease in this rate and by 48 hr after puparium formation (eye pigmentation stage) ecdysteroidogenesis is barely discernible. As viewed under the electron microscope, this is the time of the onset of apoptosis of the cells comprising the prothoracic gland portion of the ring gland, as evidenced by the appearance of hemocytes and autophagic vacuoles. Individual cells begin to separate from one another at about 72 hr after puparium formation and the process is complete just prior to adult eclosion 24 hr later. The remnants of prothoracic gland cells then enter the hemocoele where they are phagocytized by hemocytes and nephrocytes. [It is of interest that in contrast to the prothoracic glands, the corpus allatum region of the ring gland is preserved, migrates, and gives rise to the adult corpus allatum (Fig. 6), so important in the control of vitellogenesis.]

These data are of interest not only because they document prothoracic gland degeneration in this important experimental organism, but because they reveal dysfunctional gland cells at a time when previous studies on whole animal ecdysteroid titers showed a surprising ecdysteroid titer at 30–40 hr postpupariation (Bainbridge *et al.*, 1982; Pak and Gilbert, 1987). These data illustrate a reasonably recent maxim in insect endocrinology, i.e., there are extra prothoracic gland sources of ecdysteroids in immature insects, a well-known phenomenon in some adult insects where ecdysteroids play a critical role in egg maturation and the ovaries and testes of

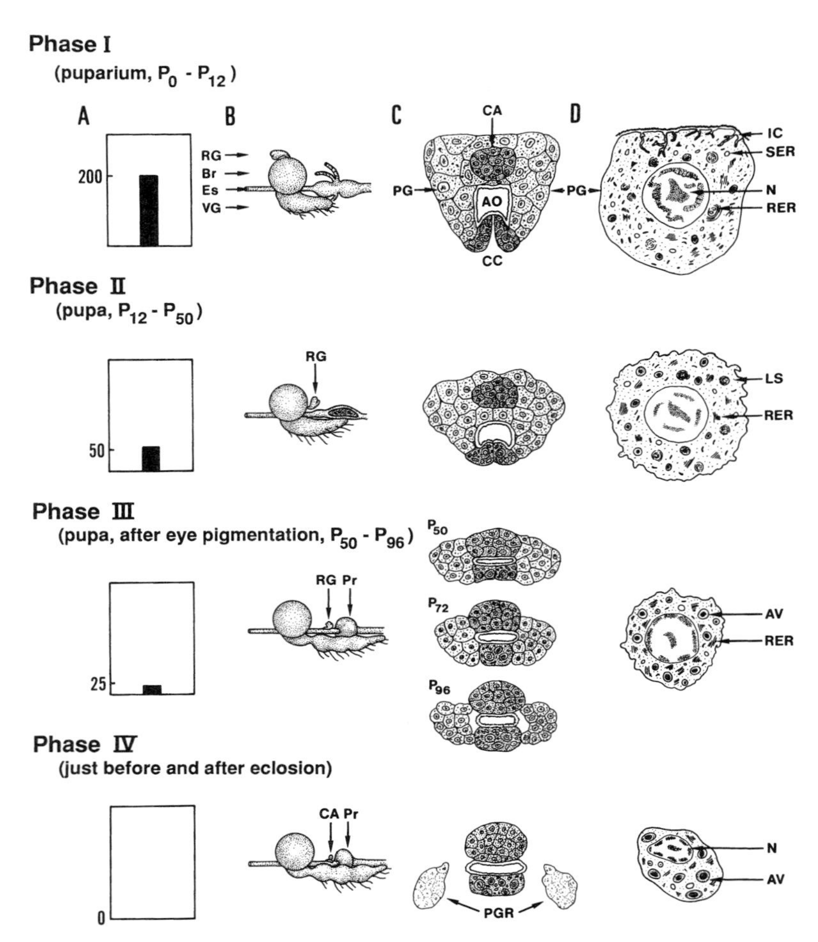

Fig. 6. Diagrammatic representation of the sequential changes in the *D. melanogaster* ring gland during pupal–adult metamorphosis. (A) Level of ecdysteroid biosynthesis by the ring gland *in vitro.* (B) Gross changes in location and structure of the ring gland. (C) Cytological changes in the ring gland. (D) Changes in the ultrastructure of the prothoracic gland cell. AO, aorta; AV, autophagic vacuole; Br, brain; CA, corpus allatum; CC, corpus cardiacum; ES, esophagus; IC, intercellular channel; LS, lysosome-like structure; N, nucleus; PG, prothoracic gland; PGR, prothoracic gland remnant; Pr, proventriculus; P_0, white puparium within 15 min after puparium formation; P_x denotes x hr after puparium formation; RG, ring gland; RER, rough endoplasmic reticulum; SER, smooth endoplasmic reticulum; VG, ventral ganglion. [Reprinted from Dai and Gilbert, 1991, with permission from Academic Press.]

several species appear to be the source of ecdysteroids (see Hagedorn, 1985).

This phenomenon of alternate sources of ecdysteroids has been recognized for many years (see Gilbert, 1974; Delbecque *et al.,* 1990) and has been reinvestigated using *Manduca* (Sakurai *et al.,* 1991). The latter showed that when abdomens devoid of prothoracic glands were isolated from the head–thorax region of the pupa, these isolated abdomens initiated development in about 2 months, but could be stimulated to accelerate the process by the implantation of pupal brains. The hemolymph ecdysteroid titer of these animals rose to 1.5 μg/ml within 5 days of receiving the brain implant. Analysis of the hemolymph ecdysteroids revealed that ecdysone was the major moiety, that the ecdysteroid composition was indistinguishable from that of normal intact pupae that had just initiated adult development, and that these observations were not due to the conversion of ecdysteroids with low affinity to the antibody used for RIA to those with high affinity. These data suggest that a brain hormone (PTTH?) can stimulate the synthesis of ecdysteroids by these as yet unidentified abdominal tissues, perhaps epidermis (see Delbecque *et al.,* 1990).

IV. JUVENILE HORMONES

A. Corpus Allatum as Site of Synthesis

As alluded to earlier, the development of structural characters that distinguish adult forms from larval forms is regulated by a complex interaction between JH and the ecdysteroids. The JHs are a unique group of sesquiterpenoid compounds that have been identified definitively only in insects (Fig. 2), although other classes of arthropods produce related sesquiterpenoids [e.g., methyl farnesoate, the unepoxidized precursor of JH III, produced by decapod crustaceans (Laufer *et al.,* 1987)]. The existence of JH was first postulated by Wigglesworth, due to its inferred inhibitory effect on metamorphosis in *Rhodnius* [originally termed "inhibitory hormone" (Wigglesworth, 1934, 1936)]; hence, the term "juvenile" hormone for its role in the retention of larval characteristics or the restraining of development toward the adult form. Wigglesworth originally demonstrated by elegant but simple transplantation experiments that JH is produced and secreted by the corpora allata, endocrine glands of ectodermal origin associated with the brain and corpora cardiaca (Figs. 1 and 6). In general, the corpora allata of larval insects appear to produce JH, although the glands of last instar larvae in the Lepidoptera, are known to produce the carboxylic acids of JH (Bhaskaran *et al.,* 1986). The corpora allata receive innervation by way of cerebral and central nervous system connections from neurosecretory cell bodies located primarily in the protocerebrum and tritocerebrum, as well as by the subesophageal ganglion. In addition, the glands themselves contain numerous neurosecretory endings which represent the site of release of allatoregulatory peptides and amines, as well as PTTH (see Section II,C).

To date, six juvenile hormones have been identified from various insect orders (Fig. 2C). JH III appears to occur in all orders and is the principal product of the corpus allatum in most, with the notable exceptions of the Lepidoptera and the Diptera. Ironically, however, JH I and JH II, the exclusively lepidopteran JHs, were the first JHs to be identified (Röller *et al.,* 1967; Meyer *et al.,* 1968). In the Lepidoptera, five JHs are produced, JH I, JH II, JH III, JH 0, and 4-methyl-JH I (the latter two are only found in *Manduca* embryos and JH 0 in *Hyalophora cecropia* males) (Bergot *et al.,* 1980; Schooley *et al.,* 1984). It is uncertain if JH 0 and 4-methyl-JH are produced by the corpora allata. In the higher flies, the bisepoxide of JH III, JHB_3, is found, in addition to JH III (Richard *et al.,* 1989a), and JHB_3 is the sole JH in some species of flies.

The absolute configurations of the epoxide group of only some of the JHs have been resolved (Fig. 2C). There are chiral centers at the 10 position of JH III and at the 10 and 11 positions of the other JHs. In addition, JHB_3 from Diptera possesses three chiral centers at positions 6, 7, and 10. At present, the absolute configurations are known only for JH I, 4-Me-JH I, JHB_3, and JH III. This is important because the unnatural enantiomers appear to be less biologically active or are degraded at different rates by esterolytic enzymes than are the natural enantiomers (Peter *et al.,* 1979; King and Tobe, 1993).

JH acids are also produced by the corpora allata of *Manduca* larvae. The glands lose their ability to methylate JH I and JH II acid during the final larval stage as a result of the disappearance or inactivation of the methyltransferase enzyme, and thus produce large quantities of these JH acids which are released into the hemolymph (Bhaskaran *et al.,* 1986; Baker *et al.,* 1987). Acid production continues into the pupal and adult stage, accompanied by high activity of the (HMG) hydroxymethylglutaryl-CoA reductase, one of the important rate-limiting enzymes in

JH and JH acid production (Bhaskaran *et al.,* 1986). These acids appear to have little biological activity but can be converted, albeit at low levels, into the JHs. It has been hypothesized that imaginal discs may be sites of methylation of these JH acids and thus the acids can be regarded as prohormones (Bhaskaran *et al.,* 1986) as with the 3-dehydroecdysone → ecdysone → 20-hydroxyecdysone cascade (see Section III,B). The biological activity of the JH acids and the changes in the abilities of imaginal tissues to methylate them may, thus, be critical to an understanding of the role of JH in metamorphosis in this species. Another acid, the unepoxidized JH III precursor, farnesoic acid, is biosynthesized and released by glands of the cockroach *Diploptera* (Yagi *et al.,* 1991) during the final larval stage. In most species studied, incubation of corpora allata with farnesoic acid or the JH acids can stimulate significant JH production at most stages, although at the time of its release by larval cockroach glands, farnesoic acid has a minimal stimulatory effect on JH production (Tobe and Stay, 1985; Yagi *et al.,* 1991). These data suggest that in metamorphosing larvae, methyltransferase activity responsible for the methylation of farnesoic acid or JH acids in the corpus allatum is dramatically reduced, with a corresponding loss in the ability of the glands to produce JH. This is important because it indicates that the regulation of JH production occurs not only at the level of early rate-limiting steps, such as HMGCoA reductase, as in vertebrate sterol biosynthesis, but also at later stages (see Tobe and Stay, 1985; Cusson *et al.,* 1991) and emphasizes the significance of multivalent feedback in the regulation of JH biosynthesis.

The release of JH from the corpora allata occurs soon after its biosynthesis as a result of physical or facilitated diffusion rather than as an active process (Tobe and Pratt, 1974) and the rapidity of this release is probably a function of the close association between the smooth endoplasmic reticulum (SER) in corpus allatum cells and the external medium. Cells of active glands are typically characterized by extensive SER and mitochondria whereas "inactive" glands producing minimal quantities of JH contain whorls of SER and an accumulation of lipid droplets but reduced SER (Dai and Gilbert, 1991). The corpus allatum does not appear to store JH since little can be extracted and isolated from the glands (Tobe and Stay, 1985). Thus, the hemolymph titer probably reflects the rate of production, although degradative enzymes exist in the hemolymph.

JHB_3 has been identified as a product of the corpus allatum portion of the ring glands of *Drosophila* larvae (Richard *et al.,* 1989a,b). However, it has not been isolated from the hemolymph to date and hence its role as a hormone remains to be clarified. Nonetheless, synthetic JHB_3 does show significant biological activity in a developmental pupal assay (Richard *et al.,* 1989a) and in a gonadotropic assay in adults (Saunders *et al.,* 1990), albeit less than JH III. This may be attributable to the racemic nature of the synthetic JHB_3 (four possible stereoisomers), since presumably only one of the enantiomers shows high bioactivity. The titer of JH III in whole body extracts of larvae of *Drosophila* has been determined using physicochemical methods (Sliter *et al.,* 1987) (Fig. 7). Although a complete correspondence does not exist between JH III and JHB_3 production, high rates of JHB_3 biosynthesis are usually accompanied by high rates of JH III biosynthesis (Richard *et al.,* 1989b), suggesting that the titer of JH III in *Drosophila* may also reflect that of JHB_3.

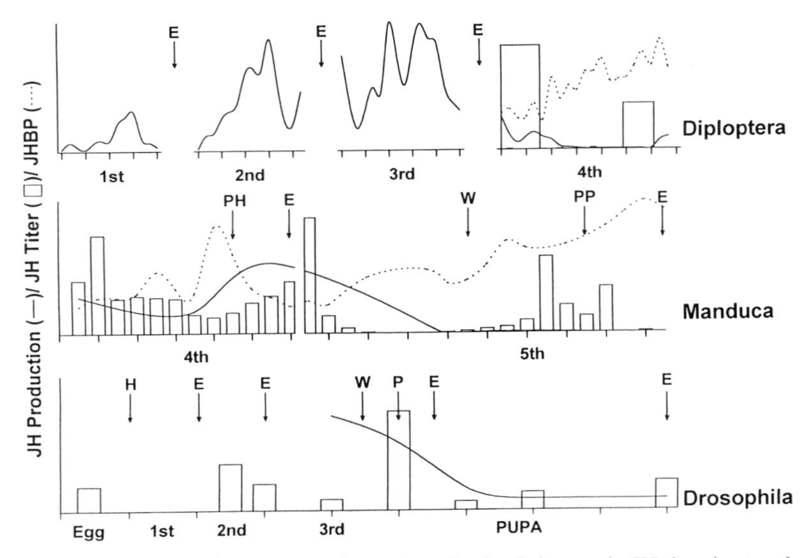

Fig. 7. Diagrammatic representation and synthesis of changes in JH titer (rectangles), JH-binding protein (JHBP) (dashed line), and JH production (solid line) during the larval life of hemimetabolous (*Diploptera*) and holometabolous (*Manduca* and *Drosophila*) insects. E, ecdysis; H, hatching; P, pupariation; PH, pharate fifth instar; PP, prepupa; W, wandering. [Data for *Diploptera* from Tobe *et al.* (1985), King and Tobe (1993), and Kikukawa and Tobe (1986a); for *Manduca sexta,* from Baker *et al.* (1987), Hidayat and Goodman (1994), Goodman (1985), Bhaskaran *et al.* (1986), and Granger *et al.* (1982); for *Drosophila melanogaster,* from Sliter *et al.* (1987), Dai and Gilbert (1991), and Richard *et al.* (1989b).]

B. Biosynthesis of Juvenile Hormones

The JHs are biosynthesized from the simple precursors acetate (JH III) and/or propionate (higher JH homologs) (Schooley and Baker, 1985). The pathway of biosynthesis in the case of JH III is identical to that for vertebrate sterol biosynthesis until the production of farnesyl pyrophosphate (Fig. 8). As noted previously, insects do not produce cholesterol and related steroids *de novo;* rather, JH is the product of this pathway. JH III occurs in the more primitive insect orders (see Tobe and Stay, 1985) of the Hemimetabola, as well as in more "primitive" members of the Holometabola, whereas the higher homologs of JH III (JH I, JH II) occur in the more advanced orders (e.g., Lepidoptera, Diptera), often in conjunction with JH III (see Sehnal *et al.,* Chapter 1, this volume). This observation, as well as its presence in all insects examined, suggests that the JH III biosynthetic pathway is more primitive. It is noteworthy that there is significant sequence similarity between the HMGCoA reductase, the enzyme responsible for the conversion of HMGCoA to mevalonate, of the insect corpus allatum and that of vertebrate liver, a principal site of *de novo* sterol biosynthesis (Chin *et al.,* 1982; Gertler *et al.,* 1988; Martinez-Gonzalez *et al.,* 1993) (Fig. 8), suggesting that this pathway to farnesyl pyrophosphate is of ancient origin.

The formation of the side chains in JH I and JH II involves differential utilization of substrates, including propionate (derived from branched chain amino acids such as isoleucine) and acetate, to give rise to both C_5 and C_6 pyrophosphate intermediates (see Fig. 8). Condensation of two C_6 units plus one C_5 unit results in the formation of JH I, whereas one C_6 unit plus two C_5 units produce JH II. The functional significance of the production of JH I and JH II remains uncertain, as does the significance of changes in the relative ratios of these compounds during larval life and metamorphosis. However, both JHs are known to have greater potency than JH III in many biological assays (see, e.g., Dahm *et al.,* 1976; Henrick *et al.,* 1976) and this may reflect different JH receptors or higher receptor affinity for these homologs. In addition, the association constants for JH I, JH II, and JH III with the JH binding protein (JHBP) (see later) differ markedly, with JH I > JH II > JH III (Hidayat and Goodman, 1994), suggesting that JH III and JH II may be more subject to degradation by esterases than JH I. Finally, the JH-specific esterases also show differences in degradative capacity for the JH homologs (Hammock, 1985; Roe and Venkatesh, 1990). These data

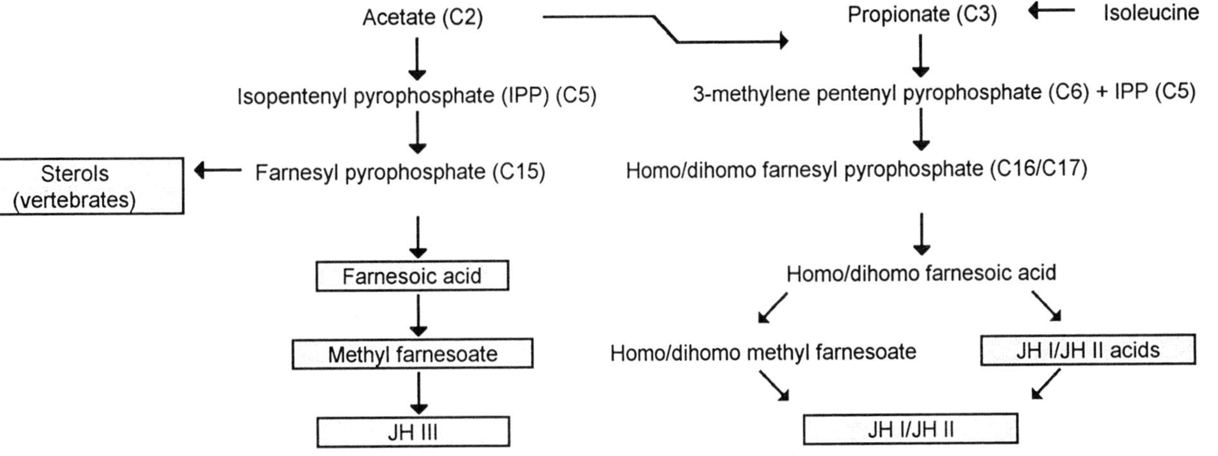

Fig. 8. Biosynthetic pathways for juvenile hormones. JH III is derived exclusively from acetate (C_2), whereas JH II is derived from both propionate (C_3) and acetate (C_2). JH I is derived from propionate (C_3). Known release products are shown in shaded boxes. The insect pathway for the production of farnesyl pyrophosphate is probably identical to that of vertebrates; insects convert farnesyl pyrophosphate into JH III, whereas in vertebrates, it is converted into sterols.

all suggest that JH I and JH II are more "persistent" and protected than JH III, perhaps due to their greater hydrophobicity, resulting in higher biological potency.

C. Regulation of Juvenile Hormone Titer

The titer of JH in the hemolymph must reflect both the rate of production and the rate of degradation. This estimate is confounded by the presence of JH-specific binding proteins in the hemolymph whose function has been hypothesized to be the protection of JH from degradation by both general and specific hemolymph esterases (see, e.g., Roe and Venkatesh, 1990). JH-specific epoxide hydrolases capable of hydrating the epoxide function to the diol also play a role in the catabolism of JH. Since the principal substrates of this enzyme, at least in lepidopteran eggs, are JH acids (Prestwich *et al.,* 1994), it may not play a major role in the direct regulation of the JH titer.

Many studies have examined JH biosynthesis and release by the corpora allata during larval development (see Tobe and Stay, 1985), although only a few have determined JH titers by precise physicochemical methods. In general, a good correlation exists between total JH biosynthesis and JH titer (see, e.g., Tobe *et al.,* 1985), further obscuring the role of esterases and hydrolases in the regulation of JH titer. It is clear that unbound JH is metabolized rapidly in hemolymph to the acid (Roe and Venkatesh, 1990), but the importance of association with the binding protein is not clear. For example, in last instar cockroaches (*Diploptera*), there are changes in the concentration of JH-binding protein (lipophorin) (JHBL) (See Fig. 7), but the relative concentration is inversely related to JH titer (King and Tobe, 1993). A similar situation is found in fourth instar *M. sexta* (Hidayat and Goodman, 1994) in which the concentration of JHBP is highest when the JH titer is minimal (Fig. 7). In both these cases, the JHBL/BP concentration far exceeds the JH titer, indicating that the majority of the BP is unloaded (Hidayat and Goodman, 1994). Therefore, virtually all JH released by the corpus allatum would be associated with JHBL/BP and thereby protected from degradation. In view of this strong association between JH and its binding proteins, how is the JH titer regulated? One possible mechanism involves the clearance of JH from the JHBP in the presence of high concentrations of the JH-specific esterase (Prestwich *et al.,* 1994). In this scheme, a high concentration of the esterase is necessary for the reduction in the JH titer in the last larval stage of Lepidoptera. The JHBP, in addition to its functions in solubilizing JH and protecting it from degradation by nonspecific esterases, plays a role in the complete degradation of JH by extracting it from lipophilic stores. This latter point is important because there are appreciable depots of JH outside of the hemolymph (see, e.g., Tobe *et al.,* 1985), attributable in large part to the lipophilic nature of JH. The high affinity of the JHBP/BL and its high concentration in the hemolymph, particularly at times when JH titers are low or decreasing (see Fig. 8), ensures an efficient removal of the JH from lipophilic stores and, in the presence of high concentrations of JH esterase, results in the hydrolysis of the hormone. Hence, at these times, JHBP contributes not only to the overall regulation of the JH titer but also to the removal, and subsequent degradation, of the hormone from the tissues, thus permitting metamorphosis. It has also been suggested that the presence of the JH esterase promotes the dissociation of bound JH from the BP and, by virtue of the high affinity of the esterase for JH, hydrolyzes "bound" JH (Abdel-Aal and Hammock, 1988). However, experimental determination of the interactions between free and sequestered JH, JHBP/BL, and JH esterase in regulating the JH titer will be necessary to establish, for example, how rapidly the JH titer can be lowered in the presence of the JHBP/BL. Given the high affinity of the BP for JH, it should be possible to establish the kinetics of JH disappearance based on rates of biosynthesis, the association constant of JHBP, and rates of degradation. It is noteworthy, however, that the JH titer cannot be reduced to zero, even in the absence of JH biosynthesis, and that rates of clearance of the hormone are not constant (Tobe *et al.,* 1985). This emphasizes the importance of a critical threshold for the JH titer in terms of metamorphic effects; however, this threshold may differ for other biological effects.

D. Juvenile Hormone Titer and Metamorphosis

The JH titer is believed to be the primary endocrine factor influencing the "quality" of developmental events during metamorphosis, (e.g., in Lepidoptera, the nature of the molt, i.e., larval–larval, larval–pupal, or pupal–adult) (Gilbert *et al.,* 1980) (Fig. 1). Thus, the JH titer directly affects metamorphosis, although in no insect species has it been possible to correlate actual JH titers with the *degree* of metamorphosis. Our present understanding of the control of metamorphosis and the role of JH rests largely on gross determinations of the JH titer in either hemolymph or whole body extracts, and its correlation with specific developmental events, or on experimental manipulation of JH titers through allatectomy and hormone replacement, or treatment of insects with JH or mimics. The classic experiments with hemimetabolous (Wigglesworth, 1934, 1936) and holometabolous insects (Williams, 1961) provided qualitative information on the role of JH in metamorphosis and, since that time, researchers have been attempting to assess the quantitative effects of JH.

It is generally assumed that the absence of JH is required for metamorphosis to the adult in holometabolous insects (see Fig. 1) (Riddiford, 1994), although the situation is less clear in the Hemimetabola. Even in holometabolous insects, the role of JH has not been fully resolved and may require reexamination because significant titers of JH are present in larvae at unexpected times (Fig. 7). For example, appreciable quantities of JH are present at the beginning of the last larval instar, as well as at the end of the instar, and JH is present in low titer at the time of commitment to the pupa. Thus, commitment to the pupa is not definable on the basis of the complete absence of JH, and experimental data on the "critical period" which dictates the interval during which the outcome of the next metamorphic or nonmetamorphic molt can be altered by experimental manipulation of JH titers (e.g., by JH treatment or allatectomy) indicate that some JH must be present at times of metamorphosis to the pupa for normal completion of the process. However, the elevated titer of JH at the beginning of the instar may control the subsequent ecdysteroid or PTTH surge responsible for the initiation of ecdysis (Hiruma, 1986; Kikukawa and Tobe, 1986b; Rountree and Bollenbacher, 1986). JH also is essential for the continued presence of the putative JH receptor in larval *Manduca* epidermis (see Riddiford, Chapter 6, this volume), particularly at times of low ecdysteroid titer. Furthermore, JH may be involved in the regulation of expression of the isoforms of the ecdysone receptor, depending again on the presence of an ecdysteroid (see Riddiford, Chapter 6, this volume). In *Manduca,* the ecdysteroid peak in the last larval stage precedes the peak in JH titer which defines the pupal molt (Baker *et al.,* 1987).

In Hemimetabola, experimental withdrawal of JH at the beginning of the last two larval stages usually results in metamorphosis. In the case of penultimate instars, this metamorphosis is premature (see, e.g., Kikukawa and Tobe, 1986a,b). As noted earlier, the presence of JH is also necessary for the normal timing of production of the ecdysteroid or PTTH peaks that ultimately elicit ecdysis (see Sections II,C and III,B,2). Conversely, experimental elevation of the JH titer at these times invariably prevents metamorphosis. Although the JH titer in the final instar is substantial at the beginning of that stage (Fig. 7) and may function not to prevent metamorphosis but rather to regulate molting through the timing of PTTH or ecdysteroid production, an elevation in the JH titer beyond this critical level at this time prevents metamorphosis. Thus, regulation of JH production, which is manifested directly in the JH titer, may be one of the primary determinants of metamorphosis in this group.

E. Regulators of Corpus Allatum Activity

It is clear that the rate of JH production is an important component in the regulation of the JH titer and that there are rapid and major changes in JH production during development and metamorphosis (Fig. 7). These changes are probably orchestrated by factors external to the corpus allatum and, to date, several peptides which have potent allatoregulatory activity have been isolated and identified in both brain and other tissues. These peptides are capable of stimulating (allatotropins) or inhibiting (allatostatins) JH production *in vitro* (Tobe and Stay, 1985; Stay *et al.,* 1994). Although these peptides have not been demonstrated conclusively to function *in vivo* in the regulation of hormone production, their potency in altering JH production *in vitro* argues strongly for a role in the control of corpus allatum activity. Only one allatotropin has been identified to date, a tridecapeptide from *Manduca* which is active only on adult female corpora allata (Kataoka *et al.,* 1989); hence, the role of allatostimulatory peptides in metamorphosis remains conjectural.

Allatostatins have been identified from several species, including *Manduca* and *Diploptera,* and both appear to be active in larval stages (Stay *et al.,* 1994). The *Manduca* peptide is a pentadecapeptide, with a pyroglutamyl-blocked N terminus whereas the *Diploptera* peptides comprise a family of 13 peptides, all with a characteristic Tyr/Phe-X-Phe-Gly-Leu-NH_2C terminus (X = Gly, Ala, Ser, Asp, Asn) (Stay *et al.,* 1994). No sequence similarity exists between the *Manduca* and *Diploptera* allatostatins. *Manduca* allatostatin (10 n*M*) inhibits JH biosynthesis by more than 80% in early last instars at times of high JH production (Kramer *et al.,* 1991), suggesting that the rapid decline in JH biosynthesis observed in early fifth instar *Manduca* (Fig. 7) is attributable in part to the action of the allatostatin. Hence, the allatostatins may be important metamorphic determinants in this species. In penultimate and last instar *Diploptera,* sensitivity to allatostatins changes dramatically during the instar with a general decline in sensitivity as the instar progresses. At the end of the penultimate instar, sensitivity is recovered whereas sensitivity is lost in the last instars (Stay *et al.,* 1994). Because JH biosynthesis is elevated at the beginning of both periods (Fig. 7), allatostatins may play a role in the reduction in JH production at these times and at the end of the penultimate stage, times when commitment to metamorphosis is occurring or has occurred. However, the suppression of JH biosynthesis in the latter half of the final stage, after the "critical" period, does not appear to be allatostatin regulated but rather is probably attributable to inactivation or turnover of strategic rate-limiting enzymes of the biosynthetic pathway as a result of metamorphosis.

In conclusion, JH defines the outcome of molts, both metamorphic and nonmetamorphic, and can therefore be regarded as the metamorphic hormone of insects. The JH titer during the "critical period," the period of sensitivity to JH, defines the nature of the molts, with a period of marked reduction in the JH titer resulting in metamorphic molts. Higher titers result in nonmetamorphic molts (larval Holometabola) or more gradual metamorphosis (larval Hemimetabola). The JH titer is determined by a complex interaction among biosynthesis, degradation, and binding to JH-specific binding proteins which may differ in hemi- and holometabolous larvae. JH-binding proteins may not only afford protection for JH in the presence of catabolic enzymes, but also may be the agents responsible for reducing JH levels within tissues. JH also regulates ecdysteroid titer and appears to be responsible in part for the ecdysteroid surge that ultimately regulates the molting process. The JH titer can be regulated by changes in the production of JH and this in turn is regulated by neuropeptides that directly influence biosynthetic activity of the corpora allata.

V. EPILOGUE

We have attempted to lay the groundwork for the contributions to follow, several of which deal with the means by which 20-hydroxyecdysone and juvenile hormone elicit their effects at the level of the gene.

It is certainly appropriate in the case of ecdysteroids since the first example of a hormone acting at the level of transcription was the classic study of Clever and Karlson (1960) on the ecdysone-elicited puffing of specific sites on the polytene chromosome of *Chironomus* (see Lezzi, Chapter 4, this volume). The groundwork for these studies was the description by Beermann (1958) of the very large alteration in the Balbiani rings of the *Chironomus* chromosome during molting. A year later it was postulated that "perhaps these chromosomal changes are, in fact, one of the more immediate actions of ecdysone, changes that reflect the elaboration of specific substances (ribonucleic acid, nucleotide coenzymes?) by the nucleus, substances that are destined to participate personally in the cytoplasmic synthesis that characterizes molting" (Schneiderman and Gilbert, 1959). And indeed they are, as the following chapters demonstrate!

ACKNOWLEDGMENTS

We thank Pat Cabarga for excellent secretarial assistance and Susan Whitfield for graphics. Research from the L. I. Gilbert Laboratory has been supported for almost 40 years by the National Institutes of Health and National Science Foundation, most recently by NIH Grants DK30118 and RR06627 and by NSF Grant IBN 9300164.

REFERENCES

Abdel-Aal, Y. A., and Hammock, B. D. (1988). Kinetics of binding and hydrolysis of juvenile hormone II in the haemolymph of *Trichoplusia ni. Insect Biochem.* **18,** 743–750.

Adachi-Yamada, T., Iwani, M., Kataoka, H., Suzuki, A., and Ishizaki, H. (1994). Structure and expression of the gene for the prothoracicotropic hormone of the silkmoth *Bombyx mori. Eur. J. Biochem.* **22,** 633–643.

Agui, N., Granger, N. A., Bollenbacher, W. E., and Gilbert, L. I. (1979). Cellular localization of the insect prothoracicotropic hormone: *In vitro* assay of a single neurosecretory cell. *Proc. Natl. Acad. Sci. U.S.A.* **76,** 5694–5698.

Agui, N., Bollenbacher, W., Granger, N., and Gilbert, L. I. (1980). Corpus allatum is release site for the insect prothoracicotropic hormone. *Nature* (*London*) **285,** 669–670.

Bainbridge, S. P., Redfern, C., and Bownes, M. (1982). Ecdysteroid titres during the stages of *Drosophila* metamorphosis. *In* "Advances in Genetics, Development and Evolution of *Drosophila*" (S. Laskavaara, ed.), pp. 165–188. Plenum, New York.

Baker, F. C., Tsai, L. W., Reuter, C. C., and Schooley, D. A. (1987). *In vivo* fluctuation of JH, JH acid, and ecdysteroid titer, and JH esterase activity, during development of fifth stadium *Manduca sexta. Insect Biochem.* **17,** 989–996.

Becker, J., and Craig, E. A. (1994). Heat-shock proteins as molecular chaperones. *Eur. J. Biochem.* **219,** 11–23.

Beermann, W. (1958). Chromosomal differentiation in insects. *In* "Developmental Cytology," Sixteenth Growth Symposium, pp. 83–103. Ronald Press, New York.

Bellés, X., Cassier, P., Cerdá, X., Pascual, N., André, M., Rosso, Y., and Piulachs, M. D. (1993). Induction of choriogenesis by 20-hydroxyecdysone in the German cockroach. *Tissue Cell* **25,** 195–204.

Bergot, B. J., Jamieson, G. C., Ratcliff, M. A., and Schooley, D. A. (1980). JH zero: New naturally occurring insect juvenile hormone from developing embryos of the tobacco hornworm. *Science* **210,** 336–338.

Bhaskaran, G., Sparagana, S. P., Barrera, P., and Dahm, K. H. (1986). Change in corpus allatum function during metamorphosis of the tobacco hornworm *Manduca sexta.* Regulation at the terminal step in juvenile hormone biosynthesis. *Arch. Insect Biochem. Physiol.* **3,** 321–338.

Bollenbacher, W. E., and Gilbert, L. I. (1980). Neuroendocrine control of postembryonic development in insects. *In* "Neurosecretion" (D. S. Farner and K. Lederis, eds.), pp. 361–370. Plenum, New York.

Bollenbacher, W. E., and Granger, N. A. (1985). Endocrinology of the prothoracicotropic hormone. *In* "Comprehensive Insect Physiology, Biochemistry and Pharmacology" (G. A. Kerkut and L. I. Gilbert, eds.), Vol. 7, pp. 109–151. Pergamon, Oxford.

Bollenbacher, W., Vedeckis, W., O'Connor, J. D., and Gilbert, L. I. (1975). Ecdysone titers and prothoracic gland activity during the larval–pupal development of *Manduca sexta. Dev. Biol.* **44,** 46–53.

Bollenbacher, W. E., O'Brien, M. A., Katahira, E. J., and Gilbert, L. I. (1983). A kinetic analysis of the action of the insect prothoracicotropic hormone. *Mol. Cell. Endocrinol.* **32,** 27–46.

Bollenbacher, W. E., Katahira, E. J., O'Brien, M. A., Gilbert, L. I., Thomas, M. K., Agui, N., and Baumhover, A. H. (1984). Insect prothoracicotropic hormone: Evidence for two molecular forms. *Science* **224,** 1243–1245.

Bowen, M. F., Bollenbacher, W. E., and Gilbert, L. I. (1984). *In vitro* studies on the role of the brain and prothoracic glands in the pupal diapause of *Manduca sexta. J. Exp. Biol.* **108,** 9–24.

Butenandt, A., and Karlson, P. (1954). Über die Isolierung eines Metamorphose-Hormons der Insekten in kristallisierter Form. *Z. Naturforsch.* **93,** 389–391.

Chin, D. J., Luskey, K. L., Faust, J. R., MacDonald, R. J., Brown, M. S., and Goldstein, J. L. (1982). Molecular cloning of 3-hydroxy-3-methylglutaryl coenzyme A reductase and evidence for regulation of its mRNA. *Proc. Natl. Acad. Sci. U.S.A.* **79,** 7704–7708.

Clever, U., and Karlson, P. (1960). Induktion von Puff-Veranderungen in den Speicheldrüsenchromosomen von *Chironomus tentans* durch ecdyson. *Exp. Cell Res.* **20,** 623–626.

Combest, W. L., and Gilbert, L. I. (1992). Polyamines modulate multiple protein phosphorylation pathways in the insect prothoracic gland. *Mol. Cell. Endocrinol.* **83,** 11–19.

Cusson, M., Yagi, K. J., Ding, Q., Duve, H., Thorpe, A., McNeil, J. N., and Tobe, S. S. (1991). Biosynthesis and release of juvenile hormone and its precursors in insects and crustaceans: The search for a unifying arthropod endocrinology. *Insect Biochem.* **21,** 1–6.

Dahm, K. H., Bhaskaran, G., Peter, M. G., Shirk, P. D., Seshan, K. R., and Röller, H. (1976). On the identity of juvenile hormone in insects. *In* "The Juvenile Hormones" (L. I. Gilbert, ed.), pp. 19–47. Plenum, New York.

Dai, J.-D., and Gilbert, L. I. (1991). Metamorphosis of the corpus allatum and degeneration of the prothoracic glands during the larval–pupal–adult transformation of *Drosophila melanogaster:* A cytophysiological analysis of the ring gland. *Dev. Biol.* **144,** 309–326.
Dai, J.-D., Costello, J., and Gilbert, L. I. (1994a). The prothoracic glands of *Manduca sexta:* A microscopic analysis of gap junctions and intercellular bridges. *Invertebr. Reprod. Dev.* **25,** 93–110.
Dai, J.-D., Mizoguchi, A., and Gilbert, L. I. (1994b). Immunoreactivity of neurosecretory granules in the brain–retrocerebral complex of *Manduca sexta* to heterologous antibodies against *Bombyx* prothoracicotropic hormone and bombyxin. *Invertebr. Reprod. Dev.* **26,** 187–196.
Dai, J. D., Mizoguchi, A., Satake, S., Ishizaki, H., and Gilbert, L. I. (1995). Developmental changes in the prothoracicotropic hormone content of the *Bombyx mori* brain-retrocerebral complex and hemolymph: analysis by immunogold electron microscopy, quantitative image analysis, and time-resolved fluoroimmunoassay. *Dev. Biol.* **171,** 212–223.
Delbecque, J.-P., Weidner, K., and Hoffmann, K. H. (1990). Alternate sites for ecdysteroid production in insects. *Invertebr. Reprod. Dev.* **18,** 29–42.
Dorn, A., Bishoff, S. T., and Gilbert, L. I. (1987). An incremental analysis of the embryonic development of the tobacco hornworm, *Manduca sexta. Invertebr. Reprod. Dev.* **11,** 137–157.
Duncan, R., and McConkey, E. H. (1982). Preferential utilization of phosphorylated 40S ribosomal subunits during initiation complex formation. *Eur. J. Biochem.* **123,** 535–538.
Gelman, D. B., Thyagaraja, B. S., Kelly, T. J., Masler, E. P., Bell, R. A., and Borkovec, A. B. (1991). The insect gut: A new source of ecdysiotropic peptides. *Experientia* **47.,** 77–80.
Gertler, F. B., Chui, C. Y., Richter-Mann, L., and Chin, D. J. (1988). Developmental and metabolic regulation of the *Drosophila melanogaster* 3-hydroxy-3-methylglutaryl coenzyme A reductase. *Mol. Cell. Biol.* **8,** 2713–2721.
Gilbert, L. I. (1964). Physiology of growth and development: Endocrine aspects. *In* "Physiology of Insecta" (M. Rockstein, ed.), Vol. 1, pp. 149–225. Academic Press, New York.
Gilbert, L. I. (1967). Lipid metabolism and function in insects. *Adv. Insect Physiol.* **4,** 69–211.
Gilbert, L. I. (1974). Endocrine action during insect growth. *Recent Prog. Horm. Res.* **30,** 347–390.
Gilbert, L. I. (1989). The endocrine control of molting: The tobacco hornworm, *Manduca sexta,* as a model system. *In* "Ecdysone From Chemistry to Mode of Action" (J. Koolman, ed.), pp. 448–471. Thieme, Stuttgart.
Gilbert, L. I., and Frieden, E., eds. (1981). "Metamorphosis," 2nd ed. Plenum, New York.
Gilbert, L. I., and King, D. S. (1973). Physiology of growth and development: Endocrine aspects. *In* "Physiology of Insects" (M. Rockstein, ed.), Vol. I, 2nd ed., pp. 249–370. Academic Press, New York.
Gilbert, L. I., and Schneiderman, H. A. (1959). Prothoracic gland stimulation by juvenile hormone extracts of insects. *Nature (London)* **184,** 171–173.
Gilbert, L. I., and Schneiderman, H. (1960). The development of a bioassay for the juvenile hormone of insects. *Trans. Am. Microsc. Soc.* **79,** 38–67.
Gilbert, L. I., Bollenbacher, W. E., Goodman, W., Smith, S. L., Agui, N., Granger, N. A., and Sedlak, B. J. (1980). Hormones controlling insect metamorphosis. *Recent Prog. Horm. Res.* **36,** 401–449.

Gilbert, L. I., Bollenbacher, W. E., Agui, N., Granger, N. A., Sedlak, B. J., Gibbs, D., and Buys, C. M. (1981). The prothoracicotropes: Source of the prothoracicotropic hormone. *Am. Zool.* **21,** 641–653.

Gilbert, L. I., Combest, W. L., Smith, W. A., Meller, V. H., and Rountree, D. B. (1988). Neuropeptides, second messengers and insect molting. *BioEssays* **8,** 153–157.

Gilbert, L. I., Warren, J. T., and Henrich, V. (1991). Emerging technologies of insect endocrinology: Analytical and molecular. *In* "Proceedings of the Centennial Meeting of the Entomological Society of America" (S. B. Vinson and R. L. Metcalf, eds.), pp. 133–160. Entomological Society of America, Lanham, Maryland.

Goodman, W. G. (1985). Relative hemolymph juvenile hormone binding protein capacity during larval–pupal and adult development of *Manduca sexta. Insect Biochem.* **15,** 557–564.

Granger, N. A., Niemiec, S. M., Gilbert, L. I., and Bollenbacher, W. E. (1982). Juvenile hormone synthesis *in vitro* by larval and pupal corpora allata of *Manduca sexta. Mol. Cell. Endocrinol.* **28,** 587–604.

Grau, V., and Lafont, R. (1994). Metabolism of ecdysone and 20-hydroxyecdysone in adult *Drosophila melanogaster. Insect Biochem. Mol. Biol.* **24,** 49–58.

Gray, R. S., Muehleisen, D. P., Katahira, E. J., and Bollenbacher, W. E. (1994). The prothoracicotropic hormone (PTTH) of the commercial silkmoth. *Bombyx mori,* in the CNS of the tobacco hornworm, *Manduca sexta. Peptides* (*N.Y.*) **15,** 777–782.

Grieneisen, M. L. (1994). Recent advances in our knowledge of ecdysteroid biosynthesis in insects and crustaceans. *Insect Biochem. Mol. Biol.* **24,** 115–132.

Grieneisen, M. L., Warren, J. T., and Gilbert, L. I. (1991). A putative route to ecdysteroids: Metabolism of cholesterol *in vitro* by mildy disrupted prothoracic glands of *Manduca sexta. Insect Biochem.* **21,** 41–51.

Grieneisen, M. L., Warren, J. T., and Gilbert, L. I. (1993). Early steps in ecdysteroid biosynthesis: Evidence for the involvement of cytochrome P-450 enzymes. *Insect Biochem. Mol. Biol.* **23,** 13–23.

Gu, S.-H., and Chou, Y.-S. (1993). Role of low ecdysteroid levels in the early last larval instar of *Bombyx mori. Experientia* **49,** 806–809.

Hagedorn, H. H. (1985). The role of ecdysteroids in reproduction. *In* "Comprehensive Insect Physiology, Biochemistry and Pharmacology" (G. A. Kerkut and L. I. Gilbert, eds.), Vol. 7, pp. 109–151. Pergamon, Oxford.

Hammock, B. D. (1985). Regulation of juvenile hormone titer. *In* "Comprehensive Insect Physiology, Biochemistry and Pharmacology" (G. A. Kerkut and L. I. Gilbert, eds.), Vol. 7, pp. 109–151. Pergamon, Oxford.

Henrich, V. C., Pak, M. D., and Gilbert, L. I. (1987). Neural factors that stimulate ecdysteroid synthesis by the larval ring gland of *Drosophila melanogaster. J. Comp. Physiol. B* **157,** 543–549.

Henrick, C. A., Staal. G. B., and Siddall, J. B. (1976). Structure activity relationships in some juvenile hormone analogs. *In* "The Juvenile Hormones" (L. I. Gilbert, ed.), pp. 48–60. Plenum, New York.

Herman, W., and Gilbert, L. I. (1966). The neuroendocrine system of *Hyalophora cecropia* (L). I. Anatomy and histology of the ecdysial glands. *Gen. Comp. Endocrinol.* **7,** 275–291.

Hetru, C., Lageaux, M., Luu, B., and Hoffmann, J. A. (1978). Adult ovaries of *Locusta migratoria* contain the sequence of biosynthetic intermediates for ecdysone. *Life Sci.* **22,** 2141–2154.

Hidayat, P., and Goodman, W. G. (1994). Juvenile hormone and hemolymph juvenile hormone binding protein titers and their interaction in the hemolymph of fourth stadium *Manduca sexta. Insect Biochem. Mol. Biol.* **24,** 709–715.

Hiruma, K. (1986). Regulation of prothoracicotropic hormone release by juvenile hormone in the penultimate and last instar larvae of *Mamestra brassicae. Gen. Comp. Endocrinol.* **63,** 201–211.

Hiruma, K., and Riddiford, L. M. (1993). Molecular mechanisms of cuticular melanization in the tobacco hornworm, *Manduca sexta* (L) (Lepidoptera: Sphingidae). *Int. J. Insect Morphol. Embryol.* **22,** 103–117.

Ichikawa, M., and Nishiitsusuji-Uwo, J. (1959). Studies on the role of the corpus allatum in the Eri-silkworm, *Philosomia cynthia ricini. Biol. Bull.* (*Woods Hole, Mass.*) **116,** 88–94.

Ishibashi, J., Kataoka, H., Isogai, A., Kawakami, A., Saegusa, H., Yagi, Y., Mizoguchi, A., Shizaki, H., and Suzuki, A. (1994). Assignment of disulfide bond location in prothoracicotropic hormone of the silkworm, *Bombyx mori:* A homodimeric protein. *Biochemistry* **33,** 5912–5919.

Ishizaki, H., and Suzuki, A. (1992). Brain secretory peptides of the silkmoth *Bombyx mori:* Prothoracicotropic hormone and bombyxin. *Prog. Brain Res.* **92,** 1–14.

Ishizaki, H., Suzuki, A., Moriya, I., Mizoguchi, A., Fujishita, M., O'Oka, H., Kataoka, H., Isogai, A., Nagasawa, H., and Suzuki, A. (1983). Prothoracicotropic hormone bioassay: Pupal–adult *Bombyx* assay. *Dev., Growth Differ.* **25,** 585–592.

Kappler, D., Hetru, C., Durst, F., and Hoffmann, J. A. (1989). Enzymes involved in ecdysone biosynthesis. *In* "Ecdysone From Chemistry to Mode of Action" (J. Koolman, ed.), pp. 161–166. Thieme, Stuttgart.

Kataoka, H., Toschi, A., Li, J. P., Carney, R. L., Schooley, D. A., and Kramer, K. J. (1989). Identification of an allatotropin from adult *Manduca sexta. Science* **243,** 1481–1483.

Kawakami, A., Kataoka, H., Oka, T., Mizoguchi, A., Kimura-Kawakami, M., Adachi, T., Iwami, M., Nagasawa, H., Suzuki, A., and Ishizaki, H. (1990). Molecular cloning of the *Bombyx mori* prothoracicotropic hormone. *Science* **247,** 1333–1335.

Keightley, D. A., Lou, K. J., and Smith, W. A. (1990). Involvement of translation and transcription in insect steroidogenesis. *Mol. Cell. Endocrinol.* **74,** 229–237.

Kikukawa, S., and Tobe, S. S. (1986a). Juvenile hormone biosynthesis in female larvae of *Diploptera punctata* and the effect of allatectomy on haemolymph ecdysteroid titre. *J. Insect Physiol.* **32,** 981–986.

Kikukawa, S., and Tobe, S. S. (1986b). Critical periods for juvenile hormone sensitivity during larval life of female *Diploptera punctata. J. Insect Physiol.* **32,** 1035–1042.

King, L. E., and Tobe, S. S. (1993). Changes in the titre of a juvenile hormone III binding lipophorin in the haemolymph of *Diploptera punctata* during development and reproduction: Functional significance. *J. Insect Physiol.* **39,** 241–251.

Kiriishi, S., Rountree, D. B., Sakurai, S., and Gilbert, L. I. (1990). Prothoracic gland synthesis of 3-dehydroecdysone and its hemolymph 3-reductase mediated conversion to ecdysone in representative insects. *Experientia* **46,** 716–721.

Kopeć, S. (1922). Studies on the necessity of the brain for the inception of insect metamorphosis. *Biol. Bull.* (*Woods Hole, Mass.*) **42,** 323–342.

Kramer, S. J., Toschi, A., Miller, C. A., Kataoka, H., Quistad, G. B., Li, J. P., Carney, R. L., and Schooley, D. A. (1991). Identification of an allatostatin from the tobacco hornworm *Manduca sexta. Proc. Natl. Acad. Sci. U.S.A.* **88,** 9458–9462.

Lafont, R. (1991). Reverse endocrinology, or "hormones" seeking functions. *Insect Biochem.* **21,** 697–721.

Laufer, H., Borst, D., Baker, F. C., Carrasco, C., Sinkus, M., Reuter, C. C., Tsai, L. W., and Schooley, D. A. (1987). The identification of a crustacean juvenile hormone. *Science* **235,** 202–205.

Martinez-Gonzalez, J., Buesa, C., Piulachs, M. D., Bellés, X., and Hegardt, G. G. (1993). Molecular cloning, developmental pattern and tissue expression of 3-hydroxy-3-methylglutaryl coenzyme A reductase of the cockroach *Blattella germanica. Eur. J. Biochem.* **213,** 233–241.

Meller, V. H., and Gilbert, L. I. (1988). Occurrence, quaternary structure and function of G protein subunits in an insect endocrine gland. *Mol. Cell. Endocrinol.* **74,** 133–141.

Meller, V. H., Combest, W. L., Smith, W. A., and Gilbert, L. I. (1988). A calmodulin-sensitive adenylate cyclase in the prothoracic glands of the tobacco hornworm, *Manduca sexta. Mol. Cell. Endocrinol.* **59,** 67–76.

Meller, V. H., Sakurai, S., and Gilbert, L. I. (1990). Developmental regulation of calmodulin-dependent adenylate cyclase in an insect endocrine gland. *Cell. Regul.* **1,** 771–780.

Mészáros, M., and Morton, D. B. (1994). Isolation and partial characterization of a gene from trachea of *Manduca sexta* that requires and is negatively regulated by ecdysteroids. *Dev. Biol.* **162,** 618–630.

Meyer, A. S., Schneiderman, H. A., Hanzman, E., and Ko, J. H. (1968). The two juvenile hormones from the cecropia silk moth. *Proc. Natl. Acad. Sci. U.S.A.* **60,** 853–860.

Mizoguchi, A., Oka, T., Kataoka, H., Nagasawa, H., Suzuki, A., and Ishizaki, H. (1990). Immunohistochemical localization of prothoracicotropic hormone-producing cells in the brain of *Bombyx mori. Dev., Growth Differ.* **32,** 591–598.

Muehleisen, D. P., Gray, R. S., Katahira, E. J., Thomas, M. K., and Bollenbacher, W. E. (1993). Immunoaffinity purification of the neuropeptide prothoracicotropic hormone from *Manduca sexta. Peptides (N.Y.)* **14,** 531–541.

Muehleisen, D. P., Katahira, E. J., Gray, R. S., and Bollenbacher, W. E. (1994). Physical characteristics of the cerebral big prothoracicotropic hormone from *Manduca sexta. Experientia* **50,** 159–163.

Nijhout, H. F., and Williams, C. M. (1974). Control of moulting and metamorphosis in the tobacco hornworm *Manduca sexta* (L): Cessation of juvenile hormone secretion as a trigger for pupation. *J. Exp. Biol.* **61,** 493–501.

O'Brien, M. A., Granger, N. A., Agui, N., Gilbert, L. I., and Bollenbacher, W. E. (1986). Prothoracicotropic hormone in the developing brain of the tobacco hornworm *Manduca sexta:* Relative amounts of two molecular forms. *J. Insect Physiol.* **32,** 719–725.

Orme-Johnson, N. R. (1990). Distinctive properties of adrenal cortex mitochrondria. *Biochim. Biophys. Acta* **1020,** 213–231.

Pak, M., and Gilbert, L. I. (1987). A developmental analysis of ecdysteroids during the metamorphosis of *Drosophila melanogaster. J. Liq. Chromatogr.* **10,** 2591–2612.

Peter, M. G., Gunawan, S., Gellissen, G., and Emmerich, H. (1979). Differences in hydrolysis and binding of homologous juvenile hormone in *Locusta migratoria* haemolymph. *Z. Naturforsch.* **34,** 588–598.

Prestwich, G. D., Touhara, K., Riddiford, L. M., and Hammock, B. D. (1994). Larva lights: A decade of photoaffinity labeling with juvenle hormone analogues. *Insect Biochem. Mol. Biol.* **24,** 747–761.

Prugh, J., Della Croce, K., and Levine, R. B. (1992). Effects of the steroid hormone, 20-hydroxyecdysone, on the growth of neurites by identified insect motoneurons *in vitro*. *Dev. Biol.* **154,** 331–345.

Rainey, W. E., Kramer, R. E., Mason, J. I., and Shay, J. W. (1985). The effects of taxol, a microtubule-stabilizing drug, on steroidogenic cells. *J. Cell. Physiol.* **123,** 17–24.

Redfern, C. P. F. (1983). Ecdysteroid synthesis by the ring gland of *Drosophila melanogaster* during late-larval, prepupal, and pupal development. *J. Insect Physiol.* **29,** 65–71.

Redfern, C. P. F. (1984). Evidence for the presence of makisterone A in *Drosophila* larvae and the secretion of 20-deoxymakisterone A by the ring glands. *Proc. Natl. Acad. Sci. U.S.A.* **81,** 5643–5647.

Rees, H. H. (1985). Biosynthesis of ecdysone. *In* "Comprehensive Insect Physiology, Biochemistry and Pharmacology" (G. A. Kerkut and L. I. Gilbert, eds.), Vol. 7, pp. 109–151. Pergamon, Oxford.

Richard, D. S., Applebaum, S. W., Sliter, T. J., Baker, F. C., Schooley, D. A., Reuter, C. C., Henrich, V. C., and Gilbert, L. I. (1989a). Juvenile hormone bisepoxide biosynthesis *in vitro* by the ring gland of *Drosophila melanogaster:* A putative juvenile hormone in the higher Diptera. *Proc. Natl. Acad. Sci. U.S.A.* **86,** 1421–1425.

Richard, D. S., Applebaum, S. W., and Gilbert, L. I. (1989b). Developmental regulation of juvenile hormone biosynthesis by the ring gland of *Drosophila melanogaster. J. Comp. Physiol. B* **159,** 383–387.

Richards, G. (1978). The relative activities of α- and ß-ecdysone and their 3-dehydro derivatives in the chromosome puffing assay. *J. Insect Physiol.* **24,** 329–335.

Riddiford, L. M. (1976a). Hormonal control of insect epidermal cell commitment *in vitro. Nature* (*London*) **250,** 115–117.

Riddiford, L. M. (1976b). Juvenile hormone control of epidermal commitment *in vivo* and *in vitro. In* "The Juvenile Hormones" (L. I. Gilbert, ed.), pp. 198–219. Plenum, New York.

Riddiford, L. M. (1994). Cellular and molecular actions of juvenile hormone. I. General considerations and premetamorphic actions. *Adv. Insect Physiol* **24,** 213–274.

Roe R. M., and Venkatesh, K. (1990). Metabolism of juvenile hormones: Degradation and titer regulation. *In* "Morphogenetic Hormones of Arthropods" (A. P. Gupta, ed.), pp. 125–179. Rutgers University Press, Piscataway, New Jersey.

Röller, H., Dahm, K. H., Sweeley, C. C., and Trost, B. M. (1967). The structure of juvenile hormone. *Angew. Chem.* **6,** 179–180.

Rountree, D. B., and Bollenbacher, W. E. (1986). The release of prothoracicotropic hormone in the tobacco hornworm, *Manduca sexta,* is controlled intrinsically by juvenile hormone. *J. Exp. Biol.* **120,** 41–58.

Rountree, D. B., Combest, W. L., and Gilbert, L. I. (1987). Protein phosphorylation in the prothoracic glands as a cellular model for juvenile hormone–prothoracicotropic hormone interactions. *Insect Biochem.* **17,** 943–948.

Rountree, D. B., Combest, W. L., and Gilbert, L. I. (1992). Prothoracicotropic hormone regulates the phosphorylation of a specific protein in the prothoracic glands of the tobacco hornworm, *Manduca sexta. Insect Biochem. Mol. Biol.* **22,** 353–362.

Rybczynski, R., and Gilbert, L. I. (1994). Changes in general and specific protein synthesis that accompany ecdysteroid synthesis in stimulated prothoracic glands of *Manduca sexta. Insect Biochem. Mol. Biol.* **24,** 175–189.

Rybczynski, R., and Gilbert, L. I. (1995a). Rapid synthesis of ß tubulin in prothoracicotropic hormone-stimulated prothoracic glands of the moth *Manduca sexta. Dev. Biol.* **169,** 15–28.

Rybczynski, R., and Gilbert, L. I. (1995b). Prothoracicotropic hormone-regulated expression of a hsp 70 cognate protein in the insect prothoracic gland. *Mol. Cell. Endocrinol.,* in press.

Sakurai, S., and Gilbert, L. I. (1990). Biosynthesis and secretion of ecdysteroids by the prothoracic glands. *In* "Molting and Metamorphosis of Insects" (E. Ohnishi and H. Ishizaki, eds.), pp. 83–106. Springer-Verlag, Berlin.

Sakurai, S., and Williams, C. M. (1989). Short-loop and positive feedback on ecdysone secretion by prothoracic gland in the tobacco hornworm, *Manduca sexta. Gen. Comp. Endocrinol.* **75,** 204–216.

Sakurai, S., Warren, J. T., and Gilbert, L. I. (1989). Mediation of ecdysone synthesis in *Manduca sexta* by a hemolymph enzyme. *Arch. Insect Biochem. Physiol.* **10,** 179–197.

Sakurai, S., Warren, J. T., and Gilbert, L. I. (1991). Ecdysteroid synthesis and molting by the tobacco hornworm, *Manduca sexta,* in the absence of prothoracic glands. *Arch. Insect Biochem. Physiol.* **18,** 13–36.

Saunders, D. S., Richard, D. S., Applebaum, S. W., Ma, M., and Gilbert, L. I. (1990). Photoperiodic diapause in *Drosophila melanogaster* involves a block to the juvenile hormone regulation of ovarian maturation. *Gen. Comp. Endocrinol.* **79,** 174–184.

Schneiderman, H. A., and Gilbert, L. I. (1959). The chemistry and physiology of insect growth hormones. *In* "Cell, Organism and Milieu" (D. Rudnick, ed.), pp. 157–187. Ronald Press, New York.

Schooley, D. A., and Baker, F. C. (1985). Juvenile hormone biosynthesis. *In* "Comprehensive Insect Physiology, Biochemistry and Pharmacology" (G. A. Kerkut and L. I. Gilbert, eds.), Vol. 7, pp. 109–151. Pergamon, Oxford.

Schooley, D. A., Baker, F. C., Tsai, L. W., Miller, C. A., and Jamieson, G. C. (1984). Juvenile hormones 0, I and II exist only in Lepidoptera. *In* "Biosynthesis, Metabolism and Mode of Action of Invertebrate Hormones" (J. A. Hoffmann and M. Porchet, eds.), pp. 373–383. Springer-Verlag, Berlin.

Shirai, Y., Aizono, Y., Iwasaki, T., Yanagida, A., Mori, H., Sumida, M., and Matsubara, F. (1993). Prothoracicotropic hormone is released five times in the 5th-larval instar of the silkworm, *Bombyx mori. J. Insect Physiol.* **39,** 83–88.

Shirai, Y., Iwasaki, T., Matsubara, F., and Aizono, Y. (1994). The carbachol-induced release of prothoracicotropic hormone from brain–corpus cardiacum–corpus allatum complex of the silkworm, *Bombyx mori. J. Insect Physiol.* **40,** 469–473.

Siew, Y. C., and Gilbert, L. I. (1971). Effects of moulting hormone and juvenile hormone on insect endocrine gland activity. *J. Insect Physiol.* **17,** 2095–2104.

Simpson, E. R., and Waterman, M. R. (1988). Regulation of the synthesis of steroidogenic enzymes in adrenal cortical cells by ACTH. *Annu. Rev. Physiol.* **50,** 427–440.

Sliter, T. J., Sedlak, B. J., Baker, F. C., and Schooley, D. A. (1987). Juvenile hormone in *Drosophila melanogaster;* identification and titer determination during development. *Insect Biochem.* **17,** 161–165.

Smith, S. L. (1985). Regulation of ecdysteroid titer: Synthesis. *In* "Comprehensive Insect Physiology, Biochemistry and Pharmacology" (G. A. Kerkut and L. I. Gilbert, eds.), Vol. 7, pp. 109–151. Pergamon, Oxford.

Smith, W. A. (1993). Second messengers and the action of prothoracicotropic hormone in *Manduca sexta. Am. Zool.* **33,** 330–339.

Smith, W. A., and Gilbert, L. I. (1989). Early events in peptide-stimulated ecdysteroid secretion by the prothoracic glands of *Manduca sexta. J. Exp. Zool.* **252,** 262–270.

Smith, W. A., and Pasquarello, T. J. (1989). Developmental changes in phosphodiesterase activity and hormonal response in the prothoracic glands of *Manduca sexta. Mol. Cell. Endocrinol.* **63,** 239–246.

Smith, S. L., Bollenbacher, W. E., and Gilbert, L. I. (1983). Ecdysone 20-monooxygenase activity during larval–pupal development of *Manduca sexta. Mol. Cell. Endocrinol.* **31,** 227–251.

Smith, W. A., Gilbert, L. I., and Bollenbacher, W. E. (1984). The role of cyclic AMP in the regulation of ecdysone synthesis. *Mol. Cell. Endocrinol.* **37,** 285–294.

Smith, W. A., Gilbert, L. I., and Bollenbacher, W. E. (1985). Calcium–cyclic AMP interactions in prothoracicotropic hormone stimulation of ecdysone synthesis. *Mol. Cell. Endocrinol.* **39,** 71–78.

Smith, W. A., Combest, W. L., and Gilbert, L. I. (1986). Involvement of cyclic AMP-dependent protein kinase in prothoracicotropic hormone-stimulated ecdysone synthesis. *Mol. Cell. Endocrinol.* **47,** 25–33.

Smith, W. A., Rountree, D. B., Bollenbacher, W. E., and Gilbert, L. I. (1987). Dissociation of the prothoracic glands of *Manduca sexta* into hormone-responsive cells. *In* "Progress in Insect Neurochemistry and Neurophysiology" (A. Borkovec and D. Gelman, eds.), pp. 319–322. Humana, Clifton, New Jersey.

Smith, W. A., Varghese, A. H., and Lou, K. J. (1993). Developmental changes in cyclic AMP-dependent protein kinase associated with increased secretory capacity of *Manduca sexta* prothoracic glands. *Mol. Cell. Endocrinol.* **90,** 187–195.

Snyder, M. J., and Feyereisen, R. (1993). A diazepam binding inhibitor (DBI) homolog from the tobacco hornworm, *Manduca sexta. Mol. Cell. Endocrinol.* **94,** R1–R4.

Song, Q., and Gilbert, L. I. (1994). S6 phosphorylation results from prothoracicotropic hormone stimulation of insect prothoracic glands: A role for S6 kinase. *Dev. Genet.* **15,** 332–338.

Song, Q., and Gilbert, L. I. (1995). Multiple phosphorylation of ribosomal protein S6 and specific protein synthesis are required for prothoracicotropic hormone-stimulated ecdysteroid biosynthesis in the prothoracic glands of *Manduca sexta. Insect Biochem. Mol. Biol.* **25,** 591–602.

Sroka, P., and Gilbert, L. I. (1971). Studies on the endocrine control of post-emergence ovarian maturation in *Manduca sexta. J. Insect Physiol.* **17,** 2409–2419.

Stay, B., Tobe, S. S., and Bendena, W. G. (1994). Allatostatins: Identification, primary structures, functions and distribution. *Adv. Insect Physiol.* **25,** 267–337.

Tanaka, Y., Asahina, M., and Takeda, S. (1994). Induction of ultranumerary larval ecdyses by ecdysone does not require an active prothoracic gland in the silkworm, *Bombyx mori. J. Insect Physiol.* **40,** 753–757.

Tobe, S. S., and Pratt, G. E. (1974). Dependence of juvenile hormone release from corpus allatum on intraglandular content. *Nature* (*London*) **252,** 474–476.

Tobe, S. S., and Stay, B. (1985). Structure and regulation of the corpus allatum. *Adv. Insect Physiol.* **18,** 305–432.

Tobe, S. S., Ruegg, R. P., Stay, B., Baker, F. C., Miller, C. A., and Schooley, D. A. (1985). Juvenile hormone titre and regulation in the cockroach *Diploptera punctata. Experientia* **41,** 1028–1034.

Tomaschko, K.-H. (1994). Defensive secretion of ecdysteroids in *Pycnogonum litorale* (Arthropoda, Pantopoda). *Z. Naturforsch.* **49C,** 367–371.

Truman, J. W. (1985). Hormonal control of ecdysis. *In* "Comprehensive Insect Physiology, Biochemistry and Pharmacology" (G. A. Kerkut and L. I. Gilbert, eds.), Vol. 7, pp. 109–151. Pergamon, Oxford.

Vedeckis, W. V., Bollenbacher, W. E., and Gilbert, L. I. (1976). Insect prothoracic glands: A role for cyclic AMP in the stimulation of α-ecdysone secretion. *Mol. Cell. Endocrinol.* **5,** 81–88.

Veldhuis, J. D., and Klase, P. A. (1982). Mechanisms by which calcium ions regulate the steroidogenic actions of luteinizing hormone in isolated ovarian cells *in vitro. Endocrinology* (*Baltimore*) **111,** 1–6.

Vogt, R. G., Rybczynski, R., Cruz, M., and Lerner, M. R. (1993). Ecdysteroid regulation of olfactory protein expression in the developing antenna of the tobacco hawk moth, *Manduca sexta. J. Neurobiol.* **24,** 581–597.

Warren, J. T., and Gilbert, L. I. (1986). Ecdysone metabolism and distribution during the pupal–adult development of *Manduca sexta. Insect Biochem.* **16,** 65–82.

Warren, J. T., Steiner, B., Dorn, A., Pak, M., and Gilbert, L. I. (1986). Metabolism of ecdysteroids during the embryogenesis of *Manduca sexta. J. Liq. Chromatogr.* **9,** 1759–1782.

Warren, J. T., Sakurai, S., Rountree, D. B., and Gilbert, L. I. (1988a). Synthesis and secretion *in vitro* of ecdysteroids by the prothoracic glands of *Manduca sexta. J. Insect Physiol.* **34,** 571–576.

Warren, J. T., Sakurai, S., Rountree, D., Gilbert, L. I., Lee, S.-S., and Nakanishi, K. (1988b). Regulation of the ecdysteroid titer of *Manduca sexta:* Reappraisal of the role of the prothoracic glands. *Proc. Natl. Acad. Sci. U.S.A.* **85,** 958–962.

Watson, R. D., Yeh, W. E., Muehleisen, D. P., Watson, C. J., and Bollenbacher, W. E. (1993). Stimulation of ecdysteroidogenesis by small prothoracicotropic hormone: Role of cyclic AMP. *Mol. Cell. Endocrinol.* **92,** 221–228.

Webb, B. A., and Summers, M. D. (1993). Stimulation of polydnavirus replication by 20-hydroxyecdysone. *Experientia* **48,** 1018–1022.

Westbrook, A. L., Regan, S. A., and Bollenbacher, W. E. (1993). Developmental expression of the prothoracicotropic hormone in the CNS of the tobacco hornworm, *Manduca sexta. J. Comp. Neurol.* **327,** 1–16.

Wigglesworth, V. B. (1934). The physiology of ecdysis in *Rhodnius prolixus* (Hemiptera). II. Factors controlling moulting and metamorphosis. *Q. J. Microsc. Sci.* **77,** 191–222.

Wigglesworth, V. B. (1936). The function of the corpus allatum in the growth and reproduction of *Rhodnius prolixus* (Hemiptera). *Q. J. Microsc. Sci.* **79,** 91–121.

Wigglesworth, V. B. (1955). The breakdown of the thoracic gland in the adult insect, *Rhodnius prolixus. J. Exp. Biol.* **32,** 485–491.

Williams, C. M. (1952). Morphogenesis and metamorphosis of insects. *Harvey Lect.* **47,** 126–155.

Williams, C. M. (1959). The juvenile hormone. I. Endocrine activity of the corpora allata of the adult Cecropia silkworm. *Biol. Bull.* (*Woods Hole, Mass.*) **116,** 323–338.

Williams, C. M. (1961). The juvenile hormone. II. Its role in the endocrine control of molting, pupation, and adult development of the Cecropia silkworm. *Biol. Bull.* (*Woods Hole, Mass.*) **121,** 572–585.

Yagi, K. J., Konz, K., Stay, B., and Tobe, S. S. (1991). Production and utilization of farnesoic acid in the juvenile hormone biosynthetic pathway by corpora allata of larval *Diploptera punctata. Gen. Comp. Endocrinol.* **81,** 284–294.

Žitňan, D., Sehnal, F., and Bryant, P. J. (1993). Neurons producing specific peptides in the central nervous system of normal and pupariation-delayed *Drosophila. Dev. Biol.* **156,** 117–135.

3

Ecdysone-Regulated Chromosome Puffing in *Drosophila melanogaster*

STEVEN RUSSELL AND MICHAEL ASHBURNER

Department of Genetics
University of Cambridge
Cambridge CB2 3EH
United Kingdom

I. INTRODUCTION

We have come a long way since Beermann (1952) first suggested that the puffs visible on the polytene chromosomes of dipteran larvae are cytological manifestations of tissue-specific and developmentally regulated gene activity. Since then, the contributions of several laboratories have confirmed this hypothesis and produced a more detailed understanding of a complex hierarchy of regulatory genes whose activities are coordinated with the early stages of metamorphosis. We discuss the way in which the study of a model system, the late larval–early pupal salivary gland chromosomes of *Drosophila melanogaster,* has led to a conceptual framework for the study of hormone-regulated developmental processes (see Cherbas and Cherbas, Chapter 5, this volume). In addition, we describe the practical uses of this system in isolating and characterizing the regulatory genes which play key roles during metamorphosis.

Chromosome puffs are diffuse swellings along the regularly banded arms of polytene chromosomes (see also Lezzi, Chapter 4, this volume). The polytene chromosomes, which are found in many species of Diptera, are the product of several rounds of DNA replication without subsequent chromosome segregation. In some tissues, such as the late larval salivary glands of *D. melanogaster* or the midge *Chironomus tentans,* the chromosomes attain a sufficient size to allow a detailed morphological study with the light microscope. It was Painter (1933) who first realized the potential for fine structure genetic mapping that polytene chromosomes offered. Detailed chromosome maps of *D. melanogaster* were produced by Bridges (1935) and these are still in use today. Puffs are reversible structural modifications to chromosomes which result in the local decondensation of, usually, a single band. Some complex puffs may result from the puffing of more than one band. The puffs increase in size, up to four-fold larger than in their unpuffed state, and accumulate high levels of RNA and protein (Berendes and Helmsing, 1974). Evidence that puffs represent gene transcription came initially from studies monitoring RNA synthesis via [^{3}H]uridine incorporation. In this way, synthesis of new RNA can be detected within minutes of puff induction (Belyaeva and Zhimulev, 1976). Furthermore, in pulse-labeling experiments, a good correlation between [^{3}H]uridine incorporation and puff size was observed (Chatterjee and Mukherjee, 1971). The advent of molecular biology and gene cloning has allowed an unequivocal demonstration that chromosome puffs do indeed represent sites of gene transcription.

Although it is now clear that puffs represent transcriptional activity, the reason why some genes are associated with puffs while others are not, despite similar levels of transcription, is unclear. What is clear, however, is that puffing is not a requirement for gene transcription nor necessarily an indicator of transcription in loci which normally puff. Germline transformants carrying larval glue genes inserted at ectopic chromosomal locations puff in some cases but not in others, with no correlation between puffing and transcript levels (for a review see Meyerowitz *et al.,* 1987). While these experiments may be interpreted as an effect of chromosomal context on puffing, a more detailed study of the regulatory sequences which mediate puffing and their relationship to transcriptional competence will be required to address the role of puffing in gene expression. The opposite is seen in some broad complex (*BR-C*) mutations [as described later, the *BR-C* is a regulator of transcription involved in many aspects of the ecdysone response (see Bayer *et al.*, Chapter 9, this volume)]; larval glue genes, which are associated with puffs in early third instar larvae, are not transcribed in certain *BR-C* mutant backgrounds, yet puffs are observed at their chromosomal locations (Crowley *et al.*, 1984). Similar observations can be made in larvae deficient in ecdysone (Hansson *et al.*, 1981). Thus, chromosome puffing is neither necessary nor sufficient for gene transcription at loci which normally puff and the significance of puffing remains an enigma.

II. PUFFS AND DEVELOPMENT

The first important study of the changes in chromosomal puffing during *Drosophila* development was the examination of the larval salivary gland by Becker (1959). Around the middle of the third larval instar, when the salivary gland chromosomes are large enough for routine analysis, only 15 or so puffs are active; these are the intermolt puffs. Toward the end of the instar, a few hours before puparium formation, the intermolt puffs regress and a new set of puffs appear. Six major puffs are rapidly induced and then regress in a 3- to 4-hr period prior to puparium formation; these are the early puffs. Concomitant with the regression of the early puffs, a set of over 100 late puffs appear and regress in a temporally specific way. This puffing sequence is continued during prepupal development where early prepupal puffs appear and finally a late set of prepupal puffs, including some of the early puffs found in late larvae, are induced before the gland is histolyzed (Fig. 1). The cycle of puff appearance and regression is so regular that Becker could conveniently define 15 puff stages where differences in the puffing pattern on chromosome arm 3L were easily identified. A more detailed subsequent analysis including other chromosome arms defined 21 puff stages, the first 15 being those of Becker (Ashburner, 1967, 1969). Some of the puff stages are correlated with particular morphological events (e.g., glue synthesis at PS4/5, puparium formation at PS10/11, and head eversion at PS21). An examination of other larval polytene tissues (which tend to have chromosomes with much poorer morphology) during this period indicates that many of the

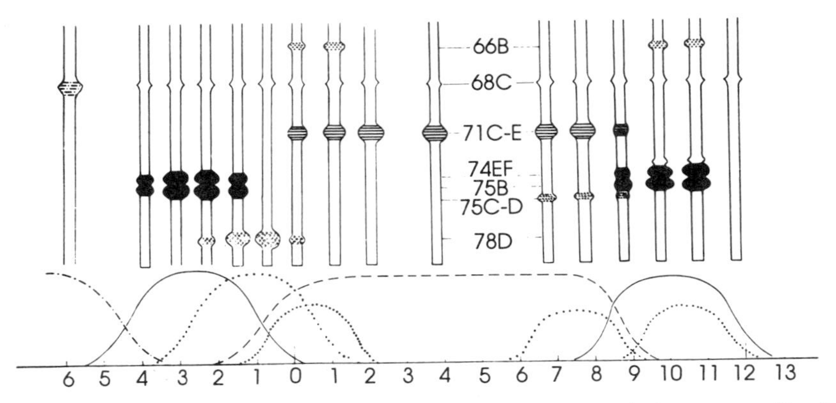

Fig. 1. The changes in puffing activity in the proximal arm of chromosome 3L of *D. melanogaster* during the late larval and prepupal stages. The abscissa represents hours before or after puparium formation (0 hr). [Reprinted with permission from Becker (1959).]

puffs are common to different tissues while some are tissue specific (Berendes, 1965; Richards, 1980). Thus the activities of a discrete set of genes which are modulated in a relatively invariant temporal sequence can be monitored by a cytological examination of the chromosomes of the animal.

It was suspected that the hormone ecdysone (see Gilbert *et al.,* Chapter 2, this volume) was responsible, at least in part, for the changes in puffing described by Becker (1959). Ecdysone, or rather 20-hydroxyecdysone, is the major molting hormone in *Drosophila* (Richards, 1978; Riddiford, 1993). [The term *ecdysone* will be used here as a generic term for those steroids that elicit molting (and puffing) while others use the term *ecdysteroid* for that class of compounds and reserve the term *ecdysone* for a specific ecdysteroid (see Gilbert *et al.,* Chapter 2, this volume).] Because of the small size and inherent developmental asynchrony of *Drosophila* larvae, ecdysone titers have been difficult to measure accurately. Many of the inferences regarding rises in hormone titers during larval development have borrowed heavily on studies in the tobacco hornworm *Manduca* (see Gilbert *et al.,* Chapter 2, this volume). Those studies which have been carried out in *Drosophila* tend to agree with much of the *Manduca* data (e.g., Richards, 1981). As discussed elsewhere in this volume, there are several rises in hormone titer which are associated with specific morphological changes in the animal. Each larval molt, for example, is preceded by a peak of hormone with the largest peaks occurring just prior to pupariation and prior to larval/pupal apolysis.

Strong support for the importance of ecdysone in chromosome puffing came from the experiments of Clever and Karlson (1960) in *Chironomus*. They were able to induce puffs typical of late fourth instar larvae by injecting purified ecdysone into younger animals, in which the puffs are not usually found. Similar experiments with *Drosophila* larvae produced comparable results (Burdette, 1964; Berendes, 1967). The elegant experiments of Becker (1962) further supported a pivotal role for ecdysone in salivary gland puff induction. Since ecdysone is produced from the ring gland, located just anterior to the larval brain, it is possible to apply a ligature behind the ring gland which bisects the salivary gland. Using this technique Becker (1962) observed that ligation prior to ecdysone release prevented the induction of early puffs in the portion of the gland posterior to the constriction, but did not affect the puffs in the anterior of the gland. If a ligature was applied after release of the hormone, then both portions of the gland exhibit a normal puffing sequene (Fig. 2). The ability to induce the normal puffing sequence in explanted salivary glands with exogenous hormone clearly demonstrates that ecdysone can be sufficient to induce puffing (Berendes, 1967). Finally, the examination of mutant larvae deficient in ecdysteroid synthesis demonstrates that most puffs fail to appear; thus, ecdysone is necessary for the correct developmental expression of puffs (Becker, 1959).

The following sections present an overview of the puffing sequence with reference to particular puffs representative of each class followed by a detailed molecular description of some of the products of these genes and their proposed functions. We will show how our current understanding of the regulation of salivary gland chromosome puffing is a synthesis of physiological studies on puff development and molecular studies of puff gene expression.

III. PHYSIOLOGY OF PUFFS

A key development in the study of chromosome puffing was the realization that the *in vivo* temporal pattern of puff appearance and regression could be faithfully mimicked in salivary glands explanted into culture medium (Ashburner, 1972). If salivary glands dissected from young third instar larvae, prior to the late larval release of ecdysone, are incu-

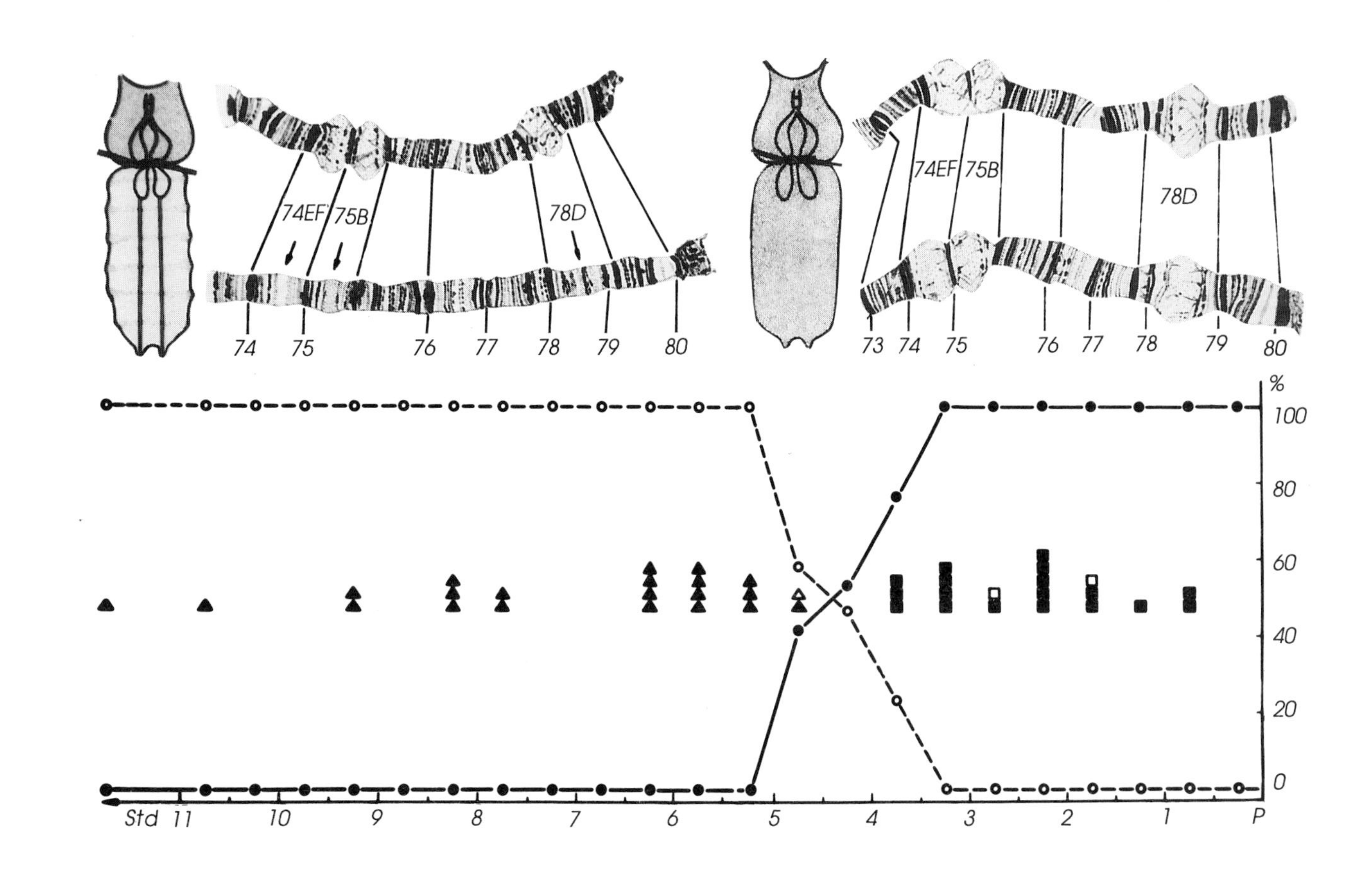

74EF
75B
78D
74
75
76
77
78
79
80
73
%
100
80
60
40
20
0
Std 11
10
9
8
7
6
5
4
3
2
1
P

bated in defined media, then the only changes are the slow regression of the intermolt puffs and the induction of a few *in vitro* puffs (these puffs are not found in the living animal). Upon the addition of purified 20 hydroxyecdysone, the intermolt puffs regress, at a rate characteristic for each puff, and a small set of early puffs is rapidly induced. These early puffs are the same as those which appear prior to pupariation in intact animals (Fig. 3). The major early puffs studied are those located at 2B5, 23E, 74EF, and 75B. The induction of the early puffs by ecdysone is extremely rapid; visible puffs are detected at 2B5 and 74EF as little as 5 min after hormone addition. In the continued presence of the hormone, each early puff reaches its maximum size within 4 hr and then regresses. It is important to note that the final size achieved and the time taken to attain this size are different for individual puffs (e.g., 23E is maximal at 1–2 hr whereas 74EF and 75B are maximal at 4 hr). Overlapping with the peak and subsequent regression of the early puffs is the induction and regression of over a hundred late puffs with a temporal profile closely resembling that observed in the intact animal. Representative late puffs which have been studied include 62E (5 hr), 78C (6 hr), 63E (8 hr), and 82F (10 hr) (the figures in parentheses refer to the time after hormone addition when maximal size is achieved). The whole sequence from PS1 to PS14 takes approximately 12 hr at 20°C. Glands incubated in the continuous presence of hormone do not progress beyond PS14. If, however, the glands are washed, transferred to a hormone-free medium for a few hours, and then returned to ecdysone-containing medium, the prepupal puffing pattern is now induced (Richards, 1976). Significantly, the hormone-free period mimics the natural *in vivo* situation where a drop in hormone titer is detected between the prepupal and pupal stages. Thus the ability to follow the changes in gene

---

Fig. 2. The ligation experiment of Becker. The abscissa represents the time in hours between the application of the ligature and puparium formation (P). The ordinate represents the percentage of animals pupariating either only anterior (○) or both anterior and posterior (●) to the ligature. The crossover point estimates the time of ecdysone release anterior to the ligature. (▲) Individuals showing the correct puffing pattern at 74EF, 75B, and 78C (formerly 78D) only anterior to the ligature as diagrammed top left; (■) individuals showing the correct puffing both anterior and posterior to the ligature (top right); and (□) animals whose chromosomes could only be analyzed in the posterior region of the gland. [Reprinted with permission from Becker (1962).]

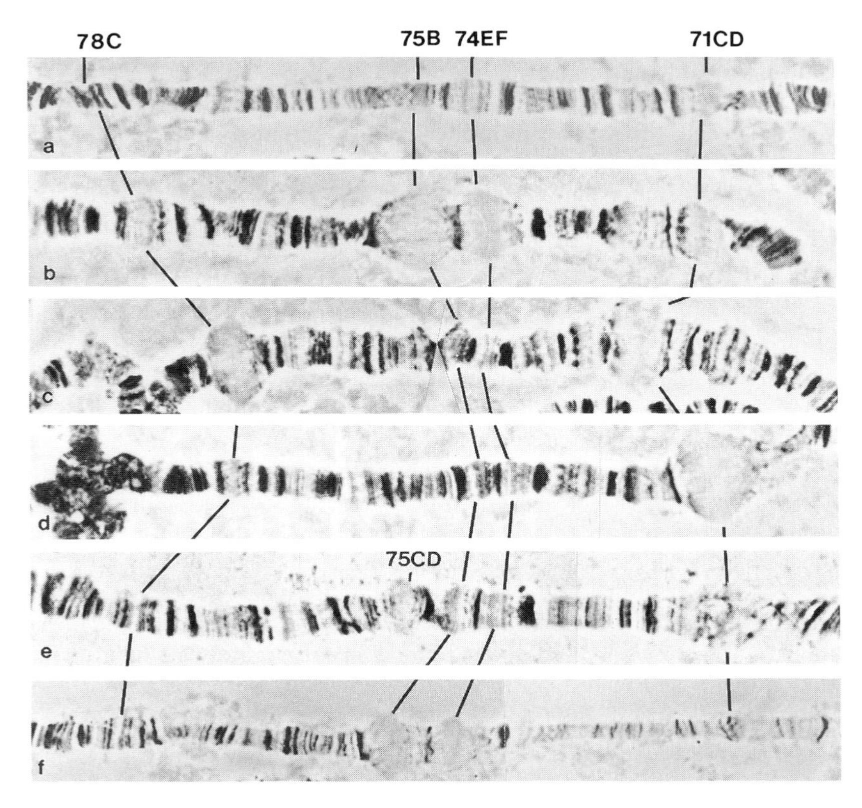

Fig. 3. The sequential induction and regression of puffs on the left arm of chromosome 3 in salivary glands dissected from larvae and prepupae: (a) PS1; (b) PS6; (c) PS10–11; (d) PS12; (e) PS18; and (f) PS19. The 74EF and 75B early puffs, the 78C early–late puff, and the 71CD and 75CD late puffs are indicated.

expression manifest as changes in chromosome puffs was now amenable to a detailed analysis *in vitro*.

The studies which examined puff induction following the injection of ecdysone into intact animals suggested that the response was dependent on the concentration of the injected hormone (Clever, 1963). With the *in vitro* system, an examination of the precise dose response of puffs to hormone became possible. The intermolt puffs, 25AC, and perhaps 3C, regress ecdysone-independently whereas the rate of 68C regression depends on hormone concentration over a 10-fold range. The early puffs are induced over a wide range of ecdysone concentrations, from 1×10^{-9} *M* to a maximal response at 5×10^{-7} *M* with a 50% response achieved at 1×10^{-7} *M*. The size of the early puffs is strongly influenced by hormone concentration: at low ecdysone levels the maximal size of the puffs is much lower than at higher doses. Furthermore, the time taken to reach maximal size for a given hormone concentration is shorter at low ecdysone doses. Hence, at low ecdysone concentrations, an early puff attains a smaller size and regresses more quickly than the same puff at a higher ecdysone concentration. It appears that the early puffs respond in a graded way to ecdysone, suggesting a rather direct response to the hormone.

The late puffs on the other hand respond over a very narrow range of hormone concentrations, with no response detected below 5×10^{-8} *M* and maximal response at 2.5×10^{-7} *M*. There is little or no size or temporal component to the response of late puffs over this concentration range. The induction of the late puffs by ecdysone may be viewed as an all or nothing response.

Other differences, in terms of response to hormone withdrawal and inhibitors of protein synthesis, are characteristic of intermolt, early, and late puffs. In the case of hormone withdrawal experiments, the intermolt puffs regress apparently normally when switched from a hormone-containing to a hormone-free medium and they are not reinducible after regression (except for 3C which is reinduced if PS1 glands are cultured without ecdysone). The early puffs regress rapidly if the hormone is withdrawn before the time of their natural regression. If the hormone is added back after withdrawal-induced regression, then the early puffs are reinduced. If, however, the early puffs are allowed to regress naturally, as is the case in the normal puffing sequence, they cannot be reinduced by adding a fresh hormone or by a reexposure to the hormone after 2 hr in a hormone-free medium. The early puffs, therefore, require the continued presence of ecdysone for their correct induction and regression. The withdrawal-induced regression is independent of protein synthesis inhibitors (see later), an observation which supports the notion that early puffs respond in a direct way to the hormone (Ashburner and Richards, 1976). As the early puffs prematurely regress on hormone withdrawal, some of the late puffs (notably 63E and 82F) are induced prematurely. Interestingly, the maximum size of ecdysone withdrawal-induced late puffs reflects the size that the early puffs had attained during their exposure to ecdysone. Furthermore, if glands in which the early puffs have been prematurely regressed are returned to the hormone-containing medium, then the prematurely induced late puffs now regress and are reinduced when the early puffs regress naturally. These data suggest that the hormone represses the appearance of late puffs and that the activity of the late puffs in some way depends on the prior activity of the early puffs. Some of the late puffs which appear earliest in the late puff sequence (62E and 78C) are not induced prematurely by hormone removal. These puffs are known as early–late puffs to indicate their requirement for the continuous presence of ecdysone.

The puffing response in the presence of protein synthesis inhibitors strongly supports the idea that early puffs respond directly to ecdysone (Ashburner, 1974). When incubated in the presence of ecdysone and cycloheximide, the intermolt puffs regress: 25AC apparently normally and 68C at a slower rate. The early puffs are induced apparently normally; however, their regression is prevented and the late puffs are not induced. Hence, in the continued presence of ecdysone and cycloheximide, only the early puffs are induced and remain active. It appears then that ecdysone-induced protein synthesis is required to repress the early puffs and to induce the late puffs. It should be noted that the early–late puffs, which require the presence of ecdysone for their induction, also appear to require the synthesis of new proteins for their induction.

Taken together, these data suggest that the early puffs are induced directly by a preexisting receptor complex which is activated in the presence of ecdysone (see Cherbas and Cherbas, Chapter 5, this volume). The early puffs require the continuous presence of this activated complex until regression, which in turn is dependent on the activity of proteins synthesized after the addition of ecdysone. After natural regression these puffs cannot be reinduced by ecdysone unless they experience a hormone-free period of at least 2 hr (see also Bayer *et al.,* Chapter 9, this volume). The late puffs do not appear to respond directly to ecdysone as they are induced only above a certain hormone threshold and cannot be induced in the presence of protein synthesis inhibitors. They thus exhibit a secondary response to the hormone. The late puffs may be induced prematurely by hormone withdrawal, but under these circumstances the maximal size they attain depends on the size reached by the early puffs prior to ecdysone withdrawal. This reinforces the suggestion that the induction of late puffs depends on early puff activity and indicates that this activity is the synthesis of protein. The regression of the intermolt puffs is more complex with individual puffs responding differently.

These observations led to the formulation of a formal model by Ashburner and colleagues for the coordination of the temporal sequence of puffing which occurs in the larval salivary gland (Fig. 4). In this model, ecdysone binds to a receptor to form an ecdysone–receptor complex (ER complex). The ER complex is envisaged to have two antagonistic effects. In the first instance it acts as a direct activator of early puffs; in the second it acts as a repressor of the late puffs and perhaps some of the intermolt puffs. The products of the early puffs are also envisaged to have a dual role: the first is to activate the late genes by lifting ecdysone repression in some way and the second to repress their own synthesis. The model explains how the early genes can react in a graded way to the concentration of hormone and react in the absence of new protein synthesis. It also accounts for the induction of late genes by premature hormone withdrawal and the fact that late genes induced in this way depend on the prior activity of the early genes (Ashburner *et al.,* 1974).

Support for aspects of the model comes from genetic experiments manipulating the dosage of early genes. In glands from larvae deficient for one copy each of 74EF and 75B, these early puffs are active for

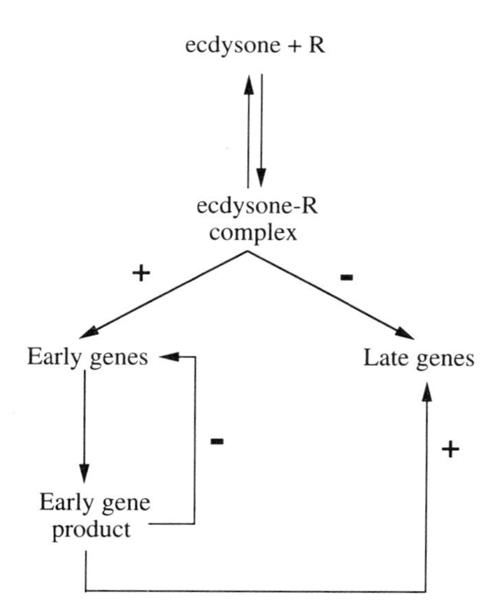

Fig. 4. A formal model for the coordinate regulation of the temporal sequence of salivary gland chromosome puffing (Ashburner *et al.,* 1974, reprinted with permission from Academic Press).

longer and the appearance of some late puffs is delayed. The opposite effect, early puffs active for a shorter period and late puffs active earlier, is observed when three copies each of 74EF and 75B are present. Interestingly, the early–late genes, noted earlier for their hormone-dependent activity, do not respond to alterations in the dosage of these early genes (Walker and Ashburner, 1981).

The model makes several key testable predictions. (1) The genes located within the early puffs contain promoters directly activated by an ecdysone receptor. (2) The products of some of the early genes are regulatory molecules which act to repress early gene expression and activate late gene promoters. (3) The late genes are transcriptionally repressed by the ecdysone receptor and are activated by an early gene product(s). In addition, we would expect that some of the intermolt genes are repressed by early gene products and that the early–late genes are activated by the ecdysone receptor and an early gene product(s). The following section describes the molecular structure of the puff genes which have been cloned and how their structure fits with the predictions of the model.

IV. MOLECULAR BIOLOGY OF PUFFS

A. Intermolt Puffs

The most thoroughly studied intermolt puffs are those containing the genes for the larval glue proteins, the *Sgs* genes. The *Sgs* gene products are components of a proteinaceous glue secreted from the salivary gland at the onset of puparium formation and used to adhere the animal to its substrate prior to metamorphosis. The *Sgs* genes have been cloned and located within the following puffs: *Sgs4* at 3C; *Sgs1* at 25B; *Sgs3, Sgs7,* and *Sgs8* clustered at 68C, and *Sgs5* at 90BC (Korge, 1975; Akam *et al.,* 1978; Muskavitch and Hogness, 1980; Velissariou and Ashburner, 1981; Meyerowitz and Hogness, 1982; Gautam, 1993; Guild, 1984). In the 68C cluster the *Sgs7* and *Sgs8* genes are divergently transcribed with their 5′ ends separated by a 475-bp intergenic region; the *Sgs3* gene is located about 2 kb downstream of the *Sgs7* 3′ end (Garfinkel *et al.,* 1983). Each gene codes for an abundant polyadenylated mRNA expressed in the salivary gland during the time the intermolt puffs are active.

Transcription of the glue genes is initiated during the mid third instar larval period as a secondary response to the low ecdysone peak occurring at this stage of development (Crowley and Meyerowitz, 1984; Hansson and Lambertsson, 1989). The role of ecdysone was first demonstrated using the temperature-sensitive mutation *l(1)su(f)*^{ts67g} which, at the restrictive temperature, is deficient in ecdysone synthesis. Individuals homozygous or hemizygous for the mutation fail to pupate when kept at the restrictive temperature and can remain in the larval stage for many days. If kept at the restrictive temperature for a period up to 70 hr after egg deposition, then *Sgs* gene transcription is not induced. Expression can be induced in these animals either by shifting to the permissive temperature or by administration of ecdysone via feeding. If the animals are shifted to the restrictive temperature after 75 hr after egg deposition, *Sgs* gene expression is induced normally (Hansson *et al.,* 1981; Hansson and Lambertsson, 1983). Similar effects are observed with the temperature-sensitive *ecdysoneless* [*l(3)ecd*^{1}] mutation, which is also deficient in ecdysterone synthesis (Garen *et al.,* 1977). In this case it was noted that although there is no transcription from the *Sgs4* gene, there is a prominent puff at 3C (Furia *et al.,* 1992). This does not necessarily indicate an uncoupling of transcription and puffing as at least five other genes, *pig1* and *ng1-4,* are in close proximity to *Sgs4* and their expression is unaffected by the *ecd*^{1} mutation. In addition, the 3C puff exhibits complex behavior in *in vitro* experiments (see later).

In addition to ecdysone, products of the *BR-C* are required for the high level expression of glue genes. Individuals carrying alleles of the *l(1)npr1* class of *BR-C* mutations fail to accumulate high levels of *Sgs* mRNAs (Kiss *et al.,* 1978). Pulse-labeling experiments demonstrate that this is due to a reduction in mRNA synthesis (Crowley *et al.,* 1984). A more detailed analysis of the effect of *BR-C* mutations on intermolt gene expression demonstrates that the *rbp* and *1(1)2Bc* functions are required for the initiation of intermolt gene transcription (Guay and Guild, 1991; Karim *et al.,* 1993; see below).

The cis-regulatory sequences required for appropriate intermolt gene expression have been determined by germline transformation and transient expression assays in which cloned fragments of glue genes are fused to reporter genes. The majority of these experiments have focused on the *Sgs4* gene and the *Sgs3, Sgs7,* and *Sgs8* group. In both of these cases it was demonstrated that *Sgs* genes inserted at ectopic sites in the genome were expressed in the same tissue-specific and temporal pattern as the native genes (Richards *et al.,* 1983; Crosby and Meyerowitz, 1986; Mc-

Nabb and Beckendorf, 1986; Hofmann *et al.,* 1987; Hofmann and Korge, 1987). Furthermore, in the majority of cases a puff was induced at the new site of insertion, concomitant with the expression of the transgene. In the case of the *Sgs3* gene, however, high level expression and puffing could be uncoupled. Transgenic animals bearing constructs containing 20 kb of DNA from the 68C puff, encompassing the *Sgs3, Sgs7,* and *Sgs8* genes, showed prominent puffs at the ectopic insertion site. Those containing a 3.5-kb construct with only the *Sgs3* gene expressed the transgene at high levels but showed no indication of an associated puff (Crosby and Meyerowitz, 1986).

Upstream of *Sgs3,* two regions, the proximal and distal regulatory elements, have been identified. These elements are functionally equivalent, as either is sufficient to confer appropriate stage- and tissue-specific transcription to a reporter gene. The elements appear to act synergistically; singly they drive expression at a low level whereas in combination full levels of expression are obtained. No evidence was found for enhancer activity (Bourouis and Richards, 1985a,b; Raghaven *et al.,* 1986; Meyerowitz *et al.,* 1987). Sequence analysis of the elements identified a conserved glue gene consensus which is also found upstream of the *Sgs3, −4, −5, −7,* and *−8* genes. The conserved sequences are also found upstream of the *Sgs3* homologs of *D. simulans, D. erecta, D. yakuba,* and *D. virilis* (Todo *et al.,* 1990). Nucleotide substitutions within the consensus sequence reduce expression levels, suggesting that the conserved elements contain binding sites for regulatory proteins. This is supported by a detailed *in vitro* and *in vivo* analysis of the distal element. This regulatory element is associated with a stage- and tissue-specific DNase I hypersensitive region which binds a stage- and tissue-specific factor, GEBF-1. The factor is apparently induced by ecdysone and is not present in larvae homozygous or hemizygous for *BR-C* mutations, indicating that GEBF-1 is either a product of the *BR-C* or that it requires a *BR-C* function for its activity (Georgel *et al.,* 1991).

In the case of the *Sgs4* gene, Korge and colleagues (Hofmann *et al.,* 1987; Hofmann and Korge, 1987) have shown that constructs containing 850 bp of upstream sequences are sufficient for correct tissue-specific and temporal expression, as well as for the induction of chromosome puffs at the site of transgene insertion. The formation of a puff was to some extent influenced by the site of insertion. Within the 850-bp upstream region, a 400-bp fragment encompassing the conserved glue gene element can confer stage- and tissue-specific enhancer activity (Shermoen *et al.,* 1987). Two specific proteins, characterized by the Bj6 and

Bx42 monoclonal antibodies, have been shown to bind to the *Sgs4* upstream region *in vivo.* Deletion of 52 bp associated with a DNase I hypersensitive site within the upstream region appears to prevent binding of the Bx42 factor (Saumweber *et al.,* 1990). This region contains sequences identified as ecdysone response elements in a heat shock gene promoter (Lawson *et al.,* 1985; Mestril *et al.,* 1986). The genes which encode the two antigens have been cloned and their structures suggest that they are involved in the regulation of chromatin structure (see Lezzi, Chapter 4, this volume). Bx42 recognizes a highly charged nuclear protein which appears to be associated with chromatin (Wieland *et al.,* 1992). The Bj6 antibody detects a basic protein product of the *nonA* gene; the protein has domains found in RNA-binding proteins (von Besser *et al.,* 1990).

In addition to the proximal region, a more distal region located between −2500 and −850 upstream of the transcription start site is able to confer stage- and tissue-specific expression at moderately high levels without the concomitant induction of puffs. The findings of McNabb and Beckendorf (1986) are contradictory to those of Korge and collaborators, as they revealed that sequences up to −850 allow appropriate expression without puffing and that further 5′ sequences are required to initiate puffing. The reasons for these differences are unclear but may represent a strain-specific variation in regulatory elements. Support for this comes from the analysis of the regulatory sequences of a number of naturally occurring *Sgs4* alleles. The Oregon-R strain expresses *Sgs4* at high levels and puffs normally, whereas the Samarkand strain expresses *Sgs4* at approximately 50% of Oregon-R levels and has a much reduced 3C puff. The Kochi-R strain only expresses *Sgs4* at 0.1% of Oregon-R levels and shows no 3C puff (Korge *et al.,* 1990). These differences are due to changes in the upstream sequences e.g., the 52-bp deletion described earlier was identified in the Kochi-R strain.

Other intermolt genes have been studied in less detail but all seem to share a common requirement for both ecdysone and the *rbp* gene product for high level stage- and tissue-specific expression. The identification of consensus sequence elements common to the upstream regulatory elements of the glue genes and the finding that constructs carrying *D. melanogaster* glue gene upstream sequences are regulated stage and tissue specifically in other *Drosophila* species (Roark *et al.,* 1990) suggest that the glue genes represent a coordinately expressed group of genes organized in a way reminiscent of the bacterial "regulon," an unlinked group of genes regulated by the same factor. This is indicated by upstream sequences conserved between different glue genes which would be expected to bind the same regulatory proteins. However, the precise arrangement of these conserved cis-acting sequences and their apparent activities (synergistic activity from *Sgs3* sequences and enhancer activity from *Sgs4* sequences) imply that additional regulatory molecules may be utilized by individual genes. The idea of common and "gene-specific" regulatory molecules within a "regulon" may reflect an underlying redundancy in the control of gene expression. For example, in *rbp* mutations, transcription of all of the glue genes is reduced but not abolished; thus, a low level of expression is maintained. The use of additional transcriptional activators not shared by members of the system may allow a level of buffering to fluctuations in the levels of activators or flexibility in response to changes in the demand for gene products.

The second feature of the intermolt genes which has been studied is their regression prior to puparium formation. As noted earlier, the intermolt puffs regress when salivary glands are exposed to ecdysone *in vitro* and, in the case of at least two puffs, 25B and 68C, the rate of regression depends on the concentration of the hormone (Ashburner, 1973). The 3C puff, which is active later in the puffing sequence, appears to regress at a rate that is independent of hormone concentration (Ashburner, 1973). *In vitro* pulse-labeling experiments demonstrate a reduction in transcription at the 68C puff within 15 min of ecdysone addition, suggesting direct repression by the ecdysone–receptor complex (Crowley and Meyerowitz, 1984). In the presence of protein synthesis inhibitors the regression of 68C is slowed, suggesting the involvement of additional, ecdysone-induced factors. Under these conditions, both 25B and 3C regress at their normal rates (Ashburner, 1974). A requirement for the *BR-C* function in intermolt gene repression is indicated by the failure of the 68C puff to regress in some *BR-C* mutations (Belyaeva *et al.*, 1981; Zhimulev *et al.*, 1982). Furthermore, it appears that the *BR-C* *l(1)2Bc* function is involved in repression, as glue genes are reinduced by the late prepupal ecdysone pulse in genotypes mutant for this function (Karim *et al.*, 1993).

Although our understanding of the induction and repression of intermolt genes is far from complete, striking similarities in both processes are apparent. Ecdysone is involved in both the induction of expression, via the low mid third instar larval peak, and the subsequent repression of glue gene transcription at the high peak prior to pupariation. In addition to ecdysone, products of the *BR-C* are required for normal induction and repression (see Bayer *et al.*, Chapter 9, this volume).

However, there are marked differences in the way in which puffs are induced and repressed. In the case of induction, as discussed earlier, the genes appear to be expressed coordinately and respond in a similar way. There appears to be very little coordinate component to their repression. This reinforces the idea that individual genes within a functional group can have regulatory sequences which respond independently of other members of the group as well as group-specific regulators.

B. Early Puffs

The genes located within the early puffs have been the focus of intensive study due to their perceived role as regulators of late gene expression. The model predicts that the early puffs contain the genes for regulatory proteins which act both as negative autoregulators and as positive activators. The cloning, sequencing, and characterization of three early puff gene products, those of the 2B, 74EF, and 75B puffs, strongly support the model and extend its applicability to other larval and imaginal tissues. Genetic evidence that these genes play a key role in regulating the ecdysone response in a variety of tissues comes from the reported lethality associated with mutations in all three genes (Belyaeva *et al.,* 1980; Segraves and Hogness, 1984; Burtis, 1985; Segraves, 1988), as well as the gene dosage experiments of Walker and Ashburner (1981). A diagram of early gene structure is shown in Fig. 5.

1. 2B

The region encompassing the 2B early puff is genetically complex; four lethal complementation groups, *broad(br), reduced bristles* on *palpus (rbp), l(1)2Bc,* and *l(1)2Bd,* and a set of *nonpupariating (npr1)* alleles which fail to complement all of these functions, are collectively designated as the *Broad-Complex* (Belvaeva *et al.,* 1980; Kiss *et al.,* 1980, 1988; see Bayer *et al.,* Chapter 9, this volume). Mutations in these functions disrupt many aspects of cell development autonomously and show many pleiotropic effects on prepupal and pupal development. Phenotypic analysis of *BR-C* mutations shows that the *BR-C* is associated with the ecdysone-induced puffing sequence. *BR-C* functions are required for intermolt puff induction, a secondary ecdysone response, and some aspects of intermolt puff repression (Crowley *et al.,* 1984; Georgel

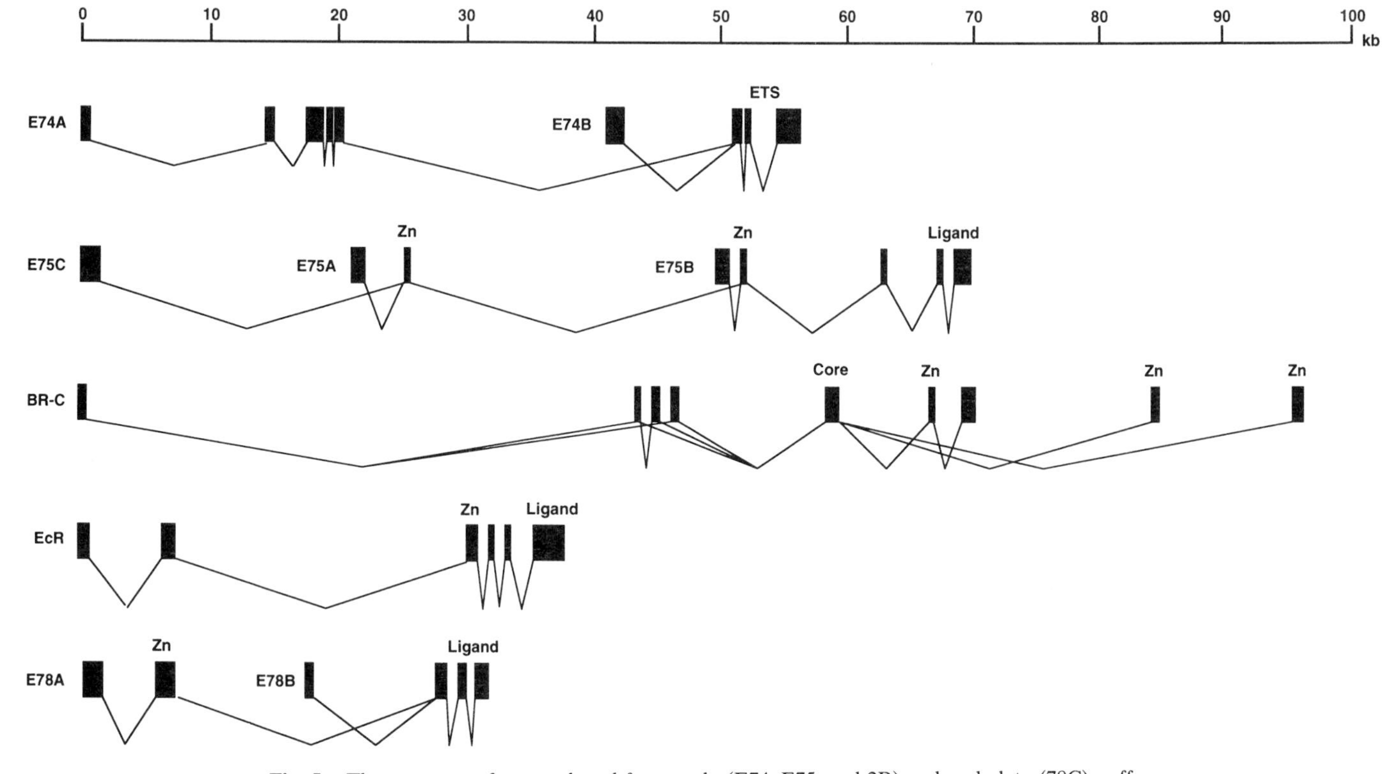

Fig. 5. The structure of genes cloned from early (E74, E75, and 2B) and early-late (78C) puffs as well as the ecdysone receptor. Black boxes represent exons and the structural features of the proteins are indicated (see text for details).

et al., 1991; Karim *et al.,* 1993). In animals deficient for *BR-C,* late puffs are never induced by ecdysone *in vivo* or *in vitro.* In addition, the early puffs 74EF and 75B respond submaximally to ecdysone and the 63E early–late puff is not induced at all. Interestingly, the 2B puff does not regress in some *BR-C* genotypes, suggesting an autoregulatory function (Belyaeva *et al.,* 1981; Guay and Guild, 1991). *BR-C* mutations cause blocks in the development of imaginal tissues, an ecdysone-dependent process, further indicating a role for the *BR-C* in the control of ecdysone-regulated gene expression (Kiss *et al.,* 1976, 1978; Zhimulev *et al.,* 1982).

BR-C covers almost 100 kb of DNA and is transcriptionally complex (Chao and Guild, 1986; Belyaeva *et al.,* 1987; Galceran *et al.,* 1990; DiBello *et al.,* 1991). The isolation and sequencing of cDNA clones identify the products of the *BR-C* as a family of related proteins sharing a common N-terminal core domain and different C-terminal domains. The C-terminal domains are characterized by the presence of zinc finger DNA-binding motifs. The complex contains three separate exons which encode three distinct sets of zinc finger pairs (DiBello *et al.,* 1991). The core domain contains a region with sequence similarity to the product of the *tramtrack* gene, a known *Drosophila* transcriptional repressor (Harrison and Travers, 1990; DiBello *et al.,* 1991; Read and Manley, 1992; Xiong and Montell, 1993). The region between the zinc finger and core domains is also variable in BR-C proteins; some classes of protein have a glutamine-rich sequence, whereas others have asparagine- and threonine-rich sequences. Similar stretches of characteristic amino acids are found in other eukaryotic transcription factors. Thus the molecular structure of *BR-C,* encoding a set of proteins with overlapping and specific domains characteristic of DNA-binding proteins and transcriptional regulators, is consistent with its proposed role in regulating aspects of the ecdysone response. The production of multiple transcription factors is also consistent with the genetic complexity of the locus and suggests that different *BR-C* functions are mediated by regulatory molecules with different specificities and functions.

The temporal and tissue distribution of *BR-C* transcripts supports the view that this gene plays a key role in the ecdysone response at different stages and in many tissues (Chao and Guild, 1986; Karim and Thummel, 1992; Andres *et al.,* 1993; Karim *et al.,* 1993). Using a core probe which detects all known *BR-C* transcripts, more than 12 different transcripts are produced from the complex. These differ in promotor, exon, and polyadenylation site usage. *In situ* hybridization studies find *BR-C* transcripts in almost all larval tissues. The expression of different transcripts is modulated in a temporal way. Transcripts can be detected in the early third instar larvae prior to the induction of glue gene synthesis, consistent with the genetic role of *BR-C* in regulating glue gene expression. *BR-C* expression appears to be upregulated, and new classes of transcript induced, at the time of the ecdysone peak prior to pupariation. Some of the transcripts are downregulated during the prepupal stage and are then further downregulated at the prepupal–pupal transition (Andres *et al.,* 1993). An examination of *BR-C* expression in organs *in vitro* has shown that mRNAs are induced rapidly over a basal level in the presence of ecdysone and that the concentration of ecdysone required for induction is very low ($\sim 2 \times 10^{-9} M$) (Karim and Thummel, 1992). In addition, there are classes of transcript which differ in their dose response to ecdysone. Although further studies will be required to precisely delineate the roles of each class of *BR-C* transcript, it is clear that given its multiplicity of gene products and complexity of temporal expression, as well as the wide variety of phenotypic consequences resulting from *BR-C* mutations, the gene plays a fundamental role in regulating the ecdysone response (Karim *et al.,* 1993).

2. *74EF*

Mutations in the *E74* locus cause lethality during the late larval–prepupal stages (Burtis, 1985). DNA associated with the 74EF puff was cloned by chromosome walking and ecdysone-induced transcription units were identified by differential hybridization with RNA from ecdysone-induced and uninduced salivary glands (Janknecht *et al.,* 1989; Burtis *et al.,* 1990). The *E74* gene is large, spanning some 60 kb of DNA, and encodes two overlapping transcription units expressed from three promoters. The *E74A* product is transcribed from a single promoter and the *E74B* product from two apparently equivalent promoters. The protein products of these transcripts have a common C-terminal region which contains a domain characteristic of members of the *ets* protooncogene superfamily. These proteins are believed to be DNA-binding regulatory proteins (Karim *et al.,* 1990). The N-terminal regions of E74A and E74B are encoded by different exons, but both are rich in acidic amino acids, inviting comparisons with the acidic "activator" domains of other transcriptional regulators. In both proteins the domains are separated by homopolymeric repeats. This structure strongly suggests that the products of the 74EF puff are involved in transcriptional regulation. The DNA-binding properties of E74A have been studied using purified protein expressed in bacteria (Urness and Thummel, 1990). Three adjacent binding sites have been identified in the middle of the *E74* gene, 11 kb downstream of the *E74A* promoter. The sites contain consensus sequences similar to those of the binding sites of other *ets* family members; this suggests that E74 proteins could autoregulate their expression. Consistent with the model, antibodies against E74A bind to a number of sites on salivary gland polytene chromosomes, including both early (22B, 63F, 74EF, and 75B) and late (at least 18 major sites including 62E, 63E, 78C, and 82F) puffs. Studies with the E74B protein indicate that it binds to the same sequences as E74A (Karim, 1992).

Initiation of transcription from the *E74A* promoter can be detected 5 min after the addition of ecdysone to incubated larval organs; the peak and subsequent decline in transcript levels after 6 hr in the presence of ecdysone parallel the appearance and regression of the 74EF puff under similar conditions. The accumulation of high levels of *E74A* transcripts in the continuous presence of ecdysone and cycloheximide also mimic the behavior of the puff. A developmental profile of *E74A* expression shows that the gene is active during most of the ecdysone pulses which occur in normal development. Developmental expression from the *E74B* promoters is similar to that of *E74A* during embryonic and early larval stages, but is induced earlier in late third instar larvae and later in pupae than *E74A*. At least some of this temporal difference may be explained by the time taken to transcribe the 60-kb *E74A* primary transcript (~1 hr) as opposed to 15–30 min for *E74B* (Karim and Thummel, 1991). The time taken to produce mature mRNA from long primary transcription units has been invoked as a possible regulatory mechanism (Thummel *et al.,* 1990). A more detailed analysis of *E74* gene expression in organs *in vitro* (Karim and Thummel, 1991) shows that *E74B* is induced by a 10-fold lower concentration of ecdysone than *E74A* ($\sim 2 \times 10^{-9}$ *M* versus $\sim 2 \times 10^{-8}$ *M*), accounting for the earlier *E74B* expression in late larvae. When incubated at ecdysone concentrations which induce maximal activity, *E74B* is only active for 2 hr and is then repressed (repression is dependent on protein synthesis). However, at suboptimal ecdysone concentrations, below 5×10^{-8} *M, E74B* is not repressed. This suggests that *E74B* repression requires a product induced by ecdysone concentrations above 5×10^{-8} *M;* an obvious candidate would be the *E74A* gene. However, a translocation breakpoint which separates the *E74* promoters and removes *E74A* function does not affect *E74B* expression (Burtis, 1985; Karim and Thummel, 1991).

The extent of *E74A* promoter sequences have been determined by *in vitro* and *in vivo* transcription assays; as little as 80 bp of upstream sequence appears to be sufficient for efficient *in vitro* expression. However, in transient expression assays, an additional 100 bp of the 5′ sequence is required for efficient expression. Sequences 43 bp downstream of the transcription start site are required for expression; a consensus TATA element is located within this transcribed region and is essential for transcription. Within the upstream sequences necessary for *in vivo* expression, binding sites for the *zeste* and GAGA transcription factors have been identified by *in vitro* footprinting assays (Thummel, 1989). It has been shown that antibodies against the GAGA factor are localized to the 74EF puff in salivary glands, supporting the association of this transcription factor with the *E74A* promoter (Tsukiyama *et al.,* 1994). Finally, evidence which suggests that these sequence elements are functionally important comes from an analysis of the *E74* gene in *D. pseudoobscura* and *D. simulans.* In both these species the *E74* gene is well conserved at the amino acid level, and in terms of the upstream sequences, nucleotide conservation is strongest over the GAGA-binding sites (Jones *et al.,* 1991). The cis-acting sequences required for ecdysone induction of *E74* expression have yet to be localized, but preliminary evidence suggests that they may be located within the first intron of *E74A* (Thummel *et al.,* 1990).

E74A RNA and protein have been localized in a variety of tissues which respond to ecdysone in both late larvae and prepupae (Thummel *et al.,* 1990; Boyd *et al.,* 1991). Cytoplasmic RNAs have been detected in larval tissues destined for histolysis (salivary gland, fat body, and gut) and imaginal tissues which develop into adult structures (eye–antennal, wing and leg disks, and the proliferation centers of the brain). Protein distribution, which is nuclear in all tissues examined, is broadly similar to that of the RNA except in the case of the brain proliferation centers where no protein has been detected in stages so far examined. Interestingly, the peak of protein expression appears to lag behind the peak of RNA expression by at least 2 hr, raising the possibility that E74A expression is regulated posttranscriptionally. In this respect, it is of interest to note that the unusually long 5′-untranslated leader of the *E74A* mRNA contains 17 short open reading frames prior to the initiation codon for E74A; these may mediate translational regulation (Burtis *et al.,* 1990).

3. 75B

Mutations which disrupt the *E75* locus fall into two genetically distinct classes, those with an early lethal phase (embryonic or early larval) and those which result in pharate adults (Segraves, 1988). The *E75* gene was cloned by chromosome walking, and the ecdysone-induced transcription units were identified by differential cDNA screening (Feigl *et al.,* 1989; Segraves and Hogness, 1990; see Cherbas and Cherbas, Chapter 5, this volume). As with the other two early genes, the *E75* locus is complex, spanning over 50 kb of DNA. Three promoters direct the production of six transcripts, each mRNA utilizing one of two polyadenylation sites located downstream of the protein-coding exons. *E75C* is the longest transcription unit, spanning approximately 55 kb; *E75A* begins around 5 kb downstream of this and *E75B* approximately 30 kb further downstream. As a consequence of this arrangement, three proteins with common C-termini and unique N termini are produced. All three products show a striking similarity with members of the steroid hormone receptor superfamily including the estrogen and retinoic acid receptors (Evans, 1988). Similarity is restricted to the C (DNA binding) and E (ligand binding/dimerization) domains with the highest similarity in the zinc finger DNA-binding domain. *E75A* and *E75C* products appear to be similar in that both contain two zinc fingers. In the case of *E75B,* however, the location of the promoter downstream of the first zinc finger exon results in the production of a protein with only one finger. Either E75B does not bind DNA or its binding differs significantly from other members of the superfamily. The *E75* genes are rapidly induced on ecdysone addition to incubated organs; expression peaks after 2 hr and regresses in the continuous presence of ecdysone and fails to regress when cycloheximide is present. The expression of *E75A* and *E75B* is similar to *E74A* in terms of their ecdysone dose response, whereas *E75C* expression is more similar to *E74B* (Karim and Thummel, 1992). As with *E74A* and *BR-C, E75* is expressed in a variety of ecdysone responsive tissues; in addition, antibodies against E75A detect the protein at early and late puff sites on polytene chromosomes (Hill *et al.,* 1993).

4. EcR

The gene for a component of the ecdysone receptor, *EcR,* has been cloned and localized to a small puff at 42A (Koelle *et al.,* 1991). As discussed by Cherbas and Cherbas (Chapter 5, this volume), the product is a member of the steroid receptor superfamily which forms an obligate heterodimer with the product of the *ultraspiricle* (*usp*) gene in order to bind ecdysone and DNA (Yao *et al.,* 1992; Oro *et al.,* 1993; Thomas *et al.,* 1993). The structure of the gene is very similar to the three early puff genes described earlier, with nested transcription units producing three related proteins. In common with the other early genes, *EcR* spans a large region of DNA, 40 kb in this case. Developmental expression of *EcR* RNA and protein has been examined; both are abundant during most of embryogenesis and during the late larval–early pupal stages. There is a particularly prominent peak of RNA expression during the late larval stage prior to pupariation (Koelle *et al.,* 1991). The EcR protein is localized in the nuclei of all late larval tissues examined, indicative of the wide role of ecdysone in controlling metamorphosis. Furthermore, it has been shown that tissues which have different fates during metamorphosis express different EcR isoforms, prompting the suggestion that different combinations of EcR isoform, and perhaps other steroid receptor molecules, mediate different tissue-specific responses (Talbot *et al.,* 1993).

A more detailed examination of *EcR* RNA expression in organs *in vitro* shows that the gene is rapidly induced above a basal level on ecdysone addition; expression peaks after approximately 1 hr and then regresses to the basal level. In addition, the induction occurs at low concentrations of ecdysone ($\sim 2 \times 10^{-9} M$); these aspects of *EcR* expression are very similar to those observed for the *E74B* gene. These data indicate that the ecdysone receptor itself behaves as an early puff gene in a way similar to *E74* and *E75*. However, the expression of *EcR* has not been measured in the presence of protein synthesis inhibitors and thus cannot be unequivocally classified as an early gene.

The molecular cloning of the early puff genes confirms the predictions of the model. The products are ecdysone-induced regulatory proteins which interact with both early and late genes. A detailed examination of the expression of these genes *in vitro* and *in vivo* indicates that the induction of these genes occurs in a coordinated way. Furthermore, the finding that these regulatory genes are expressed in a wide variety of ecdysone responsive tissues extends the applicability of the model to many ecdysone-induced processes occurring during morphogenesis. The coexpression of key regulatory molecules in a temporal- and tissue-specific way supports the belief that different combinations of regulatory molecules can modulate temporal- and tissue-specific gene expression.

C. Late Puffs

As yet, only a few of the late puff genes have been cloned and studies of their regulation by early genes are at a preliminary stage. This chapter discusses two late genes in detail: the *E78* gene, an "early–late" puff gene, and the late genes of the 71E puff. The cloning of these genes provides support for the model and, in the case of the *E78* gene, adds a hitherto unsuspected level of control to the puffing cascade.

1. 78C

As discussed earlier, the puff at 78C is classed as an early–late puff since it appears at the beginning of the late puff cycle and requires ecdysone-induced protein synthesis for its induction. The 78C puff displays an "all or nothing" response to ecdysone, a characteristic of late puffs, but unlike puffs of this class it requires ecdysone for its activity since it regresses or is not induced on hormone withdrawal (Ashburner, 1973, 1974; Ashburner and Richards, 1976). Since the size of the 78C puff is not affected by changes in the gene dosage of *E74* and *E75* and is not induced in some *BR-C* mutant backgrounds, it is probable that the protein synthesis requirement is for a *BR-C* product (Belyaeva *et al.,* 1981; Walker and Ashburner, 1981). The 78C puff gene was cloned from a chromosomal walk in the *Polycomb* region (Paro and Hogness, 1991; Stone and Thummel, 1993; G. Heimbeck *et al.,* unpublished observations). In common with the early genes, the *E78* gene is complex, spanning over 30 kb of DNA and directing the production of six transcripts from two promoters. *E78A* is the longest transcription unit and its product is a protein of the steroid receptor superfamily, as are those of EcR and E75. The *E78B* promoter is located a further 15 kb downstream and does not include the first three *E78A* exons. Two of these exons encode the E78A zinc finger domains. Both proteins share a common set of 3′ exons which encode the ligand-binding/dimerization domain. Thus the *E78B* product is a truncated member of the steroid receptor superfamily and is, presumably, unable to bind DNA.

The temporal expression of the *E78* genes proves to be surprising as the *E78A* transcripts can only be detected in 2-day-old pupae. *E78B,* on the other hand, has an expression profile which matches that of the 78C puff; the *E78B* gene is also expressed in 2- and 3-day-old pupae overlapping with *E78A* expression. Ecdysone induction of *E78B* expression in incubated organs indicates that the gene is expressed with a temporal profile similar, though slightly later, than *E74A;* there is a delay of 1 hr between hormone addition and the appearance of *E78B* mRNA (*E74A* transcripts are readily detected within 30 min of ecdysone addition). In addition, the dose response to ecdysone varies over a 100-fold range of hormone concentration in sharp contrast to the "all or nothing" response of the puff. A further surprise is the observation that *E78B* transcripts are detected in organs incubated with ecdysone and cycloheximide, although in this case the appearance of transcripts is delayed by 2 hr and does not reach the same levels as in uninhibited organs (Stone and Thummel, 1993). These data indicate that expression of the *E78B* gene and the induction of the 78C puff are not regulated in the same way. Support for this suggestion comes from the analysis of inversion chromosomes which break just upstream of the *E78B* transcription start site. In individuals homozygous for these inversions no puff is detected at 78C. However, *E78B* expression is still detectable but only at 10% of wild-type levels (S. Russell, M. Ashburner, G. Heimbeck, C. Martin, and A. Carpenter, unpublished observations).

A genetic analysis of the *E78* locus indicates that deletions of *E78A* alone, or of both *E78A* and *E78B,* have no apparent phenotypic consequences in terms of development or fertility. The *E78B* gene appears to be involved in the regulation of puffing as deletions of *E78B* result in the reduction of some late puffs (63E and 82F). This suggests that the *E78B* gene product may act as an activator of some late genes by interacting with a DNA-binding protein to enhance its activity. Support for this suggestion comes from an analysis of puffing in lines which overexpress E78B from a heat-shock promoter; in this case regression of the 74EF, 75B, and the 78C puff itself is delayed. In addition the 63E and 82F puffs which are reduced in E78B deletions are also reduced in the overexpressing lines. These effects may be due to a dominant negative effect of expressing large amounts of E78B and thus sequestering another steroid receptor molecle required for correct puffing (S. Russell, M. Ashburner, G. Heimbeck, C. Martin, and A. Carpenter, unpublished observations). Candidates for such interactions are EcR, USP, or an E75 protein. A similar modulating function has been suggested for the E75B protein (Burtis *et al.,* 1990).

The unexpected finding that one of the late genes appears to encode a regulatory molecule indicates that the control of the hierarchy is more complex than the original model envisaged. Further evidence for additional complexity comes from the finding that at least two other puff genes encode similar regulatory molecules. The product of another ecdysone-induced puff, 46F, is a steroid receptor homolog, *DHR3* (Koelle *et al.,* 1992). Another steroid receptor homolog, the *FTZ-F1* gene, appears to be a product of the prepupal 75CD puff (Lavorgna *et al.,* 1993). Thus, at least six steroid receptor superfamily members are expressed from ecdysone-induced puffs. Given that heterodimerization of class II steroid receptors appears to be widespread in eukaryotes, and indeed obligatory in the case of the ecdysone receptor, this raises the possibility that aspects of temporal- and tissue-specific gene regulation are mediated by different combinations of superfamily members (Richards, 1992). The finding that different ecdysone receptor isoforms are expressed in different tissues supports this contention. We await the *in vitro* characterization of the DNA binding and regulatory properties of different heterodimer combinations to confirm this model.

2. 71E

The 71CF region of the genome is complex, containing two ecdysone-induced late puffs, the *Sgs6* glue gene early intermolt and the genes for two proteins induced by ecdysone in tissue culture cells, Eip28 and Eip29 (Velissariou and Ashburner, 1981; Savakis *et al.,* 1984; Semeshin *et al.,* 1985). Of these, the 71E late puff has been most extensively characterized. cDNA clones representing transcripts expressed from the 71E region were isolated in a molecular screen for salivary gland genes induced by ecdysone (Wolfner, 1980). Subsequent analysis of these clones placed them into six groups representing six genes, organized as three divergently transcribed pairs, clustered within a 10-kb segment of the genome. An additional transcription unit, expressed with the temporal and tissue specificity of an intermolt gene, is located within the cluster (Restifo and White, 1989). A developmental profile of expression shows that the six clustered genes are only expressed in salivary glands during the prepupal stage, a temporal profile expected of late puff genes. These genes appear to be coordinately regulated as they appear and regress in an identical fashion over a 12-hour period from puparium formation to the prepupal–pupal transition (Andres *et al.,* 1993). The role of some early genes in the regulation of the 71E late genes has been examined and is consistent with the predictions of the model; expression is reduced dramatically in *rbp, l(1)2Bc,* or *E74A* mutant backgrounds (Guay and Guild, 1991; Karim *et al.,* 1993; J. C. Fletcher and C. S. Thummel, cited in Andres *et al.,* 1993).

A molecular analysis of the promoter regions of these and other coordinately expressed late genes, in conjunction with the further analysis of intermolt gene promoters, will provide reagents for a systematic study of the way in which the regulatory molecules described earlier act in combinatorial ways to coordinately activate and repress sets of genes.

V. MOLECULAR VIEW OF ECDYSONE-REGULATED GENE EXPRESSION

With the isolation of genes belonging to each of the major classes of late larval/prepupal puffs, a detailed comparative analysis of gene expression has become possible. Two important studies from the laboratory of C. S. Thummel have examined the expression of early and late genes by utilizing sets of identically prepared Northern blots. Early gene expression was examined in RNA extracted from organs incubated with different hormone concentrations. These data subdivided the early gene products into two classes (Karim and Thummel, 1991, 1992). The class I early products, *E74B, EcR,* and some of the *BR-C* products, are induced at a low ecdysone concentration ($\sim 2 \times 10^{-9}$ *M*). The class II early products, *E74A, E75A, E75B,* and other *BR-C* products, respond at a higher concentration ($\sim 1 \times 10^{-8}$ *M*). These differences in response *in vitro* are expected to reflect differences in response to ecdysone levels in the animal. Thus, class I genes are induced earlier in larval development in response to a low ecdysone peak and initiate the intermolt puffing sequence. The subsequent rise in hormone titer which precedes puparium formation induces the class II early genes which in turn reprogram gene expression to the prepupal late puffing sequence. This interpretation is strongly supported by an analysis of early gene expression with RNA extracted from carefully staged animals (Fig. 6) (Karim and Thummel, 1992; Andres *et al.,* 1993). In these experiments the class I early gene transcripts are detected shortly after the second to third instar molt, prior to intermolt gene expression. As hormone titers increase, initiating the onset of pupariation, the class II early genes are induced. The rise in levels of class II transcripts coincides with the disappearance of the intermolt gene transcripts and the appearance of the late gene transcripts. Some aspects of this program are repeated later in development, coinciding with the rise in hormone titer which triggers head eversion and the formation of the pupa. This type of analysis is now being extended to an analysis of gene expression in single isolated organs using sensitive polymerase chain reaction techniques (Huet *et al.,* 1993). Thus, the cascade of gene expression, initially viewed from the perspective of puffs on polytene chromosomes, can now be seen as the rise and fall in the levels of specific gene transcripts, bringing us full circle back to Beermann's (1952) suggestion that the activity of chromosome puffs is a reflection of gene activity. The availability of cloned promoter elements

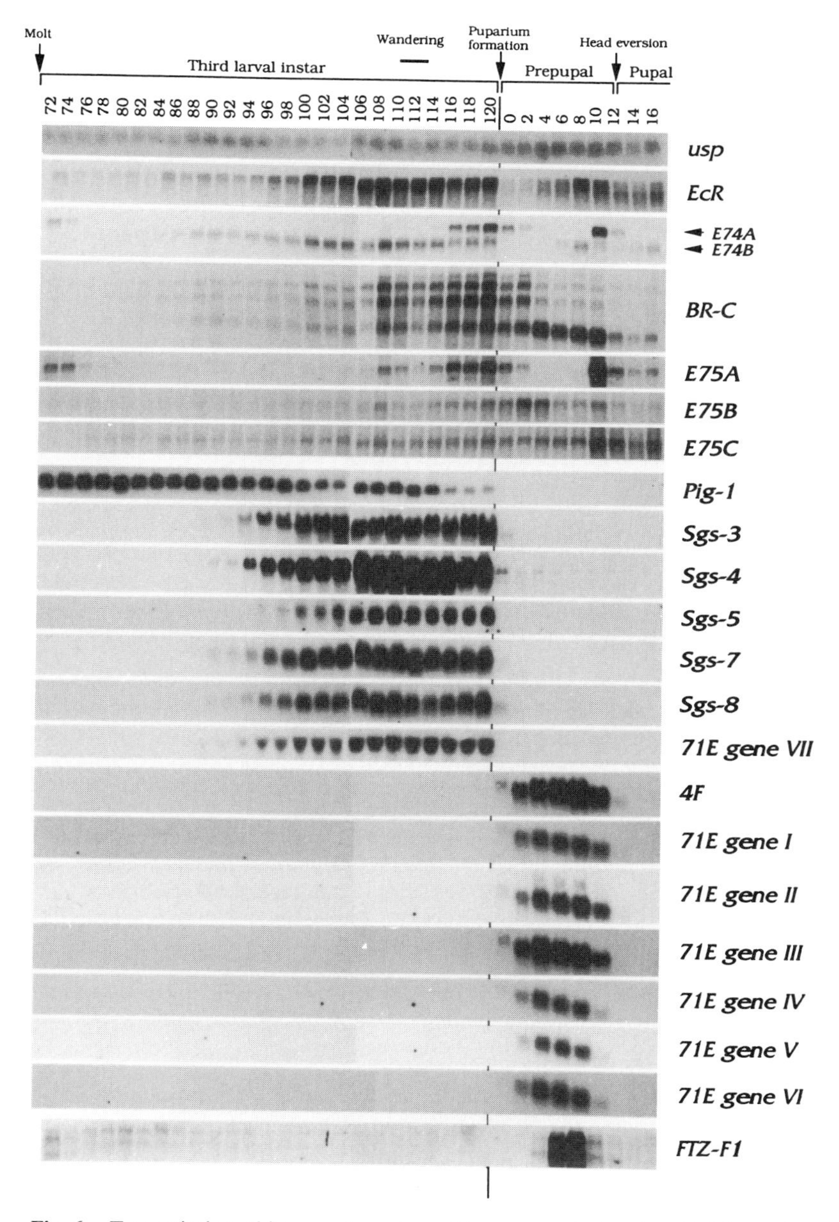

Fig. 6. Transcription of intermolt, early, and late genes during the third larval instar and prepupal stages of development. The time from egg laying is shown for third instar larvae and from puparium formation for prepupal development. Major developmental events are indicated. [Reprinted from Andres *et al.* (1993) with permission from Academic Press.]

for all three classes of puff genes and the identification of a plethora of regulatory molecules offer the prospect of understanding, at the molecular level, one way in which a complex multicellular organism can coordinately regulate sets of genes required for appropriate developmental pathways.

REFERENCES

Akam, M. E., Roberts, D. B., Richards, G. P., and Ashburner, M. (1978). *Drosophila:* The genetics of two major larval proteins. *Cell (Cambridge, Mass.)* **13,** 215–225.

Andres, A. J., Fletcher, J. C., Karim, F. D., and Thummel, C. S. (1993). Molecular analysis of the initiation of insect metamorphosis—A comparative study of *Drosophila* ecdysteroid-regulated transcription. *Dev. Biol.* **160,** 388–404.

Ashburner, M. (1967). Patterns of puffing activity in the salivary gland chromosomes of *Drosophila.* I. Autosomal puffing patterns in a laboratory stock of *Drosophila melanogaster. Chromosoma* **21,** 398–428.

Ashburner, M. (1969). Patterns of puffing activity in the salivary gland chromosomes of *Drosophila.* II. The X-chromosome puffing patterns of *Drosophila melanogaster* and *Drosophila simulans. Chromosoma* **27,** 47–63.

Ashburner, M. (1972). Patterns of puffing activity in the salivary gland chromosomes of *Drosophila.* VI. Induction by ecdysone in salivary glands of *Drosophila melanogaster* cultured in vitro. *Chromosoma* **38,** 255–281.

Ashburner, M. (1973). Sequential gene activation by ecdysone in polytene chromosomes of *Drosophila melanogaster.* I. Dependence upon ecdysone concentration. *Dev. Biol.* **35,** 47–61.

Ashburner, M. (1974). Sequential gene activation by ecdysone in polytene chromosomes of *Drosophila melanogaster.* II. The effects of inhibitors of protein synthesis. *Dev. Biol.* **39,** 141–157.

Ashburner, M., and Richards, G. (1976). Sequential gene activation by ecdysone in polytene chromosomes of *Drosophila melanogaster.* III. Consequences of ecdysone withdrawal. *Dev. Biol.* **54,** 241–255.

Ashburner, M., Chihara, C., Meltzer, P., and Richards, G. (1974). Temporal control of puffing activity in polytene chromosomes. *Cold Spring Harbor Symp. Quant. Biol.* **38,** 655–662.

Becker, H. J. (1959). Die Puffs der Spiecheldrüsenchromosomen von *Drosophila melanogaster.* I. Beobachtungen zum Verhalten des Puffmusters im Normalstamm und bei zwei Mutanten, *giant* und *lethal-giant-larvae. Chromosoma* **10,** 654–678.

Becker, H. J. (1962). Die Puffs der Speicheldrüsenchromosomen von *Drosophila melanogaster.* II. Die Auslosung der Puffbildung, ihre Spezifitat und ihre Beziehung zer Funktion der Ringdruse. *Chromosoma* **13,** 341–384.

Beermann, W. (1952). Chromomeren Konstanz und Spezifische Modifikation der Chromosomenstruktur in der Entwicklung und Organ differentzierung von *Chironomus tentans, Chromosoma* **5,** 139–198.

Belyaeva, E. S., and Zhimuley, I. F. (1976). RNA synthesis in the *Drosophila melanogaster* puffs. *Cell Differ.* **4,** 415–427.

Belyaeva, E. S., Aizenson, M. G., Semeshin, V. F., Kiss, I. I., Koczka, K., Baritcheva, E. M., Gorelova, T. V., and Zhimulev, I. F. (1980). Cytogenetic analysis of the 2B3-4–2B11 region of the X-chromosome of *Drosophila melanogaster.* I. Cytology of the region and mutant complementation groups. *Chromosoma* **81,** 281–306.

Belyaeva, E. S., Vlassova, I. E., Biyasheva, Z. M., Kakpakov, V. T., Richards, G., and Zhimulev, I. F. (1981). Cytogenetic analysis of the 2B3-4–2B11 region of the X chromosome of *Drosophila melanogaster.* II. Changes in 20-OH ecdysone puffing caused by genetic defects of puff 2B5. *Chromosoma* **84,** 207–219.

Belyaeva, E. S., Protopopov, M. O., Baricheva, E. M., Semeshin, V. F., Izquierdo, M. L., and Zhimulev, I. F. (1987). Cytogenetic analysis of region 2B3-4 to 2B11 of the X-chromosome of *Drosophila melanogaster.* VI. Molecular and cytological mapping of the *ecs* locus of the 2B puff. *Chromosoma* **92,** 351–356.

Berendes, H. D. (1965). The induction of changes in chromosomal activity in different polytene types of cell of *Drosophila hydei. Dev. Biol.* **11,** 371–384.

Berendes, H. D. (1967). The hormone ecdysone as effector of specific changes in the pattern of gene activities of *Drosophila hydei. Chromosoma* **22,** 274–293.

Berendes, H. D., and Helmsing, P. J. (1974). Nonhistone proteins of Dipteran polytene nuclei. *In* "Acidic Proteins of the Nucleus" (I. L. Cameron and J. R. Jeter, eds.), pp. 191–212. Academic Press, New York.

Bourouis, M., and Richards, G. (1985a). Hybrid genes in the study of glue gene regulation in *Drosophila. Cold Spring Harbor Symp. Quant. Biol.* **50,** 355–360.

Bourouis, M., and Richards, G. (1985b). Remote regulatory sequences of the *Drosophila* glue gene *Sgs3* as revealed by P-element transformation. *Cell (Cambridge, Mass.)* **40,** 349–357.

Boyd, L., O'Toole, E., and Thummel, C. S. (1991). Patterns of *E74A* RNA and protein expression at the onset of metamorphosis in *Drosophila. Development* **112,** 981–995.

Bridges, C. B. (1935). The structure of salivary chromosomes and the relation of the banding to the genes. *Am. Nat.* **69,** 59.

Burdette, W. J. (1964). The significance of invertebrate hormones in relation to differentiation. *Cancer Res.* **24,** 521–536.

Burtis, K. C. (1985). "Isolation and Characterization of an Ecdysone Induced Gene from *Drosophila melanogaster,*" Ph.D. thesis. Stanford University, Stanford, California.

Burtis, K., Thummel, C., Jones, C., Karim, F., and Hogness, D. (1990). The *Drosophila* 74EF early puff contains *E74,* a complex ecdysone-inducible gene that encodes two ets-related proteins. *Cell (Cambridge, Mass.)* **61,** 85–99.

Chao, A. T., and Guild, G. M. (1986). Molecular analysis of the ecdysterone-inducible 2B5 'early' puff in *Drosophila melanogaster. EMBO J.* **5,** 143–150.

Chatterjee, S. N., and Mukherjee, A. S. (1971). Chromosomal basis of dosage compensation in *Drosophila.* V. Puffwise analysis of gene activity in the X-chromosome of male and female *Drosophila hydei. Chromosoma* **36,** 46–59.

Clever, U. (1963). Von der Ecdysonkonzentration abhängige Genaktivitätsmuster in den Speicheldrüsenchromosomen von *Chironomus tentans. Dev. Biol.* **6,** 73–98.

Clever, U., and Karlson, P. (1960). Induktion von Puff-Veränderungen in den Speicheldrüsenchromosomen von *Chironomus tentans* durch Ecdyson. *Exp. Cell Res.* **20,** 623–626.

Crosby, M. A., and Meyerowitz, E. M. (1986). *Drosophila* glue gene *Sgs-3:* Sequences required for puffing and transcriptional regulation. *Dev. Biol.* **118,** 593–607.

Crowley, T. E., and Meyerowitz, E. M. (1984). Steroid regulation of RNAs transcribed from the *Drosophila* 68C polytene chromosome puff. *Dev. Biol.* **102,** 110–121.

Crowley, T. E., Mathers, P. H., and Meyerowitz, E. M. (1984). A trans-acting regulatory product necessary for expression of the *Drosophila melanogaster* 68C glue gene cluster. *Cell (Cambridge, Mass.)* **39,** 149–156.

DiBello, P. R., Withers, D. A., Bayer, C. A., Fristrom, J. W., and Guild, G. M. (1991). The *Drosophila Broad-Complex* encodes a family of related proteins containing zinc fingers. *Genetics* **129,** 385–397.

Evans, R. M. (1988). The steroid and thyroid hormone receptor superfamily. *Science* **240,** 889–895.

Feigl, G., Gram, M., and Pongs, O. (1989). A member of the steroid hormone receptor gene family is expressed in the 20-OH-ecdysone inducible puff 75B in *Drosophila melanogaster. Nucleic Acids Res.* **17,** 7167–7178.

Furia, M., D'Avino, P. P., Digilio, F. A., Crispi, S., Giordano, E., and Polito, L. C. (1992). Effect of *ecd1* mutation on the expression of genes mapped at the *Drosophila melanogaster* 3C11-12 intermoult puff. *Genet. Res.* **59,** 19–26.

Galceran, J., Llanos, J., Sampedro, J., Pongs, O., and Izquierdo, M. (1990). Transcription at the ecdysone-inducible locus-2B5 in *Drosophila. Nucleic Acids Res.* **18,** 539–545.

Garen, A., Kauvar, L., and Lepesant, J. A. (1977). Roles of ecdysone in *Drosophila* development. *Proc. Natl. Acad. Sci. U.S.A.* **74,** 5099–5103.

Garfinkel, M. D., Pruitt, R. E., and Meyerowitz, E. M. (1983). DNA sequences, gene regulation and modular protein evolution in the *Drosophila* 68C gene cluster. *J. Mol. Biol.* **169,** 765–789.

Gautam, N. (1983). Identification of 2 closely located larval salivary protein genes in *Drosophila melanogaster. Mol. Gen. Genet.* **189,** 495–500.

Georgel, P., Ramain, P., Giangrande, A., Dretzen, G., Richards, G., and Bellard, M. (1991). *Sgs-3* chromatin structure and trans-activators: Developmental and ecdysone induction of a glue enhancer-binding factor, GEBF-I, in *Drosophila* larvae. *Mol. Cell. Biol.* **11,** 523–532.

Guay, P. S., and Guild, G. M. (1991). The ecdysone-induced puffing cascade in *Drosophila* salivary glands: A *Broad-Complex* early gene regulates intermolt and late gene transcription. *Genetics* **129,** 169–175.

Guild, G. M. (1984). Molecular analysis of a developmentally regulated gene which is expressed in the larval salivary gland of *Drosophila. Dev. Biol.* **102,** 462–470.

Hansson, L., and Lambertsson, A. (1983). The role of *su-f* gene function and ecdysterone in transcription of glue polypeptide messenger RNA in *Drosophila melanogaster. Mol. Gen. Genet.* **192,** 395–401.

Hansson, L., and Lambertsson, A. (1989). Steroid regulation of glue protein genes in *Drosophila melanogaster. Hereditas* **110,** 61–67.

Hansson, L., Lineruth, K., and Lambertsson, A. (1981). Effect of the $l(1)su(f)^{ts67g}$ mutation of *Drosophila melanogaster* on glue protein synthesis. *Wilhelm Roux's Arch. Dev. Biol.* **190,** 308–312.

Harrison, S., and Travers, A. (1990). The *tramtrack* gene encodes a *Drosophila* finger protein that interacts with the *ftz* transcriptional regulatory region and shows a novel embryonic expression pattern. *EMBO J.* **9,** 207–216.

Hill, R. J., Segraves, W. A., Choi, D., Underwood, P. A., and MacAvoy, E. (1993). The reaction with polytene chromosomes of antibodies raised against *Drosophila* E75A protein. *Insect Biochem. Mol. Biol.* **23,** 99–104.

Hofmann, A., and Korge, G. (1987). Upstream sequences of dosage-compensated and non-compensated alleles of the larval secretion protein gene *Sgs-4* in *Drosophila. Chromosoma* **96,** 1–7.

Hofmann, A., Keinhorst, A., Krumm, A., and Korge, G. (1987). Regulatory sequences of the *Sgs-4* gene of *Drosophila melanogaster* analysed by P element-mediated transformation. *Chromosoma* **96,** 8–17.

Huet, F., Ruiz, C., and Richards, G. (1993). Puffs and PCR—The *in vivo* dynamics of early gene-expression during ecdysone responses in *Drosophila. Development* **118,** 613–627.

Janknecht, R., Taube, W., Ludecke, H. J., and Pongs, O. (1989). Characterization of a putative transcription factor gene expressed in the 20-OH-ecdysone inducible puff 74EF in *Drosophila melanogaster. Nucleic Acids Res.* **17,** 4455–4464.

Jones, C., Dalton, M., and Townley, L. (1991). Interspecific comparisons of the structure and regulation of the *Drosophila* ecdysone-inducible gene *E74. Genetics* **127,** 535–543.

Karim, F. D. (1992). "Regulation of Early Ecdysone Inducible Genes in *Drosophila,*" Ph.D. thesis. University of Utah, Salt Lake City.

Karim, F., and Thummel, C. (1991). Ecdysone coordinates the timing and amounts of *E74A* and *E74B* transcription in *Drosophila. Genes Dev.* **5,** 1067–1079.

Karim, F., and Thummel, C. (1992). Temporal coordination of regulatory gene expression by the steroid hormone ecdysone. *EMBO J.* **11,** 4083–4093.

Karim, F. D., Urness, L. D., Thummel, C. S., Klemsz, M. J., McKercher, S. R., Celada, A., Van Beveren, C., Maki, R. A., Gunther, C. V., Nye, J. A., and Graves, B. J. (1990). The ETS-domain: A new DNA-binding motif that recognises a purine-rich core DNA sequence. *Genes Dev.* **4,** 1451–1453.

Karim, F. D., Guild, G. M., and Thummel, C. S. (1993). The *Drosophila Broad-Complex* plays a key role in controlling ecdysone-regulated gene-expression at the onset of metamorphosis. *Development* **118,** 977–988.

Kiss, I., Benzce, G., Fodor, A., Szabad, J., and Fristrom, J. W. (1976). Prepupal larvae mosaics in *Drosophila melanogaster. Nature* (*London*) **262,** 136–138.

Kiss, I., Szabad, J., and Major, J. (1978). Genetic and developmental analysis of puparium formation in *Drosophila. Mol. Gen. Genet.* **164,** 77–83.

Kiss, I., Szabad, J., Belyaeva, E. S., Zhimulev, I. F., and Major, J. (1980). Genetic and developmental analysis of mutants in an early ecdysone-inducible puffing region in *Drosophila melanogaster. Basic Life Sci.* **16,** 163–181.

Kiss, I., Beaton, A. H., Tardiff, J., Fristrom, D., and Fristrom, J. W. (1988). Interactions and developmental effects of mutations in the *Broad-Complex* of *Drosophila melanogaster. Genetics* **118,** 247–259.

Koelle, M., Talbot, W., Segraves, W., Bender, M., Cherbas, P., and Hogness, D. (1991). The *Drosophila EcR* gene encodes an ecdysone receptor, a new member of the steroid receptor superfamily. *Cell* (*Cambridge, Mass.*) **67,** 59–77.

Koelle, M., Segraves, W., and Hogness, D. (1992). *DHR3:* A *Drosophila* steroid receptor homolog. *Proc. Natl. Acad. Sci. U.S.A.* **89,** 6167–6171.

Korge, G. (1975). Chromosome puff activity and protein synthesis in larval salivary glands of *Drosophila melanogaster. Proc. Natl. Acad. Sci. U.S.A.* **72,** 4550–4554.

Korge, G., Heide, I., Sehnert, M., and Hofmann, A. (1990). Promoter is an important determinant of developmentally regulated puffing at the *Sgs-4* locus of *Drosophila melanogaster. Dev. Biol.* **138,** 324–337.

Lavorgna, G., Karim, F., Thummel, C., and Wu, C. (1993). Potential role for a *FTZ-F1* steroid receptor superfamily member in the control of *Drosophila* metamorphosis. *Proc. Natl. Acad. Sci. U.S.A.* **90,** 3004–3008.

Lawson, R., Mestril, R., Luo, Y., and Voellmy, R. (1985). Ecdysterone selectively stimulates the expression of 23000-Da heat-shock protein–beta-galactosidase hybrid gene in cultured *Drosophila* cells. *Dev. Biol.* **110,** 321–330.

McNabb, S. L., and Beckendorf, S. K. (1986). Cis-acting sequences which regulate expression of the *Sgs-4* glue protein gene of *Drosophila. EMBO J.* **5,** 2331–2340.

Mestril, R., Schiller, P., Amin, J., Klapper, H., Ananthan, J., and Voellmy, R. (1986). Heat shock and ecdysterone activation of the *Drosophila melanogaster* hsp23 gene; a sequence element implied in developmental regulation. *EMBO J.* **5,** 1667–1673.

Meyerowitz, E., and Hogness, D. (1982). Molecular organization of a *Drosophila* puff site that responds to ecdysone. *Cell* (*Cambridge, Mass.*) **28,** 165–176.

Meyerowitz, E. M., Raghavan, K. V., Mathers, P. H., and Roark, M. (1987). How *Drosophila* larvae make glue: control of *Sgs-3* gene expression. *Trends Genet.* **3,** 288–293.

Muskavitch, M. A., and Hogness, D. S. (1980). Molecular analysis of a gene in a developmentally regulated puff of *Drosophila melanogaster. Proc. Natl. Acad. Sci. U.S.A.* **77,** 7362–7366.

Oro, A. E., Yao, T. P., and Evans, R. M. (1993). The *Drosophila* retinoid-X receptor homolog *ultraspiracle* regulates ecdysone receptor function. *J. Invest. Dermatol.* **100,** 555.

Painter, T. S. (1933). A new method for the study of chromosome rearrangements and the plotting of chromosome maps. *Science* **78,** 585–586.

Paro, R., and Hogness, D. S. (1991). The polycomb protein shares a homologous domain with a heterochromatin-associated protein of *Drosophila. Proc. Natl. Acad. Sci. U.S.A.* **8,** 263–267.

Raghavan, K. V., Crosby, M. A., Mathers, P. H., and Meyerowitz, E. M. (1986). Sequences sufficient for correct regulation of *Sgs-3* lie close to or within the gene. *EMBO J.* **5,** 3321–3326.

Read, D., and Manley, J. (1992). Alternatively spliced transcripts of the *Drosophila tramtrack* gene encode zinc finger proteins with distinct DNA binding specificities. *EMBO J.* **11,** 1035–1044.

Restifo, L. L., and White, K. (1989). An ecdysterone-regulated genetic locus is required for CNS metamorphosis in *Drosophila. Soc. Neurosci. Abstr.* **15,** 87.

Richards, G. (1976). Sequential gene activation by ecdysone in polytene chromosomes of *Drosophila melanogaster.* V. The late prepupal puffs. *Dev. Biol.* **54,** 264–275.

Richards, G. (1978). The relative biological activities of alpha- and beta-ecdysone and their dehydro-derivatives in the chromosome puffing assay. *J. Insect. Physiol.* **24,** 329–335.

Richards, G. (1980). The polytene chromosomes in the fat body nuclei of *Drosophila melanogaster. Chromosoma* **79,** 242–250.

Richards, G. (1981). The radioimmune assay of ecdysteroid titres in *Drosophila melanogaster. Mol. Cell. Endocrinol.* **21,** 181–197.

Richards, G. (1992). Switching partners? *Curr. Biol.* **2,** 657–659.

Richards, G., Cassab, A., Bourouis, M., Jarry, B., and Dissous, C. (1983). The normal developmental regulation of a cloned *Sgs-3* 'glue' gene chromosomally integrated in *Drosophila melanogaster* by P element transformation. *EMBO J.* **2,** 2137–2142.

Riddiford, L. M. (1993). Hormones and *Drosophila* development. *In* "The Development of *Drosophila melanogaster*" (M. Bate and A. Martinez-Arias, eds.), pp. 899–939. Cold Spring Harbor Laboratory, Cold Spring Harbor, New York.

Roark, M., Raghavan, K. V., Todo, T., Mayeda, C. A., and Meyerowitz, E. M. (1990). Cooperative enhancement at the *Drosophila Sgs-3* locus. *Dev. Biol.* **139,** 121–133.

Saumweber, H., Frasch, M., and Korge, G. (1990). Two puff-specific proteins bind within the 2.5 kb upstream region of the *Drosophila melanogaster Sgs-4* gene. *Chromosoma* **99,** 52–60.

Savakis, C., Koehler, M. M. D., and Cherbas, P. (1984). cDNA clones for the ecdysone-inducible polypeptide (EIP) mRNAs of *Drosophila* Kc cells. *EMBO J.* **3,** 235–243.

Segraves, W. A. (1988). "Molecular and Genetic Analysis of the *E75* Ecdysone-Responsive gene of *Drosophila melanogaster,*" Ph.D. thesis. Stanford University, Stanford, California.

Segraves, W. A., and Hogness, D. A. (1984). Molecular and genetic analysis of the 75B ecdysone inducible puff of *Drosophila melanogaster. Genetics* **107,** s96–s97.

Segraves, W., and Hogness, D. (1990). The *E75* ecdysone-inducible gene responsible for the 75B early puff in *Drosophila* encodes two new members of the steroid receptor superfamily. *Genes Dev.* **4,** 204–219.

Semeshin, V. F., Baricheva, E. M., Belyaeva, E. S., and Zhimulev, I. F. (1985). Electron microscopic analysis of *Drosophila* polytene chromosomes. II. Development of complex puffs. *Chromosoma* **91,** 210–233.

Shermoen, A. W., Jongens, J., Barnett, S. W., Flynn, K., and Beckendorf, S. K. (1987). Developmental regulation by an enhancer from the *Sgs4* gene of *Drosophila. EMBO J.* **6,** 207–214.

Stone, B. L., and Thummel, C. S. (1993). The *Drosophila* 78C early-late puff contains *E78,* an ecdysone-inducible gene that encodes a novel member of the nuclear hormone-receptor superfamily. *Cell (Cambridge, Mass.)* **75,** 307–320.

Talbot, W., Swyryd, E., and Hogness, D. (1993). *Drosophila* tissues with different metamorphic responses to ecdysone express different ecdysone receptor isoforms. *Cell (Cambridge, Mass.)* **73,** 1323–1337.

Thomas, H., Stunnenberg, H., and Stewart, A. (1993). Heterodimerization of the *Drosophila* ecdysone receptor with retinoid X receptor and ultraspiracle. *Nature (London)* **362,** 471–475.

Thummel, C. (1989). The *Drosophila E74* promoter contains essential sequences downstream from the start site of transcription. *Genes Dev.* **3,** 782–792.

Thummel, C., Burtis, K., and Hogness, D. (1990). Spatial and temporal patterns of *E74* transcription during *Drosophila* development. *Cell (Cambridge, Mass.)* **61,** 101–111.

Todo, T., Roark, M., Raghavan, K. V., Mayeda, C., and Meyerowitz, E. (1990). Fine-structure mutational analysis of a stage- and tissue-specific promoter element of the *Drosophila* glue gene *Sgs-3. Mol. Cell. Biol.* **10,** 5991–6002.

Tsukiyama, T., Becker, P. B., and Wu, C. (1994). ATP-dependent nucleosome disruption at a heat-shock promoter mediated by binding of GAGA transcription factor. *Nature (London)* **367,** 525–532.

Urness, L., and Thummel, C. (1990). Molecular interactions within the ecdysone regulatory hierarchy: DNA binding properties of the *Drosophila* ecdysone-inducible E74A protein. *Cell (Cambridge, Mass.)* **63,** 47–61.

Velissariou, V., and Ashburner, M. (1981). Cytogenetic and genetic mapping of a salivary gland secretion protein in *Drosophila melanogaster. Chromosoma* **84,** 173–185.

von Besser, H., Schnabel, P., Wieland, C., Fritz, E., Stanewsky, R., and Saumweber, H. (1990). The puff-specific *Drosophila* protein Bj6, encoded by the gene *no-on transient A,* shows homology to RNA-binding proteins. *Chromosoma* **100,** 37–47.

Walker, V. K., and Ashburner, M. (1981). The control of ecdysterone-regulated puffs in the *Drosophila* salivary glands. *Cell (Cambridge, Mass.)* **26,** 269–277.

Wieland, C., Mann, S., von Besser, H., and Saumweber, H. (1992). The *Drosophila* nuclear protein Bx42, which is found in many puffs on polytene chromosomes, is highly charged. *Chromosoma* **101,** 517–525.

Wolfner, M. (1980). "Ecdysone-Responsive Genes of the Salivary Glands of *Drosophila melanogaster,*" Ph.D. thesis. Stanford University, Stanford, California.

Xiong, W., and Montell, C. (1993). *tramtrack* is a transcriptional repressor required for cell fate determination in the *Drosophila* eye. *Genes Dev.* **7,** 1085–1096.

Yao, T., Segraves, W., Oro, A., McKeown, M., and Evans, R. (1992). *Drosophila ultraspiracle* modulates ecdysone receptor function via heterodimer formation. *Cell* (*Cambridge, Mass.*) **71,** 63–72.

Yao, T. P., Forman, B. M., Jiang, Z. Y., Cherbas, L., Chen, J. D., McKeown, M., Cherbas, P., and Evans, R. M. (1993). Functional ecdysone receptor is the product of *Ecr* and *ultraspiracle* genes. *Nature* **366,** 476–479.

Zhimulev, I. F., Vlassova, I. E., and Belyaeva, E. S. (1982). Cytogenetic analysis of the 2B3-4–2B11 region of the X chromosome of *Drosophila melanogaster.* III. Puffing disturbance in salivary gland chromosomes of homozygotes for mutation. *l(1)pp1t10. Chromosoma* **85,** 659–672.

4

Chromosome Puffing: Supramolecular Aspects of Ecdysone Action

MARKUS LEZZI

Institute for Cell Biology
Swiss Federal Institute of Technology
Hoenggerberg, CH-8093 Zurich
Switzerland

I. INTRODUCTION

Puffing is a morphological feature of giant polytene chromosomes. Because the polytenic state of interphase chromosomal material is the exception rather than the rule among metamorphosing insects, the ability of polytene chromosomes to form puffs cannot be considered a prerequisite for metamorphosis. However, the occurrence of polytene chromosomes may constitute an advantage for insects as well as for scientists investigating the mode of ecdysone action. Puffing studies revealed that ecydsone acts by controlling gene activities (Karlson, 1963), that it does so by triggering a gene activation cascade (Clever, 1964; Ashburner *et al.*, 1974), and that chromatin domains may have to open before ecdysone, together with the ecdysone–receptor (EcR) complex, finds access to gene regulatory sequences in the DNA (Lezzi and Richards, 1989). Polytene chromosomes have gained increasing importance as a substratum for molecular probes. Chromosomal localization of gene sequences as well as colocalization of known proteins may provide new insights into structural and functional interrelationships important for gene regulation by ecdysone.

II. POLYTENIC STATE OF INTERPHASE CHROMATIN

A. Formation

Generally, mitosis consists of chromosomal division, nuclear division, and cell division. Thus, omission of mitosis results in chromosomes consisting of nonsegregated chromatids residing in one nucleus of a nondividing cell. The cable-like chromosomes formed thereby may consist of up to 2^{13} threads. They are called polytene chromosomes, but have little in common with most (i.e., metaphase) chromosomes. They represent interphase chromatin of a special arrangement. Like interphase chromatin, polytene chromosomes consist of condensed and decondensed chromatin that alternates along their length, thereby producing a distinct banding pattern. Condensed regions may temporarily decondense, resulting in so-called structural modifications of the normal banding pattern (Beermann, 1962).

B. Possible Advantage for the Insect

With most diploid cells, the development of a highly specialized function and the ability to divide, and thereby to contribute to tissue growth, are mutually exclusive. In contrast, polytenic tissues and cells may grow easily and exert their specialized function at the same time since they omit the critical step of cell division. Unlike mitosis, DNA replication does not require a dedifferentiation of elaborate cytoplasmic structures and may interrupt specialized cellular functions only briefly.

The polytenic state offers a unique possibility of gene regulation that is not common to interphase chromatin. Electron microscopic studies suggest that, at a given time, not all chromosomal threads of an active polytene chromosome region are being used in transcription (Trepte, 1993, and references therein). That the number of strands being transcribed may vary is also suggested by studies involving three-dimensional reconstructions of immunofluorescent induced endogenous hybrid (IEH) signals (for method, see Lezzi and Richards, 1989) revealing the presence of nascent transcripts. These signals extend sometimes only over a portion, rather than the total cross section, of a polytene chromosome region (M. Lezzi, D. Gilligan, and M. Messerli, unpublished observations.) Variation in strand usage may greatly expand the range of expression of a given gene.

III. POLYTENE CHROMOSOME PUFFS

A puff in a native polytene chromosome is best compared with a powder puff (see, e.g., Lezzi and Robert, 1972). It is a fluffy formation, essentially constructed of loosely arranged fibrils. In light as well as electron microscopic preparations, it exhibits little contrast. Any allusion to an inflation (e.g., of a balloon) is misleading as it implies the existence of a membrane. In addition, a mere accumulation of puff-specific material cannot fully account for the puffing phenomenon since conditions causing chromosomal contraction make puffs shrink twice as much as the rest of the chromosome (unpublished observations). In the pioneering work of Beermann as well as that of Breuer and Pavan (reviewed by Beermann, 1962), it was suggested that puffs are manifestations of local gene activity. While the correlation between puff structure and gene transcription holds true in most instances, the ability to synthesize RNA must be regarded as a secondary feature of a puff.

A. Local Chromatin Decondensation: Primary Feature of Puffs

In principle, polytene chromosomes consist of the same kind of chromatin fibers as common interphase chromatin does (cf. Ananiev and Barsky, 1985; Widmer *et al.,* 1984; Trepte, 1993). Thus, information and theoretical considerations concerning the structure of these fibers in interphase chromatin and their changes are also relevant for polytene chromatin.

1. Various States of Chromatin

The basic unit of the 10-nm chromatin fiber (nucleosomal filament) is the nucleosome (Fig. 1c). In each nucleosome the DNA double helix is wound twice around an octamer of a core of histones H2A, H2B, H3, and H4, and is sealed by an externally placed histone H1. This results in a sixfold linear compaction of the DNA. A further six- to sevenfold reduction in fiber length is achieved by the coiling of the 10-nm fiber into a solenoid structure (Fig. 1b), which itself is probably further folded to yield an even higher degree of DNA packaging (Fig. 1a). Masses of densely packed (high order) and loosely packed (lower order) fibers are referred to as condensed and decondensed chromatin, respectively (for a comprehensive treatment of the topic, see van Holde, 1989).

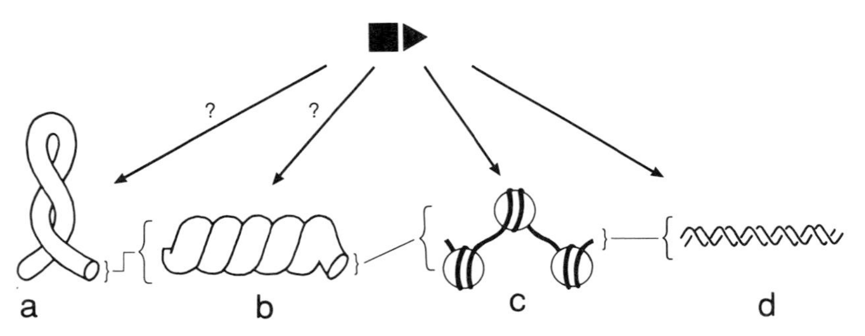

Fig. 1. Different organizational states of chromatin; effect on the binding of the EcR complex to EcREs in DNA. (a) Supercoiled loop (hypothetical structure) formed by the solenoidal fiber (b) which is formed by the nucleosomal filament (c) consisting of the DNA (d) periodically wound around histone cores (spheres). Histone H1 is not depicted. "?," can the ecdysone-activated complex of EcR (square) and USP (triangle) also bind to EcREs in chromatin of higher order structures (a, b)?

2. *State of Chromatin in Polytene Chromosome Structures*

In polytene chromosomes, condensed chromatin is generally represented by compact bands and (α-)heterochromatic regions whereas decondensed chromatin is found in the form of interbands and structural modifications which comprise puffs and active nucleolar organizer regions. In puffs, the mass of decondensed chromatin is either in line with the chromosomal body protruding laterally to some extent or arranged like a ring (Balbiani ring, BR) surrounding the chromosomal cylinder (Beermann, 1962).

a. Condensed Bands. In condensed bands, chromatin is organized in a supranucleosomal structure. Rykowski *et al.* (1988), as well as Semeshin *et al.* (1986), determined the linear packaging coefficient of DNA in specific bands of *Drosophila* salivary gland chromosomes to be 40–50 which would correspond to a solenoidal configuration. However, Frey *et al.* (1982), as well as Gruzdev and Reznik (1981), independently estimated that coefficient to be about 100–200 in *Chironomus* salivary gland chromosomes. This would require a suprasolenoidal arrangement of the chromatin fibers. These seemingly contradictory findings might be explained by observations based on nuclease sensitivity measurements indicating that there are differences in the level of chromatin compaction of bands, which are not detectable cytologically (Gilligan, 1994).

b. Balbiani Rings. Electron microscopy as well as nuclease sensitivity studies (Widmer *et al.*, 1984) clearly show that in hyperactive BRs 2 of *Chironomus tentans* the chromatin fibers carry no nucleosomes. Furthermore, the BR chromatin is much more sensitive to nuclease digestion than chromatin of condensed bands (Gilligan, 1994).

c. Puffs. The quality of the chromatin fiber present in puffs is not known. Indirect evidence supports the existence of a nucleosomal filament.

d. Interbands. Although electron microscopy studies provide evidence for the presence of nucleosomal filaments in interbands (Ananiev and Barsky, 1985), the exact nature and the functional significance of these permanently decondensed chromatin regions are still a mystery. Findings suggest that they contain DNA stretches flanking the main part of the transcription unit of the gene (Rykowski *et al.*, 1988) and that perhaps they include the transcription initiation region with stalled or idling RNA polymerase II attached (Sass, 1980). Some interbands may just represent micropuffs or small decondensed portions of a band.

e. Decondensed Bands. The existence of bands with a less than normal optical density and a somewhat greater (longitudinal) expansion is frequently overlooked. Very precise morphometric measurements revealed that the band giving rise to the early ecdysone-inducible puff I-18C in *C. tentans* salivary gland chromosomes may, under physiological conditions, expand longitudinally from a width of 1.7 to 2.3 μm without becoming transcriptionally active. During actual puffing when transcription is occurring, the I-18C region expands longitudinally to a width of 2.9 μm (unpublished observations). Theoretically, each of these steps involving a 0.6-μm length increase could be caused by the destruction of at least 10 nucleosomes, by an unwinding of at least 10 solenoidal turns, or by an unfolding of a hypothetical suprasolenoidal structure.

3. Mechanisms of Local Chromatin Decondensation

It is not known how, *in vivo,* a condensed band decondenses, forms a puff, and, in some cases, a Balbiani ring. The following parameters must be considered when trying to explain these structural changes.

a. Change in Electrostatic Interactions in Chromatin. The isolated condensed chromatin of the diploid nuclei is easily decondensed *in vitro* by either increasing or decreasing the salt concentration of the incubation buffer (Bradbury *et al.,* 1973). By lowering the ionic strength, residual negative groups in the chromatin become hydrated, resulting in decondensation. An increase in ionic strength causes a partial dissociation of ionic bonds between the negatively charged DNA and the positively charged histones, respectively, which also leads to decondensation. Histone H1, which has a pronounced propensity to form coacervates by hydrophobic interactions, seems to play a key role in this condensation/decondensation shift (Bradbury *et al.,* 1973). Ultrastructurally, the aggregated, condensed chromatin state corresponds to the solenoid (30 nm) fiber whereas the hydrated chromatin is composed of nucleosomal filaments in a zigzag configuration; the (partially) dissociated chromatin contains stretches of straight nucleosome filaments or even nucleosome-free fibers (Thoma *et al.,* 1979).

Isolated polytene chromosomes behave exactly like isolated diploid interphase chromatin when exposed to a medium having either low or high ionic strength (for references see Lezzi and Robert, 1972): in either situation the chromosomes swell. Interestingly, conditions occur under which not all bands react simultaneously, but instead some bands start decondensing earlier than others when the ionic concentration of the incubation medium is increased. Moreover, by varying the ratios of Na^{+}, K^{+}, Mg^{2+}, and Ca^{2+} in the medium, the pattern of precociously decondensing bands could be changed, while Robert devised pH conditions which do not result in only slight decondensations of individual bands, but in large puff-like structures (see Lezzi and Robert, 1972).

One may predict that any event which changes the electrostatic interaction of histone with DNA results in a change in the chromatin structure. Such events not only involve changes in the electrolytic environment but also include post-translational modifications of histones themselves and an interference by charged macromolecules, such as RNA, hyperphosphorylated RNA polymerase II (Weeks *et al.,* 1993), and other acidic proteins.

b. Action of Novel Types of Nonhistone Proteins. Multimeric protein complexes have either a destabilizing or a stabilizing effect on chromatin. The large SWI/SNF complex is able to disrupt the nucleosomal core by an ATP-dependent reaction (see the review by Lewin, 1994). In *D. melanogaster,* the Polycomb group and related proteins, which are deposited at several chromosomal regions as complexes of varying compositions, are important for maintaining genes in a repressed state. Removal of one of these proteins, E(Z), by a (temperature-sensitive) mutation impairs the deposition of other *Polycomb* group proteins and changes the polytene chromosome structure dramatically. Polytene chromosomes of these mutants are larger, thicker, and puffier than normal, exhibiting an overall decondensed banding pattern (Rastelli *et al.,* 1993).

c. Change in Topological State of DNA in Chromatin. The DNA double helix has to rotate around its axis in almost all situations in which DNA is forced to leave its normal (B-)configuration. This happens, for example, when it is wrapped around a nucleosome core or when it is transcribed. If the DNA double helix is prevented from rotating, it undergoes topological stress, i.e., it becomes supercoiled. This spring-loaded state requires a compensatory configuration which causes gross changes in DNA and chromatin fiber structure (van Holde, 1989). Preliminary *in situ* measurements with isolated polytene nuclei and chromosomes of *C. tentans* suggest that the supercoil density is considerably higher in hyperactive BRs than in bands (Gruzdev and Shurdow, 1992; A. D. Gruzdev and M. Lezzi, unpublished observations). This is most likely because the number of actively transcribing RNA polymerases II per strand is very high in hyperactive BRs (Widmer *et al.,* 1984). Obviously, topoisomerase I is not able to relax supercoiled DNA in these structures fully. This is surprising since topoisomerase I was previously shown to accumulate in BRs and to be required for maintaining their transcriptional activity (Egyhazi and Durban, 1987). The overcrossing of looping out chromosome strands (see Beermann, 1962), a typical feature of BRs, may be taken as a structural indication of an unrelaxed state of DNA in these special puff types.

B. RNA Puffs: Local Decondensations Active in RNA Synthesis

Most of the structural modifications studied in the early 1950s by Beermann represent RNA puffs (reviewed by Beermann, 1962). At that time, the significance of RNA as a messenger of genetic information was not yet established. Autoradiography, a reliable method for revealing RNA synthesis in puffs, became available only some years later (see Beermann, 1962). The IEH method, which is based on the induction of endogenous hybrids between nascent RNA and its template, is particularly useful for revealing RNA synthesis in puffs (see Lezzi and Richards, 1989). This immunohistochemical method gives a much better resolution and sensitivity than autoradiography (see, e.g., Lezzi *et al.,* 1989). All the methods have shown that most of the local decondensations in polytene chromosomes investigated by Beermann and later by other investigators are sites having high RNA synthetic activity (Beermann, 1962; see Russell and Ashburner, Chapter 3, this volume).

However, local chromatin decondensation is sometimes not strictly correlated with transcription [cf. BR1 and 2 in *Acricotopus lucidus* (Panitz *et al.,* 1972)], or, in rare cases, may occur even in its absence (for references see Lezzi and Richards, 1989). While these cases might be considered unnatural as they involve heat shock (HS) and the use of RNA synthesis inhibitors, clearly one example exists where transcription-independent local decondensation occurs under perfectly normal physiological conditions. At stage 5 of fourth instar *C. tentans* larvae, the ecdysone-inducible puff site I-18C is transcriptionally inactive as revealed by the IEH method and autoradiography. Nonetheless, it is decondensed in subitaneously developing animals of that stage in contrast to oligopausing stage 5 larvae in certain periods of the diurnal cycle (Lezzi *et al.*, 1989, 1991).

Whereas it is evident that local chromosome decondensation up to puff-like or even BR-like structures may occur in the absence of concomitant transcription, the question arises whether the reverse is true as well. RNA polymerase II can certainly transcribe a template even when nucleosomes are present (see van Holde, 1989). However, it is still under debate whether DNA organized into higher order chromatin structures, such as condensed polytene chromosome bands, is accessible to the transcription machinery. Experimental evidence exists which would suggest that it is. However, in most instances this can be traced back to limitations of the detection method used to reveal subtle changes in chromatin structure (e.g., controversy regarding the ectopically expressed *Sgs-3* gene, as discussed by Lezzi and Richards, 1989). One has to realize that any structural change in chromatin domains containing less than 3 kbp DNA falls much below the resolution of light microscopy (Note: the size of the *Sgs-3* transcription unit is 1.1 kbp). Moreover, differential strand usage (see Section II,B) may be another reason for overlooking (partial) decondensation of a band.

Several genes of ecdysone-controlled puff sites have been cloned and characterized (Table I). Many puff sites accommodate several genes, some of which have more than one transcription start site and thus promoter. Therefore, identification of that particular transcription unit which, when becoming active, would be responsible for the formation of a specific puff is hardly possible. In fact, as Table I demonstrates, a strict positive correlation between the level of a given RNA species in a tissue and the extent of puffing at its gene locus is evident only in a few cases. One might argue that the cellular steady-state level of a RNA species does not really reflect the activity state of the corresponding gene because of post-transcriptional events like RNA processing and degradation.

Therefore, attempts have been made to determine the transcriptional activity of a specific gene right at its chromosomal locus (unpublished observations). The combination of the IEH method with high-resolution *in situ* hybridization of flanking gene sequences proved that the ecdysone-inducible IEH signal in the I-18C region of *C. tentans* (Lezzi *et al.*, 1989, 1991) is in fact derived from the cloned *I-18C* gene (see Dorsch-Häsler

TABLE I

Genes in Ecdysone-Controlled Puff Sites and Their Relation to Puffing

Species	Puff site[a]	Puff type[a]	Genes	Promoters[b]	EcR/DNA binding[c]	Ec-inducible RNAs[d]	[RNA] ≈puff[e]	Protein type[f]
C. tentans	I-18C[g]	Early	*I-18C*	1	Yes	2 of 4	±Yes/no	P9/HSP18
	IV-2B[h,i]	Early	*cE75*	?	?	1 of 2	Yes/no	ST rec.
	I-17B[j]	Early–late	*sp-140*	?	?	? of 1	Yes	secr.
		Two subpuffs	*sp-420*	?	?	? of 1	No	secr.
	II-14A[h,i]	Late	*cUSP*	?	?	0 of 1	±Yes	ST rec.
R. americana	C8[k]	Late DNA puff	*C8*	?	?	? of 1	±Yes	secr.
S. coprophila	II/9A[l]	Late DNA puff	*II/9-1*	?	3EcREs	1 of 1	Yes	secr.
			II/9-2	?	No	0 of 1	Yes	secr.
D. melanogaster	74EF[m]	Early	*E74*[o]	A	Yes	1 of 1	Yes	Ets rel.
				B	?	2 of 2	No	Ets. rel.
	75B[m]	Early	*E75*[o]	A	?	2 of 2	Yes	ST rec.
				B	?	2 of 2	No	ST rec.
				C	?	2 of 2	No	ST rec.
	2B5[m]	Early	*BR-C*	4	?	3 of 4[p]	Yes/no	BR-C pr.
	78C[m]	Early–late	*E78*	A	?	? of 4	No	ST rec.
				B	?	4 of 4	Yes	ST rec.
	46F[M]	Early–late	*DHR3*	?	?	3 of 3	±Yes	ST rec.
	71C-F[m]	Mixed,[n] five subpuffs	*71 I-VI*	1	?	? of 6	Yes/no	various
			VII[o]	1	?	? of 3	Yes/no	secr.
	75CD[m]	Midprepup.	*βFTZ-F1*	?	?	? of 2	Yes	ST rec.
	3C[m]	Intermolt	*Sgs-4*	1	2EcREs[q]	1 of 1	Yes/no[r]	secr.
			Pig-1	2	?	? of 2	±No	secr.(?)
			ng-1	1	?	? of 1	±Yes	secr.(?)
	68C[m]	Intermolt	*Sgs-3*	1	?	+/−	Yes/no[r]	secr.
			Sgs-7	1	?	+/−	Yes/no[r]	secr.
			Sgs-8	1	?	+/−	Yes/no[r]	secr.

[a] Revised map positions and puff type classification.

[b] Letters specify the names of promoters, otherwise the numbers of promoters per gene are listed.

[c] Yes, binding demonstrated (e.g., by filter-binding assay). EcREs were identified by methylation interference and/or electromobility shift assays, comparison with a consensus sequence, and by functionality tests with reporter gene constructs.

[d] Number of experimentally ecdysone-inducible RNA types per total number of RNA types made by a respective gene. +/−, RNA (one species per gene) is sometimes induced, sometimes repressed by ecdysone.

[e] Does the cellular steady-state level of the RNA type change in parallel with puff size? ±, yes, to some extent. yes/no, in some situations "yes," in others "no."

[f] Gene regulatory protein types: ST rec., member of the steroid/thyroid hormone receptor superfamily; Ets rel., Ets-related protein; BR-C pr., BR-C proteins; secr., larval salivary gland specific secretion protein; various, HSP18 (HS-inducible); P9, unknown proteins.

[g] Dorsch-Häsler *et al.* (1990) and references therein; Turberg *et al.* (1992).

[h] Wegmann *et al.* (1995).

[i] M. Vögtli (personal communication).

[j] Galli and Wieslander (1993) and references therein.

[k] Frydman *et al.* (1993) and references therein.

[l] Gerbi *et al.* (1993) and references therein.

[m] See Russell and Ashburner (Chapter 3, this volume) and references therein.

[n] Complex region, difficult to analyze, but generally considered "late." Subpuffs 71C3-4, 71D/E, and 71D1.2 were reported to be "early," "early–late" (Yao *et al.*, 1993), and "constant," respectively. Genes I–VII (all "late" except one intermolt gene) apparently reside in 71E; *Sgs-6* (not listed) maps to 71C3-4. See Chapter 3 by Russell and Ashburner (this volume) and references therein.

[o] Additional less characterized genes are not listed.

[p] Up to 12 RNA species reported (Russell and Ashburner, Chapter 3, this volume).

[q] Lehmann and Korge (1995).

[r] Deviation between RNA level and puffing particularly striking in mutant or transgenic animals.

et al., 1990), and from no other gene (M. Lezzi and K. Dorsch-Häsler, unpublished observations). Even in this rather simple and very clear situation (1 puff = 1 transcription activity signal = 1 gene having 1 promoter), a strict quantitative correlation between the transcriptional activity and transcript levels of a gene and the extent of puffing at this gene's locus could not be found. Puffing (local decondensation), transcription, and cytoplasmic RNA titers have been proven to be differentially regulated events. DNA sequences which could be important in the control of local decondensation during puffing have not yet been identified.

C. DNA Puffs: RNA Puffs Containing Amplified DNA

The puffs investigated by Breuer and Pavan were DNA puffs. The major product of these puffs is the same as with RNA puffs, namely RNA. However, in contrast to RNA puffs, DNA puff sites amplify their DNA prior to, or in parallel with, puff formation and transcription. The local increase in DNA content then gives rise to a much larger expansion exhibiting a much higher RNA synthetic capacity than a RNA puff (see Lara *et al.,* 1991).

The process leading to DNA puff formation starts in the middle of the last larval instar in salivary gland chromosomes of sciarids at a time when the endogenous ecdysteroid titer begins to rise. Interestingly, the process of DNA puff formation, including DNA replication, DNA amplification, local chromatin decondensation, and RNA synthesis, can be induced prematurely by injecting ecdysone into early fourth instar larvae (Lara *et al.,* 1991). The induction of DNA puffs constitutes a late ecdysone-controlled event which is blocked by protein synthesis inhibitors (see Lara *et al.,* 1991; Gerbi *et al.,* 1993). With reference to the proposed regulation of late genes in *D. melanogaster,* it is assumed that ecdysone, together with its receptor complex, either influences DNA puff formation indirectly or acts as a repressor. However, molecular data strongly indicate that the ecdysone–ecdysone receptor (Ec-EcR) complex itself participates in the induction of DNA puffs (see Gerbi *et al.,* 1993, and references therein, also for the following).

Two genes (*II/9-1* and *II/9-2*) in the DNA puff II/9A of *Sciara coprophila,* which code for secretory proteins, were cloned and characterized. Promoter sequences of the *II/9-1* gene resembling a consensus ecdysone respone element (EcRE) were shown to confer ecdysone-inducibility to a reporter gene introduced into *D. melanogaster* and to bind EcR *in vitro* and *in vivo.* The putative origin of DNA replication lies further upstream of the transcription control region of gene *II/9-1.* Preliminary observations suggest that an EcR-binding sequence exists close to that putative origin (F. D. Urnov and S. A. Gerbi, personal communication). It will be highly interesting to see whether this sequence represents a cis-acting element which renders DNA amplification in the II/9A region ecdysone inducible. Immunohistochemical localization studies are in progress to investigate the appearance of endogenous EcR during DNA puff formation in *Sciara coprophila, Rhynchosciara americana,* and *Trichosia pubescens* (S. A. Gerbi and M. Lezzi, unpublished observations; A. J. Stocker and M. Lezzi, unpublished observations). Preliminary observations suggest that EcR binds to DNA puff sites during DNA amplification, i.e., prior to puff formation and transcription.

IV. CONTROL OF GENE ACTIVITIES IN ECDYSONE-INDUCIBLE PUFF SITES

Studies on puff site I-18C in *Chironomus* constitute the starting point of two lines of investigations focusing on two different traits of ecdysone action (Clever and Karlson, 1960; Kroeger, 1963). One line revealed the importance of chromatin structure whereas the other emphasized the role of transcription factors in ecdysone action (see Lezzi and Richards, 1989). Both, though, deal with events occurring at a supramolecular level and eventually will merge.

A. Role of Chromatin Structure in Gene Activation by Ecdysone

It is surprising that Clever and Karlson (1960) chose the chromosome region I-18C of *C. tentans* for their historical experiment, as this region is sometimes refractory to ecdysone. A very careful and quantitative analysis of the cytological structure of the I-18C region revealed that whenever this region is ecdysone-responsive it is decondensed, whereas it is refractory while being condensed (Lezzi *et al.,* 1989, 1991).

1. Accessibility Model for EcR–EcRE Interaction

These observations led to the following three-step model for the activation of the I-18C region (Lezzi and Richards, 1989) (see Fig. 2). First, the condensed chromatin state has to be converted to a decondensed state; second, the EcR complex has to bind to EcREs in the decondensed chromosome region, which requires the presence of ecdysone; and third, the bound EcR complex, then, somehow assembles and thereby sets in motion the transcription machinery. In essence, this model predicts that the chromatin structure controls the accessibility of EcREs to the EcR complex. It was described in Section III,A,3a that some chromosomal bands start decondensing prematurely when isolated chromosomes are exposed to solutions of increasing K^+ concentrations. Puff site I-18C belongs to those regions. Measurements with ion-selective microelectrodes revealed a striking parallel between the condensational state of the I-18C region attained in animals of different developmental and

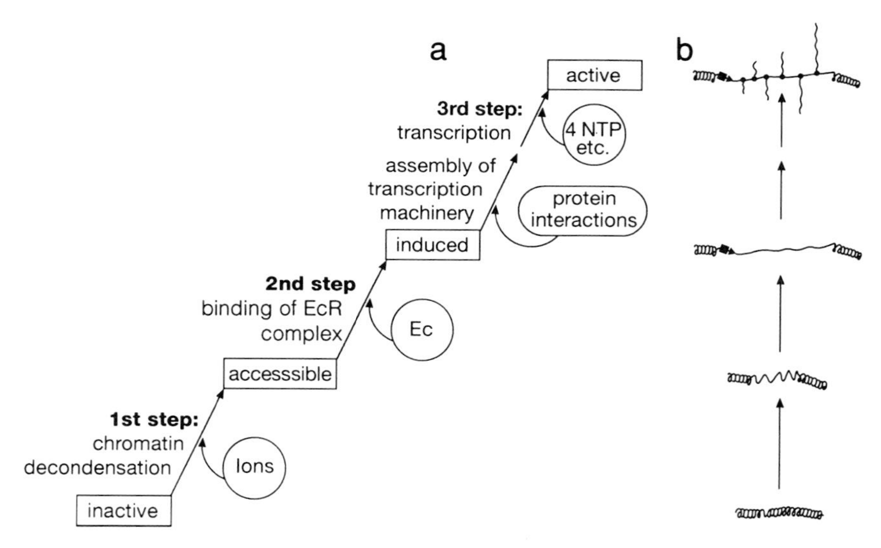

Fig. 2. Accessibility model of gene activation by ecdysone. The chromatin structure controls accessibility of EcREs to the EcR complex (square plus triangle in b) and thereby the subsequent steps (a) "Ions" stands for any condition, agent, or macromolecular modification which affects electrostatic interactions in chromatin. [Modified from Lezzi and Richards (1989). Courtesy of Georg Thieme Verlag.] (b), Different chromatin states (rectangular boxes in a).

physiological states (Lezzi *et al.,* 1989, 1991) and the intracellular K^+ activity (Lezzi *et al.,* 1991; see also Lezzi and Richards, 1989). It is therefore assumed that an increase in the nuclear K^+ concentration influences condensation of the I-18C region also *in vivo.*

The existence of a transcription factor (EcR) whose binding to DNA can be induced by a ligand (ecdysone), and the discovery of a chromosome region (I-18C) that is not permanently condensed or decondensed, may retrospectively be regarded as an exceptionally favorable situation to study the role of chromatin structure in gene regulation. With other loci, the step or accessibility model may not be recognized at all or just in parts.

Region I-19A_1 in *C. tentans,* for example, is a region which in isolated chromosomes and nuclei becomes preferably decondensed by increasing Na^+ concentrations (summarized by Lezzi and Robert, 1972). Parallel chromosomal and electrophysiological studies (with Na^+-selective microelectrodes) revealed that whenever the intracellular Na^+ activity is high, region I-19A_1 forms a transcriptionally active puff (see Lezzi and Richards, 1989). This is true for natural as well as experimental situations.

The intracellular Na^+ and K^+ activities can experimentally be caused to rise up to values two to four times higher than normal; this is achieved by a long-term incubation of larval salivary glands of *C. tentans* in tissue culture medium lacking ecdysone. Concomitantly, the lengths of chromosomes and of region I-18C increase by a factor of 1.5 and 2, respectively. With such hyperdecondensing chromosomes, the I-18C region stays transcriptionally inactive and loses its ecdysone-inducibility, much earlier in fact than other regions (see Lezzi and Richards, 1989). These observations seemingly contradict the accessibility model. They could mean, however, that an optimal rather than a maximal degree of chromatin decondensation is required for the binding and functioning of the EcR complex at specific loci. If hyperdecondensation was caused by a destruction of nucleosomes (*cf.* Section III,A,2,e), the loss of ecdysone inducibility could perhaps be explained by findings showing the importance of nucleosome formation in certain vertebrate gene promoters for steroid hormone action (see next section).

2. Chromatin Structures Controlling Interaction between Steroid Hormone Receptors and DNA

In the living cell nucleus, the EcR complex is hardly ever confronted with naked DNA, but rather with DNA which is organized into chromatin (Fig. 1). The question is whether there are chromatin states which allow or promote EcR binding to EcREs while others prohibit or impede such binding and subsequent gene activation. It is generally believed that a "closed" chromatin configuration must first be "opened" in order for transcription factors to gain access to their response elements. "Open" and "closed" are hypothetical states which may be related to condensed and decondensed chromatin conformation or, operationally, to a high or low DNase sensitivity, respectively (see van Holde, 1989).

In nuclei of primed chicken, the ovalbumin gene is situated in a DNase I-sensitive domain extending over 100 kbp. In these animals, the ovalbumin gene is transcriptionally inactive but may be easily activated by estrogen. In tissues never expressing the ovalbumin gene, this DNase-sensitive domain is missing. It is assumed that the DNase-sensitive chromatin domain has attained an "open" configuration which renders the ovalbumin gene competent for hormone induction (see van Holde, 1989).

Destabilization or unraveling of one single nucleosome (resulting in the appearance of a DNase I hypersensitive, DH, site) does not by itself result in a chromatin change which would be cytologically visible in polytene chromosomes. However, such a change may trigger structural reorganizations at higher order levels of chromatin. DH sites were found in the flanking regions of the *Sgs-4* and *Sgs-3* genes of *D. melanogaster* (Georgel *et al.*, 1991) as well as upstream of the *II/9-1* gene of *S. coprophila* (Gerbi *et al.*, 1993). Their relation to ecdysone action remains to be elucidated. They seem to be the result of an EcR interaction, a contrast to certain DH sites in the ovalbumin and the vitellogenin genes of chick which apparently have to be established prior to hormone receptor binding (discussed by van Holde, 1989).

Situations exist in which the organization of promoter DNA into nucleosomes is not inhibitory but rather beneficial for the functioning of a hormone receptor. Schild *et al.* (1993) found that reconstitution of a nucleosome at a promoter of the *Xenopus* vitellogenin B1 gene potentiates transcription induction of that gene by estrogen. With the mouse mammary tumor virus promoter, nucleosome formation is required to achieve a high specificity in gene regulation by the glucocorticoid hormone receptor (Beato, 1993).

B. Sequential Gene Activation by Ecdysone

Clever (1964, 1966) observed that when ecdysone was injected into fourth instar larvae of *C. tentans,* two puffs (I-18C and IV-2B) were formed within 30–60 min, puffs I-19A_1 and I-17B appeared 5–20 hr later, and puffs I-1A, II-4A, and III-9B formed after 48–72 hr. This sequence of early, early–late, and late ecdysone-induced puff formation reflects the natural puffing sequence unrolling in the preparatory phase of each molt. Inhibition of protein synthesis prevents the formation of those puffs which appear later than 15–20 hr after hormone administration. Furthermore, it inhibits puff IV-2B from regressing and from becoming refractory to ecdysone.

Later, Ashburner *et al.* (1974) made essentially the same observations with salivary gland chromosomes of *D. melanogaster.* Their important contribution comes from *in vitro* incubation experiments in which ecdysone was added and withdrawn in different time patterns, thereby devising conditions to induce late puffs prematurely and to break the acquired refractoriness of early ecdysone-inducible puffs. These findings and the resulting model are described in Chapter 3 by Russell and Ashburner (this volume).

Since the issue of this model is essential for a discussion of the chromosomal localization studies (described in Section VI), it will briefly be summarized. Genes in early puffs are postulated to be contacted by the Ec–EcR complex, thereby becoming activated. Products of early genes, in turn, are thought to contact and thereby activate late genes. In addition, products of early genes should contact and thereby repress the early genes themselves. Late genes, however, would be contacted and repressed by the Ec–EcR complex. It is important to keep in mind that the Ashburner model does not explain the behavior of early–late, of intermolt, or of prepupa-specific puff genes.

C. Molecular Characterization of Genes Residing in Ecdysone-Controlled Puff Sites

The relevant features of the genes which have been cloned from ecdysone-controlled puff sites are listed in Table I (for details and references see Russell and Ashburner, Chapter 3, this volume). In some aspects the available findings are consistent with the Ashburner model while in others they are not. Some early ecdysone-inducible genes may indeed code for protein-types which could serve as transcription regulators; some others, however, code for nonregulatory proteins (for examples, see Andres and Thummel, 1992). Functional EcREs conferring ecdysone inducibility to a gene have been identified in the promoters of an intermolt gene (*Sgs-4* of *D. melanogaster*) and of a late gene (*II/9-1* of *S. coprophila*). While with an intermolt gene, a direct ecdysone interaction was not necessarily to be expected, one is surprised to find the ecdysone control of a late gene by EcR to be positive. This does not seem to be an exceptional situation as will be shown later with late genes of *C. tentans*. One wonders what kind of ecdysone elements will be discovered with the late genes of *D. melanogaster* for which physiological as well as genetical findings provide strong evidence for negative control by ecdysone. A consensus-binding sequence has been identified for the product of the early gene, *E74,* within that same gene but its functional significance is not yet clear. The products of the early gene in the 2B5 region, i.e., the BR-C proteins, make contact with sequences in the promoter of an intermolt gene (*Sgs-4*), which are essential for this gene's activation (von Kalm *et al.,* 1994). An autorepressive function has not yet been demonstrated at the molecular level with any product of the early genes. Chromosomal localization studies (see later) do not give any clue to such a function. The regulatory role of ultraspiracle (USP), a late gene product in *C. tentans,* is discussed in Section V,C.

V. CHROMOSOMAL DISTRIBUTION OF PROTEINS INVOLVED IN GENE REGULATION BY ECDYSONE

The accessibility and the Ashburner models make predictions as to whether and when certain gene regions should be contacted by EcR or by early gene products. An easy and quick way to test such predictions is the immunohistochemical localization of the proteins of interest along polytene chromosomes. However, it clearly has its limitations, particularly the limited resolution of light microscopy. An assignment of a protein to a certain gene within a gene cluster or to an intragenic substructure is not easily accomplished. Nevertheless, the lack of an immunofluoresence signal most likely indicates that the respective protein has not bound to any gene in that chromosome region. In contrast, the presence of a signal makes a molecular investigation of that protein's chromosomal association most promising.

A. Ecdysone Receptor

As expected, EcR was found to be associated with early ecdysone-inducible puffs of *C. tentans* as well as *D. melanogaster* (see Table II). The early–late puffs, I-17B and I-19A_1, of *C. tentans* appear to react inconsistently with anti-EcR antibodies or not at all. In contrast, the puff sites designated by Yao *et al.* (1993) as early–late react positively with such antibodies. Late ecdysone-responsive loci were not described in the work of Yao *et al.* (1993) on *D. melanogaster*. In *C. tentans,* late ecdysone-inducible puffs are clearly stained by anti-EcR antibodies (see, e.g., locus *II-14A* in Fig. 3b and Wegmann *et al.,* 1995). At the stage investigated, these sites just start becoming puffed and transcriptionally active, as was confirmed by autoradiography, the IEH method, and colocalization of RNA polymerase II. These findings thus suggest a positive rather than a negative control of late loci in *C. tentans* by the Ec–EcR complex.

Altogether, approximately 50 loci reacted with anti-EcR antibodies in salivary gland chromosomes of *C. tentans* prepupae. These loci were grouped into strong, intermediate, and weak sites, according to the strength of their immunofluorescent signals. In glands of a stage exhibiting a low ecdysteroid titer (oligopause) or in prepupal glands subjected to an ecdysone washout, more or less the same relative signal strength distribution was observed as in untreated prepupal glands, although at a much lower intensity level. It is tempting to relate these findings to observations made by Karim and Thummel (1992) on a differential hormone dose–response relationship in the activation of different classes of ecdysone-inducible genes. Strong EcR-binding sites would correspond to class I genes. Both require a low concentration of hormone, probably because of an intrinsically high affinity to ecdysone.

B. Proteins Encoded by Genes in Early Ecdysone-Inducible Puff Sites

The products of the *BR-C* gene in the early puff 2B5 of *D. melanogaster* localize to a large number of sites (approximately 200), mainly interbands, of larval salivary glands chromosomes (Emery *et al.,* 1994). This

TABLE II

Localization of Gene Regulatory and HS Proteins at Ecdysone-Controlled Puff Sites[a]

		ECR		USP		E74	E75	FTZ-F1	HSF	HSP90	HSP70
Species and puff site	Puff type	(N)[b]	(HS)[b]	(N)[b]	(HS)[b]	(N)[b]	(N)[b]	(N)[b]	(HS)[b]	(HS)[b]	(HS)[b]
C. tentans[c]											
I-18C	Early	++[i]	++[d]	++[i]	++[d]				−[d]	++[d]	+[d]
IV-2B	Early	+++[i]	±[d]	+++[i]	+++[d]				+[d,j]	±[d]	−[d]
I-17B	Early–late	+/−[i]	±[d]	+/−[i]	±[d]				−[d]	+/−[d]	−[d]
I-1A	Late	+[i]	−[d]	+[i]	+/−[d]				−[d]	−[d]	±[d]
II-14A	Late	++[i]	++[d]	++[i]	++[d]				+[d]	+[d]	±[d]
IV-5C	HS	−[i]	++[d]	−[i]	±[d]				++[d]	++[d,h]	+[d]
I-20A	HS	+++[i]	+++[d]	+++[i]	+++[d]				+[d]	+[d,h]	−[d]
D. melanogaster[k]											
74EF	Early	++[l]		++[l]		++[e]	+++[f]	++++[g]	++[h,m]		
75B	Early	++[l]		++[l]		++[e]	+++[f]	++++[g]	+++[h,m]		
2B5	Early					−[e]	++[f]	++[g]			
78C	Early–late	++[l]		++[l]		++[e]	+[f]	++[g]			
71C-F	Mixed	+[l]		+[l]		+[e]	+[f]		+[h,m]		
63E	Late					+[e]					
93D	HS					+[e]			±[h,m]	+++[h]	±[h]

[a] Scoring: −, no signal; ±, +, ++, +++, ++++, increasing signal strength; +/−, inconsistent signal.
[b] (N) non-heat shock; (HS) heat-shock condition.
[c] Prepupae, stage 10.
[d] Colocalization studies: USP was revealed together with EcR, HSF, or HSP90, respectively (unpublished observations).
[e] Urness and Thummel (1990).
[f] Hill *et al.* (1993).
[g] Lavorgna *et al.* (1993).
[h] See Morcillo *et al.* (1993) for part of findings on *C. tentans.*
[i] Colocalization studies: EcR was revealed together with USP or RNA polymerase II. All sites occupied by EcR are also occupied by RNA polymerase II (Wegmann *et al.*, 1995).
[j] Two to three signals are already present before HS. Main signal maps to IV-2B/C.
[k] Late third instar larvae, except FTZ-F1 (prepupae).
[l] Colocalization studies by Yao *et al.* (1993).
[m] Westwood *et al.* (1991).

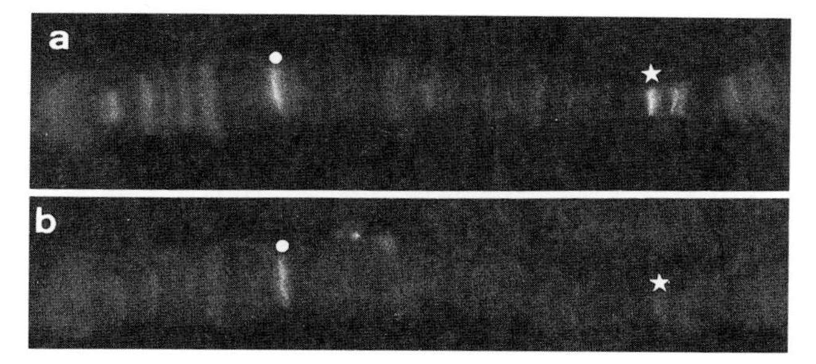

Fig. 3. Colocalization of USP (a) and EcR (b) at chromosome regions II-14A (circle) and II-17C (asterisk) harboring the chironomus *USP* and *EcR* genes, respectively. Prepupal salivary glands of *C. tentans* were incubated for 1 hr at 39° C in ecdysone-free medium. *Note:* low EcR/USP ratio at II-17C.

distribution pattern is qualitatively as well as quantitatively both surprising and very interesting.

E74 and E75, the products of genes harbored in the early ecdysone-inducible puffs 74EF and 75B, respectively, belong to two well-known DNA-binding protein groups (see Table I). Antisera raised against E74 or against E75 decorate polytene salivary gland chromosomes of *D. melanogaster* at several specific loci (see Table II), including early, early–late, and late ecdysone-inducible puff sites. Since the products of the early genes, *E74* and *E75,* were thought to have autoinhibitory effects, it is curious to find E74 and E75 accumulated at these gene loci when they are maximally puffed (see Urness and Thummel, 1990; Hill *et al.,* 1993).

C. Proteins Encoded by Genes in Late Ecdysone-Inducible Puff Sites

In *C. tentans,* the gene coding for USP maps to locus *II-14A* which is a late ecdysone-inducible puff site (see Wegmann *et al.,* 1995). This is remarkable, as one would rather expect USP to be produced very early in the ecdysone-controlled gene activation cascade since USP, obviously an indispensible partner of EcR (see Yao *et al.,* 1993; Cherbas and Cherbas, Chapter 5, this volume), would be already needed for the activation of early genes. In fact, colocalization studies carried out by Yao *et al.* (1993) as well as by Wegmann *et al.* (1995) with *D. melanogaster* and *C. tentans,* respectively, reveal that as a rule USP occupies the same puffs as EcR (for exceptions, see Section VI,D). Surprisingly, puff II-14A itself is also one of these chromosome regions to which EcR and USP bind (see Fig. 3 and Wegmann *et al.,* 1995). This raises many interesting questions concerning the autoregulation of the *USP* gene and its transcriptional and post-transcriptional control.

The gene coding for FTZ-F1 maps to locus 75 CD which is a mid prepupal puff site, responding to ecdysone in a manner similar to late ecdysone-responsive regions (cf. its sensitivity to protein synthesis inhibitors, its induction by ecdysone withdrawal, and the similarity of its activity profile with that of late puff 63E). Polyclonal antibodies directed against FTZ-F1 decorate 66 loci in prepupal salivary gland chromosomes of *D. melanogaster,* among which the strong reactions of the early ecdysone-inducible puffs and of the locus coding for FTZ-F1 itself are most striking (Lavorgna *et al.,* 1993). FITZ-F1, being a member of the steroid/thyroid hormone receptor superfamily (see Table I), is likely to play a special role in the ecdysone-controlled regulatory gene network; this role, however, is not yet clear.

D. Proteins Important in Heat-Shock Regulatory System

Heat shock results in an altered configuration of proteins. This activates the heat-shock transcription factor (HSF), which then activates genes coding for heat-shock proteins (HSPs). HSPs have the general task of correcting the abnormal situation caused by HS-damaged proteins (see Lindquist and Craig, 1988).

HSF has been localized by Westwood *et al.* (1991) not only to HS puffs but also to early ecdysone-inducible and 47 additional loci in prepupal salivary gland chromosomes of *D. melanogaster.* However, an immunofluorescent signal at these non-HS sites was detected only after HS, i.e., when preexisting puffs at these sites were supposed to have regressed. Westwood *et al.* (1991) speculate whether HSF may exert not only a positive but also a negative control on gene activities. In contrast, observations on *C. tentans* indicate that HSF may be located at some puffs also at normal temperatures. After heat shock, HSF may stay or leave these sites while appearing in a large number of new loci, mostly HS puffs (unpublished observations).

HSPs have been shown to accumulate at typical HS-inducible puff sites after heat shock (Morcillo *et al.,* 1993). A severe heat shock shifts HSP90 and HSP70 also to the early ecdysone-inducible puff I-18C of *C. tentans* (unpublished observations).

EcR appears also to have a function in HS-induced gene regulatory events. After a severe *in vitro* HS in an ecdysone-free medium, EcR is detected in the ecdysone-inducible puffs I-18C and II-14A as well as in the typical HS puffs I-20A and IV-5C (unpublished observations; for a review of HS puffs in *C. tentans,* see Lezzi *et al.,* 1984). In the HS puff IV-5C, EcR appears *de novo.* This was also observed with heat-shocked glands of oligopausing animals in which the ecdysteroid titer is very low. In most of the other regular or HS-inducible loci, the amount of EcR decreases on HS. As a rule, the USP level in puffs changes in parallel with that of EcR during HS, except for region II-17C (harboring the *Chironomus EcR* gene) in which it stays higher (see Fig. 3a) and region IV-5C (see later). Region IV-5C is not puffed or transcriptionally active under normal conditions, nor does it contain EcR, USP, HSF, HSP90, or HSP70. After HS, all these proteins assemble in the fully active IV-5C HS puff, although in the case of USP, only to a relatively small extent. Thus, the binding of EcR to IV-5C appears to be rather independent not only of its ligand but also of its heterodimerization partner. It is not known why EcR is excluded from binding to the unpuffed IV-5C region under non-HS conditions; possibly HS renders putative EcREs in this locus accessible to EcR. Region I-20A is exceptional in that it contains EcR before HS-induced puffing and activation (Wegmann *et al.,* 1995). Because the HSF is brought into that region as a transcription activator by HS, the functional significance of colocalized EcR plus USP in I-20A is not obvious. Region I-18C is also actively transcribing during HS even though HSF leaves that region under these conditions (unpublished observations). It can be speculated that, at some chromosome regions, HSP90 and HSP70 might rescue the HS-damaged EcR/USP complex [cf. the beneficial effects of HSP90 on MyoD1 (Shaknovich *et al.,* 1992)]. It seems possible that EcR becomes (re-) activated by HS even in the absence of ecdysone, which is also suggested by previous findings on purified EcR *in vitro* (Luo *et al.,* 1991).

VI. NEW LOOK AT GENE REGULATION BY ECDYSONE

A. Control of Ecdysone-Dependent Gene Expression by Multiple Factors

The activity of a gene is controlled by a variety of factors, both general and more specific. A quick look at Table II may illustrate that fact: different loci appear to accommodate different combinations of EcR, USP, E74, E75, HSF, and FTZ-F1. The available molecular data on ecdysone-controlled genes point in the same direction. The control region of the *Sgs-4* gene, for example, contains response elements not only for SEBP 2 and 3 (SEBP, secretion enhancer binding protein), BR-C proteins Z1, 3, and 4, and possibly GEBF-1 (glue enhancer binding protein) but also for the EcR/USP complex (Lehmann and Korge, 1995; von Kalm *et al.,* 1994; Georgel *et al.,* 1991). These factors appear to influence *Sgs-4* gene transcription positively in a synergistic manner. In connection with developmental studies, it might be important to know whether a gene is activated early or late by ecdysone and whether this activation is sensitive to the inhibition of protein synthesis. These parameters are mostly determined by the abundance, time course of synthesis, and degradation of some, or one, of the multitude of gene regulatory factors. This may be in contrast to investigations focusing on the molecular and supramolecular mechanisms of gene regulation by ecdysone. For those, it is more important to know whether and how these factors interact with regulatory sequences in DNA and among each other.

B. Ecdysone-Controlled System as Part of Gene Regulatory Network

In a pure cascade model (Fig. 4b), the role of ecdysone is confined to the triggering of the activation of primary genes whose products then activate the next set of genes. In an alternative model (Fig. 4a), the temporal order of gene activation is achieved by differential ecdysone sensitivities of genes responding to a gradual increase in the ecdysteroid titer. It is likely that elements of the two models are combined as illustrated by Fig. 4c. The introduction of necessary feedforward and feedback relations leads then to an actual gene regulatory network (Fig. 4d). These relations (including autoregulatory loops) may be positive,

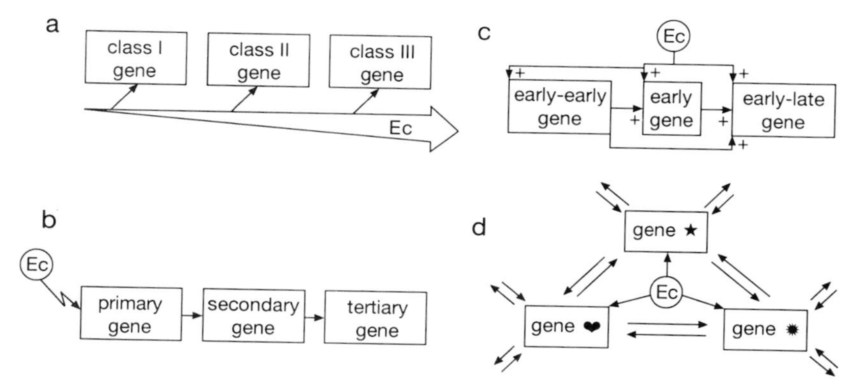

Fig. 4. Models for ecdysone-controlled gene regulation. (a) Differential ecdysone-sensitivity model. Class I, Class II, and Class III genes are thought to exhibit high, intermediate, and low affinities, resepectively, to the EcR complex (not depicted) activated by ecdysone (cf. Wegmann *et al.,* 1995). A gradual increase in the ecdysone concentration (large arrow) brings about sequential activation. For candidates of class I and II genes, see the work of Karim and Thummel (1992). (b) Model of ecdysone-triggered gene activation cascade. The Ec–EcR complex (circle) activates the primary genes which then activate the secondary ones and so on. (c) Multiple gene control model resulting from a combination of (a) with (b) and the addition of positive feedforward relations. *BR-C, E74,* and *E78* may represent members of the early–early, early, and early–late gene groups, respectively. (d) Gene network model, represented by one of its mesh. This model is derived from (c) by the addition of (hypothetical) positive and negative feedforward and -backward relations. Symbols given to each gene indicate that genes in a network do not fit into a hierarchical or temporal order. Ecdysone (circle) plays a central role in the part of the network referred to in (d). Autoregulatory loops are not depicted. *Note:* EcR and USP, forming the Ec receptor complex (not depicted), are products of genes which themselves are part of the network.

negative, or even ambivalent. Input and output signals may enter or leave the network at any point. Ecdysone, in varying concentrations, impinges more or less continuously on the changing combinations of a large number of genes.

C. Meshing of the Ecdysone- and Heat Shock-Controlled Systems of Gene Regulatory Network

It is well known that certain steroid hormone receptors in the cytoplasm of vertebrate cells require interaction with HSP90 and possibly other HSPs so that the hormone can subsequently elicit their activation and nuclear translocation (see Beato, 1993). In the promoters of small HSP genes of *D. melanogaster,* not only the HS-response but also the ecdysone-response elements were identified which appear to function independently of each other (Riddihough and Pelham, 1986). On the basis of the chromosomal localization studies reported in Section V,D, some additional points of interdigitation between ecdysone- and HS-controlled regulatory systems may be envisaged. (*i*) HSPs could contact the EcR complex, when bound to EcREs, in order to prevent or revert HS damage to EcR. (*ii*) HS might activate EcR in a ligand-independent fashion. (*iii*) HS could render EcREs accessible to EcR, possibly by chromatin changes.

Interplay between the ecdysone-controlled and the HS-controlled regulatory systems would make sense, particularly during critical developmental periods such as metamorphosis. Metamorphosis itself imposes stress on the organism. During metamorphosis, an organism would be most critically affected by internal, but also by external, stress. Ecdysone, by bringing the HS protective system to a higher level of alertness, appears to take precautionary measures against prospective stress. Heat shock, on the other hand, may in some respects potentiate the action of ecdysone in order that the metamorphic process, once set in motion, reaches its goal.

ACKNOWLEDGMENTS

I thank A. Illi for help with the preparation of the manuscript; A. J. Stocker for stimulating discussion; M. Vögtli, I. S. Wegmann, D. Gilligan, and C. Elke for valuable scientific contributions; and F. Gatzka for technical assistance in the unpublished experiments. Antibodies used in the unpublished experiments were kindly provided by F. Kafatos, R. Tanguay, and T. Westwood. The unpublished work was partially funded by the Swiss National Science Foundation. Communication of information prior to publication by M. Lehmann and G. Korge, by S. Gerbi and co-workers, and by G. Guild and co-workers is greatly appreciated. The manuscript was read by D. Gilligan. The drawings were made by U. Horstmann (graphic artist's studio).

REFERENCES

Ananiev, E. V., and Barsky, V. E. (1985). Elementary structures in polytene chromosomes of *Drosophila melanogaster. Chromosoma* **93,** 104–112.

Andres, A. J., and Thummel, C. S. (1992). Hormones, puffs and flies: The molecular control of metamorphosis by ecdysone. *Trends Genet.* **8,** 132–138.

Ashburner, M., Chihara, C., Meltzer, P., and Richards, G. (1974). Temporal control of puffing activity in polytene chromosomes. *Cold Spring Harbor Symp. Quant. Biol.* **38,** 655–662.

Beato, M. (1993). Gene regulation by steroid hormones. *In* "Gene Expression: General and Cell-Type Specific" (M. Karin, ed.), pp. 43–75. Birkhäuser Boston, Basel, Berlin.

Beermann, W. (1962). Riesenchromosomen. *Protoplasmatologia* **6D,** 1–161.

Bradbury, E. M., Carpenter, B. G., and Rattle, H. W. E. (1973). Magnetic resonance studies of deoxyribonucleoprotein. *Nature (London)* **241,** 123–126.

Clever, U. (1964). Actinomycin and puromycin: Effects on sequential gene activation by ecdysone. *Science* **146,** 794–795.

Clever, U. (1966). Induction and repression of a puff in *Chironomus tentans. Dev. Biol.* **14,** 421–438.

Clever, U., and Karlson, P. (1960). Induktion von Puff-Veränderungen in den Speicheldrüsen von *Chironomus tentans* durch Ecdyson. *Exp. Cells Res.* **20,** 623–626.

Dorsch-Häsler, K., Lutz, B., Spindler, K.-D., and Lezzi, M. (1990). Structural and developmental analysis of a gene cloned from the early ecdysterone-inducible puff site, I-18C, in *Chironomus tentans. Gene* **96,** 233–239.

Egyhazi, E., and Durban, E. (1987). Microinjection of anti-topoisomerase I immunoglobulin G into nuclei of *Chironomus tentans* salivary gland cells leads to blockage of transcription elongation. *Mol. Cell. Biol.* **7,** 4308–4316.

Emery, I. F., Bedian, V., and Guild, G. M. (1994). Differential expression of *Broad-Complex* transcription factors may forecast tissue-specific developmental fates during *Drosophila* metamorphosis. *Development* **120,** 3275–3287.

Frey, M., Koller, T., and Lezzi, M. (1982). Isolation of DNA from single microsurgically excised bands of polytene chromosomes of *Chironomus. Chromosoma* **84,** 493–503.

Frydman, H. M., Cadavid, E. O., Yokosawa, J., Silva, F. H., Navarro-Cattapan, L. D., Santelli, R. V., Jacobs-Lorena, M., Graessmann, M., Graessmann, A., Stocker, A. J., and Lara, F. J. S. (1993). Molecular characterization of the DNA puff C-8 gene of *Rhynchosciara americana. J. Mol. Biol.* **233,** 799–803.

Galli, J., and Wieslander, L. (1993). A repetitive secretory protein gene of a novel type in *Chironomus tentans* is specifically expressed in the salivary glands and exhibits extensive length polymorphism. *J. Biol. Chem.* **268,** 11888–11893.

Georgel, P., Ramain, P., Giangrande, A., Dretzen, G., Richards, G., and Bellard, M. (1991). *Sgs-3* chromatin structure and *trans*-activators: Developmental and ecdysone induction of a glue enhancer-binding factor, GEBF-I, in *Drosophila* larvae. *Mol. Cell. Biol.* **11,** 523–532.

Gerbi, S. A., Liang, C., Wu, N., DiBartolomeis, S. M., Bienz-Tadmore, B., Smith, H. S., and Urnov, F. D. (1993). DNA amplification in DNA puff II/9A of *Sciara coprophila. Cold Spring Harbor Symp. Quant. Biol.* **58,** 487–494.

Gilligan, D. (1994). "Molecular and Cytological Studies of the *BR6* Gene in *Chironomus tentans* Polytene Chromosomes," Ph.D. thesis. ETH-Zürich, Zurich.

Gruzdev, A. D., and Reznik, N. A. (1981). Evidence for the uninemy of eukaryotic chromatids. *Chromosoma* **82,** 1–8.

Gruzdev, A. D., and Shurdow, M. W. (1992). Topological state of DNA in polytene chromosomes. *Biochim. Biophys. Acta* **1131,** 35–40.

Hill, R. J., Segraves, W. A., Choi, D., Underwood, A., and Macavoy, E. (1993). The reaction with polytene chromosomes of antibodies raised against *Drosophila* E75A protein. *Insect Biochem. Mol. Biol.* **23,** 99–104.

Karim, E. D., and Thummel, C. S. (1992). Temporal coordination of regulatory gene expression by the steroid hormone ecdysone. *EMBO J.* **11,** 4083–4093.

Karlson, P. (1963). New concepts on the mode of action of hormones. *Perspect. Biol. Med.* **6,** 203–214.

Kroeger, H. (1963). Chemical nature of the system controlling gene activities in insect cells. *Nature (London)* **200,** 1234–1235.

Lara, F. J. S., Stocker, A. J., and Amabis, J. M. (1991). DNA sequence amplification in sciarid flies: Results and perspectives. *Braz. J. Med. Biol. Res.* **24,** 233–248.

Lavorgna, G., Karim, F. D., Thummel, C. S., and Wu, C. (1993). Potential role for a FTZ-F1 steroid receptor superfamily member in the control of *Drosophila* metamorphosis. *Proc. Natl. Acad. Sci. U.S.A.* **90,** 3004–3008.

Lehmann, M., and Korge, G. (1995). Ecdysone regulation of the *Drosophila Sgs-4* gene is mediated by the synergistic action of ecdysone receptor and SEBP 3. *EMBO J.* **14,** 716–726.

Lewin, B. (1994). Chromatin and gene expression: Constant questions, but changing answers. *Cell (Cambridge, Mass.)* **79,** 397–406.

Lezzi, M., and Richards, G. (1989). Salivary glands. *In* "Ecdysone" (J. Koolman, ed.), pp. 393–406. Thieme, Stuttgart.

Lezzi, M., and Robert, M. (1972). Chromosomes isolated from unfixed salivary glands of *Chironomus. Res. Probl. Cell Differ.* **4,** 35–57.

Lezzi, M., Gatzka, F., and Meyer, B. (1984). Heat-shock phenomena in *Chironomus tentans.* III. Quantitative autoradiographic studies on ^{3}H-uridine incorporaton into Balbiani ring 2 and heat-shock puff IV-5C. *Chromosoma* **90,** 204–210.

Lezzi, M., Gatzka, F., and Robert-Nicoud, M. (1989). Developmental changes in the responsiveness to ecdysterone of chromosome region I-18C of *Chironomus tentans. Chromosoma* **98,** 23–32.

Lezzi, M., Gatzka, F., Ineichen, H., and Gruzdev, A. D. (1991). Transcriptional activation of puff site I-18C of *Chironomus tentans:* Hormonal responsiveness changes in parallel with diurnal decondensation cycle. *Chromosoma* **100,** 235–241.

Lindquist, S., and Craig, E. A. (1988). The heat-shock proteins. *Annu. Rev. Genet.* **22,** 631–677.

Luo, Y., Amin, J., and Voellmy, R. (1991). Ecdysterone receptor is a sequence-specific transcription factor involved in the developmental regulation of heat shock genes. *Mol. Cell. Biol.* **11,** 3660–3675.

Morcillo, G., Diez, J. L., Carbajal, M. E., and Tanguay, R. M. (1993). HSP90 associates with specific heat shock puffs (*hsrω*). *Chromosoma* **102,** 648–659.

Panitz, R., Serfling, E., and Wobus, U. (1972). Autoradiographische Untersuchungen zur RNA-Syntheseleistung von Balbiani-Ringen. *Biol. Zentralbl.* **91,** 359–380.

Rastelli, L., Chan, C. S., and Pirrotta, V. (1993). Related chromosome binding sites for *zeste, suppressors of zeste* and *Polycomb* group proteins in *Drosophila* and their dependence on *Enhancer of zeste* function. *EMBO J.* **12,** 1513–1522.

Riddihough, G., and Pelham, H. R. B. (1986). Activation of the *Drosophila hsp27* promoter by heat shock and by ecdysone involves independent and remote regulatory sequences. *EMBO J.* **5,** 1653–1658.

Rykowski, M. C., Parmelee, S. J., Agard, D. A., and Sedat, J. W. (1988). Precise determination of the molecular limits of a polytene chromosome band: Regulatory sequences for the *Notch* gene are in the interband. *Cell (Cambridge, Mass.)* **54,** 461–472.

Sass, H. (1980). Feature of *in vitro* puffing and RNA synthesis in polytene chromosomes of *Chironomus. Chromosoma* **78,** 33–78.

Schild, C., Claret, F.-X., Wahli, W., and Wolffe, A. P. (1993). A nucleosome-dependent static loop potentiates estrogen-regulated transcription from the *Xenopus* vitellogenin B1 promoter *in vitro. EMBO J.* **12,** 423–433.

Semeshin, V. F., Belyaeva, E. S., Zhimulev, I. F., Lis, J. T., Richards, G., and Bourouis, M. (1986). Electron microscopical analysis of *Drosophila* polytene chromosomes. IV. Mapping of morphological structures appearing as a result of transformation of DNA sequences into chromosomes. *Chromosoma* **93,** 461–468.

Shaknovich, R., Shue, G., and Kohtz, D. S. (1992). Conformational activation of a basic helix–loop–helix protein (MyoD1) by the C-terminal region of murine HSP90 (HSP84). *Mol. Cell. Biol.* **12,** 5059–5068.

Thoma, F., Koller, T., and Klug, A. (1979). Involvement of histone H1 in the organization of the nucleosome and of the salt-dependent superstructures of chromatin. *J. Cell Biol.* **83,** 403–427.

Trepte, H.-H. (1993). Ultrastructural analysis of Balbiani ring genes of *Chironomus pallidivittatus* in different states of Balbiani ring activity. *Chromosoma* **102,** 433–445.

Turberg, A., Imhof, M., Lezzi, M., and Spindler, K.-D. (1992). DNA-binding properties of an "ecdysteroid receptor" from epithelial tissue culture cells of *Chironomus tentans. Insect. Biochem. Mol. Biol.* **22,** 343–351.

Urness, L. D., and Thummel, C. S. (1990). Molecular interactions within the ecdysone regulatory hierarchy: DNA binding properties of the *Drosophila* ecdysone-inducible E74A protein. *Cell (Cambridge, Mass.)* **63,** 47–61.

van Holde, K. E. (1989). "Chromatin." Springer-Verlag, New York.

von Kalm, L., Crossgrove, K., Von Seggern, D., Guild, G. M., and Beckendorf, S. K. (1994). The *Broad-Complex* directly controls a tissue-specific response to the steroid hormone ecdysone at the onset of *Drosophila* metamorphosis. *EMBO J.* **13,** 3505–3516.

Weeks, J. R., Hardin, S. E., Shen, J., Lee, J. M., and Greenleaf, A. L. (1993). Locus-specific variation in phosphorylation state of RNA polymerase II *in vivo:* Correlations with gene activity and transcript processing. *Genes Dev.* **7,** 2329–2344.

Wegmann, I. S., Quack, S., Spindler, K.-D., Dorsch-Häsler, K., Vögtli, M., and Lezzi, M. (1995). Immunological studies on the developmental and chromosomal distribution of ecdysteroid receptor protein in *Chironomus tentans. Arch. Insect. Biochem. Physiol.* **30,** 95–114.

Westwood, T. J., Clos, J., and Wu, C. (1991). Stress-induced oligomerization and chromosomal relocalization of heat-shock factor. *Nature (London)* **353,** 822–827.

Widmer, R. M., Lucchini, R., Lezzi, M., Meyer, B., Sogo, J. M., Edström, J.-E., and Koller, T. (1984). Chromatin structure of a hyperactive secretory protein gene (in Balbiani ring 2) of *Chironomus. EMBO J.* **3,** 1635–1641.

Yao, T. P., Forman, B. M., Jiang, Z. Y., Cherbas, L., Chen, J. D., McKeown, M., Cherbas, P., and Evans, R. M. (1993). Functional ecdysone receptor is the product of *EcR* and *Ultraspiracle* genes. *Nature (London)* **366,** 476–479.

5

Molecular Aspects of Ecdysteroid Hormone Action

PETER CHERBAS AND LUCY CHERBAS

Department of Biology
Indiana University
Bloomington, Indiana

I. INTRODUCTION

Ecdysteroid hormones* trigger all the events of molting and metamorphosis in arthropods. Their rapid and dramatic effects on puffing patterns in salivary gland polytene chromosomes of *Chironomus* and *Drosophila* provided one of the first clues that steroid hormones might act more or less directly to control gene expression (see Russell and Ashburner, Chapter 3, this volume; Lezzi, Chapter 4, this volume). Although the road leading from those early clues to molecular analysis of the hormone response pathway has been a long one, we now know that ecdysteroids are typical steroid hormones in the biologically significant sense that the proteins mediating their effects are members of the nuclear hormone receptor superfamily.

For the ecdysteroids, as for the vertebrate nuclear hormones, three broad questions merit investigation: (i) How, precisely, is the hormonal signal transduced into an effect on the transcription of any one responsive gene? (ii) How is it that distinct sets of genes are hormone responsive in different tissues and at different developmental stages? (iii) How is any particular tissue- and stage-specific hormone response organized, i.e., how many genes respond, how does one response lead to another, what fraction of the responsive genes are tissue and stage specific, and do developmentally distinct pathways have common features?

The third question describes the focus of much contemporary work on the sequence of events triggered by ecdysteroids in the salivary gland

* No introduction to this literature can avoid the problem of nomenclature. Two systems are in common use. Many workers, including the current authors, routinely use the word "ecdysone" in a generic sense to describe biologically active molecules, i.e., in analogy to "estrogen" or "androgen." Thus the proper name of EcR is "ecdysone receptor" and EcREs are "ecdysone response elements." In the alternative system, the word "ecdysteroid" (Goodwin *et al.*, 1978) is used for much the same purpose.

The persistence of the two systems has to do with the differing ambiguities they cause and the undoubted need for a generic biological term. The first system risks confusion where numerous hormones are being compared because "ecdysone" has become the proper name of the prohormone formerly called α-ecdysone. The second system appropriates a precisely defined chemical term ("ecdysteroid" was defined to mean compounds chemically related to 20-HE) to connote biological activity. Thus, by definition, molecules related to 2A are ecdysteroids, irrespective of biological activity, whereas the active molecule 2B is not. One consequence of which the reader should be aware is that a database should be searched using both "ecdysone" and "ecdysteroid" to find all relevant information.

The terminology based on "ecdysteroid" has been adopted in this book and we have conformed to it in this chapter. By and large we have used the abbreviations EcR and EcRE and avoided their proper names which might be confusing in this context. Where it is clear that we are speaking of the *Drosophila* proteins we refer simply to EcR (and USP). Where there is room for confusion, we have elaborated on the abbreviations to create, for example, DmEcR-B1 as an explicit reference to the B1 isoform of the *Drosophila* EcR.

Other abbreviations such as TR (thyroid receptor), VDR (vitamin D receptor), RAR (retinoic acid receptor), RXR (retinoid X receptor), TRE (thyroid response element), VDRE, and RARE are standard.

and in other responding tissues, work that is summarized in other chapters of this book. Answers to the first and second question will be prerequisite to a full understanding of those events and such answers are just beginning to become available. They are the subject of this chapter. Because much of our knowledge is preliminary, we will point out unanswered questions along the way. Our summary, like the research it describes, will draw principally on work using *Drosophila.*

Like all reviews, this chapter reflects the interests of its authors. For different perspectives, the reader is referred to the other chapters in this volume (e.g., Gilbert *et al.,* Chapter 2, this volume) and to reviews on the endocrinology of metamorphosis (Riddiford, 1993; Riddiford and Truman, 1993, 1994), insect hormone receptors (Segraves, 1991; Oro *et al.,* 1992b; Henrich and Brown, 1995), and ecdysteroid-regulated transcription (Pongs, 1988; Cherbas, 1993). No recent review of ecdysteroid receptor biochemistry has been done. Although we have been unable to remedy this in a comprehensive way, we have accorded slightly disproportionate attention to this area.

II. VERTEBRATE PRECEDENTS

As later sections will show, ecdysteroids act by a pathway that has much in common with the thyroid hormone, retinoic acid, and vitamin D response pathways. The hormone is bound by a receptor (EcR) homologous to TR, RAR, and VDR. That receptor forms a heterodimer with another member of the nuclear hormone receptor superfamily (the *ultraspiracle* gene product, USP) that is homologous to RXR. Together, they bind hormone response elements (EcREs) that are cognates of TREs, RAREs, and VDREs. Since these homologies were discovered, it has been natural to consider the ecdysteroid response pathway in the light of the far more intensively studied vertebrate models. In the end we will find that the comparison reveals many similarities but that it also highlights surprising and novel features of the ecdysteroid pathway.

To facilitate the comparison, we begin with a concise summary of thyroid hormone-mediated gene regulation (see Tata, Chapter 13, this volume; Shi, Chapter 14, this volume). Because of space limitations, we have used citations sparingly in this section, citing mostly reviews and those papers that can provide the reader with access to the literature. For the same reason, our summary deals in broad generalizations.

Ligand-regulated gene expression involves several distinguishable functions: (a) specific ligand recognition, (b) specific DNA recognition, and (c) effects on transcription. It is the distinctive feature of the members of the nuclear hormone receptor superfamily that all these functions are provided (at least to a first approximation) by a single polypeptide, the hormone receptor. Thus, the TRs—there are two genes (α and β), each generating several isoforms—are polypeptides that bind both hormone (triiodothyronine, (T_3)) and specific DNA sequences (TREs). Both the nucleotide sequences of the TREs and the polypeptide sequences of the TRs themselves indicate that they belong to the nuclear hormone receptor subfamily that also includes RARs and VDR (and EcR). For the specific descriptions that follow, we have chosen human thyroid receptor β-1 (hTRβ-1) as our model receptor.

A. Functional Domains of Human Thyroid Hormone Receptor β-1

We defer consideration of the N-terminal 106 residues (the N-terminal region or A/B domains) and begin with the DNA-binding domain (DBD, Fig. 1). The DBD (ca. residues 107–174 in hTRβ-1) is well-conserved in all members of the superfamily. It contains eight essential cysteine residues that support two zinc finger-like structures. By analogy with the glucocorticoid and estrogen DBD:DNA structures, it seems certain

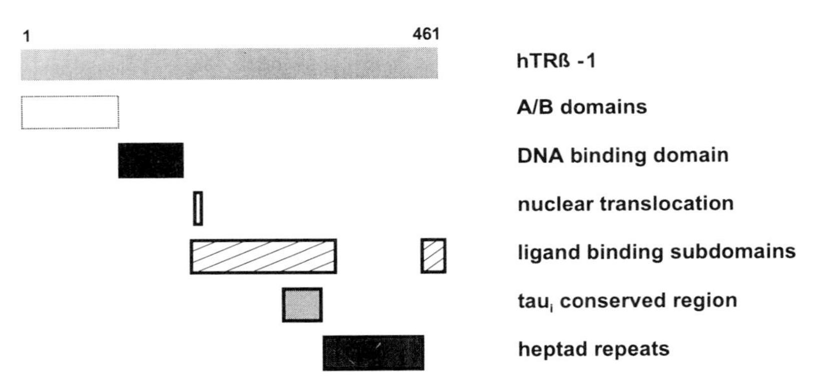

Fig. 1. Functional domains of hTRβ-1. Approximate extents of the functional domains of human thyroid hormone receptor β-1 (SWISS-PROT AC P10828) are shown.

that the regions following each Zn finger are α helices and that the interfinger helix binds the major groove of the TRE. The Zn atoms and the fingers they support organize globular regions that support the two helices in their orthogonal orientation (Freedman, 1993; Lee *et al.*, 1993). In line with these observations, three of the residues at the base of the first finger where it enters the DNA-binding helix (the "P box") are known to distinguish between receptor subfamily response elements, i.e., TR, RAR, VDR (and EcR), which share the response element half-site AGGTCA, also share an identical P box sequence (Tsai and O'Malley, 1994).

A variety of names have been given to subdivisions of the region downstream of the DBD (residues 175–461 in hTRβ1). We will call this the C-terminal region and will focus on several functional subdivisions for which information is available. TR is a nuclear protein even in the absence of ligand; the nuclear localization signal has been mapped to a cluster of basic residues just downstream of the DBD (Fig. 1) (LaCasse *et al.*, 1993). Some evidence suggests that residues in this region may also act as a hinge. Forman and Samuels (1990) first noted the existence of nine conserved hydrophobic repeats within the C-terminal region (Fig. 1) and suggested that they play an important role in dimerization. These repeating structures are highly conserved within the TR, RAR, VDR, and EcR subfamily of receptors as are the sequences of some individual repeats. There is as yet no direct evidence that repeats in these regions form a leucine zipper-like structure. However, experiments have shown that fragments containing these repeats can interfere with dimerization and there is considerable direct evidence showing the importance of the ninth heptad for both homo- and heterodimerization (Nagaya and Jameson, 1993a,b). Immediately adjacent to the heptads is the τ_i region (Fig. 1), originally recognized by its considerable conservation throughout the subfamily. Rosen *et al.* (1993) have provided evidence that sequences within τ_i are critical to TRβ interactions with at least two different dimerization partners and that the homologous regions of RAR and VDR function similarly. The syndrome of dominant thyroid hormone resistance appears to result from the presence of mutant, nonfunctional TRs which, because they are still competent to dimerize, act as poison subunits. Nagaya and Jameson (1993a) have mapped numerous such mutations and it is satisfying that they (mostly) cluster just outside the heptad repeat and τ_i regions.

TR synthesized by translation *in vitro* binds ligand (T_3) normally as do *Escherichia coli*-synthesized fragments that include the entire C-terminal region (McPhie *et al.,* 1993). Identifying the ligand-binding subdomains within the C-terminal region has been more problematic. In Figure 1 we have amalgamated all those subregions that have been implicated in hormone binding and, as the reader will observe, the resulting ligand subdomains include the entire region with the exception of the heptad repeats. Forman and Samuels (1990) noted that some C-terminal region sequences are conserved only among receptors that bind similar hormones. On that basis they identified the small subdomain shown at the C terminus as well as a larger upstream region (residues 183–272) which begins just 9 residues downstream of the DBD. The work of McPhie *et al.* (1993) was consistent with this because C-terminal region fragments lacking small numbers of residues at either end (8 residues at the C terminus or 42 residues at the N terminus) fail to bind T_3. However, in their studies of hormone-resistant mutants, Nagaya and Jameson (1993a) found mutations that affect ligand binding not only in the small C-terminal subdomain but also throughout the 310–349 region. These results may mean that residues that contact ligand are widely scattered or that a variety of defects in structure can interfere with ligand binding.

B. Thyroid Response Elements

A few T_3-responsive genes have been characterized and their TREs mapped. The TREs contain multiple copies of a sequence related to the consensus hexamer 5′-AGGTCA-3′ (Martinez and Wahli, 1991; Tsai and O'Malley, 1994). In the first examples studied the hexameric half-sites were arranged as imperfect palindromes with no intervening nucleotides. However, it has since become clear that a remarkable variety of architectures can function: (a) direct repeats with spacing 3 (DR3: 5′-AGGTCAN_3AGGTCA-3′); (b) direct repeats with spacing 2 (DR2); (c) inverted repeats with spacing 3 (5′-TGACCTN_3AGGTCA-3′); or (d) single copies of an extended (octamer) half-site element 5′-TAAGGTCA-3′ (Desvergne, 1994). [In accordance with established convention, we take the unit half-site to be a hexamer, and refer to the sequence RGGTCAN_nRGGTCA as DR*n* throughout this chapter.]

The acceptable spacing of half-sites is determined largely by DBD:DBD interactions and is reflected in the binding specificities of isolated DBDs. When palindromes are bound with dyad symmetry, the critical interaction is between D boxes of the two DBDs. The D box consists of the five residues that fall between the first two zinc-ligated cysteines of the second finger (Umesono and Evans, 1989; Dahlman-Wright *et al.,* 1991; Luisi *et al.,* 1991). The binding of direct repeats violates dyad symmetry. In this case the interaction involves the RXR D box and the "DR box" of TR or RAR. The "DR box" comprises residues in the loop of the first zinc finger (Perlmann *et al.,* 1993; Kurokawa *et al.,* 1993).

The TR complex that binds to these TREs can be a monomer, a homodimer, a heterodimer with related nuclear proteins, notably RXR (Ikeda *et al.,* 1994; Schräder and Carlberg, 1994; Williams *et al.,* 1994). Whatever the complex, its affinity for the TRE may be significantly altered by sequences adjacent to the hexamer half-site (Katz and Koenig, 1993; Schräder *et al.,* 1994a). When those sequences are especially favorable, the monomeric receptor can bind and exhibit HRE specificity. Thus, the octamer TAAGGTCA is a high-affinity binding site for monomeric TRα-1 but not for RXRs, RARs, or VDR and it mediates specifically T_3 responses (Katz and Koenig, 1993; Schräder, 1994a).

C. Dimerization

TR can homodimerize and form heterodimers with other nuclear hormone receptor family members. The predominant heterodimerization partners for TR are the RXRs. These receptors were identified originally as RAR homologs and were subsequently discovered to be dimerization partners for TR, RAR, and VDR. RXRs themselves are also capable of forming homodimers, which recognize their own specific HREs (Mangelsdorf *et al.,* 1991). In addition, RXR has a ligand of its own, 9-*cis*-retinoic acid, which appears to be widely distributed in vertebrate cells and RXR homodimers exhibit ligand-dependent effects on transcription (Subauste *et al.,* 1994).

TR–RXR heterodimerization is mediated by both the DBD and the dimerization domain in the C-terminal region. The relatively weak DBD interactions described previously are detectable only in DNA complexes; however, they are important in determining TRE specificity. The much stronger dimerization domain interactions appear to involve some of the same sequences involved in homodimerization (e.g., τ_i and the ninth heptad repeat) (Rosen *et al.,* 1993) but there is significant evidence that some mutations differentially affect homo- and heterodimerization (Nagaya and Jameson, 1993a,b).

One of the most interesting features of recent work on this branch of the superfamily is the realization that HRE binding is so extraordinarily plastic. It is surprising to observe asymmetric binding by RXR/TR heterodimers to direct repeat TREs. It is still more surprising to learn that TR can participate in interactions with isolated half-sites or octamers, palidromes, inverted palindromes, and direct repeats (of various spacings). This plasticity appears to demand unprecedented flexibility in the structures of the binding complexes. In line with that inference, evidence has been presented in favor of a model in which the DBDs of members of this family can rotate relative to their C-terminal dimerization domains (Perlmann *et al.*, 1993; Kurokawa *et al.*, 1993). Whether this particular model is correct, considerable conformational flexibility is implied by the DNA-binding data alone and such conformational flexibility must afford remarkable opportunities for variation and regulation.

Indeed, evidence is accumulating to suggest that the selection of a TRE is sensitive to many variables (Mader *et al.*, 1993; Tsai and O'Malley, 1994). On most TREs, especially DR0 and DR3, RXR/TR heterodimers bind with higher affinity than TR monomers or homodimers. However, the extent to which RXR/TR heterodimers form is sensitive not only to the concentrations of both polypeptides but to the concentrations of their individual ligands (Leng *et al.*, 1993; Cheskis and Freedman, 1994) and it may be sensitive to their post-translational modifications, e.g., phosphorylation (Tsai and O'Malley, 1994). Moreover, the affinity of the resulting heterodimer for a TRE and the magnitude (and direction) of the transcriptional consequences are affected not only by TRE architecture and sequence but also by the concentrations of the two ligands (Hallenbeck *et al.*, 1993; Schräder *et al.*, 1994b; Kurokawa *et al.*, 1994). Similar variables appear to affect homodimer binding.

D. Effects on Transcription

Transcriptional activation by members of the nuclear hormone receptor superfamily has been reviewed (Tora and Davidson, 1994) and our remarks will be succinct. Activation domains have been mapped in numerous receptors, generally by assessing the activities of receptor fragments when fused to the GAL4 DBD and tested in an appropriate system. Baniahmad *et al.* (1995) have reviewed the evidence derived from such experiments that hTRβ-1 contains four activation domains (τ1–τ4). τ1 is within the N-terminal region and is poorly defined in hTRβ-1. τ4, the strongest activation domain, is at the C terminus of the protein (residues 400–456). RARs and ER possess a similar element in a similar position; sequence conservation is limited to a clustering of prolines, hydrophobic residues, and acidic residues. τ3 (residues 399–368) is less active and τ2 (residues 207–217) has only marginal activity. Both segments involve clusters of acidic residues. For other receptors there is excellent evidence to support the idea that the receptor activation domains enhance the formation of preinitiation complexes, probably by interacting with particular TAFs (Jacq *et al.,* 1994; Chiba *et al.,* 1994; Tora and Davidson, 1994). In the case of TR, there is strong evidence that activation and silencing (see below) may involve interactions with TFIIB (Baniahmad *et al.,* 1993).

More attention has been directed to the ability of unliganded TR to inhibit transcription from TRE-containing genes (Glass *et al.,* 1989). TR silences transcription from even minimal promoters and it does so by inhibiting formation of a preinitiation complex (Fondell *et al.,* 1993; Baniahmad *et al.,* 1995). This silencing activity does not depend on the N-terminal region; if the C terminus of TR, RAR, or v-*erbA* (a modified TR) is fused to the GAL4 DBD, the fusion polypeptide silences the activity of promoters containing the GAL4-binding site. Hence, the C-terminal region contains a strong silencing domain (Baniahmad *et al.,* 1992). The τ4 activation domain appears to be an essential part of this silencing domain. Mutations in τ4 do not affect the ability of TR to bind T_3; nonetheless they convert the protein into a constitutive transcriptional silencer (Baniahmad *et al.,* 1995). Other evidence suggests that the ninth heptad repeat also plays an essential role in silencing (Flynn *et al.,* 1994; Qi *et al.,* 1995). Competition experiments have suggested that the silencing functions of TR require an as yet unidentified corepressor. Taken together, these data have suggested a model in which the τ4 domain of TR binds the corepressor and stabilizes its interaction with general transcription initiation factors (probably including TFIIB). According to this model, hormone binding alters the structure of τ4, releases the corepressor, and exposes activation domains (Baniahmad *et al.,* 1995). It is noteworthy that the oncogene v-*erbA* is a TR with the τ4 region deleted and that v-*erbA* is a constitutive silencer (Damm *et al.,* 1989).

III. ECDYSTEROID RESPONSE APPARATUS

So far as we know, all ecdysteroid effects issue from the primary effects, both positive and negative, of the hormone on the transcription of particular target genes. Our understanding of this process has advanced markedly as we have gained familiarity with the components of the molecular apparatus that mediates it. For the ecdysteroid response the critical components include response elements (EcREs) and two proteins, EcR and USP, which together comprise the heterodimeric functional receptor.

A. The EcR Protein

1. Background

Because of the complexities that arise when the functions of EcR and EcR/USP are considered, it is well to begin with a clear statement of the background ideas. The receptor of a hormone is the cellular molecule to which the hormone becomes bound to initiate a response (Ehrlich and Morgenroth, 1910). If we are fortunate, receptor occupancy will be the rate-limiting step in function. If so, candidate hormones' *in vitro* affinities for receptor will correlate with their activities. For ecdysteroids the pharmacological situation is clear and consistent with the existence of a single receptor. Bioassays that are not complicated by hormone sequestration, excretion, or metabolism (i.e., assays in many organ and cell culture systems) report consistent relative activities for various ecdysteroids. Relative to the natural hormone 20-hydroxyecdysone (20-HE, Fig. 2A), the 20-deoxy prohormone (ecdysone) is about 200-fold less effective; 25-deoxy and 14-deoxy derivatives are about one order of magnitude more active [see L. Cherbas *et al.* (1980) and P. Cherbas *et al.*, (1982) and references cited therein]. Absolute concentrations required for activity vary slightly depending on the assay. However, the summary statement that 20-HE is active only at the relatively high concentration of ca. $2\text{–}5 \times 10^{-8}$ *M* is consistent with most observations.

Experiments designed to detect a receptor with the appropriate pharmacological properties began very early using labeled 20-HE but were

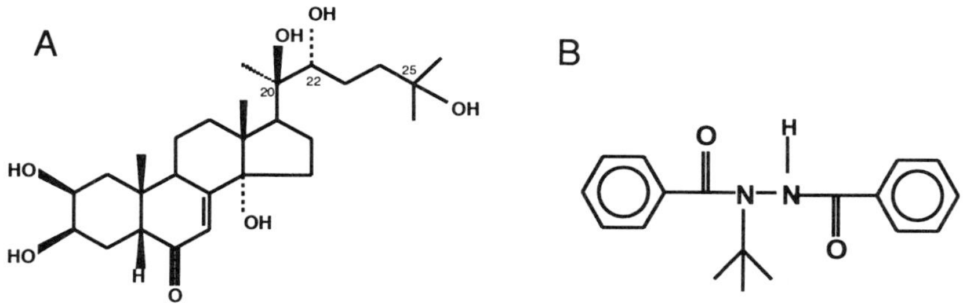

Fig. 2. Two ecdysteroid system agonists: (A) 20-hydroxyecdysone and (B) RH5849 (1,2-dibenzoyl-1-*tert*-butylhydrazine).

frustrated by that molecule's apparently low affinity for the receptor. According to the bioassay data, ponasterone A (25-deoxy 20-HE) should be a better ligand, and the use of synthetic, tritiated ponasterone A in binding studies confirmed this prediction (Maróy *et al.*, 1978; Yund *et al.*, 1978). More recently, similar logic led to the synthesis of other tritiated hormone analogs (Sage *et al.*, 1986) and to the use of the high-affinity ligand 26-[^{125}I]iodoponasterone A which can be labeled to very high specific activity (Cherbas *et al.*, 1988).

Studies using these ligands demonstrated that *Drosophila* target cells (Kc cell line, imaginal discs) do contain binding proteins with appropriately high affinities ($K_d \leq 10^{-9}\ M$) for ponasterone and iodoponasterone (Maróy *et al.*, 1978; Yund *et al.*, 1978; Beckers *et al.*, 1980; Cherbas *et al.*, 1988). These proteins are the sought-after receptors because their binding specificities (revealed by competition experiments with various ecdysteroids) matched adequately the profile of active ecdysteroids known from bioassays (see the references cited earlier and Beckers *et al.*, 1980). Receptors with similar properties were subsequently detected biochemically in a *Drosophila* tumorous blood cell line (Dinan, 1985), in a *Chironomus tentans* epithelial cell line (Turberg *et al.*, 1988), in whole larval extracts of *Calliphora vicina* (Lehmann and Koolman, 1988), and in the hypodermis and midgut of several species of crayfish (Londershausen and Spindler, 1985). They have also been detected autoradiographically in many tissues of both flies and moths (Fahrbach and Truman, 1989; Bidmon and Koolman, 1989; Bidmon *et al.*, 1991a,b, 1992; Bidmon, 1991; Dai *et al.*, 1991).

By steroid-binding assays, ecdysteroid receptors are present at about 1000 copies per cell in responsive cells (Maróy *et al.*, 1978; Yund *et al.*,

1978; Beckers *et al.*, 1980; Sage *et al.*, 1986; Cherbas *et al.*, 1988). They are nuclear even in naive cells and are not translocated following hormone treatment (see Section III,A,5). Hormone binding was observed to be sensitive not only to proteases, but also to sulfhydryl reagents (*N*-ethylmaleimide) and, with some ligands, to modest salt concentrations (Dinan, 1985; Lehmann and Koolman, 1988; Turberg *et al.*, 1988; Sage *et al.*, 1986). Working with crude nuclear extracts and at very low salt concentrations, Sage *et al.* (1982) and Dinan (1985) estimated the binding moiety to be approximately 6S. Lehmann and Koolman (1988) used gel filtration under similar conditions to estimate a size of 105 kDa. Landon *et al.* (1988) carried out a partial purification of receptor. Starting from ca. 6 g wet weight Kc cells they enriched the receptor ca. 750-fold by several chromatographic steps and used gel filtration and velocity sedimentation to estimate a molecular mass of 120 kDa (and a frictional coefficient of 1.4). Strangmann-Diekmann *et al.* (1990) used an ecdysteroid affinity label (itself ^{14}C-labeled) to detect receptor; they found peptides of 90 and 150 kDa by SDS-PAGE.

The work of Riddihough and Pelham (1987) is also relevant in this connection. Using the *hsp27* gene they identified a 23-bp footprint that includes an EcRE (see later) and showed (by UV cross-linking) that this sequence binds a protein of 80 to 90 kDa. Shortly thereafter, while studying EcREs (including the *hsp27* EcRE), Cherbas *et al.* (1991) demonstrated that the protein that binds to an EcRE includes the ecdysteroid receptor.

Luo *et al.* (1991) have purified EcRE-binding receptor complexes (from S3 cell cultures) to apparent homogeneity using traditional methods, and Ozyhar *et al.* (1992) have reported procedures for DNA affinity purification from nuclear extracts. DNA binding by receptor complexes in extracts has been characterized both for a *Calliphora* receptor (Lehmann and Koolman, 1989) and a *Chironomus* receptor (Turberg *et al.*, 1992). Turberg and Spindler (1992) have reported evidence for a cytoplasmic to nuclear translocation of receptor in a *C. tentans* epithelial cell line. Finally, Ozyhar *et al.* (1990) have observed that pyridoxal phosphate interferes with DNA binding by receptor in nuclear extracts.

Before leaving this background material, two brief asides are offered. First, the relatively poor affinity of 20-HE for fly ecdysteroid receptors is surprising given vertebrate precedents. Yund and Fristrom (1975) suggested that it is an adaptation to the very short life cycle of flies. In that connection it is interesting that, in the much longer-lived crayfish studied by Londershausen and Spindler (1985), affinities for ecdysone,

20-HE, and ponasterone A are all elevated (10- to 100-fold relative to *Drosophila* values). Since the arthropods are so diverse, these observations open the possibility that receptor ligand-binding domains may prove to be a rich source of adaptive variation in steroid binding. There is not yet sufficient sequence (or hormone binding) data to test this idea.

Second, until recently all the agonists known in these systems have been ecdysteroids (in the chemical sense). The situation has changed with the observation that RH5849 (1,2-dibenzoyl-1-*tert*-butylhydrazine, Fig. 2B) and certain molecules related to it are active inducers of ecdysteroid responses and compete for receptor binding. Moreover, RH5849-resistant cells are ecdysteroid resistant and vice versa (Wing, 1988). Obviously, these results extend the range of ligands that must be accommodated by structural models. In the present connection we mention them because RH5849 is the only agonist that binds receptor very much more weakly than predicted by its activity (Wing, 1988). Whether this anomaly reflects some further complication in receptor action remains to be seen.

2. Cloning

The ecdysteroid receptor was cloned in the laboratory of Hogness and colleagues, where a genomic library was screened at low stringency, using the orphan member of the steroid hormone receptor superfamily, *E75,* as a probe. Several genes were recovered, all encoding steroid receptor family members. cDNAs were obtained for interesting candidates and one of these, designated *EcR,* met a number of stringent criteria that identified it as the receptor (Koelle *et al.,* 1991).

1. EcR is a component of the major high-affinity ecdysteroid-binding moiety in *Drosophila* cell lines, as demonstrated by two experiments. (1) Overexpression of the EcR proteins in S2 cells leads to a large increase in the capacity of the cells to bind [^{125}I]iodoponasterone. (2) An antibody raised against bacterially expressed EcR fragments can precipitate most or all of the high-affinity [^{125}I]iodoponasterone-binding proteins in extracts of S2 or Kc cells (Koelle *et al.,* 1991; A. Mintzas and P. Cherbas, unpublished observations).
2. EcR binds specifically to EcREs. When end-labeled DNA fragments, one of which contained one or more copies of the *hsp27* EcRE (see Section III,C), were incubated with a cell extract, the EcRE-containing fragment was specifically precipitated by anti-EcR (Koelle

et al., 1991; A. Mintzas and P. Cherbas, unpublished observations). The affinity of EcR for an *hsp27* EcRE was estimated to be 850 times the affinity of EcR for nonspecific DNA (Koelle *et al.,* 1991); this is consistent with an estimate of 1000-fold specificity for the binding of the receptor complexes to EcREs in Kc cell extracts (defined by steroid binding) (Cherbas *et al.,* 1991). Similarly, gel-shift assays demonstrate specific binding of EcR to EcREs and show that antibodies against EcR interfere with the formation of the protein/EcRE complex (Koelle *et al.,* 1991).

3. EcR plays an essential role in the ecdysteroid response *in vivo.* Ecdysteroid-resistant S2 cells, selected for their ability to proliferate in the presence of ecdysteroid, display a variety of phenotypes (see later). In some cases, the cells show only weak induction of a reporter construct which is normally strongly ecdysteroid responsive; in one such line, the induction of the reporter was restored by transfection with a plasmid-expressing EcR (Koelle *et al.,* 1991). When *EcR* was mutagenized in Kc cells by a gene-targeting technique, the EcR-deficient line showed a sharp decrease in the response of an EcRE–reporter construct, and the response was restored by cotransfection with an EcR-expressing plasmid (Swevers *et al.,* 1995b).

All of these lines of evidence show that EcR is an essential component of the ecdysteroid receptor. But, as shown later, EcR itself is not sufficient to confer normal ecdysteroid responsiveness in a heterologous system. For example, EcR expressed in some mammalian cell lines (e.g., COS-1 cells) or in yeast along with a suitable EcRE–receptor construct is wholly inactive (Koelle, 1992). In other mammalian cell lines, a modest reporter induction can be observed at very high concentrations of an exceptionally active ecdysteroid (Christopherson *et al.,* 1992). We will return to these observations later in connection with the role of USP and the EcR/USP heterodimer.

3. *Structures: EcR and EcR Gene*

EcR was characterized initially as a cDNA clone. It was subsequently found that the *EcR* gene is transcriptionally complex, giving rise to at least three species of RNA that encode proteins differing in their A/B domains (Talbot *et al.,* 1993). The gene extends over approximately 70 kb of DNA at polytene position 42A. Transcripts encoding EcR-A contain three A-specific exons (A1, A2, and A3), which are spliced to a four-exon common region (exons 3–6); coding begins in exon A2. Transcripts encoding EcR-B1 initiate at a promoter approximately 35 kb downstream of the EcR-A promoter and contain two exons lacking in form A (1 and 2) which are spliced to the four-exon common region; coding begins in exon 2. Transcripts encoding EcR-B2 are identical to those encoding B1 except that exon 2 is spliced out; coding is believed to begin at a cryptic translational start site in exon 1. The DNA-binding domain is encoded in exon 3, the first exon of the common region. Thus, the proteins differ only in their A/B domains.

Isoform B1, the first version of EcR to be described, encodes a polypeptide of 878 residues with a predicted size of 94 kDa. The polypeptide, whether extracted from tissues or synthesized in bacteria, runs in SDS-PAGE as a species of ca. 105 kDa (Koelle *et al.,* 1991). It is clearly a member of the nuclear hormone receptor superfamily, retaining the usual conserved regions in the DNA-binding and C-terminal domains. Within the superfamily, its closest relatives are the members of the subfamily that includes TR, VDR, and RAR—receptors that also share the AGGTCA half-site. Figure 3 shows a number of these family members aligned with DmEcR-B1 and DmUSP and with representations of the functional domains.

DmEcR-B1 is the longest polypeptide in this set. The A/B domain of DmEcR-B1 is very long (as is that of DmEcR-A). In addition, the nonconserved upstream portion of the ligand-binding domain is quite long (matched only by hVDR). However, most of the excess length of the DmEcRs arises from the presence of a novel carboxy-terminal "tail" of >150 residues that extends the polypeptide beyond the conserved downstream portion of the ligand-binding domain. This tail is composed mostly of simple and repetitive sequences. Thus, while it might harbor activation domains, it is difficult to believe that it includes any highly structured functional elements.

The A/B domains of the nuclear hormone receptors are not conserved and those identities that are observed (e.g., those in Fig. 3 between DmEcR-B1 and hRXR) are isolated matches in long stretches of unrelated sequence. As expected, the DBDs are highly conserved. Moreover, the DmEcR P box specifies binding to AGGTCA half-site (see Section III,C). There have been no published studies of dimerization by isolated EcR DBDs, so it is unclear whether there should be a functional D box. At the D box, the DmEcRs contain the multiply charged sequence KFGRA, but this lacks acidic residues and our understanding of D box function is not adequate to predict whether it would be functional.

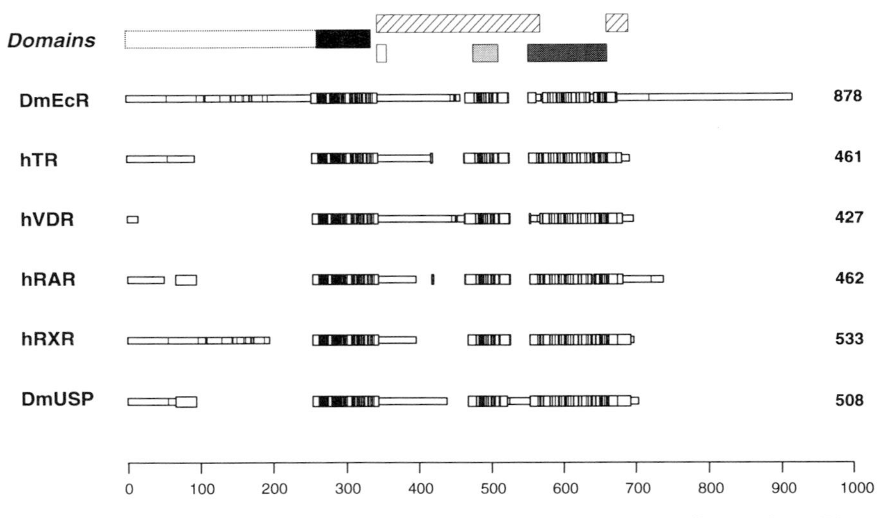

Fig. 3. An alignment of some nuclear hormone receptor superfamily members. The alignment was prepared using the MACAW multiple alignment workbench (Schuler *et al.,* (1991). Vertical lines indicate identical or conserved residues. For comparison, the functional domains of hTRβ-1 have been mapped at the top; they may be identified by reference to Fig. 1. The alignment includes the following polypeptides (SWISS-PROT ACs are given in parentheses): DmEcR-B1, *Drosophila melanogaster* ecdysone receptor, isoform B-1 (P34021); hTR, human thyroid hormone receptor β-1 (P10828); hVDR, human vitamin D receptor (P11473); hRAR, human retinoic acid receptor α-1 (P10276); hRXR, human RXRβ-2 (P28703); and DmUSP, *D. melanogaster* USP (P20153).

Nuclear transport signals are not highly conserved within the family, but the presence and the position of a highly basic region in which the signal is embedded are conserved. The DmEcRs contain that region. The τ_i subdomain is highly conserved in the DmEcRs as is the structure of heptad hydrophobic repeats. Moreover, several of the repeats are conserved in sequence. As expected, there is little sequence conservation in the ligand-binding subdomains.

It is well to mention briefly but explicitly some of the weaknesses in these comparisons. Although quite a lot is known about the structures of the conserved DBDs in this family, the same cannot be said for the other domains. Therefore it is important to note that there has been no published direct demonstration of ligand binding or dimerization by isolated EcR C-terminal regions. In preliminary experiments, we have observed that overexpression of the DmEcR C-terminal region in S2 cells increases total iodoponasterone binding without changing the K_d (L. Cherbas and S. Kim, unpublished observations). However, detailed experiments to discover how this truncated protein interacts with USP (see Section III,B) and to define the ligand-binding and dimerization domains, as well as those domains involved in nuclear transport and activation, remain to be done. Nothing is known about the presence or significance of post-translationally modified residues in EcR. Given the fact that TR and EcR are similar in their abilities to inhibit transcription when unliganded and to activate when liganded (see later), a detailed study should eventually uncover structural analogies that are refractory to sequence analysis alone.

DmEcR-B1 is encoded by a 6-kb transcript, which includes more than 1 kb of the 5′-untranslated sequence; the 5′-untranslated region includes 11 short AUG-initiated open reading frames, of which the longest is 25 codons. The DmEcR-A transcripts has 306 N of 5′-untranslated sequence, with 6 short AUG-initiated open reading frames, of which the longest is 18 codons (Koelle *et al.,* 1991; Talbot *et al.,* 1993). Long and complex 5′-untranslated regions are found frequently in steroid–receptor transcripts (though EcR seems to be an extreme case); they suggest the possibility of translational control. (For comparison, 5′-untranslated sequences from the *Drosophila* steroid receptor family members *usp, E78A, knrl, svp,* and *DHR3* contain 1, 5, 4, 3, and 1 AUGs, respectively).

4. EcR Isoforms and Their Expression

In *Drosophila* the 6-kb *EcR* transcript is detectable throughout development except in 0- to 3-hr embryos, with the highest expression just prior to pupariation. A minor 3.7-kb transcript is detected at all stages of development; its significance is not known. The expressed protein is widely distributed through the organism, both in embryos and in third instar larvae (Koelle *et al.,* 1991). Studies with carefully timed larvae suggest that *EcR* expression is roughly parallel to that of the glue genes, increasing from mid third instar until puparium formation, rapidly repressed at pupariation; unlike the glue genes, *EcR* expression rises again at approximately the time of the prepupal ecdysteroid peak (Andres *et al.,* 1993). *EcR* transcription can be rapidly induced in tissues of late third instar larvae by treatment with 20-HE *in vitro* (Karim and Thummel, 1992), an apparent contradiction to the time course of expression *in vivo.* Clearly ecdysteroid hormones play a role in the control of *EcR* expression, but that role remains unclear.

Each of the three DmEcR isoforms is a functional ecdysteroid receptor, as demonstrated by the ability to induce hormone-dependent transcription of an EcRE–reporter construct in an ecdysteroid-resistant S2 line (Talbot *et al.,* 1993). It is clear that isoforms B1 and A have independent and quite distinct spatial and temporal expression patterns during development, as shown by studies using specific monoclonal antibodies (Talbot *et al.,* 1993; Robinow *et al.,* 1993; Truman *et al.,* 1994); specific probes for isoform B2 are not available. In general, larval tissues contain more EcR-B1 than EcR-A, as do the imaginal cells located in midgut islands and in histoblast nests. Imaginal discs contain much more of isoform A than B1. For more details on the tissue and stage specificities of the EcR isoforms, see Chapter 8 by Truman (this volume).

For the functional significance of such variation in the A/B domains, we are at present mostly dependent on the vertebrate precedents. The existing evidence suggests that different isoforms of TR and RAR differ in the magnitudes of their effects on different promoters. Whether this is due to differences in the activation domains present, to effects on dimerization and/or ligand binding, or to some combination of these influences is unclear (Darling *et al.,* 1993). Eventually, genetic studies in *Drosophila* should provide a detailed answer. Work to date indicates that all three DmEcR isoforms bind both ligand and DNA as USP heterodimers (Koelle, 1992) and suggests that the three isoforms have overlapping, but not identical, functions. EMS-generated mutations in the regions common to all *EcR* isoforms are recessive embryonic lethals. Two mutations have been isolated that lead to translation termination in the B1-specific region; these mutations lead to a loss of the B1 form, but should have no effect on the expression of the other two isoforms. These B1-specific mutations, when homozygous, are lethal just prior to metamorphosis. They can be rescued by germline transformation with constructs expressing the B1 isoform and, to a lesser extent, by an A isoform-expressing construct, but virtually not at all by a B-2 expressing construct, suggesting some degree of functional differentiation (M. Bender and D. S. Hogness, personal communication.)

5. Nuclear Location of EcR

In the absence of ecdysteroids, EcR inhibits transcription (see Section III,C,3,a); hence, some EcR molecules must be nuclear even in cells unexposed to hormone. Indeed, unliganded TR has a similar inhibitory function and is nuclear (Damm *et al.,* 1989). Yund *et al.* (1978) and Sage *et al.* (1982), using biochemical approaches, first noted that EcR is nuclear in naive imaginal disk cells and Kc cells. This result has since been confirmed for all the EcR isoforms by immunocytochemistry (Koelle *et al.,* 1991; see also Truman, Chapter 8, this volume). Qualitatively, the concentration of EcR in the nucleus is unmistakable. The quantitative question of whether a small pool of cytoplasmic or cycling EcR exists has not yet been addressed.

6. EcRs of Arthropods Other Than Drosophila

EcR homologs have been isolated as cDNA clones from several insects other than *Drosophila,* using a variety of strategies. CtEcR (*C. tentans*) (Imhof *et al.,* 1993) was identified by probing a cDNA library with an oligonucleotide corresponding to a conserved region of the DBDs of related nuclear hormone receptors. AaEcR (*Aedes aegypti*) (Cho *et al.,* 1995) and MsEcR (*Manduca sexta*) (Fujiwara *et al.,* 1995) were identified by PCR on cDNA templates, using primers from conserved regions of the DBD and ligand-binding domain. BmEcR (*Bombyx mori*) (Swevers *et al.,* 1995a) was isolated by PCR using two primers from the DBD.

The impressive similarities among these molecules, extending throughout the putative ligand-binding domains, are a strong argument for their identification as EcRs. However, more direct and detailed functional tests are needed. Expressing CtEcR in *Drosophila* S2 cells decreases the ecdysteroid response of an EcRE–reporter construct (Imhof *et al.,* 1993); this is consistent with the interpretation that while CtEcR competes with DmEcR for binding to USP and/or EcREs, it is less effective than DmEcR in producing a response in these cells. A clearer functional demonstration has been provided for BmEcR, which can substitute for DmEcR in restoring the ecdysteroid response to an EcR-deficient clone of Kc cells. In addition, BmEcR and BmUSP, synthesized *in vitro,* can bind to EcREs (Swevers *et al.,* 1995b). Thus, BmEcR is a receptor, and CtEcR is at least capable of interacting with the ecdysteroid response machinery.

In fact, given that BmEcR works and given the similarities among these proteins, it appears very likely that all are EcRs. The question of whether or to what extent these polypeptides are functionally equivalent is a far more interesting one in light of their sequences. The protein sequences of DmEcR and its homologs CtEcR, AaEcR, BmEcR, and

MsEcR from various insect orders are shown aligned in Fig. 4A. The reader will note the extensive sequence conservation among them. Equally striking is the extent to which these sequences diverge. To enhance the comparison we have included (Fig. 4B) a comparable alignment of four TRβ-1 sequences from representatives of four vertebrate classes. Taken at face value, the comparison suggests that more evolution has taken place among the EcRs of the Diptera (*Drosophila, Aedes, Chironomus*) than between the TRs of primates and amphibia!

The suggestion is that sequence variation may prove to be a valuable asset in efforts to understand EcR function. Thus, if DmEcR and AaEcR

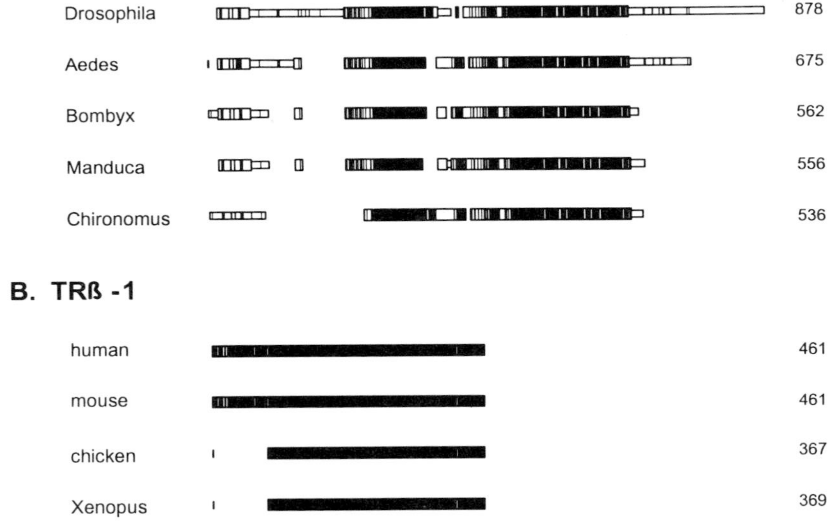

Fig. 4. Comparing sequence similarities among cognate receptors: (A) known EcRs and (B) a sample of hTRβ-1s. The alignment was prepared using MACAW. (A) and (B) have been reproduced at the same scale, but the hTRβ-1 sequences have not been aligned with the EcRs. The sequences shown include: (A) *Drosophila,* DmEcR-B1 (SWISS-PROT AC P34021); *Aedes,* AaEcR (GenBank AC U02021); *Bombyx,* BmEcR (Swevers *et al.,* 1995a); *Manduca,* MsEcR (GenBank AC U19812); and *Chironomus,* CtEcR (GenBank AC S60739). (B) Human, human TRβ-1 (SWISS-PROT AC P10828), mouse, *Mus musculis* TRβ-1 (SWISS-PROT P37242), chicken, *Gallus gallus* TRβ (SWISS-PROT P18112), *Xenopus, Xenopus laevis* TRβ-A1 (SWISS-PROT P18116).

prove to be indistinguishable in their activities, the long C-terminal tail of DmEcR can be dismissed as irrelevant. As mentioned previously, there is impressive sequence conservation, marked by long stretches of identity, among the EcRs throughout not only those domains (DBD, τ_1, heptad repeats) that are conserved throughout the family, but also in the putative ligand-binding subdomains. The basic region just downstream of DBD is basic in all the EcRs but its sequence is not conserved. In light of the comparison with the TRβ-1 sequences, it is remarkable that there is only a little more A/B domain similarity among the EcRs than among nuclear hormone family members in general (Fig. 3). Two relatively conserved segments suggest that AaEcR, BmEcR, and MsEcR correspond to DmEcR-B1 while CtEcR is more similar to DmEcR-A.

B. The *USP* Protein

1. History

Ultraspiracle (usp) was originally described as a recessive lethal *Drosophila* gene whose mutation leads to death at the L1 → L2 molt. The gene was subsequently cloned independently in three laboratories; it was isolated in Kafatos' laboratory as CF1, a transcription factor required for correct expression of the chorion gene *s15* (Shea *et al.,* 1990), and in the laboratories of Evans and Gilbert as a steroid–receptor homolog (Henrich *et al.,* 1990; Oro *et al.,* 1990).

2. Structure

USP, the protein encoded by *usp,* is a member of the steroid hormone receptor superfamily; its closest mammalian homologs are the RXRs, the "retinoid X receptors"; USP and human RXRα share 86% sequence identity in the DNA-binding domain and 49% sequence identity in the ligand-binding domain. The protein contains 507 or 508 residues, and is encoded by a single 2.4-kb mRNA, which in turn is transcribed from a single intronless gene located at position 2C1-3 on the X chromosome of *Drosophila* (Oro *et al.,* 1990; Henrich *et al.,* 1990). A *Bombyx mori* homolog has been cloned; the proteins from *Bombyx* and *Drosophila* are 96% identical in the DNA-binding domain and 56% identical in the ligand-binding domain. Thus, the DNA-binding domain is approximately as well conserved between *Drosophila* and *Bombyx* as the corresponding domain of RXRα is between human and *Xenopus,* but the ligand-binding domain is much better conserved between the two vertebrate species than between the two insect species (Tzertzinis *et al.,* 1994). USP has a short A/B domain (104 residues in *Drosophila,* 113 in *Bombyx*), which is poorly conserved between *Drosophila* and *Bombyx,* and little or no carboxy-terminal tail. Evidence that the *Bombyx* homolog is a functional USP is provided by its ability to form a heterodimer with *Bombyx* EcR and by the ability of that heterodimer to bind to EcREs (Swevers *et al.,* 1995b).

3. Binding Properties

No ligand has yet been identified for USP. RXRs, mammalian homologs of USP, bind 9-*cis*-retinoic acid (Heyman *et al.,* 1992; Levin *et al.,* 1992; Mangelsdorf *et al.,* 1992), and the presence of the ligand alters the transcriptional effects of RXR heterodimers (Kliewer *et al.,* 1992; Allegretto *et al.,* 1993; Carlberg *et al.,* 1993; Issemann *et al.,* 1993; Leng *et al.,* 1993; Macdonald *et al.,* 1993). Furthermore, 9-*cis*-retinoic acid and other RXR-binding retinoids can promote RXR homodimer formation (Lehmann *et al.,* 1992). It is likely that USP, too, has a ligand. That that ligand was not present in the *in vitro* experiments described later must be taken into account in their interpretation.

The P box of USP is identical to those in TR, VDR, RAR, RXR, and EcR. Therefore, it predicts binding to the canonical DNA half-site AGGTCA. The *s15* chorion promoter region to which it binds includes the sequence AGGTCAN_6TGTCCA (Shea *et al.,* 1990), similar in form to the palindromic sites to which other members of this family bind, including the sequence RGGTCANTGACCY to which the EcR/USP heterodimer binds (see later). However, footprinting showed that only the upstream half-site was bound to USP; mutational analysis showed that the downstream half-site was not required for binding and that truncation of the protein to prevent dimerization did not alter its DNA-binding properties. Furthermore, when bacterially expressed USP was challenged with random oligonucleotides, it bound preferentially to a single "half-site" element of the form GGGGTCAC(G/C) (Khoury Christianson *et al.,* 1992). Binding of bacterially synthesized USP to the *s15* site has an affinity of 10–20 n*M* (Khoury Christianson and Kafatos, 1993), somewhat weaker than a typical receptor/HRE interaction. However, since all of these studies of USP binding have been done with purified or bacterially expressed USP, we must also consider the possibility that USP normally functions only as a heterodimer, and hence that the binding specificity determined in these experiments does not accurately reflect the functional binding to DNA. We know this to be the case for USP function in the ecdysteroid response; we do not know whether other heterodimer partners may be involved in chorion transcription and in other USP functions (see Section III,E).

4. Genetics and Expression

Perrimon *et al.* (1985) examined the maternal effects of a series of zygotic lethals generated in the X chromosome by G. Lefevre; among these were three mutations at a locus they designated *ultraspiracle* because *usp*/Y males usually died at the L1–L2 molt, frequently going through the molt but retaining the L1 spiracle cuticular structure so that they appeared to have an extra set of posterior spiracles. *usp*/Y larvae from a *usp/usp* maternal germline clone died at hatching, often exhibiting a localized ventral defect posterior to the A8 dentical belt; *usp*/+ larvae from the same mutant germline clone were fully viable and fertile. Thus, although *usp* is a zygotic lethal, there is also a significant maternal contribution which can partially overcome the effects of a zygotic deficiency. These observations were based on two EMS-induced mutations (*usp*$^{\mathrm{VE653}}$ and *usp*$^{\mathrm{VE849}}$, now renamed *usp*^{3} and *usp*^{4}) and an X-ray induced mutation (*usp*$^{\mathrm{KA21}}$, now renamed *usp*^{2}).

Oro *et al.* (1990), having cloned a RXR homolog which was tentatively designated XR2C, showed that the breakpoint of *usp*^{2} was located in the portion of the XR2C transcription unit encoding the DNA-binding domain and that germline transformation with the XR2C transcription unit rescued all three known *usp* alleles. These experiments identified the newly cloned gene as *usp*. Since the XR2C sequence is virtually identical to the sequences of CF1, a newly cloned chorion transcription factor (Shea *et al.*, 1990), and of a newly cloned steroid hormone receptor from 2C (Henrich *et al.*, 1990), all three constitute independent clonings of *usp*. Subsequently, the mutations in *usp*^{3} and *usp*^{4} were shown to be substitutions at universally conserved Arg residues within the DNA-binding domain (Henrich *et al.*, 1994).

As predicted from the genetic studies, *usp* is expressed at a high level in nurse cells and the transcript appears to be maternally loaded into the oocyte. It is also widely transcribed in the embryo. A statistically significant failure to obtain gynandromorphs carrying *usp/usp* regions indicates that the *usp* function is required in many tissues of the embryo. Uniform expression of *usp* under control of a heat-shock promoter can provide partial rescue of the *usp* lethality, but rescue requires that the heat treatment be frequently repeated; thus, *usp* function is required throughout development, and its function seems not to require accurate temporal or spatial control (Oro *et al.,* 1992a). These observations are consistent with the very widespread spatial and temporal patterns of expression in embryos and larvae (Henrich *et al.,* 1994; Andres *et al.,* 1993). A significant role for USP in adult development is shown by the fact that flies survive to adulthood and exhibit a variety of abnormalities in the adult integument in *usp/usp/+* trisomics (Henrich *et al.,* 1994). Mosaic analysis failed to identify widespread essential *usp* functions in adult development, but the perdurance of USP protein may have been sufficient to support development in these experiments. The principal abnormalities observed in *usp/usp* patches in the adult were in the eye, where rhabdomeres were misshapen and the ventral portion of the retina appeared "sunken"; the latter phenotype is not cell autonomous. In addition, *usp* deficiency leads to decreased fertility in females, where the embryos often exhibit the ventral defect mentioned earlier and the eggs often have defective chorions (Oro *et al.,* 1992a). The chorion defects are expected, since *usp* is expressed (at a low level) in the follicular epithelium that lays down the chorion (Khoury Christianson *et al.,* 1992) and USP binds to critical elements in chorion promoters (Shea *et al.,* 1990); it is, however, puzzling that they are also observed in follicles with *usp/+* soma and *usp/usp* germline (Oro *et al.,* 1992a).

In *Bombyx,* two *usp* transcripts can be seen, apparently differing in their 3′ termini. Early analysis indicates that expression of both forms is widespread; it is not known if they are functionally distinct (Tzertzinis *et al.,* 1994).

Quantitative Western analysis indicates that in Kc cells USP is present at about 1000 copies/cell (X. Hu and P. Cherbas, unpublished observations).

C. Ecdysone Response Elements

EcREs are the *cis*-acting DNA elements that confer ecdysteroid responsiveness on a promoter. That they function implies that they are binding sites for the receptor complex but, as indicated later, the converse statement is not true. The receptor complex may bind to a sequence but still fail to mediate a response. In what follows we will attempt to distinguish carefully between EcREs and binding sites. All of the natural sequences discussed next have been recovered from *Drosophila melanogaster* genes and are shown in Table I.

1. Sequences of EcREs

A few EcREs have been identified from ecdysteroid-responsive genes; they reside within the sequences shown in Table I. The first EcRE, from *hsp27,* has been the most thoroughly studied. It was discovered when Riddihough and Pelham (1987) mapped the ecdysteroid responsiveness of *hsp27* to a 23 Np sequence. The 15 Np core of that element (Table I) has since been shown to be a fully functional EcRE, yielding strong, rapid hormone responses when tested with reporter constructs in Kc cells (Lee, 1990; Cherbas *et al.,* 1991). Oligomers based on the *hsp27* 23-mer or the core 15-mer have been used in several laboratories to generate very large (500- to 1000-fold) induction ratios (see, e.g., Koelle *et al.,* 1991). The *Eip71CD* Proximal and *Eip71CD* Distal elements were mapped by Cherbas *et al.* (1991). The sequences illustrated mediate ecdysteroid responses when tested with reporters in Kc cells, but those responses are smaller than those given by the *hsp27* 15-mer. All three of these sequences contain imperfect palindromes, suggesting a consensus EcRE of the form RRGG/TTCANTGAC/ACYY (Cherbas *et al.,* 1991). Synthetic perfect palindromes of this form yield slightly stronger responses when tested in suitable reporter constructs (Lee, 1990; Martinez *et al.,* 1991).

All three of these elements have been tested for receptor binding using nuclear extracts from Kc cells. Cherbas *et al.* (1991) showed by a blotting assay that the proteins they bound included the [^{125}I]iodoponasterone-labeled receptor. Yao *et al.* (1992) demonstrated that the bound complexes in such extracts are, in fact, EcR/USP heterodimers. This observation has since been confirmed in several laboratories and no binding of EcR monomers or homodimers has been observed using these sequences as probes. Yao *et al.* (1992) also demonstrated binding by complexes assembled from *in vitro*-translated EcR and USP.

The *Fbp1* EcRE is contained in a 70 Np region which is sufficient to confer normal *Fbp1* temporal and spatial expression, including ecdysteroid induction of *Fbp1* in the larval fat body. That 70 Np region binds the receptor complex (EcR/USP) and the 15 N segment shown is entirely

TABLE I

EcREs and Binding Sequences

Name/gene	Sequence	References
hsp27	AAGTGCATTGAACCC	Riddihough and Pelham (1987), Cherbas *et al.* (1991)
Eip71CD (Proximal)	CAAGGACAGTAAAGTGACTGACCTCAAGACG	Cherbas *et al.* (1991)
Eip71CD (Distal)	TAAAGGATCTTGACCCCAATGAACTTCTAATTTCCTGTAAGT	Cherbas *et al.* (1991)
Eip71CD (Upstream)	ATCATTACGCAGGTTAATGACCAACAATCCACACTTCAAAAGG	Cherbas *et al.* (1991)
np1 and *np2*	GCGAAAGGTCAAGAGGCCAAAGAAGGTCAGGAA	D'Avino *et al.* (1995b)
Fbp1	GGGTTGAATGAATTT	Antoniewski *et al.* (1993)
Urate oxidase	AAGTGAGAGTGAT	Wallrath *et al.* (1990), Luo *et al.* (1991)
hsp23	CAATGGCAGATAATGC	Luo *et al.* (1991)

responsible for the binding (Laval *et al.,* 1993; Antoniewski *et al.,* 1994). Therefore, although the 15 Np segment shown has not as yet been reported to be active, it is very likely to be an EcRE. This element can be aligned with the consensus EcRE only with some difficulty.

We turn now to a series of sequences whose roles are somewhat more ambiguous. Receptor complex binding sites have been detected in the regions upstream of the urate oxidase gene and *hsp23* genes. Both sites (Table I) bind purified receptor complexes (presumably EcR/USP), although with relatively low affinity compared to the *hsp27* EcRE. Neither the UO-derived element (Wallrath *et al.,* 1990; Luo *et al.,* 1991) nor the *hsp23*-derived element (Luo *et al.,* 1991) bears any obvious resemblance to the EcRE consensus. Urate oxidase transcription can be repressed by ecdysteroids and it has been suggested that the nonconsensus element is associated with repression (Wallrath *et al.,* 1990).

The coding regions of the *np1* and *np2* genes contain perfectly conserved copies of the 33 Np sequence shown in Table I. Both genes fall within an early puff (see Russell and Ashburner, Chapter 3, this volume) and are probably ecdysteroid regulated in the metamorphic salivary gland. Thus, it is interesting that the sequence shown, which contains a direct repeat of half-sites ($AGGTCAN_{12}AGGTCA$) rather than a palindrome, binds the EcR/USP heterodimer with high affinity (D'Avino *et al.,* 1995a,b). This sequence provides the first example of a putative direct repeat EcRE from a natural source. Still the evidence that it is a functional EcRE remains equivocal.

Finally, we note that the *Eip71CD* Upstream element is an especially interesting case. It is a very poor (almost undetectably so) activating element (Cherbas *et al.,* 1991). However, it mediates full inhibition of reporter expression when the hormone is absent! We will return to this element in discussing that basal inhibition.

Table II shows those nucleotides that have been observed to occur at each position in natural EcREs that are clearly related to the palindromic consensus. The *hsp27* EcRE (a close match to the consensus) has been used as the starting point for *in vitro* mutagenesis. The right hand side of Table II shows the substitutions that have been generated based on the *hsp27* consensus and indicates their ability or inability to bind receptor complexes. Clearly, considerable sequence variation is compatible with binding and it is difficult to draw any other firm conclusions from these data. The reader should be especially cautious here. (i) Not every base has been tested at every position, so positions cannot be described as variable or invariant. For example, C was acceptable at position −1 but

TABLE II

Sequence Requirements of Palindromic EcREs[a]

Position	Consensus base	Presence in known EcREs[b]				Binding by *hsp27* EcRE mutants[c]			
		G	C	A	T	G	C	A	T
−7	R	●	[b][d]	●			+[d]	○	
−6	R	●		●			+[d]	○	
−5	G	●	[b]	[b]		○			−[d]
−4	G/T	●	[b]		●	+[d]			○
−3	T	[a]	[b]	[b]	●	○			+[d]
−2	C	[b, e]	●		[b]	−[c, e]	○	−[d]	
−1	A		[b]	●			+[d]	○	−[c]
0	N	●	●	●	●	+[d]			○
+1	T				●	−[d]			○
+2	G	●				○	+[e]		−[d]
+3	A			●			+[d]	○	
+4	C/A		●	●			+[d, f]	○	
+5	C		●				○	−[d]	
+6	Y		●	[b]	●		○	−[d]	
+7	Y		●	[b]	●		○	+[d]	

[a] Positions within a hypothetical EcRE are shown and the contents of these positions in the consensus palindrome are indicated (Cherbas *et al.*, 1991). For clarity, all other sequences have been aligned with the more nearly consensus half-site on the "right."

[b] For each position, any given nonconsensus nucleotide present among the known EcREs [*hsp27, Eip71CD* (Dist) and (Prox), *Fbp1*] is indicated. The consensus nucleotide is indicated by a filled circle. A blank cell indicates that that particular substitution does not exist in this set of EcREs. A bracketed letter indicates that it does and the letter indicates the relevant reference (see below).

[c] The table shows the *hsp27* sequence and substitutions based on it and reports whether or not those substituted sequences bind the receptor complex. Open circles indicate the nucleotides present in the *hsp27* EcRE. Empty cells indicate that the substitution in question has not been tested. +, the substitution has been tested and found to bind; −, it has been tested and failed to bind.

[d] *Key to references:* [a] Riddihough and Pelham (1987); [b] Cherbas *et al.* (1991); [c] Ozyhar *et al.* (1991); [d] Ozyhar and Pongs (1993); [e] Antoniewski *et al.* (1993); [f] Martinez *et al.* (1991).

T was not when each was substituted in the *hsp27* EcRE. (ii) Whether a particular nucleotide is acceptable or unacceptable at a given position depends on context. For example, G is acceptable at position −2 in some contexts but not in others. (iii) As the number of possible architectures for EcREs increases (see later), the alignment of very imperfect palindromes becomes increasingly tendentious. (iv) So far, most sequences have been tested only for binding and only by semiquantitative assays.

We do know that the spacing between the inverted repeats of the palindrome is critical. When the intervening base is deleted, affinity for the receptor is substantially diminished (Ozyhar and Pongs, 1993, using the *hsp27* EcRE) and the ability to confer ecdysteroid responsiveness is abolished (X. Hu, S. Vigo, K. Moyse, and L. Cherbas, unpublished observations, using a consensus EcRE). Horner *et al.* (1995) reported that a palindrome with spacing 0 can bind to EcR/USP, but their data are qualitative and may not be inconsistent with the earlier results. Expanding the spacing abolishes binding [spacing 2, using the *hsp27* EcRE (Ozyhar and Pongs, 1993)] and hormone-responsiveness [spacing 3, using a consensus EcRE (Martinez *et al.*, 1991)].

As mentioned earlier, TR/RXR, RAR/RXR, and VDR/RXR heterodimers bind and function on direct repeats as well as on inverted repeats. Indeed, direct repeats are the highest affinity HREs for these hormones; the repeat spacing discriminates among the heterodimers named. Among EcREs we have noted the evidence that the *np1/np2* direct repeat (a DR12) is at least an EcR/USP-binding site. The *Eip71CD* Distal sequence contains directly repeated half-sites as well as a palindrome, and the published data (Cherbas *et al.*, 1991) can be interpreted to suggest that the direct repeat may be functional. However, there is as yet no unambiguous example of a functional direct repeat EcRE. Several laboratories have demonstrated that EcR/USP binds well *in vitro* to sequences of the form DR3, DR4, and DR5 (D'Avino *et al.*, 1995b; Horner *et al.*, 1995; X. Hu and P. Cherbas, unpublished observations). However, these sequences function poorly, if at all, when tested in appropriate reporter constructs in Kc cells (X. Hu, S. Vigo, L. Cherbas, and P. Cherbas, unpublished observations).

To conclude this section we simply note that it is still impossible here, as in the comparable vertebrate cases, to predict with certainty whether a given sequence will bind the receptor complex and, given binding, whether it will be functional as a HRE.

2. *EcREs and Basal Inhibition*

One of the striking similarities between TR and EcR is that both bind to their HREs and inhibit transcription when unliganded. With a palindromic EcRE (e.g., *hsp27*), we (Cherbas *et al.,* 1991) and others (Dobens *et al.,* 1991) have observed basal inhibitions of ca. 3-fold relative to a control reporter construct lacking an EcRE. The typical (ca. 50-fold) induction observed following hormone treatment starts from this somewhat depressed basal level. Oligomeric EcREs not only increase the magnitude of activation, but they also increase the basal inhibition. For example, two copies of the *hsp27* element give approximately 50-fold inhibition (Lee, 1990). A priori, basal inhibition might be due to unliganded EcR/USP or it might be due to another EcRE-binding transcription factor. We have used a gene-targeting procedure to manufacture Kc cells deficient in EcR. Since basal inhibition is nearly eliminated in those cells and is restored when they are transfected with an EcR expression vector, we conclude that EcR (and probably EcR/USP) is the inhibitory factor (unpublished observations).

In connection with basal inhibition we return to the *Eip71CD* Upstream EcRE. As noted before, when assayed on reporter constructs in Kc cells, Upstream is a poor activating element which, nevertheless, provides full basal inhibition. Upstream binds strongly to EcR/USP and its unusual behavior does not appear to be a consequence of local sequence context: the 15 Np imperfect palindrome at its core behaves similarly (S. Vigo and P. Cherbas, unpublished observations). Still more intriguing is the evidence that Upstream is probably a fully active EcRE, responsible for major positive and negative changes in *Eip71CD* expression in the larval epidermis (Andres and Cherbas, 1992, 1994).

The most interesting conclusion that might eventually emerge from data like these is that an EcRE is not simply a tether which can be fully described by an affinity, but is also a switch whose sequence affects the functionality of the resulting EcR/USP/DNA complex. While the data are still too sparse to support that conclusion, it is clear that at least three classes of EcR/USP-binding sites exist: (a) fully functional EcREs; (b) EcREs fully functional only for basal inhibition (Upstream); and (c) nonfunctional EcREs (e.g., DRs 3, 4, and 5).

3. EcREs and JH Function

Berger *et al.* (1992) reported that the *hsp27* EcRE can mediate a juvenile hormone-sensitive ecdysteroid response when placed before a minimal *tk* promoter with an appropriate reporter and assayed in S3 cells. Using the same EcRE and reporter but a different promoter, and conducting the assays in Kc cells, we have been unable to reproduce this effect (L. Cherbas, M. Žurovec, and P. Cherbas, unpublished observations). Since the explanation for the discrepancy is unclear, the conditions under which an EcRE can mediate juvenile hormone (JH) effects remain uncertain.

D. Ecdysteroid Resistance

The existence of ecdysteroid-resistant cell lines has been an important motivation for studies of the hormone response in those cells. In fact, resistant lines have proven to be useful (occasionally indispensable) tools at numerous stages of the work described earlier. However, the genetic bases of ecdysteroid resistance are still poorly understood. In our laboratory, the usual protocol for Kc cells has involved selection at a high hormone concentration (10^{-6} *M* 20-HE); the population that emerges has lost the normal ecdysteroid-induced block to proliferation. Fluctuation tests suggest that the events involved are quite frequent (L. Cherbas, unpublished observations). Most such populations (or clones) have reduced EcR titers as measured by steroid binding. Typically the reduction is ca. 50–80%. Ecdysteroid responses—both the cells' normal transcriptional and morphological responses and those of transfected reporter constructs—are also much diminished. However, in most cases, these defects are not rescued by transfection with a plasmid expressing EcR-B1—the principal EcR isoform of normal (ecdysteroid-sensitive) Kc cells. In short, most of these clones do not behave as though mutations have simply rendered them hypomorphic for EcR-B1.

It proved to be important in characterizing EcR (Section III,A,2) that, occasionally, the same protocol yields clones with very low EcR titers. Typically, such clones are fully rescuable by expression of EcR-B1. Unfortunately, they are also usually unstable, reverting within weeks or months to the form of resistance that cannot be rescued by EcR-B1. Indeed, the only resistant, rescuable, and stable line available is one obtained by targeting experiments directed at EcR (L. Cherbas and P. Cherbas, unpublished observations).

The events responsible for spontaneous resistance in Kc and S2 cells are probably not EcR mutations. What they are remains a mystery.

E. Other Nuclear Hormone Receptor Family Members

The list of "orphan receptors" that have been identified in *Drosophila* is long and rapidly growing (Table III). This fact is important in the

TABLE III

Members of Nuclear Hormone Receptor Superfamily in *Drosophila*

Gene	Vertebrate Homolog	Properties	Reference
EcR		EcR	Koelle *et al.* (1991)
usp	RXR	Heterodimer partner for EcR; essential component of functional receptor complex; chorion transcription factor	Henrich *et al.* (1990), Oro *et al.* (1990), Shea *et al.* (1990)
DHR3		Ecdysteroid-inducible in *Manduca* and *Drosophila*	Koelle *et al.* (1992), Palli *et al.* (1992), Horner *et al.* (1995)
DHR38	NGFI-B	Dimerizes with USP	Sutherland *et al.* (1995)
E75	ear-1	Early salivary gland puff site	Segraves and Hogness (1990)
E78		Early–late salivary gland puff site	Martín-Blanco and Kornberg (1993), Stone and Thummel (1993)
Egon		Expressed in embryonic gonad; no ligand-binding domain	Rothe *et al.* (1989)
FTZ-F1		Activates *ftz* and *Adh* transcription	Lavorgna *et al.* (1991, 1993)
FTZ-F1 β (also called *DHR39*)		Binds to *ftz* and *Adh* promoters; ecdysteroid inducible	Ohno and Petkovich (1992), Ayer *et al.* (1993), Ohno *et al.* (1994), Horner *et al.* (1995)
kni		Gap segmentation gene; no ligand-binding domain	Nauber *et al.* (1988)
knrl		Function unknown; no ligand-binding domain	Oro *et al.* (1988)
svp	COUP	Interferes with ecdysteroid response; required for embryonic development; affects identities of photoreceptor cells	Mlodzik *et al.* (1990), Zelhof *et al.* (1995)
tll	Tlx	Gap gene	Pignoni *et al.* (1990)

present context because there is ample evidence that family members can be promiscuous in their dimerization. For example, the RXRs (three species) form heterodimers with members of at least four receptor groups: the TRs, VDR, RARs, and peroxisome proliferator-activated receptor (PPAR) (Yu *et al.*, 1991; Bugge *et al.*, 1992; Hallenbeck *et al.*, 1992; Kliewer *et al.*, 1992; Leid *et al.*, 1992; Marks *et al.*, 1992; Zhang *et al.*, 1992; Issemann *et al.*, 1993; Krey *et al.*, 1993). Indeed, several of the dimerization partners of RXRs can form *inter se* heterodimers (Glass *et al.*, 1989; Berrodin *et al.*, 1992; Bogazzi *et al.*, 1994).

The significance of these observations for the ecdysteroid response is obvious. If alternatives to EcR existed (other than the known isoforms), they might have quite distinct developmental specificities. No evidence exists for alternatives to EcR. More realistically, the nature and specificity of hormone responses could be altered (a) if EcR has alternative dimerization partners that compete with USP or (b) if USP has alternative partners that compete with EcR for its services.

Alternative dimerization partners for EcR have not been detected so far. We have already noted that there is no published evidence for DNA binding by EcR monomers or homodimers. Similarly, USP homodimers have not been reported, and the binding studies described earlier suggest that USP does not bind to DNA as a homodimer. USP will form heterodimers with the mammalian RXR partners TRα, TRβ, RARα, VDR, and PPAR, but not with RXRα (Khoury Christianson *et al.*, 1992; Yao *et al.*, 1992). These, of course, are not present in ecdysteroid-responsive cells. However, a newly discovered *Drosophila* orphan receptor has been shown to form heterodimers with USP. DHR38, a *Drosophila* homolog of mammalian NGFI-B, can form heterodimers with USP *in vitro* (Sutherland *et al.*, 1995). In addition, it has been demonstrated that expression of *seven-up* (*svp*), another *Drosophila* orphan receptor, interferes with ecdysteroid responses in cells, suggesting that *svp* may pair with either EcR or USP (Zelhof *et al.*, 1995). It is not yet known whether either of these proteins is present at appropriate times, places, and concentrations to modulate the ecdysteroid response. By *in vitro* tests, USP failed to alter the DNA-binding specificities of E75A, E78A, DHR3, and Ftz-F1β; presumably it does not heterodimerize with them (Horner *et al.*, 1995).

IV. OVERVIEW OF ECDYSTEROID ACTION

As far as we know, all ecdysteroid effects follow from the ability of the hormone to induce positive and negative changes in the transcription of particular target genes. These primary events have secondary (and higher-order) consequences for the expression of other genes, leading to the kind of interregulated pathway envisioned by the Ashburner model for metamorphic gene expression in the salivary gland (see Russell and Ashburner, Chapter 3, this volume). It would not be surprising if additional, extratranscriptional effects were eventually to be discovered; there is evidence for such effects of vitamin D and progestins in some settings.

Like its closest vertebrate relatives, EcR is a nuclear protein even in naive (untreated) cells. Most likely it exists primarily as an EcR/USP heterodimer and is bound to DNA. Immunocytochemistry on polytene chromosomes demonstrates the concentration of EcR and USP on the chromosomes with very similar distributions (Yao *et al.,* 1993). Because the affinity of unliganded EcR/USP for EcREs is relatively low, we envision the naive receptor complex shuttling between neighboring DNA sequences; the occupancy of any given EcRE should be relatively low. Nonetheless, receptor complexes associated with EcREs are responsible for depressing the basal transcription of the associated genes in naive cells.

Ecdysteroids stabilize EcR/USP/EcRE complexes. One of the most striking novelties of the ecdysteroid system is that it is the heterodimer—EcR/USP—that exhibits the normal ligand-binding characteristics of the receptor (Yao *et al.,* 1993; Koelle, 1992). This contrasts sharply with the ligand-binding properties of TR, VDR, and RAR: all of these receptors bind their ligands with high affinity whether they are in the form of monomers, homodimers, or heterodimers with RXR. In contrast, *in vitro*-translated EcR by itself binds ecdysteroids very poorly, but complexes assembled from *in vitro*-translated EcR and USP accurately recreate the hormone-binding affinities and specificities previously reported for nuclear extracts. EcR/RXR heterodimers bind ecdysteroids significantly better than EcR alone but with far lower affinity than EcR/USP. In RXR-expressing mammalian cells transfected with EcR expression vectors, high doses of ecdysteroids activate transcription of an EcRE-linked reporter, presumably by virtue of the low affinity hormone interactions with EcR/RXR complexes (Christopherson *et al.,* 1992; Thomas *et al.,* 1993). Because *in vitro*-translated EcR binds ecdysteroids, however weakly, and because that binding can be stabilized—modestly—by RXR, we are comfortable with the conclusion that EcR itself is the polypeptide that contacts the steroid (Yao *et al.,* 1993). Consistent with this, the steroid dissociates from preformed EcR/RXR and EcR/USP complexes at relatively similar rates (X. Hu and P. Cherbas, unpublished observations).

If this picture is correct, then USP is an allosteric effector for ecdysteroid binding by EcR. Furthermore, this model predicts that steroid should enhance heterodimerization (observed: Yao *et al.,* 1993) and should have large effects on EcRE binding (observed: Yao *et al.,* 1992, 1993; X. Hu and P. Cherbas, unpublished observations). Because of the number of steps involved in forming functional complexes, it would not be surprising if ecdysteroids differed not only in their affinities for the complex but in their efficacies, i.e., their intrinsic abilities to provoke formation of active EcRE-binding complexes. The ecdysteroid muristerone has proven unusually active (relative to 20-HE) in experiments using mammalian cells and one interpretation of these data is that it is qualitatively more effective (at least for EcR/RXR complexes) than other ecdysteroids. However, we are not yet persuaded that this is the simplest explanation of the observations. In any event, a more quantitative picture of all these interactions should be emerging in the near future.

There is an important variable here about which we can, at this point, say virtually nothing. We do not yet know whether there are other EcR dimerization partners that play a physiological role as USP competitors. If there are, their effects on steroid binding and the converse effects of steroid on their binding to EcR will obviously be crucial. Moreover, we find it implausible that USP does not have a ligand of its own. It is attractive to suppose that that ligand will be related to 9-*cis*-retinoic acid or perhaps to juvenile hormone, but both suggestions are purely speculative. If USP has a ligand, the vertebrate precedents suggest that it will have important effects on the assembly and functions of ecdysteroid receptor complexes.

Assembled EcR/USP/steroid complexes presumably occupy EcREs with high efficiency. We do not know how many EcREs are present in the genome (or exposed in chromatin). Since active genes often have several EcREs, 300–400 available target genes would exhaust the known supplies of EcR and USP in cells even on the assumption of absolute binding specificity. The specificity that we know about (ca. three orders of magnitude preference of receptor complexes for EcREs compared any random sequence) falls far short of accounting for the localization that must occur. We assume that the biochemistry has not yet revealed all the mechanisms that account for high affinity binding to EcREs.

As surprising as the result that efficient ligand binding requires EcR/USP are the observations that EcR homodimers play no important role and that EcR/USP heterodimers function only on palindromic EcREs. Vertebrate precedents had suggested that symmetrical homodimers prefer palindromes and that heterodimers generally exhibit their largest effects on direct repeats. Moreover, it is intriguing to discover a class of sites (the DRs mentioned in previous sections) that bind the receptor complex without substantial effects on transcription. In the end we are impressed by the fluidity of our understanding of receptor complex–HRE interactions, both in this system and in the vertebrate systems. We are particularly mindful that accounting for the high affinity binding to EcREs *in vivo* will require many orders of magnitude of improvement in binding efficiency; we suspect that this will mean the discovery of better EcREs and new components of the binding machinery.

Receptor complexes localized at EcREs affect transcription. They inhibit it when they are unliganded and they activate it when liganded. So far the effects of EcREs have been tested on only a handful of promoters, but it is clear that only a minimal promoter is required (Lee, 1990; Luo *et al.,* 1991). Compared to the vertebrate situation, relatively little is known about the receptor domains involved in these processes or about the elements of the transcriptional apparatus that participate. For example, there are as yet no reports of experiments designed to map activation and inhibitory domains in EcR (or USP). Luo *et al.* (1991) reported stimulation of transcription *in vitro* by purified receptor complexes, with evidence that the limiting step is a preinitiation complex formation. Comparable results have not yet been reported by others and more details about the transcriptional components involved have not yet emerged. Nor do we know anything about the effects of EcR/USP complexes on DNA bending.

In studies of *Eip71CD* basal transcription we have discovered that the normally modest inhibition due to an EcRE becomes very substantial (ca. 50-fold) when the Inr sequence at the start of transcription is mutated or deleted. Conversely, the importance of the Inr for basal transcription of TATA-containing gene is considerably magnified by the presence of the EcRE (Cherbas and Cherbas, 1993). These effects suggest that basal inhibition, whatever its mechanism, may involve interactions with those TAFs that interact with Inr (Purnell and Gilmour, 1993; Purnell *et al.,* 1994). Some mammalian nuclear hormone receptor systems have been transferred to yeast, making possible experiments that have revealed much about the components involved in transcriptional modulation. The ecdysteroid system apparently can function in yeast (Koelle, 1992).

We began this section by emphasizing that ecdysteroid responses begin with positive and negative transcriptional effects on a limited set of primary target genes. Among those genes that are capable of responding to ecdystone, the distinction of being a primary target is not a property of a gene, but of a gene in a particular developmental setting. Thus acetylcholinesterase is not a primary target for ecdysteroids in naive Kc cells, but after brief ecdysteroid treatment it moves into that category and becomes a primary target for subsequent treatment (Cherbas and Cherbas, 1981). The distinction is perhaps clearest when one examines the range of stage- and tissue-specific effects of ecdysteroids on the *Eip71CD* gene which is affected by ecdysteroids in many (but not all) tissues and at many (but not all) times of the hormone's appearance during larval development (Andres and Cherbas, 1992).

Understanding ecdysteroid effects during development will require understanding how primary targets are selected from the set of potentially ecdysteroid-responsive genes. The specificity presumably derives, at least in part, from the effects of other *cis*-acting transcription factors that modulate either the binding or the activity of receptor complexes. As suggested by our observations on the *Eip71CD* Upstream element, it may also reflect the EcRE specificities of particular receptor complexes, complexes that may differ by EcR isoform or because of the effects of other pairing partners, ligands, or posttranslational modifications.

In unraveling these puzzles, along with the many others that have been mentioned, we expect to witness increasingly detailed biochemical analysis of the ecdysteroid response, accompanied by evolutionary analysis, taking advantage of the variation in receptor components that we have emphasized here, and by detailed genetic analysis that has long been the goal of ecdysteroid studies of *Drosophila.*

ACKNOWLEDGMENTS

We gratefully acknowledge the assistance of all our colleagues who have shared their ideas, both published and unpublished, with us and the assistance provided by the FlyBase resource for *Drosophila.*

REFERENCES

Allegretto, E. A., McClurg, M. R., Lazarchik, S. B., Clemm, D. L., Kerner, S. A., Elgort, M. C., Boehm, M. F., White, S. K., Pike, J. W., and Heyman, R. A. (1993). Transactivation properties of retinoic acid and retinoic acid and retinoid X receptors in mammalian cells and yeast. *J. Biol. Chem.* **268,** 26625–26633.

Andres, A. J., and Cherbas, P. (1992). Tissue-specific ecdysone responses: Regulation of the *Drosophila* genes *Eip28/29* and *Eip40* during larval development. *Development* **116,** 865–876.

Andres, A. J., and Cherbas, P. (1994). Tissue-specific regulation by ecdysone: Distinct patterns of *Eip28/29* expression are controlled by different ecdysone response elements. *Dev. Genet.* **15,** 320–331.

Andres, A. J., Fletcher, J. C., Karim, F. D., and Thummel, C. S. (1993). Molecular analysis of the initiation of insect metamorphosis: A comparative study of *Drosophila* ecdysteroid-regulated transcription. *Dev. Biol.* **160,** 388–404.

Antoniewski, C., Laval, M., and Lepesant, J.-A. (1993). Structural features critical to the activity of an ecdysone receptor binding site. *Insect Biochem. Mol. Biol.* **23,** 105–114.

Antoniewski, C., Laval, M., Dahan, A., and Lepesant, J.-A. (1994). The ecdysone response enhancer of the *Fbp1* gene of *Drosophila melanogaster* is a direct target for the EcR/USP nuclear receptor. *Mol. Cell. Biol.* **14,** 4465–4474.

Ayer, S., Walker, N., Mosammaparast, M., Nelson, J. P., Shilo, B., and Benyajati, C. (1993). Activation and repression of *Drosophila* alcohol dehydrogenase distal transcription by two steroid hormone receptor superfamily members binding to a common response element. *Nucleic Acids Res.* **21,** 1619–1627.

Baniahmad, A., Köhne, A. C., and Renkawitz, R. (1992). A transferable silencing domain is present in the thyroid hormone receptor, in the v-erb A oncogene product and in the retinoic acid receptor. *EMBO J.* **11,** 1015–1023.

Baniahmad, A., Ha, I., Reinberg, D., Tsai, S., Tsai, M.-J., and O'Malley, B. W. (1993). Interaction of human thyroid hormone receptor β with transcription factor TFIIB may mediate target gene derepression and activation by thyroid hormone. *Proc. Natl. Acad. Sci. U.S.A.* **90,** 8832–8836.

Baniahmad, A., Leng, X., Burris, T. P., Tsai, S. Y., Tsai, M.-J., and O'Malley, B. W. (1995). The τ 4 activation domain of the thyroid hormone receptor is required for release of a putative corepressor(s) necessary for transcriptional silencing. *Mol. Cell. Biol.* **15,** 76–86.

Beckers, C., Maróy, P., Dennis, R., O'Connor, J. D., and Emmerich, H. (1980). The uptake and release of ponasterone A by the Kc cell line of *Drosophila melanogaster. Mol. Cell. Endocrinol.* **17,** 51–59.

Berger, E. M., Goudie, K., Klieger, L., Berger, M., and DeCato, R. (1992). The juvenile hormone analogue, methoprene, inhibits ecdysterone induction of small heat shock protein gene expression. *Dev. Biol.* **151,** 410–418.

Berrodin, T. J., Marks, M. S., Ozato, K., Linney, E., and Lazar, M. A. (1992). Heterodimerization among thyroid hormone receptor, retinoic acid receptor, retinoid X receptor, chicken ovalbumin upstream promoter transcription factor, and an endogenous liver protein. *Mol. Endocrinol.* **92,** 1468–1478.

Bidmon, H.-J. (1991). Developmental changes in the presence of ecdysteroid receptors in the central nervous system of third instar larvae of *Sarcophaga bullata. Dev. Brain Res.* **63,** 121–133.

Bidmon, H.-J., and Koolman, J. (1989). Ecdysteroid receptors located in the central nervous system of an insect. *Experientia* **45,** 106–109.

Bidmon, H.-J., Granger, N. A., Cherbas, P., Maróy, P., and Stumpf, W. E. (1991a). Ecdysteroid receptors in the central nervous system of *Manduca sexta:* Their changes in distribution and quantity during larval–pupal development. *J. Comp. Neurol.* **310,** 337–355.

Bidmon, H.-J., Stumpf, W. E., and Granger, N. A. (1991b). Ecdysteroid binding sites localized by autoradiography in the central nervous system of precommitment fifth-stadium *Manduca sexta* larvae. *Cell Tissue Res.* **263,** 183–194.

Bidmon, H.-J., Stumpf, W. E., and Granger, N. A. (1992). Ecdysteroid receptors in the neuroendocrine–endocrine axis of a moth. *Experientia* **48,** 42–47.

Bogazzi, F., Hudson, L. D., and Nikodem, V. M. (1994). A novel heterodimerization partner for thyroid hormone receptor. Peroxisome proliferator-activated receptor. *J. Biol. Chem.* **269,** 11683–11686.

Bugge, T. H., Pohl, J., Lonnoy, O., and Stunnenberg, H. G. (1992). RXRα, a promiscuous partner of retinoic acid and thyroid hormone receptors. *EMBO J.* **11,** 1409–1418.

Carlberg, C., Bendik, I., Wyss, A., Meier, E., Sturzenbecker, L. J., Grippo, J. F., and Hunziger, W. (1993). Two nuclear signalling pathways for vitamin D. *Nature (London)* **361,** 657–660.

Cherbas, P. (1993). The IVth Karlson Lecture: Ecdysone-responsive genes. *Insect Biochem. Mol. Biol.* **23,** 3–11.

Cherbas, L., and Cherbas, P. (1981). The effects of ecdysteroid hormones on *Drosophila melanogaster* cell lines. *Adv. Cell Cult.* **1,** 91–124.

Cherbas, L., and Cherbas, P. (1993). The arthropod initiator: The capsite consensus plays an important role in transcription. *Insect Biochem. Mol. Biol.* **23,** 81–90.

Cherbas, L., Yonge, C. D., Cherbas, P., and Williams, C. M. (1980). The morphological response of Kc-H cells to ecdysteroids: Hormonal specificity. *Wilhelm Roux's Arch. Dev. Biol.* **189,** 1–15.

Cherbas, L., Lee, K., and Cherbas, P. (1991). Identification of ecdysone response elements by analysis of the *Drosophila Eip28/29* gene. *Genes Dev.* **5,** 120–131.

Cherbas, P., Trainor, D. A., Stonard, R. J., and Nakanishi, K. (1982). 14-Deoxymuristerone, a compound exhibiting exceptional moulting hormonal activity. *J. Chem. Soc., Chem. Commun.* **1982,** 1307–1308.

Cherbas, P., Cherbas, L., Lee, S.-S., and Nakanishi, K. (1988). 26-[^{125}I]Iodoponasterone A is a potent ecdysone and a sensitive radioligand for ecdysone receptors. *Proc. Natl. Acad. Sci. U.S.A.* **85,** 2096–2100.

Cheskis, B., and Freedman, L. P. (1994). Ligand modulates the conversion of DNA-bound vitamin D_3 receptor (VDR) homodimers into VDR–retinoid S receptor heterodimers. *Mol. Cell. Biol.* **14,** 3329–3338.

Chiba, H., Muramatsu, M., Nomoto, A., and Kato, H. (1994). Two human homologues of *Saccharomyces cerevisiae SW12/SNF2* and *Drosophila brahma* are transcriptional coactivators cooperating with the estrogen receptor and the retinoic acid receptor. *Nucleic Acids Res.* **22,** 1815–1820.

Cho, W.-L., Kapitskaya, M. Z., and Raikhel, A. S. (1995). Mosquito ecdysteroid receptor analysis of the cDNA and expression during vitellogenesis. *Insect Biochem. Mol. Biol.* **25,** 19–27.

Christopherson, K. S., Mark, M. R., Bajaj, V., and Godowski, P. J. (1992). Ecdysteroid-dependent regulation of genes in mammalian cells by a *Drosophila* ecdysone receptor and chimeric transactivators. *Proc. Natl. Acad. Sci. U.S.A.* **89,** 6314–6318.

Dahlman-Wright, K., Wright, A., Gustafsson, J.-Å., and Carlstedt-Duke, J. (1991). Interaction of the glucocorticoid receptor DNA-binding domain with DNA as a dimer is mediated by a short segment of five amino acids. *J. Biol. Chem.* **266,** 3107–3112.

Dai, J.-D., Sar, M., Warren, J. T., and Gilbert, L. I. (1991). An autoradiographic and immunocytochemical analysis of ecdysteroids and ecdysteroid binding sites in target cells of *Drosophila melanogaster. Invertebr Reprod. Dev.* **20,** 227–236.

Damm, K., Thompson, C. C., and Evans, R. M. (1989). Protein encoded by v-*erbA* functions as a thyroid-hormone receptor antagonist. *Nature* (*London*) **339,** 593–597.

Darling, D. S., Carter, R. L., Yen, P. M., Welborn, J. M., Chin, W. W., and Umeda, P. K. (1993). Different dimerization activities of α and β thyroid hormone receptor isoforms. *J. Biol. Chem.* **268,** 10221–10227.

D'Avino, P. P., Crispi, S., and Furia, M. (1995a). Hormonal regulation of the *Drosophila melanogaster ng*-gene. *Eur. J. Entomol.* **92,** 259–262.

D'Avino, P. P., Crispi, S., Cherbas, L., Cherbas, P., and Furia, M. (1995b). The moulting hormone ecdysone is able to recognize target elements composed of direct repeats. *Mol. Cell. Endocrinol.* **113,** 1–9.

Desvergne, B. (1994). How do thyroid hormone receptors bind to structurally diverse response elements. *Mol. Cell. Endocrinol.* **100,** 125–131.

Dinan, L. (1985). Ecdysteroid receptors in a tumorous blood cell line of *Drosophila melanogaster. Arch. Insect Biochem. Physiol.* **2,** 295–317.

Dobens, L., Rudolph, K., and Berger, E. M. (1991). Ecdysterone regulatory elements function as both transcriptional activators and repressors. *Mol. Cell. Biol.* **11,** 1846–1853.

Ehrlich, P., and Morgenroth, J. (1910). "Studies in Immunity." Wiley, New York.

Fahrbach, S. E., and Truman, J. W. (1989). Autoradiographic identification of ecdysteroid-binding cells in the nervous system of the moth *Manduca sexta. J. Neurobiol.* **20,** 681–702.

Flynn, T. R., Hollenberg, A. N., Cohen, O., Menke, J. B., Usala, S. J., Tollin, S., Hegarty, M. K., and Wondisford, F. E. (1994). A novel C-terminal domain in the thyroid hormone receptor selectively mediates thyroid hormone inhibition. *J. Biol. Chem.* **269,** 32713–32716.

Fondell, J. D., Roy, A. L., and Roeder, R. G. (1993). Unliganded thyroid hormone receptor inhibits formation of a functional preinitiation complex: Implications for active repression. *Genes Dev.* **7,** 1400–1410.

Forman, B. M., and Samuels, H. H. (1990). Interactions among a subfamily of nuclear hormone receptors: The regulatory zipper model. *Mol. Endocrinol.* **4,** 1293–1301.

Freedman, L. P. (1993). Structure and function of the steroid receptor zinc finger region. *In* "Steroid Hormone Action" (M. G. Parker, ed.), pp. 141–165. IRL Press, Oxford.

Fujiwara, H., Jindra, M., Newitt, R., Palli, S. R., Hiruma, K., and Riddiford, L. M. (1995). Isolation and development expression of the ecdysone receptor gene in wings of *Manduca sexta. Insect Biochem. Mol. Biol.* **25,** 845–856.

Glass, C. K., Lipkin, S. M., Devary, O. V., and Rosenfeld, M. G. (1989). Positive and negative regulation of gene transcription by a retinoic acid–thyroid hormone receptor heterodimer. *Cell* (*Cambridge, Mass.*) **59,** 697–708.

Goodwin, T. W., Horn, D. H. S., Karlson, P., Koolman, J., Nakanishi, K., Robbins, W. E., Siddall, J. B., and Takemoto, T. (1978). Ecdysteroids: A new generic term. *Nature* (*London*) **272,** 122.

Hallenbeck, P. L., Marks, M. S., Lippoldt, R. E., Ozato, K., and Nikodem, V. (1992). Heterodimerization of thyroid hormone (TH) receptor with H-2RIIBP (RXRβ) enhances DNA binding and TH-dependent transcription activation. *Proc. Natl. Acad. Sci. U.S.A.* **89,** 5572–5576.

Hallenbeck, P. L., Phyillaier, M., and Nikodem, V. M. (1993). Divergent effects of 9-*cis*-retinoic acid receptor on positive and negative thyroid hormone receptor-dependent gene expression. *J. Biol. Chem.* **268,** 3825–3828.

Henrich, V. C., and Brown, N. E. (1995). Insect hormones and their receptors: A developmental and comparative perspective. *Insect Biochem. Mol. Biol.* **25,** 881–897.

Henrich, V. C., Sliter, T. J., Lubahn, D. B., MacIntyre, A., and Gilbert, L. I. (1990). A steroid/thyroid hormone receptor superfamily member in *Drosophila melanogaster* that shares extensive sequence similarity with a mammalian homologue. *Nucleic Acids Res.* **18,** 4143–4148.

Henrich, V. C., Szekely, A. A. , Kim, S. J., Brown, N. E., Antoniewski, C., Hayden, M. A., Lepesant, J.-A., and Gilbert, L. I. (1994). Expression and function of the *ultraspiracle* (*usp*) gene during development of *Drosophila melanogaster. Dev. Biol.* **165,** 38–52.

Heyman, R. A., Mangelsdorf, D. J., Duck, J. A., Stein, R., Eichele, G., Evans, R. M., and Thaller, C. (1992). 9-*cis* Retinoic acid is a high affinity ligand for the retinoid X receptor. *Cell* (*Cambridge, Mass.*) **68,** 397–406.

Horner, M. A., Chen, T., and Thummel, C. S. (1995). Ecdysteroid regulation and DNA binding properties of *Drosophila* nuclear hormone receptor superfamily members. *Dev. Biol.* **168,** 490–502.

Ikeda, M., Rhee, M., and Chin, W. W. (1994). Thyroid hormone receptor monomer, homodimer, and heterodimer (with retinoid-X receptor) contact different nucleotide sequences in thyroid hormone response elements. *Endocrinology* (*Baltimore*) **135,** 1628–1638.

Imhof, M. O., Rusconi, S., and Lezzi, M. (1993). Cloning of a *Chironomus tentans* cDNA encoding a protein (cEcRH) homologous to the *Drosophila melanogaster* ecdysteroid receptor (dEcR). *Insect Biochem. Mol. Biol.* **23,** 115–124.

Issemann, I., Prince, R. A., Tugwood, J. D., and Green, S. (1993). The retinoid X receptor enhances the function of the peroxisome proliferator activated receptor. *Biochimie* **75,** 251–256.

Jacq, X., Brou, C., Lutz, Y., Davidson, I., Chambon, P., and Tora, L. (1994). Human $TAF_{II}30$ is present in a distinct TFIID complex and is required for transcriptional activation by the estrogen receptor. *Cell* (*Cambridge, Mass.*) **79,** 107–117.

Karim, F. D., and Thummel, C. S. (1992). Temporal coordination of regulatory gene expression by the steroid hormone ecdysone. *EMBO J.* **11,** 4083–4093.

Katz, R. W., Koenig, R. J. (1993). Nonbiased identification of DNA sequences that bind thyroid hormone receptor $\alpha 1$ with high affinity. *J. Biol. Chem.* **268,** 19392–19397.

Khoury Christianson, A. M., and Kafatos, F. C. (1993). Binding affinity of the *Drosophila melanogaster* CF1/USP protein to the chorion *s15* promoter. *Biochem. Biophys. Res. Commun.* **193,** 1318–1323.

Khoury Christianson, A. M., King, D. L., Hatzivassiliou, E., Casas, J. E., Hallenbeck, P. L., Nikodem, V. M., Mitsialis, S. A., and Kafatos, F. C. (1992). DNA binding and heteromerization of the *Drosophila* transcription factor chorion factor 1/ultraspiracle. *Proc. Natl. Acad. Sci. U.S.A.* **89,** 11503–11507.

Kliewer, S. A., Umesono, K., Noonan, D. J., Heyman, R. A., and Evans, R. M. (1992). Convergence of 9-*cis* retinoic acid and peroxisome proliferator signalling pathways through heterodimer formation of their receptors. *Nature* (*London*) **358,** 771–774.

Koelle, M. R. (1992). "Molecular Analysis of the *Drosophila* Ecdysone Receptor Complex," Ph.D. thesis. Standford University, Stanford, California.

Koelle, M. R., Talbot, W. S., Segraves, W. A., Bender, M. T., Cherbas, P., and Hogness, D. S. (1991). The *Drosophila EcR* gene encodes an ecdysone receptor, a new member of the steroid receptor superfamily. *Cell (Cambridge, Mass.)* **67,** 59–77.

Koelle, M. R., Segraves, W. A., and Hogness, D. S. (1992). DHR3: A *Drosophila* steroid receptor homolog. *Proc. Natl. Acad. Sci. U.S.A.* **89,** 6167–6171.

Krey, G., Keller, H., Mahfoudi, A., Medin, J., Ozato, K., Dreyer, C., and Wahli, W. (1993). *Xenopus* peroxisome proliferator activated receptors: Genomic organization, response element recognition, heterodimer formation with retinoid X receptor and activation by fatty acids. *J. Steroid Biochem. Mol. Biol.* **47,** 65–73.

Kurokawa, R., Yu, V. C., Naar, A., Kyakumoto, S., Han, Z., Silverman, S., Rosenfeld, M. G., and Glass, C. K. (1993). Differential orientations of the DNA-binding domain and carboxy-terminal dimerization interface regulate binding site selection by nuclear receptor heterodimers. *Genes Dev.* **7,** 1423–1435.

Kurokawa, R., DiRenzo, J., Boehm, M., Sugarman, J., Gloss, B., Rosenfeld, M., Heyman, R. A., and Glass, C. K. (1994). Regulation of retinoic signalling by receptor polarity and allosteric control of ligand binding. *Nature (London)* **371,** 528–531.

LaCasse, E. C., Lochnan, H. A., Walker, P., and Lefebvre, Y. A. (1993). Identification of binding proteins for nuclear localization signals of the glucocorticoid and thyroid hormone receptors. *Endocrinology (Baltimore)* **132,** 1017–1025.

Landon, T. M., Sage, B. A., Seeler, B. J., and O'Connor, J. D. (1988). Characterization and partial purification of the *Drosophila* Kc cell ecdysteroid receptor. *J. Biol. Chem.* **263,** 4693–4697.

Laval, M., Pourrain, F., Deutsch, J., and Lepesant, J.-A. (1993). *In vivo* functional characterization of an ecdysone response enhancer in the proximal upstream region of the *Fbp1* gene of *D. melanogaster. Mech. Dev.* **44,** 123–138.

Lavorgna, G., Ueda, H., Clos, J., and Wu, C. (1991). FTZ-F1, a steroid hormone receptor-like protein implicated in the activation of *fushi tarazu. Science* **252,** 848–851.

Lavorgna, G., Karim, F. D., Thummel, C. S., and Wu, C. (1993). Potential role for a FTZ-F1 steroid receptor superfamily member in the control of *Drosophila* metamorphosis. *Proc. Natl. Acad. Sci. U.S.A.* **90,** 3004–3008.

Lee, K. (1990). "The Identification and Characterization of Ecdysone Response Elements," Ph.D. thesis. Indiana University, Bloomington.

Lee, M. S., Kliewer, S. A., Provencal, J., Wright, P. E., and Evans, R. M. (1993). Structure of the retinoid X receptor α DNA binding domain: A helix required for homodimeric DNA binding. *Science* **260,** 1117–1121.

Lehmann, M., and Koolman, J. (1988). Ecdysteroid receptors of the blowfly *Calliphora vicinia:* Partial purification and characterization of ecdysteroid binding. *Mol. Cell. Endocrinol.* **57,** 239–249.

Lehmann, M., and Koolman, J. (1989). Ecdysteroid receptors of the blowfly *Calliphora vicinia:* Characterization of DNA binding. *Eur. J. Biochem.* **181,** 577–582.

Lehmann, J. M., Jong, L., Fanjul, A., Cameron, J. F., Lu, X. P., Haefner, P., Dawson, M. I., and Pfahl, M. (1992). Retinoids selective for retinoid X receptor response pathways. *Science* **258,** 1944–1946.

Leid, M., Kastner, P., Lyons, R., Nakshatri, H., Saunders, M., Zacharewski, T., Chen, J.-Y., Staub, A., Garnier, J. M., Mader, S., and Chambon, P. (1992). Purification, cloning, and RXR identity of the HeLa cell factor with which RAR or TR heterodimerizes to bind target sequences efficiently. *Cell (Cambridge, Mass.)* **68,** 377–395.

Leng, X., Tsai, X. Y., O'Malley, B. W., and Tsai, M.-J. (1993). Ligand-dependent conformational changes in thyroid hormone and retinoic acid receptors are potentially enhanced by heterodimerization with retinoic X receptor. *J. Steroid Biochem. Mol. Biol.* **46,** 643–661.

Levin, A. A., Sturzenbecker, L. J., Kazmer, S., Bosakowski, T., Huselton, C., Allenby, G., Speck, J., Kratzeisen, C., Rosenberger, M., Lovey, A., and Grippo, J. F. (1992). 9-*cis*-Retinoic acid stereoisomer binds and activates the nuclear receptor RXRα. *Nature (London)* **355,** 359–361.

Londershausen, M., and Spindler, K.-D. (1985). Uptake and binding of moulting hormones in crayfish. *Am. Zool.* **25,** 187–196.

Luisi, B. F., Xu, W. X., Otwinowski, Z., Freedman, L. P., Yamamoto, K. R., and Sigler, P. B. (1991). Crystallographic analysis of the interaction of the glucocorticoid receptor with DNA. *Nature (London)* **352,** 497–505.

Luo, Y., Amin, J., and Voellmy, R. (1991). Ecdysterone receptor is a sequence-specific transcription factor involved in the developmental regulation of heat shock genes. *Mol. Cell. Biol.* **11,** 3660–3675.

Macdonald, P. N., Dowd, D. R., Nakajima, S., Galligan, M. A., Reeder, M. C., Haussler, C. A., Ozato, K., and Haussler, M. R. (1993). Retinoid X receptors stimulate and 9-*cis* retinoic acid inhibits 1,25-dihydroxyvitamin D_3-activated expression of the rat osteocalcin gene. *Mol. Cell. Biol.* **13,** 5907–5917.

Mader, S., Leroy, P., Chen, J.-J., and Chambon, P. (1993). Multiple parameters control the selectivity of nuclear receptors for their response elements: Selectivity and promiscuity in response element recognition by retinoic acid receptors and retinoid X receptors. *J. Biol. Chem.* **268,** 591–600.

Mangelsdorf, D. J., Umesono, K., Kliewer, S. A., Borgmeyer, U., Ong, E. S., and Evans, R. M. (1991). A direct repeat in the cellular retinol-binding protein type II gene confers differential regulation by RXR and RAR. *Cell (Cambridge, Mass.)* **66,** 555–561.

Mangelsdorf, D. J., Borgmeyer, U., Heyman, R. A., Zhou, J. Y., Ong, E. S., Oro, A. E., Kakizuka, A., and Evans, R. M. (1992). Characterization of three RXR genes that mediate the action of 9-*cis* retinoic acid. *Genes Dev.* **6,** 329–344.

Marks, M. S., Hallenbeck, P. L., Nagata, T., Segars, J. H., Appella, E., Nikodem, V., and Ozato, K. (1992). H-2RIIBP (RXRβ) heterodimerization provides a mechanism for combinatorial diversity in the regulation of retinoic acid and thyroid hormone responsive genes. *EMBO J.* **11,** 1419–1435.

Maróy, P., Dennis, R., Beckers, C., Sage, B. A., and O'Connor, J. D. (1978). Demonstration of an ecdysteroid receptor in a cultured cell line of *Drosophila melanogaster. Proc. Natl. Acad. Sci. U.S.A.* **75,** 6035–6038.

Martín-Blanco, E., and Kornberg, T. B. (1993). DR-78, a novel *Drosophila melanogaster* genomic DNA fragment highly homologous to the DNA-binding domain of thyroid hormone–retinoic acid–vitamin D receptor subfamily. *Biochim. Biophys. Acta* **1216,** 339–341.

Martinez, E., and Wahli, W. (1991). Characterization of hormone response elements. *In* "Nuclear Hormone Receptors: Molecular Mechanisms, Cellular Functions, Clinical Abnormalities" (M. G. Parker, ed.), pp. 125–153. Academic Press, San Diego.

Martinez, E., Givel, F., and Wahli, W. (1991). A common ancestor DNA motif for invertebrate and vertebrate hormone response elements. *EMBO J.* **10,** 263–268.

McPhie, P., Parkison, C., Lee, B. K., and Cheng, S. (1993). Structure of the hormone binding domain of human β1 thyroid hormone nuclear receptor: Is it an α/β barrel? *Biochemistry* **32,** 7460–7465.

Mlodzik, M., Hiromi, Y., Weber, U., Goodman, C. S., and Rubin, G. M. (1990). The *Drosophila seven-up* gene, a member of the steroid receptor gene superfamily, controls photoreceptor cell fates. *Cell (Cambridge, Mass.)* **60,** 211–224.

Nagaya, T., and Jameson, J. L. (1993a). Thyroid hormone receptor dimerization is required for dominant negative inhibition by mutations that cause thyroid hormone resistance. *J. Biol. Chem.* **268,** 15766–15771.

Nagaya, T., and Jameson, J. L. (1993b). Distinct dimerization domains provide antagonist pathways for thyroid hormone receptor action. *J. Biol. Chem.* **268,** 24278–24282.

Nauber, U., Pankratz, J., Kienlin, A., Seifert, E., Klemm, U., and Jäckle, H. (1988). Abdominal segmentation of the *Drosophila* embryo requires a hormone receptor-like protein encoded by the gap gene *knirps. Nature (London)* **336,** 489–492.

Ohno, C. K., and Petkovich, M. (1992). *FTZ-F1β,* a novel member of the *Drosophila* nuclear receptor family. *Mech. Dev.* **40,** 13–24.

Ohno, C. K., Ueda, H., and Petkovich, M. (1994). The *Drosophila* nuclear receptors FTZ-F1α and FTZ-F1β compete as monomers for binding to a site in the *fushi tarazu* gene. *Mol. Cell. Biol.* **14,** 3166–3175.

Oro, A. E., Ong, E. S., Margolis, J. S., Posakony, J. W., McKeown, M., and Evans, R. M. (1988). The *Drosophila* gene *knirps-related* is a member of the steroid-receptor gene superfamily. *Nature (London)* **336,** 493–496.

Oro, A. E., McKeown, M., and Evans, R. M. (1990). Relationship between the product of the *Drosophila ultraspiracle* locus and the vertebrate retinoid X receptor. *Nature (London)* **347,** 298–301.

Oro, A. E., McKeown, M., and Evans, R. M. (1992a). The *Drosophila* retinoid X receptor homolog *ultraspiracle* functions in both female reproduction and eye morphogenesis. *Development* **115,** 449–462.

Oro, A. E., McKeown, M., and Evans, R. M. (1992b). The *Drosophila* nuclear receptors: New insight into the actions of nuclear receptors in development. *Curr. Opin. Genet. Dev.* **2,** 269–274.

Ozyhar, A., and Pongs, O. (1993). Mutational analysis of the interaction between ecdysteroid receptor and its response element. *J. Steroid Biochem. Mol. Biol.* **46,** 135–145.

Ozyhar, A., Kiltz, H.-H., and Pongs, O. (1990). Pyridoxal phosphate inhibits the DNA-binding activity of the ecdysteroid receptor. *Eur. J. Biochem.* **192,** 167–174.

Ozyhar, A., Strangmann-Diekmann, M., Kiltz, H.-H., and Pongs, O. (1991). Characterization of a specific ecdysteorid receptor–DNA complex reveals common properties for invertebrate and vertebrate hormone-receptor/DNA interactions. *Eur. J. Biochem.* **200,** 329–335.

Ozyhar, A., Gries, M., Kiltz, H.-H., and Pongs, O. (1992). Magnetic DNA affinity purification of ecdysteroid receptor. *J. Steroid Biochem. Mol. Biol.* **43,** 629–634.

Palli, S. R., Hiruma, K., and Riddiford, L. M. (1992). An ecdysteroid-inducible *Manduca* gene similar to the *Drosophila* DHR3 gene, a member of the steroid hormone receptor superfamily. *Dev. Biol.* **150,** 306–318.

Perlmann, T., Rangarjan, P. N., Umesono, K., and Evans, R. M. (1993). Determinants for selective RAR and TR recognition of direct repeat HREs. *Genes Dev.* **7,** 1411–1422.

Perrimon, N., Engstrom, L., and Mahowald, A. P. (1985). Developmental genetics of the 2C-D region of the *Drosophila X* chromosome. *Genetics* **111,** 23–41.

Pignoni, F., Baldarelli, R. M., Steingrimsson, E., Diaz, R. J., Patapoutian, A., Merriam, J. R., and Lengyel, J. A. (1990). The *Drosophila* gene *tailless* is expressed at the embryonic termini and is a member of the steroid receptor superfamily. *Cell (Cambridge, Mass.)* **62,** 151–163.

Pongs, O. (1988). Ecdysteroid-regulated gene expression in *Drosophila melanogaster. Eur. J. Biochem.* **175,** 199–204.

Purnell, B. A., and Gilmour, D. S. (1993). Contribution of sequences downstream of the TATA element to a protein–DNA complex containing the TATA-binding protein. *Mol. Cell. Biol.* **13,** 2593–2603.

Purnell, B. A., Emanuel, P. A., and Gilmour, D. S. (1994). TFIID sequence recognition of the initiator and sequences farther downstream in *Drosophila* class II genes. *Genes Dev.* **8,** 830–842.

Qi, J.-S., Desai-Yajnik, V., Greene, M. E., Raaka, B. M., and Samuels, H. H. (1995). The ligand-binding domains of the thyroid hormone/retinoid receptor gene subfamily function in vivo to mediate heterodimerization, gene silencing, and transactivation. *Mol. Cell. Biol.* **15,** 1817–1825.

Riddiford, L. M. (1993). Hormone receptors and the regulation of insect metamorphosis. *Receptor* **3,** 203–209.

Riddiford, L. M., and Truman, J. W. (1993). Hormone receptors and the regulation of insect metamorphosis. *Am. Zool.* **33,** 340–347.

Riddiford, L. M., and Truman, J. W. (1994). Hormone receptors and the orchestration of development during insect metamorphosis. *In* "Perspectives in Comparative Endocrinology" (K. G. Davey, R. E. Peter, and S. S. Tobe, eds.), pp. 389–394. National Research Council of Canada, Ottawa.

Riddihough, G., and Pelham, H. R. B. (1987). An ecdysone response element in the *Drosophila* hsp27 promoter. *EMBO J.* **6,** 3729–3734.

Robinow, S., Talbot, W. S., Hogness, D. S., and Truman, J. W. (1993). Programmed cell death in the *Drosophila* CNS is ecdysone-regulated and coupled with a specific ecdysone receptor isoform. *Development* **119,** 1251–1259.

Rosen, E. D., Beninghof, E. G., and Koenig, R. J. (1993). Dimerization interfaces of thyroid hormone, retinoic acid, vitamin D, and retinoid X receptors. *J. Biol. Chem.* **268,** 11534–11541.

Rothe, M., Nauber, U., and Jäckle, H. (1989). Three hormone receptor-like *Drosophila* genes encode an identical DNA-binding finger. *EMBO J.* **8,** 3087–3094.

Sage, B. A., Tanis, M. A., and O'Connor, J. D. (1982). Characterization of ecdysteroid receptors in cytosol and naive nuclear preparations of *Drosophila* Kc cells. *J. Biol. Chem.* **257,** 6373–6379.

Sage, B. A., Horn, D. H. S., Landon, T. M., and O'Connor, J. D. (1986). Alternative ligands for measurements and purification of ecdysteroid receptors in *Drosophila* Kc cells. *Arch. Insect. Biochem. Physiol.* **3 Suppl.,** 25–33.

Schräder, M. and Carlberg, C. (1994). Thyroid hormone and retinoic acid receptors form heterodimers with retinoid X receptors on direct repeats, palindromes, and inverted palindromes. *DNA Cell Biol.* **13,** 333–341.

Schräder, M., Becker-André, M., and Carlberg, C. (1994a). Thyroid hormone receptor functions as monomeric ligand-induced transcription factor on octameric half-sites. Consequences also for dimerization. *J. Biol. Chem.* **269,** 6444–6449.

Schräder, M., Muller, K. M., Nayeri, S., Kahlen, J.-P., and Carlberg, C. (1994b). Vitamin D_3–thyroid hormone receptor heterodimer polarity directs ligand sensitivity of transactivation. *Nature (London)* **370,** 382–386.

Schuler, G. D., Altschul, S. F., and Lipman, D. J. (1991). A workbench for multiple alignment construction and analysis. *Proteins: Struct. Funct. Genet.* **9,** 180–190.

Segraves, W. A. (1991). Something old, some things new: The steroid receptor superfamily in *Drosophila Cell* (*Cambridge, Mass.*) **67,** 225–228.

Segraves, W. A., and Hogness, D. S. (1990). The *E75* ecdysone-inducible gene responsible for the 75B early puff in *Drosophila* encodes two new members of the steroid receptor superfamily. *Genes Dev.* **4,** 204–219.

Shea, M. J., King, D. L., Conboy, M. J., Mariani, B. D., and Kafatos, F. C. (1990). Proteins that bind to *Drosophila* chorion *cis*-regulatory elements: A new C_2H_2 zinc finger protein and a C_2C_2 steroid receptor-like component. *Genes Dev.* **4,** 1128–1140.

Stone, B. L., and Thummel, C. S. (1993). The *Drosophila* 78C early late puff contains *E78,* an ecdysone-inducible gene that encodes a novel member of the nuclear hormone receptor superfamily. *Cell* (*Cambridge, Mass.*) **75,** 307–320.

Strangmann-Diekmann, M., Klöne, A., Ozyhar, A., Kreklau, F., Kiltz, H.-H., Hedtmann, U., Welzel, P., and Pongs, O. (1990). Affinity labelling of a partially purified ecdysteroid receptor with a bromoacetylated 20-OH-ecdysone derivative. *Eur. J. Biochem.* **189,** 137–143.

Subauste, J. S., Katz, R. W., and Koenig, R. J. (1994). DNA binding specificity and function of retinoid X receptor α. *J. Biol. Chem.* **269,** 30232–30237.

Sutherland, J. D., Kozlova, T., Tzertzinis, G., and Kafatos, F. C. (1995). DHR38: A new partner for *Drosophila* USP suggests a novel role for NGFI-B–type nuclear receptors. *Proc. Natl. Acad. Sci. U.S.A.* **92,** 7966–7970.

Swevers, L., Drevet, J. R., Lunke, M. D., and Iatrou, K. (1995a). Cloning of the ecdysone receptor of the silkmoth *Bombyx mori* and analysis of its expression during follicular cell differentiation. *Insect Biochem. Mol. Biol.* **25,** 857–866.

Swevers, L., Cherbas, L., Cherbas, P., and Iatrou, K. (1995b). *Bombyx* EcR (BmEcR) and *Bombyx* USP (BmCF1) combine to form a functional ecdysone receptor. *Insect Biochem. Mol. Biol.* in press.

Talbot, W. S., Swyryd, E. A., and Hogness, D. S. (1993). *Drosophila* tissues with different metamorphic responses to ecdysone express different ecdysone receptor isoforms. *Cell* (*Cambridge, Mass.*) **73,** 1323–1337.

Thomas, H. E., Stunnenberg, H. G., and Stewart, A. F. (1993). Heterodimerization of the *Drosophila* ecdysone receptor with retinoid X receptor and *ultraspiracle. Nature* (*London*) **362,** 471–475.

Tora, L., and Davidson, I. (1994). Transcriptional regulation by ligand-inducible nuclear receptors. *In* "Transcription: Mechanisms and Regulations" (R. C. Conaway and J. W. Conaway, eds.), Vol. 3, pp. 477–492. Raven, New York.

Truman, J. W., Talbot, W. S., Fahrbach, S. E., and Hogness, D. S. (1994). Ecdysone receptor expression in the CNS correlates with stage-specific responses to ecdysteroids during *Drosophila* and *Manduca* development. *Development* **120,** 219–234.

Tsai, M.-J., and O'Malley, B. W. (1994). Molecular mechanisms of action of steroid/thyroid receptor superfamily members. *Annu. Rev. Biochem.* **63,** 451–486.

Turberg, A., and Spindler, K.-D. (1992). Properties of nuclear and cytosolic ecdysteroid receptors from an epithelial cell line from *Chironomus tentans. J. Insect Physiol.* **38,** 81–91.

Turberg, A., Spindler-Barth, M., Lutz, B., Lezzi, M., and Spindler, K.-D. (1988). Presence of an ecdysteroid-specific binding protein ("receptor") in epithelial tissue culture cells of *Chironomus tentans. J. Insect Physiol.* **34,** 797–803.

Turberg, A., Imhof, M., Lezzi, M., and Spindler, K.-D. (1992). DNA-binding properties of an "ecdysteroid receptor" from epithelia tissue culture cells of *Chironomus tentans. Insect Biochem. Mol. Biol.* **22,** 343–351.

Tzertzinis, G., Malecki, A., and Kafatos, F. C. (1994). BmCF1, a *Bombyx mori* RXR-type receptor related to the *Drosophila ultraspiracle. J. Mol. Biol.* **238,** 479–486.

Umesono, K., and Evans, R. M. (1989). Determinants of steroid gene specificity for steroid/thyroid hormone receptors. *Cell* (*Cambridge, Mass.*) **57,** 1139–1146.

Wallrath, L. L., Burnett, J. B., and Friedman, T. B. (1990). Molecular characterization of the *Drosophila melanogaster* urate oxidase gene, an ecdysone-repressible gene expressed only in the malpighian tubules. *Mol. Cell. Biol.* **10,** 5114–5127.

Williams, G. R., Zavacki, A. M., Harney, J. W., and Brent, G. A. (1994). Thyroid hormone receptor binds with unique properties to response elements that contain hexamer domains in an inverted palindrome arrangement. *Endocrinology* (*Baltimore*) **134,** 1888–1896.

Wing, K. D. (1988). RH 5849, a nonsteroidal ecdysone agonist: Effects on a *Drosophila* cell line. *Science* **241,** 467–469.

Yao, T.-P., Segraves, W. A., Oro, A. E., McKeown, M., and Evans, R. M. (1992). *Drosophila* ultraspiracle modulates ecdysone receptor function via heterodimer formation. *Cell* (*Cambridge, Mass.*) **71,** 63–72.

Yao, T.-P., Forman, B. M., Jiang, Z., Cherbas, L., Chen, J.-D., McKeown, M., Cherbas, P., and Evans, R. M. (1993). Functional ecdysone receptor is the product of *EcR* and *Ultraspiracle* genes. *Nature* (*London*) **366,** 476–479.

Yu, V. C., Delsert, C., Andersen, B., Holloway, J. M., Devary, O. V., Naar, A. M., Kim, S. Y., Boutin, J.-M., Glass, C. K., and Rosenfeld, M. G. (1991). RXRβ: A coregulator that enhances binding of retinoic acid, thyroid hormone, and vitamin D receptors to their cognate response elements. *Cell* (*Cambridge, Mass.*) **67,** 1251–1266.

Yund, M. A., and Fristrom, J. W. (1975). Uptake and binding of β-ecdysone in imaginal discs of *Drosophila melanogaster. Dev. Biol.* **43,** 287–298.

Yund, M. A., King, D. S., and Fristrom, J. W. (1978). Ecdysteroid receptors in imaginal discs of *Drosophila melanogaster. Proc. Natl. Acad. Sci. U.S.A.* **75,** 6039–6043.

Zelhof, A. C., Yao, T.-P., Chen, J. D., Evans, R. M., and McKeown, M. (1995). Seven-up inhibits Ultraspiracle-based signaling pathways *in vitro* and *in vivo. Mol. Cell. Biol.* in press.

Zhang, X.-K., Hoffmann, B., Tran, P. B.-V., Graupner, G., and Pfahl, M. (1992). Retinoid X receptor is an auxiliary protein for thyroid hormone and retinoic acid receptors. *Nature* (*London*) **355,** 441–446.

6

Molecular Aspects of Juvenile Hormone Action in Insect Metamorphosis

LYNN M. RIDDIFORD

Department of Zoology
University of Washington
Seattle, Washington

I. INTRODUCTION

Growth in insects occurs during larval life, then metamorphosis transforms a fully grown larva into a winged, reproductive adult. The change in form is most dramatic in the Holometabola where the vermiform larva is completely unlike the adult. Yet even in the Hemimetabola where the adult body plan is similar to that of the larva, the metamorphic process involves the production of newly differentiated structures, the remodeling of some larval tissues, and the death of others. The extent of the mix of these varies with the insect.

In most insects the epidermis is polymorphic in that the same cells or daughters thereof make sequentially the larval, pupal (if holometabolous), and the adult cuticle. As detailed by Willis (Chapter 7, this volume), this cuticle is comprised mainly of proteins, some of which are stage specific and others which are characteristic of a particular kind of cuticle that may be found at any stage. These epidermal cells also are generally responsible for the external pigmentation of the animal which may change at metamorphosis. Even the specialized dermal glands which secrete a waterproofing shellac over the surface of the insect at the time of ecdysis may change the type of product secreted at metamorphosis (Horwath and Riddiford, 1988). New adult epidermal structures arise from imaginal discs or similar anlage, and their metamorphosis is discussed in detail by Bayer *et al.* (Chapter 9, this volume; see also Wilder and Perrimon, Chapter 10, this volume).

The viscera also must change at metamorphosis. The fat body is used primarily as a nutrient-processing factory during most of larval life when rapid growth is occurring. Then in the final instar of holometabolous insects this fat body must assume a second role of production and storage of proteins and nutrients for use during metamorphosis and in the cases of nonfeeding adults, also for reproductive maturation. In the adult female the fat body manufactures the yolk proteins as well as processes the nutrients to provide energy, whereas it has primarily a processing function in the adult male. Other viscera also change at metamorphosis with the most extensively studied being the nervous system (see Truman, Chapter 8, this volume).

Metamorphosis of the insect thus requires a coordinated set of changes in the various tissues and organs. This coordination is mediated by the hormones ecdysone and juvenile hormone (JH). Basically, the active metabolite of ecdysone, 20-hydroxyecdysone (20E), initiates a sequence of changes associated with molting, and the presence of JH ensures that the ensuing molt will be to another larval stage. In the absence of JH, 20E initiates the metamorphic change characteristic of a particular tissue. Thus, the action of JH is to influence in some way how the genes are regulated by 20E. The question is how?

II. OVERVIEW: "STATUS QUO" ACTION OF JUVENILE HORMONE

Since its discovery by Wigglesworth (1934), JH has both fascinated and perplexed many generations of biologists. It begins to be produced in the late embryo once the corpora allata are formed (Dorn, 1990; Gilbert *et al.,* Chapter 2, this volume), then is present in varying amounts through larval life. Its disappearance at the end of larval life is the prerequisite for metamorphosis (Riddiford, 1994). In the adult the hormone is used again in many insects to regulate reproduction (Koeppe *et al.,* 1985). The juvenile hormone also is involved in the regulation of various types of polymorphisms, ranging from cellular to behavioral (Nijhout and Wheeler, 1981). This chapter focuses on the metamorphic actions of JH.

As Wigglesworth (1934) first noted, a humoral substance from the head of the blood-sucking bug, *Rhodnius prolixus,* was necessary to prevent metamorphosis during the molt triggered by the blood meal. This "inhibitory hormone" was found to be produced by the larval corpora allata (Wigglesworth, 1936) and was then later named "juvenile hormone" (Wigglesworth, 1940b). Although these glands were inactive during the final larval instar, the tissues could still respond to the hormone and underwent supernumerary larval molts when parabiosed to earlier stage molting larvae (Wigglesworth, 1934). Conversely, decapitated first instar larvae parabiosed to molting final instar larvae underwent metamorphosis. In this hemimetabolous insect, there is a gradual progression toward adult characteristics in each larval instar with the most pronounced changes occurring at the molt to the final larval instar, including enlargement of wing pads and changes in the genitalia followed in the succeeding adult molt by full differentiation of the wings and the genitalia along with changes in epidermal bristle patterning and pigmentation (Wigglesworth, 1940a). Through numerous experiments on *Rhodnius* and based on experiments by others on different insects, Wigglesworth (1952) concluded that whether larval or adult characters are produced is "determined by the balance between two opposing hormone formulae: moulting hormone alone, and moulting hormone plus a variable amount of juvenile hormone. The final result is controlled by the relative concentration of the juvenile hormone and by the stage in moulting at which it is introduced into the system."

After Wigglesworth's initial discoveries, numerous studies showed that the larval corpora allata were responsible for preventing metamorphosis in most insects (Wigglesworth, 1972). In the 1950s Williams (1956) serendipitiously found a repository of JH in the abdomens of the wild silkmoth *Hyalophora cecropia.* He also showed that JH secreted by either larval or adult corpora allata implanted into a pupa as it was beginning development into the adult prevented formation of the adult but allowed molting of the pupa into a second pupa (Williams, 1959, 1961). Therefore, he concluded that "the role of juvenile hormone is to modify the cellular reactions to ecdyson [sic]. It appears to do so by opposing progressive differentiation without interfering with growth and molting in an unchanging state. In some unknown manner, it blocks the derepression and decoding of fresh genetic 'information' without interfering with the acting-out of information already at the disposal of the cells" (Williams, 1961). Thus, he termed it the "status quo" hormone (Williams, 1953).

The essential finding of all the studies to date is that implantation of the corpora allata or application of JH or its analogs before a critical period in either the final or late penultimate instar usually results in one or more supernumerary larval molts except for the higher Diptera (Riddiford, 1993, 1994). Conversely, the removal of the corpora allata in early instars leads to precocious metamorphosis, although whether this occurs immediately or in a succeeding molt varies. For instance, in the cockroach *Leucophaea maderae,* an intermediate molt occurs before the adult molt after removal of the corpora allata in early instar larvae (Scharrer, 1946). Similarly, after early chemical allatectomy with precocene, precocious adults are not formed after the first instar and rarely after the second instar (Staal, 1986), indicating that it may take time for the high JH characteristic of late embryos and early instar larvae to decline sufficiently within the tissues to allow metamorphosis.

Metamorphosis in the Holometabola from larva to pupa to adult was thought to depend on a declining JH titer with intermediate levels allowing transformation to the pupa and the absence of JH, the transformation to the adult (Williams, 1961). This hypothesis was based on the finding in *H. cecropia* that the corpora allata were necessary during the pupal molt to prevent precocious adult development of structures formed from imaginal discs. Interestingly, in other lepidopteran species, e.g., *Bombyx mori* (Bounhiol, 1938) and *Galleria mellonella* (Piepho, 1943), normal pupae and adults were formed after the allatectomy of penultimate instar larvae. The basis of these species differences is unknown but could be due either to a difference in tissue requirements for JH at this critical time or to differences in JH metabolism and/or storage in these different insects. It is now known that the corpora allata in *Manduca sexta* produce only the inactive JH acid during the pupal molt (Bhaskaran *et al.,* 1986), but that the eye and wing discs have the methyltransferase so convert the acid to active JH (Sparagana *et al.,* 1985). There is JH as well as JH acid in the hemolymph at this time (Baker *et al.,* 1987), so presumably it is released by the discs.

Pupal commitment of the polymorphic abdominal epidermis (Riddiford, 1976, 1978) and of the nervous system (Weeks and Truman, 1986) in Lepidoptera requires the action of low levels of ecdysteroid in the absence of JH. This ecdysteroid acts on the nervous system to cause the cessation of feeding and the onset of the wandering behavior (Dominick and Truman, 1985) and commits it to pupal development which occurs later in response to the prepupal peak of ecdysteroid. Exposure of the larval abdominal epidermis of *Manduca* to low levels of 20E *in vitro* causes its pupal commitment so that when it is next exposed to a molting concentration of ecdysteroid, it produces a pupal cuticle, irrespective of the presence or absence of JH. Importantly, in both of these systems, the presence of JH during exposure of the tissue to low ecdysteroid prevents the pupal commitment. Thus, it is critical for the larval–pupal transformation in the Lepidoptera to have tissue exposed to low ecdysteroid in the absence of JH followed by exposure to molting levels of ecdysteroid in the presence of JH.

In contrast, in the pupal–adult transformation of both the Lepidoptera and the Coleoptera, there is a steady rise of the ecdysteroid titer that at low levels commits the tissue to adult differentiation, then at higher levels elicits that differentiation (Riddiford, 1985). Second pupae can be produced by JH only if it is given at the beginning of the rise (Williams, 1956), implying that adult commitment occurs very early in response to low levels of ecdysteroid.

In higher Diptera, metamorphosis appears to be regulated by JH differently. The number of larval instars is determinate and cannot be changed by additional JH, although this additional hormone can delay pupariation (Riddiford, 1993). In *Drosophila* most larval tissues die at metamorphosis to be replaced by differentiation products of the imaginal discs (see Bayer *et al.,* Chapter 9, this volume; Wilder and Perrimon, Chapter 10, this volume), the abdominal histoblasts, and imaginal cells of the viscera. The exceptions are the nervous system (see Truman, Chapter 8, this volume) and the Malpighian tubules. Some of the larval fat body persists into the adult, but new adult fat body is also made. Imaginal discs appear immune to the actions of JH except for its apparently permissive effect on proliferation (for a review see Riddiford, 1993). Thus, experimentally one is unable to prevent the initial pupal differentiation of the imaginal discs with excess JH at any time during larval life, and adult differentiation can only be prevented by high dietary levels of a persistent JH analog (Riddiford and Ashburner, 1991). In contrast, the metamorphosis of the abdominal histoblasts can be prevented by JH application at the time of pupariation as the histoblasts begin their proliferative phase (Ashburner, 1970; Postlethwait, 1974). Preliminary studies show that the exogenous JH does not prevent the proliferation of the histoblasts or their subsequent spreading over the abdomen, but does inhibit their subsequent differentiation as well as that of the abdominal musculature (D. Currie and L. M. Riddiford, unpublished observations). Also, the remodeling of the *Drosophila* larval nervous system at metamorphosis is severely affected by high JH in the final larval instar (L. M. Riddiford and J. W. Truman, unpublished observations) as is that in Lepidoptera (Truman and Reiss, 1988).

Despite these differences in the details of the "status quo" action of JH among various insects, the essential feature is that this action is manifest only during exposure of the tissue to ecdysteroids. Therefore, to understand the molecular basis of the action of JH, we need to understand how ecdysteroids act to cause the progressive switching in gene expression that occurs at metamorphosis, whether this be for cell remodeling, new cell differentiation, or cell death.

III. MODULATION OF ECDYSTEROID ACTION BY JUVENILE HORMONE

On a molecular level, metamorphosis consists of an ecdysteroid-induced change in the pattern of gene expression in the absence of JH that includes the permanent inactivation of larval-specific genes, the activation of previously unexpressed genes, and often changes in either or both spatial and temporal expression of nonstage-specific genes. These genes include ones encoding regulatory proteins such as hormone receptors and transcription factors as well as those encoding differentiation-specific proteins. A brief review of these different types of genes in terms of how JH may affect their metamorphic change follows.

A. Ecdysone Receptor

Since the role of the ecdysone receptor (EcR) and the ecdysteroid-induced transcription factors in metamorphosis is covered in detail in Chapter 5 by Cherbas and Cherbas (this volume), only the pertinent points relevant to JH action will be covered here. The EcR is a typical member of the steroid hormone receptor superfamily with a conserved DNA-binding domain (Koelle *et al.,* 1991) and binds as a heterodimer with the *ultraspiracle* protein (USP) to activate transcription (Yao *et al.,* 1992, 1993). Although there is only one gene encoding EcR in *Drosophila* (Koelle *et al.,* 1991), it has three different isoforms which differ in their N-terminal transactivational domains and thus presumably can activate or inactivate different genes (Talbot *et al.,* 1993). Developmental studies of two of these isoforms show that both are found in the embryo, but in larval life the B1 isoform is predominant (Talbot *et al.,* 1993). Then at pupariation (the onset of metamorphosis), larval tissues that will subsequently die have primarily the B1 isoform whereas the imaginal discs and other imaginal cells that are about to begin differentiation contain primarily the A isoform. In each case, the other isoform is present in low amounts. In contrast, the histoblasts and imaginal gut cells which are about to begin proliferation contain only the B1 isoform. Thus, cell fate during metamorphosis is correlated with the combination of EcR isoforms present, and preliminary genetic studies indicate that loss of a particular isoform prevents the proper metamorphic response of only the tissues expressing that isoform. Talbot *et al.* (1993) therefore suggested that a particular EcR isoform or combination thereof is essential to initiate the specific molecular cascade involved in tissue-specific metamorphosis. Tissue-specific responses to ecdysteroid may also depend on the particular EcR binding site used, as has been found for the *Drosophila* ecdysone-induced protein gene *eip28/29* (Andres and Cherbas, 1994).

The larval nervous systems of both *Drosophila* and *Manduca* contain little EcR; then at the onset of metamorphosis all cells acquire high levels of EcR which in *Drosophila* is found to be the B1 isoform (Truman *et al.,* 1994). After pupariation, the B1 isoform rapidly disappears. During adult development, the type and quantity of the EcR isoforms present then are correlated with the fate of the cell, i.e., whether the cell arises anew or whether it is a larval cell that is to be remodeled or is to die at the end of metamorphosis (Robinow *et al.,* 1993; Truman *et al.,* 1994).

In this system where one can look at identified cells, the picture is not a simple switch of isoform at metamorphosis, but rather a changing interplay of different isoforms through development that may be involved in guiding that particular cell's response to ecdysteroid.

Importantly, the appearance of EcR in the nervous system at metamorphosis in *Manduca* is caused by 20E and is prevented by the application of JH (M. Renucci, and J. W. Truman, unpublished observations). In contrast, EcR is present in *Manduca* larval epidermis throughout larval life with higher levels occurring at the time of the larval molt and at the time of pupal commitment due to its induction by the rising ecdysteroid titer (Riddiford and Truman, 1994). Here JH does not prevent the 20E-induced increase in EcR mRNA during either the larval molt or at pupal commitment. One, however, might expect in this polymorphic cell at pupal commitment, that a new isoform of the EcR might appear or that the ratio of isoforms present might change to aid in the regulation of the new genes that are to be expressed at the pupal molt. Similarly, another such change would be expected at the pupal–adult transition. If so, then JH would be expected to prevent these changes.

In amphibians at metamorphosis, there is also a change in the pattern of receptor isoforms (encoded in this case by different genes) for the thyroid hormone which causes metamorphosis (see Tata, Chapter 13, this volume). The α thyroid hormone receptor (TRα) mRNA is present from embryonic stages onward, then at metamorphic climax as the triiodothyronine (T_3) levels increase, TRβ mRNA appears (Yaoita and Brown, 1990). But as with the EcR isoforms, there is tissue specificity in this response. The tail which is destined to be resorbed shows induction of both isoforms, whereas the hind limb which grows out at metamorphosis shows a rapid decline of TRα and little increase in TRβ (Wang and Brown, 1993). The specific role of TRβ as distinct from that of TRα in activation and/or inactivation of genes at metamorphosis is unknown.

Whereas the thyroid hormone promotes amphibian metamorphosis, prolactin has long been thought to be inhibitory and to promote growth (Kikuyama *et al.*, 1993). Yet normally the levels of the hormone are low until metamorphic climax when it increases to high levels for its functions in the adult. Although its normal role in tadpole growth is still uncertain, *in vitro* prolactin stimulates growth of the tadpole tail (reviewed by Kikuyama *et al.*, 1993) and prevents its T_3-induced regression as well as inhibits normal morphogenesis of the limbs (Tata *et al.*, 1991). In the tail it prevents the induction of TRβ by T_3 (Baker and Tata, 1992). This ability of prolactin to prevent the appearance of a new isoform that seems to be critical for metamorphosis can be likened to the possible effects of JH on the switching of the EcR isoforms discussed earlier.

B. Ecdysteroid-Induced Transcription Factors

As first demonstrated by Clever and Karlson (1960) in the salivary glands of *Chironomus tentans,* ecdysone induces rapid puffing at certain sites indicative of mRNA synthesis (for details see Russell and Ashburner, Chapter 3, this volume; Lezzi, Chapter 4, this volume). Similar studies were done in *Drosophila* by Ashburner in the early 1970s leading to a model for ecdysteroid action that postulated that the ecdysone–EcR complex directly activated early genes, the protein products of which then in turn activated late genes and inhibited early genes (Ashburner *et al.,* 1974; Russell and Ashburner, Chapter 3, this volume). These late genes were also thought to be inhibited by the presence of ecdysone. This larval puffing pattern was followed by a different sequence of puffing during the prepupal period requiring first the absence of 20E, then its presence (Richards, 1976a,b). JH had no influence on the appearance of either the early or the late larval puffs, but was able to prevent the ecdysteroid-induced appearance of the late prepupal puffs when given during the requisite ecdysteroid-free period (Richards, 1978).

Studies of the ecdysteroid-induced early genes has shown that many, but not all, are transcription factors (for a review see Andres and Thummel, 1992; see also Cherbas and Cherbas, Chapter 5, this volume). Interestingly, most of these transcription factors are not stage specific but rather appear whenever the ecdysteroid titer rises. Different isoforms due to alternative splicing and/or promoters, however, are induced, depending on concentration and/or duration of exposure to 20E and also on the tissue being studied (Karim and Thummel, 1991, 1992; Palli *et al.,* 1992; Huet *et al.,* 1993; Andres *et al.,* 1993; Jindra *et al.,* 1994). These then in combination with tissue-specific factors and in some cases with the ecdysone–EcR complex itself are thought to direct the cascade of activation/inactivation of the late tissue-specific genes that are the end result of ecdysteroid action (Thummel *et al.,* 1990).

At metamorphosis in *Drosophila,* the nature of the combination of transcription factors induced by ecdysteroid changes. First, a number of new factors encoded by the *Broad-Complex* (*Br-C*) appear in a time- and tissue-dependent manner (DiBello *et al.,* 1991; Guay and Guild,

1991; Karim *et al.,* 1993; Huet *et al.,* 1993; see Bayer *et al.,* Chapter 9, this volume). Although this gene complex is not necessary for survival to the final larval instar, it is critical for metamorphosis. The encoded factors alone or in various combinations are essential for the induction and for the later repression of the genes encoding the salivary glue proteins that fix the puparium in place, for the full induction of other ecdysteroid-induced transcription factors, and for the induction of the late genes in the salivary glands (Karim *et al.,* 1993). Clearly, the *Broad-Complex* plays a key role in orchestrating the onset of metamorphosis. Other stage-specific transcription factors may also appear at either the onset of metamorphosis or later to guide either the pupal or the adult transformation. One such example is the E78A isoform which is induced by ecdysteroid only during adult development (Stone and Thummel, 1993).

How JH influences the appearance of either the *Br-C* factors or of E78A has not been studied. *Br-C* is first activated 6–12 hr after the molt to the third instar (Andres *et al.,* 1993; Huet *et al.,* 1993) by a small rise of ecdysteroid when the JH titer has declined (Sliter *et al.,* 1987; Bownes and Rembold, 1987). First and second instar larval tissue contain a nuclear antigen that cross-reacts with the antibody to the *Manduca* nuclear JH-binding protein JP29 (see Section IV,B; L. M. Riddiford and J. W. Truman, unpublished observations). This staining disappears between 12 and 18 hr after ecdysis to the third instar. Therefore, possibly the decrease of JH and the loss of this nuclear JH-binding protein are necessary for the activation of *Br-C* by the low level of ecdysteroid. Once this complex has been activated, then JH may have little influence on its subsequent expression in analogy to what has been seen with the JH-suppressible hexameric proteins discussed later. This would explain the lack of effects of JH on the early and late larval puffs since the salivary glands that were explanted for these studies already were expressing some of the *Br-C* transcripts. In any case, one might expect that in those tissues in which JH could prevent metamorphosis, new metamorphic transcription factors would not appear if JH were given at the right time.

These findings of new transcription factors that appear during metamorphosis suggest that a new mix of transcription factors induced by ecdysteroid in the absence of JH may be necessary to activate new genes that appear at metamorphosis. Although thus far no one has found a larval-specific ecdysteroid-induced transcription factor which disappears at metamorphosis, the probability of their being one or several seems high. Its lack and/or the appearance of an inhibitory factor at this time could prevent the reactivation of larval-specific genes. A candidate for an inhibitory factor is E78B, an isoform of a member of the steroid hormone receptor superfamily which lacks the N-terminal region and the DNA-binding domain and which appears at metamorphosis (Stone and Thummel, 1993).

To cope with this new combination of transcription factors at metamorphosis, genes that are expressed in two or three different stages have to contain all possible binding sites in their promoter. Alternatively, they can have different stage-specific promoters as is found for the *Drosophila Adh* gene where the switch occurs in the final larval instar (Savakis *et al.*, 1986). Interestingly, this switch is correlated in time with the appearance of the *Broad* complex transcripts (Andres *et al.*, 1993).

C. Stage-Specific Genes: Maintenance of Ongoing Gene Expression by Juvenile Hormone

In the past few years many structural genes have been cloned and their developmental expression studied. This chapter concentrates on a few selected examples in which their response to JH has been characterized. The larval-specific genes of Lepidoptera, such as the ones for the *Manduca* larval cuticle protein LCP14 (Rebers and Riddiford, 1988; Hiruma *et al.*, 1991), the *Manduca* larval epidermal pigment insecticyanin (Riddiford *et al.*, 1990), and the arylphorins of *Manduca* (Webb and Riddiford, 1988) and *Galleria mellonella* (Memmel *et al.*, 1988, 1994), are all expressed during the larval intermolt growth phase. When ecdysteroids initiate the molt, these mRNAs disappear. This disappearance is not a direct action of ecdysteroid, at least in the case of the epidermal genes, since it can be inhibited by the presence of protein synthesis inhibitors (Hiruma *et al.*, 1991; L. M. Riddiford, unpublished observations). Thus, one or several ecdysteroid-induced early proteins apparently are necessary for this suppression. If JH is present at the outset of the ecdysteroid rise, the mRNAs reappear around the time of ecdysis when the ecdysteroid titer is again low. In the final larval instar when the tissues are exposed to a low level of ecdysteroid in the absence of JH and become pupally committed, these mRNAs also disappear. Their larval specificity is then manifest since they do not reappear when the ecdysteroid titer declines. The exception to this general picture is seen in *Galleria* where low levels of arylphorin transcripts are still found at pupation and are only completely suppressed by the ecdysteroid which initiates adult development (Kumaran *et al.,* 1993).

A cDNA for a pupal-specific gene has been isolated from the wings of the wild silkmoth, *Antheraea polyphemus* (Kumar and Sridhara, 1994). This gene is expressed at the time of pupal cuticle formation in both normal pupae and second pupae forming under the influence of exogenous JH, but not in the larva or the adult. Further studies are necessary to determine the nature of the encoded protein and its regulation by ecdysteroid and JH during normal and second pupal development.

D. Metamorphosis-Specific Genes: Prevention of New Gene Expression by Juvenile Hormone

In contrast to the larval intermolt genes described earlier, there are genes which are first expressed late during the final larval instar as a prelude to metamorphosis. Several new larval endocuticular proteins in *Manduca* are synthesized only on the final day of feeding, and their presence in the cuticle correlates with an increased flexural stiffness in preparation for metamorphosis (Wolfgang and Riddiford, 1986, 1987). The mRNAs encoding the 16- to 17-kDa proteins (LCP16/17) of this class appear in response to a low level of 20E in the absence of JH (Horodyski and Riddiford, 1989). This induction by low 20E requires concomitant protein synthesis so is not a direct action of 20E (Riddiford, 1991). Interestingly, exposure to higher concentrations of 20E equivalent to that seen at the time of pupal commitment of the epidermis causes the loss of these mRNAs as *in vivo.* These proteins are not expressed again in either the pupal or the adult abdomen.

In the lepidopteran fat body there are a number of new storage proteins that are synthesized only late in the feeding period of the final larval instar (reviewed by Riddiford, 1995). The mRNAs for these proteins first appear when the JH titer has declined, and experimental manipulations of this titer can change the timing of their appearance. However, once the mRNA appears, JH application can no longer affect their expression. In the cabbage looper, *Trichoplusia ni,* JH given at the outset of the final larval instar suppresses and delays the onset of transcription of one acidic and two basic hexameric hemolymph proteins and one trypsin-like protein that appear at this time (Jones *et al.,* 1993a–c). As the exogenous JH decays, transcript levels increase. Interestingly, after JH treatment, the transcripts for the two basic proteins that appear are not translatable *in vitro,* whereas that for the acidic protein is (Jones *et al.,* 1993a). In none of these cases has it been determined whether the unlocking of the expression of these fat body genes is simply due to the loss of JH or whether it also requires exposure to a low level of ecdysteroid as in the case of LCP16/17. These mRNAs may be expressed during the prepupal period as well. In *Galleria* the permanent cessation of transcription of LHP82 is caused by the ecdysteroid that inititates adult development (Memmel *et al.,* 1994).

In the mealworm *Tenebrio molitor,* two genes encoding proteins only found in sclerotized adult cuticle are first expressed at the onset of adult cuticle deposition during the fall of the ecdysteroid titer that initiates the adult molt (Bouhin *et al.,* 1992a,b; Charles *et al.,* 1992). Application of JH to the newly ecdysed pupa prevented their appearance 6 days later and a second pupa was formed. When JH was given 60 hr later as the ecdysteroid titer was rising, pupal–adult intermediates were formed and one of these adult cuticle genes was expressed normally (the other was not assayed) (Bouhin *et al.,* 1992b). Thus, the early application of JH is able to prevent adult commitment. Once that commitment is made, JH no longer can suppress the transcription of new adult-specific genes.

E. Other Genes Regulated by Juvenile Hormone in Larvae

One gene that is expressed in the epidermis late in every molt encodes dopa decarboxylase (DDC), the enzyme critical for the production of dopamine necessary in sclerotization of the new cuticle (Kraminsky *et al.,* 1980; Hiruma and Riddiford, 1990). Increased levels of this enzyme are required when the new cuticle also must melanize as it does in *Manduca* larvae when JH is absent during the peak of ecdysteroid for a larval molt (Hiruma and Riddiford, 1993). Consequently, maximal DDC mRNA levels are about two-fold higher in larvae lacking JH at this time. DDC regulation by ecdysteroid is complex, but expression of this gene basically requires exposure to a molting peak of ecdysteroid followed by its withdrawal (Hiruma and Riddiford, 1990). Prevention of protein synthesis during exposure to high ecdysteroid leads to the appearance of new DDC mRNA within 4 hr, indicating that an ecdysteroid-induced factor(s) is suppressing its expression (Hiruma *et al.*, 1995). How JH during the critical period of DDC induction by ecdysteroid determines the later levels of its mRNA when the ecdysteroid titers fall is not known. One possibility is that JH influences the quantity of a positive regulatory factor(s) made at this time which in turn modulates the amount of mRNA produced later (Hiruma and Riddiford, 1993).

During the final larval instar in both Lepidoptera and Diptera, JH esterase increases to eliminate JH in preparation for metamorphosis (Roe and Venkatesh, 1990; Campbell *et al.*, 1992). In *ni Trichoplusia* larval fat body this gene is inducible by JH and transcripts appear within 3 hr (Jones *et al.*, 1994). In this case, JH apparently is acting in the absence of ecdysteroid to activate a gene similarly to its effect on vitellogenin gene transcription in *Locusta* adults (Wyatt *et al.*, 1994) and in *Manduca* prepupae (Satyanarayana *et al.*, 1994). Thus, the mode of action may be different from that where it is guiding the actions of ecdysteroid.

F. Effects of Juvenile Hormone in Cell Lines

Ecdysteroid action has been extensively studied in various insect cell lines derived from embryos (for a review see Cherbas *et al.*, 1989) in which early effects include the direct induction of genes of unknown function such as Eip 28/29 (Cherbas *et al.*, 1989) as well as of transcription factors such as E75 and MHR3 (Lan *et al.*, 1996). Later effects include the cessation of proliferation and the appearance of various proteins that contribute to the morphological changes seen prior to cell death. In *Drosophila* Kc cells, the JH analog methoprene had no effect on the immediate induction of Eip 28/29 by 20E, but partially inhibited the commitment to proliferative arrest, the morphological changes, and the later induction of acetylcholinesterase (Cherbas *et al.*, 1989). Full inhibition was not attainable. Interestingly, in the presence of saturating levels of both hormones, JH limited the effects of 20E to those typically seen with low concentrations of 20E (30-fold less than saturating levels). The effect was found not to be due to competition between the two hormones.

Working with the *Drosophila* S3 cell line, Berger *et al.* (1992) showed that 20E activated transcription of a transfected reporter gene having two 23 bp heat shock protein (hsp) 27 ecdysone response elements. This transcription was prevented in a dose-responsive manner by methoprene and also by JH III, but only when the cells were pretreated with JH and the JH remained present during the exposure to 20E. They suggest that the JH–JHR complex either prevents or modifies the binding of EcR to the response element or interferes in a subsequent step involved in transcription of this reporter gene. This inhibitory action of JH is reminiscent of its prevention of ecdysteroid-induced gene switching for metamorphosis. Consequently, further analysis of this system should provide insight into the molecular basis of this action of JH.

IV. JUVENILE HORMONE DELIVERY SYSTEM

A. Hemolymph Juvenile Hormone-Binding Proteins

During larval life JH is continuously secreted by the corpora allata although there are stage-specific fluctuations in the amount released (Tobe and Stay, 1985). Then during or shortly after the molt to the final instar, the glands cease activity as a prelude to metamorphosis. The secreted lipophilic JH is immediately bound by hemolymph proteins and primarily circulates in this bound form (Goodman, 1990). In the hemimetabolous insects, the high molecular weight lipophorins serve as the primary JH carriers (Kanost *et al.,* 1990), and studies with cockroach corpora allata *in vitro* show that lipophorin both keeps the secreted JH in solution and protects it from degradation (Lanzrein *et al.,* 1993). In the holometabolous insects there are specific low molecular weight JH-binding proteins in the hemolymph (hJHBP), the best characterized of which is the one from the tobacco hornworm, *Manduca sexta* (Park *et al.,* 1993). This 32-kDa hJHBP has been cloned and found to represent a new type of hormone-binding protein (Lerro and Prestwich, 1990). Initial studies with the photoaffinity JH II analog suggest that JH binds in a hydrophobic pocket of the hJHBP (Touhara and Prestwich, 1992).

The amount of JH present in the hemolymph at a particular time depends on a balance between secretion and degradation. Degradation occurs by esterases, primarily a JH-specific esterase that appears in high amounts during the late part of the larval molts just before ecdysis and in the final instar larva (Roe and Venkatesh, 1990). The binding to the hemolymph proteins protects JH from this hydrolysis, but the increased amounts of enzyme efficiently convert all that dissociates.

There has been speculation as to whether the hJHBPs are critical for the cellular action of JH (Hidayat and Goodman, 1994). These investigators found that >99% of the JH in the hemolymph during the fourth larval instar of *Manduca* is bound by hJHBP. Since this binding is of high affinity (n*M*) (Park *et al.,* 1993), they suggest that it may also facilitate uptake into the tissues. Few experiments have been done to test this hypothesis. In one *in vitro* experimental paradigm, partially purified hJHBP of *Manduca* had no effect on the concentration of JH I necessary to inhibit the 20E-induced pupal commitment of the epidermis (Mitsui *et al.,* 1979), indicating that the hJHBP was not essential for JH action. Further experiments with pure natural or recombinant hJHBP in this or other systems have not been reported. Whether or not the JH–hJHBP complex might also interact with receptors on the cell membrane and thereby trigger second messenger action is unknown.

B. Cytosolic Juvenile Hormone-Binding Proteins

Many cellular-binding proteins for JH have been reported and partially characterized (Riddiford, 1994a). Membrane JHBPs have been found in *Rhodnius* (Ilenchuk and Davey, 1985) and *Locusta* ovaries (Sevala *et al.,* 1995). Cytosolic JHBPs (cJHBPs) are present in a wide variety of tissues in both larvae and adults (for a review see Riddiford, 1994a) and they have also been found in embryos (Touhara and Prestwich, 1994; Touhara *et al.,* 1994) and brains (King *et al.,* 1994). These cJHBPs range considerably in molecular weight from 25 kDa to over 100 kDa, depending on the insect and the tissue, and show specific binding with affinities between 1 and 67 n*M* (see Riddiford, 1994a; King *et al.,* 1994). Interestingly, the cJHBP in *Met,* the methoprene-resistant *Drosophila* mutant, has a lower affinity for JH than does the wild-type protein (Shemshedini and Wilson, 1990). The cJHBPs may help to solubilize the JH in the cytosol, and they can protect JH from cellular degradation (Touhara and Prestwich, 1994). They may also have other roles such as those of the cellular retinoid-binding proteins which regulate accessibility to the various enzymes and thus direct metabolism of the retinoids and also may act as a reservoir of ligand for the nuclear-binding proteins (Lohnes *et al.,* 1992; Ong *et al.,* 1994). Whether the cJHBPs could also mediate a cytosolic action of JH is unknown. None of the cJHBPs has been sequenced or cloned, although the cJHBP found in *Manduca* eggs is chemically and antigenically similar to the 32-kDa hJHBP (Touhara *et al.*, 1994). Presumably it entered the egg during vitellogenin uptake.

C. Nuclear Juvenile Hormone-Binding Proteins

High affinity nuclear JHBPs have been detected in larval epidermis, larval and adult fat body, and adult brain (Riddiford, 1994; Shemshedini and Wilson, 1993; King *et al.*, 1994). Studies, however, have shown that the one detected in the adult locust fat body is antigenically similar to the hJHBP made by the fat body (Wyatt *et al.*, 1994). In larval *Drosophila* fat body that remains at adult emergence, there are two nuclear proteins which specifically bind the photoaffinity JH analog, EFDA (Shemshedini and Wilson, 1993). One is 85 kDa, the same size as the EFDA-binding protein in the cytosol, whereas the other is 30 kDa and unique to the nucleus. Interestingly, the affinity of the *Diploptera punctata* (cockroach) brain nuclear protein is much higher for JH II than for the natural enantiomer of JH III, the only hormone found in the hemolymph of the cockroach (King *et al.*, 1994).

A 29-kDa nuclear protein (JP29) that binds JH in larval epidermis of *Manduca* (Palli *et al.*, 1990) has been purified and subsequently cloned (Palli *et al.*, 1994). Partially purified recombinant JP29 produced by baculovirus was reported to bind [^{3}H]JH I with an affinity of 10.7 n*M* (Palli *et al.*, 1994). Subsequent experiments have shown that this preparation also contained an esterase that released the tritiated methyl ester group on the [^{3}H]JH I, thus providing a false estimate of the amount of JH I bound (Charles, *et al.*, 1996). When purified free of esterase, no specific binding of JH I was detectable. Yet JH I and other hydrophobic isoprenoids were able to prevent binding of a photoaffinity analog of JH. Therefore, JP29 has only a very low affinity for JH and is not a JH receptor.

Immunoblotting and immunocytochemical studies show that JP29 is present both in the nucleus (Palli *et al.*, 1994) and associated with the insecticyanin pigment granules (Charles *et al.*, 1996). Its high level throughout larval life was shown to be dependent on the continued presence of JH (Palli *et al.*, 1991; Shinoda, Hiruma, and Riddiford, unpublished). At the time of pupal commitment of the epidermis, JP29 disappears as does specific binding of JH by the nucleus (Riddiford *et al.*, 1987). Its presence throughout larval life, its loss at pupal commitment coinciding with the loss of responsiveness to JH, and its dependence on the presence of JH all suggest that it may play a role in JH action in the larva. One possibility is that of an intracellular sink for JH.

V. THEORIES OF MOLECULAR ACTION OF JUVENILE HORMONE

Switches in gene expression are common in development. In viruses and prokaryotes, they may be quite simple. For example, the switch from early to late gene expression of the SV 40 (simian virus 40) virus which occurs after viral replication begins is due to the binding of all available transcriptional repressor complex produced by the host cell as the viral copy number increases to high levels (Wiley *et al.,* 1993). Interestingly, one component of this repressor complex in HeLa cells is the human estrogen-related receptor 1, an orphan member of the steroid hormone superfamily. In bacteria, switches from growth to sporulation require new sigma factors which bind to RNA polymerase and direct it to new promoters (Losick and Stragier, 1992). Similarly, in eukaryotes, the main switch may also be simple. For instance, in the slime mold, *Dictyostelium,* high continuous cAMP causes the switch from aggregation to differentiation, and one of the essential components of this switch is the production of a new transcription factor which can bind to the promoters of many of the late genes leading to their activation (Schnitzler *et al.,* 1994).

Chromatin structure is also important in gene activity in eukaryotes since it determines accessibility of the gene to the transcription apparatus. Therefore, gene switching also must include chromatin remodeling so that genes are either repressed or activated (Becker, 1994; Paranjape *et al.,* 1994). This remodeling involves proteins that function globally as activators such as the SWI/SNF proteins (Carlson and Laurent, 1994) or as repressors such as the *Polycomb* group of proteins (Pirrotta and Rastelli, 1994) as well as histones and other chromosomal proteins. Interestingly, SWI proteins are essential for glucocorticoid activation of gene expression in both yeast and *Drosophila* cells and seem to be involved in an interaction with the glucocorticoid receptor itself (Yoshinaga *et al.,* 1992). The switching of chick globin gene expression during development also involves changes in chromatin structure in addition to both qualitative and quantitative changes in the transcription factors present in a particular stage (Felsenfeld, 1993).

Despite the advances in our knowledge of ecdysteroid action in causing gene switching at metamorphosis, the molecular action of JH in preventing this switch remains an enigma. In the stimulation of vitellogenin synthesis in *Locusta,* JH is critical both for the initial preparation of the fat body for this major burst of protein synthesis (i.e., polyploidization of the cells and the increase in protein synthesis machinery) and for the increase in vitellogenin mRNA (Wyatt, 1988). Even in the secondary response where the fat body is fully prepared, the vitellogenin mRNA does not appear for at least 12 hr after exposure to JH and protein synthesis is required (Edwards *et al.,* 1993), indicating that JH may not act directly on the gene. At least one new protein that binds to the promoter region of the vitellogenin gene appears in response to JH (Braun and Wyatt, 1992), but the role of this protein is unknown.

Possibly the JH–JHR complex may not influence transcription directly by binding to DNA regulatory regions as does the Ec–EcR complex. Instead it may interact with other transcription factors important in the regulation of ongoing larval gene transcription during the intermolt. Alternatively, it could be involved in protein–protein interactions that stabilize the open chromatin configuration around these larval intermolt genes so that they can be expressed. The action of ecdysteroid during a larval molt in the presence of JH would then only cause the transient suppression of these genes. When JH was absent, the low commitment level of ecdysteroid would cause a destabilization of the chromatin to open up new regions that contain previously unexpressed genes and a reorganization of those regions around larval-specific genes so as to render them inaccessible for transcription. Therefore, new genes, including those for transcription factors as well as the stage-specific structural genes, can be expressed when the ecdysteroid next rises. The same type of chromatin change at a local level could also account for the switch from one promoter to another at metamorphosis such as in the EcR gene and in the *Adh* gene.

Other possible effects of JH could be on post-transcriptional processing or on translation. The ability of applied JH to prevent the translation of the transcripts for the basic hexameric proteins of *Trichoplusia* (Jones *et al.,* 1993a,b) after it no longer affected their gene transcription supports a possible role at this level. In *Manduca* epidermis ribosomal RNA synthesis ceases at the time of pupal commitment due to the action of 20E in the absence of JH, then new rRNA synthesis begins about a day before the prepupal rise in ecdysteroid (Shaaya and Riddiford, 1988). Is it then possible that production of a new type of ribosome is necessary for the translation of new mRNAs in a metamorphosing polymorphic cell? Stage-specific switches occur in the 18s ribosomal RNA genes of the malarial parasite *Plasmodium berghei* (Gunderson *et al.,* 1987) and in the ribosomal protein composition in *Dictyostelium* (Ramagopal, 1992), but the reasons for these switches are not understood.

All the effects of JH in directing ecdysteroid action so as to prevent metamorphosis may not be nuclear. It is now clear that in mammalian systems that extracellular signaling molecules such as peptide hormones, growth factors, and cytokines can affect transcription in the nucleus by activating membrane receptors that in turn activate various types of protein kinases which then phosphorylate transcription factors either directly or through a cascade of kinases (Karin, 1994). Prolactin is one such hormone that works this way in mammary glands through the activation of a Janus kinase which in turn tyrosine phosphorylates a cytoplasmic transcription factor and causes its translocation to the nucleus where it regulates gene expression (Campbell *et al.,* 1994). Inhibitor experiments have implicated protein kinase C in the action of prolactin on thyroxine induction of tail regression in tadpole tails (Petcoff and Platt, 1992). Since as discussed earlier, one action of prolactin is to inhibit the induction of TRβ by the thyroid hormone, it seems likely that one of these kinase signaling pathways may be involved.

JH could also be acting in this manner through a membrane receptor. In *Rhodnius* and *Locusta,* JH acts via a membrane receptor to promote vitellogenin uptake into the oocyte by causing a widening of the intercellular spaces in the follicular epithelium due to an activation of the Na^+, K^+-ATPase and the subsequent shrinkage of the cells (Ilenchuk and Davey, 1987; Davey *et al.,* 1993). Using various inhibitors and mimics, Sevala and Davey (1993) have implicated the phosphatidylinositol pathway and protein kinase C in this action of JH. Using various *Drosophila* mutants, Yamamoto *et al.* (1988) also implicated this same pathway in the stimulation of protein synthesis in the male accessory glands by JH. In this latter system, JH appears to stimulate ribosomal RNA synthesis as well (Shemshedini and Wilson, 1993). In neither system have the initial events in JH action such as the stimulation of phospholipase C or the production of protein kinase C been directly measured.

In summary, metamorphosis in insects is an ecdysteroid-mediated event which involves profound changes in gene expression to render the observed changes in form. These changes are a consequence of the new combinatorial interactions of the various EcR isoforms, the common and stage-specific transcription factors, and the tissue-specific factors that in turn regulate the specific genes. JH prevents these changes by action in the nucleus of an indeterminate nature which we hope will be resolved over the next few years.

ACKNOWLEDGMENTS

I thank Dr. James Truman for many stimulating discussions during the preparation of this chapter and I also thank both him and Dr. Kiyoshi Hiruma for a critical reading of the manuscript. The unpublished studies cited were supported by NSF and NIH.

REFERENCES

Andres, A. J., and Cherbas, P. (1994). Tissue-specific regulation by ecdysone: Distinct patterns of *Eip28/29* expression are controlled by different ecdysone response elements. *Dev. Genet.* **15,** 320–331.

Andres, A. J., and Thummel, C. S. (1992). Hormones, puffs and flies: The molecular control of metamorphosis by ecdysone. *Trends Genet.* **8,** 132–138.

Andres, A. J., Fletcher, J. C., Karim, F. D., and Thummel, C. S. (1993). Molecular analysis of the initiation of insect metamorphosis: A cooperative study of *Drosophila* ecdysone-regulated transcription. *Dev. Biol.* **160,** 388–404.

Ashburner, M. (1970). Effects of juvenile hormone on adult differentiation of *Drosophila melanogaster. Nature (London)* **227,** 187–189.

Ashburner, M., Chihara, C., Meltzer, P., and Richards, G. (1974). Temporal control of puffing activity in polytene chromosomes. *Cold Spring Harbor Symp. Quant. Biol.* **38,** 655–662.

Baker, B. S., and Tata, J. R. (1992). Prolactin prevents the autoinduction of thyroid hormone receptor mRNAs during amphibian metamorphosis. *Dev. Biol.* **149,** 463–467.

Baker, F. C., Tsai, L. W., Reuter, C. C., and Schooley, D. A. (1987). *In vivo* fluctuation of JH, JH acid, and ecdysteroid titer, and JH esterase activity during development of fifth stadium *Manduca sexta. Insect Biochem.* **17,** 989–996.

Becker, P. B. (1994). The establishment of active promoters in chromatin. *BioEssays* **16,** 541–547.

Berger, E. M., Goudie, K., Klieger, L., and DeCato, R. (1992). The juvenile hormone analogue, methoprene, inhibits ecdysterone induction of small heat shock protein gene expression. *Dev. Biol.* **151,** 410–418.

Bhaskaran, G., Sparagana, S. P., Barrera, P., and Dahm, K. H. (1986). Change in corpus allatum function during metamorphosis of the tobacco hornworn *Manduca sexta.* Regulation at the terminal step in juvenile hormone biosynthesis. *Arch. Insect Biochem. Physiol.* **3,** 321–338.

Bouhin, H., Charles, J.-P., Quennedey, B., and Delachambre, J. (1992a). Developmental profiles of epidermal mRNAs during the pupal–adult molt of *Tenebrio molitor* and isolation of a cDNA clone encoding an adult cuticular protein: Effects of a juvenile hormone analogue. *Dev. Biol.* **149,** 112–122.

Bouhin, H., Charles, J.-P., Quennedey, B., Courrent, A., and Delachambre, J. (1992b). Characterization of a cDNA clone encoding a glycine-rich cuticular protein of *Tenebrio molitor.* Developmental expression and effect of a juvenile hormone analogue. *Insect Mol. Biol.* **1,** 53–62.

Bounhiol, J. J. (1938). Recherches experimentales sur le determinisme de la metamorphose chez les Lepidopteres. *Bull. Biol. Fr. Belg., Suppl.* **24,** 1–199.

Bownes, M., and Rembold, H. (1987). The titre of juvenile hormone during the pupal and adult stages of the life cycle of *Drosophila melanogaster. Eur. J. Biochem.* **164,** 709–712.

Braun, R. P., and Wyatt, G. R. (1992). Modulation of DNA-binding proteins in *Locusta migratoria* in relation to juvenile hormone action. *Insect Mol. Biol.* **1,** 99–107.

Campbell, G. S., Argetsinger, L. S., Ihle, J. N., Kelley, P. A., Rillema, J. A., and Carter-Su, C. (1994). Activation of JAK2 tyrosine kinase by prolactin receptors in Nb2 cells and mouse mammary gland explants. *Proc. Natl. Acad. Sci. U.S.A.* **91,** 5232–5236.

Campbell, P. M., Healy, M. J., and Oakeshott, J. T. (1992). Characterisation of juvenile hormone eterase in *Drosophila melanogaster. Insect Biochem. Mol. Biol.* **22,** 665–677.

Carlson, M., and Laurent, B. C. (1994). The SNF/SWI family of global transcriptional activators. *Curr. Opin. Cell Biol.* **6,** 396–402.

Charles, J.-P.. Bouhin, H., Quennedey, B., Courrent, A., and Delachambre, J. (1992). cDNA cloning and deduced amino acid sequence of a major, glycine-rich cuticular protein from the coleopteran *Tenebrio molitor.* Temporal and spatial distribution of the transcript during metamorphosis. *Eur. J. Biochem.* **206,** 813–819.

Charles, J-P., Wojtasek, H., Lentz, A. J., Thomas, B. A., Bonning, B. C., Palli, S. R., Parker, A. G., Dorman, G., Hammock, B. D., Prestwich, G. D., and Riddiford, L. M. (1996). Purification and reassessment of ligand binding by the recombinant, putative juvenile hormone receptor of the tobacco hornworm, *Manduca sexta. Arch. Insect Biochem. Physiol.,* in press.

Cherbas, L., Koehler, M. M., and Cherbas, P. (1989). Effects of juvenile hormone on the ecdysone response of *Drosophila* Kc cells. *Dev. Genet.* **10,** 177–188.

Clever, U., and Karlson, P. (1960). Induktion von Puff-Veränderungen in den Speicheldrüsen-chromosomen von *Chironomus tentans* durch Ecdyson. *Exp. Cell Res.* **20,** 623–626.

Davey, K. G., Sevala, V. L., and Gordon, D. B. (1993). The action of juvenile hormone and antigonadotropin on the follicle cells of *Locusta migratoria. Invertebr. Reprod. Dev.* **24,** 39–46.

DiBello, P. R., Withers, D. A., Bayer, C. A., Fristom, J. W., and Guild, G. M. (1991). The *Drosophila Broad-Complex* encodes a family of related, zinc finger-containing proteins. *Genetics* **129,** 385–397.

Dominick, O. S., and Truman, J. W. (1985). The physiology of wandering behaviour in *Manduca sexta.* II. The endocrine control of wandering behaviour. *J. Exp. Biol.* **117,** 45–68.

Dorn, A. (1990). Embryonic sources of morphogenetic hormones in arthropods. *In* "Morphogenetic Hormones of Arthropods" (A. P. Gupta, ed.), Vol. 2, pp. 3–79. Rutgers University Press, New Brunswick, New Jersey.

Edwards, G. C., Braun, R. P., and Wyatt, G. R. (1993). Induction of vitellogenin synthesis in *Locusta migratoria* by the juvenile hormone analog, pyriproxifen. *J. Insect Physiol.* **39,** 609–614.

Felsenfeld, G. (1993). Chromatin structure and the expression of globin-encoding genes. *Gene* **135,** 119–124.

Goodman, W. G. (1990). Biosynthesis, titer, regulation and transport of juvenile hormones. *In* "Morphogenetic Hormones of Arthropods" (A. P. Gupta, ed.), Vol. 1, pp. 83–124. Rutgers University Press, New Brunswick, New Jersey.

Guay, P. S., and Guild, G. M. (1991). The ecdysone-induced puffing cascade in *Drosophila* salivary glands: A *Broad-Complex* early gene regulates intermolt and late gene transcription. *Genetics* **129,** 169–175.

Gunderson, J. H., Sogin, M. L., Wollett, G., Hollingdale, M., de la Cruz, V. F., Waters, A. P., and McCutchan, T. F. (1987). Structurally distinct, stage-specific ribosomes occur in *Plasmodium. Science* **238,** 933–937.

Hidayat, P., and Goodman, W. G. (1994). Juvenile hormone and hemolymph juvenile hormone binding protein titers and their interaction in the hemolymph of fourth stadium *Manduca sexta. Insect Biochem. Mol. Biol.* **24,** 709–715.

Hiruma, K., and Riddiford, L. M. (1990). Regulation of dopa decarboxylase gene expression in the larval epidermis of the tobacco hornworm by 20-hydroxyecdysone and juvenile hormone. *Dev. Biol.* **138,** 214–224.

Hiruma, K., and Riddiford, L. M. (1993). Molecular mechanisms of cuticular melanization in the tobacco hornworm, *Manduca sexta* (L) (Lepidoptera: Sphingidae). *Int. J. Embryol. Morphol.* **22,** 103–117.

Hiruma, K., Hardie, J., and Riddiford, L. M. (1991). Hormonal regulation or epidermal metamorphosis *in vitro.* Control of expression of a larval-specific cuticle gene. *Dev. Biol.* **144,** 369–378.

Hiruma, K., Carter, M. S., and Riddiford, L. M. (1995). Characterization of the dopa decarboxylase gene of *Manduca sexta* and its suppression by 20-hydroxyecdysone. *Dev. Biol.* **169,** 195–202.

Horodyski, F. M., and Riddiford, L. M. (1989). Expression and hormonal control of a new larval cuticular multigene family at the onset of metamorphosis of the tobacco hornworm. *Dev. Biol.* **132,** 292–303.

Horwath, K. L., and Riddiford, L. M. (1988). Stage and segment specificity of the secretory cell of the dermal glands of the tobacco hornworm, *Manduca sexta. Dev. Biol.* **130,** 365–373.

Huet, F., Ruiz, C., and Richards, G. (1993). Puffs and PCR: The in vivo dynamics of early gene expression during ecdysone responses in *Drosophila. Development* **118,** 613–627.

Ilenchuk, T. T., and Davey, K. G. (1985). The binding of juvenile hormone to membranes of follicle cells in the insect *Rhodnius prolixus. Can. J. Biochem. Cell Biol.* **63,** 102–106.

Ilenchuk, T. T., and Davey, K. G. (1987). Effects of various compounds on Na/K-ATPase activity, JH I binding capacity and patency response in follicles of *Rhodnius prolixus. Insect Biochem.* **17,** 1085–1088.

Jindra, M., Sehnal, F., and Riddiford, L. M. (1994). Isolation, characterization and developmental expression of the ecdysteroid-induced *E75* gene of the wax moth *Galleria mellonella. Eur. J. Biochem.* **221,** 665–675.

Jones, G., Manczak, M., and Horn, M. (1993a). Hormonal regulation and properties of a new group of basic hemolymph proteins expressed during metamorphosis. *J. Biol. Chem.* **268,** 1284–1291.

Jones, G., Venkataraman, V., Manczak, M., and Schelling, D. (1993b). Juvenile hormone action to suppress gene transcription and influence message stability. *Dev. Genet.* **14,** 323–332.

Jones, G., Venkataraman, V., and Manczak, M. (1993c). Transcriptional regulation of an unusual trypsin-related protein expressed during insect metamorphosis. *Insect Biochem. Mol. Biol.* **23,** 825–829.

Jones, G., O'Mahony, P., Schachtschabel, U., and Venkataraman, V. (1994). Juvenile hormone regulation of expression of metamorphosis-associated genes. *In* "Perspectives in Comparative Endocrinology" (K. G. Davey, R. E. Peter, and S. S. Tobe, eds.), pp. 199–201. National Research Council of Canada, Ottawa.

Kanost, M. R., Kawooya, J. K., Law, J. H., Ryan, R. O., Van Heusden, M. C., and Ziegler, R. (1990). Insect haemolymph proteins. *Adv. Insect Physiol.* **22,** 299–396.

Karim, F. D., and Thummel, C. S. (1991). Ecdysone coordinates the timing and amounts of E74A and E74B transcription in *Drosophila. Genes Dev.* **5,** 1067–1079.

Karim, F. D., and Thummel, C. S. (1992). Temporal coordination of regulatory gene expression by the steroid hormone ecdysone. *EMBO J.* **11,** 4083–4093.

Karim, F. D., Guild, G. M., and Thummel, C. S. (1993). The *Drosophila Broad-Complex* plays a key role in controlling ecdysone-regulated gene expression at the onset of metamorphosis. *Development* **118,** 977–988.

Karin, M. (1994). Signal transduction from the cell surface to the nucleus through the phosphorylation of transcription factors. *Curr. Opin. Cell Biol.* **66,** 415–424.

Kikuyama, S., Kawamura, K., Tanaka, S., and Yamamoto, K. (1993). Aspects of amphibian metamorphosis: Hormonal control. *Int. Rev. Cytol.* **1145,** 105–148.

King, L. E., Zhang, J., and Tobe, S. S. (1994). Cytosolic and nuclear juvenile hormone-binding proteins from the brain of *Diploptera punctata. Gen. Comp. Endocrinol.* **94,** 11–22.

Koelle, M. R., Talbot. W. S., Segraves, W. A., Bender, M. T., Cherbas, P., and Hogness, D. S. (1991). The *Drosophila EcR* gene encodes an ecdysone receptor, a new member of the steroid receptor superfamily. *Cell (Cambridge, Mass.)* **67,** 59–77.

Koeppe, J. K., Fuchs, M., Chen, T. T., Hunt, L.-M., Kovalick, G. E., and Briers, T. (1985). The role of juvenile hormone in reproduction. *In* "Comprehensive Insect Physiology, Biochemistry and Pharmacology" (G. A. Kerkut and L. I. Gilbert, eds.), Vol. 8, pp. 165–203. Pergamon, Oxford.

Kraminsky, G. P., Clark, W. C., Estelle, M. A., Gietz, R. D., Sage, B. A., O'Connor, J. D., and Hodgetts, R. B. (1980). Induction of translatable mRNA for dopa decarboxylase in *Drosophila:* An early response to ecdysterone. *Proc. Natl. Acad. Sci. U.S.A.* **77,** 4175–4179.

Kumar, M. N., and Sridhara, S. (1994). Characterization of four pupal wing cuticular protein genes of the silkmoth *Antheraea polyphemus. Insect Biochem. Mol. Biol.* **24,** 291–299.

Kumaran, A. K., Memmel, N. A ., Wang, C., and Trewitt, P. M. (1993). Developmental regulation of arylphorin gene activity in fat body cells and gonadal sheath cells of *Galleria mellonella. Insect Biochem. Mol. Biol.* **23,** 145–151.

Lan, Q., Wu, Z., and Riddiford, L. M. (1996). Induction of ecdysone receptor, Ultraspinacle, E75, and MHR3 mRNAs by 20-hydroxyecdysone in the GV1 cell line of *Manduca sexta. Insect Mol. Biol.,* submitted for publication.

Lanzrein, B., Wilhelm, R., and Riechsteiner, R. (1993). Differential degradation of racemic and 10R-juvenile hormone III by cockroach (*Nauphoeta cinerea*) haemolymph and the use of lipophorin for long-term culturing of corpora allata. *J. Insect Physiol.* **39,** 53–63.

Lerro, K. A., and Prestwich, G. D. (1990). Cloning and sequencing of a cDNA for the hemolymph juvenile hormone binding protein of larval *Manduca sexta. J. Biol. Chem.* **265,** 19800–19808.

Lohnes, D., Dierich, A., Ghyselinck, N., Kastner, P., Lampron, C., LeMeur, M., Lufkin, T., Mendelsohn, C., Nakshatri, H., and Chambon, P. (1992). Retinoid receptors and binding proteins. *J. Cell Sci., Suppl.* **16,** 69–76.

Losick, R., and Stragier, P. (1992). Crisscross regulation of cell-type–specific gene expression during development in *B. subtilis. Nature (London)* **355,** 601–604.

Memmel, N. A., Ray., A., and Kumaran, A. K. (1988). Role of hormones in starvation-induced delay in larval hemolymph protein gene expression in *Galleria mellonella. Wilhelm Roux's Arch. Dev. Biol.* **197,** 496–502.

Memmel, N. A., Trewitt, P. M., Grzelak, K., Rajaratnam, V. S., and Kumaran, A. K. (1994). Nucleotide sequence, structure, and developmental regulation of LHP82, a juvenile hormone-suppressible hexamerin gene from the waxmoth, *Galleria mellonella. Insect Biochem. Mol. Biol.* **24,** 133–144.

Mitsui, T., Riddford, L. M., and Bellamy, G. (1979). Metabolism of juvenile hormone by the epidermis of the tobacco hornworm, *Manduca sexta. Insect Biochem.* **9,** 637–643.

Nijhout, H. F., and Wheeler, D. E. (1982). Juvenile hormone and the physiological basis of insect polymorphism. *Q. Rev. Biol.* **57,** 109–133.

Ong, D. E., Newcomer, M. E., and Chytil, F. (1994). Cellular retinoid-binding proteins. *In* "The Retinoids: Biology, Chemistry, and Medicine" (M. B. Sporn, A. B. Roberts, and D. S. Goodman, eds.), 2nd ed., pp. 283–317. Raven, New York.

Osir, E. O., and Riddiford, L. M. (1988). Nuclear binding of juvenile hormone and its analogs in the epidermis of the tobacco hornworm. *J. Biol. Chem.* **263,** 13812–13818.

Palli, S. R., Osir, E. O., Eng, W.-S., Boehm, M. F., Edwards, M., Kulcsar, P., Ujvary, I., Hiruma, K., Prestwich, G. D,. and Riddiford, L. M. (1990). Juvenile hormone receptors in larval insect epidermis. Identification by photoaffinity labeling. *Proc. Natl. Acad. Sci. U.S.A.* **87,** 796–800.

Palli, S. R., McClelland, S., Hiruma, K., Latli, B., and Riddiford, L. M. (1991). Developmental expression and hormonal regulation of the nuclear 29 kDa juvenile hormone-binding protein in *Manduca sexta* larval epidermis. *J. Exp. Zool.* **260,** 337–344.

Palli, S. R., Hiruma, K., and Riddiford, L. M. (1992). An ecdysteroid-inducible *Manduca* gene similar to the *Drosophila* DHR3 gene, a member of the steroid hormone receptor superfamily. *Dev. Biol.* **150,** 306–318.

Palli, S. R., Touhara, K., Charles, J.-P., Bonning, B. C., Atkinson, J. K., Trowell, S. C., Hiruma, K., Goodman, W. G., Kyriakides, T., Prestwich, G. D., Hammock, B. D., and Riddiford, L. M. (1994). A nuclear juvenile hormone binding protein from larvae of *Manduca sexta:* A putative receptor for the metamorphic action of juvenile hormone. *Proc. Natl. Acad. Sci. U.S.A.* **91,** 6191–6195.

Paranjape, S. M., Kamakaka, R. T., and Kadonaga, J. T. (1994). Role of chromatin structure in the regulation of transcription by RNA polymerase II. *Annu. Rev. Biochem.* **63,** 265–297.

Park, Y. C., Tesch, M. T., Toong, Y. C., and Goodman, W. G. (1993). Affinity purification and binding analysis of the hemolymph juvenile hormone binding protein from *Manduca sexta. Biochemistry* **32,** 7909–7915.

Petcoff, D. W., and Platt, J. E. (1992). Inhibition of protein kinase C antagonizes *in vitro* tail fin regression induced by thyroxine. *Gen. Comp. Endocrinol.* **87,** 208–213.

Piepho, H. (1943). Wirkstoffe in der Metamorphose von Schmetterlingen und anderen Inseketen. *Naturwissenschaften* **31,** 329–335.
Pirrotta, V., and Rastelli, L. (1994). *white* gene expression, repressive chromatin domains and homeotic gene regulation in *Drosophila. BioEssays* **16,** 549–556.
Postlethwait, J. H. (1974). Juvenile hormone and the adult development of *Drosophila. Biol. Bull. (Woods Hole, Mass.)* **147,** 119–135.
Ramagopal, S. (1992). The *Dictyostelium* ribosome: Biochemistry, molecular biology, and developmental regulation. *Biochem. Cell Biol.* **70,** 738–750.
Rebers, J. E., and Riddiford, L. M. (1988). Structure and expression of a *Manduca sexta* larval cuticle gene homologous to *Drosophila* cuticle genes. *J. Mol. Biol.* **203,** 411–423.
Richards, G. (1976a). Sequential gene activation by ecdysone in polytene chromosomes of *Drosophila melanogaster.* IV. The mid prepupal period. *Dev. Biol.* **54,** 256–263.
Richards, G. (1976b). Sequential gene activation by ecdysone in polytene chromosomes of *Drosophila melanogaster.* V. The late prepupal period. *Dev. Biol.* **54,** 264–275.
Richards, G. (1978). Sequential gene activation by ecdysone in polytene chromosomes of *Drosophila melanogaster.* VI. Inhibition by juvenile hormones. *Dev. Biol.* **66,** 32–42.
Riddiford, L. M. (1976). Hormonal control of insect epidermal cell commitment *in vitro. Nature (London)* **259,** 115–117.
Riddiford, L. M. (1978). Ecdysone-induced change in cellular commitment of the epidermis of the tobacco hornworm, *Manduca sexta,* at the initiation of metamorphosis. *Gen. Comp. Endocrinol.* **34,** 438–446.
Riddiford, L. M. (1985). Hormone action at the cellular level. *In* "Comparative Insect Physiology, Biochemistry, and Pharmacology" (G. A. Kerkut and L. I. Gilbert, eds.). Vol. 8, pp. 37–84. Pergamon, Oxford.
Riddiford, L. M. (1991). Hormonal control of sequential gene expression in insect epidermis. *In* "Physiology of the Insect Epidermis" (K. Binnington and A. Retnakaran, eds.), pp. 46–54. Commonwealth Scientific and Industrial Research Organization, East Melbourne, Australia.
Riddiford, L. M. (1993). Hormones and *Drosophila* development. *In* "The Development of Drosophila" (M. Bate and A. Martinez Arias, eds.), pp. 899–939. Cold Spring Harbor Laboratory, Cold Spring Harbor, New York.
Riddiford, L. M. (1994). Cellular and molecular actions of juvenile hormone. I. General considerations and premetamorphic actions. *Adv. Insect Physiol.* **24,** 213–274.
Riddiford, L. M. (1995). Hormonal regulation of gene expression during lepidopteran development. *In* "The Molecular Genetics and Molecular Biology of the Lepidoptera" (M. R. Goldsmith and A. S. Wilkins, eds.), pp. 293–322. Cambridge University Press, Cambridge, England.
Riddiford, L. M., and Ashburner, M. (1991). Role of juvenile hormone in larval development and metamorphosis in *Drosophila melanogaster. Gen. Comp. Endocrinol.* **82,** 172–183.
Riddiford, L. M., and Truman, J. W. (1994). Hormone receptors and the orchestration of development during insect metamorphosis. *In* "Perspectives in Comparative Endocrinology" (K. G. Davey, R. E. Peter, and S. S. Tobe, eds.), pp. 389–394. National Research Council of Canada, Ottawa.
Riddiford, L. M., Osir, E. O., Fittinghoff, C. M., and Green, J. M. (1987). Juvenile hormone analogue binding in *Manduca* epidermis. *Insect Biochem.* **17,** 1039–1043.
Riddiford, L. M., Palli, S. R., Hiruma, K., Li, W.-C., Green, J., Hice, R. H., Wolfgang, W. J., and Webb, B. A. (1990). Developmental expression, synthesis and secretion

of insecticyanin by the epidermis of the tobacco hornworm, *Manduca sexta*. *Arch. Insect Biochem. Physiol.* **14,** 171–190.
Robinow, S., Talbot, W. S., Hogness, D. S., and Truman, J. W. (1993). Programmed cell death in the *Drosophila* CNS is ecdysone-regulated and coupled with a specific ecdysone receptor isoform. *Development* **119,** 1251–1259.
Roe, R. M., and Venkatesh, K. (1990). Metabolism of juvenile hormones: Degradation and titer regulation. *In* "Morphogenetic Hormones of Arthropods" (A. P. Gupta, ed.), Vol. 1, pp. 126–179. Rutgers University Press, New Brunswick, New Jersey.
Satyanarayana, K., Bradfield, J. Y., Bhaskaran, G., and Dahm, K. H. (1994). Stimulation of vitellogenin production by methoprene in prepupae and pupae of *Manduca sexta*. *Arch. Insect Biochem. Physiol.* **25,** 21–37.
Savakis, C., Ashburner, M., and Willis, J. H. (1986). The expression of the gene coding for alcohol dehydrogenase during the development of *Drosophila melanogaster*. *Dev. Biol.* **114,** 194–207.
Scharrer, B. (1946). The role of the corpora allata in the development of *Leucophaea maderae* (Orthoptera). *Endocrinology* (*Baltimore*) **38,** 35–45.
Schnitzler, G. R., Fischer, W. H., and Firtel, R. A. (1994). Cloning and characterization of the G-box binding factor, an essential component of the developmental switch between early and late development in *Dictyostelium*. *Genes Dev.* **8,** 502–514.
Sevala, V. L., and Davey, K. G. (1993). Juvenile hormone dependent phosphorylation of a 100 kDa polypeptide is mediated by protein kinase C in the follicle cells of *Rhodnius prolixus*. *Invertebr. Reprod. Devel.* **22,** 189–193.
Sevala, V. L., Davey, K. G., and Prestwich, G. D. (1995). Photoaffinity labeling and characterization of a juvenile hormone binding protein in the membranes of follicle cells of *Locusta migratoria*. *Insect Biochem. Molec. Biol.* **25,** 267–273.
Shaaya, E., and Riddiford, L. M. (1988). Depressed RNA synthesis in the early stages of the larval–pupal transformation of Lepidoptera: The role of 20-hydroxyecdysone. *J. Insect Physiol.* **34,** 655–659.
Shemshedini, L., and Wilson, T. G. (1990). Resistance to juvenile hormone and an insect growth regulator in *Drosophila* is associated with an altered cytosolic juvenile hormone-binding protein. *Proc. Natl. Acad. Sci. U.S.A.* **87,** 2072–2076.
Shemshedini, L., and Wilson, T. G. (1993). Juvenile hormone binding proteins in larval fat body nuclei of *Drosophila melanogaster*. *J. Insect Physiol.* **39,** 563–569.
Sliter, T. J., Sedlak, B. J., Baker, F. C., and Schooley, D. A. (1987). Juvenile hormone in *Drosophila melanogaster:* Identification and titer determination during development. *Insect Biochem.* **17,** 161–165.
Sparagana, S. P., Bhaskaran, G., and Barrera, P. (1985). Juvenile hormone acid methyltransferase activity in imaginal discs of *Manduca sexta* prepupae. *Arch. Insect Biochem. Physiol.* **2,** 191–202.
Staal, G. B. (1986). Anti juvenile hormone agents. *Annu. Rev. Entomol.* **31,** 391–429.
Stone, B. L., and Thummel, C. S. (1993). The *Drosophila* 78C early late puff contains E78, an ecdysone-inducible gene that encodes a novel member of the nuclear hormone receptor superfamily. *Cell* (*Cambridge, Mass.*) **75,** 1–20.
Talbot, W. S., Swyryd, E. A., and Hogness, D. S. (1993). *Drosophila* tissues with different metamorphic responses to ecdysone express different ecdysone receptor isoforms. *Cell* (*Cambridge, Mass.*) **73,** 1323–1337.
Tata, J. R., Kawahara, A., and Baker, B. S. (1991). Prolactin inhibits both thyroid hormone-induced morphogenesis and cell death in cultured amphibian larval tissues. *Dev. Biol.* **146,** 72–80.

Thummel, C. S., Burtis, K. C., and Hogness, D. S. (1990). Spatial and temporal patterns of *E74* transcription during *Drosophila* development. *Cell (Cambridge, Mass.)* **61,** 101–111.

Tobe, S. S., and Stay, B. (1985). Structure and regulation of the corpus allatum. *Adv. Insect Physiol.* **18,** 305–432.

Touhara, K., and Prestwich, G. D. (1992). Binding site mapping of a photoaffinity-labeled juvenile hormone binding protein. *Biochem. Biophys. Res. Commun.* **182,** 466–473.

Touhara, K., and Prestwich, G. D. (1994). Role of juvenile hormone binding protein in modulating function of JH epoxide hydrolase in eggs of *Manduca sexta. Insect Biochem. Mol. Biol.* **24,** 641–646.

Touhara, K., Soroker, V., and Prestwich, G. D. (1994). Photoaffinity labeling of juvenile hormone epoxide hydrolase and JH-binding proteins during ovarian and egg development in *Manduca sexta. Insect Biochem. Mol. Biol.* **24,** 633–640.

Truman, J. W., and Reiss, S. (1988). Hormonal regulation of the shape of identified motoneurons in the moth *Manduca sexta. J. Neurosci.* **8,** 765–775.

Truman, J. W, Talbot, W. S., Fahrbach, S. E., and Hogness, D. S. (1994). Ecdysone receptor expression in the CNS correlates with stage-specific responses to ecdysteroids during *Drosophila* and *Manduca* development. *Development* **120,** 219–234.

Wang, Z., and Brown, D. D. (1993). Thyroid hormone-induced gene expression program for amphibian tail resorption. *J. Biol. Chem.* **268,** 16270–16278.

Webb, B. A., and Riddiford, L. M. (1988). Regulation of expression of arylophorin and female-specific protein mRNAs in the tobacco hornworm, *Manduca sexta. Dev. Biol.* **130,** 682–692.

Weeks, J. C., and Truman, J. W. (1986). Hormonally mediated reprogramming of muscles and motoneurones during the larval–pupal transformation of the tobacco hornworm, *Manduca sexta. J. Exp. Biol.* **125,** 1–13.

Wigglesworth, V. B. (1934). The physiology of ecdysis in *Rhodnius prolixus.* II. Factors controlling moulting and metamorphosis. *Q. J. Microsc. Sci.* **77,** 191–222.

Wigglesworth, V. B. (1936). The function of the corpora allatum in the growth and reproduction of *Rhodnius prolixus* (Hemiptera). *Q. J. Microsc. Sci.* **79,** 91–121.

Wigglesworth, V. B. (1940a). Local and general factors in the development of "pattern" in *Rhodinus prolixus* (Hemiptera). *J. Exp. Biol.* **17,** 180–200.

Wigglesworth, V. B. (1940b). The determination of characters at metamorphosis in *Rhodnius prolixus* (Hemiptera). *J. Exp. Biol.* **17,** 201–222.

Wigglesworth, V. B. (1952). Hormone balance and the control of metamorphosis in *Rhodnius prolixus* (Hemiptera). *J. Exp. Biol.* **29,** 620–631.

Wigglesworth, V. B. (1972). "The Principles of Insect Physiology," 7th ed. Chapman & Hall, London.

Wiley, S. R., Kraus, R. J., Zuo, F., Murray, E. E., Loritz, K., and Mertz, J. E. (1993). SV40 early-to-late switch involves titration of cellular transcriptional repressors. *Genes Dev.* **7,** 2206–2219.

Williams, C. M. (1953). Morphogenesis and the metamorphosis of insects. *Harvey Lec.* **48,** 126–155.

Williams, C. M. (1956). The juvenile hormone of insects. *Nature (London)* **178,** 212–213.

Williams, C. M. (1959). The juvenile hormone. I. Endocrine activity of the corpora allata of the adult Cecropia silkworm. *Biol. Bull. (Woods Hole, Mass.)* **116,** 323–338.

Williams, C. M. (1961). The juvenile hormone. II. Its role in the endocrine control of molting, pupation, and adult development in the Cecropia silkworm. *Biol. Bull. (Woods Hole, Mass.)* **121,** 572–585.

Wolfgang, W. J., and Riddiford, L. M. (1986). Larval cuticular morphogenesis in the tobacco hornworm, *Manduca sexta,* and its hormonal regulation. *Dev. Biol.* **113,** 305–316.

Wolfgang, W. J., and Riddiford, L. M. (1987). Cuticular mechanics during larval development of the tobacco hornworm. *J. Exp. Biol.* **128,** 19–33.

Wyatt, G. R. (1988). Vitellogenin synthesis and the analysis of juvenile hormone action in locust fat body. *Can. J. Zool.* **66,** 2600–2610.

Wyatt, G. R., Braun, R. P., Edwards, G. C., Glinka, A. V., and Zhang, J. (1994). Juvenile hormone in adult insects: Two modes of action. *In* "Perspectives in Comparative Endocrinology" (K. G. Davey, R. E. Peter, and S. S. Tobe, eds.), pp. 202–208. National Research Council of Canada, Ottawa.

Yamamoto, K., Chadarevian, A., and Pelligrini, A. (1988). Juvenile hormone action mediated in male accessory glands of *Drosophila* by calcium and kinase C. *Science* **239,** 916–919.

Yao, T.-P., Segraves, W. A., Oro, A. E., McKeown, M., and Evans, R. M. (1992). *Drosophila ultraspiracle* modulates ecdysone receptor function via heterodimer formation. *Cell (Cambridge, Mass.).* **71,** 63–72.

Yao, T.-P., Forman, B. M., Jiang, Z., Cherbas, L., Chen, J.-D., McKeown, M., Cherbas, P., and Evans, R. M. (1993). Functional ecdysone receptor is the product of *EcR* and *ultraspiracle* genes. *Nature (London)* **366,** 476–479.

Yaoita, V., and Brown, D. D. (1990). A correlation of thyroid hormone receptor gene expression with amphibian metamorphosis. *Genes Dev.* **4,** 1917–1924.

Yoshinaga, S. K., Peterson, C. L., Herskowitz, I., and Yamamoto, K. R. (1992). Roles of SWI1, SWI2, and SWI3 proteins for transcriptional enhancement by steroid receptors. *Science* **258,** 1598–1603.

7

Metamorphosis of the Cuticle, Its Proteins, and Their Genes

JUDITH H. WILLIS

Department of Cellular Biology
University of Georgia
Athens, Georgia

I. INTRODUCTION

The transformation of a cryptic, sluggish, crawling caterpillar into a brilliantly decorated, vigorously flying butterfly has attracted the attention of philosophers and poets for centuries. The morphological changes in body shape, surface morphology, and texture that accompany insect metamorphosis have been subjected to serious scientific analysis for decades. It is appropriate to set the starting date for modern studies of insect metamorphosis in 1934 with Wigglesworth's discovery that a decapitated first instar nymph of the kissing bug, *Rhodnius prolixus,* will undergo premature metamorphosis if parabiosed to a fed (molt-competent) fifth stage nymph, and that a decapitated fifth (final) stage nymph will have an extra nymphal instar if it is parabiosed to a fed fourth stage nymph with its brain intact. Two years later, Wigglesworth (1936) recognized the role of the corpora allata (that lie just behind the brain) in regulating metamorphosis. Other chapters in this volume describe the action of hormones involved in regulating insect metamorphosis. This chapter focuses on one of their principal targets, integument (cuticle plus epidermis), and describes the morphological and molecular changes it undergoes during metamorphosis.

II. BACKGROUND

A. The Structure of Cuticle Changes during Metamorphosis

For many hemimetabolous insects (those with incomplete metamorphosis), the appearance of new body parts such as wings and genitalia is the most conspicuous external change that signifies that a metamorphic molt has taken place. The surface cuticle has a similar morphology, superficially and histologically, in both larva and adult; but in some hemimetabolous insects, conspicuous differences in cuticle morphology accompany metamorphosis. Indeed, such changes occur in *Rhodinus* where the cuticle goes from having distinct plaques (smooth areas) to closely spaced ripples that traverse each abdominal segment. In the holometabolous insects (those with complete metamorphosis), changes in cuticle morphology can be as dramatic as the changes in body shape that accompany metamorphosis. Instant recognition of the metamorphic stage is generally provided by a brief glimpse of the cuticle. To describe but one example, in the giant North American silk moth, *Hyalophora cecropia,* the cuticle covering the last stage larva is in most regions blue-green, flexible and smooth, studded with brightly colored tubercles that are made of rigid cuticle. Rigid cuticle is also used for the larval head capsule and mouth parts. After the first metamorphic molt, the pupa is shrouded in brown cuticle, rigid in sclerites, flexible in intersegmental membranes. The adult, which emerges after the second metamorphic molt, is covered with red, black, and white scales, embedded in a thin, amber-colored, flexible surface cuticle. Specialized regions of the adult cuticle are rigid, such as the wing hinges and genitalia.

B. A Persistent Population of Epidermal Cells Can Form Cuticle of More Than One Metamorphic Stage

Insect metamorphosis provides one of the best examples of transdifferentiation of a single population of cells. In transdifferentiation a differentiated cell changes its phenotype: first the cell makes (secretes) one product, then it transdifferentiates and forms another. [See Kafatos (1976) for a discussion of transdifferentiation and Willis (1991) for alternative interpretations of some of the Kafatos examples.]

In *H. cecropia,* as in other Lepidoptera, the entire abdomen is covered by a single layer of epidermal cells. These secrete the five larval cuticles; each instar has a distinctive coloration and/or pattern. Then these same cells and their descendants secrete both the pupal cuticle and the adult cuticle. There are no stem cells or replacement cells in the lepidopteran abdominal epidermis; one continuous cell population undergoes metamorphosis. Hence, red, ridged scale cells on the adult abdomen are progeny of the same larval epidermal cells that secreted a green, smooth larval cuticle. Structures in lepidopterans (e.g., wings, legs, eyes) arise from imaginal discs, packets of cells that do not participate in larval cuticle formation but that proliferate and differentiate to form pupal and adult cuticles. Here transdifferentiation is restricted to the pupal/adult molt.

In higher dipterans, such as *Drosophila,* all of the pupal and adult thoracic cuticles and their derivatives arise from imaginal discs (see Bayer *et al.,* Chapter 9, this volume; Wilder and Perrimon, Chapter 10, this volume). Both larval and pupal abdomens are mosaic in origin. Tiny nests of diploid epidermal cells, called histoblasts, participate in secreting larval and pupal abdominal cuticle, although most of these cuticles are formed by enormous polyploid larval epidermal cells. At the onset of adult development, the larval cells die, the histoblasts proliferate and spread, and the adult abdominal cuticle is secreted exclusively by the descendants of the histoblasts.

C. Juvenoids Can Act Directly on Epidermis to Influence Cuticle Morphology

One of the first discoveries that followed the availability of extracts of juvenile hormone (JH) (Williams, 1956) was that low doses of topically applied juvenoids (JH and synthetic analogs) would act only in a localized region. A dramatic example of this was the production of a *Rhodnius* adult with the initials "VBW" formed in larval cuticle on its abdomen. To construct this creature, Wigglesworth (1958) had taken a young final instar *Rhodnius* larva, abraded its cuticle with his initials, and treated the abraded area with a JH preparation. The level of juvenoid was insufficient to prevent metamorphosis except in the area directly treated. At the next molt, the entire epidermis secreted a new cuticle, but it was mosaic in nature, for it had "VBW" made of new larval cuticle surrounded by the normal product of the rest of the epidermal cells, adult cuticle.

Mosaicism was first interpreted as being the result of a mosaic responsiveness of the epidermis caused by gradients of sensitivity across the insect and within a segment. The gradients were postulated to reflect the cellular distribution of hormone receptors or degradative enzymes (Mitsui and Riddiford, 1976). However, studies employing microcautery of small regions of the thorax of the butterfly, *Precis coenia*, suggest that cellular responsiveness may be influenced by a signal produced by cells localized in the midline; passage of the postulated commitment factor can be blocked by wounded cells (Kremen, 1989). The nature and the mode of action of such a commitment signal add a new level of complexity to the analysis of metamorphosis.

D. Action of Juvenoids Is Not All or None

In the "engraved," mosaic *Rhodnius*, cells secreted either larval or adult cuticle. Wigglesworth (1940) had earlier performed an experiment to determine the critical period for juvenoid action which revealed that a single epidermal cell of *Rhodnius* could secrete a cuticle that had both larval and adult features. In many cuticles, the territory secreted by each underlying epidermal cell is clearly delineated on the surface because a different texture of cuticle is formed over intercellular boundaries. Willis *et al.* (1982) subsequently defined such cuticles as "composite" to distinguish them from mosaic cuticles, such as the one bearing the "VBW," where an individual cell secretes either a larval or an adult cuticle. A comprehensive analysis of such composite cuticles was carried out by Lawrence (1969) in the milkweed bug, *Oncopeltus fasciatus.* Lawrence found that distinct features of the cuticle are determined in a stepwise fashion and that injection of juvenoids blocked different features at different times. For example, it was possible to obtain cuticle with adult pigmentation and larval microsculpture, but never the reverse. Furthermore, topical application of juvenoids to the final instar larva could result in complex mosaic cuticles where some cells secreted adult cuticle while adjacent cells formed larval or composite cuticles.

Subsequently, composite cuticles were produced in another bug, *Pyrrhocoris apterus,* as well as in a variety of lepidopteran larvae. Features such as pigmentation, surface sculpturing, the shape of bristles and tubercles, and imprints of cell boundaries let one identify a cuticle as composite in character (Willis *et al.,* 1982). The implications of such composite cuticles are important because they indicate that the switch for a single cell to undergo metamorphosis does not involve a single decision, but rather a series of events. As each event is completed, some aspects of cuticle formation for the next metamorphic stage are irrevocably fixed, while other pattern elements remain sensitive to late applications of juvenoids, resulting in a composite cuticle. By the time cuticle secretion begins, application of juvenoids is without effect, even on pigmentation, a feature that is not manifest until after ecdysis occurs days later (Lawrence, 1969). A simple explanation might be that the underlying event is the synthesis of mRNAs. A pattern element would be insensitive to juvenoids once its mRNA had been synthesized. Information on how these mRNAs are used in a sequential fashion to form the cuticle is still lacking. The only molecular analysis of a composite cuticle revealed a uniform and normal distribution of the mRNA for an adult cuticular protein (Bouhin et al., 1992). Presumably there are also "pupal" proteins contributing to the composite morphology.

III. PROTEINS OF THE CUTICLE

A. Cuticles Contain Many Distinct Proteins

Until the early 1960s, analyses of changes in the cuticle during metamorphosis focused exclusively on cuticle morphology. Changes in the patterns of spots and stripes were described, and the critical periods for preventing metamorphic changes were determined for a variety of species. It gradually became apparent that a complete description of metamorphosis of the integument and its control by ecdysteroids and juvenoids must include analysis of the molecules that specify the morphology.

The cuticle consists of three major classes of molecules: cuticular lipids, chitin, and cuticular proteins. The chitin and proteins interact, in a manner still to be defined, to form cuticle. Lipids are found within and on the surface of the cuticle. The fraction of dry weight of the cuticle contributed by chitin (poly-*N*-acetylglucosamine) may change during metamorphosis, but it is assumed that the enzymes responsible for chitin synthesis remain the same. Although there have been analyses of lipid composition in different metamorphic stages, cuticular lipids represent the end product of arrays of enzymatic reactions, and elucidating the control of changes that accompany metamorphosis is apt to be complex (de Renobales *et al.,* 1991). Thus, cuticular proteins, the third major component, were, by default and because of available technology, the molecules of choice to use to examine the molecular underpinnings of metamorphic changes in cuticular morphology.

It had been known since the 1950s that there was a diversity of proteins within each type of cuticle (Hackman, 1953; for historical reviews see Willis *et al.,* 1981; Willis, 1987). Some can be extracted with water or buffer, but more are solubilized with denaturing agents such as urea, sodium dodecyl sulfate (SDS), or guanidine hydrochloride. A substantial fraction of proteins from a mature cuticle is resistant to extraction by such nondestructive agents. All protein can be removed with 2 *N* NaOH at 100°C for 2 hr (Andersen, 1973), a procedure that enables calculation of the extractable fraction. The fraction of protein that can be extracted varies with cuticle type and developmental stage (Willis *et al.,* 1981; Cox and Willis, 1985). A far greater proportion of the cuticular protein can be extracted from cuticles taken from pharate individuals prior to ecdysis or even immediately following ecdysis, before dehydration ensues or sclerotizing agents have had a chance to work (Andersen, 1973). Unfortunately, as will be discussed next, these pre-ecdysial proteins contain only a subset of the types of protein found in the mature cuticle.

Even though not all cuticular proteins can be extracted, analysis of cuticular proteins allows one to get an indication of shifting patterns of macromolecular syntheses during metamorphosis.

B. Most Cuticular Proteins Are Made by Epidermis

An appropriate goal for students of metamorphosis is to understand changes in gene action that accompany metamorphosis of the cuticle. Hence the source of each cuticular protein must be known so that changes in gene activity can be studied in the appropriate tissue. While most of the cuticular proteins are made by the epidermis, epidermal tissue is not the sole source of cuticular proteins.

Early analyses of cuticular proteins by one-dimensional (1D) electrophoresis revealed many bands with identical electrophoretic mobilities in cuticle extracts and hemolymph. Some proteins with electrophoretic mobility identical to cuticular proteins were even shown to be produced by the fat body, the major source of hemolymph proteins (Palli and Locke, 1988). Even as higher resolution electrophoretic techniques were used, proteins common to both hemolymph and cuticle remained.

Several different experimental techniques have been used to learn which tissues synthesize cuticular proteins. Insect integument was placed in tissue culture with a labeled amino acid and the resulting labeled proteins were extracted from cleaned cuticle and displayed on gels. An early application of this technique using [^{3}H]leucine showed that almost all of the cuticular proteins were synthesized by the epidermis (Willis *et al.*, 1981). Some workers used [^{35}S]methionine as the labeled amino acid and found most of the cuticular proteins labeled. This was an unexpected result. Although methionine is used as the initiating amino acid and may occur elsewhere in the signal peptide, most mature (processed) cuticular proteins lack both of the sulfur-containing amino acids (for details see Andersen *et al.*, 1995). Possibly the finding (Kalinich and McClain, 1992; Browder *et al.*, 1992) that [^{35}S]methionine can donate its label to a variety of amino acids in preformed proteins explains its appearance and certainly casts doubt on all studies done with [^{35}S]methionine, unless the labeled proteins have been hydrolyzed and the label is shown to reside exclusively in methionine.

The origin of cuticular proteins continued to plague workers who knew that there were proteins with identical electrophoretic mobility in hemolymph and cuticular protein extracts. Publications from Locke and associates have clarified the site of synthesis of cuticular proteins and have revealed unexpected complexity (Palli and Locke, 1987; Leung *et al.*, 1989; Sass *et al.*, 1993, 1994).

The key reagents in their work were a battery of specific polyclonal antibodies raised against several individual cuticular protein bands isolated from electrophoretic gels. These antibodies were used in conjunction with immunogold particles to stain ultrathin sections of tissues in order to discover which cells contained the protein within Golgi vesicles, a clear indication that the cells were indeed the site of synthesis. Furthermore, the antibody–immunogold system was used to localize individual proteins within cuticle and tracheae. The antibodies were also used to identify proteins synthesized by isolated tissues and following *in vitro* translation of mRNAs.

By these methods, three broad classes of cuticular proteins were identified. Class C proteins are traditional cuticular proteins, made by the epidermis and secreted across its apical surface directly into the cuticle. Class BD proteins are secreted in both directions, across both apical and basal surfaces of the epidermis. Hence the epidermis can be the source for hemolymph proteins common to both hemolymph and cuticle. Transport from hemolymph to cuticle need not be evoked to explain the joint occurrence. Two well-studied proteins belong to this BD class, arylphorin (Leung *et al.*, 1989) and insecticyanin (Riddiford *et al.*, 1990), a blue pigment found in the epidermis, hemolymph, and cuticle of *Manduca sexta.* Class T proteins appear to be transported to the cuticle for they are not made by the epidermis. The site of synthesis of one of these cuticular proteins has been shown to be spherulocytes, a particular class of hemocyte. Several other proteins found in the cuticle are synthesized by hemocytes as well as by epidermis, and some of these have not been found in the hemolymph (Sass *et al.*, 1994). Sass *et al.* (1994) speculated that some of these "cuticular" proteins made by hemocytes may participate in encapsulation of foreign material within the hemocoel and not be destined for integumentary cuticle. The conclusion to be drawn from these studies is that while most cuticular proteins are synthesized by the epidermis, a few may arise from other sources and be transported via hemolymph to the cuticle.

An additional indication of the complex origin of cuticular proteins is revealed by cuticles formed under pathological conditions (juvenoids or a variety of drugs). They have been shown to contain proteins with identical electrophoretic mobility to some of the hemolymph proteins (Roberts and Willis, 1980c).

Cox and Willis (1985) analyzed cuticular proteins of *H. cecropia* with isoelectric-focusing slab gels. Cuticle and tracheae shared proteins with identical electrophoretic mobility. Sass *et al.* (1993) confirmed this with their more precise immunological identification of particular proteins; 11 of 14 cuticular proteins subjected to detailed analysis were also found in tracheae, these proteins came from all three classes. Since tracheae are associated with all insect tissues, one must be cautious in interpreting the significance of the presence of mRNAs or cuticular proteins from nonintegumental sources.

C. Early and Late Cuticular Proteins Are Secreted within a Metamorphic Stage

Given that the cuticle is composed of morphologically distinct layers secreted in sequence after apolysis (separation of the epidermis from the old cuticle), it was not surprising to learn that even within a metamorphic stage, there is a temporal pattern of cuticular protein synthesis. Distinct pre- and postecdysial proteins have been identified by electrophoresis in *Tenebrio* (Roberts and Willis, 1980a; Andersen, 1975; Lemoine and Delachambre, 1986). Furthermore, when cuticular protein synthesis is measured one sees that some proteins are synthesized before others (Roberts and Willis, 1980b). This program of synthesis occurs within the epidermis; synthesis of specific cuticular proteins has been followed in isolated integuments and imaginal discs at various times in the molt cycle (Doctor *et al.,* 1985; Roter *et al.,* 1985; Hiruma *et al.,* 1991). Also, mRNAs for individual cuticular proteins appear in the epidermis in a temporal pattern (Kimbrell *et al.,* 1988; Fechtel *et al.,* 1989; Riddiford, 1994). In *Rhodnius,* the cuticle ultrastructure changes after the blood meal that initiates molting and a new protein can be extracted from the cuticle (Hillerton, 1978). In *Manduca,* the cuticle changes its morphology and physical properties late in the final instar, on the final day of feeding (Wolfgang and Riddiford, 1986, 1987). Accompanying the decrease in spacing of cuticular lamellae, a new set of cuticular proteins is added to the final larval cuticle (Wolfgang and Riddiford, 1986; Horodyski and Riddiford, 1989). At this time, the cuticle becomes stiffer, presumably as an adaptation for burrowing into the soil. Such a late instar shift in cuticular protein synthesis has not been observed in *H. cecropia* in which the final instar larva surrounds itself with a cocoon.

D. Juvenoids Cause Repetition of Temporal Sequence of Syntheses

As seen earlier, juvenoids cause an insect to make a new cuticle that is a replica of the cuticle of the previous stage. It is this observation that led to calling JH a "status quo" hormone (Williams, 1953). Analyses of cuticular proteins and their patterns of synthesis enable one to learn more about the nature of how completely a new cuticle mimics the previous one. The mealworm, *Tenebrio molitor,* is a useful subject for such studies. Juvenoids applied to young pupae cause the formation of a new, second pupal cuticle. Roberts and Willis (1980b) found that such second pupae synthesize first early, and then late, cuticular proteins, thereby mimicking the pattern of synthesis of a normal pupa. Bouhin *et al.* (1992) showed that an application of juvenoids blocks the appearance of mRNA for an adult-specific cuticular protein. Sridhara has routinely used pupae of *Antheraea polyphemus* molting in the presence of juvenoids as a source of pupal tissue active in synthesizing cuticular proteins. He has confirmed that proteins produced *in vitro* using mRNAs from pharate second pupae are comparable to those obtained from normal pharate pupae (Sridhara, 1993).

E. Cuticular Protein Sequences Reveal Several Multigene Families

Complete sequences are now available for about 40 different cuticular proteins, isolated from seven species belonging to four different orders. Data come from direct sequencing and conceptual translations of cDNAs. The majority of cuticular proteins from all orders sequenced to date show evidence of a motif within a region of hydrophilic amino acids. This motif was first recognized by Rebers and Riddiford (1988) and consists of the pattern $GX_8GX_6YXAXEGYX_7PX_2P$ (X is any amino acid; A, alanine; E, glutamine; G, glycine; P, proline; Y, tyrosine). The consensus is a loose one: phenylanine is frequently substituted for one or both tyrosine residues, the terminal prolines may be absent, and minor differences in spacing for the central AXE may exist. Additional conserved amino acids in this region have been identified. The position of the motif within a protein varies; in some it is central, while in others it lies closer to the amino or carboxy terminus. Once these variations were appreciated, it was possible to recognize the motif in proteins extracted from both flexible and rigid cuticles and from all metamorphic stages of a single species (Lampe and Willis, 1994; Andersen *et al.,* 1995). The existence of this motif suggests that this major group of cuticular proteins shared a common ancestor long before the divergence of insect orders. Its function remains unknown.

Other cuticular proteins also have properties that suggest that they belong to families. There are cuticular proteins with high glycine contents and others with high alanine contents. There are cuticular proteins with diverse internal repeats and specific motifs. Not surprisingly, considering how few cuticular protein sequences are known, there are cuticular proteins that so far have sequences that are unique. A review by Anderson *et al.* (1995) organizes all available sequence information and speculates constructively on the possible significance of various motifs.

F. Protein Composition Reflects Physical Properties of Cuticle More Than Metamorphic Stage

An early hypothesis concerning the regulation of insect metamorphosis postulated that larval, pupal, and adult gene sets exist (Wigglesworth, 1959). Certainly, if such gene sets existed, they would be expected to be found in the cuticular proteins. The available data do not support this hypothesis.

1. Stage-Specific Cuticular Proteins

Numerous reports of cuticular proteins have been found in only a single metamorphic stage, as postulated by the gene set hypothesis. The coleopteran *Pachynoda epphipiata* has a grub-like larva with soft cuticle and a pupa with hard cuticle. Extracts of these cuticles have proteins with different electrophoretic mobilities (Andersen, 1975). The proteins of adult *Tenebrio* differ in electrophoretic mobility from those found in either larvae or pupae (Roberts and Willis, 1980a; Andersen, 1975; Lemoine and Delachambre, 1986). Analyses of several dipterans revealed stage-specific cuticular proteins, including differences among larval instars. For *Drosophila melanogaster,* electrophoretic banding patterns displayed four sets of cuticular proteins: one set is found in first and second instar larvae, a second in the third instar, another in the pupa, and yet another in the adult (Chihara *et al.,* 1982). Similar results come from another dipteran, *Dacus oleae* (Souliotis *et al.,* 1988), and some new proteins appeared in the final larval cuticle of *Lucilia cuprina* (Skelly and Howells, 1987).

There are data for stage-specific cuticular proteins in lepidopterans. In an extensive analysis of the cuticular proteins from *H. cecropia,* Cox and Willis (1985) scored 152 bands on 1D isoelectrofocusing electrophoretic gels. Of these, 7% were unique to the larva, 15% were found only in pupae, and 9% were seen only in adults.

More precise measures of protein distribution than comparison of electrophoretic bands/spots have been employed. Analysis of message distribution for specific cuticular protein genes or cDNAs revealed larval-specific cuticular proteins in *M. sexta* [LCP14 (Rebers and Riddiford, 1988), LCP16/17 (Horodyski and Riddiford, 1989), and insecticyanin (Riddiford *et al.,* 1990)], two adult specific cuticular proteins in *Tenebrio* (Bouhin *et al.,* 1992; Charles *et al.,* 1992), and larval- and pupal-restricted mRNAs in *Drosophila* (Snyder *et al.,* 1981; Apple and Fristrom, 1991).

2. *Cuticular Proteins Used in More Than One Metamorphic Stage*

Although there is now considerable compelling evidence for stage-specific cuticular proteins, many proteins are used in cuticles of more than a single metamorphic stage. This presents a challenge to the gene set hypothesis. Andersen (1973) published similar electropherograms for the SDS-soluble proteins from femurs of freshly ecdysed locust larvae and adults. Also in 1973, it was reported that the amino acid compositions of cuticles from *Tenebrio* larvae and pupae were almost identical (Andersen *et al.,* 1973). While the morphology of larval and pupal *Tenebrio* is distinctive, only subtle differences in lamellar spacing were found in the ultrastructure of their cuticles (Caveney, 1970). Nonetheless it was surprising to learn by this gross analysis of cuticular protein composition that larval and pupal cuticles were so similar. A subsequent examination of the cuticular proteins of *Tenebrio* by SDS electrophoresis was carried out in three different laboratories. The consensus was that almost all bands found in larval cuticles were also present in pupal cuticles (Andersen, 1975; Roberts and Willis, 1980a; Lemoine and Delachambre, 1986).

The identity of cuticular proteins in pharate larval and adult *Tenebrio* cuticles has been verified. Andersen *et al.* (1995) used two-dimensional (2D) electrophoresis and ion-exchange chromatography to compare cuticular proteins, and found nearly identical patterns. They purified several individual proteins and sequenced one (E1) from pupae and part of its larval counterpart. Both larval and pupal E1 had identical molecular masses and the same 47 amino acid residues that were sequenced in both. Thirteen additional larval/pupal protein pairs had identical molecular masses as determined by electrospray ionization mass spectrometry.

Studies with *H. cecropia* revealed that physical properties of the cuticle were a better indicator of what proteins it would contain than the metamorphic stage (Willis and Cox, 1984; Cox and Willis, 1985, 1987). Many of the cuticular proteins isolated from pupal fore-wings and sclerites (rigid cuticles) are identical in electrophoretic mobility on 2D gels to cuticular proteins isolated from larval tubercles or head capsules (also rigid cuticles). No differences were found among cuticular proteins from the flexible larval sclerites and intersegmental membranes. In addition, many of these were identical in electrophoretic mobility to the proteins from pupal and adult intersegmental membranes, and even to some of the proteins of the flexible adult sclerites.

Now that it is known that cuticles with different physical properties have different sets of extractable proteins, proof of stage-specific cuticular proteins requires examination of minor cuticles, with appropriate physical properties, of another stage.

On the other hand, given that proteins with identical electrophoretic mobility on 1D and even 2D gels may still represent products of different genes, definitive proof that a single cuticular protein gene is used in more than a single metamorphic stage requires analysis of genes and messages. Data confirming the results of these electrophoretic analyses are given in Section IV,C.

IV. MESSAGES AND GENES THAT CODE FOR CUTICULAR PROTEINS

A. Distribution of Cuticular Protein Messages Parallels Cuticular Protein Synthesis

The changing pattern of cuticular proteins within a metamorphic stage, as well as differences in protein solubility in different regions and at different times within an instar, prompted some workers to worry that differences in protein distribution reflected differential sclerotization rather than differential synthesis. Hence they postulated that a protein might be restricted to flexible cuticles because, although present in hard cuticles, it was not extractable.

Examination of mRNA distribution eliminated this concern. Binger and Willis (1990) studied the translation products of total mRNAs isolated from different anatomical regions and different metamorphic stages of *H. cecropia.* Protein products were produced in a manner that both translated and processed the protein, allowing mature proteins to be visualized on 2D gels. The distribution of proteins HCCP12 and HCCP66 in the cuticles coincides with the distribution of translation products from mRNA. A tantalizing result not yet resolved is that most of the processed protein products failed to comigrate with cuticular proteins. Perhaps some of these unknown spots represent that fraction of proteins which cannot be extracted.

Subsequent data from Northern analyses have mainly confirmed the restricted appearance of message in epidermis underlying cuticle that contains a particular protein. Messages for HCCP12 were found in epidermis underlying flexible cuticles (such as larval dorsal abdomen and pupal and adult intersegmental membranes), but not in pupal sclerites or larval tubercles (Binger and Willis, 1994). As expected, the message for HCCP66 was found in larval tubercles and pupal sclerites but not in epidermis underlying flexible cuticles (Lampe and Willis, 1994).

There are, however, two exceptions to the generalization of colocalization of epidermal messages and proteins in the overlying cuticle. HCCP66 mRNA has been found to be abundant in pupal hind-wings (Lampe and Willis, 1994). Although a narrow rim of rigid cuticle is made by the hindwing, most of the epidermis secretes a thin cuticle that contains HCCP12, a major protein of flexible cuticles. This thin cuticle is not visible on the surface. It is not yet known whether the appearance of the mRNA for a protein not found in the overlying cuticle represents a case of translational control or if tracheal cells that also secrete HCCP66 are the source. The second exception involves a *Bombyx* cuticular protein found only in pupal integuments, yet the mRNA is abundant in integuments on the first day of the penultimate and final larval instars (Nakato *et al.,* 1992).

Several groups have used *in situ* hybridization to follow the distribution of mRNAs in different regions of the epidermis. The conclusions from these studies are that individual messages for cuticular proteins can have very precise distributions and that the boundary between cells with and without a particular message is as abrupt as the change in the type of cuticle overlying the cells (Horodyski and Riddiford, 1989; Apple and Fristrom, 1991; Bouhin *et al.,* 1992).

B. Common Features Are Found in Cuticular Protein Genes from Diverse Species

In 1981, genomic level analysis of cuticular proteins began with the discovery that a 7.9-kb region from the *Drosophila* genomic library contained the sequences that coded for four of the five major cuticular proteins that had been isolated from late third instar larval cuticles (Snyder *et al.,* 1981, 1982). Forty thousand λ clones with *Drosophila* genomic DNA were screened with late third instar larval integumentary cDNAs and positive clones were then rescreened for strong binding to those cDNAs rather than to embryonic or pupal cDNAs. The resulting positive phage were used in R-looping studies to determine the number of major transcripts. Then *in vitro* translation products from mRNAs that had been hybrid selected with one of these clones and its subfragments were examined by electrophoresis and by precipitation with antibodies that had been raised against *Drosophila* larval cuticular proteins. Final confirmation that the coding sequences for the cuticular proteins were indeed present was obtained by comparing conceptual translations with the N-terminal amino acid sequences of the cuticular proteins. *In situ* hybridization revealed that the gene cluster hybridized to region 44D on chromosome II, precisely where genetic variants of two of the cuticular proteins had previously been mapped. With these comprehensive findings, molecular analysis of cuticular protein genes was launched.

Components and modifications of this pioneering study have been used to isolated additional cuticular protein genes from *Drosophila, Manduca, Hyalophora, Tenebrio,* and *Antheraea.* In some cases, cDNA libraries were prepared from epidermal mRNA isolated at the time of intense cuticular protein synthesis. Other workers purified proteins, obtained N-terminal and sometimes internal peptide sequences, and used these to construct probes or primers from which cDNAs and subsequently genes were obtained. Verification that one had a cuticular protein gene or cDNA involved matching of known sequence data with the conceptual translation of the isolated material or showing that translation products of the cDNA or hybrid selected messages could be precipitated by antibodies raised against cuticular proteins. Sometimes identification relied on finding that conceptual translations matched what are now known as sequences typical of cuticular proteins.

Sequences of genes for about 25 cuticular proteins (plus some homologs in closely related species) have been published (for a comprehensive review see Andersen *et al.,* 1995). Two common features have emerged: linkage and intron position. Some of the genes are found in close proximity to one another. This was first noted for the four larval cuticular protein genes isolated from *Drosophila;* all reside within 7.9 kb that maps to chromosomal region 44D (Snyder *et al.,* 1981, 1982). The *Drosophila* cuticular protein gene *EDG84* appears to be "part of a cluster of genes with related sequences" (Apple and Fristrom, 1991). In *Manduca, LCP16/17* and two relatives have been recognized within 12 kb on one fragment of genomic DNA whereas two other relatives lie on another fragment within 8 kb of one another (Horodyski and Riddiford, 1989). A 10-kb region from *A. polyphemus* codes for two cuticular proteins, not yet sequenced (Kumar and Sridhara, 1994).

Another of the cuticular protein genes in *Drosophila* has been found in an unexpected location, within the first intron of a complex locus that encodes three constitutive enzymes in the purine biosynthetic pathway, the *Gart* locus (Henikoff *et al.,* 1986). A major transcript for a cuticular protein, now called *Gart intron* gene, is derived from the opposite strand to that which codes for the enzymes. Its conceptual translation reveals a protein with a perfect Rebers–Riddiford amino acid consensus. A homologous gene has been identified in the same location in *Drosophila pseudoobscura* (Henikoff and Eghtedarzadeh, 1987).

Another common feature of cuticular protein genes is an intron that interrupts the signal peptide. Genes for all proteins with at least a remnant of the Rebers–Riddiford consensus have an intron in this location. Even some of the cuticular protein genes that show no evidence of belonging to this family, such as the *Drosophila* genes *EDG84* and *EDG91* and the *Bombyx* genes *PCP* and *LCP30,* have an intron interrupting their signal peptide (Apple and Fristrom, 1991; Nakato *et al.,* 1992, 1994). The intron can range from about 60 bp in several *Drosophila* genes (Snyder *et al.,* 1982; Apple and Fristrom, 1991) to over 5.8 kb in *Bombyx PCP* (Nakato *et al.,* 1992). The 3.5-kb intron in *HCCP66* has three partial short interspersed repeated DNA elements, each about 450 nucleotides in length (Lampe and Willis, 1994). It is tantalizing to speculate that this conservatively placed intron contains some critical regulatory sequences. The extensive sequence conservation found in the *Gart* cuticular protein genes from *D. melanogaster* and *D. pseudoobscura* supports this hypothesis (Henikoff and Eghtedarzadeh, 1987). Unfortunately, computer-assisted sequence comparisons have not yet revealed any potential regulatory sequences.

Genes for two cuticular proteins lack an intron interrupting the signal peptide, insecticyanin from *Manduca* (Li and Riddiford, 1992) and *yellow* from *Drosophila* (Geyer *et al.,* 1986; Chia *et al.,* 1986).

Obviously, the justification of using cuticular proteins as indicators of molecular events that accompany metamorphosis implies that we will be able to compare the regulation of their genes during metamorphic and nonmetamorphic molts. The similarities in gene structure, already identified for so many of the cuticular proteins, hold out promise that as regulatory mechanisms are elucidated, the results will be applicable to diverse genes and species.

C. A Single Cuticular Protein Gene Can Be Expressed in More Than One Metamorphic Stage

Several cuticular proteins have been shown to be used to construct a cuticle of more than a single metamorphic stage. The evidence for an absence of stage specificity of a gene product was originally based on identical electrophoretic mobility of pairs of cuticular proteins. Identical mobility could be due to coincidence or a gene duplication event, with each copy used in a different stage. Data discussed here are based on genomic sequence data and analysis of messages, and provide proof that a gene need not be duplicated in order to be expressed in more than one metamorphic stage. Hence segregating genes into stage-specific sets is not necessary for metamorphosis.

Several examples of genes expressed in more than one metamorphic stage come from *Drosophila.* Cuticular protein transcripts from the *Gart intron* gene are prominent in late third instar larvae and in late prepupae; lesser amounts are detected in the early third instar, early prepupa, and pupa (Henikoff and Eghtedarzadeh, 1987). It may be significant that the maximum activity of this cuticular protein gene coincides with maximal transcription of the other *Gart* messages.

The most complex pattern of cuticular protein gene expression analyzed to date is for the *yellow* gene of *Drosophila.* The wild-type allele is essential for normal pigmentation of a variety of cuticles, from larval mouth hooks to diverse adult structures, ranging from the general body surface to individual bristles and hairs. "At least 40 different structures of the adult cuticle can independently express the normal or mutant *y* phenotype" (Nash, 1976). The *yellow* gene has now been cloned and sequenced, and various mutants have been analyzed. The protein sequence of *yellow,* based on conceptual translation of the cloned gene, is not that of a typical structural cuticular protein, for it contains methionine and cysteine residues, amino acids lacking in most other cuticular proteins. The evidence is consistent with a single gene being expressed in both larvae and adults (Geyer *et al.*, 1986; Chia *et al.*, 1986). The basis for the complex expression pattern revealed by mutagenesis has been explained by postulating a diversity of cis-acting factors clustered in the 5′ region of the gene that are controlled by elements that act in a combinatorial fashion (Chia *et al.*, 1986). Mapping of these sites is under way (Geyer and Corces, 1992).

A cDNA has been isolated for a 14.6-kDa *M. sexta* cuticular protein. *In situ* hybridization patterns reveal that the corresponding gene is expressed throughout sclerites and intersegmental membranes in the larva and then only in a narrow zone in intersegmental membranes of pupae and adults (Riddiford, 1991).

Three of the four cuticular proteins genes isolated from *A. polyphemus* are expressed in both pharate pupae and pharate adults (Kumar and Sridhara, 1994).

The isolation of cuticular protein genes in *H. cecropia* was directed toward two proteins that appeared in more than one metamorphic stage. Protein HCCP66 is found in rigid cuticles of larvae and pupae; protein HCCP12 is found in flexible cuticles of all three stages. Genes for both of these proteins have now been isolated and transcripts analyzed. For each, only a single gene is expressed irrespective of the metamorphic stage (Binger and Willis, 1994; Lampe and Willis, 1994).

A gene for *Bombyx* cuticular protein LCP30 was isolated by screening a cDNA expression library with antibody against the protein and using the cDNA to isolate the gene. Immunoblotting and mRNA analyses reveal that the appearance/expression of *Bombyx LCP30* occurs in larval and adult abdomens, but not in pupae or adult wings (Nakato *et al.*, 1994).

Hence, Wigglesworth's (1959) postulate that each metamorphic stage is orchestrated by activity of a separate set of genes is incorrect. Some cuticular proteins are stage specific whereas others are not. Thus, the evolution of novel structures that accompanied the evolution of new developmental stages did not require the duplication of genes in order to package them into stage-specific sets. Metamorphosis of the cuticle, the prime indicator of stage, is not dependent on discrete, stage-specific sets of genes coding for the cuticular structural proteins.

D. Metamorphic Gene Regulation Does Not Require Different Promoters

Williams and Kafatos (1971) presented a prescient model for insect metamorphosis that modified Wigglesworth's gene set concept. They suggested that if stage-specific genes did not exist, then at least there should be stage-specific promoters. Now that gene regulation is beginning to be understood, this idea is not without merit. Stage-specific trans-acting regulatory factors could be envisioned as interacting with discrete cis-acting regulatory regions of cuticular protein genes. A gene could be modeled that used three specific transcription start sites, each controlled by its stage-specific regulatory factors. Indeed, just such a metamorphic-associated regulation is known for several insect genes. In the *Drosophila* fat body there is a switch in promoters for the alcohol dehydrogenase (*ADH*) gene late in the last larval instar (Savakis *et al.,* 1986). Cis-acting regulatory regions for ADH have been defined (for details see Abel *et al.,* 1993). Multiple promoters are used for the early genes in the ecdysteroid cascade (Karim and Thummel, 1992; see Cherbas and Cherbas, Chapter 5, this volume).

All evidence to date suggests that the same transcription start site for cuticular protein genes is used in different metamorphic stages. Most of the evidence is indirect; the same size message is found in different stages (Chia *et al.,* 1986; Nakato *et al.,* 1990; Henikoff and Eghtedarzadeh, 1987; Kumar and Sridhara, 1994). Precise mapping of the transcription start site at different metamorphic stages has only been carried out for three cuticular protein genes. Geyer *et al.* (1986) found identical transcription start sites for *yellow* with mRNA isolated from larvae or pharate adults. For *HCCP12* and *HCCP66,* only a single promoter is used for each gene, irrespective of the metamorphic stage (Binger and Willis, 1994; Lampe and Willis, 1994).

If the same transcription start site can be used in different metamorphic stages, all of the stage-specific regulatory information and all of the tissue-specifying factors must ultimately impact on a single site for initiation of transcription. Sorting out how genes are regulated during metamorphosis will be more complicated than if there were stage-specific promoters and we only needed to isolate the corresponding trans-acting factors.

E. Cuticular Protein Genes Have Ecdysteroid Response Elements and Bind Putative Juvenoid Receptors

An important first step in elucidating the hormonal regulation of metamorphosis will entail learning whether the genes for cuticular proteins are direct or indirect targets for juvenoids and ecdysteroids. While one might anticipate that cuticular protein genes would lie far downstream in the ecdysteroid action cascade (see Cherbas and Cherbas, Chapter 5, this volume), it must be remembered that Natzle (1993) has shown that genes for other structural proteins are primary targets of ecdysteroid action.

Drosophila ecdysteroid response elements (EcREs) have been identified in several of the genes of the ecdysteroid cascade and also in the small heat shock genes that are regulated by ecdysteroids; consensus elements have been described (Riddihough and Pelham, 1987; Cherbas *et al.,* 1991). Complete or partial elements that resemble these EcREs have been located on cuticular protein genes. Two of the pupal cuticular protein genes from *Drosophila* (*EDG78* and *EDG84*) have imperfect matches to an EcRE, including a region in *EDG78* that surrounds the TATA box. These *Drosophila* genes are activated by a pulse of ecdysteroid; if exposed to continuous hormone, no message appears. Accordingly, Apple and Fristrom (1991) speculated that the putative EcREs they have identified may play a repressive role in regulating *EDG78* and *EDG84. H. cecropia* cuticular protein genes *HCCP66* and *HCCP12* also have regions close to the transcription start site with recognizable similarity to EcREs (Lampe and Willis, 1994; Binger and Willis, 1994). Preliminary experiments with extracts of wings (taken from day 2 pharate adults), demonstrated to contain active receptors for ecdysteroids, failed to retard the mobility of a nucleotide segment containing the *HCCP66* putative EcRE (Lampe and Willis, 1994).

Hence there is not yet any direct evidence that EcRE-like regions regulate genes for cuticular proteins. Obviously, further experimentation will be necessary before the level at which ecdysteroids influence the activity of cuticular proteins genes can be understood.

The mode of JH action in regulating metamorphosis is even less clear than with ecdysteroids (for reviews see Riddiford, 1994, and Chapter 6, this volume). The sequence for a putative juvenile hormone receptor has no obvious DNA-binding motifs (Palli *et al.,* 1994), yet a protein with the same molecular mass (29 kDa) as this putative receptor was shown to bind to genes for two *Manduca* cuticular proteins (Palli *et al.,* 1990). Data are not yet available to indicate whether the recombinant protein made from the putative receptor gene also has this binding ability.

Thus although both ecdysteroids and juvenoids obviously play a major role in regulating cuticular protein synthesis, it is not yet known if they exert their action directly on cuticular protein genes. Even if hormone–receptor complexes are shown to bind to cuticular protein genes and drive appropriate reporter constructs, we still need to learn whether this regulation is important in the metamorphic transitions of cuticular protein gene expression. Possibly these hormones are regulating levels of gene expression after metamorphic decisions have been made (for further discussion see Willis, 1990).

In this regard, it is of interest that *Bombyx* larvae exposed to methoprene (a JH analog) still had an unprocessed transcript for *LCP30* (indicative of ongoing transcription) 8 days after ecdysis to the final instar, whereas transcription was low on day 4 and absent on day 8 in normal larvae (Nakato *et al.,* 1994). Here, while juvenoids are obviously sustaining activity of the gene (either directly or indirectly), the juvenoid-preventable decline in mRNA may not be related to metamorphosis since the message also disappears at the end of the penultimate instar.

The distribution of cuticular proteins indicates that gene regulation must involve the use of tissue-specific trans-acting regulatory factors. Thus it was not surprising to find that both a gene from *H. cecropia* (*HCCP66*) (Lampe and Willis, 1994) and *Bombyx* (*PCP*) (Nakato *et al.,* 1992) have response elements for members of the POU family of regulators. These proteins are transcription factors implicated in tissue-specific gene regulation in mammals (Scholer, 1991). That the POU response element "works" was demonstrated for *HCCP66* with gel retardation assays. Active extracts were obtained from a variety of epidermal tissues, cells that were secreting that protein (such as fresh pupal wings), and cells that were not (such as wing discs from larvae). Binding was competed with DNA containing the octamer implicated in binding (Lampe and Willis, 1994). Whether this binding factor may be translocated into the nucleus as part of the scheme for gene regulation or, in fact, whether it plays a role in regulating *HCCP66* is not yet known.

Amazingly, it appears that adult cuticular patterns may be controlled by the same regulatory molecules that work early in embryonic development (Carroll *et al.*, 1994).

V. PERSPECTIVES GAINED FROM ANALYSIS OF CUTICULAR PROTEINS

A. Two Models Are Proposed for Regulation of Cuticular Protein Genes

Analyses of cuticular proteins provide a framework for understanding both the evolution and the regulation of insect metamorphosis. Among the few cuticular protein genes isolated to date, we have already learned that expression patterns for some are stage specific, most are region dependent, some are transcribed in more than one stage, and some are transcribed but not translated. This complex pattern of cuticular protein gene expression might cause the reader to despair of ever sorting out the signals that control appropriate spatial and temporal patterns of activity. Yet insects themselves are able to control metamorphosis with but a single compound, the juvenile hormone. A molt in the presence of juvenoids results in the reformation of the same type of cuticle that had last been made. When ecdysteroids act in the absence of juvenoids, the epidermis becomes "reprogrammed" and forms a cuticle characteristic of the next metamorphic stage. Since juvenoids can prevent a molting insect from undergoing metamorphosis, then "all" we as investigators have to do to understand how the insect moves from the patterns of gene expression of one metamorphic stage to the patterns of the next is to work out the mechanism of action of JH.

There are two major schools of thought for explaining the antimetamorphic action of juvenoids: the direct action school and the memory system school (for an earlier discussion of this dichotomy, see Willis, 1990). All cognoscenti appreciate that the expression of a particular gene requires the input of a diversity of trans-acting regulatory factors that impinge on the control region of that gene. They also anticipate that these factors will act in some sort of combinational manner so that each gene will not require a unique set of factors. The key difference then between the two views (direct action and memory system) is whether juvenoids act to make available specific transcription factors for genes that code for structural proteins or whether they work in a more generic sense to render such genes capable of responding to factors.

1. Direct Action School

Adherents of the direct action school envision that JH regulates metamorphosis directly (in conjunction with a receptor) or indirectly (via a cascade of regulatory proteins that directs the activity of each cuticular protein gene). Possible points for juvenoid regulation might be the production of specific transcription factors. The recognition that the genes in the *Broad-Complex* [early genes in the ecdysteroid cascade (see Bayer *et al.,* Chapter 9, this volume)] are not expressed until a metamorphic molt (Karim *et al.,* 1993) provides a clue as to how such a mechanism might work. The *Broad-Complex* uses nested promoters and differential splicing to produce a group of zinc finger transcription factors (Karim *et al.,* 1993). Devotes of the direct action school might postulate that proteins from the *Broad-Complex* are needed to activate genes for pupal cuticular proteins. These factors might also serve as negative regulators of larval gene expression.

Yet, as shown in this chapter, cuticular protein genes are not segregated into distinct larval, pupal, and adult sets. Although the appearance of some cuticular proteins is restricted to a single metamorphic stage, many are expressed in more than one stage. In *H. cecropia,* a situation has been identified where one cell may express HCCP66, while an immediately adjacent cell expresses HCCP12. At metamorphosis, descendants of both express HCCP66. For most of the sclerite, HCCP66 is a pupal protein. But for cells from the tubercle, easily recognizable in the pupa because they retain the orange, yellow, or blue color they had in the larva, HCCP66 is both a larval and a pupal protein. Now that we know that HCCP66 is made by a single gene, using a single promoter, it is difficult to see how a simple change in the availability of transcription factors can be at the heart of metamorphic regulation.

2. Memory System School

Adherents of the memory system school emphasize that JH acts as a status quo agent, preventing cells from changing their pattern of synthesis. Such action might well occur at a higher level of nuclear organization than the individual gene. Action of juvenoids, then, is postulated to be on the memory system of the cell, the way the cell knows that it last secreted a larval cuticle. Hence, the tubercle cells know that at the last molt they secreted HCCP66, whereas the surrounding cells of the sclerite "remember" that they had secreted HCCP12.

What might that memory mechanism be? We know that the memory signal does not reside in the cuticle, for epidermal cells stripped from the cuticle of diapausing pupae and cultured *in vitro* make a second pupal cuticle in the presence of ecdysteroids and juvenoids and an adult cuticle when just ecdysteroids are present (Willis and Hollowell, 1976). The memory might reside in the state of methylation of genes for cuticular proteins, just as plants and vertebrates use methylation to stabilize gene activity patterns (Bird, 1992). Yet insects have only trace levels of methylated DNA and, with one exception (Field *et al.*, 1989), do not appear to use methylation to regulate the activity of their genes.

Memory might involve other parameters that stabilize gene activity such as chromatin structure. Precedents exist in vertebrates where hemoglobin genes reside in more compacted chromatin in tissues that will never express them (Weintraub and Groudine, 1976; for a discussion of this possibility, see Willis, 1990).

The location of a gene within the nucleus can also affect its availability to transcription factors. For example, several homeotic genes in *Drosophila* remain active in cells only when a protein made by the *polycomb* gene is present. If the *polycomb* gene product is absent, cells first correctly express the homeotic products, but later show a mosaic and inappropriate pattern of expression, and the development of abnormal adults ensues (Paro, 1990). Polycomb stabilization of gene activity is mediated by nuclear compartmentalization, not chromatin organziation (Schlossherr *et al.*, 1994).

If JH prevented genes from moving from one chromatin state to another or one nuclear compartment to another (active to inactive or vice versa), then molts in its presence would be for the status quo. Cells could repeat the pattern of syntheses they had carried out in the previous instar, but would not express genes that had not been previously expressed.

Another possibility exists. Gurdon (1992) suggested that cells might use mRNAs to stabilize a determined state. If molts in the presence of JH prevented the disappearance of such indicator mRNAs, a molt for the status quo would result.

B. What Is Next?

An immediate goal should be to learn how insects can regulate events within an epidermal cell so that a constellation of genes is expressed at each metamorphic stage: some stage specific and some common to one or more metamorphic stages. If the lessons already learned from analyzing cuticular proteins and reviewed in this chapter are kept in mind, then we should be able to design and interpret experiments so that we can, at long last, break the code the insects evolved in the past and are using at present to regulate metamorphosis.

ACKNOWLEDGMENTS

I am grateful to M. K. Farmer, A. Gutkina, A. E. Sluder, and J. S. Willis for their perceptive comments and editorial assistance. Research from my laboratory has been supported by grants from the National Science Foundation and the National Institutes of Health, most recently GM 44511.

REFERENCES

Abel, T., Michelson, A. M., and Maniatis, T. (1993). A *Drosophila* GATA family member that binds to *Adh* regulatory sequences is expressed in the developing fat body. *Development* **119,** 623–633.

Andersen, S. O. (1973). Comparison between the sclerotization of adult and larval cuticle in *Schistocerca gregaria. J. Insect Physiol.* **19,** 1603–1614.

Andersen, S. O. (1975). Cuticular sclerotization in the beetles *Pachynoda epphipiata* and *Tenebrio molitor. J. Insect Physiol.* **21,** 1225–1232.

Andersen, S. O., Chase, A. M., and Willis, J. H. (1973). The amino-acid composition of cuticles from *Tenebrio molitor* with special reference to the action of juvenile hormone. *Insect Biochem.* **3,** 171–180.

Andersen, S. O., Hojrup, P., and Roepstorff, P. (1995). Insect cuticular proteins. *Insect Biochem. Mol. Biol.* **25,** 153–176.

Andersen, S. O., Rafn, K., Krogh, T. N., Hojrup, P., and Roepstorff, P. (1995). Comparison of larval and pupal cuticular proteins in *Tenebrio molitor. Insect Biochem. Mol. Biol.* **25,** 177–187.

Apple, R. T., and Fristrom, J. W. (1991). 20-Hydroxyecdysone is required for, and negatively regulates, transcription of *Drosophila* pupal cuticle protein genes. *Dev. Biol.* **146,** 569–582.

Binger, L. C., and Willis, J. H. (1990). *In vitro* translation of epidermal RNAs from different anatomical regions and metamorphic stages of *Hyalophora cecropia. Insect Biochem.* **20,** 573–583.

Binger, L. C., and Willis, J. H. (1994). Identification of the cDNA, gene and promoter for a major protein from flexible cuticles of the giant silkmoth *Hyalophora cecropia. Insect Biochem. Mol. Biol.* **24,** 989–1000.

Bird, A. (1992). The essentials of DNA methylation. *Cell (Cambridge, Mass.)* **70,** 5–8.

Bouhin, H., Charles, J.-P., Quennedey, B., Courrent, A., and Delachambre, J. (1992). Characterization of a cDNA clone encoding a glycine-rich cuticular protein of *Tenebrio molitor:* Developmental expression and effect of a juvenile hormone analogue. *Insect Mol. Biol.* **1,** 53–62.

Browder, L. W., Wilkes, J., and Rodenhiser, D. I. (1992). Preparative labeling of proteins with [^{35}S]methionine. *Anal. Biochem.* **204,** 85–89.

Carroll, S. B., Gates, J., Keys, D. N., Paddock, S. W., Panganiban, G. E. F., Selegue, J. E., and Williams, J. A. (1994). Pattern formation and eyespot determination in butterfly wings. *Science* **265,** 109–114.

Caveney, S. (1970). Juvenile hormone and wound modelling of *Tenebrio* cuticle architecture. *J. Insect Physiol.* **16,** 1087–1107.

Charles, J.-P., Bouhin, H., Quennedey, B., Courrent, A., and Delachambre, J. (1992). cDNA cloning and deduced amino acid sequence of a major, glycine-rich cuticular protein from the coleopteran *Tenebrio molitor.* Temporal and spatial distribution of the transcript during metamorphosis. *Eur. J. Biochm.* **206,** 813–819.

Cherbas, L., Lee, K., and Cherbas, P. (1991). Identification of ecdysone response elements by analysis of the *Drosophila Eip28/29* gene. *Genes Dev.* **5,** 120–131.

Chia, W., Howes, G., Martin, M., Meng, Y. B., Moses, K., and Tsubota, S. (1986). Molecular analysis of the *yellow* locus of *Drosophila. EMBO J.* **5,** 3597–3605.

Chihara, C. J., Silvert, D. J., and Fristrom, J. W. (1982). The cuticle proteins of *Drosophila melanogaster:* Stage specificity. *Dev. Biol.* **89,** 379–388.

Cox, D. L., and Willis, J. H. (1985). The cuticular proteins of *Hyalophora cecropia* from different anatomical regions and metamorphic stages. *Insect Biochem.* **15,** 349–362.

Cox, D. L., and Willis, J. H. (1987). Analysis of the cuticular proteins of *Hyalophora cecropia* with two dimensional electrophoresis. *Insect Biochem.* **17,** 457–468.

de Renobales, M., Nelson, D. R., and Blomquist, G. J. (1991). Cuticular lipids. *In* "Physiology of the Insect Epidermis" (K. Binnington and A. Retnakaran, eds.), pp. 240–251. Commonwealth Scientific and Industrial Research Organization, East Melbourne, Australia.

Doctor, J., Fristrom, D., and Fristrom, J. W. (1985). The pupal cuticle of *Drosophila:* Biphasic synthesis of pupal cuticle proteins in vivo and in vitro in response to 20-hydroxyecdysone. *J. Cell Biol.* **101,** 189–200.

Fechtel, K., Fristrom, D. K., and Fristrom, J. W. (1989). Prepupal differentiation in *Drosophila:* Distinct cell types elaborate a shared structure, the pupal cuticle, but accumulate transcripts in unique patterns. *Development* **106,** 649–656.

Field, L. M., Devonshire, A. L., Ffrench-Constant, R. H., and Forde, B. G. (1989). Changes in DNA methylation are associated with loss of insecticide resistance in the peach-potato aphid *Myzus persicae* (Sulz.). *FEBS Lett.* **243,** 323–327.

Geyer, P. K., and Corces, V. G., (1992). DNA position-specific repression of transcription by a *Drosophila* zinc finger protein. *Genes Dev.* **6,** 1865–1873.

Geyer, P. K., Spana, C., and Corces, V. G. (1986). On the molecular mechanism of gypsy-induced mutations at the *yellow* locus of *Drosophila melanogaster. EMBO J.* **5,** 2657–2662.

Gurdon, J. B. (1992). The generation of diversity and pattern in animal development. *Cell (Cambridge, Mass.)* **68,** 185–199.

Hackman, R. H. (1953). Chemistry of insect cuticle. 1. The water-soluble proteins. *Biochem. J.* **54,** 362–367.

Henikoff, S., and Eghtedarzadeh, M. K. (1987). Conserved arrangement of nested genes at the *Drosophila Gart* locus. *Genetics* **117,** 711–725.

Henikoff, S., Keene, M. A., Fechtel, K., and Fristrom, J. W. (1986). Gene within a gene: Nested Drosophila genes encode unrelated proteins on opposite DNA strands. *Cell (Cambridge, Mass.)* **44,** 33–42.

Hillerton, J. E. (1978). Changes in the structure and composition of the extensible cuticle of *Rhodnius prolixus* through the 5th larval instar. *J. Insect. Physiol.* **24,** 399–412.

Hiruma, K., Hardie, J., and Riddiford, L. M. (1991). Hormonal regulation of epidermal metamorphosis *in vitro:* Control of expression of a larval-specific cuticle gene. *Dev. Biol.* **144,** 369–378.

Horodyski, F. M., and Riddiford, L. M. (1989). Expression and hormonal control of a new larval cuticular multigene family at the onset of metamorphosis of the tobacco hornworm. *Dev. Biol.* **132,** 292–303.

Kafatos, F. C. (1976). Sequential cell polymorphism: A fundamental concept in developmental biology. *Adv. Insect Physiol.* **12,** 1–15.

Kalinich, J. F., and McClain, D. E. (1992). An *in vitro* method for radiolabeling proteins with ^{35}S. *Anal. Biochem.* **205,** 208–212.

Karim, F. D., and Thummel, C. S. (1992). Temporal coordination of regulatory gene expression by the steroid hormone ecdysone. *EMBO J.* **11,** 4083–4093.

Karim, F. D., Guild, G. M., Thummel, C. S. (1993). The *Drosophila Broad-Complex* plays a key role in controlling ecdysone-regulated gene expression at the onset of metamorphosis. *Development* **118,** 977–988.

Kimbrell, D. A., Berger, E., King, D. S., Wolfgang, W. J., and Fristrom, J. W. (1988). Cuticle protein gene expression during the third instar of *Drosophila melanogaster. Insect Biochem.* **18,** 229–235.

Kremen, C. (1989). Patterning during pupal commitment of the epidermis in the butterfly, *Precis coenia:* The role of intercellular communication. *Dev. Biol.* **133,** 336–347.

Kumar, M. N., and Sridhara, S. (1994). Characterization of four pupal wing cuticular protein genes of the silkmoth *Antheraea polyphemus. Insect Biochem. Mol. Biol.* **24,** 291–299.

Lampe, D. J., and Willis, J. H. (1994). Characterization of a cDNA and gene encoding a cuticular protein from rigid cuticles of the giant silkmoth, *Hyalophora cecropia. Insect Biochem. Mol. Biol.* **24,** 419–435.

Lawrence, P. A. (1969). Cellular differentiation and pattern formation during metamorphosis of the milkweed bug *Oncopeltus. Dev. Biol.* **19,** 12–40.

Lemoine, A., and Delachambre, J. (1986). A water-soluble protein specific to the adult cuticle in *Tenebrio.* Its use as a marker of a new programme expressed by epidermal cells. *Insect Biochem.* **16,** 483–489.

Leung, H., Palli, S. R., and Locke, M. (1989). The localization of arylphorin in an insect, *Calpodes ethlius. J. Insect Physiol.* **35,** 223–231.

Li, W., and Riddiford, L. M. (1992). Two distinct genes encode two major isoelectric forms of insecticyanin in the tobacco hornworm, *Manduca sexta*. *Eur. J. Biochem.* **205,** 491–499.

Mitsui, T., and Riddiford, L. M. (1976). Pupal cuticle formation by *Manduca sexta* epidermis *in vitro:* Patterns of ecdysone sensitivity. *Dev. Biol.* **54,** 172–186.

Nakato, H., Toriyama, M., Izumi, S., and Tomino, S. (1990). Structure and expression of mRNA for a pupal cuticle protein of the silkworm, *Bombyx mori*. *Insect Biochem.* **7,** 667–678.

Nakato, H., Izumi, S., and Tomino, S. (1992). Structure and expression of gene coding for a pupal cuticle protein of *Bombyx mori*. *Biochim. Biophys. Acta* **1132,** 161–167.

Nakato, H., Shofuda, K., Izumi, S., and Tomino, S. (1994). Structure and developmental expression of a larval cuticle protein gene of the silkworm, *Bombyx mori*. *Biochim. Biophys. Acta* **1218,** 64–74.

Nash, W. G. (1976). Patterns of pigmentation color states regulated by the *y* locus in *Drosophila melanogaster*. *Dev. Biol.* **48,** 336–343.

Natzle, J. E. (1993). Temporal regulation of *Drosophila* imaginal disc morphogenesis: A hierarchy of primary and secondary 20-hydroxyecdysone–responsive loci. *Dev. Biol.* **155,** 516–532.

Palli, S. R., and Locke, M. (1987). The synthesis of hemolymph proteins by the larval epidermis of an insect *Calpodes ethlius* (Lepidoptera: Hesperiidae). *Insect Biochem.* **17,** 711–722.

Palli, S. R., and Locke, M. (1988). The synthesis of hemolymph proteins by the larval fat body of an insect *Calpodes ethlius* (Lepidoptera: Hesperiidae). *Insect Biochem.* **18,** 405–413.

Palli, S. R., Osir, E. O., Eng, W.-S., Boehm, M. F., Edwards, M., Kulcsar, P., Ujvary, I., Hiruma, K., Prestwich, G. D., and Riddiford, L. M. (1990). Juvenile hormone receptors in insect larval epidermis: Identification by photoaffinity labeling. *Proc. Natl. Acad. Sci. U.S.A.* **87,** 796–800.

Palli, S. R., Touhara, K., Charles, J.-P., Bonning, B. C., Atkinson, J. K., Trowell, S. C., Hiruma, K., Goodman, W. G., Kyriakides, T., Prestwich, G. D., Hammock, B. D., and Riddiford, L. M. (1994). A nuclear juvenile hormone-binding protein from larvae of *Manduca sexta:* A putative receptor for the metamorphic action of juvenile hormone. *Proc. Natl. Acad. Sci. U.S.A.* **91,** 6191–6195.

Paro, R. (1990). Imprinting a determined state into the chromatin of *Drosophila*. *Trends Genet.* **6,** 416–421.

Rebers, J. F., and Riddiford, L. M. (1988). Structure and expression of a *Manduca sexta* larval cuticle gene homologous to *Drosophila* cuticle genes. *J. Mol. Biol.* **203,** 411–423.

Riddiford, L. M. (1991). Hormonal control of sequential gene expression in insect epidermis. *In* "Physiology of the Insect Epidermis" (K. Binnington and A. Retnakaran, eds.), pp. 46–54. Commonwealth Scientific and Industrial Research Organization, East Melbourne, Australia.

Riddiford, L. M. (1994). Cellular and molecular actions of juvenile hormone I. General considerations and premetamorphic actions. *Adv. Insect Physiol.* **24,** 213–274.

Riddiford, L. M., Palli, S. R., Hiruma, K., Li, W.-C., Green, J., Hice, R. H., Wolfgang, W. J., and Webb, B. A. (1990). Developmental expression, synthesis and secretion of insecticyanin by the epidermis of the tobacco hornworm, *Manduca sexta*. *Arch. Insect Biochem. Physiol.* **14,** 171–190.

Riddihough, G., and Pelham, H. R. B. (1987). An ecdysone response element in the *Drosophila* hsp27 promoter. *EMBO J.* **6,** 3729–3734.

Roberts, P. E., and Willis, J. H. (1980a). The cuticular proteins of *Tenebrio molitor.* I. Electrophoretic banding patterns during postembryonic development. *Dev. Biol.* **75,** 59–69.

Roberts, P. E., and Willis, J. H. (1980b). The cuticular proteins of *Tenebrio molitor.* II. Patterns of synthesis during postembryonic development. *Dev. Biol.* **75,** 70–77.

Roberts, P. E., and Willis, J. H. (1980c). Effects of juvenile hormone, ecdysterone, actinomycin D, and mitomycin C on the cuticular proteins of *Tenebrio molitor. J. Embryol. Exp. Morphol.* **56,** 107–123.

Roter, A. H., Spofford, J. B., and Swift, H. (1985). Synthesis of the major cuticle proteins of *Drosophila melanogaster* during hypoderm differentiation. *Dev. Biol.* **107,** 420–431.

Sass, M., Kiss, A., and Locke, M. (1993). Classes of integument peptides. *Insect Biochem. Mol. Biol.* **23,** 845–857.

Sass, M., Kiss, A., and Locke, M. (1994). Integument and hemocyte peptides. *J. Insect Physiol.* **40,** 407–421.

Savakis, C. M., Ashburner, M., and Willis, J. H. (1986). The expression of the gene coding for alcohol dehydrogenase during the development of *Drosophila melanogaster. Dev. Biol.* **114,** 194–207.

Schlossherr, J., Eggert, H., Paro, R., Cremer, S., and Jack, R. S. (1994). Gene inactivation in *Drosophila* mediated by the *Polycomb* gene product or by position-effect variegation does not involve major changes in the accessibility of the chromatin fibre. *Mol. Gen. Genet.* **243,** 453–462.

Scholer, H. R. (1991). Octamania: The POU factors in murine development. *Trends Genet.* **7,** 323–328.

Skelly, P. J., and Howells, A. J. (1987). Larval cuticle proteins of *Lucilia cuprina:* Electrophoretic separation, quantitation and developmental changes. *Insect Biochem.* **17,** 625–633.

Snyder, M., Hirsh, J., and Davidson, N. (1981). The cuticle genes of *Drosophila:* A developmentally regulated gene cluster. *Cell (Cambridge, Mass.)* **25,** 165–177.

Snyder, M., Hunkapiller, M., Yuen, D., Silvert, D., Fristrom, J., and Davidson, N. (1982). Cuticle protein genes of *Drosophila:* Structure, organization and evolution of four clustered genes. *Cell (Cambridge, Mass.)* **29,** 1027–1040.

Souliotis, V., Patriou-Georgoula, M., Zongza, V., and Dimitriadis, G. J. (1988). Cuticle proteins during the development of *Dacus oleae. Insect Biochem.* **18,** 485–492.

Sridhara, S. (1993). mRNAs of wing epidermis of the oak silkmoth *Antheraea polyphemus* and other silkmoths during pupal development. *Insect Biochem. Mol. Biol.* **23,** 631–641.

Weintraub, H., and Groudine, M. (1976). Chromosomal subunits in active genes have an altered conformation. *Science* **193,** 848–856.

Wigglesworth, V. B. (1934). The physiology of ecdysis in *Rhodnius prolixus* (Hemiptera). II. Factors controlling moulting and 'metamorphosis.' *Q. J. Microsc. Sci.* **77,** 191–222.

Wigglesworth, V. B. (1936). The function of the corpus allatum in the growth and reproduction of *Rhodnius prolixus* (Hemiptera). *Q. J. Microsc. Sci.* **79,** 91–121.

Wigglesworth, V. B. (1940). The determination of characters at metamorphosis in *Rhodnius prolixus* (Hemiptera). *J. Exp. Biol.* **17,** 201–222.

Wigglesworth, V. B. (1958). Some methods for assaying extracts of the juvenile hormone in insects. *J. Insect Physiol.* **2,** 73–84.

Wigglesworth, V. B. (1959). Metamorphosis, polymorphism, differentiation. *Sci. Am.* **200**(2), 100–106.

Williams, C. M. (1953). Morphogenesis and the metamorphosis of insects. *Harvey Lect.* **47,** 126–155.

Williams, C. M. (1956). The juvenile hormone of insects. *Nature* (*London*) **178,** 212–213.

Williams, C. M., and Kafatos, F. C. (1971). Theoretical aspects of the action of juvenile hormone. *Mitt. Schweiz. Entomol. Ges.* **44,** 151–162.

Willis, J. H. (1987). Cuticular proteins: The neglected component. *Arch. Insect. Biochem. Physiol.* **6,** 203–215.

Willis, J. H. (1990). Regulating genes for metamorphosis: Concepts and results. *In* "Molecular Insect Science" (H. H. Hagedorn, J. G. Hildebrand, M. G. Kidwell, and J. H. Law, eds.), pp. 91–98. Plenum, New York.

Willis, J. H. (1991). The epidermis and metamorphosis. *In* "Physiology of the Insect Epidermis" (K. Binnington and A. Retnakaran, eds.), pp. 36–45. Commonwealth Scientific and Industrial Research Organization, East Melbourne, Australia.

Willis, J. H., Cox, D. L. (1984). Defining the anti-metamorphic action of juvenile hormone. *In* "Biosynthesis, Metabolism and Mode of Action of Invertebrate Hormones" (J. A. Hoffmann and M. Porchet, eds.), pp. 466–474. Springer-Verlag, Heidelberg.

Willis, J. H., and Hollowell, M. P. (1976). The interaction of juvenile hormone and ecdysone: Antagonistic, synergistic, or permissive? *In* "The Juvenile Hormones" (L. I. Gilbert, ed.), pp. 270–287. Plenum, New York.

Willis, J. H., Regier, J. C., and Debrunner, B. A. (1981). The metamorphosis of arthropodin. *In* "Current Topics in Insect Endocrinology and Nutrition" (G. Bhaskaran, S. Friedman, and J. G. Rodriguez, eds.), pp. 27–46. Plenum, New York.

Willis, J. H., Rezaur, R., and Sehnal, F. (1982). Juvenoids cause some insects to form composite cuticles. *J. Embryol. Exp. Morphol.* **71,** 25–40.

Wolfgang, W. J., and Riddiford, L. M. (1986). Larval cuticular morphogenesis in the tobacco hornworm, *Manduca sexta,* and its hormonal regulation. *Dev. Biol.* **113,** 305–316.

Wolfgang, W. J., and Riddiford, L. M. (1987). Cuticular mechanics during larval development of the tobacco hornworm, *Manduca sexta. J. Exp. Biol.* **128,** 19–33.

8

Metamorphosis of the Insect Nervous System

JAMES W. TRUMAN

Department of Zoology
University of Washington
Seattle, Washington

I. INTRODUCTION

During metamorphosis, larval structures are discarded and new adult structures are formed, but the extent of this tissue changeover varies among insects. It reaches its extreme in higher Diptera such as *Drosophila* in which almost all larval cells are discarded and adult tissues are built up of new cells from imaginal discs or imaginal proliferation zones (see Bayer *et al.,* Chapter 9, this volume; Wilder and Perrimon, Chapter 10, this volume). Even in higher flies, the central nervous system (CNS) is an exception to this rule of cellular changeover. Many neurons in the CNS persist through metamorphosis so that the same cell functions in both the larva and the adult.

Many neurons, then, are faced with the challenge of controlling behaviors of two radically different animals. Metamorphosis is typically associated with a dramatic change in the habitat of an insect, usually from an aquatic or land-bound larva to an aerial adult (see Sehnal *et al.,* Chapter 1, this volume). Both larval and adult stages show morphological and behavioral adaptations that are molded by their respective environments with little regard for that of the other stage. For example, the larval stage, which is adapted for feeding and growth, is relatively sedentary with a veniform shape that becomes more extreme in forms that live in aquatic or semiaquatic habitats. The larval abdomen has assumed a major role in locomotion. Indeed, many larvae have dispensed with legs altogether and are adapted for burrowing through the substratum. The sensory world of holometabolous larvae similarly reflects this sedentary life-style. Their sensory systems for vision, olfaction, and taste are composed of only a small number of receptors that are tuned to detection and discrimination of proximal stimuli (Zacharuk and Shields, 1991). The greatest numbers of larval sensory neurons are mechanoreceptors that provide tactile information about the immediate environment of the larva.

In contrast to the larva, the adult stage is typically a highly active animal devoted to reproduction and dispersal. The thorax has completely taken over locomotor functions both for walking and for the uniquely adult behavior of flight. Associated with this active life-style is one of the most sophisticated sensory systems seen outside of the vertebrates. Mechanoreception is no longer the dominant sensory modality. The adult visual and olfactory systems may include thousands of sensory receptors as compared with the dozens of receptors for the corresponding modalities in the larva. The compound eyes provide rapid and detailed discrimination of distant objects while elaborate antennal systems allow long distance detection of chemical cues from hosts or mates.

The profound changes in morphology and behavior that occur during metamorphosis are accommodated by equally dramatic changes in the

CNS of the insect. The analysis of changes in the CNS is aided by the fact that many insect neurons are large cells that have unique morphology, biochemistry, and patterns of synaptic connections. The same cell can be reliably found from individual to individual and, hence, its fate can be tracked through metamorphosis. Numerous physiological, anatomical, and immunocytochemical studies have focused on the metamorphosis of insect neurons. What are the fates of the larval neurons? Are they discarded or are they preserved for use in the adult stage? Also, are there new neurons that are unique to the adult stage, and what are their origins and their function in the adult CNS?

II. PATTERNS OF DEVELOPMENT OF THE INSECT CENTRAL NERVOUS SYSTEM

To understand the metamorphosis of the CNS of holometabolous insects one also has to understand how the CNS forms during embryogenesis. Our modern concepts of CNS development are rooted in the studies of embryogenesis in the grasshopper, a hemimetabolous insect. These studies serve as an appropriate starting point since the Holometabola arose from hemimetabolous ancestors.

A. Hemimetabolous Pattern of Development

The insect CNS is composed of the brain and a chain of ventral segmental ganglia that are linked via paired connectives. The embryogenesis of the segmental ganglia is relatively stereotyped and has been the object of intensive study. The formation of each ganglion begins when a set of neuronal stem cells, the neuroblasts (NBs), segregate from the ventral ectoderm (Fig. 1). Eventually 61 NBs (a single median NB and 30 pairs of lateral NBs) move internally from the overlying ectoderm. The NBs are arranged in a stereotyped set of rows, which allows a given stem cell to be found reproducibly from segment to segment and from embryo to embryo (Bate, 1976; Doe and Goodman, 1985). In addition to the NBs, a small number of midline precursor cells also contribute a few neurons in each ganglion (Bate and Grunewald, 1981). The NB divides asymmetrically; the smaller product is a ganglion mother cell (GMC),

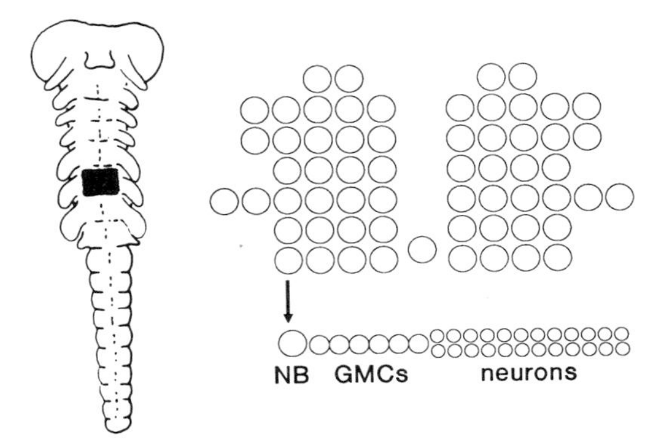

Fig. 1. Ventral view of an early stage grasshopper embryo showing (in black) the position of the 61 neuroblasts (NBs) that produce the neurons in the second thoracic ganglion. The enlargement on the right shows the arrangement of these NBs into 30 bilateral pairs with a median unpaired NB. Each NB undergoes asymmetric divisions, the smaller product of which is a ganglion mother cell (GMC), which, in turn, divides to produce two daughter neurons. In the lineage depicted the oldest neurons are to the right, the youngest are nearest the NB.

which in turn divides once to produce two daughter neurons. Each NB undergoes repeated divisions; the result is a series of neurons that are produced two by two through time (Fig. 1). The first neurons produced in a lineage are typically large neurons, the motoneurons and many interganglionic interneurons, whereas the later born cells are small intersegmental and local interneurons (see, e.g., Shepherd and Bate, 1990; Thompson and Siegler, 1993). The identity of a neuron is dependent on the identity of the NB that produced it. Consequently, the ablation of a specific NB results in a ganglion lacking the neurons normally derived from that stem cell (Taghert and Goodman, 1984). Hence, once the NBs have segregated from the ectoderm, establishment of neuronal identities appears to be strictly lineage dependent with no regulation occurring between lineages.

In the case of grasshoppers (Shepherd and Bate, 1990), and probably for hemimetabolous insects in general, neurogenesis is completed by the end of embryogenesis so that the newly hatched insect has all of the neurons that it needs for the rest of its life. Two notable exceptions occur in the brain; the optic lobes continue to add visual interneurons to accommodate the new ommatidia that are added to the compound eye at each larval molt (Anderson, 1978). Also, the mushroom bodies, areas of the brain associated with learning and memory (Davis, 1993), continue to add neurons throughout larval life and even into the adult stage (Cayre *et al.*, 1994).

Not only do newly hatched grasshoppers have their adult complement of neurons, but these cells already show the basic morphology that will characterize them in the adult (Altman and Tyrer, 1974). Indeed, the neural circuits for adult-specific behaviors such as flight (Stevenson and Kutsch, 1988) or oviposition (Thompson, 1993) are already in a recognizable form around the time of hatching. From such observations, though, one should not conclude that the CNS of hemimetabolous insects is static after hatching. To the contrary, it undergoes marked changes during larval life to accommodate growth and the addition of new sensory elements (see, e.g., Murphey and Chiba, 1990). Nevertheless, the essence of the adult nervous system, in terms of the number of CNS neurons and their main pattern of connections, is clearly evident in hemimetabolous larva by the end of embryogenesis.

The sensory systems of hemimetabolous larvae show both qualitative and quantitative changes during postembryonic life. During each larval molt there is the progressive addition of more receptors as structures such as the compound eyes, antennae, and cerci increase in size. A qualitative change is typically seen during the final molt to the adult stage as the insect adds novel types of receptors required for adult behavior. For example, in cockroaches the antennal sensilla that detect the sex pheromone only differentiate during the adult molt (Schaffer and Sanchez, 1973). Likewise, many of the receptors associated with the wings and genitalia appear at this time when the mature structures become manifest.

B. Holometabolous Pattern of Development

Unlike the situation in grasshoppers, the CNS of a newly hatched holometabolous larva has little in common with that of the adult (Fig. 2). On a gross level, areas of the brain involved with the processing of sensory information from the larval eyes and antennae are poorly developed in accord with the relative simplicity of larval sensory systems. This reduction is in stark contrast to the subsequent hypertrophy of these brain regions in the adult. In most holometabolous groups the ventral CNS of the larva is a chain of discrete ganglia with only the last few fused to form a compound

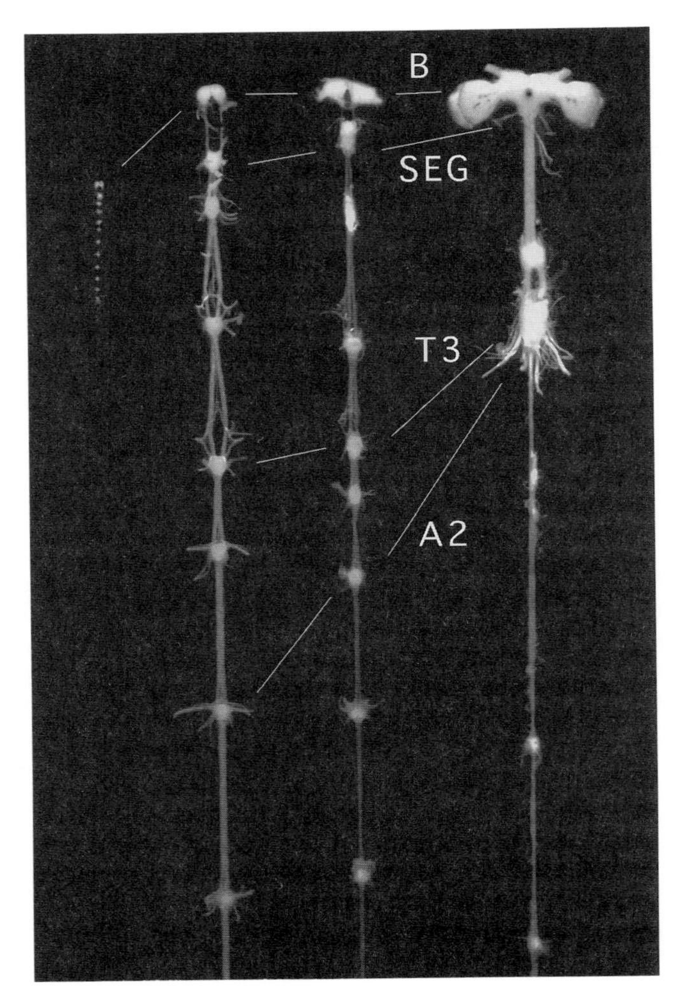

Fig. 2. Changes in the size and morphology of the CNS of *Manduca sexta* during its life history. The stages from left to right are freshly hatched first stage larva, wandering fifth stage larva, pupa, and adult. The lines indicate the location of the same ganglia in subsequent stages. A2, second abdominal ganglion; B, brain; SEG, subesophageal ganglion; T3, third thoracic ganglion. In the adult ganglia, T2 to A2 fuse to form a compound ganglion.

terminal ganglion. Metamorphosis often includes migration and fusion of some segmental ganglia; typically the first few abdominal ganglia with the last one or two thoracic ganglia (see, e.g., Pipa, 1967). In this compound "pterothoracic" ganglion, the fusion shortens interganglionic axons and, accordingly, the nerve impulse conduction time between neurons in different ganglia. Presumably this arrangement facilitates the use of abdominal neurons for the rapid, coordinated activity needed in flight control (Robertson and Pearson, 1983).

At a cellular level, larval neurons may bear little resemblence to their adult counterparts. Individual neurons that can be identified in the two stages remain in their main functional class (i.e., larval motoneurons are also motoneurons in the adult), but their morphology and synaptic connections may change dramatically during metamorphosis. Also, no one has yet found latent circuits for adult behaviors such as flight or reproductive movements in the larval CNS. Indeed, in moths the opposite situation is seen for ecdysis behavior in which most of the larval neural circuitry is carried through metamorphosis, more or less intact, to provide the basis of the adult motor program (Mesce and Truman, 1988).

Although hemimetabolous and holometabolous nervous systems differ markedly in their postembryonic changes, they are strikingly similar in their early embryonic development. The neurons in the segmental ganglia of *Drosophila* and *Manduca* embryos are generated from essentially the same array of neuroblasts as seen in grasshoppers (Thomas *et al.*, 1984; Doe, 1992). Also, similar neurons are produced by corresponding NBs in these diverse species, underscoring the conservative nature of insect neuronal development (Thomas *et al.*, 1984; Goodman and Doe, 1993).

Two fundamental differences separate the embryogenesis of a holometabolous CNS from that of a grasshopper (Fig. 3). First, as indicated earlier, the CNS of a newly hatched grasshopper is a miniature version of the adult CNS, a similarity that extends even to the level of individual neurons. In contrast, the neurons in the CNS of a newly hatched holometabolous larva are adapted to the needs of the larva and may not remotely resemble their eventual form in the adult. During metamorphosis these cells have to remove these larval specializations as well as acquire their adult features. A second difference relates to neurogenesis. In addition the embryonic neurogenic period, holometabolous insects have a second neurogenic period that occurs during larval life. NBs in the brain and segmental ganglia typically become mitotically active in the early larval instars and produce new neurons (the "imaginal" neurons) up until the larval–pupal transition (White and Kankel, 1978; Booker and Truman,

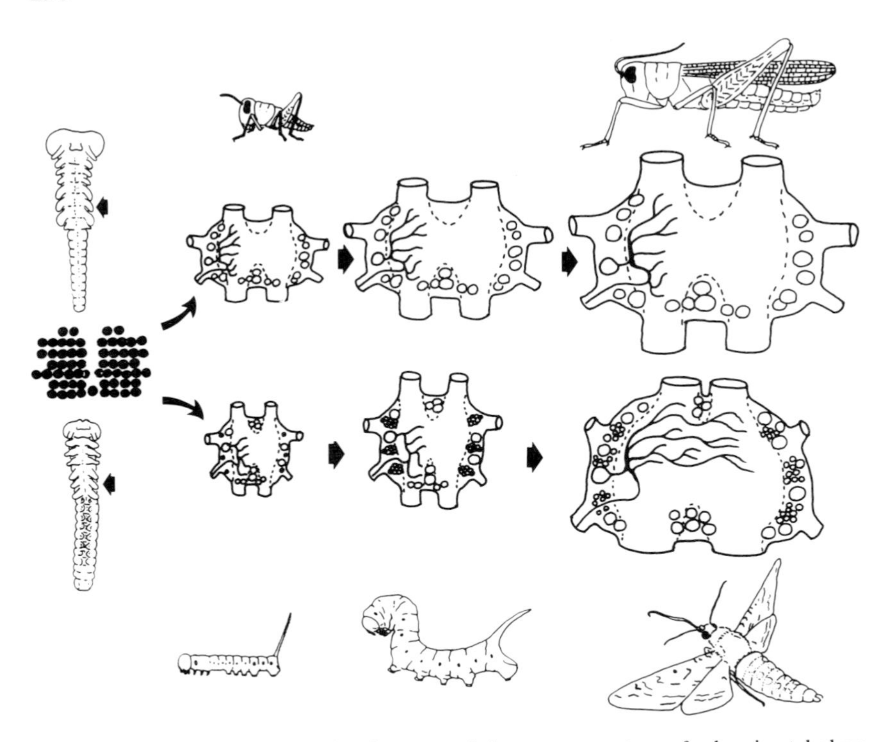

Fig. 3. Comparison of the development of the nervous system of a hemimetabolous insect such as a grasshopper (top) with a holometabolous insect such as a moth (bottom). During embryogenesis of both, the neurons in a thoracic segment (arrow) arise from identical sets of NBs (black circles). The postembryonic development of the two groups differs in two respects. In holometabolous insects some of the embryonic NBs persist into larval life where they make new imaginal neurons that mature at metamorphosis. Also, larval neurons are remodeled at metamorphosis for new adult functions.

1987a; Truman and Bate, 1988). These new neurons extend an axon into the neuropil but then arrest their development until metamorphosis when they mature into functional neurons (Booker and Truman, 1987a).

Comparative studies on the tsetse fly *Glossina pallidipes* (Truman, 1990) as well as an elegant series of cell transplantation experiments in *Drosophila* (Prokop and Technau, 1991) show that the NBs that make the imaginal neurons are persistent embryonic NBs. The NBs in the abdomen typically complete their program of divisions during embryogenesis and presumably die prior to hatching. In contrast, most of those in the brain and thoracic ganglia stop dividing before the completion of their lineages and persist, in a dormant state, into larval life. Because of this premature mitotic arrest, the brain and thoracic ganglia of holometabolous larvae have many fewer neurons than do the corresponding regions of a young grasshopper. The dormant NBs then reactivate during larval life to produce the remainder of their lineages (Fig. 3). Therefore, the shift from a hemimetabolous to a holometabolous life-style has involved a heterochronic shift in which a portion of embryonic neurogenesis is delayed into postembryonic life.

Holometabolous larvae differ in the time during embryogenesis when their NBs arrest. The timing of this arrest is significant since, as indicated earlier, the first neurons to be born are motoneurons and large intersegmental interneurons whereas smaller interneurons are born later. Consequently, all holometabolous larvae have a complete set of motoneurons but, depending on when the arrest occurs, they vary widely in the number of interneurons they have available for sensory processing and for motor control. In the most primitive Holometabola, the Neuroptera and primitive beetles, the NB arrest happens relatively late and these larvae hatch with reasonable numbers of central neurons. Such larvae are often capable of sophisticated behaviors as indicated by many of them being active hunters. In contrast, the larvae of more advanced holometabolic groups show earlier and earlier neurogenic arrests. In *Drosophila,* for example, the thoracic NBs undergo an average of only about 7 embryonic divisions (Hartenstein *et al.,* 1987) as compared to 25 or more for embryonic grasshoppers. This reduction in numbers of neurons present in the CNS of advanced holometabolous larvae is reflected in their simplified behavior. Indeed, in the higher Hymenoptera (the bees, ants, and wasps), larval behavior is reduced to the extent that most larvae are totally dependent on the adult for their care and feeding. They are cared for either by direct provisioning or by their being endoparasites in plants or other insects.

When the arrested embryonic NBs subsequently reactivate in the larva, they do not restart from the beginning of their lineage with the production of a new set of motoneurons. Instead, they appear to make neurons that are appropriate for the time at which they stopped in the embryo. As summarized in Fig. 4, this relationship results in the neurons of the sensory and motor sides of the adult CNS having different developmental origins. The only source of motoneurons for the adult are the larval cells that must then be remodeled at metamorphosis to accommodate the needs of the adult. The imaginal neurons, by comparison, are almost exclusively interneurons, most apparently dedicated to the processing of information provided by the new adult sensory systems.

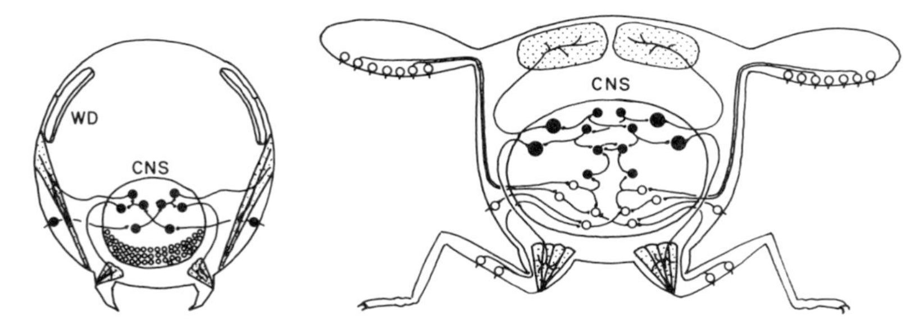

Fig. 4. Schematic section through the thorax of a holometabolous larva and adult. Black neurons arise during embryogenesis whereas the white neurons arise during larval life or metamorphosis. The motor side of the adult nervous system is predominantly made from embryonically derived neurons whereas most sensory neurons and sensory interneurons are made during postembryonic life. Also, in the latter, larval neurons change shape and connections during their transition to the adult. WD, wing imaginal disc.

III. CELLULAR CHANGES DURING METAMORPHOSIS OF THE CENTRAL NERVOUS SYSTEM

The cellular changes that occur during metamorphosis are involved with two major issues: the fate of larval neurons and the maturation of imaginal neurons. Larval neurons have three principle fates which are illustrated for *Manduca* abdominal motoneurons in Fig. 5. Some neurons, such as motoneuron PPR which innervates a larval proleg muscle, have a larval-specific function and die after the molt to the pupal stage. Other neurons, such as the D-IV motoneurons that innervate the ventral internal muscles, persist largely unchanged through metamorphosis, are used

---

Fig. 5. Relationship of the fate of particular motoneurons in *Manduca* to the ecdysteroid titer during the last portion of postembryonic life. Drawings show the appearance of the central dendritic arbors of various identified neurons through metamorphosis. MN-3 undergoes dendritic loss just after pupal ecdysis followed by dendritic sprouting during adult differentiation; the D-IV motoneuron remains essentially unchanged during metamorphosis but dies after adult emergence; and PPR undergoes dendritic reduction during the larval–pupal transition but then dies soon thereafter. [Ecdysteroid titer based on data from Bollenbacher *et al.* (1981).]

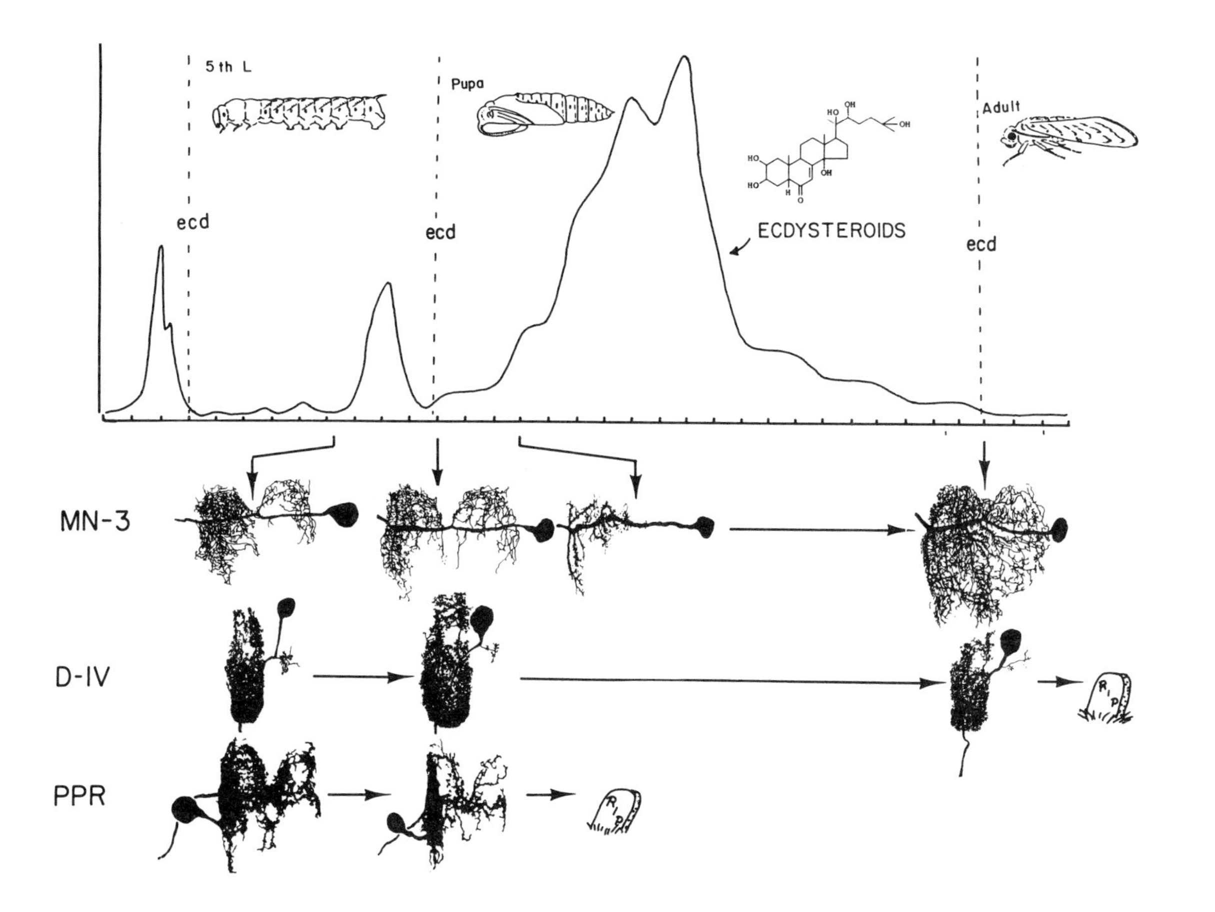
5 th L
Pupa
Adult
ecd
ecd
ecd
ECDYSTEROIDS
MN-3
D-IV
PPR

during the emergence of the adult, and then die soon thereafter. Most larval neurons (e.g., MN-3) are remodeled during metamorphosis and are subsequently reused in the adult.

A. Programmed Death of Larval Neurons

Metamorphosis is accompanied by the death of larval neurons throughout the CNS but it is most pronounced in the abdominal ganglia, in accord with the shift in locomotor function from the abdomen of the larva to the thorax of the adult. In *Manduca,* for example, the number of muscle groups in an abdominal segment drops from about 50 pairs in the larva to about 10 pairs in a mature adult. The number of motoneurons accordingly drops from over 35 pairs to about 10 pairs, respectively (Taylor and Truman, 1974). Indeed, an abdominal ganglion in the adult contains only about 350 neurons as compared to 700–800 in the larva. The larval neurons die in two waves. The first wave of death, which includes the proleg motoneurons PPR and APR, occurs about 2 days after pupal ecdysis. Some of these deaths show segment specificity. For example, the APR motoneurons innervate larval proleg muscles in segments A3 to A6. These neurons die in segments A5 and A6 whereas in A3 and A4 they survive to function in the pupa and adult (Weeks and Ernst-Utzschneider, 1989). Thus, selective cell death provides a mechanism to generate segmental specializations in the adult CNS that were not present in the larva.

The second wave of neuronal death occurs after the emergence of the adult. Many of the neurons that die at this time are involved in maintaining the behavior during the pupal–adult transition or in the performance of ecdysis and associated behaviors. Because the adult is a terminal stage that does not molt again, neurons and muscles dedicated to ecdysial behaviors undergo programmed degeneration after adult emergence (Truman, 1983; Kimura and Truman, 1990). In both *Manduca* and *Drosophila,* though, conditions that prolong ecdysis and associated behaviors can delay the demise of these neurons.

B. Remodeling of Larval Neurons

One of the most striking aspects of CNS metamorphosis is the remodeling of larval neurons. These changes have been followed principally in large identifiable motoneurons by dye fills of the same cell at intervals through the transition from the larval to the adult (Truman and Reiss, 1976; Casaday and Camhi, 1976; Levine and Truman, 1985). Alternatively, larval motoneurons have been labeled *in vivo* by injecting a dye into their target muscles which was taken up by the axon terminals and transported back to the cell body. After metamorphosis, the presence of dye in the cell body unequivocally marked the neurons which were then refilled with dye using microelectrodes to assess their adult morphology (Kent and Levine, 1988).

The changes to both the central and the peripheral parts of these motoneurons is a two stage process involving an initial loss of larval-specific processes followed by the outgrowth of adult-specific structures

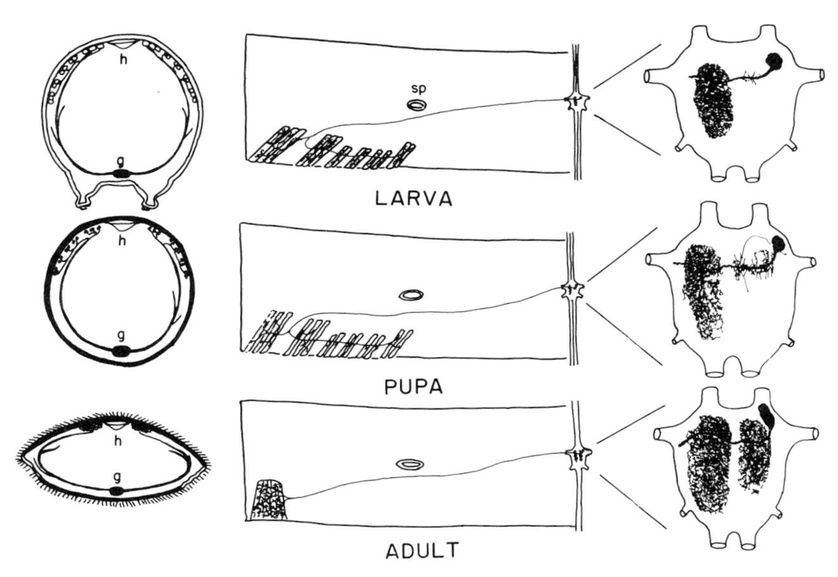

Fig. 6. Changes in motoneuron MN-1 during metamorphosis. (Right) The central, dendritic arbor of the cell in the larva, pupa, and adult. (Center) Left abdominal hemisegment showing the position of MN-1's target muscle that extends along the posterior margin of the segment from the lateral spiracle (sp) to the dorsal midline in the larva. The muscles degenerate in the pupa but the most dorsal fibers serve as a scaffold for the growth of the adult muscle. (Left) Cross section of the abdomen showing the position of target muscles relative to the dorsal heart (h) and the ventral ganglion (g). [Reprinted from Truman (1992) with permission from John Wiley and Sons.]

(Fig. 6). In the periphery, the larval muscles controlled by these cells degenerate, but, in moths, their remains are used as a scaffolding for the growth of the adult muscle. The axons of the motoneurons remain in contact with their targets throughout this remodeling (Heinertz, 1976; Truman and Reiss, 1995). In contrast, the axons of the motoneurons in *Drosophila* appear to be pruned back severely. As they regrow out in the periphery, myoblasts migrate along them to the sites where they fuse to form the new adult muscles (Currie and Bate, 1991). The first few days around pupal ecdysis in *Manduca* are devoted to the removal of larval processes (e.g., MN-3 in Fig. 4). The sprouting of new processes then begins at the start of adult differentiation and is finished by about midway through the pupal–adult transition. During the rest of metamorphosis the cell forms and matures its central and peripheral connections.

The leg motoneurons in *Drosophila* may present an extreme form of remodeling. Although these cells are apparently born during embryogenesis, they do not send axons into the periphery until the onset of metamorphosis, when they rapidly extend axons into the evaginating leg imaginal discs to innervate the forming leg musculature (M. Bate, unpublished results). It is not known whether these cells function as interneurons or stay as arrested immature motoneurons during larval life.

The motor system has to deal with two major problems as it is remodeled from its larval to its adult configuration. One problem is the shift from a hydraulic skeleton that is maintained by muscle tone and blood pressure in the larva to the rigid exoskeleton of the pupa and the adult. In addition to maintaining larval body shape, blood pressure is also used for extending certain structures such as the abdominal prolegs. Relatively little is known about the neuronal control of the hydraulic skeleton in insect larvae, and the muscle groups responsible for maintaining body tone are not well defined. Most likely, they are muscles in the external muscle layer. The larval CNS shows a large number of tonically active motor units that presumably provide the drive to these muscles. This tonic motor activity is seen throughout larval life and into the early phases of the larval–pupal transition, but it ceases at the time the pupal cuticle is forming and the dehydrating larval cuticle becomes stiff enough to maintain body form (see, e.g., Jacobs and Weeks, 1990). Whether the loss of tonic activity results from changes in the properties of the motoneurons and/or from the loss of interneuronal inputs to these cells is not known.

The second problem facing the motor system is dealing with the changes in the functional relationships of target muscle groups. For example, in *Manduca* the leg motoneuron FeFlx/FeExt causes flexion of the femur in the larva but femur extension in the adult (Kent and Levine, 1988, 1993). At metamorphosis the larval dendritic arbor is almost completely removed and is replaced by the extensive outgrowth of adult-specific branches. The extreme remodeling of this neuron is likely associated with its very different behavioral roles in the two stages. The complexity of leg movements and the leg neuropils make it difficult to associate particular anatomical changes with new functional roles.

The simple anatomy of the abdomen in *Manduca* makes it easier to relate changes in form of abdominal motoneurons with their new functions. For example, motoneuron MN-1 innervates a set of dorsal oblique muscles (DEO2/DEO3) that extend from the spiracle to the heart in the larva. Its adult target is a small longitudinal muscle (DE4) just lateral of the heart tube (Figs. 6 and 7). In the larva, the right and left DEO2/DEO3s function as *antagonists* during lateral bending of the abdomen. In contrast, the adult abdomen is dorsoventrally flattened with little capacity for lateral movement. Movements are confined to the vertical plane, during which the right and left DE4s function as *synergists.*

The change in functional relationship of MN-1's targets is reflected in changes in the neuron's dendritic morphology (Fig. 6) and synaptic connections (Levine and Truman, 1982). In the larva, MN-1 has a unilateral dendritic tree, and the right and left MN-1s have nonoverlapping dendritic fields. Inputs that excite one MN-1, such as from the lateral stretch receptor (Fig. 7), tend to inhibit its contralateral homolog and vice versa. Such as arrangement is typical for motor units that control antagonistic muscles. After metamorphosis, however, the two neurons now appear to share common synaptic inputs and those that excite one MN-1 also excite the other. This sharing of inputs is associated with the right MN-1 growing a new dendritic tree that overlaps the larval dendritic tree of the left cell and vice versa. Hence, in its adult form, MN-1 has a bilateral dendritic tree with extensive branching in both right and left neuropils. This extensive overlap of the dendrites of the two cells presumably facilitates their sharing common inputs and, hence, their function as synergists (Fig. 7). Most abdominal motoneurons show a similar shift from a unilateral to a bilateral dendritic tree as they change from their larval to their adult form (Levine and Truman, 1985; Weeks and Ernst-Utzschneider, 1989). Consequently, metamorphosis brings about a widespread reorganization of the motor architecture of the abdominal CNS to accommodate the morphological constraints of the adult abdomen.

In general, the transmitter content of a neuron is more stable than its morphology. There are no known examples of shifts through metamor-

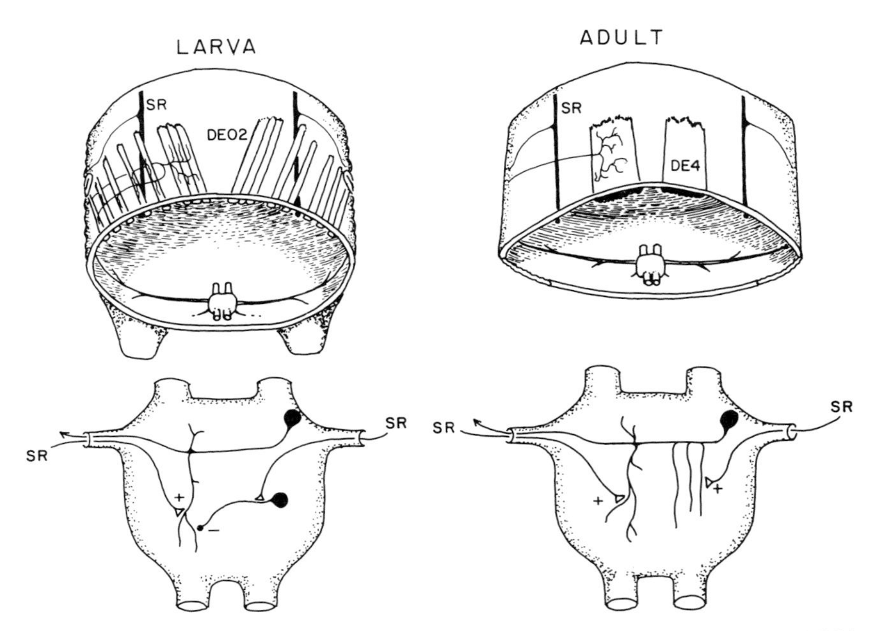

Fig. 7. Interpretation of the changes in the connectivity between motoneuron MN-1 and the segmental stretch receptors (SRs) during metamorphosis. The larval and adult stages depicted are the peripheral (top) and central (bottom) branching patterns of the MN-1 that innervates muscles (DEO2 or DE4) on the left side of an abdominal segment. In the larva, this MN-1 is excited by the left SR but is inhibited by the right SR via an inhibitory interneuron. After metamorphosis, the right SR now also excites MN-1 through contacts on the new adult-specific portion of its dendritic tree. [Based on data from Levine and Truman (1982).]

phosis in small molecule transmitters such as acetylcholine, glutamate, γ-aminobutyric acid, or biogenic amines. Neuropeptide expression, though, appears to be more plastic. For example, some lateral abdominal neurosecretory cells in *Manduca* shift from producing a cardioacceleratory peptide during larval life and to making bursicon in the adult (Tublitz and Sylwester, 1990). Also, many *Manduca* motoneurons contain a FMLFamide-like peptide as a cotransmitter in the larva but not in the adult (Witten and Truman, 1990). In *Drosophila* the TVA neurons express FMRFamide in the adult whereas they do not do so in the larva (O'Brien *et al.,* 1991). The impact of these shifts in neuromodulatory substances on the functioning of the CNS in the respective stages is not known.

C. Growth and Metamorphosis of Sensory Systems

Sensory neurons arise in the periphery in association with the epidermis. In some larvae, such as those of higher flies, no new sensory neurons are added after hatching. In contrast, *Manduca* represents the more usual case, with mechanosensory neurons being progressively added during the course of larval growth. For instance, between the first and the fifth larval stage the number of sensory hairs on an abdominal hemisegment increases from 8 to about 600 (Levine *et al.,* 1985). About 30 of these sensory neurons then persist into the pupal stage to supply the sensory hairs in the pupal-specific gin trap as well as the general body surface (Bate, 1973; Levine *et al.,* 1985). Large numbers of new mechanoreceptors are then formed during adult differentiation, although not to the numbers that were present in the last stage larva. It is not known whether the persisting larval sensory neurons come to innervate some of these adult hairs. The only mechanosensory neurons that are known to persist from the larva into the adult are the segmental stretch receptors. Although there is a general reduction in numbers of mechanoreceptors over the body surface of the adult, large numbers of new mechanoreceptors appear on specialized structures like the legs, wings, and genitalia.

Outside of mechanoreception, metamorphosis brings dramatic increases in the number of receptors for other sensory modalities. For example, a larval antenna, such as in *Manduca,* bears about a dozen sensilla whereas the antenna of the adult male is covered with over 100,000 sensilla (Sanes and Hildebrand, 1976). Similar increases are evident in the visual system. In some insects such as mosquitoes (Pflugfelder, 1937) the larval photoreceptors persist through metamorphosis and are associated with the adult compound eye. In contrast, the larval photoreceptors in *Drosophila* are thought to be lost relatively early in metamorphosis (Tix *et al.,* 1989).

The larval sensory cells may have a developmental function of providing pathways for the ingrowth of axons from the adult receptors. The best known case is in *Drosophila* in which Bowig's nerve, the nerve that carries the axons of the larval photoreceptors to the brain, serves as a pathway for the subsequent ingrowth of adult photoreceptor axons into the brain (see Meinertzhagen and Hanson, 1993).

Details of the development of complex sensory structures such as the antenna in *Manduca* (Hildebrand, 1985) and the compound eye in *Drosophila* (Wolff and Ready, 1993) are well described and will not be treated in detail here.

D. Production and Maturation of Imaginal Neurons

In *Manduca* the ventral NBs reactivate late in the second larval instar and produce neurons through the remainder of larval life and up to 1 to 2 days after pupal ecdysis (Booker and Truman, 1987a). The newborn cells arrest after they grow out an initial axon and are stock-piled until metamorphosis (Fig. 8). With the termination of feeding by the last (fifth) larval stage, these cells begin to enlarge (Booker and Truman, 1987b). Their maturation is completed during the pupal–adult transition with transmitter phenotypes appearing by about 30–40% of the pupal–adult transition (Fig. 9) (Witten and Truman, 1991). In *Manduca,* 3000–4000 new imaginal neurons are added to each thoracic ganglion and about 60–70 new neurons are added to each abdominal ganglion (Booker and Truman, 1987a,b). The extensive death of larval neurons in the abdomen and the large numbers of new imaginal neurons added in the thorax result in the great disparity in the number of neurons found in thoracic versus abdominal regions of the adult CNS.

The imaginal neurons of the mushroom bodies are the only imaginal neurons that appear to function in the larval stage. At the outset of metamorphosis they show a dramatic pruning of their larval axons and subsequently regrow new axons during adult differentiation (Technau and Heisenberg, 1982).

Many imaginal neurons are apparently used for the processing of inputs from the new adult sensory structures. Hence, ingrowing axons from the developing eye and antenna have a profound impact on the development of their associated brain areas as detailed in the following discussions. The roles of other sensory projections on the metamorphosis and development on other parts of the brain or the ventral CNS are poorly understood. In *Sarcophaga,* for example, the extirpation of a mesothoracic leg imaginal disc prior to metamorphosis results in the loss of small cells from the cortex of the adult thoracic ganglion (Chiarodo, 1963). Whether these missing cells are glial cells or imaginal neurons involved in processing information from the adult leg is not known.

1. Development of Optic Lobes

The most dramatic metamorphic change in the insect CNS is the appearance of the adult optic lobes (Fig. 10). The replacement of the simple lateral eyes (the stemmata) of the larva by the compound eyes

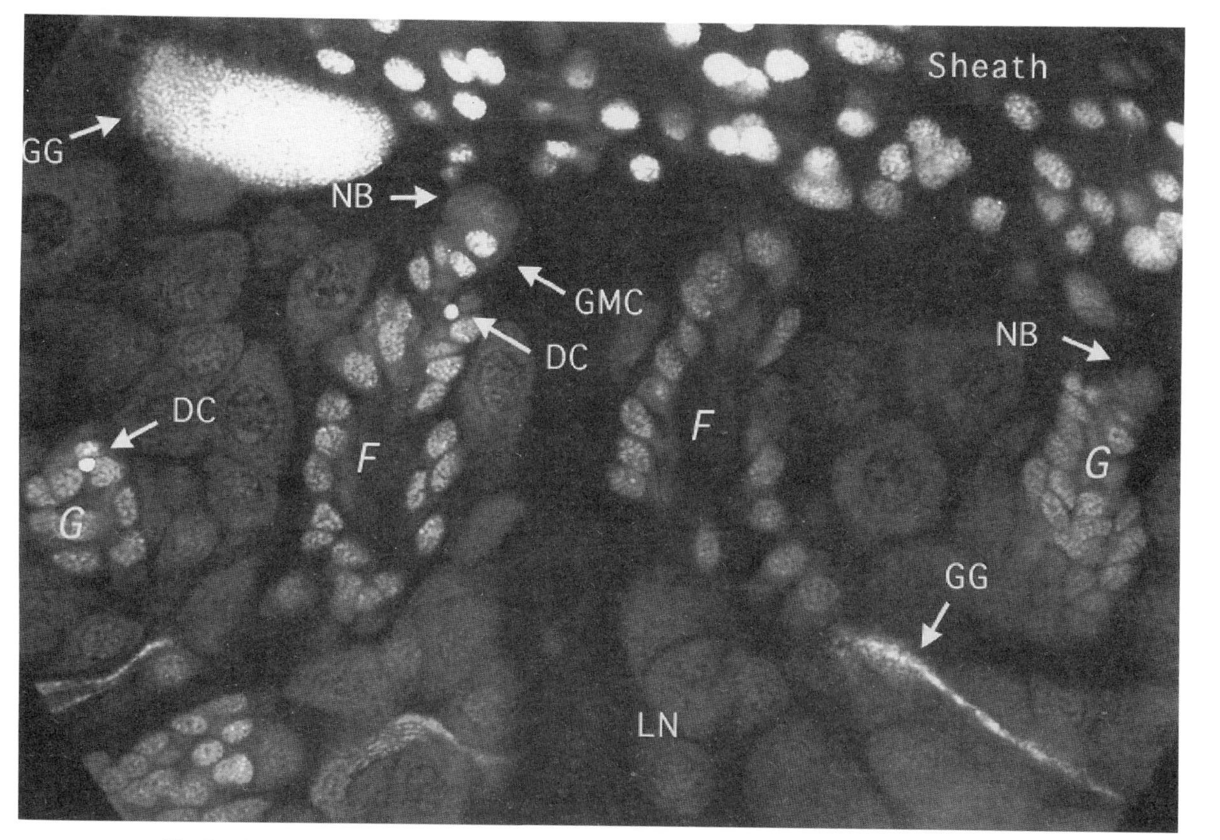

Fig. 8. Optical section of the anterior ventral region of a thoracic ganglion from a fifth instar *Manduca* larva showing four clusters of immature imaginal neurons (the two "F" and two "G" clusters) that are situated among the larval neurons (LN). DC, dying cell; GG, giant glia, GMC, ganglion mother cell; NB, neuroblast.

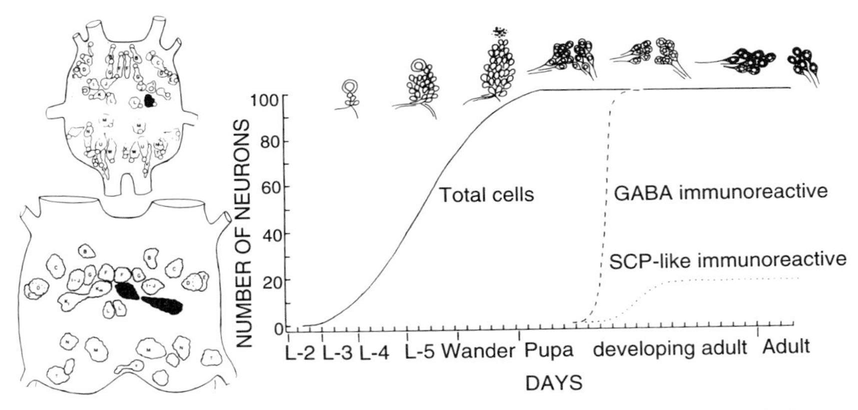

Fig. 9. Spatial and temporal aspects of the production of imaginal neurons in larvae of *Manduca.* (Left) Ventral view of ganglion T2 from a larva (top) and an adult (bottom) showing the position of the NBs (circles) and their associated clusters of progeny (irregular outlines). The NBs are only present in the larva. (Right) The time course of production of neurons by the K neuroblast (in black at left) and their subsequent acquisition of their transmitter phenotypes; all eventually express γ-aminobutyric acid (GABA) and 15 also express a SCP-like peptide. [Reprinted with permission from Witten and Truman *J. Neurosci.* **11,** 1980–89 (1991).]

of the adult is accompanied by the development of this complex brain area for visual integration. Each optic lobe has three primary integration areas, the lamina, medulla, and lobula/lobula plate, that are ordered from the compound eye to the central brain. The patterns of connections from the photoreceptors to the interneurons in the lamina and medulla are well known for flies (Meinertzhagen and Hanson, 1993).

The formation of the optic lobes begins early in larval life as two bands of NBs, the inner and outer proliferation zones, begin dividing. Initially these cells divide symmetrically as the population of NBs expands. As metamorphosis approaches, however, the NBs switch to producing immature neurons according to the pattern of asymmetic divisions characteristic of neuronal stem cells (Nordlander and Edwards, 1969). The NBs of the outer proliferation zone produce neurons of the lamina and the medulla. The neurons of the lobula/lobula plate come from the inner zone.

The development of the optic lobes is closely linked to that of the compound eye. In *Drosophila,* as the eye imaginal disc grows in the late last larval instar, a morphogenetic furrow passes from posterior to

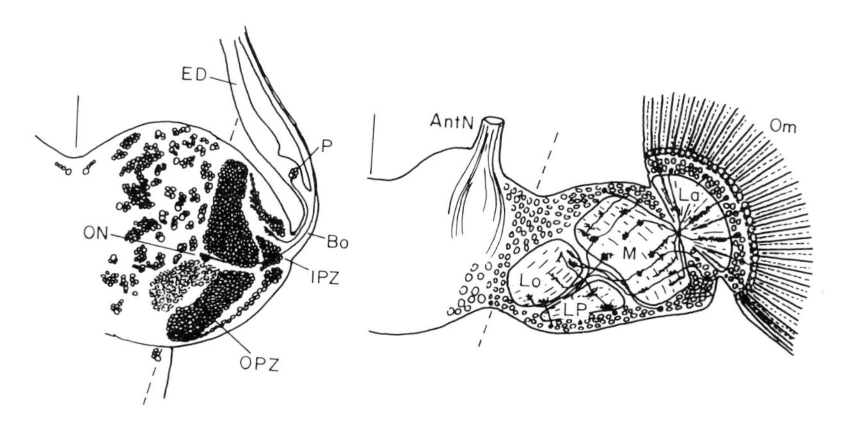

Fig. 10. Metamorphosis of the optic lobe of *Drosophila.* (Left) Dorsal view of the right brain hemisphere of a last stage larva showing cells having undergone DNA synthesis during the past 6 hr. These include NBs and their lineages in the medial brain and cells in the inner (IPZ) and outer proliferation zones (OPZ) in the optic lobes. Also shown is the eye imaginal disc (ED) indicating one set of forming photoreceptors (P) whose axons project into the forming lamina. The axons of the larval photoreceptors (Bo) and their termination in the optic association neuropil (ON) are also shown. (Right) The same region of the adult brain showing the main optic neuropils, the lamina (La), medulla (M), lobula (Lo), and lobula plate (LP). Representative optic lobe neurons have been included. AntN, antennal nerve; Om, base of the ommatidia of the compound eye.

anterior across the surface of the disc (Wolff and Ready, 1993). Cells in the wake of the furrow begin to organize into ommatidial units and differentiate into various cell types, including photoreceptors. This wave of ommatidia production results in a wave of ingrowth of photoreceptor axons with the first ones arriving during the middle of the last larval stage and the last ones by about 10 hr after puparium formation (APF). These ingrowing axons have an important organizational action on the forming optic lobe. First, the retinal axons provide a signal needed by lamina NBs in order for them to divide (Selleck and Steller, 1991). Consequently, the wave of photoreceptor production over the surface of the eye produces a wave of proliferation that sweeps over the surface of the lamina and finishes by about 25 hr APF. The subsequent survival of these lamina neurons is then dependent on their establishing connections with the receptor axons, and a wave of neuronal death then follows the proliferation wave and is finished by about 45 hr APF (Hofbauer and

Campos-Ortega, 1990). During the first half of metamorphosis, this asynchrony of development across the eye and lamina is also reflected in the deeper layers of the optic lobes. These developmental gradients are not evident during the last half of metamorphosis.

2. *Development of Antennal Lobes*

The metamorphosis of the antennal lobes has been most extensively studied in Lepidoptera such as *Manduca sexta* (Hildebrand, 1985). The receptor axons from the adult antenna project to an association area in the deutocerebrum, the antennal lobes. Each lobe is organized into about 60 discrete glomeruli. The antennal lobe is sexually dimorphic in that males possess an additional large glomerulus, the macroglomerular complex (MGC), that receives input from the sex pheromone-sensitive hairs that are specific to the male antenna. Two major classes of interneurons have processes in the antennal lobes: local interneurons that have no output axon and typically branch in most or all of the glomeruli and output interneurons that have a dendritic tuft in a single glomerulus and an output axon that projects to the protocerebrum.

The antennal afferents play an important role in the organization of the antennal lobes (see Hildebrand, 1985; Oland *et al.,* 1990). Unlike in the eye, though, the ingrowth of antennal afferents occurs too late to have any effect on the production of the antennal lobe interneurons nor do afferents seem necessary for the survival of these interneurons. In the absence of afferent input, however, the organization of the antennal lobe remains in a protoglomerular stage. The local interneurons show sparser dendritic arbors than do their normal counterparts and the output neurons exhibit diffuse dendrites in contrast to their normal uniglomerular tufts. This organizing effect of the afferents seems to be mediated through glial cells (Tolbert and Oland, 1989; Oland *et al.,* 1990). The ingrowing afferents cause changes in the distribution of glia in a fashion that prefigures the organization of the future glomeruli. Early removal of these glial cells by a variety of means prevents the organization of the glomeruli even in the presence of antennal axons.

IV. SEX-SPECIFIC DIFFERENTIATION WITHIN THE CENTRAL NERVOUS SYSTEM

An interesting issue that arises during metamorphosis is the transformation of a monomorphic larval CNS into a dimorphic adult structure.

Although it is functionally monomorphic, the larval CNS already has covert dimorphisms due to sex-specific patterns of imaginal neuron production. In the abdominal ganglia of *Manduca,* for example, male larvae have an additional pair of NBs, the I-J NB, that is not found in female larvae (Booker and Truman, 1987a). Male larvae also produce more midline neurons in their terminal ganglia than do female larvae (Thorn and Truman, 1994a). Similarly, in *Drosophila* the terminal abdominal NBs generate more cells in male larvae than they do in females (Taylor and Truman, 1992).

In addition to sex-specific neurogenesis, dimorphisms are also established through cell death. This is especially evident in the motor system innervating the skeletal musculature of the genitalia (Fig. 11). Larvae of both sexes have identical sets of motoneurons that innervate identical sets of body wall muscles. During metamorphosis the larval muscles degenerate and then redifferentiate in a sex-specific manner to supply the male or female genitalia. Within the CNS, cell death during the pupal–adult transition acts on this common set of larval motoneurons to select subsets of cells for use in males or in females (Giebultowicz and Truman, 1984). A few motoneurons survive in both sexes but come to innervate muscles that have regrown in a sex-specific pattern. The subsequent remodeling of these motoneurons is similarly dimorphic (Thorn and Truman, 1989).

While most aspects of sex determination in insects are thought to be cell autonomous, cell–cell interactions seem to be involved in at least some aspects of sexual differentiation that involves the CNS. A striking example is the development of the MGC in the antennal lobe of male *Manduca.* The formation of the MGC requires the axons of the pheromone-sensitive afferents. Consequently, the removal of the antennal disc from male larvae at the start of metamorphosis results in brains with poorly developed antennal lobes that lack a MGC. If the removed male antennal disc is replaced by one from a female, a normal antennal lobe is formed but it has a female appearance because it lacks a MGC. The reciprocal experiment of transplanting a male antennal disc onto a female larvae results in the formation of a MGC in the brain of the female (Schneiderman *et al.,* 1982)! Not only are gross aspects of the antennal lobe changed in such animals, but the female olfactory interneurons show male-like differentiation with dendritic branches that extend into the MGC to contact the pheromone-sensitive afferents. Moreover, while females normally do not respond to the sex pheromone because they lack pheromone-sensitive receptors on their antennae, females bearing grafts of male antennae show behavioral orientation to pheromone

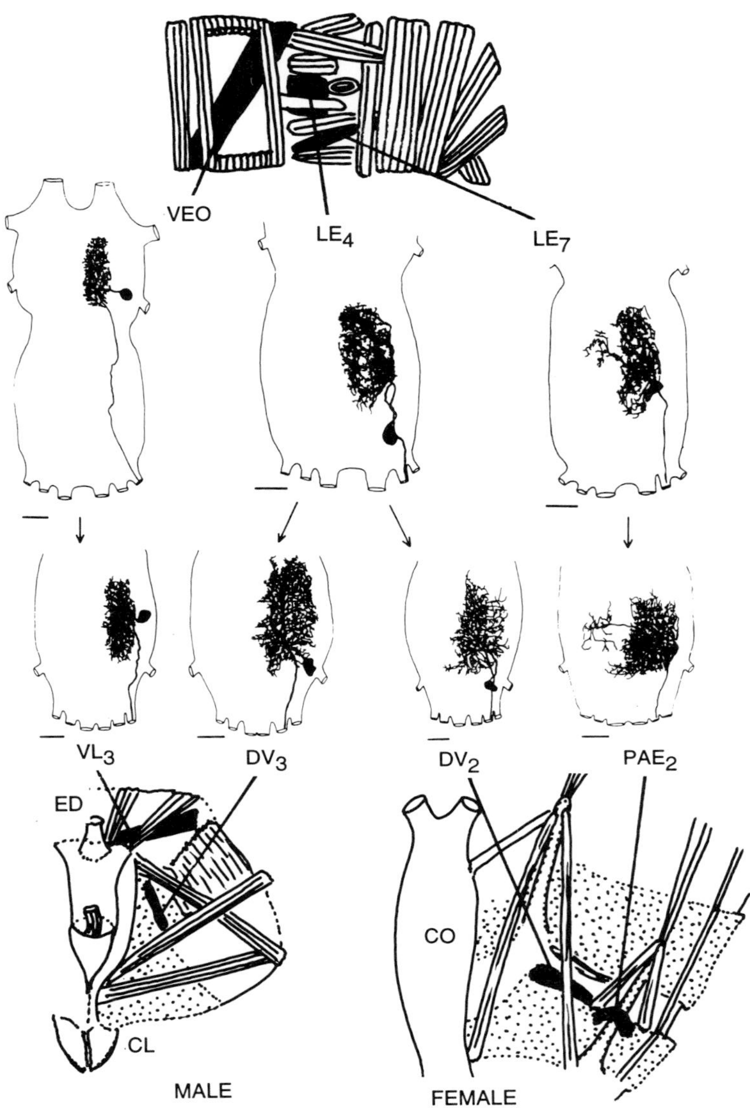

Fig. 11. Cobalt backfills of single motoneurons in *Manduca* depicting sex-related fates associated with the metamorphosis of the terminal abdominal segments. The top row shows the larval neurons and their muscles and the bottom row shows the corresponding neurons and the muscles that they innervate in the adult. The motoneuron to larval muscle VEO dies in females but survives and innervates muscle LV_3 in males; the motoneuron to larval muscle LE_7 dies in males but innervates muscle PAE_2 in females. [Reprinted from Thorn and Truman (1989) with permission from John Wiley and Sons.]

presented in airstreams (Schneidermann *et al.,* 1986). Thus, the sex of the antenna influences the sexual maturation of antennal lobe interneurons and perhaps even deeper brain structures.

Inductive interactions are also seen on the motor side. In *Drosophila,* males have a specially modified muscle, the male-specific muscle, on segment A5. Analysis of mosaic animals produced by embryonic cell transplantations showed that the sexual differentiation of this muscle depends on the sex of the CNS and not that of the epidermis or of the muscle itself (Lawrence and Johnson, 1986).

The reverse situation of the sexual differentiation of the motor system being influenced by the sex of the periphery is evident in the differentiation and survival of midline reproductive tract neurons in *Manduca* (Thorn and Truman, 1994a,b). These midline cells are born in the A7 and A9 region of the terminal ganglion and innervate the reproductive ducts arising from the female or the male genital disc. The A7 cells innervate the oviducts whereas the A9 set innervate the sperm ducts. Accordingly, the A7 cells persist in females but die in males whereas the A9 cells live in males but die in females. To determine whether a target was actually required for the survival of the cells in the appropriate sex, the forming sperm ducts were reduced or removed in male pupae by surgical or endocrine means. The number of A9 neurons that survived and matured in these males was directly related to the amount of sperm duct tissue the animal retained (Thorn and Truman, 1994b). Therefore, the presence of a target is necessary for the survival of these sexually dimorphic neurons. This type of target dependence is not seen in motoneurons that innervate nondimorphic muscles.

V. HORMONAL REGULATION OF METAMORPHIC CHANGES

A. Larval–Pupal Transition

The decline and disappearance of juvenile hormone (JH) early in the last stage allows metamorphosis to occur (see Gilbert *et al.,* Chapter 2, this volume). In the absence of JH, the appearance of ecdysteroids, which formerly caused molting from one larval stage to the next, now causes a metamorphic molt resulting in the formation of the pupal stage. There are two peaks of ecdysteroids during the larval–pupal transformation (Fig. 5). The initial commitment peak is relatively small; it turns off feeding behavior and induces behaviors directed toward finding an appropriate pupation site. It also "commits" larval tissues such as the epidermis for a pupal response when they next see ecdysteroids (Riddiford, 1978). There then follows the prepupal peak that actually causes the molt to the pupal stage.

Larval neurons show no overt developmental changes when ecdysteroids acting in the presence of JH cause the molt from one larval stage to the next. The commitment peak is the first time that they see ecdysteroids in the absence of JH. This peak also has no obvious effect on their synaptic connections or morphology, but it has a profound effect on their subsequent response to ecdysteroids. It renders these neurons insensitive to JH and it increases their sensitivity to ecdysteroids (Weeks and Truman, 1986). When next exposed to ecdysteroids at the prepupal peak, many larval neurons respond by pruning back their dendritic processes, irrespective of the presence or the absence of JH.

The most detailed studies of the relationship of the prepupal ecdysteroid peak to dendritic regression are for motoneuron PPR (Weeks and Truman, 1985). As seen in Fig. 12, this motoneuron shows dendritic regression 2 to 3 days after wandering, at the time that its muscle is degenerating. Prevention of the prepupal peak by abdominal ligation blocks both PPR's dendritic regression and the muscle loss while infusion of 20-hydroxyecdysone (20-HE) into ligated abdomens restores both responses. The correlation of the neuron's dendritic regression with the death of its target muscle does not mean that muscle loss triggers the pruning response. Indeed, the surgical removal of PPR's target in the larval stage does not induce precocious dendritic loss in PPR nor does it block the ability of the neuron to respond subsequently to the appropriate steroid signals. Therefore, both neuron and muscle seem to have coadapted to respond to the same endocrine cues.

PPR subsequently degenerates about 2 days after pupal ecdysis. Degeneration is also a response to the prepupal ecdysteroid peak even though the death of the neuron is delayed until a few days later. Interestingly, the regression and degeneration responses of PPR can be separated from one another by timed ligation experiments that restrict the amount of the prepupal peak to which the neuron is exposed (Weeks, 1987). A longer exposure to ecdysteroid is required for the degeneration response as compared to dendritic regression. This ability to separate the two responses underscores that dendritic regression is a discrete response of the cell and not a transient phase in the course of the cell's demise. This

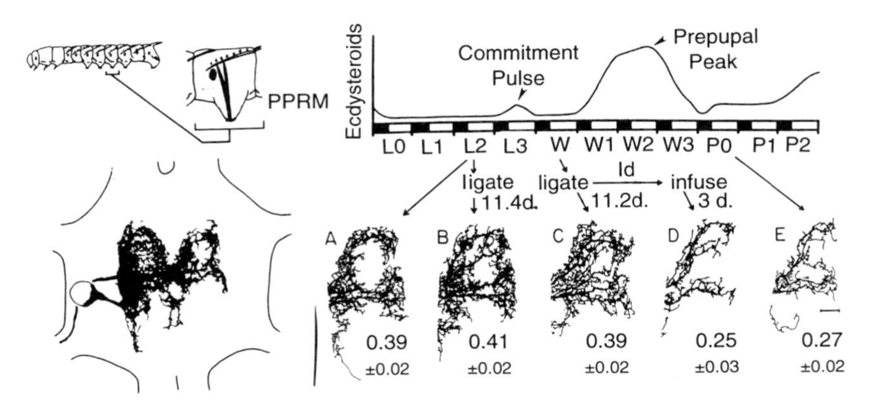

Fig. 12. Effect of steroid manipulations on dendrite loss in motoneuron PPR in *Manduca*. (Left) The central morphology of PPR and the location of its target muscle (PPRM). (Right) The blood ecdysteroid titer during the larval–pupal transformation and a summary of the effects of various manipulations on the dendritic arbor of PPR. Drawings are of the portion of the dendritic tree contralateral to the cell body and numbers give a measure of the density of dendritic branching (mean ±SE). A ligature placed between the thorax and abdomen deprives the abdomen of ecdysteroids; infusions were with 5.6 μg of 20-HE/hr for 12 hr. [Reprinted with permission from Weeks and Truman *J. Neurosci* **5,** 2290–2300 (1985).]

conclusion is also supported by the fact that many neurons that show similar patterns of dendritic regression survive to subsequently grow adult-specific arbors (e.g., MN-3 in Fig. 5).

The death of neurons after pupal ecdysis appears to be a direct response to ecdysteroid. Experiments involving isolating and maintaining the APR motoneurons *in vitro* have shown that these neurons survive in culture for extended periods if maintained without ecdysteroids but that addition of 20-HE to the culture results in their eventual death (Streichert and Weeks, 1994). Importantly, the response of APR to 20-HE depends on the abdominal ganglion from which the neurons was isolated. APRs from A6 subsequently die when cultured with 20-HE, whereas those from A4 continue to live when subjected to the same treatment. Thus, the segment-specific fates of the APRs that are evident *in vivo* are maintained when the cells are isolated *in vitro*.

While larval neurons respond to the prepupal ecdysteroid peak by pruning back dendritic and axonal arbors, the situation is different for the imaginal neurons. Since the latter are arrested immature neurons, they have no larval specializations to remove and they respond to the prepupal peak by initiating maturation. In *Manduca* the resumption of maturation is first evident as an increase in the size of the cell body (Booker and Truman, 1987b). Studies in *Drosophila* indicate that this size increase is preceeded by an up-regulation of regulatory genes, such as the homeotic gene *ultrabithorax* (Glicksman and Truman, 1990). This up-regulation is presumably in preparation for the cells' rapid maturation that will come during the pupal–adult transition.

B. Pupal–Adult Transition

A prolonged release of ecdysteroids causes the transformation of the pupa into the adult (Warren and Gilbert, 1986). If this steroid surge is averted, as in the case of pupal diapause, the larval neurons regress but then remain in this regressed condition. Infusion of 20-HE then induces dendritic sprouting within 36 hr (Truman and Reiss, 1988). Application of JH mimics can block this maturation but only if given during the early stages of the outgrowth response.

An intriguing aspect of the response of these larval neurons to ecdysteroid is that the same cell shows different responses (dendritic regression versus sprouting) depending on whether it is in the larval or the pupal stage. Insight into the basis of this stage specificity comes from the culture experiments of Prugh *et al.* (1992), in which they maintained identified leg motoneurons from *Manduca* in the presence or absence of 20-HE. When taken from the pupal stage, these motoneurons showed good survival in culture with or without ecdysteroid but the addition of 20-HE resulted in profuse dendritic growth, a response that parallels the situation seen *in vivo*. Importantly, this 20-HE-induced sprouting was suppressed by the addition of JH mimics to the medium. Moreover, the same leg motoneurons taken from the larval stage also showed good survival in culture but showed no steroid-dependent growth. Indeed, there was a tendency for these cells to have fewer processes in the presence of 20-HE. Hence, the stage specificity of response to 20-HE seen *in vivo* is preserved when these cells are challenged with steroid *in vitro*. These results argue that the stage-specific alteration in a neuron's response to 20-HE is due to changes that are intrinsic to the neuron itself and are not dependent on alterations in the central or peripheral environment of the cell.

The final metamorphic change in the CNS is the wave of neuronal death that occurs after adult emergence. In both *Manduca* and *Drosophila* the withdrawal of ecdysteroids at the end of metamorphosis is essential for these cells to die. Treatment with 20-HE either *in vivo* (Truman and Schwartz, 1984; Robinow *et al.*, 1993) or to cultured ganglia (Bennett and Truman, 1985) delays or prevents the death of these neurons. Although the steroid decline is *necessary* for cell death, it alone may not be *sufficient* for triggering the death of some of these cells. This is indicated by the effectiveness of behavioral manipulations in delaying the degeneration response in newly emerged insects (Truman, 1983; Kimura and Truman, 1990). Moreover, nerve cord transection experiments can permanently save some of the "doomed" neurons, indicating that descending a neural signal is needed to trigger their death (Fahrbach and Truman, 1989).

VI. RELATIONSHIP OF ECDYSONE RECEPTOR TO METAMORPHIC CHANGES IN THE CENTRAL NERVOUS SYSTEM

An intriguing aspect of the metamorphosis of the CNS is that a single hormone, 20-HE, regulates a complex temporal and spatial pattern of neuronal responses. Work on *Drosophila* has focused on the ecdysone receptor (EcR) as a potential site for controlling this diversity.

The *ecr* gene encodes three forms of the ecdysone receptor (EcR-A, EcR-B1, and EcR-B2) (Koelle *et al.*, 1991; Talbot *et al.*, 1993; see Cherbas and Cherbas, Chapter 5, this volume). These forms of EcR then combine with another member of the nuclear hormone receptor superfamily, *ultraspiracle,* to make the active heterodimer receptor (Yao *et al.*, 1993). Studies with antibodies to regions common to all EcR isoforms as well as antibodies specific for EcR-A and for EcR-B1 (Talbot *et al.*, 1993) show that neurons exhibit complex temporal and spatial patterns of EcR expression which correlate with the different types of steroid responses (Truman *et al.*, 1994).

Through most of larval life, larval neurons lack detectable EcR. Similarly, they fail to show developmental responses to the ecdysteroid surges that cause larval molting. EcR-A and EcR-B1 both appear in these cells midway through the last larval instar (Figs. 13 and 14A). Initially, EcR-B1 levels are much higher than EcR-A and it is the major isoform present

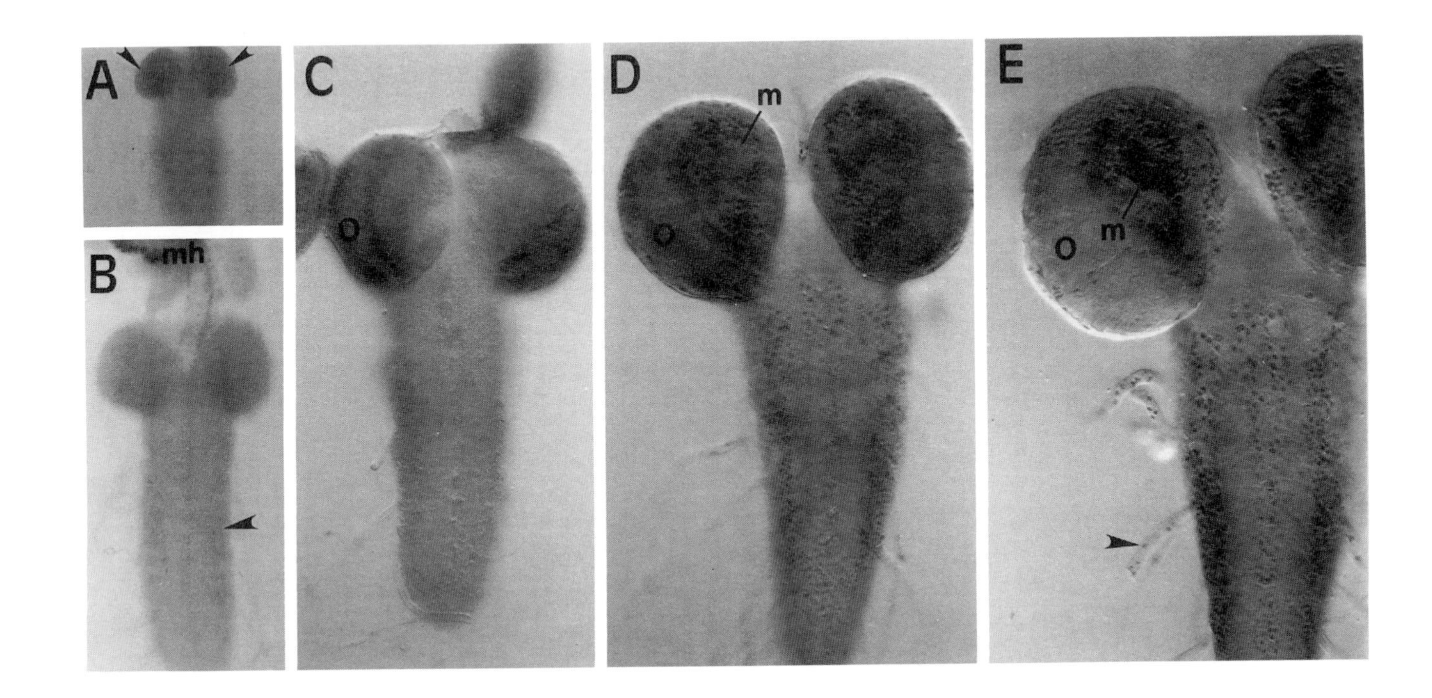
A
B
mh
C
o
D
m
o
E
o
m

during the ecdysteroid peak that causes the larval–pupal transition. EcR-B1 levels then drop and EcR-A becomes predominant during the transition from pupa to adult. The presence of different EcR isoforms during the two major peaks of ecdysteroids is intriguing since the first ecdysteroid peak causes synapse loss and neurite regression whereas the second induces sprouting and synaptogenesis.

The relationship of a particular EcR isoform to a particular type of neuronal response was consistent for all classes of central neurons. For example, during the larval–pupal transition when most larval neurons express high levels of EcR-B1, most imaginal neurons express only EcR-A (Fig. 14B). Since imaginal neurons have no larval specializations to lose, they show maturational responses during the larval–pupal transition, in association with this early expression of EcR-A.

There are two sets of imaginal neurons that are exceptional in that they show high levels of EcR-B1 expression. The mushroom bodies in the central brain show prominent EcR-B1 expression at the onset of metamorphosis (Fig. 14C). As discussed earlier, they are the only imaginal neurons known to prune back their larval processes at the start of metamorphosis (Technau and Heisenberg, 1982) and the time of highest EcR-B1 expression coincides with the time of the pruning.

The optic lobes neurons show very strong (EcR-B1 expression from 10 to 55 hr APF (Fig. 14D). This temporal window begins at the time the last retinal axons arrive in the optic lobes and continues through the period when waves of differentiation sweep across the optic lobes. Through this interval the optic lobes have the problem of maintaining a gradient of differentiation in an environment of high ecdysteroids that promotes the rapid maturation of other imaginal neurons. It has been speculated that EcR-B1 expression in the optic lobes through this period may allow neurons to remain plastic and to participate in inductive interactions even in the face of these high ecdysteroid titers (Truman *et*

Fig. 13. CNS from larval *Drosophila* immunostained to show the acquisition of EcR-B1 in preparation from metamorphosis. (A) Ten-hour first instar larva; arrowheads: the position of a pair of weakly staining, brain neurons. (B) Twelve-hour second instar larva with weak staining in tracheal nuclei (arrow); mouth hooks (mh) show strong nuclear staining. (C–E) third instar larvae at 14 (C), 30 (D), and 40 (E) hr postecdysis. As larvae aged, B1 staining was lost from the OL proliferation zones (O) but appeared in larval neurons (D,E) and mushroom body neurons (E; m). Arrowhead: peripheral glia. [Reprinted with permission from Truman *et al.* (1994).]

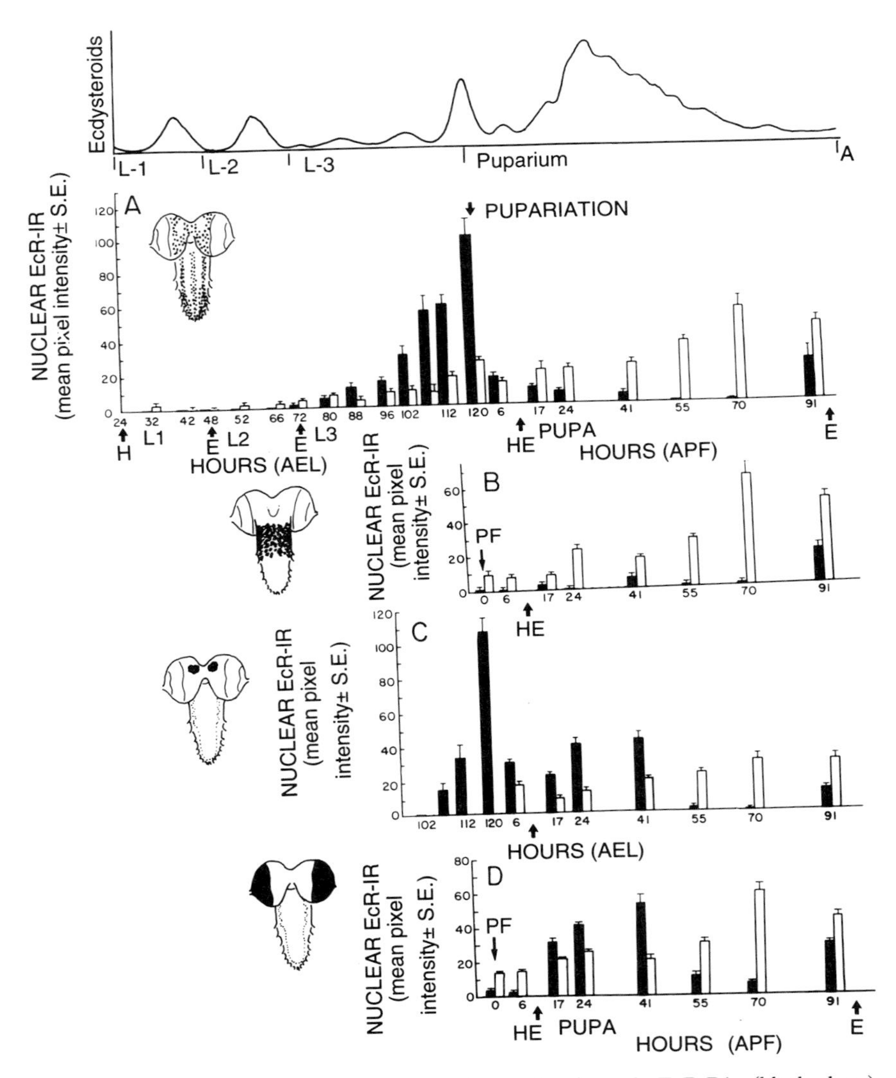

Fig. 14. Relative levels of EcR-A (open bars) and EcR-B1 (black bars) immunofluorescence in the nuclei of of various types of neurons during larval life and metamorphosis in *Drosophila.* (A) Larval neurons in ventral CNS; (B) imaginal neurons in the thorax; (C) imaginal neurons of the mushroom bodies; and (D) imaginal neurons of the optic lobes (bars give mean ±SE with background subtracted). AEL, after egg laying; APF, after puparium formation; E, ecdysis; H, hatching; HE, head eversion; L1, first larval stage; *etc.* [Data are from Truman *et al.* (1994). The ecdysteroid titer is modified from Riddiford (1993).]

al., 1994). The EcR-B1 fades away at the time that these gradients disappear.

Although most variations in EcR expression are evident at the start of metamorphosis, a few neurons acquire altered patterns of EcR that are maintained throughout this transition. Starting at 12–18 hr APF a group of about 300 larval neurons begin to express levels of EcR-A that are about 10 times higher than their neighbors. Their receptor levels remain high through the remainder of metamorphosis and into adult life. Between 4 and 10 hr of adult life, all of these neurons then undergo programmed cell death (Robinow *et al.,* 1993). Similar elevated levels of ecdysteroid binding are also seen in doomed neurons in *Manduca* (Fahrbach and Truman, 1989). Thus, neurons that are going to die are set apart from other cells in their receptor expression from the start of the pupal–adult transition. Whether elevated levels of EcR-A is the switch that puts these cells on the pathway that results in their eventual death has yet to be determined.

Although there are strong correlations between EcR isoform expression and the nature of a neuron's response to ecdysteroids, there is not yet direct evidence that the relationship is causal. Most of the variation in EcR expression does occur at the outset of metamorphosis. Presumably, this early heterogeneity reflects the different needs of the various neuronal types as the ecdysteroid titer begins to knit the larval and imaginal neurons into a unified CNS. The ability to work with individual insect neurons as unique cells with defined fates will undoubtedly aid in establishing the role of EcR isoforms in directing these diverse responses.

REFERENCES

Altman, J. S., and Tyrer, N. M. (1974). Insect flight as a system for the study of neuronal connections. *In* "Experimental Analysis of Insect Behaviour" (L. Barton Browne, ed.), pp. 159–179. Springer-Verlag, Berlin.

Anderson, H. (1978). Postembryonic development of the visual system of the locust *Schistocerca gregaria.* I. Patterns of growth and developmental interactions in the retina and optic lobes. *J. Embryol. Exp. Morphol.* **45,** 55–83.

Bate, C. M. (1973). The mechanism of the pupal gin-trap. I. Segmental gradients and the connections of the triggering sensilla. *J. Exp. Biol.* **59,** 95–107.

Bate, C. M. (1976). Embryogenesis of an insect nervous system. I. A map of the thoracic and abdominal neuroblasts in *Locusta migratoria. J. Embryol. Exp. Morphol.* **35,** 107–123.

Bate, C. M., and Grunewald, E. G. (1981). Embryogenesis of an insect nervous system. II. A second class of precursor cells and the origin of the intersegmental connectives. *J. Embryol. Exp. Morphol.* **61,** 317–330.

Bennett, K. L., and Truman, J. W. (1985). Steroid-dependent survival of identifiable neurons in cultured ganglia of the moth *Manduca sexta. Science* **229,** 58–60.

Bollenbacher, W. E., Smith, S. L., Goodman, W., and Gilbert, L. I. (1981). Ecdysteroid titer during the larval–pupal–adult development of the tobacco hornworm, *Manduca sexta. Gen. Comp. Endocrinol.* **44,** 302–306.

Booker, R., and Truman, J. W. (1987a). Postembryonic neurogenesis in the CNS of the tobacco hornworm, *Manduca sexta.* I. Neuroblast arrays and the fate of their progeny during metamorphosis. *J. Comp. Neurol.* **255,** 548–559.

Booker, R., and Truman, J. W. (1987b). Postembryonic neurogenesis in the CNS of the tobacco hornworm, *Manduca sexta.* II. Hormonal control of imaginal nest cell degeneration and differentiation during metamorphosis. *J. Neurosci.* **7,** 4107–4114.

Casaday, G. B., and Camhi, J. M. (1976). Metamorphosis of the flight motor neurons in the moth *Manduca sexta. J. Comp. Physiol.* **112,** 143–158.

Cayre, M., Strambi, C., and Strambi, A. (1994). Neurogenesis in an adult insect brain and its hormonal control. *Nature* (*London*) **368,** 57–59.

Chiarodo, A. J. (1963). The effects of mesothoracic leg extirpation on the postembryonic development of the nervous system of the blowfly, *Sarcophaga bullata. J. Exp. Zool.* **153,** 263–277.

Currie, D., and Bate, M. (1991). Development of adult abdominal muscles in *Drosophila:* Adult myoblasts express *twist* and are associated with nerves. *Development* **113,** 91–102.

Davis, R. L. (1993). Mushroom bodies and *Drosophila* learning. *Neuron* **11,** 1–14.

Doe, C. Q. (1992). Molecular markers for identified neuroblasts and ganglion mother cells in the *Drosophila* nervous system. Development **16,** 855–864.

Doe, C. Q., and Goodman, C. S. (1985). Early events in insect neurogenesis. I. Development and segmental differences in the pattern of neuronal precursor cells. *Dev. Biol.* **111,** 193–205.

Fahrbach, S. E., and Truman, J. W. (1987). Possible interactions of a steroid hormone and neuronal inputs in controlling the death of an identified neuron in the moth *Manduca sexta. J. Neurobiol.* **18,** 497–508.

Fahrbach, S. E., and Truman, J. W. (1989). Autoradiographic identification of ecdysteroid binding cells in the nervous system of the moth *Manduca sexta. J. Neurobiol.* **20,** 681–702.

Giebultowicz, J. M., and Truman, J. W. (1984). Sexual differentiation in the terminal ganglion of the moth *Manduca sexta:* Role of sex-specific neuronal death. *J. Comp. Neurol.* **226,** 87–95.

Glicksman, M. A., and Truman, J. W. (1990). Regulation of the homeotic gene *Ultrabithorax* by ecdysone in the *Drosophila* larval central nervous system. *Soc. Neurosci. Abstr.* **16,** 324.

Goodman, C. S., and Doe, C. Q. (1993). Embryonic development of the *Drosophila* central nervous system. *In* "The Development of *Drosophila melanogaster*" (M. Bate and A. Martinez Arias, eds.), pp. 1131–1206. Cold Spring Harbor Laboratory, Cold Spring Harbor, New York.

Hartenstein, V., Rudloff, E., and Campos-Ortega, J. A. (1987). The pattern of proliferation of the neuroblasts in the wild-type embryo of *Drosophila melanogaster. Wilhelm Roux's Arch. Dev. Biol.* **196,** 473–485.

Heinertz, R. (1976). Untersuchungen am thorakalen Nervensystem von *Antheraea polyphemus* Cr. (Lepidoptera) unter besonderer Berücksichtigung der Metamorphose. *Rev. Suisse Zool.* **83,** 215–242.

Hildebrand, J. G. (1985). Metamorphosis of the insect nervous system. Influences of the periphery on the postembryonic development of the antennal sensory pathway in the brain of *Manduca sexta. In* "Model Neural Networks and Behavior" (A. I. Selverston, ed.), pp. 129–148. Plenum, New York.

Hofbauer, A., and Campos-Ortega, J. A. (1990). Proliferation pattern and early differentiation of the optic lobes in *Drosophila melanogaster. Wilhelm Roux's Arch. Dev. Biol.* **198,** 264–274.

Jacobs, G. A., and Weeks, J. C. (1990). Postsynaptic changes at a sensory-to-motor neuron synapse contribute to the developmental loss of a reflex behavior during insect metamorphosis. *J. Neurosci.* **10,** 1341–1356.

Kent, K. S., and Levine, R. B. (1988). Neural control of leg movements in a metamorphic insect: Persistence of the larval leg motor neurons to innervate the adult legs of *Manduca sexta. J. Comp. Neurol.* **276,** 303.

Kent, K. S., and Levine, R. B. (1993). Dendritic reorganization of an identified neuron during metamorphosis of the moth *Manduca sexta:* The influence of interactions with the periphery. *J. Neurobiol.* **24,** 1–22.

Kimura, K.-I., and Truman, J. W. (1990). Postmetamorphic cell death in the nervous and muscular systems of *Drosophila melanogaster. J. Neurosci.* **10,** 403–411.

Koelle, M. R., Talbot, W. S., Segraves, W. A., Bender, M. T., Cherbas, P., and Hogness, D. S. (1991). The *Drosophila EcR* gene encodes an ecdysone receptor, a new member of the steroid receptor superfamily. *Cell (Cambridge, Mass.)* **67,** 59–77.

Lawrence, P. A., and Johnston, P. (1986). The muscle pattern of a segment of *Drosophila* may be determined by neurons and not by contributing myoblasts. *Cell (Cambridge, Mass.)* **45,** 505–513.

Levine, R. B., and Truman, J. W. (1982). Metamorphosis of the insect nervous system: Changes in the morphology and synaptic interactions of identified cells. *Nature (London)* **299,** 250–252.

Levine, R. B., and Truman, J. W. (1985). Dendritic reorganization of abdominal motoneurons during metamorphosis of the moth, *Manduca sexta. J. Neurosci.* **5,** 2424–2431.

Levine, R. B., Pak, C., and Linn, D. (1985). The structure, function and metamorphic reorganization of somatotopically projecting sensory neurons in *Manduca sexta* larvae. *J. Comp. Physiol. A* **157,** 1–13.

Meinertzhagen, I. A., and Hanson, T. E. (1993). The development of the optic lobes. *In* "The Development of *Drosophila melanogaster*" (M. Bate and A. Martinez Arias, eds.), pp. 1363–1491. Cold Spring Harbor Laboratory, Cold Spring Harbor, New York.

Mesce, K. A., and Truman, J. W. (1988). Metamorphosis of the ecdysis motor pattern in the hawkmoth, *Manduca sexta. J. Comp. Physiol. A* **163,** 287–299.

Murphey, R. K., and Chiba, A. (1990). Assembly of the cricket cercal sensory system: Genetic and epigenetic control. *J. Neurobiol.* **21,** 120–137.

Nordlander, R. H., and Edwards, J. S. (1969). Postembryonic brain development in the monarch butterfly, *Danaus plexippus plexippus,* L. II. The optic lobes. *Wilhelm Roux's Arch. Dev. Biol.* **163,** 197–200.

O'Brien, M. A., Schneider, L. E., and Taghert, P. H. (1991). *In situ* hybridization analysis of the FMRFamide neuropeptide gene in *Drosophila.* II. Constancy in the cellular pattern of expression during metamorphosis. *J. Comp. Neurol.* **304,** 623–638.

Oland, L. A., Orr, G., and Tolbert, L. P. (1990). Construction of a protoglomerular template by olfactory axons initiates the formation of olfactory glomeruli in the insect brain. *J. Neurosci.* **10,** 2096–2112.

Pflugfelder, O. (1937). Die Entwicklung der optischen Ganglien von *Culex pipiens. Zool. Anz.* **117,** 31–36.

Pipa, R. L. (1967). Insect neurometamorphosis—III. Nerve cord shortening in a moth, *Galleria mellonella* (L.), may be accomplished by hormonal potential of neuroglia motility. *J. Exp. Zool.* **164,** 47–60.

Prokop, A., and Technau, G. M. (1991). The origin of postembryonic neuroblasts in the ventral nerve cord of *Drosophila melanogaster. Development* **111,** 79–88.

Prugh, J., Della Croce, K., and Levine, R. B. (1992). Effects of the steroid hormone, 20-hydroxyecdysone, on the growth of neurites by identified insect motoneurons *in vitro. Dev. Biol.* **154,** 331–347.

Riddiford, L. M. (1978). Ecdysone-induced change in cellular commitment of the epidermis of the tobacco hornworm, *Manduca sexta,* at the initiation of metamorphosis. *Gen. Comp. Endocrinol.* **34,** 438–446.

Riddiford, L. M. (1993). Hormones and *Drosophila* development. *In* "The Development of *Drosophila melanogaster*" (M. Bate and A. Martinez Arias, eds.), pp. 899–939. Cold Spring Harbor Laboratory, Cold Spring Harbor, New York.

Robertson, R. M., and Pearson, K. G. (1983). Interneurons in the flight system of the locust: Distributions, connections and resulting properties. *J. Comp. Neurol.* **215,** 33–50.

Robinow, S., Talbot, W. S., Hogness, D. S., and Truman, J. W. (1993). Programmed cell death in the *Drosophila* CNS is ecdysone-regulated and coupled with a specific ecdysone receptor isoform. *Development* **119,** 1251–1259.

Sanes, J. R., and Hildebrand, J. G. (1976). Structure and development of antennae in a moth, *Manduca sexta. Dev. Biol.* **51,** 282–299.

Schaffer, R., and Sanchez, T. V. (1973). Antennal sensory system of the cockroach *Periplanta americana:* Postembryonic development and morphology of sense organs. *J. Comp. Neurol.* **149,** 335–354.

Schneiderman, A. M., Matsumoto, S. G., and Hildebrand, J. G. (1982). Trans-sexually grafted antennae influence development of sexually dimorphic neurones in moth brain. *Nature* (*London*) **298,** 844–846.

Schneiderman, A. M., Hildebrand, J. G., Brennan, M. M., and Tumlinson, J. H. (1986). Trans-sexually grafted antennae alter pheromone-directed behavior in a moth. *Nature* (*London*) **823,** 801–803.

Selleck, S. B., and Steller, H. (1991). The influence of retinal innervation on neurogenesis in the first optic ganglion of *Drosophila. Neuron* **6,** 83–99.

Shepherd, D., and Bate, C. M. (1990). Spatial and temporal patterns of neurogenesis in the embryo of the locust (*Schistocerca gregaria*). *Development* **108,** 83–96.

Stevenson, P. A., and Kutsch, W. (1988). Demonstration of functional connectivity of the flight motor system in all stages of the locust. *J. Comp. Physiol. A* **162,** 247–259.

Streichert, L. C., and Weeks, J. C. (1994). Selective death of identified *Manduca sexta* motoneurons is induced by ecdysteroids *in vitro. Soc. Neurosci. Abstr.* **20,** 461.

Taghert, P. H., and Goodman, C. S. (1984). Cell determination and differentiation of identified serotonin-containing neurons in the grasshopper embryo. *J. Neurosci.* **4,** 989–1000.

Talbot, W. S., Swyryd, E. A., and Hogness, D. S. (1993). *Drosophila* tissues with different metamorphic responses to ecdysone express different ecdysone receptor isoforms. *Cell* (*Cambridge, Mass.*) **73,** 1323–1337.

Taylor, H. M., and Truman, J. W. (1974). Metamorphosis of the abdominal ganglia of the tobacco hornworm, *Manduca sexta:* Changes in populations of identified motor neurons. *J. Comp. Physiol.* **90,** 367–388.

Taylor, B. J., and Truman, J. W. (1992). Commitment of abdominal neuroblasts in *Drosophila* to a male or female fate is dependent on genes of the sex-determining hierarchy. *Development* **114,** 625–642.

Technau, G., and Heisenberg, M. (1982). Neural reorganization during metamorphosis of the corpora pedunculata in *Drosophila melanogaster. Nature (London)* **295,** 405–407.

Thomas, J. B., Bastiani, M. J., Bate, M., and Goodman, C. S. (1984). From grasshopper to *Drosophila:* A common plan for neuronal development. *Nature (London)* **310,** 203–207.

Thompson, K. J. (1993). Embryonic neural activity and serial homology underlying grasshopper oviposition. *Soc. Neurosci. Abstr.* **19,** 348.

Thompson, K. J., and Siegler, M. V. S. (1993). Development of segment specificity in identified lineages of the grasshopper CNS. *J. Neurosci.* **13,** 3309–3318.

Thorn, R. S., and Truman, J. W. (1989). Neuronal respecification during the metamorphosis of the genital segments in the tobacco hornworm moth, *Manduca sexta. J. Comp. Neurol.* **284,** 489–503.

Thorn, R. S., and Truman, J. W. (1994a). Sexual differentiation in the CNS of the moth, *Manduca sexta.* I. Sex and segment-specificity in production, differentiation, and survival of the imaginal midline neurons. *J. Neurobiol.* **25,** 1039–1053.

Thorn, R. S., and Truman, J. W. (1994b). Sexual differentiation in the CNS of the moth, *Manduca sexta.* II. Target dependence for the survival of the imaginal midline neurons. *J. Neurobiol.* **25,** 1054–1066.

Tix, S., Minden, J. S., and Technau, G. M. (1989). Pre-existing neuronal pathways in the developing optic lobes of *Drosophila. Development* **105,** 739–746.

Tolbert, L. P., and Oland, L. A. (1989). A role for glia in the development of organized neuropilar structures. *Trends NeuroSci.* **12,** 70–75.

Truman, J. W. (1983). Programmed cell death in the nervous system of an adult insect. *J. Comp. Neurol.* **216,** 445–452.

Truman, J. W. (1990). Metamorphosis of the central nervous system of *Drosophila. J. Neurobiol.* **21,** 1072–1084.

Truman, J. W. (1992). Developmental neuroethology of insect metamorphosis. *J. Neurobiol.* **23,** 1404–1422.

Truman, J. W., and Bate, M. (1988). Spatial and temporal patterns of neurogenesis in the central nervous system of *Drosophila melanogaster. Dev. Biol.* **125,** 145–157.

Truman, J. W., and Reiss, S. E. (1976). Dendritic reorganization of an identified motoneuron during metamorphosis of the tobacco hornworm moth. *Science* **192,** 477–479.

Truman, J. W., and Reiss, S. E. (1988). Hormonal regulation of the shape of identified motoneurons in the moth. *Manduca sexta. J. Neurosci.* **8,** 765–775.

Truman, J. W., and Reiss, S. E. (1995). Neuromuscular metamorphosis in the moth *Manduca sexta:* Hormonal regulation of synapse loss and remodeling. *J. Neurosci.* **15,** 4815–4826.

Truman, J. W., and Schwartz, L. M. (1984). Steroid regulation of neuronal death in the moth nervous system. *J. Neurosci.* **4,** 274–280.

Truman, J. W., Taylor, B. J., and Awad, T. (1993). Formation of the adult nervous system. *In* "The Development of *Drosophila melanogaster*" (M. Bate and A. Martinez Arias, eds.), pp. 1245–1275. Cold Spring Harbor Laboratory, Cold Spring Harbor, New York.

Truman, J. W., Talbot, W. S., Fahrbach, S. E., and Hogness, D. S. (1994). Ecdysone receptor expression in the CNS correlates with stage-specific responses to ecdysteroids during *Drosophila* and *Manduca* development. *Development* **120,** 219–234.

Tublitz, N. J., and Sylwester, A. W. (1990). Postembryonic alteration of transmitter phenotype in individually identified peptidergic neurons. *J. Neurosci.* **10,** 161–168.

Warren, J. T., and Gilbert, L. I. (1986). Ecdysone metabolism and distribution during the pupal–adult development of *Manduca sexta. Insect Biochem.* **16,** 65–82.

Weeks, J. C. (1987). Time course of hormonal independence for developmental events in neurons and other cell types during insect metamorphosis. *Dev. Biol.* **124,** 163–176.

Weeks, J. C., and Ernst-Utzschneider, K. (1989). Respecification of larval proleg motor neurons during metamorphosis of the tobacco hornworm, *Manduca sexta:* Segmental dependence and hormonal regulation. *J. Neurobiol.* **20,** 569–592.

Weeks, J. C., and Truman, J. W. (1985). Independent steroid control of the fates of motoneurons and their muscles during insect metamorphosis. *J. Neurosci.* **5,** 2290–2300.

Weeks, J. C., and Truman, J. W. (1986). Hormonally mediated reprogramming of muscles and motoneurones during the larval–pupal transformation of the tobacco hornworm, *Manduca sexta. J. Exp. Biol.* **125,** 1–13.

White, K., and Kankel, D. R. (1978). Patterns of cell division and cell movement in the formation of the imaginal nervous system of *Drosophila melanogaster. Dev. Biol.* **65,** 296–321.

Witten, J. L., and Truman, J. W. (1990). Stage specific expression of FMRFamide-like immunoreactivity in motoneurons of the tobacco hornworm, *Manduca sexta,* is mediated by steroid hormones. *Soc. Neurosci. Abstr.* **16,** 633.

Witten, J. L., and Truman, J. W. (1991). The regulation of transmitter expression in adult-specific lineages in the moth, *Manduca sexta.* I. Transmitter identification and developmental acquisition of expression. *J. Neurosci.* **11,** 1980–1989.

Wolff, T., and Ready, D. F. (1993). Pattern formation in the *Drosophila* retina. *In* "The Development of *Drosophila melanogaster*" (M. Bate and A. Martinez Arias, eds.), pp. 1277–1325. Cold Spring Harbor Laboratory, Cold Spring Harbor, New York.

Yao, T.-P., Forman, B. M., Jiang, Z., Cherbas, L., Chen, J.-D., McKeown, M., Cherbas, P., and Evans, R. M. (1993). Functional ecdysone receptor is the product of *EcR* and *ultraspiracle* genes. *Nature* (*London*) **366,** 476–479.

Zacharuk, R. Y., and Shields, V. D. (1991). Sensilla of immature insects. *Annu. Rev. Entomol.* **36,** 331–354.

9

Gene Regulation in Imaginal Disc and Salivary Gland Development during *Drosophila* Metamorphosis

CYNTHIA BAYER, LAURENCE VON KALM, AND JAMES W. FRISTROM

Department of Molecular and Cell Biology
University of California
Berkeley, California

Complete metamorphosis of insects is remarkable for many reasons. Among these is the dichotomy between the developmental programs of larval and imaginal tissues in *Drosophila.* The former, e.g., salivary glands, are programmed to die, whereas the latter, e.g., imaginal discs, are destined to form adult structures. An important function of the polytene larval salivary glands is glue secretion at the end of larval development (see Russell and Ashburner, Chapter 3, this volume). This is followed by progressive waves of gene expression in the early stages of metamorphosis before the glands finally succumb to histolysis in the early pupal period. In contrast, the diploid imaginal discs undergo rapid morphogenesis during the prepupal period to form the basic epidermal structures of the adult thorax and head. Wing discs, for example, elongate, evert, and fold into a bilayered wing. Despite the disparity in developmental fate, glands and discs depend on common regulatory molecules to effect development. Both tissues respond to the molting hormone 20-hydroxyecdysone (20-HE), both contain 20-HE receptors, and both express a set of "early ecdysone genes" that regulate their respective developmental programs (see Russell and Ashburner, Chapter 3, this volume). The receptors and master regulators, although similar in the two tissues, are not identical. An understanding of the similarities and differences will lead ultimately to a better understanding of the elaboration of the developmental programs in both tissues.

One 20-HE-responsive master regulatory gene is the *Broad-Complex* (*BR-C*), located at the "early" puff site at chromosomal position 2B5 on the X chromosome (Ashburner *et al.,* 1974; Zhimulev *et al.,* 1982). In contrast to other early 20-HE-responsive genes, the *BR-C* is well defined genetically (Stewart *et al.,* 1972; Belyaeva *et al.,* 1980, 1981, 1987; Kiss *et al.,* 1988). The genetics of the locus provide important insights into its developmental and metamorphic functions. For example, complete null mutations (e.g., deficiencies) stop development in a cell autonomous manner at the end of larval life. Mutant cells do not enter into metamorphosis even in the midst of wild-type tissues in genetically mosaic animals (Kiss *et al.,* 1976). The role of the *BR-C* in metamorphosis, particularly its role in imaginal discs and in salivary glands, is a focus of this chapter.

I. METAMORPHIC DEVELOPMENT OF IMAGINAL DISCS

A. Overview

Drosophila metamorphosis takes place over approximately 4 days at 25°C and includes first the 12-hr "prepupal period" (between pupariation and pupation) and then the 84-hr "pupal period" (between pupation and eclosion) (Fig. 1). Imaginal disc development during the prepupal period presages the pupal development of discs, providing a convenient model for the study of metamorphosis.

At the end of the third instar, approximately 6 hr after the initial rise in 20-HE titer (Fig. 1), the larva stops crawling, everts its spiracles, and forms the puparium (pupariation) that surrounds the organism for the duration of metamorphosis. While the hormone level is high the imaginal discs undergo dramatic morphogenetic changes to form the basic shape of the adult head, thorax, and appendages (Poodry and Schneiderman, 1970; Fristrom and Fristrom, 1975). The hormone level returns to low (intermolt) levels approximately 4 hr after pupariation (AP). By the end of the prepupal period, the developing adult is surrounded by a continuous exoskeleton or cuticle, the pupal cuticle, whose elaboration continues into the early pupal period (Wolfgang *et al.,* 1986). Pupal cuticle secretion occurs mainly while the concentration of 20-HE is low.

B. Larval Imaginal Disc Morphology

Each of 21 imaginal discs forms a specific part of the adult epidermis (Fristrom and Fristrom, 1993). Determined during embryogenesis, these adult precursors are separated from their larval counterparts as invaginations of the embryonic epidermis at anterior–posterior compartment boundaries (reviewed by Cohen, 1993). Discs increase in size by cell division during larval development. A mature, late third instar disc contains as many as 50,000 cells in a folded, single-cell-thick epithelium. Each disc has a characteristic size, shape, and pattern of folds that is a consequence of the number, size, and shape of its component cells. Typically, mature imaginal discs are sac-like epithelial organs composed of a folded columnar epithelium (the disc proper) on one side that grades into a squamous epithelium (peripodial epithelium) on the other. Most discs are directly continuous with the larval epidermis by way of attachment stalks. Junctions characteristic of insect epithelia attach disc cells to each other. The disc epithelium has apical–basal polarity with a microvillar apical surface facing the disc lumen (interior of the sac) and a basal surface resting on a basal lamina. At the apical surface a contractile actin–myosin cytoskeleton—the apical contractile belt—follows the cell boundaries (Condic *et al.,* 1991). During the second half of the third instar and at the beginning

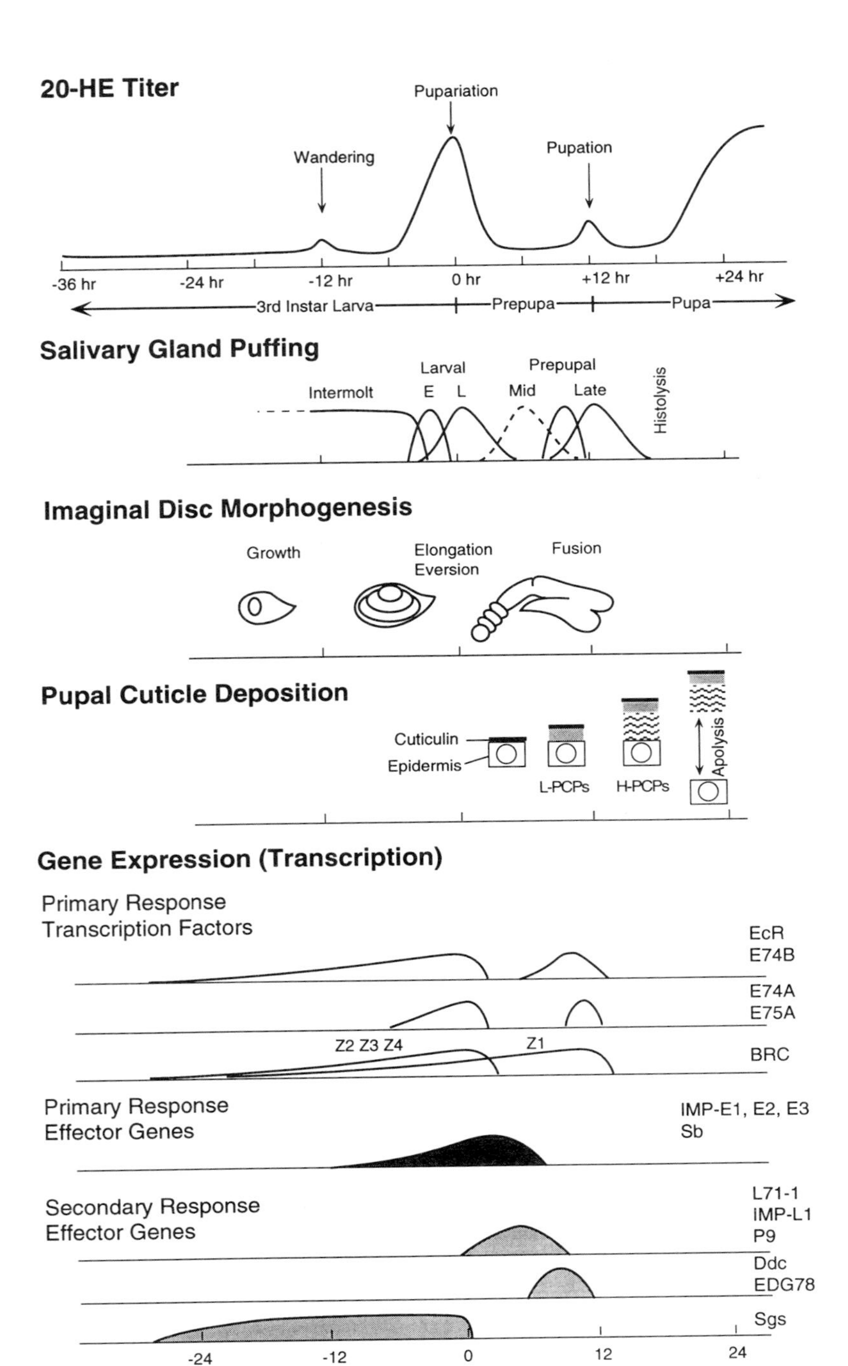

20-HE Titer
Pupariation
Wandering
Pupation
-36 hr
-24 hr
-12 hr
0 hr
+12 hr
+24 hr
3rd Instar Larva
Prepupa
Pupa
Salivary Gland Puffing
Larval
Prepupal
Intermolt
E
L
Mid
Late
Histolysis
Imaginal Disc Morphogenesis
Growth
Elongation
Eversion
Fusion
Pupal Cuticle Deposition
Cuticulin
Epidermis
L-PCPs
H-PCPs
Apolysis
Gene Expression (Transcription)
Primary Response
Transcription Factors
EcR
E74B
E74A
E75A
Z2 Z3 Z4
Z1
BRC
Primary Response
Effector Genes
IMP-E1, E2, E3
Sb
Secondary Response
Effector Genes
L71-1
IMP-L1
P9
Ddc
EDG78
Sgs
-24
-12
0
12
24

of the prepupal period, disc cells secrete a poorly characterized, tenuous nonchitinous apical extracellular matrix. So, a mature imaginal disc is a single-cell-thick apically contractile epithelium whose cells are encased within extracellular matrices and attached by junctions.

C. Imaginal Disc Morphogenesis

Imaginal discs that form appendages such as legs, wings, and antennae undergo similar morphogenetic changes in response to 20-HE (Fig. 2). In the late third instar and early prepupal period the appendages elongate and the discs evert to the surface of the organism (the apical surface is now outside). As eversion occurs, the periphery of the disc expands until it meets and fuses with adjacent discs to form a continuous adult epithelium (Fristrom and Fristrom, 1993). Confocal microscopy, pharmacology, and genetical analyses indicate that elongation and eversion

---

Fig. 1. 20-HE regulation of changes in tissue metamorphosis and gene expression. The peaks in 20-HE titer at the onset of metamorphosis are shown on a time scale relative to 0 hr prepupae (white prepupae). The progression of three tissue-specific hormone-regulated developmental events—salivary gland puffing, imaginal disc morphogenesis, and cuticle deposition—are shown below the hormone titer profile. The sets of salivary gland puffs—intermolt puffs, early (E) and late (L) larval puffs, and mid and late prepupal puffs—become active in a strict temporal sequence in response to 20-HE at the end of larval development and throughout the prepupal period. (The mid prepupal puffing profile is indicated by a dotted line to distinguish it from the late prepupal puff profiles). The morphogenetic changes in imaginal discs beginning in −6-hr larvae occur via elongation, eversion, and fusion of the discs, and are complete by +12 hr AP in prepupae. Beginning in the prepupal period, pupal cuticle proteins (PCPs) are secreted by imaginal disc epidermal cells in a hormone-dependent manner to form the procuticle beneath the cuticulin layer. L-PCPs, comprising the exocuticle region, begin to be deposited by +8 to +9 hr AP. H-PCPs, comprising the endocuticle region, are deposited after pupation at +12 hr AP. The entire cuticle is apolysed from the epidermis at +18 to +20 hr AP in response to the pupal rise in 20-HE titer. Transcript accumulation profiles for some of the 20-HE-responsive genes described in the text are also shown relative to these time points. These curves describe profiles of gene expression in whole animals: *EcR, E74, E75, IMP-E1, -E2, -E3, -L1, L71-1, Sgs* genes and *EDG78* (Andres *et al.*, 1993), *BR-C* (C. Bayer and J. W. Fristrom, unpublished observations), *Sb* (Appel *et al.*, 1993), and *P9* (J. Emery and G. M. Guild, unpublished observations). The curve for *Ddc* represents only imaginal disc-specific expression (Clark *et al.*, 1986).

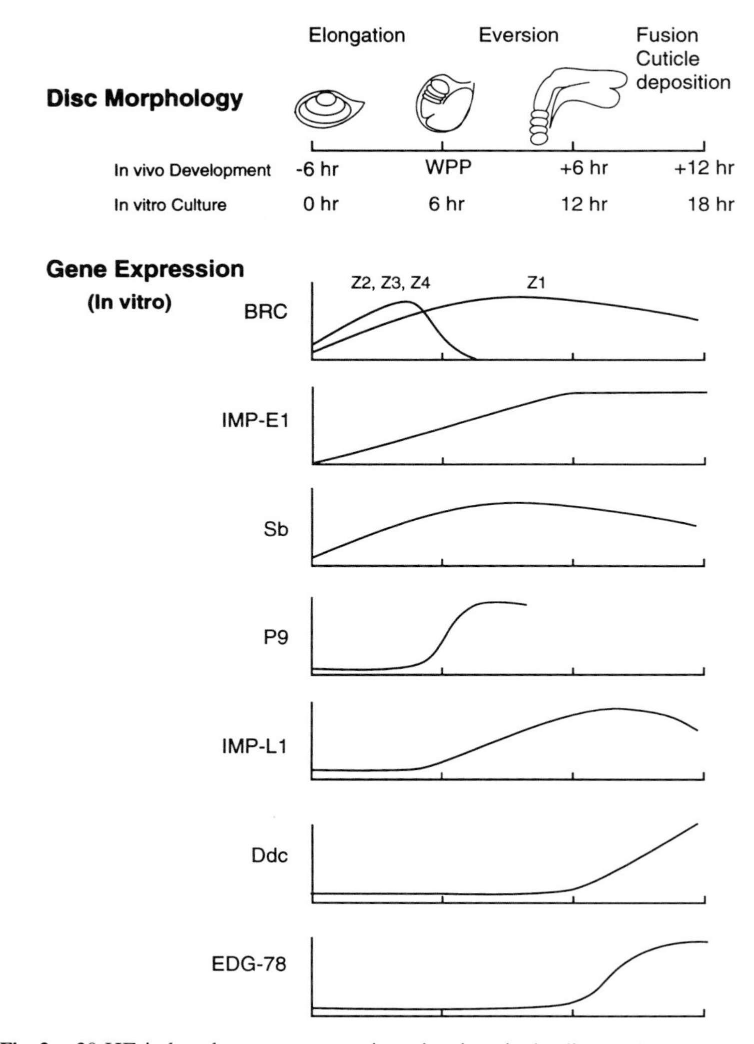

Fig. 2. 20-HE-induced gene expression in imaginal disc culture. Transcript accumulation profiles during 18 hr of imaginal disc culture in the continous presence of 1 μg/ml 20-HE is shown for *BR-C* (C. Bayer, and J. W. Fristrom, unpublished observations), *IMP-E1* and *IMP-L1* (Natzle, 1993) *Sb* (R. Abu-Shumays and J. W. Fristrom, unpublished observations), *P9* (J. Emery and G. M. Guild, unpublished observations), *Ddc* (Clark *et al.*, 1986), and *EDG-78* (Fechtel *et al.*, 1988). Accumulation of *Ddc* and *EDG-78* transcripts requires the removal of hormone after 6 hr of culture in its presence. For reference, changes in morphology of a leg disc during 18 hr of culture in the continuous presence of 20-HE are shown at the top. Also shown is the correspondence of the timing of these morphogenetic changes in culture to the −6-hr to +12-hr AP metamorphic period *in vivo*.

depend on cell shape changes mediated by contraction of the actin–myosin apical contractile belt (Fekete *et al.,* 1975; Fristrom and Fristrom, 1975; Condic *et al.,* 1991). Coupled with cell shape changes are modifications in the attachments of cells to the extracellular matrices. Some of these modifications probably depend on proteases that are synthesized in response to 20-HE (Pino-Heiss and Schubiger, 1989; Appel *et al.,* 1993).

D. Imaginal Disc Differentiation: Cuticle Formation

Cuticle deposition in prepupal discs is typical of that of insects (Locke, 1966; Fristrom *et al.,* 1986; see Willis, Chapter 7, this volume). The pupal cuticle is composed of an outer nonchitinous cuticulin layer and an inner procuticle. The outer cuticulin layer forms first but is not complete until prepupal appendage morphogenesis is finished (4 to 5 hr AP) (Fig. 1). Deposition of the chitin-containing, lamellate pupal procuticle begins at 8 to 9 hr AP. [Between the cuticulin layer and the procuticle there is a structurally amorphous inner epicuticle containing chitin, or a chitin-like substance, and two L-PCPs (Fristrom *et al.,* 1986) that may represent a structural transition zone between the chitin-free cuticulin layer and the lamellate procuticle.] Before pupation, only L-PCPs (low molecular weight pupal cuticle proteins) are synthesized and secreted, and only two or three lamellae of the procuticle are formed (Doctor *et al.,* 1985; Wolfgang *et al.,* 1986). Following pupation, H-PCPs (high molecular weight pupal cuticle proteins) are synthesized and secreted, and up to 18 more lamellae are formed. The slight rise in 20-HE titer just before pupation (Fig. 1) may be critical for the transition between synthesis of low and high molecular weight PCPs. The exocuticle, the region of the procuticle deposited before pupation, is structurally distinct from the endocuticle, which forms after pupation. L-PCPs are limited to the exocuticle; H-PCPs to the endocuticle (Wolfgang *et al.,* 1986). Apolysis of the pupal cuticle, the separation of the cuticle from the underlying disc epidermis, occurs around 18 hr AP in response to a drop in 20-HE titer.

II. DEVELOPMENT OF LARVAL SALIVARY GLANDS

Larval salivary glands are large secretory epithelial tubes connected to the exterior by narrow, nonsecretory ducts. The glands per se first appear as lateroventral plates or placodes in the labial region of 6- to 7-hr embryos (Campos-Ortega and Hartenstein, 1985; Panzer *et al.*, 1992). Shortly after, the placode cells invaginate to form the glands. Next, the cells immediately ventral to the placodes invaginate to form the ducts and fuse with the anterior of each developing gland (B. Zhou and S. K. Beckendorf, personal communication). Cells at the boundary between the ducts and the glands form imaginal rings, the precursors of the adult salivary glands (reviewed by Berendes and Ashburner, 1978).

Salivary glands have completed mitosis and attained their full complement of 100–120 cells by the placode stage (Berendes and Ashburner, 1978). Thereafter, gland cells continue to increase in both nuclear and total volumes because their DNA is endoreplicated, beginning immediately after and perhaps even during invagination (Smith and Orr-Weaver, 1991; Y. Kuo and S. K. Beckendorf, personal communication). Salivary gland cells are the first of the larval tissues to undergo endoreplication which, in the absence of mitosis, leads to polyteny. These cells endoreplicate 8 to 10 times (n = 512 to 2048) by the end of third instar. An anterior to posterior gradient of increasing cell size and nuclear volume possibly reflects differential rates of polytenization (Berendes, 1965).

Secretion into the lumen of the gland is first observed in 10 hr-old-embryos (Campos-Ortega and Hartenstein, 1985). The function of this embryonic secretion is unknown. By mid-third instar, highly refractile secretory granules appear in the cytoplasm of the salivary gland cells in a posterior to anterior progression (Berendes and Ashburner, 1978). By the end of the third instar all but the 15–30 cells located next to the imaginal ring are engorged with this material. These granules, composed of the salivary gland glue (Fraenkel, 1952; Korge, 1975; Beckendorf and Kafatos, 1976), are produced in a 20-HE-dependent manner during the third instar (see Section III,A). Shortly before puparium formation, the granules migrate to the cell apex and are secreted into the gland lumen. At the time of puparium formation the glue is expelled to attach the pupa to a dry substrate (Fraenkel and Brookes, 1953). At this point the lumen is reduced in size, but shortly thereafter increases in diameter again as new 20-HE-induced secretion commences. The functions of prepupal secretions are unknown. Larval salivary gland development ends with histolysis. Beginning approximately 2 hr after pupation, again in a posterior to anterior gradient, the nuclei become pycnotic and nuclear and basement membranes break down (Berendes and Ashburner, 1978). Histolysis, which is dependent on 20-HE, proceeds for several hours.

Cells within the imaginal rings proliferate throughout third instar and prepupal development (Berendes and Ashburner, 1978). Differentiation of the adult salivary glands occurs as the larval glands are histolysing and is dependent on 20-HE. Although the adult gland precursors are in contact with histolysing larval cells and hemocytes, they are protected from the 20-HE-induced histolysis in larval gland cells. Adult glands reach maximum extension approximately 36 hr AP (Robertson, 1936). The function of the adult salivary glands is unknown; however, one might logically presume a role in feeding or digestion.

III. 20-HYDROXYECDYSONE-REGULATED GENE EXPRESSION

Changes in the titer of the steroid hormone 20-HE coordinate the metamorphic development of both salivary glands and imaginal discs. The sharp rise in 20-HE titer at the end of third instar (Fig. 1) initiates a rapid increase in expression of a set of transcriptional regulators (see Cherbas and Cherbas, Chapter 5, this volume). These regulators exert subsequent temporal control over the expression of tissue-specific effector genes that mediate the morphogenetic processes unique to each hormone-responsive tissue.

A. Ecdysone Gene Hierarchy in Salivary Glands

The hierarchy of ecdysone-triggered gene expression during metamorphosis was first described by Ashburner and colleagues (Ashburner *et al.*, 1974; Russell and Ashburner, Chapter 3, this volume). By using explanted *Drosophila* salivary glands, they were able to reproduce in culture the polytene chromosome puffing sequence that occurs during the last several hours of larval development in response to 20-HE (Fig. 1). They showed that "intermolt" puffs, already active in mid-third instar larvae, regress rapidly in response to the addition of 20-HE. At the same time a new set of six "early" puffs (including chromosomal sites 2B5, 74EF, and 75E) forms within 5 min of 20-HE addition, without requiring protein synthesis (primary response). After reaching maximal size, the early puffs regress and ~100 "late" puffs form in a strict temporal sequence. The regression of the early puffs and induction of the late puffs are both dependent on 20-HE addition and protein synthesis (secondary response). These observations of a temporally defined sequence of gene activity gave rise to a formal model of an ecdysone-triggered, self-regulated gene hierarchy. In this model, 20-HE complexed with its receptor initially represses late gene activity while it activates early genes. Secondarily, accumulation of these early gene products activates tissue-specific late gene expression, while repressing transcription of the early genes.

Two other sets of temporally defined puffs become active later in development (Richards 1976a,b). The first is the set of "mid prepupal" puffs that require a reduction in the hormone titer to intermolt levels for their activity. Their activity also requires protein synthesis, suggesting a dependence on early puff gene activity. The second is the set of "late prepupal" puffs that are induced by the rise in hormone titer preceding pupation (Fig. 1). The set of late prepupal puffs includes some of the early and late larval puff sites that were active previously and appear to be regulated in a similar manner at both stages of development.

The tight temporal regulation of salivary gland gene expression during metamorphosis presumably reflects a requirement for different protein products in this tissue as development proceeds. During third instar, glue proteins (the products of intermolt puffs) are synthesized. Once the glue is secreted, the products of the early puffs repress intermolt puff activity, and soon after initiate the late puffing sequence. Consistent with their regulatory roles in metamorphosis, early gene products have been shown to bind to many sites on polytene chromosomes, including the late puff sites (Urness and Thummel, 1990; Hill *et al.*, 1993; Emery *et al.*, 1994). The roles of the late puff and mid prepupal puff gene products are unknown; however, products of the late prepupal puffs are likely to be important in triggering programmed cell death in the early pupa.

The genes encoding *EcR* and *ultraspiracle* (*usp*), the components of the 20-HE receptor (see Cherbas and Cherbas, Chapter 5, this volume), and the early ecdysone genes *E74, E75,* and *BR-C* have been characterized at a molecular level. The *EcR* gene encodes a member of the steroid receptor superfamily (Koelle *et al.*, 1991) and produces a family of three *EcR* isoforms which share DNA- and ligand-binding domains (Talbot *et al.*, 1993). EcR and USP protein (a retinoid X receptor homolog) heterodimerize to form an active ecdysone receptor complex (Yao *et al.*, 1992, 1993; Thomas *et al.*, 1993; for a more detailed discussion see Cherbas and Cherbas, Chapter 5, this volume). Each of the early ecdysone genes, *E74, E75,* and *BR-C,* encodes a family of transcription factors

(Feigl *et al.*, 1989; Janknecht *et al.*, 1989; Burtis *et al.*, 1990; Segraves and Hogness, 1990; DiBello *et al.*, 1991) that are coordinately expressed in many tissues, including salivary glands and imaginal discs (Chao and Guild, 1986; Karim and Thummel, 1992; Huet *et al.*, 1993). The isoforms of each of these gene families are differentially regulated by 20-HE at the end of the third instar and throughout the prepupal period (Karim and Thummel, 1992; Andres *et al.*, 1993) (Fig. 1).

B. Ecdysone Gene Hierarchy in Imaginal Discs

Gene activity in the form of chromosome puffing cannot be observed in diploid imaginal disc cells. Nonetheless, a similar hierarchy of 20-HE-triggered gene expression occurs in discs in the form of primary and secondary response genes. Imaginal discs express all isoforms of *EcR* and each of the known primary response genes *E74, E75,* and *BR-C* (Fig. 1). The relative levels of isoforms, however, produced by each gene differ in imaginal discs and salivary glands. For example, Huet *et al.* (1993) found differences between these two tissues in the temporal profile of transcript isoforms specific to the A and B promoters of both the *E74* and *E75* genes. These differences may result from either different hormone titers or receptor concentrations within tissues. In contrast to these relatively subtle differences, antibodies to two *EcR* protein isoforms detect very different patterns in these tissues (Talbot *et al.*, 1993). The *EcRB* protein isoform is the predominant form in salivary gland cells (as well as most other larval tissues). The *EcRA* isoform predominates in imaginal disc cells (with the exception of peripodial epithelium cells that express high levels of *EcRB* along with *EcRA*). Such differences suggest that *EcRA/USP* and *EcRB/USP* heterodimer complexes direct the expression of different regulatory hierarchies in imaginal disc and salivary gland cells, respectively. Consistent with this view, the relative abundance of *BR-C* RNA isoforms differs significantly between these two tissues. Of the four alternatively spliced *BR-C* zinc finger isoforms produced in third instar larvae (see Section V,B), the Z1 and Z3 forms predominate in salivary glands, whereas Z2 and Z3 are the major forms in third instar imaginal discs (Huet *et al.*, 1993; von Kalm *et al.*, 1994; C. Bayer and J. W. Fristrom, unpublished observations). Consequently, differential regulation of the *BR-C* by different ecdysone receptor complexes may be one means of mediating tissue-specific responses to ecdysone.

Although the 20-HE responses of imaginal discs and salivary glands are different in form, the timing of major developmental changes in both tissues is coordinated by the hormone. For example, at the time when mid prepupal puffs are induced in salivary glands, the pupal cuticle is synthesized by differentiating discs. Transcription of L-PCP genes (e.g., *EDG-78*) and *Dopa decarboxylase* (*Ddc*) *in vitro* requires exposure of larval imaginal discs to 20-HE for 6 hr followed by 6 hr of culture in its absence (Clark *et al.,* 1986; Fechtel *et al.,* 1988; Apple and Fristrom, 1991) (Fig. 2). This hormone regimen emulates the rise and fall of hormone titer *in vivo* during the larval/prepupal transition (−6-hr larvae to +10-hr AP) and is similar to the culture conditions required to replicate mid prepupal puffing *in vitro* (Richards, 1976a). The continuous presence of hormone in culture prevents expression of both the pupal cuticle genes and the mid prepupal salivary gland puffs. Thus, sets of genes that appear to depend on the presence of early gene products, but are repressed by 20-HE, are expressed in both salivary glands and imaginal discs during the mid prepupal period. In the case of imaginal discs, this temporal coordination of 20-HE-regulated gene expression is essential for restricting the deposition and sclerotization of pupal cuticle to the end of the prepupal period, after disc morphogenesis is completed.

Another phase of gene regulation common to salivary glands and imaginal discs occurs in late prepupae. As described in Section III,A, the late prepupal puffs are activated in salivary glands in response to the rise in 20-HE at 10 hr AP. This puffing sequence occurs just before histolysis of the glands in early pupae. In imaginal discs, this same hormone peak represses transcription of L-PCPs (Apple and Fristrom, 1991) and induces expression of H-PCPs (Doctor *et al.,* 1985) (Fig. 1). This shift in pupal cuticle protein synthesis completes the innermost layer of the pupal cuticle. After the 20-HE titer returns to intermolt levels, the pupal cuticle is apolysed from the epidermis, in anticipation of the pupa's next phase of epidermal morphogenesis, e.g., bristle and hair extrusion.

C. Ecdysone-Responsive Effector Molecules in Imaginal Discs

Although little is known about the nature of effector molecules in salivary glands, many effector genes have been identified in imaginal discs (Table I). Some of these imaginal disc effector genes exhibit a primary response to hormone (Fig. 2). For example, some *IMP* (inducible

TABLE I

20-HE-Responsive Genes Expressed in Imaginal Discs

Role in Metamorphosis	Gene	Locus	Nature of Product and/or function	Response to 20-HE in imaginal discs	References[c]
Regulatory	*BRC*	2B5	Four zinc finger protein isoforms essential for metamorphosis	Primary response gene.[a] All four isoforms respond to 20-HE within 1 hr	*(1)*
	EcR	42A	Ecdysone receptor; complexes with *USP* protein (retinoid X receptor homolog).	*EcR* transcription is unchanged (*USP* response to 20-HE not known)	*(2)*
	E74	74B	Members of *ets* family of DNA-binding proteins.	Switch from *E74B* to *E74A* isoform expression.	*(3)*
	E75	75EF	Members of steroid receptor superfamily	*E75A* isoform is induced by 20-HE	*(4)*
Pupal cuticle formation	*EDG-78*	78E	L-PCP. Member of cutin cuticle protein family.	Secondary response gene.[b] Expressed after 4–6 hr exposure to 20-HE. Transcripts accumulate within 6 hr after withdrawal of 20-HE.	*(5)*
	EDG-84	84A1	L-PCP. Sequence similarity to *Locusta* and *Hyalophora* cuticle proteins.	Like EDG-78.	*(6)*
	EDG-91	91A	L-PCP. Glycine rich, related to chorion proteins	Like EDG-78	*(7)*
	EDG-64	64CD	3.0; 3.9; 5.2; 5.4-kb transcripts also expressed in all larval instars. Not a L-PCP.	Like EDG-78.	*(8)*
	EDG-43	43A	3.0-kb transcript expressed only in prepupae. Presumably not a L-PCP.	Like EDG-78 but transcripts accumulate 6–9 hr after removal of 20-HE.	*(9)*
	Ddc	37C1	Dopa decarboxylase. Provides substrate for cross-linking cuticle components.	Transcript accumulation in discs like that of EDG-43	*(10)*

(*continues*)

TABLE I

Continued

Role in Metamorphosis	Gene	Locus	Nature of Product and/or function	Response to 20-HE in imaginal discs	References[c]
Morphogenesis	*IMP-E1*	66C	Apparent secreted protein with three EGF-like repeats followed by a mucin-like domain (probable site of O-glycosylation). Also expressed in neurolemma. May be essential gene.	Primary response gene.[a] Transcript levels increase within 30 min of exposure to 20-HE. Peak expression at pupariation.	*(11)*
	IMP-E2	63E	Apically secreted protein with 16 EIK-like repeats (potential protease cleavage sites). No known homologs. Function in disc morphogenesis unknown.	Primary response gene.[a] Transcript levels increase within 0–3 hr of exposure to 20-HE. Peak expression at pupariation.	*(12)*
	IMP-E3	84E	Apparently an apically secreted protein. No known homologs. Function in disc morphogenesis unknown.	Primary response gene.[a] Transcript levels increase within 1 hr of exposure to 20-HE. Peak expression at pupariation.	*(13)*
	IMP-L1	70A	Apparent 38-kDa secreted protein with no known homologs. Role in disc morphogenesis unknown.	Secondary response gene.[b] Transcript levels increase 4–6 hr after exposure to 20-HE.	*(14)*
	IMP-L2	64B	Immunoglobulin family member implicated in spreading of imaginal tissues in prepupae (discs) and early pupae (abdominal histoblast nests).	Transcript levels increases within 1 hr of exposure to 20-HE. Peak expression in histoblast nests after pupation.	*(15)*
	Stubble	89B9	Transmembrane serine protease.	Primary response gene.[a] Transcript levels increase within 1 hr of exposure to 20-HE. Peak expression at pupariation.	*(16)*
	P9	29F	Putative serine-protease inhibitor. Gene is regulated by the *BRC*.	Secondary response gene.[b] Transcript levels increase after a 6-hr exposure to 20-HE. Peak expression 4 hr AP.	*(17)*

	zipper	60E	Nonmuscle myosin heavy chain essential for disc morphogenesis (elongation eversion).	Not 20-HE responsive in discs at pupariation.	(*18*)
	actin	5C 42	Cytoplasmic actin is essential for disc morphogenesis (elongation eversion)	No or minor 20-HE responsiveness in discs at pupariation.	(*19*)
	β_3-tubulin	60D	β_3-tubulin. Role in disc morphogenesis unknown.	Expressed in discs at pupariation. Only β-tubulin gene with 20-HE-induced expression in discs.	(*20*)
	collagen type IV	25C	Type IV collagen. Located in basal lamina. Specific role in disc morphogenesis unknown.	Proteolytically cleaved in response to 20-HE. Transcriptional response not known.	(*21*)
Metabolic	*IMP-L3*	65B	Lactate dehydrogenase	Transcript levels increase by 6 hr of exposure to 20-HE	(*22*)
	rRNA (bobbed)	20E-F	18S and 20S ribosomal RNA	Approximately twofold increase in synthesis after 5 hr of culture with 20-HE.	(*23*)

[a] Primary response gene: transcription occurs in the presence of protein synthesis inhibitors.

[b] Secondary response gene: transcription is blocked in the presence of protein synthesis inhibitors.

[c] Key to references: (*1*) Chao and Guild (1986), Kiss *et al.* (1988), Huet *et al.* (1993), Karim *et al.* (1993), DiBello *et al.* (1991), Emery *et al.* (1994), C. B. and J. W. Fristrom (unpublished observations). (*2*) Koelle *et al.* (1991), Huet *et al.* (1993), Talbot *et al.* (1993). (*3*) Burtis *et al.* (1990), Thummel *et al.* (1990), Urness and Thummel (1990), Huet *et al.* (1993). (*4*) Segraves and Hogness (1990), Huet *et al.* (1993). (*5*) Fechtel *et al.* (1988, 1989), Apple and Fristrom (1991). (*6*) Fechtel *et al.* (1988, 1989), Apple and Fristrom (1991). (*7*) Fechtel *et al.* (1988), Apple and Fristrom (1991). (*8*) Fechtel *et al.* (1988, 1989). (*9*) Fechtel *et al.* (1988, 1989). (*10*) Clark *et al.* (1986). (*11*) Natzle *et al.* (1988), Natzle (1993). J. E. Natzle (personal communication). (*12*) Paine-Saunders *et al.* (1990). (*13*) Moore *et al.* (1990). (*14*) Natzle *et al.* (1992), Natzle (1993). (*15*) Osterbur *et al.* (1988), Garbe *et al.* (1993). (*16*) Beaton *et al.* (1988), Appel, *et al.* (1993). (*17*) J. Emery and G. M. Guild (personal communication). (*18*) Kiehart (1991), Young *et al.* (1993), J. W. Fristrom, A. Hammonds, P. Gotwals, and D. P. Kiehart (unpublished observation). (*19*) Fyrberg *et al.* (1980), Tobin *et al.* (1980), Fyrberg and Goldstein (1990), S. L. Tobin (personal communication). (*20*) Gasch *et al.* (1988), Kimble *et al.* (1989), Sobrier *et al.* (1989), P. Gotwals and J. W. Fristrom (unpublished observations). (*21*) Birr *et al.* (1990), Fessler *et al.* (1993). (*22*) R. Abu-Shumays and J. W. Fristrom (unpublished observations). (*23*) Petri *et al.* (1971), Raikow and Fristrom (1971).

membrane-associated protein) genes respond directly to 20-HE in imaginal disc culture. Among these, *IMP-E1* encodes an apparent secreted protein that contains three EGF domains and a mucin domain (site of possible O-linked glycosylation) (J. E. Natzle, personal communication). Early *IMP-E1* expression and a nonuniform distribution in imaginal discs and the neuralemma suggest that it may have a role in epithelial morphogenesis in these two tissues (Natzle *et al.,* 1988).

The primary 20-OH-responsive *Stubble* (*Sb*) locus encodes a transmembrane serine protease that is required for leg and wing morphogenesis and apparently localizes proteolytic activity at the apical cell surface (Appel *et al.,* 1993; D. Fristrom, personal communication). Prepupal imaginal discs of *Sb* mutants are unable to unfold or elongate properly; a phenotype that can be alleviated *in vitro* by treatment with trypsin, another serine protease that has been shown to accelerate the elongation of wild-type discs cultured *in vitro* (Poodry and Schneiderman, 1971; Fekete *et al.,* 1975). The "malformed" disc phenotype of *Sb* mutants—short, thick, adult legs and short, broad and blistered adult wings—is similar to that of the *br* group of *BR-C* mutants. The malformed phenotype also arises through genetic interactions in combinations of *BR-C* and *Sb* mutations (Beaton *et al.,* 1988). These genetic interactions suggest that either the *BR-C* regulates *Sb* expression or that a target of the *BR-C* is involved in the function of the *Sb* protease in discs (see Section V,D,2).

Other serine proteases as well as serine protease inhibitors accumulate in discs in response to 20-HE (Pino-Heiss and Schubiger, 1989). These serine proteases are active during elongation and eversion, and their activity is controlled by inhibitors, presumably by regulating zymogen activation or by direct inhibition of the active protease. One function of serine proteases in imaginal discs may be to modify the extracellular matrix encasing disc cells to facilitate morphogenesis. For example, the proteolytic cleavage of type IV collagen in discs occurs in response to 20-HE (Birr *et al.,* 1990; Fessler *et al.,* 1993).

Mutations in the nonmuscle myosin heavy chain gene, *zipper* (Young *et al.,* 1993), also produce the malformed disc phenotype (J. W. Fristrom, unpublished observations), demonstrating the role of myosin and of contractility in disc morphogenesis. As described in Section I,C, morphogenesis in imaginal discs (elongation and eversion) results from changes in cell shape (Condic *et al.,* 1991; Fristrom and Fristrom, 1993) that depend on the contractile activity of an actin–myosin-based cytoskeleton. Specifically, an apical belt of actin filaments constricts in response to 20-HE, causing the sheet of cells to change shape, elongating (and constricting in the case of the leg disc) the appendage (Fristrom and Fristrom, 1975). The nonmuscle myosin heavy chain encoded by *zipper,* the only myosin found in discs, is associated with this apical actin belt (D. Fristrom and D. P. Kiehart, personal communication). So partial loss of function *zipper* mutations apparently lead to a failure in contractility and to the malformed phenotype. The malformed phenotype also arises through dominant genetic interactions between *zipper* mutations and *BR-C* or *Sb* mutations. Transcription of *zipper* in imaginal discs does not depend on 20-HE (A. Hammonds and J. W. Fristrom, unpublished observations); therefore the genetic interactions between *zipper* and the *BR-C* do not occur at the level of transcription.

The *zipper* gene is an example of a gene that plays a key role in metamorphosis but is not ecdysone responsive. Actin cytoskeletal contractility, however, is 20-HE responsive. Thus, it appears that the actin–myosin cytoskeleton is the scaffold upon which the effectors of cell shape change act. It is likely that some of these effectors are other 20-HE-regulated cytoskeletal components that either signal these cells or actively drive their contraction and shape changes.

The set of secondary responsive genes expressed in imaginal discs include other *IMP* genes. *IMP-L1* expression is limited to imaginal discs and requires protein synthesis in order to respond to 20-HE (Natzle, 1993) (Fig. 2). It encodes a secreted or transmembrane protein that is expressed just before puparium formation and may be involved in mediating the reshaping of imaginal discs into adult appendages (Natzle *et al.,* 1992). Another 20-HE-inducible late *IMP* gene expressed in discs, *IMP-L3,* encodes lactate dehydrogenase (*LDH*) (R. Abu-Shumays and J. W. Fristrom, unpublished observations). LDH protein is abundantly expressed in the developing larval musculature, where it has a role in facilitating muscle contraction by providing NAD for glycolysis (Rechsteiner, 1970). *LDH* RNA levels in the whole animal decrease in the third instar after the completion of development of the larval musculature. In contrast, expression of the *LDH* gene in discs is induced after 6 hr of culture in the presence of hormone and *in vivo* in prepupal discs. This tissue-specific induction suggests that LDH and presumably other metabolic enzymes have specific roles in disc morphogenesis. It is likely, for example, that discs undergoing metamorphosis have high requirements for energy. Indeed, given that discs emulate muscle (imaginal disc morphogenesis depends on contraction of the actin—myosin contractile belt), it is not surprising that LDH is abundantly expressed in discs during this period where it could help maintain levels of ATP production necessary for contractile activity.

IV. ROLE OF PATTERN IN IMAGINAL DISC MORPHOGENESIS

Genes involved in specifying the body pattern are also essential for successful metamorphosis of the insect larva into an adult (see Wilder and Perrimon, Chapter 10, this volume). At some level the spatial determinants of body pattern, established in the embryo and ultimately manifested in the adult, must intersect with the ecdysone-responsive temporal regulators. During larval development, different homeotic selector gene products are required in different imaginal disc types, e.g., leg discs versus wing discs, to maintain the segmental identities first set up in the embryo. It is possible that unique morphogenetic changes induced in each disc type at the end of larval development are mediated by a set of ecdysone-dependent effector gene products that are common to all imaginal discs. The action of these common effector gene products could, in the context of different intracellular and extracellular structures previously determined by the patterning genes, induce different morphogenetic processes specific to each disc type (Talbot *et al.,* 1993). Alternatively, it is possible that different sets of ecdysone-responsive effector genes may be induced in different disc types. This may result from an interaction of spatial and temporal regulators at the level of the 20-HE receptor complex itself. For example, leg disc-specific expression of the homeotic genes *Ultrabithorax* and *Antennapedia* may lead to synthesis of factors that affect the function of the 20-HE receptor complex, possibly by multimerization, and thus direct a leg-specific hierarchy of metamorphic gene expression.

Evidence for the intersection of spatial and temporal regulators at a third level has been suggested. Restifo and Merrill (1994) noted that mutants of the homeotic selector gene *Deformed* exhibit defects in the development of ventral head structures (derived from the antennal disc) and the central nervous system (CNS) that are similar to the phenotypes of these structures in *BR-C* mutants. The head phenotypes are enhanced in *BR-C;Dfd* double mutants. *Deformed* expression in the CNS is insensitive to ecdysone and is unaffected in *BR-C* mutant backgrounds (Restifo *et al.,* 1995). In contrast to the CNS, *Dfd* expression in the antennal disc appears to be upregulated by 20-HE (L. L. Restifo, personal communication). These results suggest the intriguing possibility that the patterning genes are temporally regulated by the ecdysone hierarchy, at least in imaginal discs. Further investigation should reveal whether other regulators of the adult body pattern are regulated by 20-HE as a way to coordinate their action during metamorphosis.

V. THE *BROAD-COMPLEX*: A KEY REGULATOR OF *DROSOPHILA* METAMORPHOSIS

The *Broad-Complex* is a multifunctional gene essential for metamorphic development in *Drosophila.* Mutants affecting the function of this locus alter virtually all major aspects of development during the larval to prepupal transition. Animals lacking all *BR-C* functions stop developing as late third instar larvae and are unable to initiate the metamorphic program. Mutations affecting specific subfunctions of the *BR-C* have lethal phases at various stages of development ranging from early prepupae to late pupae. Viable mutants exhibit phenotypes in adults such as a broad wing or a reduced number of bristles on the palpus or abdomen. *BR-C* mutations affect metamorphic progression of both larval tissue and imaginal discs. Indeed, as described next, the *BR-C* has a central coordinating regulatory role in directing tissue-specific hormone responses during metamorphosis.

A. Genetics and Structure of the *Broad-Complex*

1. Genetics

Three complementing genetic functions, *broad* (*br*), *reduced bristles on palpus* (*rbp*), and *2Bc,* have been identified at the *BR-C* (Belyaeva *et al.,* 1980; Kiss *et al.,* 1988; Lindsley and Zimm, 1992) (Fig. 3). Evidence that each of these complementation groups represents *BR-C* mutations comes

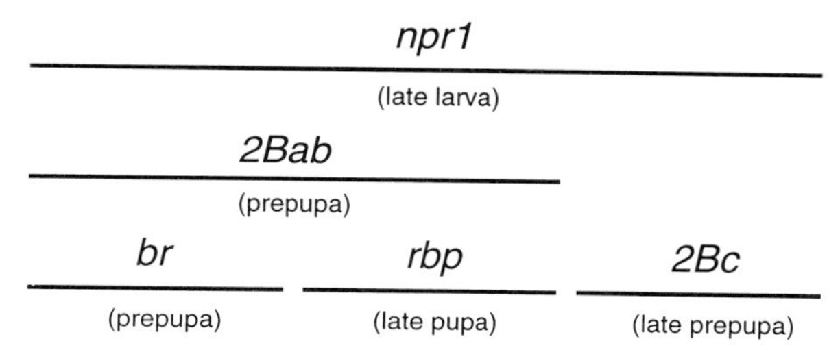

Fig. 3. *BR-C* lethal complementation groups. The lethal phase of each group is indicated beneath the bar.

from an additional group of *nonpupariating 1* (*npr1*) alleles that fail to complement *br, rbp,* or *2Bc* mutants. Additionally, cytologically nonvisible *npr1* mutants are phenotypically identical to *npr1* mutants carrying deficiencies for the chromosomal region containing the *BR-C*. Thus, *npr1* mutants appear to have lost all genetic function at the locus. The phenotype of *npr1* mutants emphasizes the essential role of the *BR-C* in directing metamorphosis. Mutant animals leave the food as late third instar larvae and wander in search of a place to pupariate. However, despite this behavioral inclination to metamorphose, *npr1* larvae are unable to proceed with the metamorphic program, remaining as wandering larvae for several days before they die (Stewart *et al.*, 1972; Kiss *et al.*, 1978).

In addition to *npr1*, another class of mutations, *2Bab*, fails to complement the lethality of both *br* and *rbp* alleles (Belyaeva *et al.*, 1980; Kiss *et al.*, 1988). Although *2Bab* mutants complement the lethality of *2Bc* alleles when placed in heterozygous combinations over *2Bc* mutants, a highly penetrant minor defect in leg development is exhibited. Thus, *2Bab* alleles appear to be at least partially defective for all *BR-C* genetic functions and therefore may represent a hypomorphic class of *npr1* alleles.

2. *Molecular Structure*

The *BR-C* was cloned using a chromosomal walk (Chao and Guild, 1986; Belyaeva *et al.*, 1987). By determining the breakpoints of chromosomal rearrangements associated with *BR-C* mutations, the genetic limits of the *BR-C* was defined to a 100-kb region of DNA. 20-HE-inducible transcripts within this region were identified, and a number of cDNA clones were isolated (Chao and Guild, 1986; DiBello *et al.*, 1991). Consistent with its role as an early response gene of the ecdysone hierarchy and its cell autonomous mode of action, analysis of the cDNA clones revealed that the *BR-C* encodes a family of zinc finger DNA-binding proteins related to the C_2H_2 TFIIIA-like class of zinc finger proteins (discussed in more detail in Section V,B).

The physical organization of the *BR-C* is perversely complex with transcripts originating from at least two promoters (P_{distal} and $P_{proximal}$) separated by 45 kb of DNA (DiBello *et al.*, 1991; C. Bayer and J. W. Fristrom, unpublished observations) (Fig. 4). In addition to variability at the 5′ ends of transcripts, at least four different polyadenylation sites are also utilized. All of the transcripts share a 1680 nucleotide core region containing amino-terminal coding sequences common to all *BR-C* proteins. The core region is alternatively spliced to one of four zinc

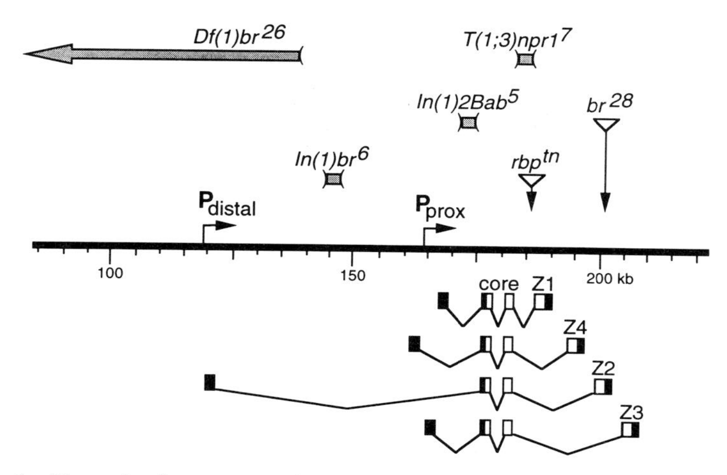

Fig. 4. The molecular structure of the *BR-C* and location of genetic rearrangements. Coordinates are taken from a genomic walk in the 2B5 region (Chao and Guild, 1986). The locations of the two *BR-C* promoters are indicated (P_{distal} and P_{prox}). It is not known if multiple transcription start sites are used at each of these promoters. Representative cDNAs encoding different zinc finger protein isoforms are shown below. Filled black boxes represent untranslated exons. Protein-coding sequences are shown as open boxes. Some inversion, translocation, and deficiency breakpoints associated with *BR-C* mutants are indicated above. In the case of $Df(1)br^{26}$, genomic sequences to the right of the arrow are retained in the mutant. The locations of two P element insertions associated with *BR-C* mutants are indicated by triangles.

finger domains that define the four classes of protein isoforms that make up the *BR-C* family (Fig. 5). Located between the coding portion of the core region and the zinc fingers of each protein is a "linker" region of varible length. The choice of linker region and polyadenylation site strictly correlate with zinc finger domain usage. In contrast, there is no demonstrated correlation between the choice of promoter and the choice of the zinc finger DNA-binding domain (Fig. 4). However, the majority of transcripts identified originate from the promoter most proximal to the core region.

3. *Correlations between Genetics and Molecular Genetics*

A number of *BR-C* mutants are associated with cytologically visible rearrangements within the chromosomal region delimiting the *BR-C*

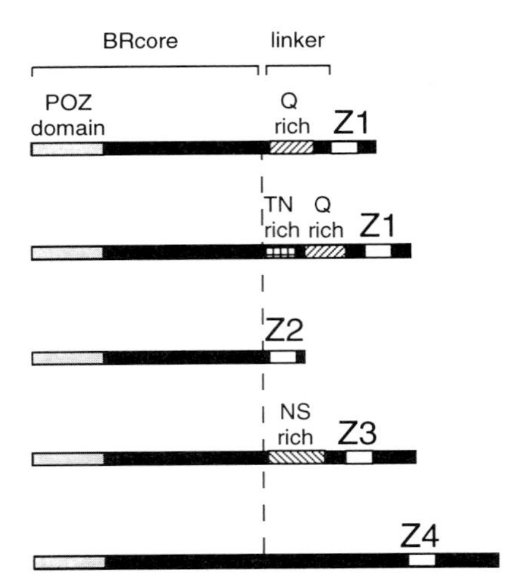

Fig. 5. Structure of the *BR-C* protein isoforms. The BRcore, found in all *BR-C* proteins, contains the amino-terminal POZ domain motif. Linker regions between the Z1 and Z3 BRcore and zinc finger domains contain runs rich in the indicated amino acids (Q, glutamine; T, threonine; N, asparagine; S, serine).

(Fig. 4). Mapping of these rearrangements to the chromosomal walk and analyses of RNA and protein expression patterns in mutants have provided new insights into *BR-C* regulation and the molecular nature of mutant defects.

Other than deficiencies that remove the entire locus, the only cytologically visible *npr1* allele, *npr1*^{7}, is a translocation that separates the core from all the zinc finger domains (DiBello *et al.,* 1991). No stable *BR-C* RNA products are detected in these mutants (C. Bayer and J. W. Fristrom, unpublished observations), a result consistent with previous conclusions that *npr1* alleles are deficient in all genetic functions at the *BR-C*. Protein analysis using antibodies directed against the core region further supports the notion that *nprl* alleles are deficient in all *BR-C* functions (Emery *et al.,* 1994). Only a small amount of an aberrantly sized protein is detected in *npr1*^{7} and *nprl*^{6} mutants, whereas no protein is detected in *npr1*^{1} and *npr1*^{3} animals.

Structural mutations affecting *rbp* and *br* function are independent, i.e., they fully complement each other. There are, however, three classes of mutations that affect both *rbp* and *br* functions (Kiss *et al.,* 1988). These mutations all affect *BR-C* transcription (C. Bayer and J. W. Fristrom, unpublished observations). The first class consists of mutants that are lethal with *br* mutants, but exhibit the *rbp* phenotype in combination with *rbp* mutants. Two mutants in this class, In(1)br^6 and Df(1)br^{26}, both separate the distal promoter from the rest of the gene, suggesting that the distal promoter is more important for *br* than for *rbp* function (DiBello *et al.*, 1991). A second class consists of lethal *rbp* mutants that exhibit the *br* phenotype in combination with *br* mutants (Kiss *et al.*, 1988). Two mutations in this class, rbp^1 and rbp^4, are uncharacterized at a molecular level, but by inference must affect *BR-C* transcription in a manner different from *br* lethal–*rbp* viable mutations (perhaps by altering the activity of the proximal promoter). The third class, *2Bab* mutants, is lethal with both *br* and *rbp* mutants. A key mutation in this class, $2Bab^5$, is genetically null for *br* and *rbp*. In $2Bab^5$ mutants, *BR-C* coding sequence are separated from both promoters by an inversion (Fig. 4), and transcription from the locus now occurs using an adventitious promoter (C. Bayer and J. W. Fristrom, unpublished observations).

Of these three classes of mutants in which both *br* and *rbp* functions are affected, only the *2Bab* class affects *2Bc* function, producing a weak viable *2Bc* phenotype (kinked leg) (Kiss *et al.*, 1988). Because only *2Bab* mutations affect *2Bc* function, we propose that normal *2Bc* function has a lower threshold requirement for *BR-C* activity than do *br* and *rbp* functions. Consistent with this view, *BR-C* transcript levels are lower in $2Bab^5$ mutants than in the other mutant classes that effect both *br* and *rbp,* but not *2Bc,* function. In addition, the remaining *2Bc* function in $2Bab^5$ mutants does not appear to depend on the activity of either the distal or the proximal promoter. This suggests that a separate regulatory region specific for *2Bc* function may lie downstream of the proximal promoter.

B. *Broad-Complex* Proteins

Each of the four *BR-C* zinc finger protein isoforms contains at least three physically distinct domains (DiBello *et al.,* 1991; C. Bayer and J. W. Fristrom, unpublished observations). All share an amino-terminal domain of 425 amino acids (BRcore, Fig. 5). In addition to the BRcore each isoform ecodes a unique pair of C_2H_2 zinc finger domains. Located between the BRcore and the zinc fingers is a linker region of variable length, ranging from 275 amino acids in the case of the Z4 protein to only 8 amino acids in the case of Z2 isoform. Z1 proteins contain one of three different linker domains. Two of these have alternate versions of a glutamine-rich sequence. The other has a glutamine domain next to a region rich in the amino acids threonine and asparagine. The Z3 class of *BR-C* protein also has a distinctive linker domain in which asparagine and serine are unusually abundant. The role of these linker domains in the function of *BR-C* proteins is unclear. Glutamine- and serine/threonine-rich repeats have been implicated in protein–protein contacts in the transcriptional activation domains of several transcription factors (Courey and Tjian, 1988; Seipel *et al.,* 1992). Alternatively, the threonine- and serine-rich domains in Z1 and Z3 proteins may serve as phosphorylation sites used to differentially regulate their activities (Tanaka and Herr, 1990; O'Neill *et al.,* 1994).

The presence of pairs of C_2H_2 zinc fingers in the *BR-C* proteins immediately suggests that they regulate metamorphosis by binding to DNA and modulating transcription. Binding to DNA occurs *in vitro* to the regulatory region of a *BR-C* target, the salivary glue secretion gene, *Sgs4* (von Kalm *et al.,* 1994). *BR-C* proteins bind to *Sgs4* sequences known to be required for the proper regulation of this gene *in vivo.* The pattern of binding of *BR-C* proteins is complex, with 6 to 12 binding sites for each zinc finger isoform in the 838 nucleotide *Sgs4* regulatory region. However, there is considerable specificity in the choice of binding sites for each protein isoform. An inspection of the sequences bound by *BR-C* proteins upstream of *Sgs4* suggests a basis for this specificity. Each *BR-C* protein appears to bind to a different consensus sequence. Most striking is the presence within each consensus of an invariant trinucleotide that is unique to each zinc finger isoform. This finding is of particular interest as trinucleotide sequences are important in the binding of three other C_2H_2 zinc finger proteins: tramtrack, Zif268, and the chorion transcription factor CF2 (Pavletich and Pabo, 1991; Gogos *et al.,* 1992; Fairall *et al.,* 1992, 1993). The co-crystal structures of tramtrack and Zif268 bound to their binding sites have been resolved. In both cases, amino acid residues in the α-helix of each finger make contact with a core trinucleotide sequence. A comparison of the pairs of zinc finger domains in *BR-C* proteins further suggests a basis for specificity in binding to DNA. Several amino acids adjacent to the first finger of the tramtrack 69-kDa protein (TTK69) zinc finger domain are essential for full binding activity (Fairall *et al.,* 1992). The equivalent region is divergent among all *BR-C* proteins (Fig. 6) and suggests that their DNA-binding specifici-

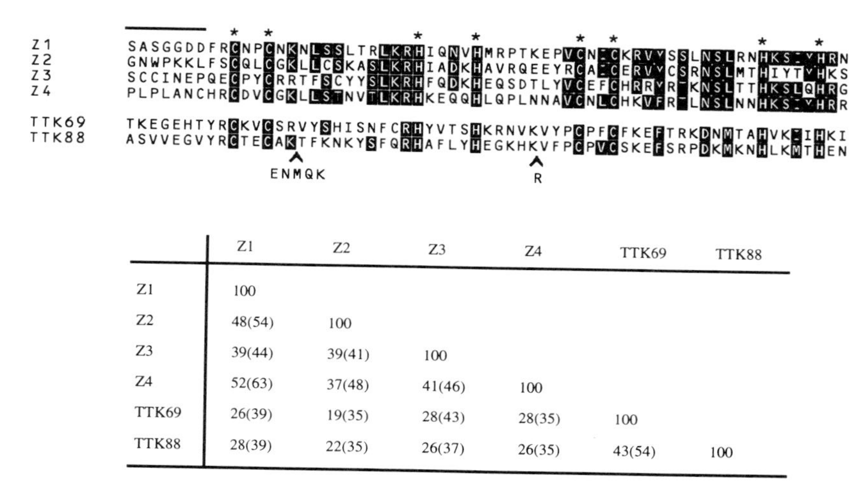

	Z1	Z2	Z3	Z4	TTK69	TTK88
Z1	100					
Z2	48(54)	100				
Z3	39(44)	39(41)	100			
Z4	52(63)	37(48)	41(46)	100		
TTK69	26(39)	19(35)	28(43)	28(35)	100	
TTK88	28(39)	22(35)	26(37)	26(35)	43(54)	100

Fig. 6. Comparison of the amino acid sequences of each *BR-C* zinc finger pair. For comparison, the zinc finger pairs contained in the related tramtrack proteins (TTK69 and TTK88) are also shown (Harrison and Travers, 1990; Read and Manley, 1992) (the locations of additional amino acids found within the TTK88 domain are indicated with an open arrowhead). White letters in black boxes indicate amino acid positions within each *BR-C* zinc finger pair that are conserved (identities or conservative substitutions) in at least three of the proteins. Amino acid positions in the tramtrack protein matching these conserved regions of the *BR-C* zinc fingers are also shown. An asterisk indicates the position of the zinc coordinating cysteine and histidine residues. The bar shows the location of seven divergent amino acids immediately N-terminal to the first finger of the tramtrack and *BR-C* zinc finger domains that may influence binding specificity. Similarities between the zinc finger domains are shown in the table; percentages identities (percentage similarities including identities and conservative substitutions).

ties may depend in part on the sequence N-terminal to the first zinc finger of each pair.

The first 120 N-terminal residues of all *BR-C* proteins represent a conserved motif (the POZ domain) present in more than 30 proteins in a variety of organisms (DiBello *et al.,* 1991; Bardwell and Treisman, 1994). About one-third of these proteins, like the *BR-C* proteins, bind to DNA and have a C_2H_2 zinc finger domains. The others appear to be structural proteins or have unknown functions. In the C_2H_2 class of proteins, the POZ domain mediates specific protein–protein interactions that inhibit binding to DNA by a factor of 5- to 10-fold (Bardwell and Treisman, 1994). Inhibition of binding by the POZ domain is not limited to zinc finger proteins and includes other DNA-binding motifs such as POU- and homeodomains. Not surprisingly, the POZ domain does not interact directly with zinc fingers to repress binding. Rather, interactions between POZ domains occur. However, these interactions are very specific, and not all combinations of POZ domains interact. This raises the possibility that additional cellular factors interact with *BR-C* proteins to facilitate their binding to DNA. Alternatively, DNA binding by *BR-C* proteins may be enhanced in particular heterodimeric or multimeric combinations and may be repressed in others. These possibilities are exciting given that all isoforms of the *BR-C* are ubiquitously expressed, but in varying ratios in specific tissues (see Section V,D) Perhaps the relative ratio of *BR-C* zinc finger isoforms produced in a particular tissue determines which isoforms bind to DNA and activate the appropriate target genes in that tissue.

C. Relationships between *Broad-Complex* Protein Products and Genetic Functions

Thus far, the various functions provided by the *BR-C* during metamorphosis have been defined at the level of mutant phenotypes. Ultimately, however, an understanding of the function of the *BR-C* and its role in mediating the 20-HE signal requires identification of the relationships between *BR-C* genetic functions and the zinc finger proteins encoded by this locus.

Two complementary approaches have been used to correlate *BR-C* protein products with specific genetic functions. First, structural changes at the *BR-C* and truncated proteins have been identified in mutants defective in specific genetic functions. Second, transgenic animals carrying cDNAs encoding individual *BR-C* proteins and driven by an *hsp70* promoter have been used to test the ability of *BR-C* proteins to rescue mutant lethality or visible defects. As an extension of this analysis, we have also found that these transgenes can rescue the transcriptional activity of genes that are downregulated in *BR-C* mutant backgrounds. Although many aspects of these studies are still in progress, a picture of the relationships between zinc finger proteins and genetic functions is emerging.

1. *br* Function

The Z2 protein isoform provides *br* function. An amorphic *br* allele, br^{28}, is associated with a P element insertion into the Z2 domain (DiBello *et al.*, 1991) (Fig. 4). Revertants of br^{28}, resulting from precise excision of the P element, are wild type for all *BR-C* functions, demonstrating that the P element insertion is responsible for the loss of *br* function (C. Bayer and J. W. Fristrom, unpublished observations). As expected, the wild-type Z2 protein is absent in br^{28} mutants, and a truncated Z2 protein is present instead (Emery *et al.*, 1994).

The ability of the *BR-C* transgenes to rescue lethal *br* mutant phenotypes has been tested. In the null br^5 and br^{28} mutations, imaginal discs fail to elongate and development arrests at the white prepupal stage (Kiss *et al.*, 1988). Expression of a Z2 transgene during larval development rescues both disc morphogenesis and subsequent development through eclosion in both of these mutants (C. Bayer and J. W. Fristrom, unpublished observations). In contrast to Z2, expression of Z1, Z3, or Z4 proteins are unable to rescue br^5 or br^{28} prepupal lethality.

2. *rbp* Function

The Z1 protein isoform provides *rbp* function. In the lethal rbp^5 mutant, all of the Z1-containing proteins are truncated (Emery *et al.*, 1994). They appear to be prematurely terminated between the conserved core region and the zinc finger domain. Thus, rbp^5 animals produce no Z1 proteins with DNA-binding activity. The homozygous viable *rbp* mutant, rbp^{tm}, contains a P element insertion a few kilobases upstream from the Z1 zinc finger domain (Moran and Torkamanzehi, 1990; C. Bayer and J. W. Fristrom, unpublished observations) (Fig. 4). A novel transcript containing both Z1 and Z4 sequences is found in this mutant whereas Z2 and Z3 transcripts are unaffected. The bristle defect in rbp^{tm} can be rescued by expression of the Z1 transgene during prepupal development, but not by any of the other *BR-C* proteins. Consistent with a primary role for Z1 in *rbp* function, Z1 transgene expression can also rescue the pupal lethality of rbp^1 mutants.

A correlation between Z1 and *rbp* function is also indicated by experiments aimed at rescuing the transcriptional activity of genes dependent on *BR-C* function. Five salivary gland genes, *Sgs4* (Guay and Guild, 1991; Karim *et al.*, 1993; von Kalm *et al.*, 1994) and the late puff genes

L71-1,-3,-6, and *-9* (Guay and Guild, 1991; Karim *et al.*, 1993), are strongly underexpressed in *rbp*^{5} mutants. The transcriptional activity of all of these genes is restored to approximately wild-type levels in *rbp*^{5} animals when a Z1 transgene is expressed (K. Crossgrove and G. M. Guild, personal communication; L. von Kalm, C. Bayer, and J. W. Fristrom, unpublished observations). Expression of a Z2 transgene cannot rescue the transcriptional activity of these target genes. It is not yet known if a Z3 or Z4 transgene can also rescue the activity of these genes.

3. *2Bc* Function

The relationship between *BR-C* proteins and *2Bc* function is unclear. The correlation between the centromere proximal genetic map position of the *2Bc*^{1} mutation and the physical position of the Z3 zinc finger domain has led to the suggestion that the Z3 domain may mediate *2Bc* function (Aizenon and Belyaeva, 1982; DiBello *et al.,* 1991). However, expression of either a Z2 or a Z3 transgene (but not a Z1 or a Z4 transgene) during larval development can rescue the prepupal lethality of *2Bc* mutants (C. Bayer and J. W. Fristrom, unpublished observations). While these experiments demonstrate that either the Z2 or the Z3 protein can provide *2Bc* function, they do not address whether they both are actually required for *2Bc* function. Genetic tests indicate that the Z2-defective mutant *br*^{28} is *2Bc*$^+$ (because it fully complements *2Bc* mutants) and thus suggest that Z3 provides the *2Bc* function in this animal. The relative roles of the Z2 and Z3 proteins in *2Bc* function can be resolved once a Z3-defective mutant is isolated and genetically analyzed.

Unlike *br*^{28} and *rbp*^{5}, no structurally abnormal *BR-C* proteins or RNAs have been identified in *2Bc* mutants (Karim *et al.,* 1993; Emery *et al.,* 1994). However, a transcript analysis of *2Bc* mutants suggests that the underlying molecular basis of this class of mutations is complex. *BR-C* transcript levels are greatly reduced in *2Bc* whole animals or third instar salivary glands (Karim *et al.,* 1993; L. von Kalm, C. Bayer, J. W. Fristrom, and S. K. Beckendorf, unpublished observations). Thus *2Bc* mutations appear to be regulatory, and *2Bc* function may even be part of an autoregulatory loop used to modulate *BR-C* expression. A striking feature of the transcriptional defects in *2Bc* mutant larvae is that accumulation of *BR-C* RNA (and *BR-C* target gene RNA) is delayed rather than permanently reduced (Karim *et al.,* 1993; L. von Kalm, C. Bayer, J. W. Fristrom, and S. K. Beckendorf, unpublished observations). Together with the rescue results described earlier, this suggests that although the activities of all four *BR-C* proteins may be reduced in *2Bc* mutants during third instar, it is only the lack of sufficient levels of Z2 or Z3 protein that prevents successful advancement through metamorphosis.

4. *Z4 Protein Function*

Z4-specific defects have not been identified in any *BR-C* mutant, and Z4 transgenes have not been implicated unequivocally in any *BR-C* function. One possible explanation is that Z4 may encode a nonessential and/or redundant function at the *BR-C*. A role for the Z4 protein in *BR-C* function can best be studied by specifically mutating the Z4 zinc finger domain so that the transcription of other *BR-C* RNAs is not affected. Phenotypic and complementation analysis of such Z4 mutants may then reveal the function of this zinc finger domain.

Structurally, the Z4 protein is most similar to Z1 and most divergent from Z2 (Fig. 6). Z1 and Z4 proteins also appear to share functional properties. High levels of Z1 or Z4 transgene expression during third instar can induce larval death. In contrast, larval lethality does not result from equivalently high levels of Z2 or Z3 transgene expression during third instar. These results suggest that excessive Z1 or Z4 expression during third instar may prematurely activate cell death pathways in larval tissues.

D. Regulatory Roles of the *Broad-Complex*

The regulatory roles of the *BR-C* are best known in imaginal discs and salivary glands. Genetic evidence also suggests a role for the *BR-C* in gut, muscle, and CNS metamorphosis, indicating that the *BR-C* is a global regulator of this whole developmental transition (Restifo and White, 1991, 1992). In salivary glands, *BR-C* action is understood through its effects on the expression of target genes. In contrast, in imaginal discs and other tissues, mutant phenotypes indicate a requirement for *BR-C* functions for specific morphogenetic processes, whereas relatively little is known about specific *BR-C* targets.

1. *Salivary Glands*

Salivary gland development during metamorphosis is characterized by progressive waves of gene expression, each wave depending on proper regulation by the preceding one. *BR-C* activity is required for each wave of gene expression in salivary glands, emphasizing its role as a key regulator of metamorphosis in this tissue (Karim *et al.,* 1993).

In salivary glands, *BR-C* RNA is first detected in early third instar larvae (von Kalm *et al.,* 1994). Defects in salivary gland gene expression patterns are apparent in *BR-C* mutants by mid third instar. Induction of intermolt genes fails in *npr1* mutants and is delayed in *rbp* and *2Bc* mutants (Crowley *et al.,* 1984; Guay and Guild, 1991; Karim *et al.,* 1993). In wild-type animals, Z1 and Z3 RNA isoforms accumulate in salivary glands at this time, and Z1 and Z3 proteins bind directly to sequences required to induce one of the intermolt genes, *Sgs4* (von Kalm *et al.,* 1994). Thus, the timing of glue gene induction appears to depend on when *BR-C* protein products reach a threshold level. Delayed induction in *rbp* and *2Bc* mutants can be viewed as a consequence of the failure of the corresponding *BR-C* protein products to accumulate to sufficient levels at the proper time in the third instar. *2Bc* function is also required to repress glue gene transcription at the end of larval development. In *2Bc* mutants, expression of these target genes persists well into the prepupal period, consistent with the observation that the regression of intermolt puffs is delayed (Zhimulev *et al.,* 1982; Karim *et al.,* 1993).

Data from puffing and transcriptional analysis indicate that the *BR-C* also regulates the activity of both early puff and late puff genes at the end of larval development. In $npr1^6$ mutants the early 74EF and 75E puffs are submaximally induced, whereas another early puff, at 63F, does not form (Belyaeva *et al.,* 1981). *E74A* and *E75A* RNAs are underexpressed in *2Bc* mutants in late larvae and early prepupae (Karim *et al.,* 1993). (It is important to note that while puffing studies specifically reflect events occurring in salivary glands, transcriptional data for early genes come from whole animals and mixed organ culture and may not accurately reflect salivary gland gene regulation.) As described in Section V,C,3, the *BR-C* itself is also underexpressed in *2Bc* mutants, suggesting that it may autoregulate its expression at this time.

Late puff gene transcription is abnormal in *rbp* and *2Bc* mutants (Guay and Guild, 1991; Karim *et al.,* 1993). The *rbp* function appears to be crucial for all late genes tested so far (the *L71* cluster). However, a variety of effects, ranging from wild-type expression to severely reduced transcript accumulation, are seen in *2Bc* mutants. Consistent with these observations, puffing at some, but not all, of the late gene sites fails to occur in *2Bc* mutants (Zhimulev *et al.,* 1982). In the cases where late gene expression is affected in *2Bc* mutants, induction is delayed and repression occurs several hours earlier than normal (Karim *et al.,* 1993). Thus, the *BR-C* may regulate intermolt and late expression in similar ways, i.e., by controlling the timing of induction and repression.

In contrast to the early and late puffs, the mid prepupal puffing pattern is normal in *2Bc* mutants (Zhimulev *et al.,* 1982). Although *2Bc* function is not required for this phase of salivary gland gene expression, one of the mid prepupal puff, 75C-D, is prematurely induced in $npr1^6$ mutants (Belyaeva *et al.,* 1981). Therefore it is possible that another *BR-C* function (*br* or *rbp*) may have a role in repressing this mid prepupal puff at the larval to prepupal transition.

Like the early and late puffs, the activities of late prepupal puffs are aberrant in *2Bc* mutants (Zhimulev *et al.,* 1982). Here, some puffs are underdeveloped whereas others fail to form. Again transcriptional studies in whole animals support a role for the *BR-C* in late prepupal puff gene expression. For example, in a manner reminiscent of the late larval response, *E74A, E75A,* and the *BR-C* itself are underexpressed in *2Bc* mutants in late prepupae (Karim *et al.,* 1993).

Finally, the last step of larval salivary gland metamorphosis—histolysis—fails to occur in *rbp* mutants (Restifo and White, 1992). The adult salivary glands fail to develop properly in these mutants as well. Abnormalities in adult salivary gland morphogenesis are not limited to *rbp* mutants, but occur also in lethal *br* and *2Bc* mutants. This suggests that the *BR-C* controls programmed salivary gland cell death in early pupae.

In summary, the *BR-C* appears to be a crucial regulator of the timing of almost every aspect of gene expression in salivary glands. Clearly, two of the three *BR-C* gene functions, *rbp* and *2Bc,* are essential for the normal progression of gene expression during larval salivary gland metamorphosis. Because Z1 and Z3 RNAs are the predominant isoforms expressed in salivary glands during late larval and prepupal development, it is likely that they encode the *BR-C* proteins essential for salivary gland metamorphosis (Huet *et al.,* 1993; von Kalm *et al.,* 1994). Surprisingly, although the Z1 (*rbp*) protein is abundant in white prepupae, the Z3 protein cannot be detected at this time (Emery *et al.,* 1994), once again suggesting that *2Bc* function may be complex and potentially involves more than one *BR-C* zinc finger isoform.

2. *Imaginal Discs*

Imaginal disc morphogenesis begins at different times during the third instar depending on the fate of the specific disc. For example, eye discs begin to differentiate in mid third instar, whereas wing and leg imaginal disc morphogenesis begins in late third instar in response to 20-HE. *BR-C* mutants have strong effects on the morphogenesis of discs that form appendages. As the 20-HE titer rises, wing and leg discs follow a well-defined developmental pattern (reviewed by Fristrom and Fristrom, 1993). Briefly, elongation and shaping of the appendages are followed by eversion of the appendages to the outside of the animal. Finally, the discs fuse with one another to form a continuous head and thoracic epidermis. Two *BR-C* genetic functions, *br* and *2Bc,* are required for wing and leg disc morphogenesis (Kiss *et al.,* 1988). In *br* mutants, leg and wing discs fail to both elongate and evert. Thus *br* function is required in multiple disc subdomains since the centers of elongation and eversion are the disc proper and peripodial epithelium, respectively. In contrast, in *2Bc* mutants, discs fail to fuse properly, leaving gaps in the epidermis (Kiss *et al.,* 1988).

Wing and leg disc morphogenesis begins almost immediately after the late third instar rise in hormone titer (6 hr before puparium formation; −6-hr larvae) with elongation and eversion essentially completed by 6 hr AP (Fristrom and Fristrom, 1993) (Fig. 2). Fusion is completed by around 8 hr AP. When *in vitro* cultured imaginal discs are exposed to 20-HE (approximating the rise in hormone titer at the end of third instar), all four *BR-C* zinc finger RNAs are rapidly induced (C. Bayer and J. W. Fristrom, unpublished observations). However, whereas Z1 and Z3 RNAs predominate in salivary glands, Z2 RNA and protein are most abundant in discs both *in vitro* and *in vivo.* This is expected, given the relationship between Z2 and *br* function and the requirement for *br* activity in the early stages of disc morphogenesis.

In contrast to the wealth of information concerning the role of *BR-C* genetic functions in imaginal disc morphogenesis, almost nothing is known about the nature of *BR-C* target genes in this tissue. There is, however, evidence that the products of the *BR-C* and *Stubble* (*Sb*) genes interact to control appendage elongation. The *Sb* locus interacts with *br* mutants in heteroallelic combinations to produce a severe malformation of appendages (Beaton *et al.,* 1988).

The *Sb* serine protease is required for proper morphogenesis of imaginal discs. *Sb* expression is rapidly induced by 20-HE in imaginal discs prior to appendage elongation and eversion (Appel *et al.,* 1993; R. Abu-Shumays and J. W. Fristrom, unpublished observations) (Fig. 2). A likely role for the *Sb* product at this time could be to cleave a component of the extracellular matrix that is necessary for disc elongation. Several 20-

OH-inducible serine protease inhibitors have also been implicated in imaginal disc morphogenesis (Pino-Heiss and Schubiger, 1989), and some may be required to modulate or downregulate *Sb* activity after elongation is complete.

One can imagine several ways that 20-HE could regulate the timing of *Sb* action in mediating disc elongation. One of these could involve *BR-C* regulation of *Sb* expression. Alternatively, 20-HE may directly upregulate *Sb* expression, whereas the *BR-C* may modulate activity of the *Sb* protease by regulating the expression of a protease inhibitor. The latter mechanism is suggested by the identification of a *BR-C* target gene, *P9*, that contains a serine protease inhibitor domain of the kunitz family (J. Emery and G. M. Guild, personal communication). The *P9* gene is induced by 20-HE as disc elongation nears completion (Fig. 2), and its expression is positively regulated by *rbp* function and negatively regulated by *br* function (J. Emery and G. M. Guild, personal communication). Although there is no evidence demonstrating *P9* regulation of *Sb* activity, the timing of *Sb* and *P9* induction in imaginal discs and their relationships to the *BR-C* suggest a general hypothetical model by which 20-HE acts through the *BR-C* to modulate *Sb* activity and control elongation of imaginal discs (Fig. 7).

Shortly after the late third instar rise in 20-HE titer, the *BR-C* is transcriptionally upregulated in imaginal discs and acts to repress tran-

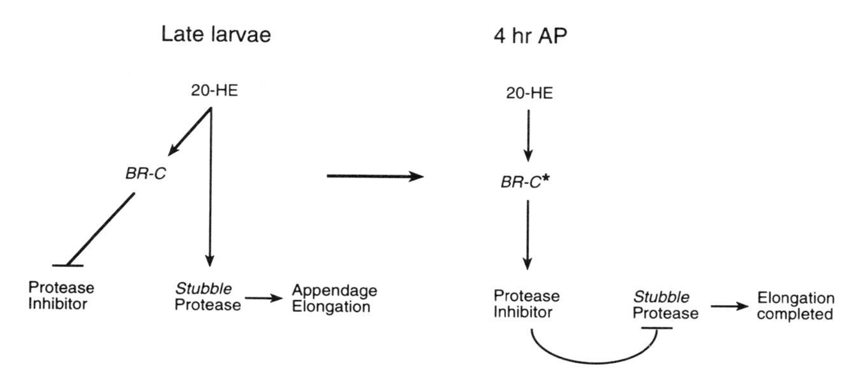

Fig. 7. A hypothetical model showing how *Stubble* serine protease activity might be regulated during imaginal disc morphogenesis. A key component of the model is a protein isoform switch at the *BR-C* (*BR-C* to *BR-C**) at 4 hr AP that induces expression of an inhibitor of the *Stubble* protease, preventing further elongation (see text for details).

scription of a *Sb* serine protease inhibitor. *Sb* expression is also induced by the increase in hormone titer, and its product mediates elongation. At approximately 3 to 4 hr AP, *BR-C* protein expression undergoes an isoform switch, leading to up-regulation of a *Sb* inhibitor and consequent down-regulation of *Sb* protein activity. Indeed, a *BR-C* protein isoform switch (from Z2 to Z1 protein) does occur in discs around 4 hr AP, the time when elongation nears completion (Emery *et al.,* 1994 (Fig. 2). Most aspects of this model can be tested experimentally once the key step of identifying an inhibitor of the *Sb* serine protease has been achieved.

E. Future Directions

The coupling of genetical and molecular developmental studies of the *BR-C* has revealed much new information about the metamorphic functions of this complex gene, as well as its regulation. As one might expect for a key regulator of development, each advance in our knowledge raises new questions.

Our understanding of the relationships between *BR-C* genetic functions and zinc finger proteins has improved significantly. *rbp* function depends on the Z1 protein whereas *br* function depends on the Z2 protein. Yet important questions remain. Most notably, *2Bc* function has been linked to both the Z2 and the Z3 protein isoforms. Perhaps the Z1 and Z2 isoforms provide all essential *BR-C* function, and the Z3 and Z4 isoforms are not essential for metamorphosis. Determining whether Z3 isoforms are defective in *2Bc* mutants will help to address this possibility. However, identifying defects in *2Bc* mutants may not be easy, particularly if they are regulatory. An alternative approach to understanding the role of Z3 and Z4 protein isoforms is to mutagenize each zinc finger domain individually and directly test the effect of these mutations.

Although much is known about the role of the *BR-C* in directing specific morphogenetic processes in imaginal discs, and to a lesser degree in other tissues, few downstream effector target genes have been identified. The isolation and characterization of such target genes are crucial for a full understanding of metamorphosis. Different ratios of *BR-C* protein isoforms appear to be utilized in a tissue-specific manner to regulate these target genes. In order to understand why particular target genes are expressed in specific tissues it will be necessary to determine how *BR-C* protein isoform activities are regulated. Regulation of *BR-C* isoform activity could occur at multiple levels. Some target genes may contain binding sites for only one isoform. Alternatively, two or more isoforms may compete for, or cooperate at, binding sites in a given target. The mode of binding may depend on the relative concentration of each protein. In addition, binding activities may be post-translationally regulated. The POZ domain, which presumably regulates the affinity of *BR-C* proteins for their binding sites, is found in all *BR-C* isoforms. It seems likely that the regulatory effects of this domain are modulated by cellular factors in an isoform- and tissue-specific manner.

Perhaps the most critical and challenging question to be faced is the transcriptional mechanism by which different tissue-specific ratios of *BR-C* isoforms are actually produced. This is likely to be a key factor in determining the resulting tissue-specific patterns of *BR-C* protein activity. As with the control of protein isoform activity, one can imagine multiple levels of regulation. For example, is alternative splicing of exons downstream from the BRcore regulated by the presence or absence of particular tissue-specific splicing factors? How does the choice of promoter influence which processed transcripts will be produced? The *BR-C* may also be differentially regulated by combinations of *EcR* isoforms, early gene products, and tissue-specific transcription factors. As a prerequisite to identifying and understanding such interactions, further studies of *BR-C* regulatory sequences are necessary. Given the large size of this locus, this is a daunting task. However, in the end, understanding how the *BR-C* exerts its control over the complex process of *Drosophila* metamorphosis will be a rich reward.

ACKNOWLEDGMENTS

We thank Greg Guild, Kirsten Crossgrove, John Emery, Steve Beckendorf, Yien Kuo, Bing Zhou, Linda Restifo, Dianne Fristrom, Robin Abu-Shumays, and Ann Hammonds for sharing unpublished results. We are grateful to Dianne Fristrom for assistance with the figures. This review was significantly improved by critical comments from Robin Abu-Shumays, Dianne Fristrom, Greg Guild, Ann Hammonds, and Mary Prout.

REFERENCES

Aizenzon, M. G., and Belyaeva, E. S. (1982). Genetic loci in the X chromosome region 2AB of *Drosophila melanogaster. Drosophila Inf. Serv.* **58,** 3–7.

Andres, A. A., Fletcher, J. C., Karim, F. D., and Thummel, C. S. (1993). Molecular analysis of the initiation of insect metamorphosis: A comparative study of *Drosophila* ecdysteroid-regulated transcription. *Dev. Biol.* **160,** 388–404.
Appel, L. F., Prout, M., Abu-Shumays, R., Hammonds, A., Garbe, J. C., Fristrom, D., and Fristrom, J. (1993). The *Drosophila Stubble–stubbloid* gene encodes an apparent transmembrane serine protease required for epithelial morphogenesis. *Proc. Natl. Acad. Sci. U.S.A.* **90,** 4937–4941.
Apple, R. T., and Fristrom, J. W. (1991). 20-Hydroxyecdysone is required for, and negatively regulates, transcription of *Drosophila* pupal cuticle protein genes. *Dev. Biol.* **146,** 569–582.
Ashburner, M., Chihara, C., Meltzer, P., and Richards, G. (1974). Temporal control of puffing activity in polytene chromosomes. *Cold Spring Harbor Symp. Quant. Biol.* **38,** 655–662.
Bardwell, V. J., and Treisman, R. (1994). The POZ domain: A conserved protein–protein interaction motif. *Genes Dev.* **8,** 1664–1677.
Beaton, A. H., Kiss, I., Fristrom, D., and Fristom, J. W. (1988). Interaction of the *Stubble–stubbloid* locus and the *Broad-Complex* of *Drosophila melanogaster. Genetics* **120,** 453–464.
Beckendorf, S. K., and Kafatos, F. C. (1976). Differentiation in the salivary glands of *Drosophila melanogaster:* Characterization of the glue proteins and their developmental appearance. *Cell* (*Cambridge, Mass.*) **9,** 365–373.
Belyaeva, E. S., Aizenzon, M. G., Semeshin, V. F., Kiss, I., Koczka, K., Baricheva, E. M., Gorelova, T. D., and Zhimulev, I. F. (1980). Cytogenetic analysis of the 2B3-4–2B11 region of the X chromosome of *Drosophila melanogaster* I: Cytology of the region and mutant complementation groups. *Chromosoma* **81,** 281–306.
Belyaeva, E. S., Vlassova, I. E., Biyasheva, Z. M., Kakpakov, V. T., Richards, G., and Zhimulev, I. F. (1981). Cytogenetic analysis of the 2B3/4–2B11 region of the X chromosome of *Drosophila melanogaster* II. Changes in 20-OH ecdysone puffing caused by genetic defects of puff 2B5. *Chromosoma* **84,** 207–219.
Belyaeva, E. S., Protopopov, M. O., Baricheva, E. M., Semeshin, V. F., Izquierdo, M. L., and Zhimulev, I. F. (1987). Cytogenetic analysis of the region 2B3-4–2B11 of the X chromosome of *Drosophila melanogaster* VI: Molecular and cytological mapping of the *ecs* locus and the 2B puff. *Chromosoma* **95,** 295–310.
Berendes, H. D. (1965). Salivary gland function and chromosomal puffing patterns in *Drosophila hydei. Chromosoma* **17,** 35–77.
Berendes, H. D., and Ashburner, M. (1978). The salivary glands. *In* "The Genetics and Biology of *Drosophila*" (M. Ashburner and T. F. Wright, eds.), Vol. 2b, pp. 453–498. Academic Press, New York.
Birr, C. A., Fristrom, D., King, D. S., and Fristrom, J. W. (1990). Ecdysone-dependent proteolysis of an apical surface glycoprotein may play a role in imaginal disc morphogenesis in *Drosophila. Development* **110,** 239–248.
Burtis, K. C., Thummel, C. S., Jones, C. W., Karim, F. D., and Hogness, D. S. (1990). The *Drosophila* 74EF early puff contains *E74,* a complex ecdysone-inducible gene that encodes two *ets*-related proteins. *Cell* (*Cambridge, Mass.*) **61,** 85–99.
Campos-Ortega, J. A., and Hartenstein, V. (1985). "The Embryonic Development of *Drosophila melanogaster.*" Springer-Verlag, Berlin.
Chao, A. T., and Guild, G. M. (1986). Molecular analysis of the ecdysterone-inducible 2B5 "early" puff in *Drosophila melanogaster. EMBO J.* **5,** 143–150.

Clark, W. C., Doctor, J., Fristrom, J. W., and Hodgetts, R. B. (1986). Differential responses of the dopa decarboxylase gene to 20-OH-ecdysone in *Drosophila melanogaster. Dev. Biol.* **114,** 141–150.

Cohen, S. M. (1993). Imaginal disc development. *In* "The Development of *Drosophila melanogaster*" (M. Bate and A. Martinez Arias, eds.), Vol. II, pp. 747–841. Cold Spring Harbor Laboratory, Cold Spring Harbor, New York.

Condic, M. L., Fristrom, D., and Fristrom, J. W. (1991). Apical cell shape changes during *Drosophila* imaginal leg disc elongation: A novel morphogenetic mechanism. *Development* **111,** 23–33.

Courey, A. J., and Tjian, R. (1988). Analysis of SP1 *in vivo* reveals multiple transcriptional domains, including a novel glutamine-rich activation motif. *Cell (Cambridge, Mass.)* **55,** 887–898.

Crowley, T. E., Mathers, P. H., and Meyerowitz, E. M. (1984). A *trans*-acting regulatory product necessary for expression of the *Drosophila melanogaster* 68C glue gene cluster. *Cell (Cambridge, Mass.)* **39,** 149–156.

DiBello, P. R., Withers, D. A., Bayer, C. B., Fristrom, J. W., and Guild, G. M. (1991). The *Drosophila Broad-Complex* encodes a family of related proteins containing zinc fingers. *Genetics* **129,** 385–397.

Doctor, J., Fristrom, D., and Fristrom, J. W. (1985). The pupal cuticle of *Drosophila:* Biphasic synthesis of pupal cuticle proteins *in vivo* and *in vitro* in response to 20-hydroxyecdysone. *J. Cell Biol.* **101,** 189–200.

Emery, I. F., Bedian, V., and Guild, G. M. (1994). Differential expression of *Broad-Complex* transcription factors may forecast distinct development tissue fates during *Drosophila* metamorphosis. *Development* **120,** 3275–3287.

Fairall, L., Harrison, S. D., Travers, A. A., and Rhodes, D. (1992). Sequence-specific DNA binding by a two zinc-finger peptide from the *Drosophila melanogaster* tramtrack protein. *J. Mol. Biol.* **226,** 349–366.

Fairall, L., Schwabe, J. W. R., Chapman, L., Finch, J. T., and Rhodes, D. (1993). The crystal structure of a two zinc-finger peptide reveals an extension to the rules for zinc-finger/DNA recognition. *Nature (London)* **366,** 483–487.

Fechtel, K., Natzle, J. E., Brown, E. E., and Fristrom, J. W. (1988). Prepupal differentiation of *Drosophila* imaginal discs: Identification of four genes whose transcripts accumulate in response to a pulse of 20-hydroxyecdysone. *Genetics* **120,** 465–474.

Fechtel, K., Fristrom, D., and Fristrom, J. W. (1989). Prepupal differentiation in *Drosophila:* Distinct cell types elaborate a shared structure, the pupal cuticle, but accumulate transcripts in unique patterns. *Development* **106,** 649–656.

Feigl, G., Gram, M., and Pongs, O. (1989). A member of the steroid hormone receptor gene family is expressed in the 20-OH-ecdysone inducible puff 75B in *Drosophila melanogaster. Nucleic Acids Res.* **17,** 7167–7178.

Fekete, E., Fristrom, D., Kiss, I., and Fristrom, J. W. (1975). The mechanism of evagination of imaginal discs of *Drosophila melanogaster.* II Studies on trypsin-accelerated evagination. *Wilhelm Roux's Arch. Dev. Biol.* **173,** 123–138.

Fessler, L. I., Condic, M., Nelson, R., Fessler, J., and Fristrom, J. (1993). Site-specific cleavage of basement membrane collagen IV during *Drosophila* metamorphosis. *Development* **117,** 1061–1070.

Fraenkel, G. (1952). A function of the salivary glands of the larvae of *Drosophila* and other flies. *Biol. Bull. (Woods Hole, Mass.)* **103,** 285–286.

Fraenkel, G., and Brookes, V. J. (1953). The process by which the puparia of many species of flies become fixed to substrate. *Biol. Bull. (Woods Hole, Mass.)* **105,** 442–449.

Fristrom, D., and Fristrom, J. W. (1975). The mechanism of evagination of imaginal discs of *Drosophila melanogaster.* I. General considerations. *Dev. Biol.* **43,** 1–23.

Fristrom, D., and Fristrom, J. W. (1993). The metamorphic development of the adult epidermis. *In* "The Development of *Drosophila melanogaster*" (M. Bate and A. Martinez Arias, eds.), Vol. II, pp. 843–897. Cold Spring Harbor Laboratory, Cold Spring Harbor, New York.

Fristrom, D., Doctor, J., and Fristrom, J. W. (1986). Procuticle proteins and chitin-like material in the inner epicuticle of the *Drosophila* pupal cuticle. *Tissue Cell* **18,** 531–543.

Fyrberg, E. A., and Goldstein, L. S. B. (1990). The *Drosophila* cytoskeleton. *Annu. Rev. Cell Biol.* **6,** 559–596.

Fyrberg, E. A., Kindle, K. L., Davidson, N., and Sodja, A. (1980). The actin genes of *Drosophila:* A dispersed multigene family. *Cell (Cambridge, Mass.)* **19,** 365–378.

Garbe, J. C., Yang, E., and Fristrom, J. W. (1993). *IMP-L2:* An essential secreted immunoglobulin family member implicated in neural and ectodermal development in *Drosophila. Development* **119,** 1237–1250.

Gasch, A., Hinz, U., Leiss, D., and Renkawitz-Pohl, R. (1988). The expression of β_1 and β_3 tubulin genes of *Drosophila melanogaster* is spatially regulated during embryogenesis. Development **211,** 8–16.

Gogos, J. A., Hsu, T., Bolton, J., and Kafatos, F. C. (1992). Sequence discrimination by alternatively spliced isoforms of a DNA binding zinc finger protein. *Science* **257,** 1951–1955.

Guay, P. S., and Guild, G. M. (1991). The ecdysone-induced puffing cascade in *Drosophila* salivary glands: A *Broad-Complex* early gene regulates intermolt and late gene transcription. *Genetics* **129,** 169–175.

Harrison, S. D., and Travers, A. A. (1990). The *tramtrack* gene encodes a *Drosophila* finger protein that interacts with the *ftz* transcriptional regulatory region and shows a novel embryonic expression pattern. *EMBO J.* **9,** 207–216.

Hill, R. J., Segraves, W. A., Choi, D., Underwood, P. A., and Macavoy, E. (1993). The reaction with polytene chromosomes of antibodies raised against *Drosophila* E75A protein. *Insect Biochem.* **23,** 99–104.

Huet, F., Ruiz, C., and Richards, G. (1993). Puffs and PCR: The in vivo dynamics of early gene expression during ecdysone responses in *Drosophila. Development* **118,** 613–627.

Janknecht, R., Taube, W., Ludecke, H.-J., and Pongs, O. (1989). Characterization of a putative transcription factor gene expressed in the 20-OH-ecdysone inducible puff 74EF in *Drosophila melanogaster. Nucleic Acids Res.* **17,** 4455–4464.

Karim, F. D., and Thummel, C. S. (1992). Temporal coordination of regulatory gene expression by the steroid hormone ecdysone. *EMBO J.* **11,** 4083–4093.

Karim, F. D., Guild, G. M., and Thummel, C. S. (1993). The *Drosophila Broad-Complex* plays a key role in controlling ecdysone-regulated gene expression at the onset of metamorphosis. *Development* **118,** 977–988.

Kiehart, D. P. (1991). Contractile and cytoskeletal proteins in *Drosophila* embryogenesis. *Curr. Top. Membr. Transp.* **38,** 79–97.

Kimble, M., Incardona, J. P., and Raff, E. C. (1989). A variant β-tubulin isoform of *Drosophila melanogaster* (β_3) is expressed primarily in tissues of mesodermal origin in embryos and pupae, and is utilized in populations of transient microtubules. *Dev. Biol.* **131,** 415–429.

Kiss, I., Bencze, G., Fodor, A., Szabad, J., and Fristrom, J. (1976). Prepupal larval mosaics in *Drosophila melanogaster. Nature* (*London*) **262,** 136–138.

Kiss, I., Szabad, J., and Major, J. (1978). Genetic and developmental analysis of puparium formation in *Drosophila. Mol. Gen. Gent.* **164,** 77–83.

Kiss, I., Beaton, A. H., Tardiff, J., Fristrom, D., and Fristrom, J. W. (1988). Interactions and developmental effects of mutations in the *Broad-Complex* of *Drosophila melanogaster. Genetics* **118,** 247–259.

Koelle, M. R., Talbot, W. S., Segraves, W. A., Bender, M. T., Cherbas, P., and Hogness, D. S. (1991). The *Drosophila EcR* gene encodes an ecdysone receptor, a new member of the steroid receptor family. *Cell* (*Cambridge, Mass.*) **67,** 59–77.

Korge, G. (1975). Chromosome puff activity and protein synthesis in the larval salivary glands of *Drosophila melanogaster. Proc. Natl. Acad. Sci. U.S.A.* **72,** 4550–4554.

Lindsley, D. L., and Zimm, G. G. (1992). "The Genome of *Drosophila melanogaster.*" Academic Press, San Diego.

Locke, M. (1966). The structure and formation of the cuticulin layer in the epicuticle of an insect, *Calpodes ethlius* (Lepidoptera, Hesperiidae). *J. Morphol.* **118,** 461–494.

Moore, J., Fristrom, D., Hammonds, A., and Fristrom, J. W. (1990). Characterization of *IMP-E3,* a gene active during imaginal disc morphogenesis in *Drosophila melanogaster. Dev. Genet.* **11,** 299–309.

Moran, C., and Torkamanzehi, A. (1990). P-elements and quantitative variation in *Drosophila. In* "Ecological and Evolutionary Genetics of *Drosophila*" (J. S. F. Barker, W. T. Starmer, and R. J. MacIntyre, eds.), Plenum, New York.

Natzle, J. (1993). Temporal regulation of *Drosophila* imaginal disc morphogenesis: A hierarchy of primary and secondary 20-hydroxyecdysone–responsive loci. *Dev. Biol.* **155,** 516–532.

Natzle, J. E., Fristrom, D. K., and Fristrom, J. W. (1988). Genes expressed during imaginal disc morphogenesis: *IMP-E1,* a gene associated with epithelial cell rearrangement. *Dev. Biol.* **129,** 428–438.

Natzle, J. E., Robertson, J. P., Majumdar, A., Vesenka, G. D., Enlow, B., and Clark, K. E. (1992). Sequence and expression of *IMP-L1,* an ecdysone-inducible gene expressed during *Drosophila* imaginal disc morphogenesis. *Dev. Genet.* **13,** 331–344.

O'Neill, E. M., Rebay, I., Tjian, R., and Rubin, G. M. (1994). The activities of two ETS-related transcription factors required for *Drosophila* eye development are modulated by the Ras/MAPK pathway. *Cell* (*Cambridge, Mass.*) **78,** 137–147.

Osterbur, D., Fristrom, D., Natzle, J., and Fristrom, J. W. (1988). Genes expressed during imaginal disc morphogenesis: *IMP-L2,* a gene implicated in epithelial fusion. *Dev. Biol.* **129,** 439–448.

Paine-Saunders, S., Fristrom, D., and Fristrom, J. W. (1990). The *Drosophila IMP-E2* gene encodes an apically secreted protein expressed during imaginal disc morphogenesis. *Dev. Biol.* **140,** 337–351.

Panzer, S., Weigel, D., and Beckendorf, S. K. (1992). Organogenesis in *Drosophila melanogaster:* Embryonic salivary gland determination is controlled by homeotic and dorsoventral patterning genes. *Development* **114,** 49–57.

Pavletich, N. P., and Pabo, C. O. (1991). Zinc finger–DNA recognition: Crystal structure of a Zif268–DNA complex at 2.1 A. *Science* **252,** 809–817.

Petri, W., Fristrom, J., Stewart, D., and Hanly, E. W. (1971). The *in vitro* synthesis and characteristics of ribosomal RNA in imaginal discs of *Drosophila melanogaster. Mol. Gen. Genet.* **110,** 245–262.

Pino-Heiss, S., and Schubiger, G. (1989). Extracellular protease production by *Drosophila* imaginal discs. *Dev. Biol.* **132,** 282–291.

Poodry, C. A., and Schneiderman, H. A. (1970). The ultrastructure of the developing leg of *Drosophila melanogaster. Wilhelm Roux's Arch. Dev. Biol.* **168,** 464–477.

Poodry, C. A., and Schneiderman, H. A. (1971). Intercellular adhesivity and pupal morphogenesis in *Drosophila melanogaster. Wilhelm Roux's Arch. Dev. Biol.* **168,** 1–9.

Raikow, R., and Fristrom, J. (1971). Effects of β-ecdysone on RNa metabolism of imaginal discs of *Drosophila melanogaster. J. Insect Physiol.* **17,** 1599–1614.

Read, D., and Manley, J. L. (1992). Alternatively spliced transcripts of the *Drosophila tramtrack* gene encode zinc finger proteins with distinct DNA binding specificities. *EMBO J.* **11,** 1035–1044.

Rechsteiner, M. C. (1970). *Drosophila* lactate dehydrogenase and α-glycerolphosphate dehydrogenase: Distribution and change in activity during development. *J. Insect Physiol.* **16,** 1179–1192.

Restifo, L. L., and Merrill, V. K. L. (1994). Two *Drosophila* regulatory genes, *Deformed* and the *Broad-Complex,* share common functions in development of adult CNS, head and salivary glands. *Dev. Biol.* **162,** 465–485.

Restifo, L. L., and White, K. (1991). Mutations in a steroid hormone-regulated gene disrupt the metamorphosis of the central nervous system in *Drosophila. Dev. Biol.* **148,** 174–194.

Restifo, L. L., and White, K. (1992). Mutations in a steroid hormone-regulated gene disrupt the metamorphosis of internal tissues in *Drosophila:* Salivary glands, muscle and gut. *Wilhelm Roux's Arch. Dev. Biol.* **201,** 221–234.

Restifo, L. L., Estes, P. S., and Dello Russo, C. (1995). Genetics of ecdysteroid-regulated CNS metamorphosis. *Eur. J. Entomol.* **92,** 169–187.

Richards, G. (1976a). Sequential gene activation by ecdysone in polytene chromosomes of *Drosophila melanogaster.* IV. The mid prepupal period. *Dev. Biol.* **54,** 256–263.

Richards, G. (1976b). Sequential gene activation by ecdysone in polytene chromosomes of *Drosophila melanogaster.* V. The late prepupal puffs. *Dev. Biol.* **54,** 264–275.

Robertson, C. W. (1936). The metamorphosis of *Drosophila melanogaster,* including an accurately timed account of the principal morphological changes. *J. Morphol.* **59,** 351–399.

Segraves, W., and Hogness, D. S. (1990). The *E75* ecdysone-inducible gene responsible for the 75B early puff in *Drosophila* encodes two new members of the steroid receptor superfamily. *Genes Dev.* **4,** 204–219.

Seipel, K., Georgiev, O., and Schaffner, W. (1992). Different activation domains stimulate transcription from remote ('enhancer') and proximal ('promoter') positions. *EMBO J.* **11,** 4961–4968.

Smith, A. V., and Orr-Weaver, T. L. (1991). The regulation of the cell cycle during *Drosophila* embryogenesis: The transition to polyteny. *Development* **112,** 998–1008.

Sobrier, M. L., Chapel, S., Couderc, J. L., Micard, D., Lecher, P., Somme-Martin, G., and Dastugue, B. (1989). 20-OH-Ecdysone regulates 60C β-tubulin gene expression in Kc cells and during *Drosophila* development. *Exp. Cell Res.* **184,** 241–249.

Stewart, M., Murphy, C., and Fristrom, J. W. (1972). The recovery and preliminary characterization of X chromosome mutants affecting imaginal discs of *Drosophila melanogaster. Dev. Biol.* **27,** 71–83.

Talbot, W. S., Swyryd, E. A., and Hogness, D. S. (1993). *Drosophila* tissues with different metamorphic responses to ecdysone express different ecdysone receptor isoforms. *Cell (Cambridge, Mass.)* **73,** 1323–1337.

Tanaka, M., and Herr, W. (1990). Differential transcriptional activation by Oct-1 and Oct-2: Interdependent activation domains induce Oct-2 phosphorylation. *Cell (Cambridge, Mass.)* **60,** 375–386.

Thomas, H. E., Stunnenberg, H. G., and Stewart, A. F. (1993). Heterodimerization of the *Drosophila* ecdysone receptor with retinoid X receptor and *ultraspiracle. Nature (London)* **362,** 471–475.

Thummel, C. S., Burtis, K. C., and Hogness, D. S. (1990). Spatial and temporal patterns of *E74* transcription during *Drosophila* development. *Cell (Cambridge, Mass.)* **61,** 101–111.

Tobin, S. L., Zulauf, E., Sanchez, F., Craig, E. A., and McCarthy, B. J. (1980). Multiple actin-related sequences in the *Drosophila* genome. *Cell (Cambridge, Mass.)* **19,** 121–131.

Urness, L. D., and Thummel, C. S. (1990). Molecular interactions within the ecdysone regulatory hierarchy: DNA binding properties of the *Drosophila* ecdysone-inducible *E74A* protein. *Cell (Cambridge, Mass.)* **63,** 47–61.

von Kalm, L., Crossgrove, K., Von Seggern, D., Guild, G. M., and Beckendorf, S. K. (1994). The *Broad-Complex* directly controls a tissue-specific response to the steroid hormone ecdysone at the onset of metamorphosis. *EMBO J.* **13,** 3505–3516.

Wolfgang, W., Fristrom, D., and Fristrom, J. W. (1986). The pupal cuticle of *Drosophila:* Differential ultrastructural immuno-localization of cuticle proteins. *J. Cell Biol.* **102,** 306–311.

Yao, T.-P., Segraves, W. A., Oro, A. E., McKeown, M., and Evans, R. M. (1992). *Drosophila* ultraspiracle modulates ecdysone receptor function via heterodimer formation. *Cell (Cambridge, Mass.)* **71,** 63–72.

Yao, T.-P., Forman, B. M., Jlang, Z., Cherbas, L., Chen, J.-D., McKeown, M., Cherbas, P., and Evans, R. M. (1993). Functional ecdysone receptor is the product of *EcR* and *Ultraspiracle* genes. *Nature (London)* **366,** 476–479.

Young, P. E., Richman, A. M., Ketchum, A. S., and Kiehart, D. P. (1993). Morphogenesis in *Drosophila* requires nonmuscle myosin heavy chain function. *Genes Dev.* **7,** 29–41.

Zhimulev, I. F., Vlassova, I. E., and Belyaeva, E. S. (1982). Cytogenetic analysis of the 2B3-4–2B11 region of the X chromosome of *Drosophila melanogaster* III. Puffing disturbance in salivary gland chromosomes of homozygotes for mutation $l(1)pp1^{t10}$. *Chromosoma* **85,** 659–672.

10

Genes Involved in Postembryonic Cell Proliferation in *Drosophila*

ELIZABETH L. WILDER AND NORBERT PERRIMON

Department of Genetics
Howard Hughes Medical Institute
Harvard Medical School
Boston, Massachusetts

I. INTRODUCTION

During the development of multicellular organisms, the determination of cell fates usually occurs concomitantly with cellular divisions. How

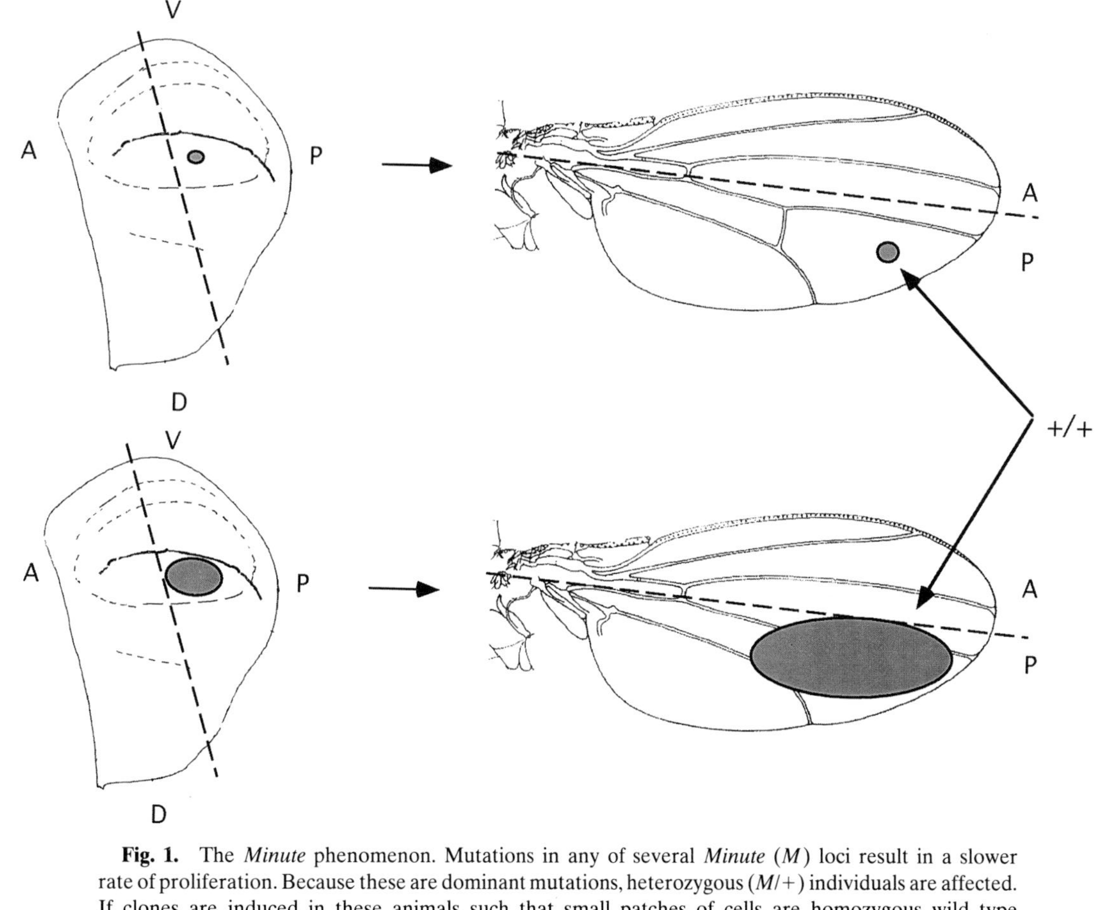

Fig. 1. The *Minute* phenomenon. Mutations in any of several *Minute* (*M*) loci result in a slower rate of proliferation. Because these are dominant mutations, heterozygous (*M*/+) individuals are affected. If clones are induced in these animals such that small patches of cells are homozygous wild type

cell proliferation and cell determination events are interrelated remains a poorly understood process. The primordia of the adult structures in *Drosophila,* the imaginal discs, provide an attractive model system to characterize the relationship between these two developmental processes. Unlike the embryo which undergoes few cell divisions, imaginal discs undergo extensive growth during larval development. A unique property of imaginal discs is their accessibility to direct manipulation. They can be transplanted from animal to animal allowing one to conduct classic embryological analysis not available with other tissues. These experimental features in combination with genetic approaches available in *Drosophila,* i.e., analysis of mutations and techniques of ectopic expression, allow investigators to address how pattern is established in a proliferating tissue.

A striking example of the intricate integration between cellular proliferation and the establishment of pattern is provided by the classic studies of the effect of *Minute* mutations on development (Garcia-Bellido *et al.,* 1973; Morata and Ripoll, 1975; Simpson, 1981; Simpson and Morata, 1981) (Fig. 1). In this experiment, clones of wild-type cells are generated in animals that carry a dominant *Minute* mutation that reduces the growth rate of heterozygous animals. Clones of wild-type cells have a growth advantages such that they finally occupy a large portion of the appendage. Interestingly, the size of the appendage and overall pattern is unaffected, indicating that events that control growth and patterning are tightly regulated and interdependent. This phenomenon is rather reminiscent of regulative events observed in vertebrate embryos, thus raising the possibility that some of the underlying molecular processes are similar.

This chapter reviews the literature describing the control of cell proliferation in the larva, with particular emphasis on the way in which cell

---

(+/+) or homozygous mutant (*M/M*), the effects on altered growth rate can be analyzed. When +/+ clones are induced in a *M*/+ background, the wild-type cells have a growth advantage and come to occupy a relatively large portion of the disc and the resulting wing. The final size and shape of the wing, however, are unchanged, indicating that the heterozygous population undergoes fewer divisions to compensate for the quickly dividing wild-type cells. Anterior, posterior, dorsal, and ventral compartments are fixed regions of the wing disc with positional information that prevents mixing of cells between compartments. The +/+ cells do not cross these borders. The mechanism that determines the completion of pattern and the subsequent induction of quiescence is not understood.

division is linked to pattern formation in the imaginal discs. A number of genes whose function in the control of proliferation is somewhat understood have been selected.

II. BACKGROUND ON *DROSOPHILA* DEVELOPMENT

Drosophila development from egg to imago takes about 10 days at 25°C (described by Bodenstein, 1965). Embryogenesis occupies the first 24 hr and ends with the hatching of a first instar larva. One of the features of the *Drosophila* embryo is that the first 13 nuclear divisions occur without cytokinesis. As these divisions proceed, the nuclei migrate to the periphery of the egg where the plasma membrane extends between the nuclei. Following completion of cellularization at the cellular blastoderm stage (3 hr after egg laying, AEL) (Zalokar and Erk, 1976), the germ layers form, and pattern is established through a series of cell movements with relatively few cell divisions (Foe, 1989).

Following hatching from the egg case, the larva undergoes two molts, one at the end of the first instar, at about 48 hr AEL, and the other at the end of the second instar, at about 72 hr AEL. The third instar lasts approximately 2 days, after which the puparium forms and the 5-day pupal period begins. The imaginal tissues are distinct from the larval tissues in their ploidy (Pearson, 1974). Cells of the larva that do not give rise to adult tissues do not divide, but become increasingly polyploid and undergo histolysis during morphogenesis. The imaginal tissues, however, remain diploid and undergo mitotic divisions rather than the endomitotic divisions characteristic of polyploid cells. Among the cells that will form the adult tissues are the imaginal discs, which are distinguishable from larval cells on the basis of morphology and gene expression at 9–10 hr of embryogenesis (Bate and Martinez Arias, 1991). Other diploid cells in the larvae include the abdominal histoblasts that form the epidermis of the adult abdomen, the gonads that comprise both germ cells and somatic cells, the imaginal rings of the foregut, hindgut, anus, trachea, and salivary glands located in discrete regions of the larval organs, muscles and neural precursors, and blood cells. In the early first instar larva, just after hatching, the discs can be identified as clusters of approximately 40 cells (Madhavan and Schneiderman, 1977). The cells in these clusters proliferate throughout larval life and form folded epithelial sheets that differentiate and evert during morphogenesis. The pattern of the appendages is established entirely in the larval period; the discs are thus somewhat unique in the clear temporal separation between the events of pattern determination and differentiation.

III. CELL PROLIFERATION IN LARVAL TISSUES

Genetic analysis of imaginal precursors has provided a large amount of information about the mechanisms underlying their determination and how their proliferation and differentiation are regulated. This section describes a few examples that illustrate the role of specific genes in these processes.

A. *Escargot* Required for Maintenance of Diploidy in Quiescent Cells

The adult abdomen is formed from segmentally repeated nests of cells in the larva called histoblasts. There are four histoblast nests located dorsally in each segment (Roseland and Schniederman, 1979). Unlike most imaginal tissues, these cells do not proliferate during larval life, but remain quiescent until the onset of morphogenesis. At this point, they begin to proliferate, undergo approximately nine cell divisions, and migrate to form the epidermis of the adult abdomen (Roseland and Schneiderman, 1979). The factors that maintain quiescence of histoblast cells during larval life and that trigger proliferation at puparium formation have not been defined, but are likely to be regulated by the hormones that initiate morphogenesis (see Gilbert *et al.,* Chapter 2, this volume). Progress has been made in understanding how ploidy is controlled in imaginal versus larval cells. The gene *escargot (esg)* encodes a zinc finger-containing protein that appears to be required for the maintenance of diploidy of imaginal cells (Hayashi *et al.,* 1993). This gene was identified through a P element enhancer detection insertion and is expressed in all imaginal tissues (Hayashi *et al.,* 1993; Whiteley *et al.,* 1992). Mutations at this locus often result in embryonic or early larval lethality, but individuals that survive until adulthood exhibit a variety of defects, including poorly differentiated abdomens (Ashburner *et al.,* 1990). Analysis of the histoblasts of mutant larvae revealed that these cells undergo DNA replication without cell division and appear similar to larval polyploid cells (Hayashi *et al.,* 1993). Subsequently, during the pupal period, the cells cannot divide and differentiate. This effect is not observed in imaginal disc cells, which divide throughout larval life. However, if their proliferation is inhibited genetically using a mutation in the *D-raf* serine/threonine kinase (see below), disc cells also become polyploid in an *esg* mutant background. Thus *esg* appears to be required for the maintenance of diploidy in quiescent cells, and this requirement appears to be reduced if the cells are actively dividing. Analysis of the basis for the earlier lethality associated with *esg* mutants will be required to determine additional roles for *esg* during embryogenesis.

B. Role of *twist* in Proliferation of Muscle Precursors

The precursors of the adult abdominal muscles form clusters of cells that proliferate in the larva and are associated with specific nerves (Bate *et al.,* 1991). In the thorax, the muscle precursors make up the adepithelial cells that are associated with the imaginal discs. These cells are established as muscle precursors along with the larval muscle precursors in the embryo, but do not undergo differentiation with their larval counterparts. The mechanism that separates those precursors that differentiate to form larval muscles from those that proliferate appears to involve the gene *twist* (*twi*). The proliferating muscle precursors express *twi,* which encodes a DNA-binding protein of the HLH class (Thisse *et al.,* 1988). Expression of *twi* is initiated at the blastoderm stage (Thisse *et al.,* 1988), persists in the adult muscle precursors (Bate *et al.,* 1991), and is continued as long as these cells proliferate. The initiation of muscle fusion coincides with the cessation of *twi* expression at morphogenesis (Currie and Bate, 1991). Although *twi* expression correlates with proliferation of muscle precursors, a causal relationship has not been established. To date, the mechanism that permits *twi*-expressing cells to remain undifferentiated and proliferating while larval muscles fuse has not been elucidated.

C. Control of Cell Proliferation in Neural Cells

Generation of the adult central nervous system follows a schedule similar to that of myoblasts in that a few neuroblasts proliferate during larval life and differentiate as morphogenesis begins (Truman and Bate, 1988; see Truman, Chapter 8, this volume). Little is known about the regulatory mechanism governing this process. Mutations in a membrane-associated protein with apparent kinase activity encoded by the *lethal giant larvae (lgl)* locus (Strand *et al.,* 1991) cause overproliferation of larval neuroblasts (Gateff, 1978) and produce tissue that can be transplanted to form invasive malignant neuroblastomas (Gateff and Schneiderman, 1974). In addition, studies of the *anachronism* (*ana*) locus (Ebens *et al.,* 1993) have shown that the glia play an important role. The *ana* gene encodes a glycoprotein that is secreted by glial cells and functions to maintain a quiescent phase prior to postembryonic neuroblast proliferation. This gene was identified in a screen for genes that affect the organization of the adult optic lobes. In *ana* mutants, neuroblast proliferation begins precociously, resulting in disorganized brain and optic lobes. The expression of *ana* does not change with the onset of neuroblast proliferation in wild-type animals so the mechanism controlling stem cell proliferation is still unclear.

D. Control of Cell Proliferation in Larval Blood System

A number of mutations have been isolated that result in the production of dark masses of tissue in the animal body cavity. These masses, which are referred to as melanotic tumors (Fig. 2; Table I), have been correlated with the overproliferation of blood cells which leads to invasion of normal larval tissues causing subsequent encapsulation and melanization (Gateff, 1978). Relatively little is known about the role of these genes in the control of proliferation during hematopoiesis since few of the corresponding genes have been cloned.

Watson *et al.* (1991) isolated more than 20 genes on the X chromosome that are associated with the production of tumors. These mutations have been grouped into two classes based on their mutant phenotypes. Class 1 mutations produce tumors through an autoimmune response in which a normal immune system responds to abnormal target tissues. Class 2 mutations result in overgrowth of the hematopoietic organs or hemocytes. One of them, *Aberrant immune response 8 (air8),* a member of the second class, encodes the *Drosophila* homolog of the ribosomal protein, S6 (Watson *et al.,* 1992). Mutations in *air8* inhibit the growth of most larval organs, but cause overproliferation in the lymph gland

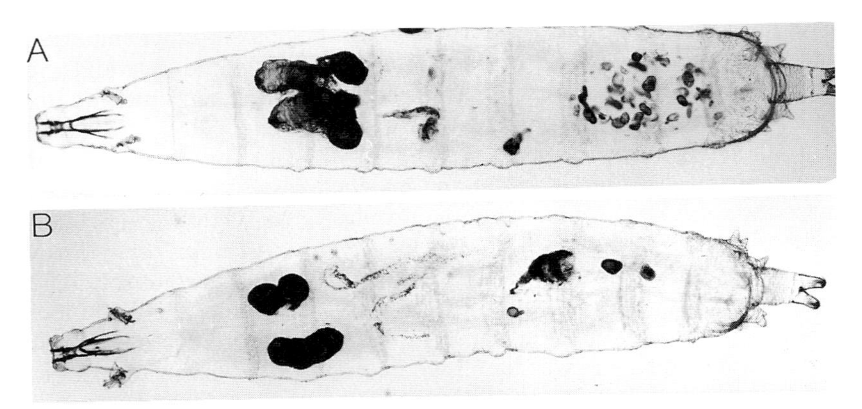

Fig. 2. Melanotic tumors in *Drosophila* larvae. Many mutations induce the onset of tumors in the blood system of *Drosophila*. These tumors are blackened from melanotic pigment. They can arise either through an abnormal proliferation of cells of the blood system or from the response of the immune system to other abnormal cells. (A) A *Tumorous lethal* larva, carrying a dominant mutation in *hopscotch* which results in the hyperproliferation of blood cells. (B) Tumors produced by exposure to ubiquitous ectopic *hopscotch*. The phenotypes produced from the Tum^L lesion and from ectopic *hop* activity are very similar, indicating that the Tum^L mutation results in a gain of *hop* activity. (Photographs courtesy of D. A. Harrison.)

(K. L. Watson, personal communication). In addition, blood cells are overproduced and plasmatocytes prematurely differentiate into lamellocytes (Bryan *et al.*, 1993). Hemocytes from *air8* mutants proliferate in-

TABLE I

Cloned Hematopoietic Tumor Genes

Gene	Homology	Tumor	Other tissues	Reference
air8	Ribosomal S6	Loss of function	Undergrowth	Watson *et al.* (1992)
hopscotch	Janus kinase (JAK)	Gain of function	Undergrowth of discs with loss of function	Perrimon and Mahowald (1986) Harrison *et al.* (1995)
l(3)mbn-1	Cytokeratin domains; novel	Loss of function	None reported	Konrad *et al.* (1994)

definitely *in vitro*. A possible role for S6 in cell proliferation has been sugested by analysis of this protein in mammalian cells. Mammalian S6 is phosphorylated in a developmentally controlled and mitogen-induced manner, and has been proposed to be involved in the selective translation of mRNAs (Mutoh *et al.*, 1992; Sturgill and Wu, 1991; Traugh and Pendergast, 1986).

Another gene that is associated with melanotic tumors, *l(3)malignant blood neoplasm −1 (l(3) mbn-1)* has been shown to encode a protein with a region of homology to cytokeratin, but the majority of the coding sequence bares no homology to known proteins (Konrad *et al.*, 1994). Through its cytokeratin–homology domains, *l(3)mbn-1* has been postulated to bind specific proteins destined for nuclear transport (Konrad *et al.*, 1994).

Perhaps a more obvious candidate for a regulatory molecule is encoded by the gene *hopscotch* (*hop*). This gene encodes a nonreceptor tyrosine kinase of the Jak family (Binari and Perrimon, 1994) and produces melanotic tumors when overexpressed (Harrison *et al.*, 1995). In addition, an allele of *hop, Tumorous lethal* (*Tum-l*) (Hanratty and Dearolf, 1993), encodes a mutant protein in which a single amino acid change produces tumors in a dominant fashion (Harrison *et al.*, 1995). Although mutations that disrupt the control of proliferation in the larval blood system result in similar phenotypes, i.e., melanotic tumors, the extent to which the genes interact in a common pathway is unclear.

E. Genes Involved in Imaginal Disc Cell Growth

Mutations that perturb imaginal disc cell growth can either have a global effect on the growth of the discs or have more restricted effects on the pattern with which proliferation proceeds within the disc. The first class has been grossly divided into two groups, undergrowth and overgrowth, based on their effects on the overall apparent growth of the imaginal discs. It should be pointed out, however, that a further analysis of mutations in this group may reveal a more restricted effect on cell proliferation.

1. Genes Associated with Undergrowth

A number of genes have been identified that are required generally for proliferation of the discs. Mutations in these genes result in a "small disc" or "discless" phenotype in which the cells of the imaginal discs fail to proliferate (Table II). The difference between these two categories has been suggested to be a matter of degree rather than a fundamental difference between the two groups (Szabad and Bryant, 1982). All contain some disc tissue, so the defects do not interfere with the process of establishing the disc primordia. With the exception of *minidiscs,* mutants in this class are able to support the growth of discs transplanted from wild-type hosts, indicating that the proliferation defects are not due to growth factors present in the circulating hemolymph (Shearn and Garen,

TABLE II

Mutants Resulting in Small Discs

Gene	Affected discs[a]	Mitotic defect	Reference
torpedo (D-EGF receptor)	All		Clifford and Schupbach (1989)
l(1)pole hole (D-raf)	All		Nishida *et al.* (1988)
hopscotch	All		Binari and Perrimon (1994)
defective dorsal discs	W, H, DP		Simcox *et al.* (1987)
minidiscs	All		Shearn and Garen (1974)
many discs missing	All		Cohen (1993)
diminished discs	All		Shearn and Garen (1974)
quartet	All		Shearn and Garen (1974)
discs small (2)	All		Cohen (1993)
discs small (3)	All		Cohen (1993)
all discs missing	All		Cohen (1993)
discless	W-missing L-small	No	Kiss *et al.* (1976), Szabad and Bryant (1982)
l(3)discless-1	All	Yes	Shearn *et al.* (1971), Szabad and Bryant (1982)
l(3)discless-3	All	Yes	Shearn *et al.* (1971), Shearn and Garen (1974), Szabad and Bryant (1982), Gatti and Baker (1989)
l(3)discless-5	All	Yes	Shearn *et al.* (1971), Shearn and Garen (1974), Szabad and Bryant (1982)

[a] W, wing; L, leg; H, haltere; DP, dorsal prothoracic.

1974). Among mutations of the "small disc" group is *hop* (Perrimon and Mahowald, 1986), which, as mentioned earlier can result in blood tumors when overexpressed in wild-type animals. In addition, the *Drosophila* homologs of the vertebrate protooncogenes *raf* and EGF-receptor (D-raf and DER, respectively) belong to this group of genes (Clifford and Schupbach, 1989; Nishida *et al.,* 1988). Some genes of this group appear to encode general cell division functions. This is suggested by an analysis of mutations in five of the discless genes which showed that proliferation in the maternal germline was also inhibited by a loss of these functions (Taubert and Szabad, 1987).

A small number of mutations have been isolated in which the defects in cell proliferation are restricted to a subset of the discs. The *defective dorsal discs* (*ddd*) gene is only required for the growth of the humerus, wing, and haltere discs (Simcox *et al.,* 1987). The gene functions nonautonomously within a given disc, such that mosaic discs in which only a few cells are wild type develop normally. However, the wild-type *ddd* product cannot rescue *ddd* mutant cells in adjacent discs, indicating that each must contain at least some wild-type cells for normal growth to occur. Dorsal disc primordia can be observed in *ddd* mutant embryos, indicating that *ddd* is not required for the establishment of the discs. Altogether, these results suggest that *ddd* is likely to encode a growth factor that is specific for dorsal discs and that is produced by the discs themselves. A second mutation, *hyperplastic discs,* can result in either overgrowth or undergrowth of wing discs but has no effect on leg discs (Cohen, 1993). These mutations suggest that different mechanisms govern proliferation in wing and leg discs.

2. *Genes Associated with Overgrowth*

Mutants that result in overgrowth of the imaginal discs can be categorized as either hyperplastic or neoplastic, depending on the effect on the histological structure of the discs (Bryant and Schmidt, 1990) (Table III). In hyperplastic mutants, discs show abnormal folding patterns and they grow to several times their normal size during an extended larval period. They maintain their epithelial structures and the ability to differentiate, but the overproliferation results in additional lobes. Hypomorphic alleles of hyperplastic mutants are sometimes viable until the pharate adult stage, just before eclosion from the pupal case. The structures that are produced are abnormal and vary with the particular lesion.

TABLE III

Genes Resulting in Tumorous Discs

Gene	Homology	Category	Adhesion effects	Reference
fat	Cadherin	Hyperplastic	Yes	Mahoney *et al.* (1991)
giant discs	nd[a]	Hyperplastic	No	Buratovich and Bryant (1995)
tumorous discs	nd	Hyperplastic	nd	Gateff and Mechler (1989)
c43	nd	Hyperplastic	Yes	Martin *et al.* (1977), Jursnich *et al.* (1990)
discs overgrown	nd	Hyperplastic	Yes	Jursnich *et al.* (1990)
discs large	Guanylate kinase	Neoplastic	Yes	Stewart *et al.* (1972), Murphy (1974), Perrimon (1988), Woods and Bryant (1991)
giant larvae	Ser/Thr kinase; cadherin	Neoplastic	Yes	Klambt *et al.* (1989), Strand *et al.* (1991)
warts	nd	Neoplastic	Yes	Bryant *et al.* (1993)

[a] Not determined.

Individuals homozygous for weak *fat* alleles are characterized by invaginations and evaginations of cuticle from the body surface, separated cuticle vesicles, and bristle polarity alterations (Bryant *et al.*, 1988). These defects suggest problems with cell adhesion within the epithelium. Clones of cells that are homozygous for strong alleles produce outgrowths in the adult (Mahoney *et al.*, 1991). Defective cell adhesion could produce abnormalities both in morphogenesis and in the cell communication needed for the control of cell proliferation. Molecular analysis of the *fat* locus revealed that, consistent with its mutant phenotype, *fat* encodes a transmembrane protein with homology to vertebrate cadherins (Mahoney *et al.*, 1991). These proteins are present in adherens junctions and are involved in homophilic cell recognition (Geiger and Avalon, 1992; Takeichi, 1990).

Mutations in *discs overgrown* (*dco*) produce a phenotype similar to that observed in *fat* pharate adults (Jursnich *et al.*, 1990). Mutations in both *dco* and *fat,* as well as lesions in *c43,* produce widening and disruption of the segmentation of the tarsus, the most distal segments of the leg (Bryant, 1987). Discs lacking *dco* and *c43* activity also show reductions in gap-junctional transmission, a common trait of transformed cells (Jursnich *et al.*, 1990). In *lethal giant discs* (*lgd*), other diploid tissues in the larva overproliferate as well. In *lgd* mutants the number of cuticular sense organs is reduced, the number of tarsal segments is reduced, and the leg discs often duplicate (Bryant and Schubiger, 1971). An analysis of this phenotype has been initiated which further underscores the link between the control of proliferation and the establishment of pattern in the discs (Buratovich and Bryant, 1995; see below).

In addition to causing overproliferation, neoplastic mutations disrupt the epithelial structure of the discs, resulting in their transformation into sponge-like masses (Bryant and Schmidt, 1990). These mutations result in larval lethality. The cells become cuboidal and lose the ability to differentiate when transplanted into wild-type hosts. Since loss of functions of these genes produces tumorous discs, these genes are believed to encode tumor suppressor genes. Three genes have been reported to have this function: *warts* (*wts*), *giant larvae* (*lgl*), and *discs large* (*dlg*). *wts* was isolated in a screen in which mutagenized chromosomes were made homozygous in clones of the F_1 individual through the use of FRT-induced mitotic recombination (Bryant *et al.*, 1993). Clones of wild-type cells form long clusters of cells that are oriented parallel to the proximodistal axis. This reflects a preferential orientation for cell division in wild-type discs. In contrast, *wts* mutant clones from large round clumps in the adult legs, suggesting that the orientation of cell division is disturbed. In addition, each cell is outlined by a thick cuticle, which has been suggested to be a result of failure in cell adhesion.

As mentioned earlier, *lgl* causes overproliferation in the larval brain (Gateff, 1978), but is primarily expressed in the imaginal discs (Klambt *et al.*, 1989; Klambt and Schmidt, 1986). The *lgl* protein is present in tissues that are soon to cease proliferation and begin differentiation (Klambt and Schmidt, 1986; Klambt *et al.*, 1989). Cell adhesion is also apparently disrupted in homozygous mutants at this locus, as disc cells show reduced contact with each other and are columnar rather than cuboidal. The identification of the *lgl* gene product as a possible membrane-associated kinase (Strand *et al.*, 1991) that is required for cell adhesion (Gateff, 1978; Klambt *et al.*, 1989) has interesting implications. Intracellular factors are apparently required for cell adhesion in addition to extracellular-binding functions (Takeichi, 1990). Such a factor could conceivably be supplied by the *lgl* gene product.

The *dlg* gene encodes a SH3 domain-containing protein with homology to vertebrate guanylate kinase, implicating it in the regulation of the GTP : GDP ratio (Woods and Bryant, 1991). The protein is localized to septate junctions which appear to provide a selective permeability barrier (Skaer and Maddrell, 1987) and may play a role in cell adhesion (Tepass and Hartenstein, 1994). The recurring finding that adhesion-related molecules are associated with overgrowth strongly suggests that proper cell contacts are crucial in the regulation of proliferation.

3. Genes That Control Pattern of Proliferation in Imaginal Discs

Specific patterns of proliferation within the discs have been most carefully studied in the eye–antennal disc and the wing disc. Cell division in the eye disc is tightly coupled to neuronal differentiation (Ready *et al.,* 1976). Differentiation initiates posteriorly and proceeds anteriorly in a wave called the morphogenetic furrow. Anterior to the furrow, cells divide asynchronously. At the posterior edge, cells form regularly spaced clusters, each containing postmitotic precursors of five of the eight photoreceptor neurons. The remaining cells enter S phase synchronously and divide to form the three remaining photoreceptor cells and nonneuronal accessory cells (Ready *et al.,* 1976). The eye disc thus offers a tissue in which to study the synchronous control of cell division. Thomas *et al.* (1994) have shown that the establishment of synchrony in G_1 at the furrow itself is critical for the proper pattern of cell determination to occur and that the gene *roughex* (*rux*) is required to establish this synchrony. It is not required for proliferation per se, however, as cells anterior to the furrow divide in *rux* mutants, and the discs are normal in size. Since differentiation and proliferation are both associated with the morphogenetic furrow, the signals that control furrow progression are likely to coordinate both processes. One such signal is encoded by the gene *hedgehog* (*hh*), the product of which is thought to function as a secreted signal in the induction of furrow progression through induction of a second secreted protein encoded by the *decapentaplegic* (*dpp*) gene (Heberlein *et al.,* 1993; Ma *et al.,* 1993). Dpp is a member of the transforming growth factor B (TGF-β) family of secreted proteins (Padgett *et al.,* 1987), the members of which perform a variety of functions in cell proliferation and differentiation in vertebrate systems.

In the other large discs, the leg and wing discs, synchronous control of cell divisions has not been observed. However, in the wing disc, cells do not appear to proliferate homogeneously across the disc. Cell proliferation stops in the wing disc in a reproducible manner during the late third instar and early pupal phase (Schubiger and Palka, 1987). The first cells to stop proliferating lie at the boundary between the presumptive dorsal and ventral surfaces of the wing blade (O'Brochta and Bryant, 1985). Formation of this zone of nonproliferating cells (ZNC) requires the activity of the gene *wingless* (*wg*) (Phillips and Whittle, 1993). Gonzalez-Gaitan *et al.* (1994) have suggested that proliferation in the remaining part of the wing disc occurs in waves that derive from proliferation centers within the disc. These centers are restricted to the areas between the veins, indicating that the presumptive vein regions represent proliferation boundaries (see Fig. 4).

The induction of *Notch* (*N*) mutant clones early in larval development produces cells that are unable to proliferate (de Celis and Garcia-Bellido, 1994). Although this gene is associated at other times in development with the transformation of epidermal cells to neural cells, the loss of N mutant epidermal cells in the wing blade cannot be explained by this phenomenon. If the clones are induced in a *Minute* background such that their induction leads simultaneously to a loss of the dominant *Minute* mutation, a growth advantage is obtained relative to the remainder of the disc and *N* mutant clones can proliferate. This indicates that *N* activity provides a growth stimulus but is not required for cell viability. Moreover, *N* mutant clones that are located near the presumptive veins can proliferate in a wild-type background, indicating that *N* activity is required to a lesser extent in this region (de Celis and Garcia-Bellido, 1994).

Understanding *N* function is made more difficult by its pleiotropy. As with many genes, it is used in many aspects of development, including the specification of both embryonic and adult pattern. Because pattern cannot be established in the absence of cell proliferation, these two processes are always linked. Thus if a particular gene product functions in pattern specification, it inevitably influences proliferation and vice versa. Such is the case with *N:* it overtly influences pattern, so to what extent can we say that it regulates the control of cell proliferation? Ultimately, clarification of this issue for *N* and other pattern-specifying genes will require that assays be developed that distinguish between pattern forming functions and functions controlling proliferation, if such distinctions exist. Most of the evidence to date, as described in the following section, suggests that the control of proliferation in the imaginal discs may be entirely dependent on the establishment of pattern.

IV. MUTUAL CONTROL OF PATTERN AND PROLIFERATION

Although imaginal discs are apparently homogeneous epithelial sheets, by late in the third instar, the pattern of the adult structures has been completely determined within the discs. Thus different pieces of an individual disc will differentiate as different pattern elements if transplanted to a permissive environment. This has allowed the establishment of fate maps of third instar discs (Bryant, 1975; Haynie and Bryant, 1986; Schubiger, 1968; see also Bryant, 1978) (Figs. 3 and 4). The observation that discs from which fragments have been removed initiate proliferation to regenerate the missing fragment [or duplicate the existing one (see Bryant, 1978)] suggests that proliferation is driven by an incomplete pattern. Bryant and Simpson (1984) proposed that cell proliferation is controlled by a develop-

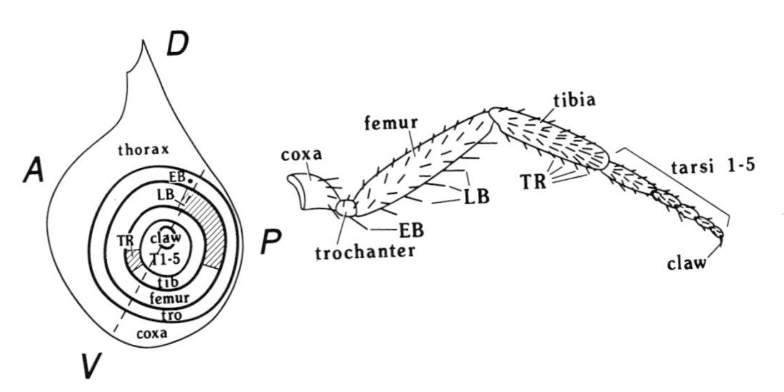

Fig. 3. Fate map of a leg imaginal disc (Schubiger, 1968; Steiner, 1976). The leg imaginal disc consists of a folded epithelial sheet, in which the folds occur in a series of concentric rings. During morphogenesis, the leg everts from the center outward, such that the most distal element (the claw) is specified in the innermost ring of the disc. Increasingly proximal structures are mapped to the increasingly large circles. Elements of the thorax and body wall are specified in the outermost regions of the disc. Anterior and posterior compartments are fixed in the leg disc, but the dorsoventral boundary is not so clearly defined. Some of the most characteristic bristles along the proximal–distal axis are illustrated T1-5, five tarsal segments; TR, transverse row bristles of the tibia; LB, large bristles of the femur; EB, edge bristle of the trochanter.

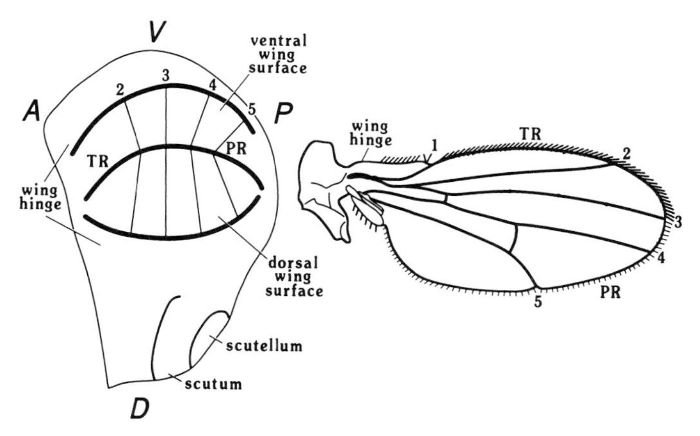

Fig. 4. Fate map of a wing imaginal disc (Bryant, 1975). The wing disc gives rise to the wing as well as the dorsal thorax and thus consists of a folded region that everts outward in a manner similar to the leg disc to produce the wing and an unfolded region that produces the dorsal thorax (the scutum and the scutellum). The presumptive wing blade is dissimilar to the leg in that it is organized around a stripe rather than a point. Thus, eversion produces the two opposed sheets of the wing rather than a cylindrical leg. The stripe corresponds to the boundary between the dorsal and ventral surfaces of the wing. This boundary, the wing margin, is characterized by bristles that mark the anterior (triple row, TR) and posterior surfaces (posterior row, PR). Both dorsal and ventral surfaces contain regions of condensed cells, the veins, that form hollow passageways for nerves and hemolymph. These veins (1–5) occur in stereotyped positions that can be marked in the wing disc by expression of the gene *rhomboid* (Sturtevant *et al.,* 1993). The first vein is localized at the wing margin.

ing map of positional information (Wolpert, 1971) in the disc and that proliferation stops when the map is complete. This hypothesis raises the question: How do the discs "know" when the pattern is complete? The cessation of proliferation in late third instar coincides with the release of the hormone ecdysone, which when converted to 2-hydroxyecdysone, initiates the onset of morphogenesis (see Gilbert *et al.,* Chapter 2, this volume; Russell and Ashburner, Chapter 3, this volume). However, 2-hydroxyecdysone does not trigger the change in the proliferative state. When discs from young larvae are transplanted into the growth-permissive environment of adult abdomens, they grow until they reach their normal size at which point proliferation ceases (Bryant and Levinson, 1985). Thus, the control of proliferation is a disc-intrinsic property.

A. Polar Coordinate Model

Analyses of regenerating or duplicating disc fragments indicate that proliferation is controlled by interactions between cells. Wounding, by surgical manipulation, X-rays, or induction of a temperature-sensitive cell lethal mutant, causes local proliferation at the wound edge (Abbott *et al.,* 1981; French *et al.,* 1976; Postlethwait and Schneiderman, 1973; Russell, 1974). If fragments of the disc are removed, the disc compensates by reprogramming the cells with reference to the identities of the cells at the wound edge so that the missing fragment is regenerated, or the existing piece is duplicated, through a process of proliferation and intercalation, or filling in, of positional values. French *et al.* (1976) proposed that the position within the disc can be described in terms of a circumferential map, with different positions on a circle representing different positional identities, or values, in the imaginal disc (see Fig. 5). This model of positional specification, the polar coordinate model, suggests that the disc ceases proliferation when all of the values around the circumference are present. Meinhardt (1983) extended this model by proposing a mechanism through which pattern might be established during the normal course of development. He proposed that the initial disparity of positional fate arose from positional signals acquired from the embryo in sectors (Fig. 5B). The intersection of these sectors was proposed to establish the imaginal primordium. Differential gene expression would define each sector, and a minimum of three would be required since their intersection should define a point around which the disc grows. The expression of specific genes in each sector would establish their positional identities and lead to the proliferation and pattern intercalation that was described by the polar coordinate model.

B. Molecular Definition of Positional Values

The basic proposition of the polar coordinate model and Meinhardt's model that cell proliferation and, consequently, axis formation are induced by the juxtaposition of cells with disparate positional values has not been refuted. Indeed, experiments continue to support the basic tenets of this model and of Meinhardt's (1983) modifications. A number of genes that function during embryonic pattern formation have been

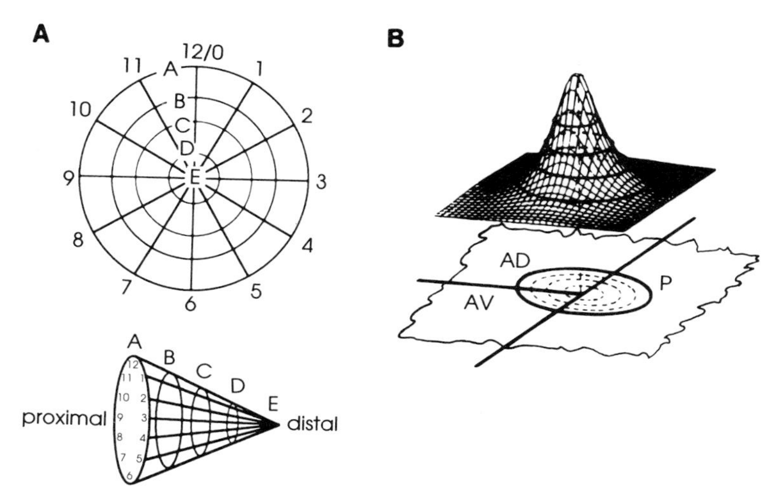

Fig. 5. Models for positional fate specification in the imaginal discs. (A) The polar coordinate model. This model suggests that position is specified circumferentially and radially within the disc. Anterior, posterior, dorsal, and ventral positions can be assigned numbers around the circumference (typically depicted as 0–12) whereas proximal–distal values are assigned radial values (A–E). During wound healing, missing values are intercalated by the "shortest distance" which results in either regeneration of the missing values or duplication of the existing ones. For example, excision of a piece containing values 1–4 would be expected to result in regeneration in the large fragment and duplication of the small one. [Adapted from French *et al.* (1976).] (B) Meinhardt's model. Meinhardt extrapolated from the polar coordinate model to predict a mechanism by which the disc is initially established. He postulated that the initial positional information within the disc occurs in sectors, the intersection of which created a source of an organizing element for the proximodistal axis. Pattern intercalation occurs as result of the interaction between the sectors to promote growth of the developing disc or regeneration in a wounded disc. [Adapted from Meinhardt (1983).]

found to be expressed in sectors of the imaginal discs and to be required for the specification of cell fate in their respective sectors (see Fig. 6).

The *wingless* (*wg*) gene is expressed in the anterior ventral region of the leg disc and is required for the specification of ventral fate (Baker, 1988; Couso *et al.,* 1993). It is expressed in a dynamic pattern in the wing disc and plays multiple roles there (Couso *et al.,* 1993). It is expressed in a ventral region early, and loss of Wg activity during this time results in loss of the wing. *wg* encodes a secreted glycoprotein which is homologous to the vertebrate protooncogene *Wnt-1* (Gonzalez *et al.,* 1991; Nusse

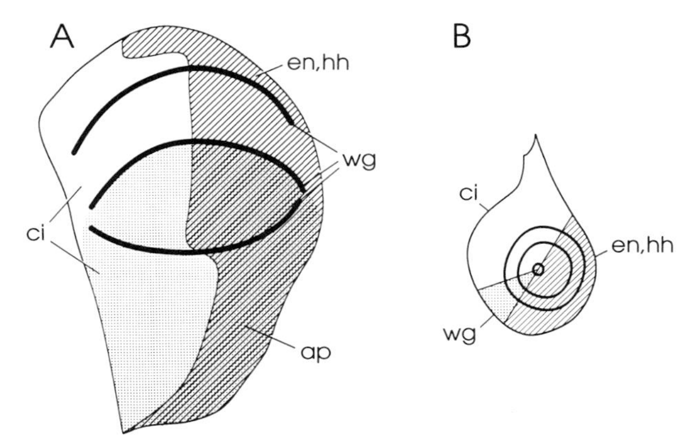

Fig. 6. Sectorial gene expression: (A) wing disc and (B) leg disc. In both leg and wing discs, *en* and *hh* are expressed in the posterior compartment. Loss of *en* function in these cells results in a transformation to anterior fate. The anterior compartment is marked by expression of the gene *ci.* In the wing, the dorsal compartment expresses *ap.* Early in wing development, *wg* is expressed ventrally, but by mid-third instar, its expression becomes restricted to the presumptive dorsoventral boundary and to stripes along both the dorsal and ventral hinge regions. In the leg, *wg* continues to be expressed ventrally throughout larval development. The intersection of sectors in both discs occurs at a point around which the proximal–distal axis has been proposed to be organized (Meinhardt, 1983; Campbell *et al.*, 1993). However, in the wing, the dorsoventral boundary appears to be the organizing region (Diaz-Benjumea and Cohen, 1993; Williams *et al.*, 1994), although an interaction with the anterior–posterior compartment is evidently required for full pattern specification (Adapted with permission from *Nature,* Basler and Struhl, Copyright (1994) Macmillan Magazines Ltd.) (see Fig. 6).

and Varmus, 1992; van den Heuvel *et al.,* 1989). Also expressed in this region, but extending throughout the anterior dorsal region as well, is the gene *cubitus interruptus (ci).* This gene encodes a protein with homology to the vertebrate zinc finger containing GLI-1 and is expected to regulate transcription (Orenic *et al.,* 1990).

In the posterior part of the disc, *engrailed* (*en*) is expressed and defines a compartment, such that *en*-expressing cells do not mix with those of the anterior compartment (Morata and Lawrence, 1975). This gene encodes a homeobox-containing protein which indicates that it functions as a transcription factor. It apparently regulates the transcription of a second gene found in the posterior compartment, *hedgehog.* This gene encodes a protein that is likely to be secreted and is expressed in *en*-expressing cells in both the embryo and in the imaginal discs (Lee *et al.,*

1992; Mohler and Vani, 1992; Tabata and Kornberg, 1994; Tabata *et al.*, 1992). In vertebrate limbs, *hh* has been shown to provide the polarizing activity present in the posterior region of the limb bud and has been postulated to specify pattern in a graded fashion (Riddle *et al.,* 1993). In both vertebrates and *Drosophila,* members of the (TGF-β) family of secreted peptides are induced in response to *hh* (Basler and Struhl, 1994; Tabata and Kornberg, 1994).

The specification of the dorsoventral axis in the leg and wing discs appears to occur via different mechanisms (see later). This is made obvious by the expression of the gene *apterous (ap)* in dorsal cells of the wing but not the leg disc (Cohen *et al.,* 1992). This gene encodes a protein with homology to the LIM class of transcription factors and is required for the specification of dorsal fate in the wing disc (see later).

The *wg, ci, en, hh,* and *ap* genes define anterior, posterior, dorsal, and ventral sectors of the discs. Although complexities arise when leg and wing discs are compared, the fact that sectors in the discs have distinct patterns of gene expression corroborates the ideas of French *et al.* (1976) and Meinhardt (1983). How these sectors might interact to stimulate cell proliferation and pattern in the discs is discussed next.

C. Evidence That Proliferation Is Induced By Sector Juxtaposition

In addition to several surgical experiments in which cells were inappropriately juxtaposed, resulting in proliferation and axial duplications, genetic maniplation of the imaginal discs has revealed that altering the positional information within the disc can also induce pattern intercalation. This has been accomplished either by removing the gene products mentioned earlier through the use of mutants or by ectopically expressing the genes. Three sets of experiments that lend support to the notion that proliferation is induced by the juxtaposition of cells with disparate positional identities are reviewed.

1. Wingless Function in Leg Disc

Loss of Wg activity in the imaginal discs can be achieved either through the use of a pupal–lethal mutation (Baker, 1988) or through the use of a temperature-sensitive allele (Couso *et al.,* 1993). This produces a loss of ventral pattern elements in the legs and axial duplications. Since *wg* is expressed in the anterior ventral sector of the disc, these experiments have been interpreted as evidence that the mechanisms described in the polar coordinate model for regenerating discs operate during normal development (Couso *et al.,* 1993). The disc responds to the loss of fate in the *wg* sector by duplicating the dorsal fate. Moreover, because the duplications occur in a mirror image direction, this provided evidence that pattern was intercalated between the unaffected edges. Interestingly, distal structures are also lost in the absence of *wg.* Failure to specify the distal-most elements of the leg is also seen in certain regeneration experiments and has been suggested to result from a lack of a complete circumference of positional values (French *et al.,* 1976). Thus, distalization of the axis has been thought to require the entire circumference to be specified. However, this is likely to represent an oversimplification of Wg function in the generation of distal fate since the loss of function of the gene *dishevelled* (*dsh*), which is required for Wg function in the embryo, results in the loss of ventral fate in a manner similar to the loss of Wg, but the legs often retain their distal structures (Klingensmith *et al.,* 1994, Thiessen *et al.,* 1994). This suggests that Wg may have a separate function in the proximodistal axis that is independent of ventralization.

Ectopic expression of Wg in clones has also suggested that an inappropriate juxtaposition of cells can lead to extra proliferation and production of a separate axis. Clones of cells that express Wg ectopically have been marked with a cuticular marker and are randomly generated throughout the leg disc (Struhl and Basler, 1993). When these clones are generated on the ventral side of the leg, no aberrant phenotype is produced. However, when these clones fall on the dorsal side of the leg, dorsal cells are transformed to ventral fate and, in some cases, bifurcated legs form. The random generation of the clones precludes the possibility of analyzing for any one clone both the disc and the leg it gives rise to. However, the location of the marked cells indicates that these bifurcations probably arise from clones that were induced at the dorsal anterior–posterior border of the disc. Clones found at the anterior–posterior border of the disc express the gene *aristaless (al)* which is also expressed in the center of the disc (Campbell *et al.,* 1993). Thus *wg* has been proposed to interact with a factor(s) at the anterior–posterior border to establish the proximodistal axis through the activity of *al.* A candidate for the anterior–posterior border-interacting factor was proposed to be *decapentaplegic (dpp)* (Campbell *et al.,* 1993), a member of the TGF-β family of secreted proteins (Padgett *et al.,* 1987). Interestingly, analysis of a tumor suppressor gene *l* (2) *giant discs* has revealed that *wg* and *dpp* are ectopically expressed in mutant discs (Buratovich and Bryant, 1995). These discs fail to cease proliferation, often duplication, and has led to the suggestion that hyperplasia is a result of misexpression of genes that specify pattern (Buratovich and Bryant, 1995).

The expanded ventral structures and bifurcated axes produced from ectopic Wg activity in clones included not only those cells that expressed the ectopic product, but also many wild-type cells (Struhl and Basler, 1993). This led to the hypothesis that Wg specified ventral fate and organized the proximodistal axis by virtue of its secreted nature. However, it has been proposed that the organization of the proximodistal axis is not dependent on the ability of Wg to be secreted but only on its ability to specify ventral fate (Diaz-Benjumea and Cohen, 1994). Analysis of the Wg signaling pathway in the embryo has identified the gene *zeste-white 3 (zw3)* that functions antagonistically to Wg (Siegfried *et al.*, 1992). Wg has been proposed to function through the repression of Zw3 activity (Siegfried *et al.*, 1992), and the ectopic expression of Wg mimics the loss of Zw3 (Noordermeer *et al.*, 1992). Consistently, when clones of *zw3* mutant tissue are induced in the leg discs, ventral structures are observed in ectopic positions. Interestingly, although Zw3 functions only within the cell that expresses it (i.e., it functions cell autonomously), a proximodistal pattern involving wild-type cells is organized around mutant clones in a manner similar to that observed in clones expressing ectopic Wg (Diaz-Benjumea and Cohen, 1994). Thus the establishment of the axis was proposed to result from juxtaposition of dorsal and ventral cells near the anterior–posterior border followed by the repatterning of wild-type cells to intercalate the missing positional identities. This is in agreement with the mechanism proposed by Meinhardt (1983) in which the interaction between sectors produces the growth and patterning of the proximodistal axis.

2. *apterous* Function in Wing Disc

Dorsoventral identities in the wing disc appear to be established during the second instar as marked by the expression of the gene *apterous* in the dorsal part of the disc (Williams *et al.*, 1993). This gene appears to be necessary to specify dorsal fate in the wing disc since dorsal clones mutant at this locus are ventralized in a cell autonomous fashion (Diaz-Benjumea and Cohen, 1993). Interestingly, these clones are often associated with an ectopic wing which forms around the mutant clone (Fig. 7). The dorsoventral boundary of the ectopic wing blade is marked by characteristic bristles that define the wing margin. All elements of the wing margin form, indicating that the juxtaposition of dorsal and ventral cells is sufficient to induce the outgrowth of an ectopic wing.

The *ap* gene encodes a putative transcription factor (Cohen *et al.*, 1992), so other molecules are presumably required for the cell interactions

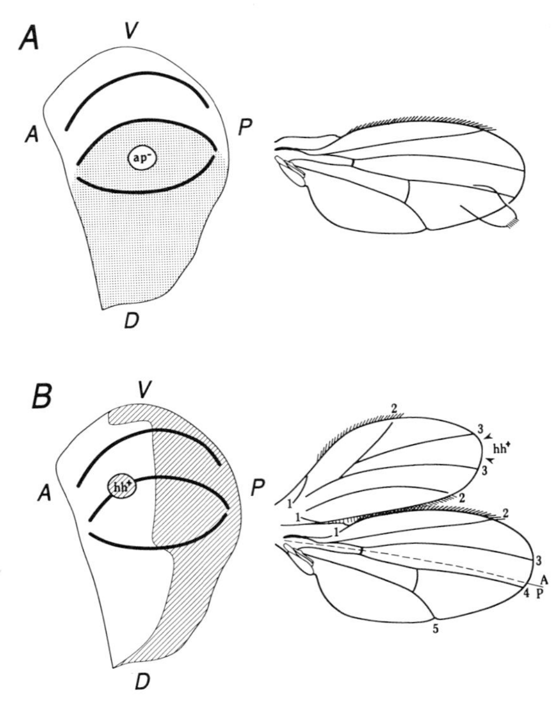

Fig. 7. Induction of wing bifurcations. (A) Dorsal clones of cells lacking *ap* function interact with wild-type cells to produce ectopic marginal structures and outgrowths from the wing blade (Adapted with permission from Diaz-Benjumea and Cohen, 1993). The juxtaposition of dorsal and nondorsal cells creates an ectopic margin at the periphery of the clone, and wild-type cells proliferate to form the majority of the outgrowth. This effect is independent of the anterior–posterior border. (B) Clones of cells ectopically expressing *hh* in the anterior compartment induce wing bifurcations when they arise on the dorsoventral border (Adapted with permission from *Nature,* Basler and Struhl, Copyright (1994) Macmillan Magazines Ltd.). Ectopic anterior–posterior borders are established on either side of the clone, creating a mirror-image symmetric ectopic wing with two anterior halves. This was proposed to be a result of *hh* directing the anterior pattern through the induction of *dpp,* but an interaction with factors at the dorsoventral boundary is also apparently required.

that trigger these events. The gene that is induced first following the interaction of dorsal and ventral cells of the wing disc appears to be *vestigial* (*vg*) (Williams *et al.,* 1994). The *vg* gene is initially transcribed in a stripe that abuts the dorsoventral boundary. It subsequently expands to fill the entire wing blade, where it has been proposed to be required for growth of the wing. Consistently, a loss of *vg* function results in an increasing loss of the wing blade with the increasing severity of the mutation. The analysis of *vg* has led to the hypothesis that dorsoventral interactions lead to the induction of *vg* expression and margin specification and that the expansion of *vg* expression is required for cell proliferation. The early expression of *wg* in the ventral part of the wing disc and the requirement for *wg* function in the formation of the wing has led to the suggestion that Wg provides the signal from the ventral cells. Couso *et al.* (1995) have suggested that the gene *Serrate* (*Ser*), which encodes a transmembrane protein with homology to EGF, is responsible for the dorsal signal. An interaction between Ser and Wg has been proposed to trigger the induction of *vg* and the subsequent outgrowth of the wing.

3. hedgehog Function in Legs and Wings

In wild-type wing and leg discs, *hh* is expressed in the posterior compartment (see Fig. 6). The ectopic expression of *hh* in the anterior compartment of the wing disc on or near the dorsoventral boundary results in transformation to an anterior fate and to the reorganization of pattern such that the wing bifurcates (Basler and Struhl, 1994) (Fig. 7). As a secreted peptide, Hh is thought to act over some distance, although the mechanism of this action is not well understood. *dpp* is normally expressed in anterior cells that abut the posterior compartment. The induction of ectopic *dpp* in anterior cells following ectopic *hh* in the wing disc led to the suggestion that Hh elicits its effects through induction of this member of the TGF-β family (Basler and Struhl, 1994). However, an interaction with a molecule(s) at the dorsoventral boundary is also apparently required to produce axis bifurcation. In the leg disc, *wg* has also been suggested to be downstream of the *hh* function. In the ventral half of the leg disc, *wg* expression is expanded following ectopic expression of *hh,* whereas in the dorsal half, *dpp* expression is expanded. Thus, *hh* was proposed to act through the induction of other secreted factors.

What the ectopic *hh* experiment also suggests is a confirmation of the model of Meinhardt (1983). Ectopic expression of *hh* induces axis bifurcation only when it occurs in the anterior compartment at the dorsoventral boundary. Therefore an interaction between patterning proteins of multiple sectors is required for axis formation. This apparently results in the intercalation of intervening positional values in a manner similar to that following loss of *ap* activity. In this case, the anterior–posterior axis bifurcates instead of the dorsoventral axis. A similar requirement for an interaction between sectors was observed for *wg:* dorsal clones expressing ectopic Wg have to be near the anterior–posterior boundary to induce ectopic legs (Campbell *et al.,* 1993). Meinhardt's model of a requirement for three sectors to intersect is met in both cases. What these experiments add to Meinhardt's model is the identification of those genes that are sufficient to determine sectorial position and the identification of *dpp* as a possible mitogen that induces the proliferation associated with sector juxtaposition.

D. Role of Cell Adhesion in Control of Proliferation

As discussed earlier, the specification of pattern appears to be the driving force behind proliferation in the discs. The fact that many of the tumor suppressor genes encode molecules associated with cell adhesion (see Table III) provides strong evidence that proper adhesion is critical for the control of proliferation. At least three models for the mechanism through which the molecules that specify pattern and those that mediate cell adhesion interact in growth control are obvious (see Fig. 8).

Model 1. The acquisition of stable adhesion partners as specified by patterning molecules such as *ap* and *hh* may provide a signal to cease proliferation. Cell adhesion may form the signal to cells in the disc that pattern is complete. In 1966, Garcia-Bellido showed that cells in different regions of the disc have differential adhesion properties. This provides a mechanism by which cell proliferation might be induced by the juxtaposition of cells with disparate positional information: cell adhesion might be disrupted, resulting in suppression of quiescence. Acquisition of adhesive properties would be expected to be downstream of pattern specification, and loss of adhesive properties might be expected to mimic the disruption of pattern. Such a role for cell adhesion might also explain the finding of Schubiger and co-workers (Abbot *et al.,* 1981; Karpen and Schubiger, 1981; Kiehle and Schubiger, 1985) that the initiation of proliferation following wounding occurs prior to the healing of the two

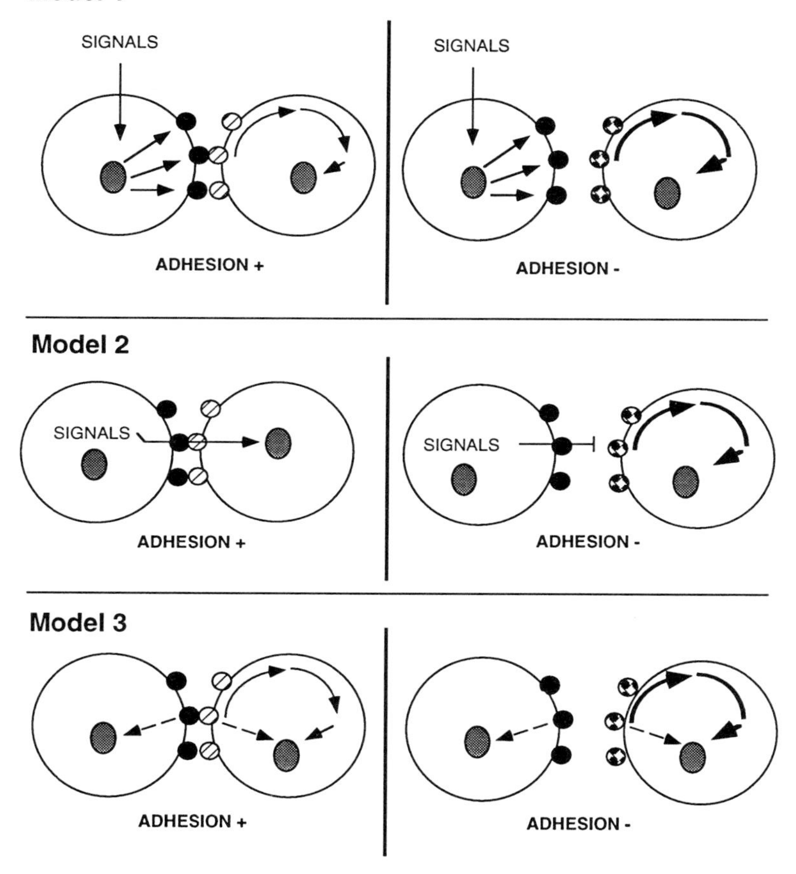

Fig. 8. Models for the role of cell adhesion in control of proliferation. Model 1. Signals from genes that specify pattern, such as *wg* and *hh,* are received by cells which respond by expressing position-specific cell adhesion molecules. These proteins interact with cell adhesion molecules on neighboring cells. This interaction provides a signal. If the interaction is appropriate, cell proliferation is discontinued. If this interaction is inappropriate or if adhesion is disrupted by wounding, proliferation continues. Model 2. The transmission of signals from cell to cell occurs via cell junctions. In the presence of intact junctions, signals pass between cells to control proliferation. If these junctions are disrupted, signaling is perturbed and proliferation continues unchecked. Model 3. Proteins that mediate pattern specification through intercellular interactions may also mediate cell adhesion. Instead of controlling nuclear activities through a series of kinases, proteins that mediate both pattern specification and cell adhesion may regulate cellular processes via the cytoarchitecture. Different adhesive contexts may provide different cellular signals. In all cells, the nucleus is indicated in gray.

surfaces. Loss of adhesion rather than the contact of particular cell surface determinants may provide a proliferative stimulus.

Model 2. Cell adhesion may govern cell proliferation and pattern by facilitating signal transduction. In this model, pattern signals that control proliferation might require cell junctions to be transduced. Adhesion molecules, such as cadherins and integrins, and transmembrane signaling proteins, such as Sevenless (Tomlinson *et al.,* 1987), are both localized to adherens junctions, supporting the notion that these junctions are involved in transmitting signals as well as adhesion (Hynes, 1992).

Model 3. Molecules that specify pattern may directly function in cell adhesion as well. A signaling protein may mediate cell adhesion and lead to changes in cell fate and/or proliferative status via the cytoarchitecture. Such a mechanism is suggested by proteins such as *Notch* and *Delta* which mediate cell adhesion but also provide cellular signals (Artavanis-Tsakonas *et al.,* 1991). Interaction of these proteins could conceivably provide stable adhesive properties to constitute a signal for the completion of pattern as well as a signal to influence the pattern. The extracellular domain of *N* consists of 36 repeats of a domain that is homologous to mammalian epidermal growth factor (Artavanis-Tsakonas *et al.,* 1991). Three of these repeats mediate heterophilic cell adhesion through binding to *Dl* (Rebay *et al.,* 1991). The function of the remaining repeats is unknown, but they offer the possibility of binding to several different proteins. This suggests a mechanism whereby a series of adhesive affinities as well as a series of cell signals may be achieved.

V. ADDITIONAL ROLE FOR *WINGLESS*

The data described thus far are consistent with the notion that cell proliferation in the discs is controlled by cell interactions that determine the completion of pattern specification, perhaps through differential cell adhesion, although cell adhesion may also serve to facilitate signaling between cells. A significant part of these data involve *wg* and rest on the assumption that the only function of *wg* is to specify ventral fate. Evidence indicates that this is likely to be an oversimplification of the role that *wg* plays in the discs.

A. Expression of *wingless* Induced at Wound Edge

In an effort to understand the processes controlling wound healing and pattern intercalation, Russell and co-workers examined the expression of *wg* following surgical wounding or induction of cell death through a temperature-sensitive cell lethal mutation (W. J. Brook, S. Scanga, A. Manoukian, and M. A. Russell, personal communication). The expression of *wg* is induced early following wounding, suggesting a function of this gene in wound healing. Brook *et al.* (1993) provided evidence that *dpp* is also induced early during wound healing. These events precede the activation of genes such as *en* that mark the establishment of newly generated pattern elements. Thus *wg* and *dpp* may play a role in the events that are a prerequisite for pattern intercalation. The ectopic activation of these genes is reminiscent of the expression observed in *gd* mutant discs (Buratovich and Bryant, 1995) and suggests that loss of activity of this tumor suppressor gene may mimic the generation of a wound.

B. Interference of Dorsal Differentiation by *wingless*

The expression of *wg* at the wound edge prior to the healing of the two edges indicates that *wg* functions in generating the proliferating cells that have the capacity to regenerate, i.e., the blastema. Evidence that *wg* has this capacity is provided through directed ectopic expression of *wg* in dorsal cells (Wilder and Perrimon, 1995). These experiments have revealed that *wg* can interfere with the determination of dorsal cells inasmuch as gene expression that marks the commitment of cells as tarsal cells is not observed dorsally. This observation suggests that *wg* expression in dorsal cells induces events that are also necessary for blastema formation. This effect is independent of the function of *wg* in the specification of ventral fate since it can be observed without concomitant ectopic expression of a gene that marks ventral fate. Interestingly, an anterior–posterior difference in response to *wg* has been observed. While anterior cells appear to be able to respond to *wg* either by adopting a ventral fate or by losing markers of determination, posterior dorsal cells cannot be ventralized. This may possibly explain the differential capacity of cells for regeneration (Schubiger, 1971).

It also allows the mechanism of ventral specification to be analyzed separately from the mechanism through which dorsal differentiation is inhibited.

Since *wg* is expressed ventrally, can it be expected to influence the process of determination in dorsal cells during the normal course of development? The secreted nature of Wg protein indicates that the answer is probably "yes." It is generally observed to be secreted over short distances and so is likely to be available to dorsal cells at the disc center. Perhaps Wg at the disc center maintains a blastema-like state that is essential for disc growth.

C. Does *wingless* Function through Cell Adhesion?

The question of the mechanism through which *wg* achieves its functions both in ventral cells and at the disc center is a matter of speculation, but several data address the issue and offer mechanisms by which specification of pattern and cell adhesion might be linked by *wg*. First, analysis of *wg* signaling in the embryo has revealed that the *Drosophila* homolog of the adherens junction protein β-catenin, *armadillo* (*arm*), is required for transmission of the *wg* signal (Peifer *et al.,* 1991; Peifer and Wieschaus, 1990). This at first seems consistent with Model 2 (see Fig. 8) for the role of cell adhesion in signaling and proliferation. The adherens junction, including *arm,* might facilitate the uptake or transmission of the *wg* signal into adjoining cells. However, epistasis experiments revealed that *arm* functions downstream of other genes in the pathway (Noordermeer *et al.,* 1994; Siegfried *et al.,* 1994). Moreover, Peifer *et al.* (1994) have provided evidence that the cell adhesion function of *arm* in the adherens junction is separate from the function of *arm* in Wg signaling. Thus it seems unlikely that Wg signals through adherens junctions. However, a second adhesion-mediated mechanism of action is suggested by the biochemical properties of Wg and other members of the Wnt gene family: *Wg* has been found to tightly adhere to the extracellular matrix (Nusse and Varmus, 1992). This raises the possibility that Wg may elicit part of its function through changes in the adhesive quality of cells, although no data to date have confirmed such a functional significance to the interaction between Wg and the extracellular matrix.

VI. CONCLUDING REMARKS

The control of cell proliferation is likely to involve the integration of several regulatory components. In the imaginal discs these include genes that might be predicted to control proliferation due to their homology to vertebrate protooncogenes. In addition, a number of genes that are involved in cell adhesion have been shown to affect proliferation. Increasingly, cell adhesion molecules in vertebrate systems are being implicated in the control of proliferation as well. The analysis of the mechanisms through which these genes function in this process in *Drosophila* is therefore likely to add to the understanding of growth control generally. Finally, the way in which proliferation and pattern are coupled such that the discs cease growth when the pattern is completely specified is reminiscent of growth control in developing vertebrate systems and implicates genes that specify pattern as growth mediators as well. The degree to which they function directly in the control of proliferation is at this point unclear but may be higher than has been previously thought. The unresolved issue of how these regulatory components are integrated is very much an open question, but the ease of genetic manipulation in *Drosophila* makes the imaginal discs an attractive arena in which to pursue the problem.

REFERENCES

Abbott, L. C., Karpen, G. H., and Schubiger, G. (1981). Compartmental restrictions and blastema formation during pattern regulation in *Drosophila* imaginal leg discs. *Dev. Biol.* **87,** 64–75.

Artavanis-Tsakonas, S., Delidakis, C., and Fehon, R. G. (1991). The *Notch* locus and the cell biology of neuroblast segregation. *Annu. Rev. Cell Biol.* **7,** 427–452.

Ashburner, M., Thompson, P., Roote, J., Lasko, P. F., Grau, Y., El Messal, M., Roth, S., and Simpson, P. (1990). The genetics of a small autosomal region of *Drosophila melanogaster* containing the structural gene for alcohol dehydrogenase. VII. Characterization of the region around the *snail* and the *cactus* loci. *Genetics* **126,** 679–694.

Baker, N. E. (1988). Transcription of the segment-polarity gene wingless in the imaginal discs of *Drosophila,* and the phenotype of a pupal-lethal wg mutation. *Development* **102,** 489–497.

Basler, K., and Struhl, G. (1994). Compartment boundaries and the control of *Drosophila* limb pattern by Hedgehog protein. *Nature* (*London*) **368,** 208–214.

Bate, M., and Martinez Arias, A. (1991). The embryonic origin of imaginal discs in *Drosophila. Development* **112,** 755–761.

Bate, M., Rushton, E., and Currie, D. A. (1991). Cells with persistent *twist* expression are the embryonic precursors of adult muscles of *Drosophila. Development* **113,** 79–89.

Binari, R., and Perrimon, N. (1994). Stripe-specific regulation of pair-rule genes by *hopscotch,* a putative Jak family tyrosine kinase in *Drosophila. Genes Dev.* **8,** 300–312.

Bodenstein, D. (1965). The postembryonic development of *Drosophila. In* "The Biology of *Drosophila*" (M. Demerec, ed.), pp. 275–367. Hafner, New York.

Brook, W. J., Ostafichuk, L. M., Piorecky, J., Wilkinson, M. O., Hodgetts, D. J., and Russell, M. A. (1993). Gene expression during imaginal disc regeneration detected using enhancer-sensitive P-elements. *Development* **117,** 1287–1297.

Bryant, P. J. (1975). Pattern formation in the imaginal wing disc of *Drosophila melanogaster:* Fate map, regeneration, and duplication. *J. Exp. Zool.* **193,** 49–78.

Bryant, P. J. (1978). Pattern formation in imaginal discs. *In* "The Genetics and Biology of *Drosophila*" (M. Ashburner and T. R. F. Wright, eds.), Vol. 2c, pp. 229–335. Academic Press, New York.

Bryant, P. J. (1987). Experimental and genetic analysis of growth and cell proliferation in *Drosophila* imaginal discs. *In* "Genetic Regulation of Development" (W. F. Loomis, ed.), pp. 339–372. Liss, New York.

Bryant, P. J., and Levinson, P. (1985). Intrinsic growth control in the imaginal primordia of *Drosophila,* and the autonomous action of a lethal mutation causing overgrowth. *Dev. Biol.* **107,** 355–363.

Bryant, P. J., and Schmidt, O. (1990). The genetic control of cell proliferation in *Drosophila* imaginal discs. *J. Cell Sci., Suppl.* **13,** 169–189.

Bryant, P. J., and Schubiger, G. (1971). Giant and duplicated imaginal discs in a new lethal mutant of *Drosophila melanogaster. Dev. Biol.* **24,** 233–263.

Bryant, P. J., Huettner, B., Held, L. J., Ryerse, J., and Szidonya, J. (1988). Mutations at the fat locus interfere with cell proliferation control and epithelial morphogenesis in *Drosophila. Dev. Biol.* **129,** 541–554.

Bryant, P. J., and Simpson, P. (1984). Intrinsic and extrinsic control of growth in developing organs. *Q. Rev. Biol.* **59,** 387–415.

Bryant, P. J., Watson, K. L., Justice, R. W., and Woods, D. F. (1993). Tumor suppressor genes encoding proteins required for cell interactions and signal transduction in *Drosophila. Development, Suppl.* 239–249.

Buratovich, M. A., and Bryant, P. J. (1995). Duplication of *l(2)giant discs* imaginal discs in *Drosophila* is mediated by ectopic expression of *wingless* and *decapentaplegic. Dev. Biol.* **168,** 452–463.

Campbell, G., Weaver, T., and Tomlinson, A. (1993). Axis specification in the developing *Drosophila* appendage: The role of wingless, decapentaplegic, and the homeobox gene aristaless. *Cell (Cambridge, Mass.)* **74,** 1113–1123.

Clifford, R. J., and Schupbach, T. (1989). Coordinately and differentially mutable activities of torpedo, the *Drosophila melanogaster* homolog of the vertebrate EGF receptor gene. *Genetics* **123,** 771–787.

Cohen, S. M. (1993). Imaginal disc development. *In* "The Development of *Drosophila melanogaster*" (M. Bate and A. Martinez Arias, eds.), Vol. 2, pp. 747–841. Cold Spring Harbor Laboratory, Cold Spring Harbor, New York.

Cohen, B., McGuffin, M. E., Pfeifle, C., Segal, D., and Cohen, S. M. (1992). *apterous,* a gene required for imaginal disc development in *Drosophila* encodes a member of the LIM family of developmental regulatory proteins. *Genes Dev.* **6,** 715–729.

Couso, J. P., Bate, M., and Martinez Arias, A. (1993). A wingless-dependent polar coordinate system in *Drosophila* imaginal discs. *Science* **259,** 484–489.

Couso, J. P., Carroll, S. B., Knust, E., and Martinez Arias, A. (1994). Serrate and Wingless signaling induce *vestigial* gene expression and wing formation in *Drosophila. Current Biol.* in press.

Currie, D. A., and Bate, M. (1991). The development of adult abdominal muscles in *Drosophila:* Myoblasts express *twist* and are associated with nerves. *Development* **113,** 91–102.

de Celis, J. F., and Garcia-Bellido, A. (1994). Roles of the *Notch* gene in *Drosophila* wing morphogenesis. *Mech. Dev.* **46,** 109–122.

Diaz-Benjumea, F. J., and Cohen, S. M. (1993). Interaction between dorsal and ventral cells in the imaginal disc directs wing development in *Drosophila. Cell (Cambridge, Mass.)* **75,** 741–752.

Diaz-Benjumea, F. J., and Cohen, S. M. (1994). *wingless* acts through the *shaggy/zeste-white 3* kinase to direct dorsal–ventral axis formation in the *Drosophila* leg. *Development* **120,** 1661–1670.

Ebens, A. J., Garren, H., Cheyette, B. N. R., and Zipursky, S. L. (1993). The *Drosophila anachronism* locus: A glycoprotein secreted by glia inhibits neuroblast proliferation. *Cell (Cambridge, Mass.)* **74,** 15–28.

Foe, V. E. (1989). Mitotic domains reveal early commitment of cells in *Drosophila* embryos. *Development* **107,** 1–22.

French, V., Bryant, P. J., and Bryant, S. V. (1976). Pattern regulation in epimorphic fields. *Science* **193,** 969–981.

Garcia-Bellido, A. (1966). Pattern reconstruction by dissociated imaginal disc cells of *Drosophila melanogaster. Dev. Biol.* **14,** 278–306.

Garcia-Bellido, A., Ripoll, P., and Morata, G. (1973). Developmental compartmentalization of the wing disc of *Drosophila. Nature (London) New Biol.* **245,** 251–253.

Gateff, E. (1978). Malignant neoplasms of genetic origin in *Drosophila melanogaster. Science* **200,** 1448–1459.

Gateff, E., and Mechler, B. (1989). Tumor suppressor genes of *Drosophila melanogaster. CRC Crit. Rev. Oncogenesis* **1,** 221–245.

Gateff, E., and Schneiderman, H. A. (1974). Developmental capacities of benign and malignant neoplasms of *Drosophila. Wilhelm Roux Arch. Entwicklungsmech. Org.* **176,** 23–65.

Gatti, M., and Baker, B. (1989). Genes controlling essential cell cycle functions in *Drosophila melanogaster. Genes Dev.* **3,** 438–453.

Geiger, B., and Ayalon, O. (1992). Cadherins. *Annu. Rev. Cell Biol.* **8,** 307–332.

Gonzalez, F., Swales, L., Bejsovec, A., Skaer, H., and Martinez Arias, A. (1991). Secretion and movement of wingless protein in the epidermis of the *Drosophila* embryo. *Mech. Dev.* **35,** 43–54.

Gonzalez-Gaitan, M., Capdevila, M., and Garcia-Bellido, A. (1994). Cell proliferation patterns in the wing imaginal disc of *Drosophila. Mech. Dev.* **40,** 183–200.

Hanratty, W. P., and Dearolf, C. R. (1993). The *Drosophila Tumorous-lethal* hematopoietic oncogene is a dominant mutant in the *hopscotch* locus. *Mol. Gen. Genet.* **238,** 33–37.

Harrison, D. A., Binari, R., Stines-Nahreini, T., Gelman, M., and Perrimon, N. (1995). Activation of a *Drosophila* Janus kinase (JAK) causes hematopoietic neoplasia and developmental defects. *EMBO J.* **14,** 2857–2865.

Hayashi, S., Hirose, S., Metcalfe, T., and Shirras, A. D. (1993). Control of imaginal cell development by the *escargot* gene of *Drosophila. Development* **118,** 105–115.

Haynie, J. L., and Bryant, P. J. (1986). Development of the eye-antennal imaginal disc and morphogenesis of the adult head in *Drosophila melanogaster. J. Exp. Zool.* **237,** 293–308.

Heberlein, U., Wolff, T., and Rubin, G. M. (1993). The TGF-β homolog *decapentaplegic* and the segment polarity gene *hedgehog* are required for propagation of a morphogenetic wave in the *Drosophila* retina. *Cell (Cambridge, Mass.)* **75,** 913–926.

Hynes, R. O. (1992). Integrins: Versatility, modulation, and signaling in cell adhesion. *Cell (Cambridge, Mass.)* **69,** 11–25.

Jursnich, V. A., Fraser, S. E., Held, L. J., Ryerse, J., and Bryant, P. J. (1990). Defective gap-junctional communication associated with imaginal disc overgrowth and degeneration caused by mutations of the dco gene in *Drosophila. Dev. Biol.* **140,** 413–429.

Karpen, G. H., and Schubiger, G. (1981). Extensive regulatory capabilities of a *Drosophila* imaginal disk blastema. *Nature (London)* **294,** 744–747.

Kiehle, C. P., and Schubiger, G. (1985). Cell proliferation changes during pattern regulation in imaginal leg discs of *Drosophila melanogaster. Dev. Biol.* **109,** 336–346.

Kiss, I., Bencze, G., Fekete, E., Fodor, A., Gausz, J., Maroy, P., Szabad, J., and Szidonya, J. (1976). Isolation and characterization of x-linked lethal mutants affecting differentiation of the imaginal discs in *Drosophila melanogaster. Theoret. App. Genet.* **48,** 217–226.

Klambt, C., and Schmidt, O. (1986). Developmental expression and tissue distribution of the *lethal (2) giant larvae* protein of *Drosophila melanogaster. EMBO J.* **5,** 2955–2961.

Klambt, C., Muller, S., Lutzelschwab, R., Rossa, R., Totzke, F., and Schmidt, O. (1989). *l(2) giant larvae* encodes a protein homologous to the cadherin cell-adhesion molecule family. *Dev. Biol.* **133,** 425–436.

Klingensmith, J., Nusse, R., and Perrimon, N. (1994). The *Drosophila* segment polarity gene dishevelled encodes a novel protein required for response to the wingless signal. *Genes Dev.* **8,** 118–130.

Konrad, L., Becker, G., Schmidt, A., Klockner, T., Kaufer-Stillger, G., Dreschers, S., Edstrom, J.-E., and Gateff, E. (1994). Cloning, structure, cellular localization, and possible function of the tumor suppressor gene *lethal (3) malignant blood neoplasm-1* of *Drosophila melanogaster. Dev. Biol.* **163,** 98–111.

Lee, J. J., von Kessler, D., Parks, S., and Beachy, P. A. (1992). Secretion and localized transcription suggest a role in positional signaling for products of the segmentation gene hedgehog. *Cell (Cambridge, Mass.)* **71,** 33–50.

Ma, C., Zhou, Y., Beachy, P. A., and Moses, K. (1993). The segment polarity gene *hedgehog* is required for progression of the morphogenetic furrow in the developing *Drosophila* eye. *Cell (Cambridge, Mass.)* **75,** 927–938.

Madhavan, M. M., and Schneiderman, H. A. (1977). Histological analysis of the dynamics of growth of imaginal discs and histoblast nests during the larval development of *Drosophila melanogaster. Wilhelm Roux's Arch. Dev. Biol.* **183,** 269–305.

Mahoney, P. A., Weber, U., Onofrechuk, P., Biessmann, H., Bryant, P. J., and Goodman, C. S. (1991). The fat tumor suppressor gene in *Drosophila* encodes a novel member of the cadherin gene superfamily. *Cell (Cambridge, Mass.)* **67,** 853–868.

Martin, P., Martin, A., and Shearn, A. (1977). Studies of l(3)C43^{hs1} a polyphasic, temperature sensitive mutant of *Drosophila melanogaster* with a variety of imaginal disc defects. *Dev. Biol.* **55,** 213–232.

Meinhardt, H. (1983). Cell determination boundaries as organizing regions for secondary embryonic fields. *Dev. Biol.* **96,** 375–385.

Mohler, J., and Vani, K. (1992). Molecular organization and embryonic expression of the hedgehog gene involved in cell–cell communication in segmental patterning of *Drosophila. Development* **115,** 957–971.

Morata, G., and Lawrence, P. A. (1975). Control of compartment development by the *engrailed* gene in *Drosophila. Nature* (*London*) **255,** 614–617.

Morata, G., and Ripoll, P. (1975). Minutes: Mutants of *Drosophila* autonomously affecting cell division rate. *Dev. Biol.* **42,** 211–221.

Murphy, C. (1974). Cell death and autonomous gene action in lethals affecting imaginal discs in *Drosophila melanogaster. Dev. Biol.* **39,** 23–36.

Mutoh, T., Rudkin, B. B., and Guroff, G. (1992). Differential responses of the phosphorylation of ribosomal protein S6 to nerve growth factor and epidermal growth factor in PC12 cells. *J. Neurochem.* **58,** 175–185.

Nishida, Y., Hata, M., Ayaki, T., Ryo, H., Yamagarta, M., Shimuza, K., and Nishizuka, Y. (1988). Proliferation of both somatic and germ cells is affected in the *Drosophila* mutants of *raf* proto-oncogene. *EMBO J.* **7,** 775–781.

Noordermeer, J., Johnston, P., Rijsewijk, F., Nusse, R., and Lawrence, P. A. (1992). The consequences of ubiquitous expression of the wingless gene in the *Drosophila* embryo. *Development* **116,** 711–719.

Noordermeer, J., Klingensmith, J., Perrimon, N., and Nusse, R. (1994). *dishevelled* and *armadillo* act in the Wingless signaling pathway in *Drosophila. Nature* (*London*) **367,** 80–83.

Nusse, R., and Varmus, H. E. (1992). Wnt genes. *Cell* (*Cambridge, Mass.*) **69,** 1073–1087.

O'Brochta, D. A., and Bryant, P. J. (1985). A zone of non-proliferating cells at a lineage restriction boundary in *Drosophila. Nature* (*London*) **313,** 138–141.

Orenic, T. V., Slusarski, D. C., Kroll, K. L., and Holmgren, R. A. (1990). Cloning and characterization of the segment polarity gene cubitus interruptus dominant of *Drosophila. Genes Dev.* **4,** 1053–1067.

Padgett, R. W., St. Johnston, D., and Gelbart, W. M. (1987). A transcript from a *Drosophila* pattern gene predicts a protein homologous to the transforming growth factor-beta family. *Nature* (*London*) **325,** 81–84.

Pearson, M. J. (1974). The abdominal epidermis of *Calliphora erythrocephala* (Diptera). I. Polyteny and growth in the larval cells. *J. Cell Sci.* **16,** 113–131.

Peifer, M., and Wieschaus, E. (1990). The segment polarity gene armadillo encodes a functionally modular protein that is the *Drosophila* homolog of human plakoglobin. *Cell* (*Cambridge, Mass.*) **63,** 1167–1176.

Peifer, M., Rauskolb, C., Williams, M., Riggleman, B., and Wieschaus, E. (1991). The segment polarity gene armadillo interacts with the wingless signaling pathway in both embryonic and adult pattern formation. *Development* **111,** 1029–1043.

Peifer, M., Sweeton, D., Casey, M., and Wieschaus, E. (1994). *Wingless* signal and Zeste-white3 kinase trigger opposing changes in the intracellular distribution of Armadillo. *Devel.* **120,** 369–380.

Perrimon, N. (1988). The maternal effect of *lethal* (*1*)*discs large-1:* a recessive oncogene of *Drosophila melanogaster. Dev. Biol.* **119,** 137–142.

Perrimon, N., and Mahowald, A. P. (1986). *Lethal(1) hopscotch,* a larval–pupal zygotic lethal with a specific maternal effect on segmentation in *Drosophila. Dev. Biol.* **118,** 28–41.

Phillips, R. G., and Whittle, J. R. S. (1993). wingless expression mediates determination of peripheral nervous system elements in late stages of *Drosophila* wing disc development. *Development* **118,** 427–438.

Postlethwait, J. H., and Schneiderman, H. A. (1973). Pattern formation in imaginal discs of *Drosophila melanogaster* after irradiation of embryos and young larvae. *Dev. Biol.* **32,** 345–360.

Ready, D. F., Hanson, T. E., and Benzer, S. (1976). Development of the *Drosophila* retina, a neurocrystalline lattice. *Dev. Biol.* **53,** 217–240.

Rebay, I., Fleming, R. J., Fehon, R. G., Cherbas, L., Cherbas, P., and Artavanis-Tsakonas, S. (1991). Specific EGF repeats of *Notch* mediate interactions with *Delta* and *Serrate:* Implications for *Notch* as a multifunctional receptor. *Cell* (*Cambridge, Mass.*) **67,** 687–699.

Riddle, R. D., Johnson, R. L., Laufer, E., and Tabin, C. (1993). Sonic hedgehog mediates the polarizing activity of the ZPA. *Cell* (*Cambridge, Mass.*) **75,** 1401–1416.

Roseland, C. R., and Schneiderman, H. A. (1979). Regulation and metamorphosis of the abdominal histoblasts of *Drosophila melanogaster. Wilhelm Roux's Arch. Dev. Biol.* **186,** 235–265.

Russell, M. A. (1974). Pattern formation in the imaginal discs of a temperature-sensitive cell lethal mutation of *Drosophila melanogaster. Dev. Biol.* **40,** 24–39.

Schubiger, G. (1968). Anlageplan, determinationszustand, und transdeterminationsleistungen der mannlichen Vorderbeinscheibe von *Drosophila melanogaster. Wilhelm Roux Arch. Entwicklungs mech. Org.* **160,** 9–40.

Schubiger, G. (1971). Regeneration, duplication, and transdetermination in fragments of the leg disc of *Drosophila melanogaster. Dev. Biol.* **26,** 277–295.

Schubiger, G., and Palka, J. (1987). Changing spatial patterns of DNA replication in the developing wing of *Drosophila. Dev. Biol.* **123,** 145–153.

Shearn, A., Rice, T., Garen, A., and Gehring, W. (1971). Imaginal disc abnormalities in lethal mutants of *Drosophila. Proc. Natl. Acad. Sci.* **68,** 2594–2598.

Shearn, A., and Garen, A. (1974). Genetic control of imaginal disc development. *Proc. Natl. Acad. Sci. U.S.A.* **71,** 1393–1397.

Siegfried, E., Chou, T. B., and Perrimon, N. (1992). wingless signaling acts through zeste-white 3, the *Drosophila* homolog of glycogen synthase kinase-3, to regulate engrailed and establish cell fate. *Cell* (*Cambridge, Mass.*) **71,** 1167–1179.

Siegfried, E., Wilder, E. L., and Perrimon, N. (1994). Components of Wingless signaling in *Drosophila. Nature* (*London*) **367,** 76–80.

Simcox, A. A., Wurst, G., Hersperger, E., and Shearn, A. (1987). The *defective dorsal discs* gene of *Drosophila* is required for the growth of specific imaginal discs. *Dev. Biol.* **122,** 169–189.

Simpson, P. (1981). Growth and cell competition in *Drosophila. J. Embryol. Exp. Morphol., Suppl.* **65,** 77–88.

Simpson, P., and Morata, G. (1981). Differential mitotic rates and patterns of growth in compartments in the *Drosophila* wing. *Dev. Biol.* **85,** 299–308.

Skaer, H., and Maddrell, S. H. P. (1987). How are invertebrate epithelia made tight? *J. Cell Sci.* **88,** 139–141.

Steiner, E. (1976). Establishment of compartments in the developing leg imaginal discs of *Drosophila melanogaster. Wilhelm Roux's Arch. Dev. Biol.* **180,** 9–30.

Strand, D., Kalmes, A., Walther, H.-P., Schwinn-Arnold, T., and Mechler, B. M. (1991). The *l(2)giant larvae* tumor suppressor p127 protein accumulates in the inner surface of the plasma membrane at regions of cell to cell contact. *Eur. Drosophila Res. Conf., 12th* p. 43 (abstract).

Struhl, G., and Basler, K. (1993). Organizing activity of wingless protein in *Drosophila. (Cambridge, Mass.)* **72,** 527–540.

Sturgill, T. W., and Wu, J. (1991). Recent progress in characterization of protein kinase cascades for phosphorylation of ribosomal protein S6. *Biochem. Biophys. Acta* **1092,** 350–357.

Sturtevant, M. A., Roark, M., and Bier, E. (1993). The *Drosophila* rhomboid gene mediates the localized formation of wing veins and interacts genetically with components of the epidermal growth factor receptor signalling pathway. *Genes Dev.* **7,** 961–973.

Szabad, J., and Bryant, P. J. (1982). The mode of action of "discless" mutations in *Drosophila melanogaster. Dev. Biol.* **93,** 240–256.

Tabata, T., and Kornberg, T. (1994). Hedgehog is a signaling protein with a key role in patterning *Drosophila* imaginal discs. Cell *(Cambridge, Mass.)* **76,** 89–102.

Tabata, T., Eaton, S., and Kornberg, T. B. (1992). The *Drosophila hedgehog* gene is expressed specifically in posterior compartment cells and is a target of *engrailed* regulation. *Genes Dev.* **6,** 2635–2645.

Takeichi, M. (1990). Cadherins: A molecular family important in selective cell–cell adhesion. *Annu. Rev. Biochem.* **59,** 237–252.

Taubert, H., and Szabad, J. (1987). Genetic control of cell proliferation in female germ line cells of *Drosophila:* Mosaic analysis of five *discless* mutations. *Mol. Gen. Genet.* **209,** 545–551.

Tepass, U., and Hartenstein, V. (1994). The development of cellular junctions in the *Drosophila* embryo. *Dev. Biol.* **161,** 563–596.

Thiessen, H., Purcell, J., Bennett, M., Kansagara, D., Syed, A., and Marsh, J. L. (1994). *dishevelled* is required during *wingless* signaling to establish both cell polarity and cell identity. *Development* **120,** 347–360.

Thisse, B., Stoetzce, C., Gorostiza, T. C., and Schmidt, P. F. (1988). Sequence of the *twist* gene and nuclear localization of its protein in endomesodermal cells of early *Drosophila* embryos. *EMBO J.* **7,** 2175–2183.

Thomas, B. J., Gunning, D. A., Cho, J., and Zipursky, S. L. (1994). Cell cycle progression in the developing *Drosophila* eye: *roughex* encodes a novel protein required for the establishment of G1. *Cell (Cambridge, Mass.)* **77,** 1003–1014.

Tomlinson, A., Bowtell, D. D., Hafen, E., and Rubin, G. M. (1987). Localization of the sevenless protein, a putative receptor for positional information, in the eye imaginal disc of *Drosophila. Cell (Cambridge, Mass.)* **51,** 143–150.

Traugh, J. A., and Pendergast, A. M. (1986). Regulation of protein synthesis by phosphorylation of ribosomal protein S6 and aminoacyl-tRNA synthetases. *Prog. Nucleic Acid Res. Mol. Biol.* **33,** 195–230.

Truman, J. W., and Bate, M. (1988). Spatial and temporal patterns of neurogenesis in the central nervous system of *Drosophila melanogaster. Dev. Biol.* **125,** 145–157.

van den Heuvel, M., Nusse, R., Johnston, P., and Lawrence, P. A. (1989). Distribution of the wingless gene product in *Drosophila* embryos: A protein involved in cell–cell comunication. *Cell (Cambridge, Mass.)* **59,** 739–749.

Watson, K. L., Johnson, T. K., and Denell, R. E. (1991). *Lethal (1) Aberrant Immune Response* mutations leading to melanotic tumor formation in *Drosophila melanogaster. Dev. Genet.* **12,** 173–187.

Watson, K. L., Konrad, K. D., Woods, D. F., and Bryant, P. J. (1992). *Drosophila* homolog of the human S6 ribosomal protein is required for tumor suppression in the hematopoietic system. *Proc. Natl. Acad. Sci. U.S.A.* **89,** 11302–11306.

Whiteley, M., Noguchi, P. D., Sensabaugh, S. M., Odenwald, W. F., and Kassis, J. A. (1992). The *Drosophila* gene *escargot* encodes a zinc finger motif found in *snail*-related genes. *Mech. Dev.* **36,** 117–127.

Wilder, E. L., and Perrimon, N. (1994). Dual functions of *wingless* in *Drosophila* leg imaginal discs. *Development* **121,** 477–488.

Williams, J. A., Paddock, S. W., and Carroll, S. B. (1993). Pattern formation in a secondary field: A hierarchy of regulatory genes subdivides the developing *Drosophila* wing disc into discrete subregions. *Development* **117,** 571–584.

Williams, J. A., Paddock, S. W., Vorwerk K., and Carroll, S. B. (1994). Organization of wing formation and induction of a wing-patterning gene at the dorsal/ventral compartment boundary. *Nature* (*London*) **368,** 299–305.

Wolpert, L. (1971). Positional information and pattern formation. *Curr. Top. Dev. Biol.* **6,** 183–224.

Woods, D. F., and Bryant, P. J. (1991). The discs–large tumor suppressor gene of *Drosophila* encodes a guanylate kinase homolog localized at septate junctions. *Cell* (*Cambridge, Mass.*) **66,** 451–464.

Zalokar, M., and Erk, I. (1976). Division and migration of nuclei during early embryogenesis of *Drosophila melanogaster. J. Microsc. Biol. Cell.* **25,** 97–106.

II

Amphibians

11

Endocrinology of Amphibian Metamorphosis

JANE C. KALTENBACH

Department of Biological Sciences
Mount Holyoke College
South Hadley, Massachusetts

I. INTRODUCTION

This chapter introduces recent studies on hormonal controls of metamorphic events in amphibians, particularly in anurans. The wealth of information built up since 1912 has established the prime importance of the thyroid hormone in metamorphic processes, enabling aquatic larvae to adapt to terrestrial life-styles of adult frogs.

Accuracy in comparing data from different laboratories has been made possible by the division of larval development into stages based on external characteristics and events. Larval development of *Rana pipiens* was divided into three phases by Etkin (1964, 1968): premetamorphosis (larval growth), prometamorphosis (limb growth and differentiation), and metamorphic climax [foreleg emergence (fle) and very rapid changes such as tail resorption]. This terminology is commonly referred to and has been applied to other species including *Rana catesbeiana* and *Xenopus laevis,* which are now frequently used in experiments. However, when more precise stages are called for, those of Taylor and Kollros (1946) for *Rana* and those of Nieuwkoop and Faber (e.g., 1967) for *Xenopus* are generally used. These types of staging have been compared in detail with the terms used by Etkin (for reviews see, e.g., Dodd and Dodd, 1976; Kikuyama *et al.,* 1993) and are briefly summarized in Fig. 1. With respect to the development of larvae of different species, consideration should be given to the duration of the stages (which varies depending on lengths of the various larval life spans) and also to differences in the timing of tail resorption as compared to fle (especially with respect to *Xenopus* and *Rana*).

Early reports focused on morphological changes initiated by thyroid hormone and later studies focused on accompanying biochemical changes. Thus, a firm basis has been established for experiments in which new methods are being used to probe basic mechanisms and molecular aspects of endocrine regulation of metamorphosis. The many advances in these areas clearly have increased our understanding of the critical role of hormones in developmental processes. Details of thyroid hormone effects in specific target organs, even in the immune system, have been clarified. Enormous strides have been made with respect to the mechanisms of hormone action in peripheral tissues, e.g., (1) hormone-induced alterations in gene expression resulting in specific metamorphic changes and (2) the nature of thyroid hormone receptors and the genes which encode them. In addition, the importance of hormonal regulation and fine-tuning of thyroid hormone-induced metamorphic changes is now appreciated with respect to neurohormones as well as hormones from specific cells in endocrine glands, such as the pituitary and adrenal (interrenal) glands. Examples of the role of hormones in amphibian metamorphosis are presented in this chapter and are discussed in detail by other authors in the following chapters. Amphibian metamorphosis has become an exciting and fast-moving field.

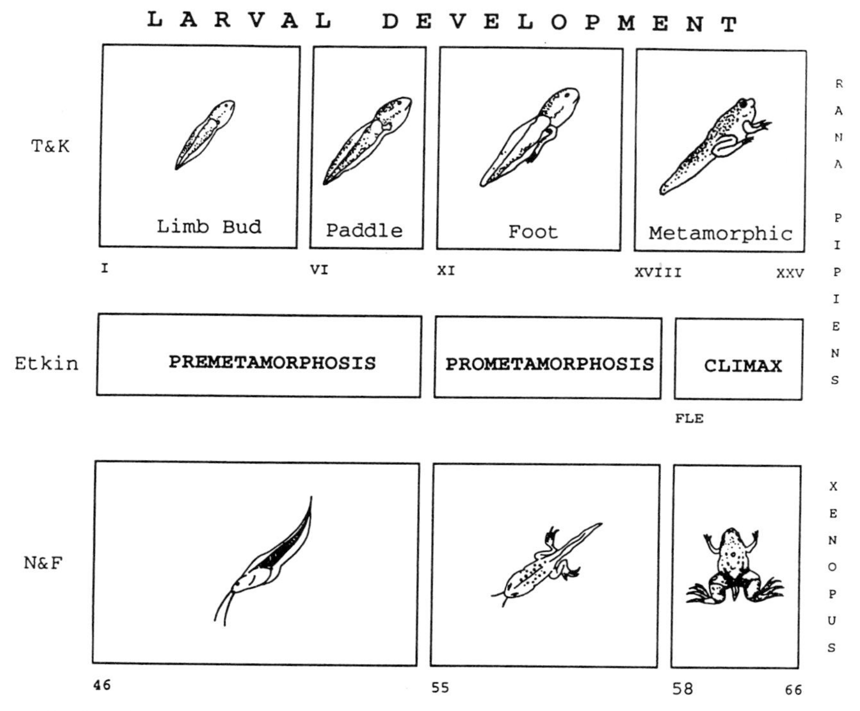

Fig. 1. Phases of larval development (*Rana pipiens*) according to Etkin (1964, 1968) as compared to the stages of Taylor and Kollros (1946) for *R. pipiens* and to the stages of Nieuwkoop and Faber (e.g., 1967) for *Xenopus laevis.* Etkin's larval phases and their characteristics are premetamorphosis, larval growth; prometamorphosis, limb growth and differentiation; and metamorphic climax, starting with foreleg emergence (fle) followed by rapid changes in the mouth and tail, resulting in a small frog. [Embryonic stages prior to larval stages are referred to as 1–25 for *Rana* (Shumway, 1940) and 1–45 for *Xenopus* (Nieuwkoop and Faber, 1967).]

II. THYROID HORMONE

A. Induction of Metamorphic Events

Although thyroid hormone has long been known to induce precocious metamorphosis (Gudernatsch, 1912), questions are still being addressed with respect to the role of the hormone in mediating morphological changes in specific tissues and organs, as for example in the development of specific cells in the nervous system, repatterning of the larval cranium, and epithelial degeneration and shortening of the gastrointestinal (GI) tract.

Among examples of thyroid hormone effects upon specific neurons and regions of the larval nervous system is the response in *Rana pipiens* larvae of cells in the mesencephalic V nucleus (proprioceptors of the jaw musculature). With thyroid hormone exposure cell size and number increased, but at high hormone concentrations cell loss occurred as it did during the climax of spontaneously metamorphosing tadpoles (for reviews see Kollros, 1981, 1984; Kollros and Bovbjerg, 1990). Another example is the precocious induction of ipsilateral retinothalamic projections from the retina of thyroxine-injected eyes but not from contralateral control eyes (Hoskins and Grobstein, 1984, 1985), indicating a direct, local hormone action on specific cells in the eye and induction of a "neural pathway to a distant target." [For a review of other metamorphic changes in the eye and the role of thyroid hormone in bringing them about, see Hoskins (1990).]

Metamorphosis of the cartilaginous larval skull to the predominantly bony adult skull is another event mediated by thyroid hormone (Hanken and Hall, 1988; Hanken and Summers, 1988; Hanken *et al.,* 1989). Both cartilage and bone were found to respond precociously and independently to thyroid hormone. However, because of the more rapid and extensive response of cartilage, the normal morphological integration of the two types of tissue was not effected by the concentrations of hormone used in the experiments.

As the GI tract shortens during metamorphosis, the larval epithelial lining degenerates (and is replaced by an adult-type epithelium) (for a review see, e.g., Dauça and Hourdry, 1985). *In vivo* and *in vitro* studies showed that thyroid hormone-induced epithelial degeneration involved both apoptosis of the epithelial cells and heterolysis by macrophages which proliferate, differentiate, and migrate through the epithelium into the lumen (Ishizuya-Oka and Shimozawa, 1992). To determine where in the tract metamorphic shortening actually occurs, Pretty *et al.* (1995) used an ingenious method. By marking and mapping specific locations in the intestinal coils and stimulating metamorphosis with exogenous thyroid hormone, they demonstrated that shortening takes place uniformly, not regionally, along the length of the tract.

Thyroid hormone initiates metamorphic events by acting directly on peripheral tissues. This has been demonstrated (1) by hormone implants causing adjacent, localized metamorphic responses (for a review see Kaltenbach, 1968) and (2) by culture of larval organs in hormone-containing media resulting in morphological and biochemical changes. Remarkable progress has been made in determining the mechanism of such hormone action. For example, many genes have been isolated and characterized and changes in their expression found to be induced by thyroid hormone. Gene expression programs have been investigated in such larval organs as intestine, pancreas, liver, red blood cells, skin, tail, and hind limbs.

However, in order for thyroid hormone to alter the expression of specific genes in peripheral tissues, the hormone must first react with nuclear receptors which bind to specific DNA sequences (thyroid hormone response elements). These interactions alter the transcription of specific genes, as indicated by changes in levels of mRNA, followed by translation to protein, and finally to striking biochemical and morphological metamorphic changes. By binding both to hormone and to DNA, the nuclear receptors become intermediates in the hormonal alteration of gene expression and thus play a critical role in regulating those genes which encode both for specific metamorphic events and for the hormone receptors themselves. Mechanisms of thyroid hormone action are covered extensively in the following chapters.

B. Deiodination in Tissues and Hormone Levels in Circulation

1. Deiodination

Although T_4 (thyroxine or tetraiodothyronine) is the main form of thyroid hormone secreted from the thyroid gland into the circulation, T_3 (triiodothyronine, sometimes referred to as another thyroid hormone) is more active and is the predominant form which interacts with receptors in target organs (Fig. 2). Deiodinative reactions within peripheral tissues of tadpoles have been clarified, especially by the work of Buscaglia and Leloup on *Xenopus* and of Galton on *R. catesbeiana* (for reviews see Buscaglia *et al.,* 1985; Galton, 1983, 1988, 1992; Kikuyama *et al.,* 1993; Galton *et al.,* 1994). In early stages (premetamorphosis), tadpoles are unable to change T_4 to T_3. However, activity of the enzyme 5′-deiodinase, which converts T_4 to T_3, increases (e.g., in skin, gut, tail) during both spontaneous and hormone-induced metamorphic climax. On the other hand, activity of 5-deiodinase, which degrades both T_4 and T_3 to relatively inactive compounds, rT_3 (reverse T_3) and T_2 (3,3′-diiodothyronine), is present in prometamorphic tadpoles but is very low during climax. The balance existing between the two enzyme systems may both protect the tadpole from excessive T_3 levels prior to climax and contribute to their rise during climax. Thus, they play a very important role in regulating metamorphic progress.

2. *Circulating Hormone Levels*

As suggested in the preceding sections and discussed in detail in the following chapters, the metamorphic action of thyroid hormone depends on molecular reactions in the peripheral tissues, such as increases in (1) the activity of the enzymes in the deiodinative system leading to higher levels of T_3 within cells, (2) the number and affinity of thyroid hormone receptors in the cell nuclei, and (3) the rate at which the hormone–receptor complexes alter gene transcription and mRNA levels.

However, not only responses in the target tissues, but also activity of the thyroid glands, must be taken into account. The rate of hormone synthesis, as well as the release of hormone from the glands into the bloodstream, has been shown, e.g., by uptake of radioactive iodine into the glands and by electron microscopy of the thyroid cells (e.g., development of rough endoplasmic reticulum indicative of thyroglobulin synthesis; formation of large lysosomes, indicative of hormone release) (for a review see Regard, 1978). Changes in the structure and activity of the glands are responsible for the levels of circulating hormone, which in turn have been correlated with metamorphic progress (for reviews see Fox, 1981; White and Nicoll, 1981; Kaltenbach, 1982; Galton, 1983; Rosenkilde and Ussing, 1990).

Concentrations of T_4 and T_3 in blood plasma (or serum) have been determined by radioimmunoassay (RIA) in tadpoles of a number of anuran species. *R. catesbeiana* tadpoles have been favorites for such studies, their large size allowing collection of sufficient blood for assay on samples from individual animals (Miyauchi *et al.,* 1977; Regard *et al.,* 1978; Mondou and Kaltenbach, 1979; Suzuki and Suzuki, 1981). Hormone concentrations have also been reported for blood from *Rana clamitans* (fairly large tadpoles) (Weil, 1986) and from *Rana esculenta* tadpoles (Schultheiss, 1980), as well as from four species of small *Xenopus* larvae (requiring pooled plasma samples) (Leloup and Buscaglia, 1977; Schultheiss, 1980; Buscaglia *et al.,* 1985). Hormone values have been obtained for toad tadpoles, *Bufo japonicus* (Niinuma *et al.*, 1991a) and *Bufo marinus* (Weber *et al.*, 1994); however, because of their size, whole body extracts rather than plasma samples were assayed.

T_4 values are generally higher than those of T_3, but overall patterns of hormone concentrations are similar throughout larval development despite some individual and species variations. T_4 and T_3, although usually undetectable in early stages (premetamorphosis), rise during prometamorphosis, peak dramatically during climax, and then decrease to the lower levels of the frog. However, T_4 was detected (but not quantified) in plasma of premetamorphic tadpoles by means of immunofluorescent staining of blood smears (Piotrowski and Kaltenbach, 1985). Moreover, the wide variations reported for hormone levels of individual tadpoles in climax (Miyauchi *et al.*, 1977; Mondou and Kaltenbach, 1979; Suzuki and Suzuki, 1981) might well be due to bursts of hormone release into the circulation by different animals at different times, as suggested by Rosenkilde and Ussing (1990).

Precocious induction of metamorphic events by exogenous thyroid hormone, as well as correlation of metamorphic progress with endogenous hormone concentrations, indicates that control of metamorphosis occurs not only at the level of the target organs (as pointed out earlier) but also at the level of the thyroid glands and the circulating hormone levels.

A developmental pattern of serum thyroglobulin (TG), similar to that of thyroid hormone, was reported for *R. catesbeiana* tadpoles by Suzuki and Fujikura (1992). Although bovine thyroid-stimulating hormone (TSH) did not augment TG release from thyroid glands in culture, possible stimulatory effects of the more potent frog TSH would be of interest. The significance of circulating TG remains to be determined. Might it be an extrathyroidal source of T_4 and T_3, as reported for mammals (Brix and Herzog, 1994)?

III. HYPOTHALAMUS–PITUITARY–THYROID HORMONE INTERACTIONS

A. Pituitary–Thyroid Axis

It is well established that thyroid gland activity, and consequently metamorphic progress, is controlled by hormonal stimulation from the pituitary gland. The classical vertebrate pituitary–thyroid axis starts to function in amphibians in early larval stages (for a review see White and Nicoll, 1981). TSH (thyroid-stimulating hormone or thyrotropin*) from TSH cells (thyrotropes*) in the pars distalis stimulates the synthesis and release of thyroid hormone, which in turn not only induces metamorphic changes, but also acts back upon the pituitary to decrease TSH release—a negative feedback system (Fig. 2).

TSH, FSH (follicle-stimulating hormone), and LH (luteinizing hormone) are highly glycosylated polypeptides consisting of α and β subunits; both peptide chains (but not the sugars) are needed for hormone–receptor interaction. The β subunit is unique to each hormone; the α subunit is the same in the three hormones and its carbohydrate components are critical for the activation of intracellular events (for a review see Sairam, 1989). Purification of amphibian TSH from bullfrog pituitaries resulted in a preparation four times as potent as bovine TSH in releasing T4 from cultured larval thyroid glands (Sakai *et al.,* 1991). Later an amphibian TSH of even greater potency was obtained, with one subunit showing notable sequence homology with mammalian TSH_{β} and the other with the α subunit of bullfrog gonadotropins (for a review see Kikuyama *et al.,* 1993). Despite these advances in amphibian hormone purification, plasma levels of TSH in metamorphosing tadpoles have not yet been measured (an appropriate RIA for amphibian TSH is not yet available).

However, the TSH content of pituitary glands has been assessed by bioassay (for a review see White and Nicoll, 1981) and identification and activity of TSH cells by various histological techniques. For example, electron microscopy showed that size and distribution of secretory granules within the cells reflected cellular activity and thus were correlated with thyroid function and metamorphic progress (for a review see Rosenkilde and Ussing, 1990). Moreover, immunohistochemical staining (antibodies against mammalian TSHβ) of larval pituitaries resulted in the accurate identification of thyrotropes in a number of species (for a review see Kikuyama *et al.,* 1993). Such cells, generally located ventrocentrally in the pars distalis, increased in size and number during metamorphosis (Moriceau-Hay *et al.,* 1982; Kar and Naik, 1986). Moreover, staining with several antibodies (against TSHβ, LHβ, FSHβ, and its α subunit) indicated that bullfrog thyrotropes appear earlier in development than

* In the literature the "tropic" pituitary hormones are sometimes spelled "trophic," e.g., thyrotropin vs thyrotrophin. Similarly, the pituitary cells secreting such hormones are referred to either as "tropes" or trophs," e.g., thyrotrophs.

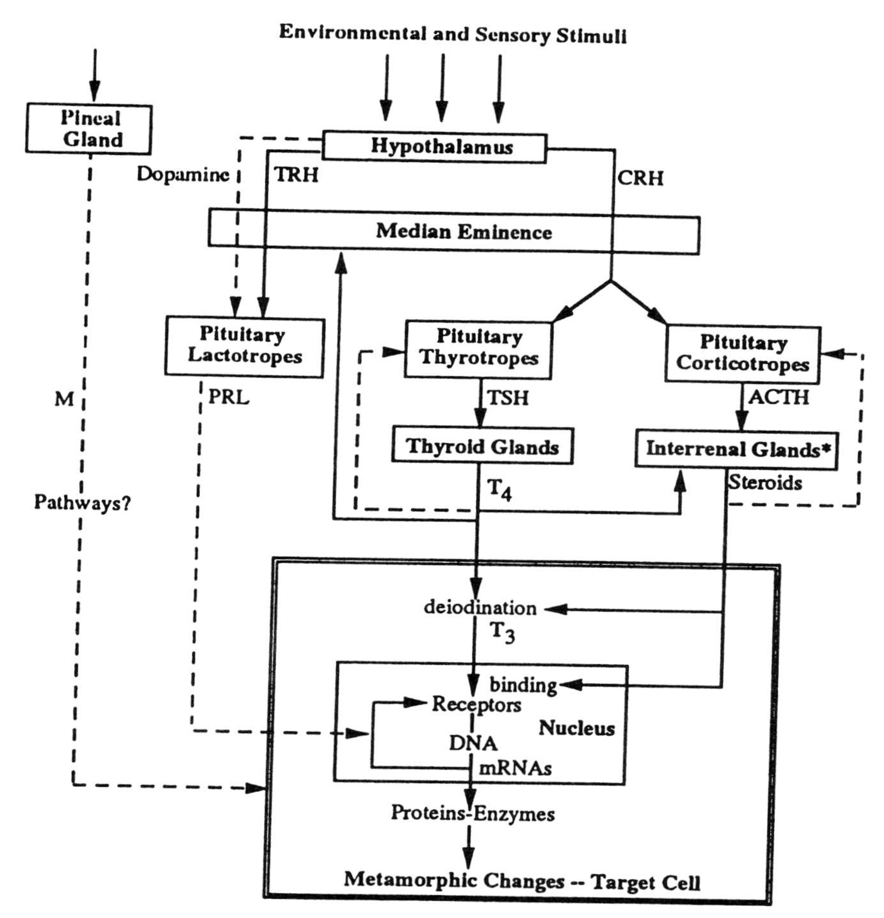

Fig. 2. Proposed hormonal controls and fine-tuning of thyroid hormone-induced metamorphosis; these help assure appropriate timing and rate of metamorphic changes in specific tissues of amphibian tadpoles. Inhibition of metamorphosis by prolactin (PRL) and melatonin (M) and stimulation by adrenal steroids (corticosterone, aldosterone, cortisol) are included in the diagram, but possible interactions of other hormones and exact mechanisms of hormone actions have been omitted. ---->, inhibitory action; →, stimulatory action. In amphibians, adrenal cortical tissue is interspersed among kidney tubules and is referred to as interrenal glands (*).

do gonadotropes (Tanaka *et al.*, 1991). Immunostaining coupled with morphometric techniques led García-Navarro *et al.* (1988) to conclude that storage of TSH within the thyrotropes predominates during pre- and prometamorphosis and that release of TSH occurs early in climax, causing high serum T_4 and T_3 levels and thus metamorphic changes typical of climax stages.

cDNAs encoding each of the subunits of *Xenopus* TSH have been cloned; moreover, the developmental pattern of the TSH α and β mRNAs is similar to that of endogenous thyroid hormone levels, although the peak occurs for the mRNAs somewhat earlier than the thyroid hormone peak (Buckbinder and Brown, 1993).

In light of the importance of carbohydrate moieties of glycoprotein hormones in general (for a review see Sairam, 1989), as well as purification of TSH, the time has come to focus on the structure and role of the carbohydrate components of amphibian TSH (and other pituitary glycoproteins), including possible differences in carbohydrates of intrapituitary and circulating TSH, as well as the glycosylation process itself. If specific steps in the overall process were found to be hormonally controlled, that would add to the complexity of endocrine regulation in larval amphibians.

B. Hypothalamic Control

Early classical experiments involving hypothalectomy, pituitary transplants, etc. indicated that the pituitary–thyroid axis is regulated by the hypothalamus (for reviews see White and Nicoll, 1981; Kikuyama *et al.*, 1993). Neurohormones released from nerve endings travel in the circulation of the median eminence and portal vessels to the anterior pituitary, where they stimulate specific cells such as lactotropes, thyrotropes, and corticotropes. Interestingly, growth of axons from specific nuclei in the hypothalamus and development of the median eminence (thickening and internalization of capillaries) are stimulated by thyroid hormone—reactions considered as metamorphic events and as examples of positive thyroid hormone feedback on the hypothalamus and median eminence.

In the late 1960s Guillemin and Schally independently isolated thyrotropin-releasing hormone (TRH) from mammalian hypothalami; it was identified as a tripeptide and named for its ability to stimulate

TSH release in mammals. In a survey of TRH levels in various tissues of the frog *R. pipiens,* high concentrations were found in the hypothalamus and other parts of the brain, although the highest concentration was in the skin (Jackson and Reichlin, 1977). A later study on skin of *Xenopus* indicated that TRH is derived from a large precursor molecule containing four copies of the hormone (Richter *et al.,* 1984). In the larval brain (*Xenopus*), concentration of TRH increased during prometamorphic and climax stages (King and Millar, 1981; Bray and Sicard, 1982; Millar *et al.,* 1983; Balls *et al.,* 1985). In addition, with immunohistochemical staining, TRH was identified within specific nuclei and tracts of the *R. catesbeiana* tadpole brain, including the hypothalamus and median eminence, as well as the pars nervosa of the pituitary; staining became more extensive during climax stages (Mimnagh *et al.,* 1987; Taniguchi *et al.,* 1990).

Such experiments provided a basis for considering TRH as a neuropeptide stimulator of the larval pituitary–thyroid axis, yet exogenous TRH usually failed to promote precocious metamorphosis (as shown in numerous studies with differences in TRH concentrations and methods of TRH administration as well as different species and stage of tadpoles, etc.). Pros and cons of a metamorphic role for TRH are listed by Balls *et al.* (1985). *In vitro* and *in vivo* studies have demonstrated that TRH is ineffective in stimulating the larval pituitary–thyroid axis or metamorphosis, even though it does promote TSH release from pituitaries of adult frogs (Denver, 1988; 1993; Denver and Licht, 1989).

The latest research points to CRH (corticotropin-releasing hormone) rather than TRH as the main neuroendocrine regulator of thyroid activity and hence metamorphosis in anuran larvae (Fig. 2); in contrast, both neuropeptides (possibly, more) stimulate the pituitary–thyroid axis of adult frogs (Denver, 1988; 1993; Denver and Licht, 1989; Gancedo *et al.,* 1992). Questions which come to mind include the following. Is there an advantage to different neuroendocrine controls for thyroid hormone in larval and adult frogs? Is CRH released in a pulsatile fashion and thereby indirectly responsible for possible bursts of thyroid hormone release during metamorphic climax? Does thyroid hormone have a negative feedback effect on the synthesis/release of CRH from hypothalamic neurons? [For a detailed discussion of molecular mechanisms, neurohormones, and the significance of the role which CRH plays in tadpole metamorphosis, see Chapter 12 by Denver (this volume).]

IV. HORMONAL FINE-TUNING

The dependency of metamorphic changes on the action of thyroid hormone, along with its stimulation by the hypothalamic–pituitary (TSH) axis, provides an excellent system, not only for probing molecular events responsible for such changes, but also for exploring the role of other hormones in modulating or fine-tuning the rate of metamorphic progress. For example, prolactin serves as a brake to slow metamorphosis, whereas specific corticoids from the adrenal glands have an opposite effect. However, such actions may vary depending on thyroxine concentrations (Wright *et al.,* 1994). The nature of such hormones and their mechanisms of action have been clarified by use of molecular techniques.

A. Prolactin

It was in the 1960s (50 years after the metamorphosing role of thyroid hormone was first reported) that prolactin was proposed as a growth and antimetamorphic hormone (Fig. 2) (e.g., by Etkin and Lehrer, 1960; Bern *et al.,* 1967; Etkin and Gona, 1967); this has been well documented over the years (for reviews see White and Nicoll, 1981; Kikuyama *et al.,* 1993). Growth hormone, on the other hand, appears to be more important in stimulating the growth of juvenile frogs than of tadpoles, although the amphibian growth hormone (not amphibian prolactin) was reported to promote hind limb growth in *Bufo* larvae (Kikuyama *et al.,* 1993). Examples of exogenous prolactin effects include an increase in larval growth and weight gain, as well as changes in certain organs, such as increases in height and length of the tail and in length of the GI tract. Similar effects and, for the first time in anuran larvae, an increase in weight of the gills (arches and filaments as a unit) have also been found (J. C. Kaltenbach and J. Amaral, unpublished observations). Decreases in thyroid hormone-induced degenerative changes (tail resorption, GI tract shortening, and gill resorption) reflect the antimetamorphic actions of prolactin. Moreover, prolactin inhibits the growth of hind limb buds *in vivo* (Wright *et al.,* 1979) and T_3-stimulated limb buds (and tails) in culture (Tata *et al.,* 1991). Some variation in results from different laboratories may be related to differences in species and stages of tadpoles, as well as in hormone concentrations, ratios, and lengths of treatment.

Prolactin may induce its antimetamorphic effects by an inhibitory (goitrogenic) action on the thyroid gland (Gona, 1967; for a review see Dent, 1988) and, also, as established later with *in vitro* experiments, by a direct action on thyroid-stimulated peripheral tissues (e.g., tail and hind limbs) (Tata *et al.,* 1991). Morever, as indicated by the work in Nicoll's laboratory, prolactin may also stimulate secretion of a liver factor, synlactin, which in turn acts synergistically to enhance the effect of prolactin on peripheral tissues (e.g., Delidow *et al.,* 1988)—an interesting example of fine-tuning of hormone action by a secondary or intermediate factor. Future studies may indicate if such synergism can be demonstrated in cultured organs. What secondary factors may be involved in actions of other hormones?

Antimetamorphic actions of prolactin should be considered in the context of receptor regulation. Specific binding of prolactin has been reported for various tadpole organs (e.g., tail, kidney) (for a review see Kikuyama *et al.,* 1993); in mammals, prolactin receptors have been grouped along with growth hormone receptors into a large cytokine receptor superfamily (for a review see Doppler, 1994). Prolactin binding in the tadpole kidney increases dramatically during climax or in response to exogenous T_4, suggesting that renal prolactin receptors can be induced by thyroid hormone (White and Nicoll, 1979). Moreover, prolactin not only reduces the T_4 induction of its own renal-binding proteins (White *et al.,* 1981), but also prevents autoinduction (T_3-induced up-regulation) of thyroid hormone receptor mRNAs (as shown in *Xenopus* in tail cultures) (Baker and Tata, 1992). Intriguing reactions between two hormones and their receptors!

Although mammalian prolactin was used in early experiments, amphibian prolactin from *R. catesbeiana* pituitaries has now been purified and sequenced (Yamamoto and Kikuyama, 1981; Takahashi *et al.,* 1990; Yasuda *et al.,* 1991). Prolactins have also been purified and characterized from other anurans (e.g., *Bufo* and *Xenopus*) (Yamamoto *et al.,* 1986b; Yamashita *et al.,* 1993). Such studies paved the way for the production of antiserum and its use in immunofluorescent staining of prolactin cells in the larval pituitary (distribution and developmental changes) and in RIA (circulating levels of prolactin) (for reviews see Rosenkilde and Ussing, 1990; Kikuyama *et al.,* 1993; Kikuyama, 1994). Interestingly, endogenous plasma prolactin levels were found to peak late in climax, a time when they may have an osmoregulatory role

(Clemons and Nicoll, 1977; Yamamoto and Kikuyama, 1982; Niinuma *et al.,* 1991b).

cDNAs which encode bullfrog and *Xenopus* prolactins have been cloned, expression of the gene has been found to be inducible by thyroid hormone, and developmental patterns of prolactin mRNAs in the pituitary have been determined (Takahashi *et al.,* 1990; Buckbinder and Brown, 1993). The latter were similar to levels of prolactin in the pituitary (and in the plasma), i.e., low in premetamorphosis and high in climax (Yamamoto *et al.,* 1986a).

Prolactin release is under both inhibitory and stimulatory control from the hypothalamus (for reviews see Kikuyama *et al.,* 1993; Kikuyama, 1994). Evidence points to dopamine as a main inhibitory factor (Fig. 2). During prometamorphosis, it is probably released from monoaminergic nerve fibers that end in the pars distalis, but disappear from that location during spontaneous or thyroxine-induced climax (by which time the median eminence has become developed) (Aronsson, 1978). However, for endogenous plasma prolactin levels to peak during climax, not only is the inhibitory control probably decreased, but hypothalamic stimulation is needed as well (Kawamura *et al.,* 1986). The neuropeptide TRH (although not a TSH stimulator in tadpoles) is now well known as a major prolactin-releasing peptide (Fig. 2). During prometamorphosis the pattern of TRH nerve fibers in the pars distalis is similar to that of the monoaminergic fibers and is present only for a limited time, i.e., until climax (Taniguchi *et al.,* 1990). Thus, evidence suggests that the pathways by which these two regulatory factors reach the pituitary differ with metamorphic stage, i.e., (1) during prometamorphosis from nerve fibers within the pars distalis and (2) during climax from nerve endings near capillaries in the median eminence followed by transport in the portal circulation to the pars distalis. Other neuroendocrine factors may also be involved in prolactin regulation. [For further discussion of prolactin, see Chapter 12 by Denver and Chapter 13 by Tata (this volume).]

B. Adrenal Steroid Hormones

In contrast to the inhibitory action of prolactin on thyroid hormone-induced metamorphosis, adrenal corticoids synergize with thyroid hormone in promoting metamorphosis, a role which has become well established (Fig. 2) (for reviews see Dent, 1988; Rosenkilde and Ussing, 1990; Kikuyama *et al.,* 1993). Early reports on steroid action in tadpoles were inconsistent, most probably due to differences in tadpoles stages and hormone concentrations used. Later evidence showed that thyroxine-induced metamorphosis (as indicated by rate of overall shortening) was, indeed, enhanced by specific steroids, e.g., hydrocortisone (Frieden and Naile, 1955) and deoxycorticosterone (Kobayashi, 1958). Tail resorption then became a favorite criterion for investigations on hormone interactions, although other metamorphic criteria have also been used. For example, activity of a hepatic enzyme (carbamoyl-phosphate synthase) was greater and plasma levels of T_4 (or T_3) were higher with a combination of corticosterone and T_4 or T_3 than with thyroid hormone alone (Galton, 1990).

Corticoid–thyroid hormone synergism occurs *directly* within peripheral tissues as demonstrated (1) by localized tail fin resorption in the region of implants containing a steroid plus a subthreshold amount of thyroxine (Kaltenbach, 1958) and (2) by *in vitro* studies in which tail shortening was accelerated in medium containing both steroid and thyroid hormone as compared to thyroid hormone alone (see e.g., Kikuyama *et al.,* 1982, 1983).

Cultured tail tips have also been used to investigate the mechanism by which adrenal steroids enhance T_4 and T_3 action. For example, the lag period before the start of T_4-induced tail resorption was reduced by steroid hormone action (Kaltenbach, 1985). Moreover, aldosterone and corticosterone were reported to increase the maximal nuclear binding capacity of T_3 in cultured tails of both the toad (*B. japonicus*) and bullfrog (*R. catesbeiana*) (Niki *et al.,* 1981; Suzuki and Kikuyama, 1983). Yet, results from an *in vivo* study indicated that the binding characteristics of T_3 nuclear receptors in tail (and liver and red blood cells) were not affected by corticosterone (Galton, 1990). On the other hand, the activity of 5′-deiodinase (in skin) was increased and the activity of 5-deiodinase (in gut and liver) was decreased in tadpoles exposed to both corticosterone and T_4 compared to T_4 alone; thus, T_3 formation from T_4 was increased and T_3 degradation decreased (Galton, 1990). *In vitro* experiments suggest that hydrocortisone acts synergistically on T_3-induced 63-kDa keratin gene expression—the first example of a corticoid and thyroid hormone synergistic action at the mRNA level in cultured epidermal cells (Shimizu-Nishikawa and Miller, 1992). These examples suggest that adrenal corticoids may enhance the action of thyroid hormone by more than one mechanism within the target tissues.

The patterns of several adrenal steroids in plasma during spontaneous metamorphosis have been determined. Radioimmunoassay indicated that levels of aldosterone and corticosterone (the main larval corticoids) and cortisol are elevated during climax (when thyroid hormone levels are high). Each steroid has its own unique developmental pattern, but variations, even within a species (e.g., *R. catesbeiana*), have been reported by different laboratories (e.g., Jaffe, 1981; Krug *et al.,* 1983; Kikuyama *et al.,* 1986). Plasma levels of the main corticoids have also been determined in *X. laevis* (Jaudet and Hatey, 1984) and *B. japonicus* (Niinuma *et al.,* 1989). Relatively high levels of prolactin and of adrenal corticoids during climax enable these hormones to fine-tune the metamorphic actions of thyroid hormone and thereby ensure that metamorphic processes can proceed in a controlled, orderly manner.

Thyroid hormone, adrenocorticotropin (ACTH), and CRH are involved directly or indirectly in the control of biosynthesis and secretion of adrenalcortical steroids (Fig. 2) (for reviews see Dent, 1988; Kikuyama *et al.,* 1993). In addition to accelerating the many well-known degenerative metamorphic changes (e.g., tail, gills), thyroid hormone also stimulates development of the median eminence and adrenal cortical tissue (interrenal glands). When thyroid hormone levels are high (climax) or after exposure of tadpoles to exogenous thyroid hormone, activity of a key enzyme involved in steroid biosynthesis reaches a peak (Hsu *et al.,* 1980, 1984); one might then expect a rise in levels of plasma steroids. These in turn synergize with thyroid hormone to promote metamorphic changes in peripheral tissues. Moreover, enzyme activity in the larval adrenal gland as well as plasma steroid levels was increased not only by thyroxine, but also by ACTH; the two hormones together are more effective than either one alone (Jaffe, 1981; Krug *et al.,* 1983; Hsu *et al.,* 1984; Kikuyama *et al.,* 1986).

By the beginning of climax, immunostaining of CRH has increased in the hypothalamus and well-developed median eminence (Carr and Norris, 1990), allowing CRH to reach the pituitary and act on the thyrotropes and corticotropes causing release of both TSH and ACTH (Denver and Licht, 1989). Thus, it is possible that a single hypothalamic factor may be indirectly responsible for regulation of both thyroxine and adrenal corticoids—two chemically different types of hormones which interact to ensure appropriate metamorphic progress. Moreover, by acting through such a neurohormonal system, environmental factors might be able to influence metamorphosis. [For further discussion see Chapter 12 by Denver and Chapter 13 by Tata, (this volume).]

C. Melatonin

The pineal gland and its main hormone, melatonin, have long been studied with respect to effects on tadpole growth and metamorphosis. Although numerous contradictory results have been reported, melatonin generally retards those processes (Fig. 2) (see, e.g., Delgado *et al.,* 1987). However, modulating effects of various lighting regimens, melatonin dosages, and species should be taken into consideration, e.g., metamorphic rate of larvae given melatonin varied depending on length of the photoperiod; moreover, the metamorphic rate was accelerated by low melatonin concentrations but slowed by high concentrations (Edwards and Pivorum, 1991). Most such experiments involved melatonin effects on spontaneous metamorphosis, but exogenous melatonin has also been reported to inhibit T_4-induced metamorphosis of *R. pipiens* tadpoles; this effect depends on timing of the melatonin injections in relation to the light/dark (L/D) cycle (Wright *et al.,* 1991). Moreover, tail tips in culture regress more slowly in a medium containing both T_4 and melatonin than T_4 alone, indicating a direct antagonistic action of melatonin at the level of peripheral tissues (Wright *et al.,* 1991). Additional evidence suggests that melatonin may also affect metamorphosis by depressing growth of the larval thyroid gland (Wright *et al.,* 1992).

The mechanism by which melatonin acts at the cellular and molecular levels is still unclear. However, melatonin receptor cDNA has been cloned from *Xenopus* skin pigment cells and the receptor protein sequenced and assigned to the G-protein-linked receptor superfamily (Ebisawa *et al.,* 1994).

In the adult frog *Rana perezi,* endogenous melatonin concentrations peaked in the dark, were depressed by low temperature, and were higher in the eye and plasma than in the pineal gland (Delgado and Vivien-Roels, 1989). Changes in endogenous levels of melatonin during tadpole development and the importance of the eye as a source of melatonin in the tadpole remain to be determined.

These experimental examples suggest that the pineal gland, by transducing environmental light into a hormonal signal, melatonin, may play a role in fine-tuning the rate of metamorphosis.

V. ENVIRONMENTAL INFLUENCE

Environmental factors known to influence metamorphosis may do so by interactions with the endocrine system (as implied in the preceding section on melatonin). Two examples of such factors, i.e., light and temperature, and their effects on hormonal and resultant metamorphic changes are described briefly below.

A. Light

Effects of changes in the L/D cycle indicate that, in general, when light phases are increased (in duration or frequency), both spontaneous and T_4-induced metamorphosis are accelerated (see, e.g., Wright *et al.*, 1988, 1990a). Moreover, the effectiveness of exogenous T_4 in stimulating metamorphosis depends on the *time* in the L/D cycle in which the hormone is administered. For example, in tadpoles of *R. pipiens* (Wright *et al.*, 1986) and *X. laevis* (Burns *et al.*, 1987), metamorphosis is faster if tadpoles are exposed to T_4 during light, rather than dark, phases. Such an effect may be correlated with patterns of tadpole melatonin secretion (Wright *et al.*, 1990a), i.e., during light phases melatonin levels may be too low to appreciably slow down the rate of T_4-induced metamorphic changes.

Moreover, not only the metamorphic rate, but also rhythms of cell proliferation in limb epidermis vary with changes in the L/D cycle; for example, cell division was delayed when T_4 was administered in the dark (although DNA synthesis was not affected) (Wright *et al.*, 1990b). It is possible that effects of lighting changes on cell proliferation are mediated by the pineal gland and melatonin and are partially responsible for the "occurrence" of spontaneous metamorphosis when conditions in nature are favorable (Wright *et al.*, 1993).

B. Temperature

The influence of temperature on metamorphosis has been known for some time, yet only a few detailed experiments on the relationship of temperature and hormones have been reported. For example, *in vivo* experiments of Frieden *et al.* (1965) demonstrated that in response to an injection of T_3, the rate of tail shortening decreased at successively lower temperatures—no T_3 response occurred at 5°C. However, tadpoles injected with T_3 at 25°C but maintained at 5°C resumed metamorphosis upon return to the higher temperature. *In vitro* experiments with tail tips cultured at different temperatures in a thyroxine-containing medium indicated that temperature influences hormone action directly at the level of peripheral tissues such as the tail (Fry, 1972). Such *in vivo* and *in vitro* systems may become useful models for determining molecular mechanisms responsible for the temperature modification of thyroid hormone-stimulated metamorphosis. Does temperature influence T_4-stimulated gene expression (mRNAs)? Are number of T_4/T_3 nuclear receptors altered by changes in temperature?

In addition to the inhibiting effects of low temperature on tadpole metamorphosis of commonly studied species (e.g., *X. laevis, Rana grylio, R. pipiens, R. temporaria*), effects of temperature and T_4 on metamorphosis have been determined in *Ascaphus truei* tadpoles, which are adapted to cold mountain streams. In the laboratory, low temperatures (5°C) retarded T_4 metamorphosis, suggesting that in nature "cold" may be one of the factors contributing to the unusually long (2–4 years) life span of these larvae (Brown, 1990). Length of larval life spans vary tremendously in different species (e.g., 2–3 weeks in *Ceratophrys ornata* to as many as 4 years in *A. truei*), but controlling factors await investigation.

VI. CONCLUDING REMARKS

Amphibian metamorphosis is a complex process, with exquisite hormonal control of gene expression as its basis. As more is learned about the molecular mechanisms involved, it becomes evident that metamorphosis has become a model system for exploring the mechanisms of gene expression during postembryonic development (for reviews see Tata,

1993; Atkinson, 1994). Interest has focused on thyroid hormone stimulation of genes responsible for thyroid hormone receptors and for proteins involved in the initiation of specific metamorphic events. In addition, genes for transcription factors are now being identified and the role of such factors in thyroid hormone-induced events is being examined (e.g., involvement of RcC/EBP-1 factor in expression of urea cycle enzymes in the liver and AP-2 in keratin gene expression in the skin) (Xu and Tata, 1992; Xu *et al.,* 1993; Chen *et al.,* 1994; French *et al.,* 1994).

The complexity of hormonal regulation of metamorphosis may also be appreciated by considering fine-tuning of specific T_4/T_3-stimulated metamorphic events, not only by prolactin, corticosteroids, and melatonin, but also by possible direct or indirect involvement of other hormones. For example, exogenous gonadal steroids (17β-estradiol and testosterone) inhibited T_3-induced metamorphosis *in vivo* (but not *in vitro*) (see, e.g., Gray and Janssens, 1990), and the amphibian neurohypophyseal hormone mesotocin (MT), but not arginine vasotocin (AVT), promoted T_4-induced resorption of tadpole tails in culture (Iwamura *et al.,* 1985).

On the other hand, MT and especially AVT (as well as T_4/T_3, prolactin, and corticosteroids) affect osmoregulation, i.e., water and ion movement. Studies on the osmoregulatory effects of these hormones on such organs as larval gills, epidermis, and kidneys suggest that hormonal controls of osmoregulation are similar to those of metamorphosis. The two processes appear to be closely associated as well as appropriate for change of an animal from an aquatic to a generally more terrestrial environment (for reviews see White and Nicoll, 1981; Burggren and Just, 1992; Kikuyama *et al.,* 1993).

To add to the complexity of the endocrinological picture associated with metamorphosis, hormones, produced elsewhere than in classical endocrine glands, have now been found in larval and adult anurans. For example, atrial natriuretic peptide (ANP) has been localized in cardiac muscle cells throughout larval development, yet little is known about its control or its function in the tadpole (in mammals, it regulates blood pressure, renal function, and salt balance) (Hirohama *et al.,* 1989).

Also to consider is the possibility of one hormone having different functions, depending on the site of its synthesis/secretion. The neuropeptide hormone TRH released from hypothalamic neurons indirectly influences metamorphosis by stimulating prolactin secretion from the pituitary, but the function of the same peptide secreted from granular glands in the skin is unknown (e.g., Stroz and Kaltenbach, 1986). One might add that such exocrine glands release a host of bioactive peptides and other compounds onto the surface of the skin, some with known functions (e.g., antimicrobial activity, defense against predators, etc.) (for a review see Bevens and Zasloff, 1990). Since glandular development depends on thyroid hormone, it appear that cutaneous TRH is a product rather than a regulator of metamorphosis.

In future hormonal investigations, it might be well to keep in mind the importance of carbohydrate moieties on glycosylated hormones, of ratios of exogenous interacting hormones, as well as hormone concentrations relative to larval size, and of the time in the L/D cycle in which hormones are administered, as well as species and length of larval life spans. Do endogenous hormone levels differ in larvae with very long or very short life spans?

Interest in determining evolutionary relationships will undoubtedly continue. Similarity between specific amino acid sequences in peptide hormones and in receptors of amphibians and other vertebrates suggests that they have changed very little over the long course of evolution, i.e., they have been highly conserved. In fact, the tripeptide TRH has even been found in alfalfa plants (Roth and LeRoith, 1987).

The cytokines are an intensely studied major group of peptide regulatory factors. Broadly speaking, this group includes growth factors (GF), interferons, colony-stimulating factors, and interleukins (IL) (for reviews see Nathan and Sporn, 1991; Lee, 1992). Some of these (e.g., IL-1, IL-2, and TGF-β5), and possibly classical hormones as well, influence metamorphic change in the immune system of amphibians and represent a very promising area of research. [For details see Chapter 18 by Rollins-Smith and Cohen (this volume).]

The advances in endocrinology and molecular biology have made these extraordinary times for expanding our knowledge and gaining valuable insights into the endocrinology of amphibian metamorphosis.

REFERENCES

Aronsson, S. (1978). The pars distalis nerves and metamorphosis in *Rana temporaria*. *Gen. Comp. Endocrinol.* **36,** 497–501.

Atkinson, B. G. (1994). Metamorphosis: Model systems for studying gene expression in postembryonic development. *Dev. Genet.* **15,** 313–319.

Baker, B. S., and Tata, J. R. (1992). Prolactin prevents the autoinduction of thyroid hormone receptor mRNAs during amphibian metamorphosis. *Dev. Biol.* **149,** 463–467.

Balls, M., Clothier, R. H., Rowles, J. M., Kiteley, N. A., and Bennett, G. W. (1985). TRH distribution, levels, and significance during the development of *Xenopus laevis*. *In* "Metamorphosis" (M. Balls and M. Bownes, eds.), pp. 260–272. Clarendon, Oxford.

Bern, H. A., Nicoll, C. S., and Strohman, R. C. (1967). Prolactin and tadpole growth. *Proc. Soc. Exp. Biol. Med.* **126,** 518–520.

Bevins, C. L., and Zasloff, M. (1990). Peptides from frog skin. *Annu. Rev. Biochem.* **59,** 395–414.

Bray, T., and Sicard, R. E. (1982). Correlation among the changes in the levels of thyroid hormones, thyrotropin and thyrotropin-releasing hormone during the development of *Xenopus laevis. Exp. Cell Biol.* **50,** 101–107.

Brix, K., and Herzog, V. (1994). Extrathyroidal release of thyroidal hormones from thyroglobulin by J774 mouse macrophages. *J. Clin. Invest.* **93,** 1388–1396.

Brown, H. A. (1990). Temperature, thyroxine, and induced metamorphosis in tadpoles of a primitive frog, *Ascaphus truei. Gen. Comp. Endocrinol.* **79,** 136–146.

Buckbinder, L., and Brown, D. D. (1993). Expression of the *Xenopus laevis* prolactin and thyrotropin genes during metamorphosis. *Proc. Natl. Acad. Sci. U.S.A.* **90,** 3820–3824.

Burggren, W. W., and Just, J. J. (1992). Developmental changes in physiological systems. *In* "Environmental Physiology of the Amphibians" (M. E. Feder and W. W. Burggren, eds.), pp. 467–530. University of Chicago Press, Chicago.

Burns, J. T., Patyna, R., and Ryland, S. (1987). A circadian rhythm in the effect of thyroxine in the stimulation of metamorphosis in the African clawed frog, *Xenopus laevis. J. Interdiscip. Cycle Res.* **18,** 293–296.

Buscaglia, M., Leloup, J., and De Luze, A. (1985). The role and regulation of monodeiododination of thyroxine to 3,5,3′-triiodothyronine during amphibian metamorphosis. *In* "Metamorphosis" (M. Balls and M. Bownes, eds.), pp. 273–293. Clarendon, Oxford.

Carr, J. A., and Norris, D. O. (1990). Immunochemical localization of corticotropic-releasing factor– and arginine vasotocin-like immunoreactivities in the brain and pituitary of the American bullfrog (*Rana catesbeiana*) during development and metamorphosis. *Gen. Comp. Endocrinol.* **78,** 180–188.

Chen, Y., Hu, H., and Atkinson, B. G. (1994). Characterization and expression of C/EBP-like genes in the liver of *Rana catesbeiana* tadpoles during spontaneous and thyroid hormone-induced metamorphosis. *Dev. Genet.* **15,** 366–377.

Clemons, G. K., and Nicoll, C. S. (1977). Development and preliminary application of a homologous radioimmunoassay for bullfrog prolactin. *Gen. Comp. Endocrinol.* **32,** 531–535.

Dauça, M., and Hourdry, J. (1985). Transformations in the intestinal epithelium during anuran metamorphosis. *In* "Metamorphosis" (M. Balls and M. Bownes, eds.), pp. 36–58. Clarendon, Oxford.

Delgado, M. J., and Vivien-Roels, B. (1989). Effect of environmental temperature and photoperiod on the melatonin levels in the pineal, lateral eye, and plasma of the frog, *Rana perezi:* Importance of ocular melatonin. *Gen. Comp. Endocrinol.* **75,** 46–53.

Delgado, M. J., Gutiérrez, P., and Alonso-Bedate, M. (1987). Melatonin and photoperiod alter growth and larval development in *Xenopus laevis* tadpoles. *Comp. Biochem. Physiol. A* **86A,** 417–421.

Delidow, B. C., Baldocchi, R. A., and Nicoll, C. S. (1988). Evidence for hepatic involvement in the regulation of amphibian development by prolactin. *Gen. Comp. Endocrinol.* **70,** 418–424.

Dent, J. N. (1988). Hormonal interaction in amphibian metamorphosis. *Am. Zool.* **28,** 297–308.
Denver, R. J. (1988). Several hypothalamic peptides stimulate *in vitro* thyrotropin secretion by pituitaries of anuran amphibians. *Gen. Comp. Endocrinol.* **72,** 383–393.
Denver, R. J. (1993). Acceleration of anuran amphibian metamorphosis by corticotropin-releasing hormone-like peptides. *Gen. Comp. Endocrinol.* **91,** 38–51.
Denver, R. J., and Licht, P. (1989). Neuropeptide stimulation of thyrotropin secretion in the larval bullfrog: Evidence for a common neuroregulator of thyroid and interrenal activity in metamorphosis. *J. Exp. Zool.* **252,** 101–104.
Dodd, M. H. I., and Dodd, J. M. (1976). The biology of metamorphosis. *In* "Physiology of the Amphibia" (B. Lofts, ed.), Vol. 3, pp. 467–599. Academic Press, New York.
Doppler W. (1994). Regulation of gene expression by prolactin. *Rev. Physiol. Biochem. Pharmacol.* **124,** 93–130.
Ebisawa, T., Karne, S., Lerner, M. R., and Reppert, S. M. (1994). Expression cloning of a high-affinity melatonin receptor from *Xenopus* dermal melanophores. *Proc Natl. Acad. Sci. U.S.A.* **91,** 6133–6137.
Edwards, M. L. O., and Pivorum, E. B. (1991). The effects of photoperiod and different dosages of melatonin on metamorphic rate and weight gain in *Xenopus laevis* tadpoles. *Gen. Comp. Endocrinol.* **81,** 28–38.
Etkin, W. (1964). Metamorphosis. *In* "Physiology of the Amphibia" (J. A. Moore, ed.), pp. 427–468. Academic Press, New York.
Etkin, W. (1968). Hormonal control of amphibian metamorphosis. *In* "Metamorphosis: A Problem in Developmental Biology" (W. Etkin and L. I. Gilbert, eds.), pp 313–348. Appleton-Century-Crofts, New York.
Etkin, W., and Gona, A. G. (1967). Antagonism between prolactin and thyroid hormone in amphibian development. *J. Exp. Zool.* **165,** 249–258.
Etkin, W., and Lehrer, R. (1960). Excess growth in tadpoles after transplantation of the adenohypophysis. *Endocrinology* (*Baltimore*) **67,** 457–466.
Fox, H. (1981). Cytological and morphological changes during amphibian metamorphosis. *In* "Metamorphosis: A Problem in Developmental Biology" (L. I. Gilbert and E. Frieden, eds.), 2nd ed., pp. 327–362. Plenum, New York.
French, R. P., Warshawsky, D., Tybor, L., Mylniczenko, N. D., and Miller, L. (1994). Upregulation of AP-2 in the skin of *Xenopus laevis* during thyroid hormone-induced metamorphosis. *Dev. Genet.* **15,** 356–365.
Frieden, E., and Naile, B. (1955). Biochemistry of amphibian metamorphosis: I. Enhancement of induced metamorphosis by gluco-corticoids. *Science* **121,** 37–40.
Frieden, E., Wahlborg, A., and Howard, E. (1965). Temperature control of the response of tadpoles to triiodothyronine. *Nature* (*London*) **205,** 1173–1176.
Fry, A. E. (1972). Effects of temperature on shortening of isolated *Rana pipiens* tadpole tail tips. *J. Exp. Zool.* **180,** 197–208.
Galton, V. A. (1983). Thyroid hormone action in amphibian metamorphosis. *In* "Molecular Basis of Thyroid Hormone Action" (J. H. Oppenheimer and H. H. Samuels, eds.), pp. 445–483. Academic Press, New York.
Galton, V. A. (1988). The role of thyroid hormone in amphibian development. *Am. Zool.* **28,** 309–318.
Galton, V. (1990). Mechanisms underlying the acceleration of thyroid hormone-induced tadpole metamorphosis by corticosterone. *Endocrinology* (*Baltimore*) **127,** 2997–3002.

Galton, V. A. (1992). The role of thyroid hormone in amphibian metamorphosis. *Trends Endocrinol. Metab.* **3,** 96–100.

Galton, V. A., Davey, J. C., and Schneider, M. J. (1994). Mechanisms of thyroid hormone action in developing *Rana catesbeiana* tadpoles. *In* "Perspectives in Comparative Endocrinology" (K. G. Davey, R. E. Peter, and S. S. Tobe, eds.), pp. 412–415. National Research Council of Canada, Ottawa.

Gancedo, B., Corpas, I., Alonso-Gómez, A. L., Delgado, M. J., Morreale de Escobar, G., and Alonso-Bedate, M. (1992). Corticotropin-releasing factor stimulates metamorphosis and increases thyroid hormone concentration in prometamorphic *Rana perezi* larvae. *Gen. Comp. Endocrinol.* **87,** 6–13.

García-Navarro, S., Malagón, M. M., and Gracia-Navarro, F. (1988). Immunohistochemical localization of thyrotropic cells during amphibian morphogenesis: A stereological study. *Gen. Comp. Endocrinol.* **71,** 116–123.

Gona, A. G. (1967). Prolactin as a goitrogenic agent in Amphibia. *Endocrinology (Baltimore)* **81,** 748–754.

Gray, K. M., and Janssens, P. A. (1990). Gonadal hormones inhibit the induction of metamorphosis by thyroid hormones in *Xenopus laevis* tadpoles *in vivo,* but not *in vitro. Gen. Comp. Endocrinol.* **77,** 202–211.

Gudernatsch, J. F. (1912). Feeding experiments on tadpoles. I. The influence of specific organs given as food on growth and differentiation. A contribution to the knowledge of organs with internal secretion. *Wilhelm Roux Arch. Entwicklungsmech. Org.* **35,** 457–483.

Hanken, J., and Hall, B. K. (1988). Skull development during anuran metamorphosis. II. Role of thyroid hormone in osteogenesis. *Anat. Embryol.* **178,** 219–227.

Hanken, J., and Summers, C. H. (1988). Skull development during anuran metamorphosis. III. Role of thyroid hormone in chondrogenesis. *J. Exp. Zool.* **246,** 156–170.

Hanken, J., Summers, C. H., and Hall, B. K. (1989). Morphological integration in the cranium during anuran metamorphosis. *Experientia* **45,** 872–875.

Hirohama, T., Uemura, H., Nakamura, S., Naruse, M., and Aota, T. (1989). Ultrastructure and atrial natriuretic peptide (ANP)-like immunoreactivity of cardiocytes in the larval, metamorphosing and adult specimens of the Japanese toad, *Bufo japonicus formosus. Dev., Growth Differ.* **31,** 113–121.

Hoskins, S. G. (1990). Metamorphosis of the amphibian eye. *J. Neurobiol.* **21,** 970–989.

Hoskins, S. G., and Grobstein, P. (1984). Induction of the ipsilateral retinothalamic projection in *Xenopus laevis* by thyroxine. *Nature (London)* **307,** 730–733.

Hoskins, S. G., and Grobstein, P. (1985). Development of the ipsilateral retinothalamic projection in the frog *Xenopus laevis.* III. The role of thyroxine. *J. Neurosci.* **5,** 930–940.

Hsu, C.-Y., Yu, N.-W., and Chen, S.-J. (1980). Development of $\Delta^5 - 3\beta$-hydroxysteroid activity in the interrenal gland of *Rana catesbeiana. Gen. Comp. Endocrinol.* **42,** 167–170.

Hsu, C.-Y., Yu, N.-W., Pi, C.-M., Chen, S.-J., and Ruan, C.-C. (1984). Hormonal regulation of development of the interrenal activity of $\Delta^5 - 3\beta$-hydroxysteroid dehydrogenase in bullfrog tadpoles. *J. Exp. Zool.* **232,** 73–78.

Ishizuya-Oka, A., and Shimozawa, A. (1992). Programmed cell death and heterolysis of larval epithelial cells by macrophage-like cells in the anuran small intestine *in vivo* and *in vitro. J. Morphol.* **213,** 185–195.

Iwamuro, S., Yamaguchi, Y., Kobayashi, T., and Kikuyama, S. (1985). Effect of neurohypophyseal hormone on thyroxine-induced resorption of tadpole tail *in vitro. Proc. Jpn. Acad., Ser. B* **61,** 441–443.

Jackson, I. M. D., and Reichlin, S. (1977). Thyrotropin-releasing hormone: Abundance in the skin of the frog, *Rana pipiens. Science* **198,** 414–415.

Jaffe, R. C. (1981). Plasma concentration of corticosterone during *Rana catesbeiana* tadpole metamorphosis. *Gen. Comp. Endocrinol.* **44,** 314–318.

Jaudet, G. J., and Hatey, J. L. (1984). Variations in aldosterone and corticosterone plasma levels during metamorphosis in *Xenopus laevis* tadpoles. *Gen. Comp. Endocrinol.* **56,** 59–65.

Kaltenbach, J. C. (1958). Direct steroid enhancement of induced metamorphosis in peripheral tissues. *Anat. Rec.* **131,** 569–570 (abstract).

Kaltenbach, J. C. (1968). Nature of hormone action in amphibian metamorphosis. *In* "Metamorphosis: A Problem in Developmental Biology" (W. Etkin and L. I. Gilbert, eds.), pp. 399–441. Appleton-Century-Crofts, New York.

Kaltenbach, J. C. (1982). Circulating thyroid hormone levels in Amphibia. *Gunma Symp. Endocrinol.* **19,** 63–74.

Kaltenbach, J. C. (1985). Amphibian metamorphosis: Influence of thyroid and steroid hormones. *In* "Current Trends in Comparative Endocrinology" (B. Lofts and W. N. Holmes, eds.), pp. 533–534. Hong Kong University Press, Hong Kong.

Kar, S., and Naik, D. R. (1986). Cytodifferentiation and immunocharacteristics of adenohypophyseal cells in the toad, *Bufo melanosticus. Anat. Embryol.* **175,** 137–146.

Kawamura, K., Yamamoto, K., and Kikuyama, S. (1986). Effects of thyroid hormone, stalk section, and transplantation of the pituitary gland on plasma prolactin levels at metamorphic climax in *Rana catesbeiana. Gen. Comp. Endocrinol.* **64,** 129–135.

Kikuyama, S. (1994). Amphibian prolactin and control of its secretion. *In* "Perspectives in Comparative Endocrinology" (K. G. Davey, R. E. Peter, and S. S. Tobe, eds.), pp. 237–242. National Research Council of Canada, Ottawa.

Kikuyama, S., Suzuki, M. R., Niki, K., and Yoshizato, K. (1982). Thyroid hormone–adrenal corticoid interaction in tadpole tail. *Int. Congr. Ser.—Excerpta Med.* **605,** 202–205.

Kikuyama, S., Niki, K., Mayumi, M., Shibayama, R., Nishikawa, M., and Shintake, N. (1983). Studies on corticoid action on the toad tadpole tail *in vitro. Gen. Comp. Endocrinol.* **52,** 395–399.

Kikuyama, S., Suzuki, M. R., and Iwamuro, S. (1986). Elevation of plasma aldosterone levels of tadpoles at metamorphic climax. *Gen. Comp. Endocrinol.* **63,** 186–190.

Kikuyama, S., Kawamura, K., Tanaka, S., and Yamamoto, K (1993). Aspects of amphibian metamorphosis: Hormonal control. *Int. Rev. Cytol.* **145,** 105–148.

King, J. A., and Miller, R. P. (1981). TRH, GH-RIH, and LH-RH in metamorphosing *Xenopus laevis. Gen. Comp. Endocrinol.* **44,** 20–27.

Kobayashi, H. (1958). Effect of desoxycorticosterone acetate on metamorphosis induced by thyroxine in anuran tadpoles. *Endocrinology* (*Baltimore*) **62,** 371–377.

Kollros, J. J. (1981). Transitions in the nervous system during amphibian metamorphosis. *In* "Metamorphosis: A Problem in Developmental Biology" (L. I. Gilbert and E. Frieden, eds.), 2nd ed., pp. 445–459. Plenum, New York.

Kollros, J. J. (1984). Growth and death of cells of the mesencephalic fifth nucleus in *Rana pipiens* larvae. *J. Comp. Neurol.* **224,** 386–394.

Kollros, J. J., and Bovbjerg, A. M. (1990). Mesencephalic fifth nucleus cell responses to thyroid hormone: One population or two? *J. Neurobiol.* **21,** 1002–1010.

Krug, E. C., Horn, K. V., Battista, J., and Nicoll, C. S. (1983). Corticosteroids in serum of *Rana catesbeiana* during development and metamorphosis. *Gen. Comp. Endocrinol.* **52,** 232–241.

Lee, F. D. (1992). The role of interleukin-6 in development. *Dev. Biol.* **151,** 331–338.
Leloup, J., and Buscaglia, M. (1977). La triiodothyronine, hormone de la métamorphose des Amphibiens. *C. R. Hebd. Seances Acad. Sci., Ser. D* **284,** 2261–2263.
Millar, R. P., Nicolson, S., King, J. A., and Louw, G. N. (1983). Functional significance of TRH in metamorphosing and adult anurans. *In* "Thyrotropin-Releasing Hormone" (E. C. Griffiths and G. W. Bennett, eds.), pp. 217–227. Raven, New York.
Mimnagh, K. M., Bolaffi, J. L., Montgomery, N. M., and Kaltenbach, J. C. (1987). Thyrotropin-releasing hormone (TRH): Immunohistochemical distribution in tadpole and frog brain. *Gen. Comp. Endocrinol.* **66,** 394–404.
Miyauchi, H., LaRochelle, F. T., Jr., Suzuki, M., Freeman, M., and Frieden, E. (1977). Studies on thyroid hormones and their binding in bullfrog tadpole plasma during metamorphosis. *Gen. Comp. Endocrinol.* **33,** 254–266.
Mondou, P. M., and Kaltenbach, J. C. (1979). Thyroxine concentrations in blood serum and pericardial fluid of metamorphosing tadpoles and of adult frogs. *Gen. Comp. Endocrinol.* **39,** 343–349.
Moriceau-Hay, D., Doerr-Schott, J., and Dubois, M. (1982). Immunohistochemical demonstration of TSH-, LH-, and ACTH-cells in the hypophysis of tadpoles of *Xenopus laevis* D. *Cell Tissue Res.* **225,** 57–64.
Nathan, C., and Sporn, M. (1991). Cytokines in context. *J. Cell Biol.* **113,** 981–986.
Nieuwkoop, P. D., and Faber, J. (1967). "Normal Table of *Xenopus laevis* (Daudin)," 2nd ed. North-Holland, Amsterdam.
Niinuma, K., Mamiya, N., Yamamoto, K., Iwamuro, S., Vaudry, H., and Kikuyama, S. (1989). Plasma concentrations of aldosterone and prolactin in metamorphosing toad tadpoles. *Bull. Sci. Eng. Res. Lab., Waseda Univ.* **122,** 17–21.
Niinuma, K., Tagawa, M., Hirano, T., and Kikuyama, S. (1991a). Changes in tissue concentrations of thyroid hormones in metamorphosing toad larvae. *Zool. Sci.* **8,** 345–350.
Niinuma, K., Yamamoto, K., and Kikuyama, S. (1991b). Changes in plasma and pituitary prolactin levels in toad (*Bufo japonicus*) larvae during metamorphosis. *Zool. Sci.* **8,** 97–101.
Niki, K., Yoshizato, K., and Kikuyama, K. (1981). Augmentation of nuclear binding capacity for triiodothyronine by aldosterone in tadpole tail. *Proc. Jpn. Acad., Ser. B* **57,** 271–275.
Piotrowski, D. C., and Kaltenbach, J. C. (1985). Immunofluorescent detection and localization of thyroxine in blood of *Rana catesbeiana* from early larval through metamorphic stages. *Gen. Comp. Endocrinol.* **59,** 82–90.
Pretty, R., Naitoh, T., and Wassersug, R. J. (1995). Metamorphic shortening of the alimentary tract in anuran larvae (*Rana catesbeiana*). *Anat. Rec.* **242,** 417–423.
Regard, E. (1978). Cytophysiology of the amphibian thyroid gland through larval development and metamorphosis. *Int. Rev. Cytol.* **52,** 81–118.
Regard, E., Taurog, A., and Nakashima, T. (1978). Plasma thyroxine and triiodothyronine levels in spontaneously metamorphosing *Rana catesbeiana* tadpoles and in adult anuran Amphibia. *Endocrinology* (*Baltimore*) **102,** 674–684.
Richter, K., Kawashima, E., Egger, R., and Kreil, G. (1984). Biosynthesis of thyrotropin releasing hormone in the skin of *Xenopus laevis:* Partial sequence of the precursor deduced from cloned cDNA. *EMBO J.* **3,** 617–621.
Rosenkilde, P., and Ussing, A. P. (1990). Regulation of metamorphosis. *Fortschr. Zool.* **38,** 125–138.
Roth, J., and LeRoith, D. (1987). Chemical cross talk. *Sciences* (*N.Y.*) **May/June,** 51–54.

Sairam, M. R. (1989). Role of carbohydrates in glycoprotein hormone signal transduction. *FASEB J.* **3,** 1915–1926.

Sakai, M., Hanaoka, Y., Tanaka, S., Hayashi, H., and Kikuyama, S. (1991). Thyrotropic activity of various adenohypophyseal hormones of the bullfrog. *Zool. Sci.* **8,** 929–934.

Schultheiss, H. (1980). T_3 and T_4 concentrations during metamorphosis of *Xenopus laevis* and *Rana esculenta* and in the neotenic Mexican axolotl. *Gen. Comp. Endocrinol.* **40,** 372 (abstract).

Shimizu-Nishikawa, K., and Miller, L. (1992). Hormonal regulation of adult type keratin gene expression in larval epidermal cells of the frog *Xenopus laevis. Differentiation (Berlin)* **49,** 77–83.

Shumway, W. (1940). Stages in the normal development of *Rana pipiens.* I. External form. *Anat. Rec.* **78,** 139–147.

Stroz, D.-A., and Kaltenbach, J. C. (1986). Thyrotropin-releasing hormone (TRH): Immunohistochemical localization in skin of larval and adult *Rana pipiens* and *Rana catesbeiana. J. Am. Osteopath. Assoc.* **86,** 680 (abstract).

Suzuki, M. R., and Kikuyama, S. (1983). Corticoids augment nuclear binding capacity for triidothyronine in bullfrog tadpole tail fins. *Gen. Comp. Endocrinol.* **52,** 272–278.

Suzuki, S., and Fujikura, K. (1992). Thyroglobulin and thyroid hormone secretion in the tadpoles and adult frogs. *Proc. Int. Symp. Amphib. Endocrinol., Waseda Univ.* p. 6 (abstract).

Suzuki, S., and Suzuki, M. (1981). Changes in thyroidal and plasma iodine compounds during and after metamorphosis of the bullfrog, *Rana catesbeiana. Gen. Comp. Endocrinol.* **45,** 74–81.

Takahashi, N., Yoshihama, K., Kikuyama, S., Yamamoto, K., Wakabayashi, K., and Kato, Y. (1990). Molecular cloning and nucleotide sequence analysis of complementary DNA for bullfrog prolactin. *J. Mol. Endocrinol.* **5,** 281–287.

Tanaka, S., Sakai, M., Park, M. K., and Kurosumi, K. (1991). Differential appearance of the subunits of glycoprotein hormones (LH, FSH, and TSH) in the pituitary of bullfrog (*Rana catesbeiana*) larvae during metamorphosis. *Gen. Comp. Endocrinol.* **84,** 318–327.

Taniguchi, Y., Tanaka, S., and Kurosumi, K. (1990). Distribution of immunoreactive thyrotropin-releasing hormone in the brain and hypophysis of larval bullfrogs with special reference to nerve fibers in the pars distalis. *Zool. Sci.* **7,** 427–433.

Tata, J. R. (1993). Gene expression during metamorphosis: An ideal model for post-embryonic development. *BioEssays* **15,** 239–248.

Tata, J. R., Kawahara, A., and Baker, B. S. (1991). Prolactin inhibits both thyroid hormone-induced morphogenesis and cell death in cultured amphibian larval tissues. *Dev. Biol.* **146,** 72–80.

Taylor, A. C., and Kollros, J. J. (1946). Stages in the normal development of *Rana pipiens* larvae. *Anat. Rec.* **94,** 7–24.

Weber, G. M., Farrar, E. S., Tom, C. K. F., and Grau, E. G. (1994). Changes in whole-body thyroxine and triiodothyronine concentrations and total content during early development and metamorphosis of the toad *Bufo marinus. Gen. Comp. Endocrinol.* **94,** 62–71.

Weil, M. R. (1986). Changes in plasma thyroxine levels during and after spontaneous metamorphosis in a natural population of the green frog, *Rana clamitans. Gen. Comp. Endocrinol.* **62,** 8–12.

White, B. A., and Nicoll, C. S. (1979). Prolactin receptors in *Rana catesbeiana* during development and metamorphosis. *Science* **204,** 851–853.

White, B. A., and Nicoll, C. S. (1981). Hormonal control of amphibian metamorphosis. *In* "Metamorphosis: A Problem in Developmental Biology" (L. I. Gilbert and E. Frieden, eds.), 2nd ed., pp. 363–396. Plenum, New York.

White, B. A., Lebovic, G. S., and Nicoll, C. S. (1981). Prolactin inhibits the induction of its own renal receptors in *Rana catesbeiana* tadpoles. *Gen. Comp. Endocrinol.* **43,** 30–38.

Wright, M. L., Majerowski, M. A., Lukas, S. M., and Pike, P. A. (1979). Effect of prolactin on growth, development, and epidermal cell proliferation in the hindlimb of the *Rana pipiens* tadpole. *Gen. Comp. Endocrinol.* **39,** 53–62.

Wright, M. L., Frim, E. K., Bonak, V. A., and Baril, C. (1986). Metamorphic rate in *Rana pipiens* larvae treated with thyroxine or prolactin at different times in the light/dark cycle. *Gen. Comp. Endocrinol.* **63,** 51–61.

Wright, M. L., Jorey, S. T., Myers, Y. M., Fieldstad, M. L., Paquette, C. M., and Clark, M. B. (1988). Influence of photoperiod, daylength, and feeding schedule on tadpole growth and development. *Dev., Growth Differ.* **30,** 315–323.

Wright, M. L., Blanchard, L. S., Jorey, S. T., Basso, C. A., Myers, Y. M., and Paquette, C. M. (1990a). Metamorphic rate as a function of the light/dark cycle in *Rana pipiens* larvae. *Comp. Biochem. Physiol. A* **96A,** 215–220.

Wright, M. L., Pathammavong, N., and Basso, C. A. (1990b). DNA synthesis is unaffected but subsequent cell division is delayed in tadpole hindlimb epidermis when thyroxine is given in the dark. *Gen. Comp. Endocrinol.* **79,** 89–94.

Wright, M. L., Cykowski, L. J., Mayrand, S. M., Blanchard, L. S., Kraszewska, A. A., Gonzales, T. M., and Patnaude, M. (1991). Influence of melatonin on the rate of *Rana pipiens* tadpole metamorphosis *in vivo* and regression of thyroxine-treated tail tips *in vitro. Dev., Growth Differ.* **33,** 243–249.

Wright, M. L., Cykowski, L. J., and Blanchard, L. S. (1992). Studies of cell proliferation and follicle characteristics in the thyroid of the *Rana pipiens* tadpole at various metamorphic stages. *Proc. Int. Symp. Amphib. Endocrinol., Waseda Univ.* p. 3 (abstract).

Wright, M. L., Jorey, S. T., Blanchard, L. S., Garatti, L., Mayrand, S. M., and Kraszewska, A. A. (1993). Modulation of cell cycle phases by various circadian and noncircadian light/dark schedules and evidence implicating melatonin. *J. Interdiscip. Cycle Res.* **24,** 101–117.

Wright, M. L., Cykowski, L. J., Lundrigan, L., Hemond, K. L., Kochan, D. M. Faszewski, E. E., and Anuszewski, C. M. (1994). Anterior pituitary and adrenal cortical hormones accelerate or inhibit tadpole hindlimb growth and development depending on stage of spontaneous development or thyroxine concentration in induced metamorphosis. *J. Exp. Zool.* **270,** 175–188.

Xu, Q., and Tata, J. R. (1992). Characterization and developmental expression of *Xenopus* C/EBP gene. *Mech. Dev.* **38,** 69–81.

Xu, Q., Baker, B. S., and Tata, J. R. (1993). Developmental and hormonal regulation of the *Xenopus* liver-type arginase gene. *Eur. J. Biochem.* **211,** 891–898.

Yamamoto, K., and Kikuyama, S. (1981). Purification and properties of bullfrog prolactin. *Endocrinol. Jpn.* **28,** 59–64.

Yamamoto, K., and Kikuyama, S. (1982). Radioimmunoassay of prolactin in plasma of bullfrog tadpoles. *Endocrinol. Jpn.* **29,** 159–167.

Yamamoto, K., Niinuma, K., and Kikuyama, S. (1986a). Synthesis and storage of prolactin in the pituitary gland of bullfrog tadpoles during metamorphosis. *Gen. Comp. Endocrinol.* **62,** 247–253.

Yamamoto, K., Kobayashi, T., and Kikuyama, S. (1986b). Purification and characterization of toad prolactin. *Gen. Comp. Endocrinol.* **63,** 104–109.

Yamashita, K., Matsuda, K., Hayashi, H., Hanaoka, Y., Tanaka, S., Yamamoto, K., and Kikuyama, S. (1993). Isolation and characterization of two forms of *Xenopus* prolactin. *Gen. Comp. Endocrinol.* **91,** 307–317.

Yasuda, A., Yamaguchi, K., Kobayashi, T., Yamamoto, K., Kikuyama, S., and Kawauchi, H. (1991). The complete amino acid sequence of prolactin from the bullfrog, *Rana catesbeiana. Gen. Comp. Endocrinol.* **83,** 218–226.

12

Neuroendocrine Control of Amphibian Metamorphosis

ROBERT J. DENVER

Department of Biology
University of Michigan
Ann Arbor, Michigan

I. INTRODUCTION

Since the early 18th century, amphibians have been important models for developmental biology. The typical amphibian life history is unique among terrestrial vertebrates in the presence of an aquatic fish-like larval stage. This characteristic makes amphibian larvae ideal for the study of postembryonic development (Tata, 1993; Atkinson, 1994). Early studies on morphogenesis of the amphibian tadpole were the first to show a role for a hormone in directing development. The investigation of amphibian metamorphosis continues to provide valuable insights into the mechanisms of action of hormones in development (Tata, 1993; Atkinson, 1994; Shi, 1994). The central role of the thyroid gland in metamorphosis has become a classic model for the study of thyroid hormone (TH) action (Shi, 1994). Several excellent reviews have covered the endocrinology of metamorphosis (Dodd and Dodd, 1976; White and Nicoll, 1981; Norris, 1989; Kikuyama *et al.,* 1993a). The scope of this chapter is to describe recent findings on the development of the amphibian neuroendocrine system and the identification of the neurohormone(s) responsible for controlling metamorphosis. Emphasis will be placed, where appropriate, on what is known of the molecular mechanisms underlying each of these aspects of amphibian development. Since most information in this area has come from studies of anurans (frogs), this chapter focuses on this amphibian order, with only limited treatment of urodeles. [For coverage of the neuroendocrinology of urodele metamorphosis, see Norris (1989).]

Etkin (1968) proposed a model for changes in endocrine activity during amphibian metamorphosis, and he coined the terms for describing the stages of anuran development: "premetamorphosis," during which time considerable larval growth but little morphological change occurs and TH levels are low; "prometamorphosis," when hind limb growth is accelerating and plasma titers of TH are rising; and "metamorphic climax," the final and most rapid phase of morphological change when thyroid activity is at its highest (White and Nicoll, 1981). Elevations in TH production during metamorphosis as predicted by Etkin (1968) have been confirmed by many investigators (White and Nicoll, 1981; Kikuyama *et al.,* 1993a). While TH is the major stimulus for metamorphosis, pituitary prolactin (PRL) is thought to function as a larval growth hormone and to have antimetamorphic actions (Dodd and Dodd, 1976; White and Nicoll, 1981; Kikuyama *et al.,* 1993a). Etkin (1968) proposed that PRL production would be high during larval life and then decline at metamorphic climax. However, recent findings have shown that plasma PRL concentrations and pituitary PRL mRNA are very low during premetamorphosis but rise late in prometamorphosis and exhibit a peak at metamorphic climax (see Section III,C) (Fig. 1).

Corticosteroids produced by the interrenal glands (amphibian homolog of adrenal cortex) accelerate thyroxine (T_4)-induced tadpole metamorphosis (see Dodd and Dodd, 1976). Work in this area supports an important role for corticosteroids in controlling morphogenesis. Radio-

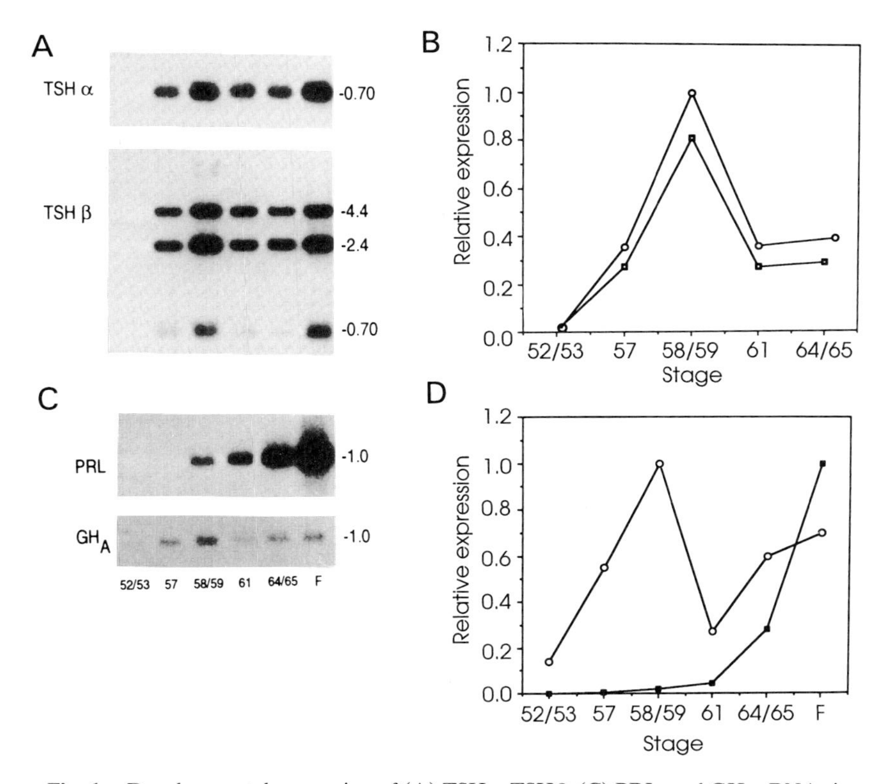

Fig. 1. Developmental expression of (A) TSHα, TSHβ, (C) PRL, and GH mRNAs in pituitary glands of *Xenopus* tadpoles. Total RNA was analyzed by Northern blot. Nieuwkoop–Faber stages are indicated (F, frog) and molecular sizes are at the right of the autoradiograms. The quantitative expression (based on laser densitometry) of (B) TSHβ (□) and TSHβ (○), and (D) PRL (■) and GH (○) is given relative to the highest value which was arbitrarily defined as 1.0. [Adapted with permission from Buckbinder and Brown (1993).]

immunoassay (RIA) data for several corticosteroid hormones have shown marked elevations in blood hormone levels during prometamorphosis (reviewed by White and Nicoll, 1981; Kikuyama *et al.,* 1993a).

In vertebrates, secretions of T_4 by the thyroid and corticosteroids by the interrenals are regulated by pituitary thyrotropin (TSH) and adrenocorticotropin (ACTH), respectively; the biosynthesis and secretion of TSH and ACTH (and also PRL) are controlled by neurohormones produced in the hypothalamus. The neuroendocrine system is central to the control of endocrine changes which occur during metamorphosis and it is at the level of the hypothalamus that certain types of environmental influences which alter the rate of development are transduced into an endocrine response.

II. DEVELOPMENT OF HYPOTHALAMOHYPOPHYSEAL SYSTEM

A. Embryonic Origins of Pituitary Gland and Hypothalamus

The first specific account of the development of the vertebrate pituitary gland was that of Rathke (1838) for the snake *Natrix natrix.* Rathke described a solid bud of tissue on the dorsal side of the stomodeal invagination which was ectodermal in origin and which appeared to give rise to the epithelial hypophysis (pars distalis, pars intermedia, pars tuberalis). Until recently (see later), there has been general agreement that the adenohypophysis forms from buccal ectoderm. This pouch was thought to make contact with the hypothalamus where it differentiates into the pituitary gland and then loses contact with the buccal cavity. The neurohypophysis forms from the floor of the primitive diencephalon (neuroectoderm) as an evagination (infundibulum). The pars intermedia was thought to derive from buccal ectoderm that is in close contact with the neurohypophysis (Nyholm and Doerr-Schott, 1977).

Contrasting views of the development of the pituitary gland have been put forth by several investigators (for a discussion see Kawamura and Kikuyama, 1992). Eagleson *et al.* (1986) showed that certain cells of the frog (*Xenopus laevis*) adenohypophysis and the brain arise from part of the anterior neural ridge (ANR; called ventral neural ridge by the authors based on the nomenclature for avian embryos). They transplanted ^{125}I-labeled ANR tissue into *Xenopus* embryos and showed that it gives rise to the preoptic region of the hypothalamus and the most anterior portion of the adenohypophysis containing ACTH-immunoreactive cells. Transplantation of melanin-rich tissues from open neurula stage toad embryos (*Bufo japonicus*) to the same region of albino embryos showed that the ANR gives rise to both the adenohypophysis and the anterior hypothalamus [preoptic region (Kawamura and Kikuyama, 1992)]. These authors suggest that the entire adenohypophysis might be placodal rather than stomodeal, arising from neuroectoderm, not buccal ectoderm. Although it cannot be concluded from their data that the adenohypophysis is of neuroectodermal origin due to the presence of mesodermal tissue in their ANR grafts, the authors present strong evidence that the anlage of the adenohypophysis is very closely related to the neuronal primordium. Further evidence for an ontogenetic relationship between the hypothalamus and the pituitary comes from the demonstration by Van Noorden *et al.* (1984) that the neural marker neuron-specific enolase is present in the anterior pituitary gland.

B. Reciprocal Interactions between Brain and Pituitary during Development

Kawamura and Kikuyama (1992) showed that the anlagen of the adenohypophysis and the anterior hypothalamus are closely apposed at the open neurula stage. While these two tissues have well-known functional relationships later in development, several lines of evidence suggest the possibility that inductive and perhaps regulatory interactions occur between amphibian pituitary and hypothalamic cells very early in development. For instance, posterior hypothalectomy of *Rana pipiens* or *Bufo japonicus* tadpoles by removal of the anlage of the posterior hypothalamus and infundibulum (central part of anterior neural plate) at the open neurula stage results in development of the pituitary without contact with the brain (Chang, 1957; Hanoaka, 1967; Kawamura and Kikuyama, 1989). All cells of the adenohypophysis develop normally except cells producing proopiomelanocortin (POMC)-derived peptides such as ACTH, MSH (melanocyte stimulating hormone), and endorphins (Kawamura and Kikuyama, 1995); however, these animals do not proceed through metamorphosis (Chang, 1957; Hanoaka, 1967) or, in the case of *B. japonicus,* complete metamorphosis later than intact animals (Kawamura and Kikuyama, 1995). Early studies suggested that the developing brain induces intermediate lobe development in amphibians [based on the appearance of α-MSH bioactivity (Blount, 1945; Driscoll and Eakin, 1955; Eakin, 1956; Thurmond and Eakin, 1959; Pehlemann, 1962; Etkin, 1967)]. The work of Hayes and Loh (1990), using *in situ* hybridization histochemistry, support the view that brain–pituitary interactions may be responsible for the simultaneous onset of POMC gene expression in hypothalamus and pituitary.

Direct morphological contact between the pituitary primordium and infundibulum appears to be required for the differentiation of POMC-producing cells in the pituitary (see Kawamura and Kikuyama, 1995); these are the first pituitary cell types to develop, both in amphibians (Moriceau-Hay *et al.,* 1982; Eagleson *et al.,* 1986; Hayes and Loh, 1990) and in mammals (Begeot *et al.,* 1982). Although other pituitary cells develop slightly later than the POMC cells (see Kikuyama *et al.,* 1993a), they may require contact with primordial hypothalamic tissue earlier in development before the infundibulum forms. Kikuyama *et al.* (1993b) showed that when ANR of open neurula-stage embryos is transplanted to the tail of *B. japonicus* embryos (tailbud stage), no pituitary cells develop, while the cells (only PRL and MSH cells were monitored) developed normally when the pituitary primordium of tailbud stage embryos was grafted. They interpret this to mean that pituitary cells other than the POMC-producing cells also require surrounding brain tissue to develop normally, at least early in development (before the tailbud stage). Thus, some type of inductive interaction, perhaps reciprocal between the brain and pituitary, may be required at some stage of development for the normal differentiation of each class of pituitary cell in amphibians. Similar interactions between brain and pituitary primordium during rat development are suggested by studies employing cocultures of diencephalon and pituitary primordium (Watanabe, 1982a,b, 1985). These studies suggest that there are different critical periods for brain–pituitary contact during the development of the different pituitary cell types. For instance, corticotropes require contact with the diencephalon sometime before day 13.5 of gestation; separation of brain and pituitary anlagen after that time does not result in impaired corticotrope development (Watanabe, 1982a). In contrast, the differentiation of luteinizing hormone (LH) cells requires brain–pituitary contact before day 11.5 (Watanabe, 1985).

In tadpoles, the presence of the pituitary gland is required for the development of hypothalamic nerve terminals in the median eminence (Etkin *et al.,* 1965; Kikuyama *et al.,* 1979). Kikuyama *et al.* (1979) showed that the development of monoaminergic neurons of the preoptic nucleus (PON) is dependent on TH. However, when tadpoles are hypophysectomized at the tail bud stage, thyroxine is not capable of inducing the development of these nerve terminals nor of the capillaries in the median eminence, perhaps due to the absence of target (i.e., pituitary) cell surface markers or secreted factors.

While classical transplant studies have suggested that the amphibian pituitary gland and hypothalamus are both derived from neuronal primordium, the molecular mechanisms for the specification of cells toward these distinct lineages in amphibians are unknown. Organ culture studies with mammalian tissues have shown that a variety of hormones (i.e., neuropeptides, steroids) can influence the differentiation of different pituitary cell types; however, these studies have not defined whether the hormones are responsible for specifying the different cell types, enhancing the expression of genes encoding trophic factors, or triggering the expansion of previously committed cells (for a review see Voss and Rosenfeld, 1992). As discussed earlier, development of the amphibian hypothalamus and pituitary likely depends on reciprocal inductive interactions, perhaps involving the activation of tissue-specific homeobox genes. At least one example exists of a hypothalamic factor which can induce differentiation of a pituitary cell type: gonadotropin-releasing hormone (GnRH) induced the differentiation of gonadotropes in organ cultures of rat pituitary primordium (Begeot *et al.,* 1984).

The very earliest events in the commitment of ectoderm to the pituitary anlage may result from the expression of homeotic gene(s). Greater than 80 homeotic genes have been identified in animals; most are responsible for directing the development of specific structures and tissues (Scott *et al.,* 1989). A member of the LIM class homeotic genes, *Xlim-3,* is expressed very early in the pituitary primordium of *Xenopus* embryos and it is the earliest marker for pituitary primordium thus far identified (Taira *et al.,* 1993). Taira *et al.* (1993) have suggested that *Xlim-3* is involved in the initial specification of cells toward the pituitary lineage. Later specification of distinct pituitary cell lineages may result from the expression of other cell-specific transcription factors whose expression is controlled by inductive stimuli from neural tissue (Voss and Rosenfeld, 1992). For instance, in rats, expression of the thyrotrope embryonic factor (TEF), a basic leucine repeat transcription factor, correlates with the onset of TSHβ expression (Drolet *et al.,* 1991). The pituitary-specific Pit-1, which is a POU homeodomain transcription factor, is thought to be responsible for the specification and maintenance of somatotropes and lactotropes in the rat pituitary (Voss and Rosenfeld, 1992). No comparable data are now available for the molecular mechanisms involved in amphibian pituitary development.

The median eminence does not develop until prometamorphosis in *R. pipiens* (Etkin, 1968). This later development also occurs in *Xenopus* (see Goos, 1978). However, this does not preclude the earlier existence of functional regulatory relationships between the hypothalamus and the pituitary since brain and pituitary are closely apposed at very early stages of development (Kawamura and Kikuyama, 1992) and neurosecretory fibers terminate near the pituitary cells (Goos, 1978). The development of the median eminence is correlated with the dramatic increase in thyroid activity seen during late prometamorphosis. Etkin (1968) suggested that the development of this tissue may be one mechanism by which thyroid activity increases through an increase in the supply of TSH-releasing factor to the pituitary at this time.

C. Development of Neurosecretory Neurons and Role of Thyroid Hormone in Differentiation of Neuroendocrine Centers

The presence of neurosecretory cells in the preoptic region of the tadpole (*Xenopus*) hypothalamus was shown in very early stages of premetamorphosis [Nieuwkoop and Faber, 1956 (NF) stage 50] using pseudoisocyanin (PIC) staining (Goos *et al.,* 1968a). Cells first appear in the dorsal part of the preoptic region and continue to develop throughout pre- and prometamorphosis; cells in the ventral region develop later (NF stage 54).

Thyroid hormone plays a central role in the neural development of amphibians (see Kollros, 1981; Gona *et al.,* 1988). Several investigators have shown that chemical or surgical thyroidectomy results in a failure in development of hypothalamic neurosecretory centers (reviewed by Goos, 1978). Neurosecretory cells in the PON develop in *Xenopus* tadpoles in parallel with the development of the thyroid follicles (Goos, 1978). Treatment of *Xenopus* tadpoles with the goitrogen propylthiouracil (PTU) at NF stage 45 results in the failure of PIC-positive cells to develop in the ventral region of the PON, but the dorsal cells do develop (Goos *et al.,* 1968a); these effects can be reversed by treatment with thyroxine (Goos *et al.,* 1968b). The dependence of the development of hypothalamic monoaminergic neurons on TH has been shown by Kikuyama *et al.* (1979). In neotenic tiger salamanders, injection of T_4

into the hypothalamus induced metamorphosis while injection of the same dose intraperitoneally did not (Norris and Gern, 1976), thus supporting a role for TH in the maturation of the hypothalamohypophyseal axis.

Several investigators have shown that the median eminence is dependent on TH for its development (Voitkevich, 1962; Etkin, 1965a,b; Srebro, 1962). Data of Etkin (1965a,b) suggest that the sensitivity of this tissue to TH does not develop until prometamorphosis, long after other tissues have acquired sensitivity. However, this may reflect a lag in the system resulting from the need for neurons in the PON to project axons to the region of the future median eminence. It is noteworthy that neurosecretory neurons in the PON that are dependent on TH for their development appear earlier than the median eminence (see Goos *et al.*, 1968a). Thus, TH not only stimulates the growth and differentiation of the hypothalamic neurosecretory centers controlling TSH production, but also the structure which conveys thyrotropin-releasing factor (TRF) to the hypophysis.

III. ROLES OF PITUITARY HORMONES IN METAMORPHOSIS

A large body of data now supports a key role for the pituitary gland in controlling the progression of metamorphosis and metamorphic climax in most amphibian species (see Dodd and Dodd, 1976; White and Nicoll, 1981; Kikuyama *et al.*, 1993a). For instance, hypophysectomy induces metamorphic stasis, which is reversed by administering pituitary extract (see Dodd and Dodd, 1976).

A. Thyrotropin

The primary pituitary hormone which regulates thyroid gland secretion is TSH, a glycoprotein composed of two subunits, the α and β. The α subunit is common among the glycoprotein hormones, whereas the β subunit confers hormonal specificity on the molecule (Pierce and Parsons, 1981). Measures of TSH bioactivity (see Dodd and Dodd, 1976) and immunocytochemical analyses of pituitaries using antiserum to human TSHβ (this antiserum cross-reacts with amphibian TSH; Garcia-Navarro *et al.,* 1988; Tanaka *et al.,* 1991) have shown its presence in the pituitary during early stages of premetamorphosis. However, because of the lack of an amphibian TSH radioimmunoassay, no direct measures of circulating levels of TSH or description of secretory dynamics have been possible. The TSH α and β subunits of *Xenopus* were cloned and found to have 60 to 70% sequence similarity to other vertebrate TSH subunits (Buckbinder and Brown, 1993). As expected from changes in thyroid activity during metamorphosis and bioassay results of changes in thyrotropic activity in the pituitary gland of metamorphosing tadpoles (see Dodd and Dodd, 1976), the expression of mRNAs of both subunits increased up to metamorphic climax and then declined (Buckbinder and Brown, 1993) (see Fig. 1).

B. Adrenocorticotropin

The amphibian pituitary gland also produces ACTH, a 39 amino acid peptide that controls the production of interrenal steroids (Yasuda *et al.,* 1989; see Kikuyama *et al.,* 1993a). While interrenal activity increases during metamorphosis and injections of ACTH increase serum corticosteroids and accelerate T_4-induced metamorphosis (see White and Nicoll, 1981; Kikuyama *et al.,* 1993a), there is at present no direct evidence for increased ACTH synthesis and secretion during metamorphosis or information on hormonal modulators of ACTH secretion in tadpoles.

C. Prolactin and Growth Hormone

Pituitary hormones of the PRL/GH family have been implicated in the negative control of amphibian development (through antagonism of the actions of TH) and in the positive control of growth. The effects of GH and PRL on amphibian growth and development have been extensively reviewed (Dodd and Dodd, 1976; White and Nicoll, 1981; Kikuyama *et al.,* 1993a). Treatment of tadpoles or cultured tissues with mammalian PRL/GHs blocks the actions of endogenous and exogenous TH and promotes the growth of larval tissues (Dodd and Dodd, 1976; White and Nicoll, 1981; Tata, 1993; Kikuyama *et al.,* 1993). Mammalian GH has some similar and some dissimilar actions to mammalian PRL, generally being less potent in stimulating the growth of larval structures and blocking metamorphosis, but being more potent in stimulating the growth of adult structures (i.e., limb growth) (Dodd and Dodd, 1976; White and Nicoll, 1981; Delidow, 1989; Kikuyama *et al.,* 1993a). Because of the possibility of heterologous hormone effects of the mammalian hormone preparations and the lack of sufficient quantities of homologous amphibian GHs and PRLs to conduct similar tests, clarification of the different roles that these two hormones might play during development has been difficult. However, based on limited studies with amphibian PRL preparations, it appears that the homologous PRL has similar effects to the mammalian hormones (Kikuyama *et al.,* 1993a); purified frog GH has not been available in sufficient quantities to do similar tests on tadpole growth, although GH preparations from amphibian pituitaries do stimulate the growth of postmetamorphic frogs (Kikuyama *et al.,* 1993a). A possible mechanism for the antimetamorphic action of PRL/GH may be through the blockade of the autoinduction of TH receptor gene expression (Tata, 1993).

While the administration of exogenous PRL and GH may be criticized as having a pharmacological rather than a physiological effect, several investigators have attempted to explain the role of PRL/GH in amphibian development by blocking the secretion or activity of the endogenous hormone. Since the ergot alkaloid drugs (which function as dopamine agonists) inhibit PRL release from cultured adult bullfrog pituitaries (Seki and Kikuyama, 1982), some investigators have argued that the stimulatory effects on metamorphosis of these compounds support a role for endogenous PRL as an antimetamorphic hormone (Platt, 1976; White and Nicoll, 1981). However, the effects of these drugs on PRL production in amphibian larvae have not been shown, and other explanations for the observed effects besides an action on PRL secretion are possible [i.e., central nervous system (CNS) activation of thyrotropic centers, direct stimulation of TSH release, direct activation of TH biosynthesis, etc.]. A more specific way of blocking endogenous PRL bioactivity is by injecting antiserum to the animal's PRL. Three groups have accelerated spontaneous or T_4-induced metamorphic climax by treating tadpoles with anti-PRL serum (anti-frog PRL: Clemons and Nicoll, 1977; Yamamoto and Kikuyama, 1982a; anti-ovine PRL: Eddy and Lipner, 1975); anti-frog GH serum did not influence development rate (Clemens and Nicoll, 1977). While these studies support a physiological role for the endogenous hormone, it is not certain that these antisera blocked endogenous PRL bioactivity (see Clemons and Nicoll, 1977) or that normal circulating levels of PRL are responsible for larval growth (see Yamamoto and Kikuyama, 1982a).

As mentioned earlier, Etkin (1968) proposed that larval growth and metamorphosis are controlled by a balance between TH and PRL, and that the two should show an inverse relationship in their blood concentrations at metamorphic climax. However, while circulating concentrations of TH rise during prometamorphosis and peak at climax (Dodd and Dodd, 1976; White and Nicoll, 1981; Kikuyama *et al.,* 1993a), circulating levels of PRL and pituitary PRL mRNA levels have also been found to rise at late prometamorphosis or at metamorphic climax depending on the species (Clemons and Nicoll, 1977; Yamamoto and Kikuyama, 1982b; Takahashi *et al.,* 1990; Niinuma *et al.,* 1991; Buckbinder and Brown, 1993) (see Fig. 1), thus contradicting the earlier hypothesis of an inverse relationship of the two hormones (Etkin, 1968). The rise in PRL production tends to occur slightly later than the rise in TSH subunit expression and circulating TH (see Buckbinder and Brown, 1993). White and Nicoll (1979) showed that renal PRL binding (receptors?) was low in premetamorphic bullfrog tadpoles and increased during metamorphic climax. These findings have led Buckbinder and Brown (1993) to question the accepted view that PRL functions as an amphibian "juvenile hormone" since the hormone is not expressed at high levels during the time when it should be exerting its biological action on larval growth and metamorphosis. However, the possibility remains that low levels of PRL protein and mRNA that are undetectable by radioimmunoassay and Northern blot, respectively, could be sufficient for maintaining an antimetamorphic/larval growth-promoting activity during premetamorphosis.

The rise in PRL production during metamorphic climax appears to be regulated by TH. Treatment of thyroidectomized bullfrog tadpoles with T_4 resulted in a rise in plasma PRL levels at late climax, and pituitary stalk transection experiments support an influence of T_4 on PRL production at the level of the hypothalamus (Kawamura *et al.,* 1986). Buckbinder and Brown (1993) also showed a rise in pituitary PRL mRNA following the treatment of early prometamorphic *Xenopus* tadpoles with 3,5,3′-triiodothyronine (T_3) and the virtual absence of PRL mRNA following treatment with the goitrogen methimazole. Giant tadpoles following thyroidectomy or pituitary transplantation were thought to result from the elevated production of PRL (see White and Nicoll, 1981; Kawamura *et al.,* 1986). However, Kawamura *et al.* (1986) showed that thyroidectomy did not result in hypersecretion of PRL. Given the dependence of PRL on TH for expression, PRL may not be responsible for the abnormal growth of these animals. In this regard, it is noteworthy that Buckbinder and Brown (1993) demonstrated negative regulation by TH of pituitary GH mRNA in *Xenopus* tadpoles [negative feedback on GH production has also been shown in reptiles and birds (Licht and Denver, 1990; Denver and Harvey, 1991)]. Also, GH mRNA is expressed earlier in *Xenopus* development than PRL mRNA [see Fig. 1; however, in ranid frogs, GH production is low during premetamorphosis and gradually rises up to the time of metamorphic climax (see Kikuyama *et al.*, 1993a)], and these findings led Buckbinder and Brown (1993) to suggest that GH might be the larval growth factor in *Xenopus* and that PRL might play a role in frogs. This hypothesis has not yet been tested by physiological means in *Xenopus.*

The function of the late prometamorphic rise in PRL production has yet to be explained, but some have suggested that it may function in osmoregulation (see Norris, 1989). While some have proposed that the low level of circulating PRL during early premetamorphosis may be sufficient to maintain larval growth (see Kikuyama *et al.,* 1993a), the function of the surge in PRL production at climax may be to modulate the rate of metamorphosis in the face of elevated levels of TH, e.g., to act as a "brake" on the action of high levels of TH. Early treatment with high concentrations of TH results in the precocious development of hind limbs and the reorganization of larval structures into adult, but the process is not completely normal and death generally ensues (R. J. Denver, unpublished observations; see Gudernatsch, 1912). As was shown by Etkin (1935), only very low levels of TH can accelerate development without causing death. Although T_3 induces PRL production, the level of endogenous PRL induced by T_3 treatment (which causes death) may not be sufficient to counteract the effects of high T_3 levels. This hypothesis awaits a rigorous test.

IV. ROLE OF HYPOTHALAMUS IN CONTROLLING PITUITARY SECRETION DURING METAMORPHOSIS

The involvement of the hypothalamus in controlling amphibian metamorphosis is well established; autografting the pituitary gland to the tail, posterior hypothalectomy, or the placement of an impermeable barrier between the hypothalamus and the pituitary results in metamorphic stasis (see Dodd and Dodd, 1976; Goos, 1978; White and Nicoll, 1981;

Norris, 1989; Kikuyama *et al.*, 1993a). The only species where the involvement of hypothalamic control is less clear-cut is *B. japonicus* (for a discussion see Kikuyama *et al.*, 1993a).

Location of Thyrotropin-Releasing Factor Producing Cells in Tadpole Hypothalamus

Classical surgical ablation and lesioning studies, with metamorphic stasis as the indicator of the removal of TRF nuclei, have localized the TRF-producing regions of the hypothalamus to the dorsal magnocellular neurons of the PON (Voitkevich, 1962; see Goos, 1978; Dodd and Dodd, 1976) (see Fig. 2). Lesions just posterior to the PON can also block metamorphosis, presumably by disrupting the preoptic–hypophyseal nerve tract (see Dodd *et al.*, 1971). Goos (1978) has pointed out that the amount of neurosecretory material (PIC-positive) in the PON is highly correlated with the activity of the pituitary thyrotropes and thyroid gland (based on histological observations). The neurons of the PON first appear at the time the thyroid follicles differentiate and continue to increase in size up to the time of metamorphic climax (see Goos, 1978). Finally, the PON is the major site of production of neurohormones identified as having thyrotropic and metamorphosis-accelerating activity (see Section V,B).

V. NEUROHORMONES CONTROLLING PITUITARY SECRETION DURING METAMORPHOSIS

The secretion of hormones by the adenohypophysis during metamorphosis is controlled by specific release and release-inhibiting factors produced by modified neurons in the hypothalamus and released into the pituitary portal circulation. Several attempts have been made since the 1970s to identify candidate molecules for the amphibian thyrotropin-releasing factor. Only recently has this search produced a candidate peptide with TSH-releasing activity in tadpoles.

A. Thyrotropin-Releasing Hormone

In mammals the principal stimulatory hypothalamic regulator of TSH synthesis and secretion is the tripeptide amide, thyrotropin-releasing

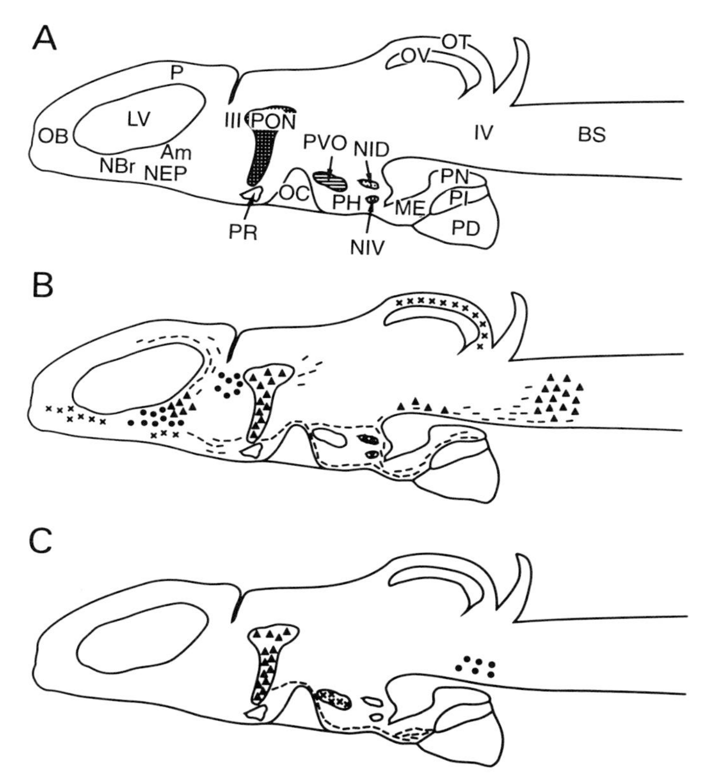

Fig. 2. Distribution of TRH and CRF neurons in tadpole brain. (A) Location of major nuclei and brain regions in generalized tadpole brain. Am, amygdala; BS, brain stem; LV, lateral ventricle; ME, median eminence; NBr, nucleus of diagonal band of Broca; NEP, nucleus entopedunclularis; NID, nucleus infundibularis dorsalis; NIV, nucleus infundibularis ventralis; OB, olfactory bulb; OC, optic chiasma; OT, otic tectum; OV, otic tectum ventricle; P, dorsal medial pallium; PD, pars distalis; PH, preoptic-hypophyseal tract; PI, pars intermedia; PON, preoptic nucleus; PN, pars nervosa; PR, preoptic recess; PVO, paraventricular organ; III, third ventricle; IV, fourth ventricle: (B) Distribution of TRH in tadpole brain. (C) Distribution of CRF in tadpole brain. (Perikarya located by immunocytochemistry (IC) only (●), perikarya located by *in situ* hybridization histochemistry (ISHH) only (×), perikarya located by both IC and ISHH (▲), and nerve fibers located by IC (−). [Based on data from Carr and Norris (1990), Gonzalez and Lederis (1988), Mimnagh *et al.* (1987), Olivereau *et al.* (1987), Seki *et al.* (1983), Stenzel-Poore *et al.* (1992), Verhaert *et al.* (1984), and Zoeller and Conway (1989). Drawings by Leif Saul.]

hormone (TRH) (reviewed by Morley, 1981). This neuropeptide was the first hypophysiotropin to be isolated and later synthesized, and its role as a physiological regulator of mammalian TSH secretion has been established. TRH binds to specific receptors on pituitary thyrotrope cells, stimulates the *in vitro* and *in vivo* synthesis and release of TSH, and passive immunization with TRH antiserum greatly reduces blood levels of TSH (see Morley, 1981). TRH has also been shown to regulate the thyroid system in birds (Sharp and Klandorf, 1985).

Although the TRH molecule has been localized in the hypothalamus of representatives of all vertebrate classes, its role as a TSH-releasing hormone has been controversial. TRH was thought to be inactive in stimulating the thyroid system in lower vertebrates (see Licht and Denver, 1990). On the other hand, TRH is widely distributed in the nervous system of all vertebrates studied, and so several investigators suggested that its primitive function was as a neurotransmitter (Jackson and Bolaffi, 1983) and that its role as a TSH-releasing hormone arose in parallel and convergently with the evolution of homeothermy (Sawin *et al.,* 1981; Millar *et al.,* 1983). Later work has contradicted this hypothesis by showing that the pituitary of reptiles (Licht and Denver, 1990) and amphibians (summarized later) can respond to TRH by releasing TSH and that this action in amphibians varies with the stage of development.

The earlier conclusion that TRH did not control the thyroid axis of amphibians was based on the failure of exogenous TRH to elicit metamorphosis in anuran tadpoles. Etkin and Gona (1968) injected large quantities of TRH (1 mg/animal/day) into tadpoles but failed to accelerate metamorphic climax. Since that time, many have tried and failed to stimulate tadpole metamorphosis with TRH (see Ball, 1981; Jackson and Bolaffi, 1983; Norris, 1983; Millar *et al.,* 1983) (Fig. 3) despite demonstrations of elevations in brain content of the hormone during metamorphosis (Millar *et al.,* 1983). The lack of effect of TRH on metamorphosis is not because of rapid clearance of the hormone since larval blood does not degrade TRH, but blood from metamorphosed animals does (Taurog *et al.,* 1974). Later studies by Darras and Kuhn (1982) showed that an intravenous infusion of TRH in adult frogs elevated circulating TH levels, and Denver (1988) showed that TRH can act directly on the adult frog pituitary to stimulate the release of thyrotropic bioactivity (TSH). Histological analyses of the adult frog pituitary gland following exposure to TRH *in vivo* or *in vitro* showed clear activation of the thyrotropes (Malagon *et al.,* 1991; Castano *et al.,* 1992). However, *in vitro* studies by Denver and Licht (1989) showed that TRH is not active

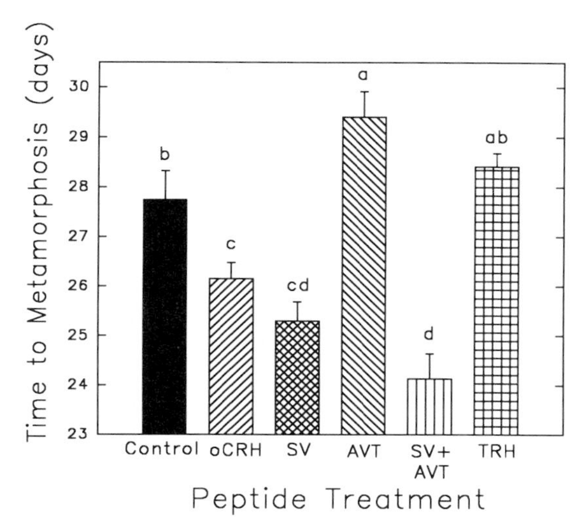

Fig. 3. Mean time to metamorphosis (front limb emergence; Gosner stage 42) in *Scaphiopus hammondii* tadpoles injected with neuropeptides (2 μg/animal/day every other day for 3 weeks). Vertical lines represent SEM and the letters separate the means based on Duncan's multiple range test ($P < 0.05$; $n = 8$ to 22/treatment). [Reprinted from Denver (1993) with permission from Academic Press.]

in stimulating TSH release from tadpole pituitaries, thus confirming the lack of effect of TRH on TSH in tadpole metamorphosis. These data suggest that there are developmental changes in TSH responsiveness to TRH in amphibians, with the tripeptide becoming active as a TRF after metamorphic climax (see Denver and Licht, 1989). TRH elevates circulating T_4 when injected into metamorphosed but not neotenic axolotls, suggesting a developmental switch in TRH action in urodeles also (Darras and Kuhn, 1983; Jacobs and Kuhn, 1987). This lack of effect of TRH may be due to the down-regulation or inactivation of TRH receptors on the larval thyrotropes; TRH receptors are downregulated by TH in both mammals and birds (Hinkle and Goh, 1982; Harvey and Baidwan, 1990), and it may be that the high plasma levels of TH at late prometamorphosis are responsible for the down-regulation of TRH receptors.

Besides regulating TSH in adult amphibians, TRH is a potent stimulator of pituitary PRL secretion (see White and Nicoll, 1981; Norris, 1989; Kikuyama *et al.,* 1993a). Work from the laboratory of Kikuyama has shown that TRH is the major PRL-releasing factor in the adult bullfrog hypothalamus (Seki *et al.,* 1988; Nakajima *et al.,* 1993). Norris (1989) suggested that the primary endocrine function of TRH in metamorphosis may be to induce the surge in circulating PRL levels that occur at climax. At present, pituitary PRL responses to TRH in larval amphibians have not been studied, despite the availability of homologous PRL RIAs (see Kikuyama *et al.,* 1993a).

Immunocytochemical and RIA studies have shown that TRH immunoreactivity is widely distributed in the tadpole and adult amphibian brain (Jackson and Bolaffi, 1983; Mimnagh *et al.,* 1987; Seki *et al.,* 1983) (see Fig. 2). TRH levels in the frog (*R. pipiens*) hypothalamus were 6–12 times greater than in the rat; quantitatively, much more TRH lies in other brain regions outside of the hypothalamic area of the frog, supporting the view that TRH functions as a central neurotransmitter (Jackson and Bolaffi, 1983). The area of highest concentration of TRH perikarya in the tadpole and frog brain is the preoptic region (Mimnagh *et al.,* 1987; Seki *et al.,* 1983). The distribution of TRH mRNA in adult *Xenopus* brain as determined by *in situ* hybridization histochemistry more or less parallels that of the peptide as determined by immunocytochemistry in ranid species except for certain brain areas such as the otic tectum, where only TRH mRNA has been detected (Zoeller and Conway, 1989) (see Fig. 2). This could be due to species differences, as the anatomical distribution of TRH immunoreactivity has not been described in *Xenopus* brain. Concentrations of TRH in whole brain extracts of *Xenopus* tadpoles increased progressively during prometamorphosis up to the time of metamorphic climax (Millar *et al.,* 1983). The onset of the rise in brain TRH levels was correlated with the first rise in plasma TH levels (Millar *et al.,* 1983; Leloup and Buscaglia, 1977); however, as discussed earlier, data suggest that TRH is not responsible for controlling the thyroid axis of tadpoles.

B. Corticotropin-Releasing Factor

Corticotropin-releasing factor (CRF) is a 41 amino acid straight chain polypeptide that was first isolated from ovine hypothalamus and shown to be a potent stimulator of pituitary adrenocorticotropin (ACTH) secretion in mammals (Vale *et al.,* 1981); this is the primary regulatory action of CRF on the pituitary gland of mammals (see Vale, 1992). CRF also stimulates ACTH secretion by adult frog pituitaries in organ culture

(Tonon *et al.*, 1986; Kikuyama *et al.*, 1993a) and by dispersed frog pituitary cells (Gracia-Navarro *et al.*, 1992).

Studies on TSH secretory control in adult frogs have shown that CRF acts directly on the pituitary to stimulate the release of TSH [as measured by changes in levels of thyrotropic bioactivity released by pituitaries exposed to CRF *in vitro* (Denver, 1988; Jacobs and Kuhn, 1992)]. Also, intravenous injections of CRF elevate circulating TH levels in adult frogs (Jacobs and Kuhn, 1989). Morphometric studies of the adult frog pituitary gland showed clear activation of the thyrotrope cells by CRF (Malagon *et al.*, 1991). Pituitaries of bullfrog tadpoles respond to CRF by releasing TSH *in vitro* whereas TRH is inactive (Denver and Licht, 1989). Similar tests in other nonmammalian vertebrates for which a TSH RIA is available showed that CRF is active in stimulating TSH secretion (turtles: see Licht and Denver, 1990; salmon: D. Larsen and P. Swanson, personal communication). Injections of CRF also elevate TH levels in chick embryos (Meeuwis *et al.*, 1989). Thus, it appears that a primitive role of the CRF peptide may be as both a thyrotropin- and a corticotropin-releasing factor.

Two forms of the *Xenopus* CRF gene have been cloned and shown to share more than 93% sequence similarity with the mammalian CRFs but only 50% similarity with sauvagine (SV; a 39 amino acid CRF-like peptide first isolated from the skin of the frog *Phylomedusa sauvageii*) (Montecucchi and Henschen, 1981). Using heterologous antisera, CRF immunoreactive perikarya and nerve fibers have been identified in the frog brain (Verhaert *et al.*, 1984; Olivereau *et al.*, 1987; Gonzalez and Lederis, 1988; Carr and Norris, 1990). Sauvagine immunoreactivity has also been shown in the adult frog hypothalamus (Gonzalez and Lederis, 1988). Most of the CRF and SV immunoreactivity is localized in the preoptic area, with very few CRF perikarya in extrahypothalamic regions (see Fig. 2). A similar distribution pattern was found by *in situ* hybridization histochemistry of adult *Xenopus* brain using a fragment of the cloned *Xenopus* CRF gene as a probe (Stenzel-Poore *et al.*, 1992). The pattern of CRF distribution contrasts sharply with that of TRH; the latter is much more widely distributed in the amphibian brain (see Fig. 2). The presence of CRF immunoreactivity in the median eminence increases significantly during prometamorphosis and peaks at metamorphic climax (Carr and Norris, 1990), suggesting that the peptide is available to function as a hypophysiotropin during metamorphosis.

The demonstration of the stimulation of tadpole TSH secretion by Denver and Licht (1989) was the first disovery of a hypophysiotropin that was capable of stimulating TSH in the tadpole. Since CRF is known to stimulate ACTH secretion in frogs, Denver and Licht (1989) suggested that CRF may act as a common neuroregulator of the thyroid and interrenal axes in metamorphosis. The next logical step was to test whether administration of CRF to tadpoles could accelerate metamorphosis. The first attempt to accelerate metamorphosis by injecting CRF was reported by Gancedo *et al.* (1992). They showed that injections of either ovine (oCRF) or human CRF could accelerate metamorphosis of *R. perezi* larvae. They also showed that both acute responses to a single injection (2 or 8 hr) and chronic treatment with CRF resulted in significant elevations of whole body T_3 and T_4. These results were corroborated and extended by Denver (1993) who showed that injections of both oCRF and SV accelerated metamorphosis in two anuran species, *Scaphiopus hammondii* (Western spadefoot toad; see Fig. 3) and *R. catesbeiana* (North American bullfrog). In contrast to the CRF peptides, TRH had no effect on development (see Fig. 3). In *S. hammondii* tadpoles, injections of both oCRF and SV produced acute (2 or 6 hr) rises in whole body thyroxine, whereas TRH injections tended to reduce whole body T_4. Moreover, Denver (1993) showed that passive immunization of bullfrog tadpoles with anti-CRF serum decelerated metamorphosis, whereas normal serum had no effect. Similar results have been obtained in *S. hammondii* with several anti-CRF sera and a CRF peptide antagonist (α-helical CRF) (Rivier *et al.*, 1984; R. J. Denver, unpublished observations). Taken together, these findings suggest the possibility that the endogenous CRF peptide functions as a larval amphibian TRF and plays a central role in the control of metamorphosis. Given the sensitivity of CRF neuronal activity to external environmental influences (e.g., stress) (Rivier *et al.,* 1986), the secretion of CRF may be one mechanism by which tadpoles accelerate development in response to a deteriorating larval habitat (i.e., an environmental stress) (Newman, 1992; Denver, 1993).

C. Other Neuropeptides

1. Gonadotropin-Releasing Hormone

In addition to regulating the gonadal axis in amphibians, GnRH has been shown to elevate circulating T_4 levels in axolotls and frogs (Jacobs and Kuhn, 1988; Jacobs *et al.,* 1988) and to act by directly stimulating the release of thyrotropic bioactivity from the pituitary gland of adult frogs (Denver, 1988). This action of GnRH was presumed to be through the stimulation of TSH release since highly purified bullfrog luteinizing hormone (LH) and follicle-stimulating hormone (FSH) had very low thyrotropic activity in anuran species compared to purified bullfrog TSH (MacKenzie *et al.*, 1978). MacKenzie *et al.* (1978) claimed that the minimal thyrotropic activity present in the LH and FSH preparations could be accounted for by contamination with TSH. LH and FSH exhibited high potency in gonadotropic bioassays (Licht and Papkoff, 1974). The conclusion that GnRH is stimulating TSH release in the frog has been questioned by Sakai *et al.* (1991) who showed that their bullfrog LH preparation, which they claim is of a high degree of purity, exhibited 10 to 40% of the thyrotropic activity of their bullfrog TSH. This group is confident that their TSH preparation will be suitable for the development of a TSH radioimmunoassay, and measurement of TSH release from frog pituitaries stimulated with GnRH should resolve this issue.

Gonadotropin-releasing hormone immunoreactive fibers and radioimmunoassayable GnRH appear in the median eminence just before metamorphic climax (ranid frogs: Crim, 1984; Muske and Moore, 1990; *Xenopus:* Millar *et al.,* 1983). Using *in situ* hybridization histochemistry and immunocytochemistry for the mammalian form of GnRH, Hayes *et al.* (1994) confirmed the earlier reports that GnRH is not expressed until late in metamorphosis. While GnRH is active in stimulating the thyroid axis in adult frogs (see earlier discussion), it has not yet been tested for activity in stimulating the tadpole thyroid axis or in accelerating metamorphosis. However, because of its late expression pattern in the tadpole hypothalamus, it is unlikely that GnRH is responsible for stimulating the steady rise in circulating TH seen during metamorphosis, although it could possibly contribute to the peak in thyroid activity seen at climax.

2. *Somatostatin*

Inhibitory control of TSH secretion by the neuropeptide somatostatin (SRIF) has been shown in mammals (reviewed by Reichlin, 1986) and reptiles (reviewed by Licht and Denver, 1990). Brain concentrations of radioimmunoassayable SRIF increased in *Xenopus* tadpoles during metamorphosis, suggesting a potential role for this peptide in modulating endocrine activity during development (Millar *et al.,* 1983). However, no data exist on the actions of SRIF on metamorphosis nor on the biosynthesis and secretion of pituitary hormones in amphibians.

3. Arginine Vasotocin

Arginine vasopressin (AVP) has been shown to stimulate TSH secretion and to interact with CRF to stimulate ACTH secretion in mammals (Rivier and Plotsky, 1986; Reichlin, 1986); AVP is also a potent stimulator of ACTH secretion by frog pituitaries *in vitro* (Tonon *et al.,* 1986). The major amphibian neurohypophyseal nonapeptide is arginine vasotocin (AVT; Acher, 1974), and immunoreactive AVT in the median eminence of bullfrog tadpoles increases during metamorphosis (Carr and Norris, 1990). AVT was tested for effects on metamorphosis in spadefoot toad tadpoles, either alone or in combination with CRF-like peptides; however, this peptide had no stimulatory effect on metamorphosis and if anything, given alone, it slowed metamorphosis (see Fig. 3).

VI. NEGATIVE FEEDBACK ON HYPOTHALAMOHYPOPHYSEAL AXIS

Negative feedback of TH on larval amphibian pituitary thyrotropes is suggested by data derived from classical histological techniques. Treatment of tadpoles with goitrogens induced enlargement of the thyroid and degranulation of thyrotropes (Goos *et al.,* 1968a; Goos, 1968; see Dodd and Dodd, 1976), while replacement with T_4 restored granules to the TSH cells, suggesting negative feedback of T_4 on TSH (Goos *et al.,* 1968b; Kurabuchi *et al.,* 1992). Kaye (1961) showed that negative feedback can operate from a very early stage of development—tadpoles of *R. pipiens* treated with TH at TK stage III exhibited depressed ^{131}I uptake by their thyroids. Buckbinder and Brown (1993) demonstrated elevated TSHβ mRNA in the pituitary of a single tadpole made hypothyroid with methimazole, suggesting negative feedback of TH on amphibian TSHβ gene expression; TSHα was not affected by thyroid status. Direct negative feedback of TH on TSH secretion has been shown for two species of adult frogs using the *in vitro* pituitary culture and thyroid bioassay system described by Denver (1988; *R. pipiens:* S. Pavgi, personal communication; *R. ridibunda:* Jacobs and Kuhn, 1992). No data supporting direct negative feedback of TH on tadpole TSH production are now available.

While TH induces growth and differentiation of neurosecretory structures in the tadpole hypothalamus, it can also exert a negative feedback action on hypothalamic neurons. Negative feedback of TH on CRF neurons is suggested by the reduction of CRF mRNA levels in the hypothalamus of prometamorphic *Xenopus* tadpoles following treatment with T_3 for 24 hr; the level of TH receptor β (TRβ) mRNA was elevated by this same treatment (see Fig. 4). These data lend further support to the hypothesis that CRF functions as a larval amphibian TRF (see earlier discussion), since CRF neurons appear to be part of the negative feedback loop of the tadpole hypothalamohypophyseal–thyroid axis.

VII. SUMMARY AND FUTURE DIRECTIONS FOR RESEARCH

The hypothalamus is undoubtedly the central controller of the hormonal changes that drive amphibian metamorphosis. However, whether the hypothalamus controls the initiation of morphogenesis or the rate of morphogenesis once it is proceeding or both has yet to be resolved.

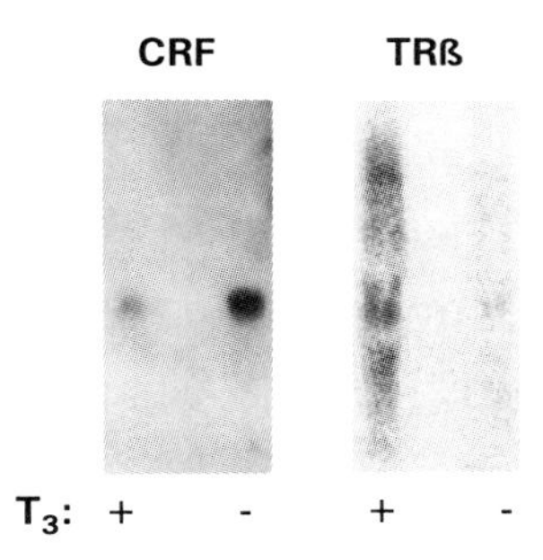

Fig. 4. Thyroid hormone-regulated expression of CRF and thyroid hormone receptor (TR) β mRNAs in *Xenopus* tadpole hypothalamus. Tadpoles were treated with T_3 (5 n*M* dissolved in water) for 24 hr before tissue collection. Total RNA (15 μg/lane) was analyzed by Northern blot using a *Xenopus* ^{32}P-labeled CRF probe (Stenzel-Poore *et al.*, 1992) and a *Xenopus* ^{32}P-labeled TRβ probe (Yaoita *et al.*, 1990). Blots were stripped and reprobed with a ^{32}P-labeled probe for the *Xenopus* ribosomal protein L8 gene as a control for RNA loading (Shi and Liang, 1994); this demonstrated equal amounts of total RNA loaded in each lane.

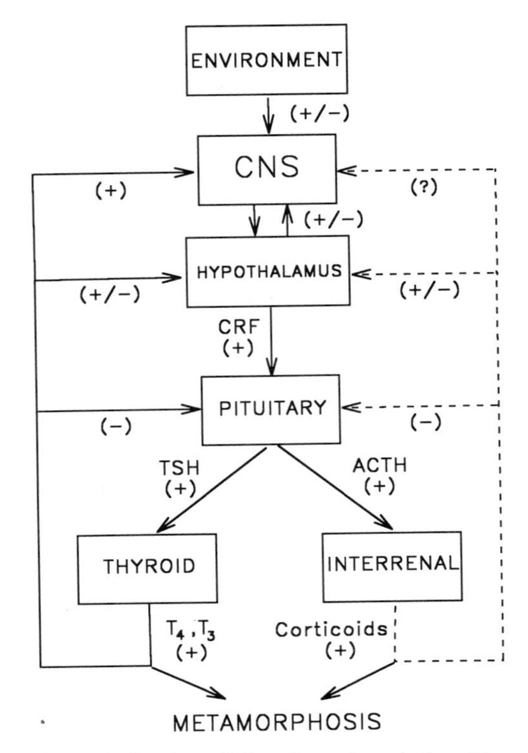

Fig. 5. Integration of the hypothalamohypophyseal–thyroid and hypothalamo-hypophyseal–interrenal axes during amphibian metamorphosis. Solid lines and arrows designate pathways for which substantial evidence exists whereas the dashed lines and arrows are for pathways for which little or no data exist in tadpoles. A plus sign indicates a positive or stimulatory action and a minus sign indicates a negative or inhibitory action. T_4, T_3, and corticoids can exert positive effects on the development of neurons in the hypothalamus and other brain areas and a negative feedback on the HHT and HHI axes; these effects may occur simultaneously (see text). CRF has been shown to be a positive regulator of the thyroid and interrenal axes but other stimulatory neurohormones and perhaps inhibitory factors may exist. Prolactin is not shown, which may control metamorphosis in a negative or inhibitory way (may block actions of thyroid hormone; see text) and which may be controlled by thyrotropin-releasing hormone and possible other neurohormones. CNS, central nervous system; CRF, corticotropin-releasing factor; TSH, thyrotropin; ACTH, adrenocorticotropin; T_4, thyroxine; T_3, 3,5,3′-triiodothyronine.

While it has been known for some time that there are likely to be inductive interactions between the brain and pituitary during development, only recently have the molecular mechanisms of these interactions begun to be explored. Future work using molecular genetic methods should shed light on this issue.

In tadpoles, TH can exert both positive and negative actions on neuroendocrine functions: positive in that they are required for development of the median eminence and hypothalamic neurosecretory neurons and negative in that they inhibit the production of TSH and the neurohormones which control TSH secretion (see Fig. 5). In order to understand the mechanisms by which TH controls the differentiation of hypothalamic neurosecretory centers, the genes that are regulated by TH must be cloned, identified, and their functions explained.

The failure of exogenous TRH to accelerate metamorphosis has promoted several investigators to suggest that the molecule responsible for controlling the thyroid axis of tadpoles is different from the tripeptide TRH. Data from several laboratories support the hypothesis that a CRF-like peptide plays a central role in controlling both the thyroid and the interrenal axes during metamorphosis (Fig. 5). The CRF neuron may also function as a link between the external environment and the endocrine system, and this may allow tadpoles to monitor the quality of the larval habitat and respond accordingly by altering their rate of development. Other neurohormones produced in the tadpole hypothalamus which might potentially regulate TSH and ACTH must be identified in order to form a complete picture of the regulation of pituitary secretion during metamorphosis.

ACKNOWLEDGMENTS

I thank Dr. Sushama Pavgi and the editors for critical comments on the manuscript. Dr. Mary P. Stenzel-Poore and Dr. Yun-Bo Shi kindly supplied the *Xenopus* CRF and TRβ clones, respectively. Drs. Jean Rivier and Wylie Vale very generously supplied synthetic neuropeptides and antipeptide sera used in some of the studies described in this chapter. Illustrations were prepared by Dr. Leif Saul. This work was supported in part by NSF Grant IBN 92-19211.

REFERENCES

Acher, R. (1974). Chemistry of the neurohypophysial hormones: An example of molecular evolution. *In* "Handbook of Physiology," (J. Field, ed) Vol. 16, pp. 119–130. Williams & Wilkins, Baltimore.

Atkinson, B. G. (1994). Metamorphosis: Model systems for studying gene expression in postembryonic development. *Dev. Genet.* **15,** 313–319.

Ball, J. N. (1981). Hypothalamic control of the pars distalis in fishes, amphibians, and reptiles. *Gen. Comp. Endocrinol.* **44,** 135–170.

Begeot, M., Dubois, M. P., and Dubois, P. M. (1982). Comparative study *in vivo* and *in vitro* of the differentiation of immunoreactive corticotropic cells in fetal rat anterior pituitary. *Neuroendocrinology* **35,** 255–264.

Begeot, M., Morel, G., Rivest, R. W., Aubert, M. L., Dubois, M. P., and Dubois, P. M. (1984). Influence of gonadoliberin on the differentiation of rat gonadotrophs: An *in vivo* and *in vitro* study. *Neuroendocrinology* **38,** 217–225.

Blount, R. F. (1945). The interrelationship of the parts of the hypophysis in development. *J. Exp. Zool.* **100,** 79–101.

Buckbinder, L., and Brown, D. D. (1993). Expression of the *Xenopus laevis* prolactin and thyrotropin genes during metamorphosis. *Proc. Natl. Acad. Sci. U.S.A.* **90,** 3820–3824.

Carr, J. A., and Norris, D. O. (1990). Immunohistochemical localization of corticotropin-releasing factor–and arginine vasotocin-like immunoreactivities in the brain and pituitary of the American bullfrog (*Rana catesbeiana*) during development and metamorphosis. *Gen. Comp. Endocrinol.* **78,** 180–188.

Castano, J.-P., Ramirez, J.-L., Malagon, M. M., and Gracia-Navarro, F. (1992). Differential response to amphibian PRL and TSH pituitary cells to *in vitro* TRH treatment. *Gen. Comp. Endocrinol.* **88,** 178–187.

Chang, C. R. (1957). Hypothalectomy in *Rana pipiens* neurulae. *Anat. Rec.* **128,** 531.

Clemons, G. K., and Nicoll, C. S. (1977). Effects of antisera to bullfrog prolactin and growth hormone on metamorphosis of *Rana catesbeiana* tadpoles. *Gen. Comp. Endocrinol.* **31,** 495–497.

Crim, J. W. (1984). Immunocytochemistry of luteinizing hormone-releasing hormone in brains of bullfrogs (*Rana catesbeiana*) during spontaneous metamorphosis. *J. Exp. Zool.* **229,** 327–337.

Darras, V. M., and Kuhn, E. R. (1982). Increased plasma levels of thyroid hormones in a frog *Rana ridibunda* following intravenous administration of TRH. *Gen. Comp. Endocrinol.* **48,** 469–475.

Darras, V. M., and Kuhn, E. R. (1983). Effects of TRH, bovine TSH, and pituitary extracts on thyroidal T4 release in *Ambystoma mexicanum. Gen. Comp. Endocrinol.* **51,** 286–291.

Delidow, B. C. (1989). Reevaluation of the effects of growth hormone and prolactin on anuran tadpole growth and development. *J. Exp. Zool.* **249,** 279–283.

Denver, R. J. (1988). Several hypothalamic peptides stimulate *in vitro* thyrotropin secretion by pituitaries of anuran amphibians. *Gen. Comp. Endocrinol.* **72,** 383–393.

Denver, R. J. (1993). Acceleration of anuran amphibian metamorphosis by corticotropin-releasing hormone–like peptides. *Gen. Comp. Endocrinol.* **91,** 38–51.

Denver, R. J., and Harvey, S. (1991). Thyroidal inhibition of chicken pituitary growth hormone: Alterations in the secretion and accumulation of newly synthesized hormone. *J. Endocrinol.* **131,** 39–48.

Denver, R. J., and Licht, P. (1989). Neuropeptide stimulation of thyrotropin secretion in the larval bullfrog: Evidence for a common neuroregulator of thyroid and interrenal activity during metamorphosis. *J. Exp. Zool.* **252,** 101–104.

Dodd, M. H. I., and Dodd, J. M. (1976). The biology of metamorphosis. *In* "Physiology of the Amphibia" (B. Lofts, ed.), Vol. 3, pp. 467–599. Academic Press, New York.

Dodd, J. M., Follett, B. K., and Sharp, P. J. (1971). Hypothalamic control of pituitary function in submammalian vertebrates. *In* "Advances in Comparative Physiology and Biochemistry" (O. Lowenstein, ed.), Vol. 4, pp. 113–223. Academic Press, New York.

Driscoll, W. T., and Eakin, R. M. (1955). The effects of sucrose on amphibian development with special reference to the pituitary body. *J. Exp. Zool.* **129,** 149–175.

Drolet, D. W., Scully, K. M., Simmons, D. M., Wegner, M., Chu, K. T., Swanson, L. W., and Rosenfeld, M. G. (1991). TEF, a transcription factor expressed specifically in the anterior pituitary during embryogenesis, defines a new class of leucine zipper proteins. *Genes Dev.* **5,** 1739–1753.

Eagleson, G. W., Jenks, B. G., and van Overbeeke, P. (1986). The pituitary adrenocorticotropes originate from neural ridge tissue in *Xenopus laevis. J. Embryol. Exp. Morphol.* **95,** 1–14.

Eakin, R. M. (1956). Differentiation of the transplanted and explanted hypophysis of amphibian embryo. *J. Exp. Zool.* **131,** 263–296.

Eddy, L., and Lipner, H. (1975). Acceleration of thyroxine-induced metamorphosis by prolactin antiserum. *Gen. Comp. Endocrinol.* **25,** 462–466.

Etkin, W. (1935). The mechanisms of anuran metamorphosis. I. Thyroxine concentration and the metamorphic pattern. *J. Exp. Zool.* **71,** 317–340.

Etkin, W. (1965a). The phenomenon of amphibian metamorphosis. IV. The development of the median eminence. *J. Morphol.* **116,** 371–378.

Etkin, W. (1965b). Hypothalamic sensitivity to thyroid feedback in the tadpole. *Neuroendocrinology* **1,** 293–302.

Etkin, W. (1967). Relation of the pars intermedia to the hypothalamus. *In* "Neuroendocrinology" (L. Martini and W. F. Ganong, eds.), Vol. 2, pp. 261–282. Academic Press, New York.

Etkin, W. (1968). Hormonal control of amphibian metamorphosis. *In* "Metamorphosis: A Problem in Developmental Biology" (W. Etkin and L. I. Gilbert, eds.), pp. 313–348. Appleton-Century-Crofts, New York.

Etkin, W., and Gona, A. G. (1968). Failure of mammalian thyrotropin releasing factor preparation to elicit metamorphic response in tadpoles. *Endocrinology* (*Baltimore*) **82,** 1067–1068.

Etkin, W., Kikuyama, S., and Rosenbluth, J. (1965). Thyroid feedback to the hypothalamic neurosecretory system in frog larvae. *Neuroendocrinology* **1,** 45–64.

Gancedo, B., Corpas, I., Alonso-Gomez, A. L., Delgado, M. J., Morreale de Escobar, G., and Alonso-Bedate, M. (1992). Corticotropin-releasing factor stimulates metamorphosis and increases thyroid hormone concentration in prometamorphic *Rana perezi* larvae. *Gen. Comp. Endocrinol.* **87,** 6–13.

Garcia-Navarro, S., Malgon, M. M., and Gracia-Navarro, F. (1988). Immunohistochemical localization of thyrotropic cells during amphibian morphogenesis: A stereological study. *Gen. Comp. Endocrinol.* **71,** 116–123.

Gona, A. G., Uray, N. J., and Hauser, K. F. (1988). Neurogenesis of the frog cerebellum. *In* "Developmental Neurobiology of the Frog" (E. D. Pollack and H. D. Bibb, eds.), pp. 255–276. Liss, New York.

Gonzalez, G. C., and Lederis, K. (1988). Sauvagine-like and corticotropin-releasing factor–like immunoreactivity in the brain of the bullfrog (*Rana catesbeiana*). *Cell Tissue Res.* **253,** 29–37.

Goos, H. J. T. (1968). Hypothalamic neurosecretion and metamorphosis in *Xenopus laevis.* III. The effect of an interruption of thyroid hormone synthesis. *Z. Zellforsch.* **92,** 583–587.

Goos, H. J. T. (1978). Hypophysiotropic centers in the brain of amphibians and fish. *Am. Zool.* **18,** 401–410.

Goos, H. J. T., De Knecht, A.-M., and De Vries, J. (1968a). Hypothalamic neurosecretion and metamorphosis in *Xenopus laevis.* I. The effect of propylthiouracil. *Z. Zellforsch.* **86,** 384–392.

Goos, H. J. T., Zwanebeek, H. C. M., and van Oordt, P. G. W. J. (1968b). Hypothalamic neurosecretion and metamorphosis. II. The effect of thyroxine following treatment with propylthiouracil. *Arch. Anat. Embryol.* **51,** 268–274.

Gracia-Navarro, F., Lamacz, M., Tonon, M.-C., and Vaudry, H. (1992). Pituitary adenylate cyclase-activating polypeptide stimulates calcium mobilization in amphibian pituitary cells. *Endocrinology (Baltimore)* **131,** 1069–1074.

Gudernatsch, J. F. (1912). Feeding experiments on tadpoles. I. The influence of specific organs given as food on growth and differentiation. A contribution to the knowledge of organs with internal secretion. *Wilhelm Roux Arch. Entwicklungsmech. Org.* **35,** 457–483.

Hanoaka, Y. (1967). Effect of phenylthiourea upon posterior hypothalectomized tadpoles of *Rana pipiens. Gen. Comp. Endocrinol.* **9,** 24–30.

Harvey, S., and Baidwan, J. S. (1990). Thyroidal inhibition of growth hormone secretion in fowl: Tri-iodothyronine–induced down-regulation of thyrotrophin-releasing hormone–binding sites on pituitary membranes. *J. Mol. Endocrinol* **4,** 127–134.

Hayes, W. P., and Loh, Y. P. (1990). Correlated onset and patterning of propiomelanocortin gene expression in embryonic *Xenopus* brain and pituitary. *Development* **110,** 747–757.

Hayes, W. P., Wray, S., and Battey, J. F. (1994). The frog gonadotropin-releasing hormone–I (GnRH-I) gene has a mammalian-like expression pattern and conserved domains in GnRH-associated peptide, but brain onset is delayed until metamorphosis. *Endocrinology (Baltimore)* **134,** 1835–1845.

Hinkle, P. M., and Goh, K. B. C. (1982). Regulation of thyrotropin-releasing hormone receptors and responses by L-triiodothyronine in dispersed rat pituitary cell cultures. *Endocrinology (Baltimore)* **110,** 1725–1731.

Jackson, I. M. D., and Bolaffi, J. L. (1983). Phylogenetic distribution of TRH: Significance and function. *In* "Thyrotropin-Releasing Hormone." (E. C. Griffiths and G. W. Bennett, eds.), pp. 191–202. Raven, New York.

Jacobs, G. F. M., and Kuhn, E. R. (1987). TRH injection induces thyroxine release in the metamorphosed but not in the neotenic axolotl *Ambystoma mexicanum. Gen. Comp. Endocrinol.* **66,** 40.

Jacobs, G. F. M., and Kuhn, E. R. (1988). Luteinizing hormone-releasing hormone induces thyroxine release together with testosterone in the neotenic axolotl *Ambystoma mexicanum. Gen. Comp. Endocrinol.* **71,** 502–505.

Jacobs, G. F. M., and Kuhn, E. R. (1989). Thyroid function may be controlled by several hypothalamic factors in frogs and at least by one in the neotenic axolotl. *Abstr. Int. Symp. Comp. Endocrinol., 11th* p. 174.

Jacobs, G. F. M., and Kuhn, E. R. (1992). Thyroid hormone feedback regulation of the secretion of bioactive thyrotropin in the frog. *Gen. Comp. Endocrinol.* **88,** 415–423.

Jacobs, G. F. M., Goyvaerts, M. P., Vandorpe, G., Quaghebeur, A. M. L., and Kuhn, E. R. (1988). Luteinizing hormone-releasing hormone as a potent stimulator of the thyroidal axis in ranid frogs. *Gen. Comp. Endocrinol.* **70,** 274–283.

Kawamura, K., and Kikuyama, S. (1995). Induction from posterior hypothalamus is essential for the development of the pituitary proopiomelanocortin (POMC) cells of the toad (*Bufo japonicus*). *Cell Tissue Res.* **279,** 233–239.

Kawamura, K., and Kikuyama, S. (1992). Evidence that hypophysis and hypothalamus constitute a single entity from the primary stage of histogenesis. *Development* **115,** 1–9.
Kawamura, K., Yamamoto, K., and Kikuyama, S. (1986). Effects of thyroid hormone, stalk section, and transplantation of the pituitary gland on plasma prolactin levels at metamorphic climax in *Rana catesbeiana. Gen. Comp. Endocrinol.* **64,** 129–135.
Kaye, N. W. (1961). Interrelationships of the thyroid and pituitary in embryonic and premetamorphic stages of the frog. *Rana pipiens. Gen. Comp. Endocrinol.* **1,** 1–19.
Kikuyama, S., Miyakawa, M., and Arai, Y. (1979). Influence of thyroid hormone on the development of preoptic–hypothalamic monoaminergic neurons in tadpoles of *Bufo bufo japonicus. Cell Tissue Res.* **198,** 27–33.
Kikuyama, S., Kawamura, K., Tanaka, S., and Yamamoto, K. (1993a). Aspects of amphibian metamorphosis: Hormonal control. *Int. Rev. Cytol.* **145,** 105–148.
Kikuyama, S., Inaco, H., Jenks, B. G., and Kawamura, K. (1993b). Development of the ectopically transplanted primordium of epithelial hypophysis (anterior neural ridge) in *Bufo japonicus* embryos. *J. Exp. Zool.* **266,** 216–220.
Kollros, J. J. (1981). Transitions in the nervous system during amphibian metamorphosis. *In* "Metamorphosis: A Problem in Developmental Biology" (L. I. Gilbert and E. Frieden, eds.), 2nd ed, pp. 445–459. Plenum, New York.
Kurabuchi, S., Tanaka, S., and Kikuyama, S. (1992). Development of TSH cells in larval toad, *bufo japonicus* during metamorphosis and their response to thyroxine and thiourea. *Proc. Int. Symp. Amphib. Endocrinol. Waseda Univ.* p. 26 (abstract).
Leloup, J., and Buscaglia, M. (1977). La triiodothyronine, hormone de la métamorphose des Amphibiens. *C. R. Acad. Sci. Ser. D.* **284,** 2261–2263.
Licht, P., and Denver, R. J. (1990). Regulation of thyrotropin secretion. *In* "Progress in Comparative Endocrinology" (A. Epple, C. G. Scanes, and M. H. Stetson, eds.), pp. 427–432. Wiley–Liss, New York.
Licht, P., and Papkoff, H. (1974). Separation of two distinct gonadotropins from the pituitary gland of the bullfrog *Rana catesbeiana. Endocrinology* (*Baltimore*) **94,** 1587–1594.
MacKenzie, D. S., Licht, P., and Papkoff, H. (1978). Thyrotropin from amphibian (*Rana catesbeiana*) pituitaries and evidence for heterothyrotropic activity of bullfrog luteinizing hormone in reptiles. *Gen. Comp. Endocrinol.* **36,** 566–574.
Malagon, M. M., Ruiz-Navarro, A., Torronteras, R., and Gracia-Navarro, F. (1991). Effects of ovine CRF on amphibian pituitary ACTH and TSH cells *in vivo:* A quantitative ultrastructural study. *Gen. Comp. Endocrinol.* **83,** 487–497.
Meeuwis, R., Michielsen, R., Decuypere, E., and Kuhn, E. R. (1989). Thyrotropic activity of the ovine corticotropin-releasing factor in the chick embryo. *Gen. Comp. Endocrinol.* **76,** 357–363.
Millar, R. P., Nicolson, S., King, J. A., and Louw, G. N. (1983). Functional significance of TRH in metamorphosing and adult anurans. *In* "Thyrotropin-Releasing Hormone" (E. C. Griffiths and G. W. Bennett, eds.), pp. 217–227. Raven, New York.
Mimnagh, K. M., Bolaffi, J. L., Montgomery, N. M., and Kaltenbach, J. C. (1987). Thyrotropin-releasing hormone (TRH): Immunohistochemical distribution in tadpole and frog brain. *Gen. Comp. Endocrinol.* **66,** 394–404.
Montecucchi, P. C., and Henschen, A. (1981). Amino acid composition and sequence analysis of sauvagine, a new active peptide from the skin of *Phyllomedusa sauvagei. Int. J. Pept. Protein Res.* **18,** 113–120.
Moriceau-Hay, D., Doerr-Schott, J., and Dubois, M. P. (1982). Immunohistochemical demonstration of TSH-, LH- and ACTH-cells in the hypophysis of tadpoles of *Xenopus laevis* Daudin. *Cell Tissue Res.* **225,** 57–64.

Morley, J. E. (1981). Neuroendocrine control of thyrotropin secretion. *Endocr. Rev.* **2,** 396–436.

Muske, L. E., and Moore, F. L. (1990). Ontogeny of immunoreactive GnRH neuronal systems in amphibians. *Brain Res.* **534,** 177–187.

Nakajima, K., Uchida, D., Sakai, M., Takahashi, N., Yanagisawa, T., Yamamoto, K., and Kikuyama, S. (1993). Thyrotropin-releasing hormone (TRH) is the major prolactin-releasing factor in the bullfrog hypothalamus. *Gen. Comp. Endocrinol.* **89,** 11–16.

Newman, R. A. (1992). Adaptive plasticity in amphibian metamorphosis. *Biosciences* **42,** 671–678.

Nieuwkoop, P. D., and Faber, J. (1956). *In* "Normal Table of *Xenopus laevis* Daudin." North Holland Publishers, Amsterdam.

Niinuma, K., Yamamoto, K., and Kikuyama, S. (1991). Changes in plasma and pituitary prolactin levels in toad (*bufo japonicus*) larvae during metamorphosis. *Zool. Sci.* **8,** 97–101.

Norris, D. O. (1983). Evolution of endocrine regulation of metamorphosis in lower vertebrates. *Am. Zool.* **23,** 709–718.

Norris, D. O. (1989). Neuroendocrine aspects of amphibian metamorphosis. *In* "Development, Maturation and Senescence of Neuroendocrine Systems: A Comparative Approach" (C. G. Scanes and M. P. Schreibman, eds.), pp. 63–90. Academic Press, San Diego.

Norris, D. O., and Gern, W. A. (1976). Thyroxine-induced activation of hypothalamo-hypophysial axis in neotenic salamander larvae. *Science* **194,** 525–527.

Nyholm, N. E. I., and Doerr-Schott, J. (1977). Developmental immunohistology of melanotrophs in *Xenopus laevis* tadpoles. *Cell Tissue Res.* **180,** 231–239.

Olivereau, M., Vandesande, F., Boucique, E., Ollevier, F., and Oliveau, J. M. (1987). Immunocytochemical localization and spatial relation to the adenohypophysis of a somatostatin-like and a corticotropin-releasing factor–like peptide in the brain of four amphibian species. *Cell Tiss. Res.* **247,** 317–324.

Pehlemann, F.-W. (1962). Experimentelle Untersuchungen zur Determination und Differenzierung der Hypophyse bei Anuren (*Pelobates fuscus, Rana esculenta*). *Wilhelm Roux Arch. Entwicklungsmech. Org.* **153,** 551–602.

Pierce, J. G., and Parsons, T. F. (1981). Glycoprotein hormones: Structure and function. *Annu. Rev. Physiol.* **50,** 465–495.

Platt, J. E. (1976). The effects of ergocornine on tail height, spontaneous and T4-induced metamorphosis and thyroidal uptake of radioiodide in neotenic *Ambystoma tigrinum*. *Gen. Comp. Endocrinol.* **28,** 71–81.

Rathke, H. (1838). Uber die Entstehung der Glandula Pituitaria. *Arch. Anat. Physiol. Wiss. Med.* 482–485.

Reichlin, S. (1986). Neuroendocrine control of thyrotropin secretion. *In* "Werner's The Thyroid: A Fundamental and Clinical Text" (S. H. Ingbar and L. E. Braverman, eds.), 5th ed., pp. 241–266. Lippincott, Philadelphia.

Rivier, C. L., and Plotsky, P. M. (1986). Mediation by corticotropin releasing factor (CRF) of adenohypophysial hormone secretion. *Annu. Rev. Physiol.* **48,** 475–494.

Rivier, J., Rivier, C., and Vale, W. (1984). Synthetic competitive antagonists of corticotropin-releasing factor: Effect on ACTH secretion in the rat. *Science* **224,** 889–891.

Rivier, C., Rivier, J., and Vale, W. (1986). Stress-induced inhibition of reproductive functions: Role of endogenous corticotropin-releasing factor. *Science* **231,** 607–609.

Sakai, M., Hanaoka, Y., Tanaka, S., Hayashi, H., and Kikuyama, S. (1991). Thyrotropic activity of various adenohypophyseal hormones of the bullfrog. *Zool. Sci.* **8,** 929–934.

Sawin, C. T., Bacharach, P., and Lance, V. (1981). Thyrotropin-releasing hormone and thyrotropin in the control of thyroid function in the turtle, *Chrysemys picta*. *Gen. Comp. Endocrinol.* **45,** 7–11.

Scott, M. P., Tamkun, J. W., and Hartzell, G. W., III (1989). The structure and function of the homeodomain. *Biochim. Biophys. Acta* **989,** 25–48.

Seki, T., and Kikuyama, S. (1982). *In vitro* studies on the regulation of prolactin secretion in the bullfrog pituitary gland. *Gen. Comp. Endocrinol.* **46,** 473–479.

Seki, T., Nakai, Y., Shioda, S., Mitsuma, T., and Kikuyama, S. (1983). Distribution of immunoreactive thyrotropin-releasing hormone in the forebrain and hypophysis of the bullfrog, *Rana catesbeiana*. *Cell Tissue Res.* **233,** 507–516.

Seki, T., Kikuyama, S., and Suzuki, M. (1988). Effect of hypothalamic extract on the prolactin release from the bullfrog pituitary gland with special reference to thyrotropin-releasing hormone (TRH). *Zool. Sci.* **5,** 407–413.

Sharp, P. J., and Klandorf, H. (1985). Environmental and physiological factors controlling thyroid function in galliformes. *In* "The Endocrine System and the Environment" (B. K. Follett, S. Ishii, and A. Chandola, eds.), pp. 175–188. Japan Scientific Society Press, Tokyo/Springer-Verlag, Berlin.

Shi, Y.-B. (1994). Molecular biology of amphibian metamorphosis. A new approach to an old problem. *Trends Endocrinol. Metab.* **5,** 14–20.

Shi, Y.-B., and Liang, V. C.-T. (1994). Cloning and characterization of the ribosomal protein L8 gene from *Xenopus laevis*. *Biochim. Biophys. Acta* **1217,** 227–228.

Srebro, Z. (1962). Neurosecretory activity in the brain of adult *Xenopus laevis* and during metamorphosis. *Folia Biol.* (*Krakow*) **10,** 93–111.

Stenzel-Poore, M. P., Heldwein, K. A., Stenzel, P., Lee, S., and Vale, W. W. (1992). Characterization of the genomic corticotropin-releasing factor (CRF) gene from *Xenopus laevis:* Two members of the CRF family exist in amphibians. *Mol. Endocrinol.* **6,** 1716–1724.

Taira, M., Hayes, W. P., Otani, H., and Dawid, I. B. (1993). Expression of LIM class homeobox gene *Xlim-3* in *Xenopus* development is limited to neural and neuroendocrine tissues. *Dev. Biol.* **159,** 245–256.

Takahashi, N., Yoshihama, K., Kikuyama, S., Yamamoto, K., Wakabayashi, K., and Kato, Y. (1990). Molecular cloning and nucleotide sequence analysis of complementary DNA for bullfrog prolactin. *J. Mol. Endocrinol.* **5,** 281–287.

Tanaka, S., Sakai, M., Park, M. K., and Kurosumi, K. (1991). Differential appearance of the subunits of glycoprotein hormones (LH, FSH, and TSH) in the pituitary of bullfrog (*Rana catesbeiana*) larvae during metamorphosis. *Gen. Comp. Endocrinol.* **84,** 318–327.

Tata, J. R. (1993). Gene expression during metamorphosis: An ideal model for post-embryonic development. *BioEssays* **15,** 239–248.

Taurog, A., Oliver, C., Eskay, R. L., Porter, J. C., and McKenzie, J. M. (1974). The role of TRH in the neoteny of the Mexican axolotl (*Ambystoma mexicanum*). *Gen. Comp. Endocrinol.* **24,** 267–279.

Thurmond, W., and Eakin, R. M. (1959). Implantation of the amphibian adenohypophysial anlage into albino larvae. *J. Exp. Zool.* **140,** 145–167.

Tonon, M.-C., Cuet, P., Lamacz, M., Jegou, S., Cote, J., Gouteux, L., Ling, N., Pelletier, G., and Vaudry, H. (1986). Comparative effects of corticotropin-releasing factor, arginien vasopressin, and related neuropeptides on the secretion of ACTH and α-

MSH by frog anterior pituitary cells and neurointermediate lobes *in vitro. Gen. Comp. Endocrinol.* **61,** 438–445.

Vale, W. W. (1992). Introduction to the symposium. *In* "Corticotropin-Releasing Factor" (D. J. Chadwick, J. Marsh, and K. Ackrill, eds.), pp. 1–5. Wiley, New York.

Vale, W., Speiss, J., Rivier, C., and Rivier, J. (1981). Characterization of a 41-residue ovine hypothalamic peptide that stimulates corticotropin and β-endorphin. *Science* **213,** 1394–1397.

Van Noorden, S., Polak, J. M., Robinson, M., Pearse, A. G. E., and Marangos, P. J. (1984). Neuron-specific enolase in the pituitary gland. *Neuroendocrinology* **38,** 309–316.

Verhaert, P., Marivoet, S., Vandesande, F., and De Loof, A. (1984). Localization of CRF immunoreactivity in the central nervous system of three vertebrate and one insect species. *Cell Tissue Res.* **238,** 49–53.

Voitkevich, A. A. (1962). Neurosecretory control of the amphibian metamorphosis. *Gen. Comp. Endocrinol., Suppl.* **1,** 133–147.

Voss, J. W., and Rosenfeld, M. G. (1992). Anterior pituitary development: Short tales from dwarf mice. *Cell (Cambridge, Mass.)* **70,** 527–530.

Watanabe, Y. G. (1982a). Effects of brain and mesenchyme upon the cytogenesis of rat adenohypophysis *in vitro.* I. Differentiation of adrenocorticotropes. *Cell Tissue Res.* **227,** 257–266.

Watanabe, Y. G. (1982b). An organ culture study on the site of determination of ACTH and LH cells in the rat adenohypophysis. *Cell Tissue Res.* **227,** 267–275.

Watanabe, Y. G. (1985). Effects of brain and mesenchyme upon the cytogenesis of rat adenohypophysis *in vitro.* II. Differentiation of LH cells. *Cell Tissue Res.* **242,** 49–55.

White, B. A., and Nicoll, C. S. (1979). Prolactin receptors in *Rana catesbeiana* during development and metamorphosis. *Science* **204,** 851–853.

White, B. A., and Nicoll, C. S. (1981). Hormonal control of amphibian metamorphosis. *In* "Metamorphosis: A Problem in Developmental Biology" (L. I. Gilbert and E. Frieden, eds.), 2nd ed., pp. 363–396. Plenum, New York.

Yamamoto, K., and Kikuyama, S. (1982a). Radioimmunoassay of prolactin in plasma of bullfrog tadpoles. *Endocrinol. Jpn.* **29,** 159–167.

Yamamoto, K., and Kikuyama, S. (1982b). Effects of prolactin antiserum on growth and resorption of tadpole tail. *Endocrinol. Jpn.* **29,** 81–85.

Yaoita, Y., Shi, Y. B., and Brown, D. D. (1990). *Xenopus laevis* α and β thyroid hormone receptors. *Proc. Natl. Acad. Sci. U.S.A.* **87,** 7090–7094.

Yasuda, A., Kawauchi, H., and Kikuyama, S. (1989). Isolation and characterization of proopiomelanocortin-related hormone from an amphibian, the bullfrog (*Rana catesbeiana*). *Abstr. Int. Symp. Comp. Endocrinol.,* 11th p. 368.

Zoeller, R. T., and Conway, K. M. (1989). Neurons expressing thyrotropin releasing hormone-like messenger ribonucleic acid are widely distributed in *Xenopus laevis* brain. *Gen. Comp. Endocrinol.* **76,** 139–146.

13

Hormonal Interplay and Thyroid Hormone Receptor Expression during Amphibian Metamorphosis

JAMSHED R. TATA

Laboratory of Developmental Biochemistry
National Institute for Medical Research
The Ridgeway, Mill Hill
London NW7 1AA, United Kingdom

I. INTRODUCTION

In both vertebrates and invertebrates metamorphosis is perhaps the most dramatic example of postembryonic development. Ever since the discovery by Gudernatsch (1912) that feeding thyroid glands to frog tadpoles precociously induced metamorphosis, the importance of thyroid hormones (THs) in vertebrate postembryonic development has been fully realized. With the availability of synthetic thyroid hormones, first L-thyroxine (T_4) and later 3,3′,5-triiodo-L-thyronine (T_3), a detailed picture has emerged since 1945 of the various physiological and biochemical responses of amphibian larval tissues to these hormones during normal and induced metamorphosis (Weber, 1967; Frieden and Just, 1970; Cohen, 1970; Beckingham Smith and Tata, 1976a; White and Nicoll, 1981; Galton, 1983; Atkinson, 1994; see Kaltenbach, Chapter 11, this volume; Atkinson *et al.,* Chapter 15, this volume).

With the recognition of the hypothalamus–pituitary–thyroid axis, the link between the central nervous system (CNS) and the initiation of metamorphosis could be traced to environmental signals operating via hypothalamic and pituitary hormones. To explain the unique kinetics of amphibian metamorphosis, namely a slow initial phase accelerating to a metamorphic climax, Etkin (1964) put forward the novel concept of positive feedback between T_4 and thyrotropic hormone (TSH). This was in marked contrast to the more accepted, general notion of negative feedback loops operating between the pituitary/hypothalamus and the peripheral endocrine glands. Although it explained the kinetics of the process, once initiated, a positive T_4–TSH feedback loop did not explain the timing of onset of metamorphosis or how the progression of metamorphosis was controlled. The discovery of juvenile hormones in invertebrates, their role in regulating metamorphosis, and their modification of responses of larval and pupal tissues to ecdysteroids explained many features of the dynamics of metamorphosis in invertebrates (see Riddiford and Truman, 1978; Gilbert and Frieden, 1981; Gilbert *et al.,* Chapter 2, this volume). An analogous hormonal interplay was not known for amphibian metamorphosis until the discovery in the early 1970s that the pituitary hormone prolactin (PRL) could modulate, or even totally abolish, thyroid hormone-induced metamorphosis in many amphibian species (Nicoll, 1974; White and Nicoll, 1981; see Denver, Chapter 12, this volume). Although amphibian prolactins have been purified (Kikuyama *et al.,* 1993), much of the just-mentioned evidence was based on exogenous mammalian prolactin. Whether endogenous prolactin can be considered as equivalent to juvenile hormone in invertebrates is debatable, it can be a useful tool in exploring the mechanism of action of TH in regulating amphibian metamorphosis.

Since the 1940s, many investigators, attracted by the various aspects of amphibian metamorphosis as a model for unraveling the mechanism of action of thyroid hormone, had built up an impressive repository of information on the morphological and biochemical responses of anuran larval tissues to the hormone (see Cohen, 1970; Weber, 1967; Beckingham Smith and Tata, 1976a; Frieden and Just, 1970; see Kaltenbach, Chapter 11, this volume; Atkinson *et al.,* Chapter 15, this volume). There was, however, a missing element in these studies—the role of thyroid hormone receptors. Since 1980 it has become amply clear that receptors play a central role in hormone action: their genes have been cloned, pure recombinant receptors have been isolated, and their functions analyzed in great detail. These studies have made it possible to study the developmental and hormonal expression of nuclear hormone receptor genes and revealed their importance in the regulatory action of the relevant hormone (Green and Chambon, 1988; Evans, 1988; Parker, 1991, 1993). This chapter focuses attention on the significance of auto- and cross-regulation of thyroid hormone receptor genes in an attempt to understand how these hormones induce morphological and biochemical responses in larval tissues and thus regulate amphibian metamorphosis.

II. MORPHOLOGICAL AND BIOCHEMICAL RESPONSES OF AMPHIBIAN LARVAE TO THYROID HORMONES

A survey of the major morphological and biochemical changes occurring during metamorphosis is helpful in assessing the importance of thyroid hormone receptors and expression in considering their role in metamorphosis.

A. Diversity of Responses

Virtually every tissue in the metamorphosing amphibian larva, especially in anurans, exhibits profound morphological and biochemical changes. All these changes could be induced precociously by exogenous thyroid hormones or, conversely, be prevented from occurring if the thyroid gland was surgically or chemically ablated, in premetamorphic tadpoles (Etkin, 1964; Weber, 1967; Frieden and Just, 1970; Beckingham Smith and Tata, 1976a; Atkinson, 1981, 1994; Tata, 1993; see Kaltenbach, Chapter 11, this volume). In this context, it is important to consider that the biochemical and physiological responses of amphibian larval tissues undergoing metamorphosis do not represent adaptation to environmental changes but occur in anticipation of new demands.

Table I lists some of the well-known responses of anuran tadpole tissues to thyroid hormones during natural and hormonally induced metamorphosis, which emphasizes both their diversity and multiplicity. Indeed, no two tissues exhibit the same hormonal responses, which range from *de novo* morphogenesis to functional reprogramming to total regression. For example, the larval CNS and eyes are major hormonal targets, with a wide variety of changes in its organization and function (Kollros, 1984; Burd, 1990; Hoskins, 1990). During metamorphosis the liver and skin undergo genetic reprogramming that leads to the acquisition of new biochemical characteristics such as the appearance of urea cycle enzymes and synthesis of serum albumin in the liver and collagen deposition and keratinization of the skin (see Atkinson *et al.,* Chapter 15, this volume; Miller, Chapter 17, this volume; Yoshizato, Chapter 19, this volume). Among the most dramatic morphological and biochemical changes are the emergence of limbs and lungs, the loss of tail and gills and the substantial regression, followed by regeneration, of the tadpole intestine (see chapters by Shi, Chapter 14, this volume; Yoshizato, Chapter 19, this volume). It is important to note that a similar wide range of responses is provoked by ecdysteroids in different larval and pupal tissues of invertebrates (Gilbert and Frieden, 1981; see Gilbert *et al.,* Chapter 2, this volume; Willis, Chapter 7, this volume; Truman, Chapter 8, this volume).

B. Thyroid Hormones Acting Directly and Locally

Being lipophilic molecules, thyroid hormones can be applied topically to individual tissues or to different parts of a given tissue of the amphibian tadpole, thus highlighting an important feature of the action of thyroid hormones, namely, that it is not systemic, but direct and local (see Kaltenbach, Chapter 11, this volume).

TABLE I

Diversity of Morphological and Biochemical Responses during Thyroid Hormone-Induced Amphibian (Anuran) Metamorphosis

Tissue	Response	
	Morphological	Biochemical
Brain	Restructuring, axon guidance, axon growth, cell proliferation, and death	Cell division, apoptosis, and new protein synthesis
Liver	Restructuring, functional differentiation	Induction of urea cycle enzymes and albumin; larval → adult hemoglobin gene switching
Eye	Repositioning; new retinal neurons and connections; lens structure	Visual pigment transformation (porphyropsin → rhodopsin); β-crystallin induction
Skin	Restructuring; skin granular gland formation; keratinization and hardening	Induction of collagen, 63-kDa (adult) keratin and magainin genes
Limb bud, lung	*De novo* formation of bone, skin, muscle, nerves, etc.	Cell proliferation and differentiation; chondrogenesis
Tail, gills	Complete regression	Programmed cell death; induction and activation of lytic enzymes (collagenase, nucleases, phosphatases); lysosome proliferation
Pancreas, intestine	Major tissue restructuring	Reprogramming of phenotype acquisition of new digestive functions
Immune system	Redistribution of cell populations	Altered immune system and appearance of new immunocompetent components

More convincing evidence for a nonsystemic action of TH is provided by organ culture experiments, whereby T_4 and T_3 act directly on tadpole tissues such as liver, skin, intestine, limb buds, and tail to initiate the same metamorphic changes as in whole larvae (Tata, 1966; Weber, 1969; Tata *et al.,* 1991; Ishizuya-Oka and Shimozawa, 1991; Helbing and Atkinson, 1994). The culture approach reinforced the view that the same hormonal signal can trigger off quite different postembryonic developmental programs. Thus, exposure of tadpole organ cultures to T_3 causes a genetic reprogramming to induce serum proteins and urea cycle enzymes in liver, keratinization of the skin, morphogenesis of limb buds, and major or total cell loss in intestine and tail. Figure 1 illustrates an example of T_3 acting directly on organ cultures of *Xenopus* tadpole tails and hind limb buds removed from the same animals. Clearly, the hormone has initiated changes in these two larval tissues that are diametrically opposite in their ultimate developmental fate. Another advantage of the culture system is that it allows one to answer such questions as the involvement of protein synthesis, or a particular metabolic pathway, in bringing about a programmed developmental change by using specific inhibitors which would otherwise be toxic to the intact tadpole. This feature is discussed in Section II,D.

C. Thyroid Hormones Setting into Motion Predetermined Genetic Program

Although these culture studies confirmed the direct action of the hormone on individual tissues, one has to reconcile them with the fact that *in vivo* neighboring tissues may possibly influence the response of a given tissue to the hormone. It is therefore significant that early transplantation studies had clearly demonstrated that changing the location of a given organ or tissue did not alter the response to endogenous or exogenous thyroid hormone. For example, the tadpole eye or limb bud underwent the same morphological and biochemical differentiation to the adult form when transplanted into the tail as in their normal location during natural or induced metamorphosis (Etkin, 1964; Kaltenbach, 1953), thus proving that thyroid hormone does not determine the developmental program but only acts to initiate it.

From early studies on the activation of the tadpole's thyroid gland via the pituitary, it emerged that metamorphosis would occur soon after the

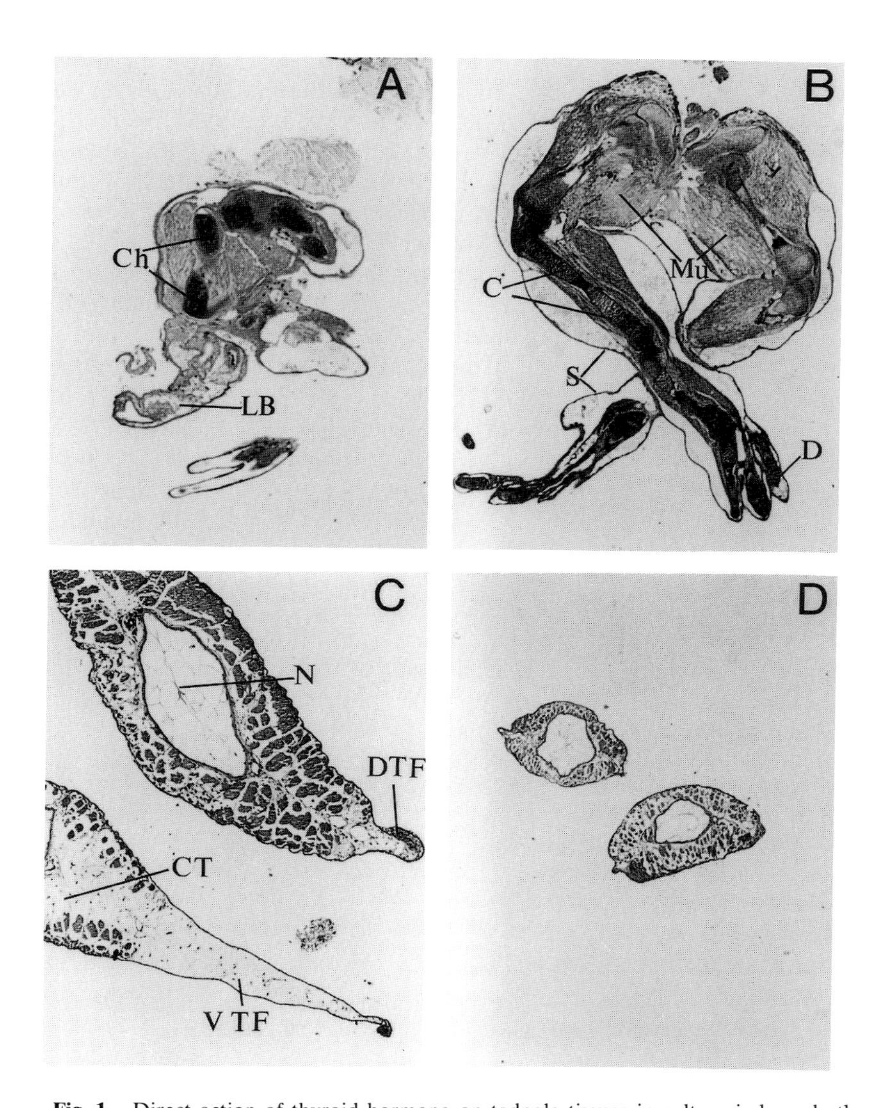

Fig. 1. Direct action of thyroid hormone on tadpole tissues in culture induces both morphogenesis and cell death. Histological sections of hind limb buds (or limbs) and tails, from the same stage 54/55 *Xenopus* tadpoles, after 8 days in culture and treated with 2×10^{-9} *M* T_3. Longitudinal sections through limb buds or limbs: (A) control limb buds and (B) T_3 added after 1 day in culture. Transverse sections through tails; (C) control and (D) T_3 added after 1 day. The limb bud and tail sections are different magnifications, but the control and T_3-treated samples for each are at the same magnification. LB, limb bud; Ch, early chondrogenic cells; C, cartilage; D, digits; Mu, muscle; S, skin; DTF, dorsal tail fin; VTF, ventral tail fin; N, notochord; CT, connective tissue. [For other details see Tata *et al.* (1991).]

secretion of endogenous thyroid hormones (Etkin, 1964; Leloup and Buscaglia, 1977; Galton, 1983). At the same time, ablation of the thyroid gland of premetamorphic tadpoles, or administration of anti-thyroid drugs, followed by exogenous TH, clearly showed that the individual larval tissues were competent to undergo metamorphosis well before the thyroid gland was activated. This raised the question as to how early in development would the different tissues be "ready" to respond to thyroid hormones, a question highly relevant to the developmental expression of thyroid hormone receptors (see Section IV,A).

A systematic study in which different stages of *Xenopus* embryos and tadpoles were exposed to TH, and a number of biochemical and morphological responses measured, showed that the competence to respond to T_3 was established by as early as stage 44 or 45, while clearly exhibiting several responses to T_3 before they reached stage 47 or 48 a few days later (Tata, 1968). The acquisition of this response (Fig. 2) occurs some weeks before the first detection of T_4 or T_3 in blood in stages 53/54 tadpoles (Leloup and Buscaglia, 1977; Tata *et al.,* 1993). The rapid buildup and decline in circulating T_3 temporally matches the onset and completion of natural metamorphosis (Fig. 3). With the application of highly sensitive techniques of molecular biology, such as PCR (polymerase chain reaction) amplification of RNA and *in situ* hybridization techniques, it has now been possible to establish the early competence of response to T_3 in terms of well-defined gene products.

Fig. 2. (i) Major visible morphological changes induced in very early stages of *Xenopus* tadpoles by T_3. (A and C) Control animals on days 10 and 25 after fertilization, respectively. (B and D) Tadpoles at 6 and 21 days after fertilization, respectively, treated with 4×10^{-9} *M* T_3 and photographed 4 days later. Arrows indicate the regression of the tail, the appearance of a beak-like structure, the altered positioning of eyes, and swelling of hind limb buds. (ii) Summary of several responses to T_3 showing the early acquisition of metamorphic competence of *Xenopus* tadpoles. The time on the abscissa refers to be developmental stage when larvae were first exposed to $2\text{–}5 \times 10^{-9}$ *M* T_3 and the measurements were made 4 days later. The responses include the increased rate of synthesis of RNA, DNA, and membrane phospholipids (PL) and net loss of weight (Wt) and ability to concentrate inorganic phosphate $[^{32}P]P_i$. [Reprinted from Tata (1968) with permission from Academic Press.]

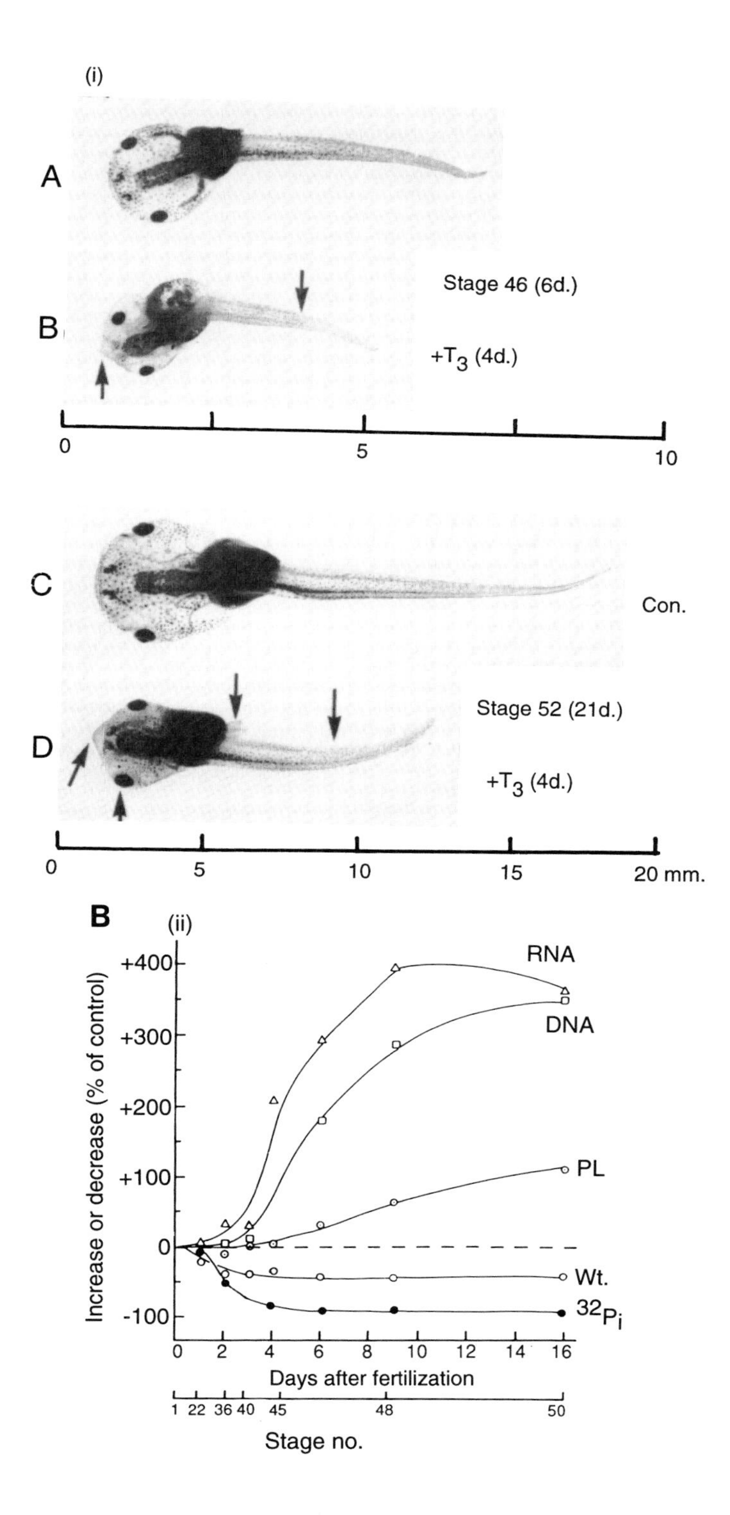

(i)
A
B
Stage 46 (6d.)
+T_3 (4d.)
0
5
10
C
Con.
D
Stage 52 (21d.)
+T_3 (4d.)
0
5
10
15
20 mm.
B
(ii)
Increase or decrease (% of control)
+400
+300
+200
+100
0
-100
RNA
DNA
PL
Wt.
$^{32}P_i$
0 2 4 6 8 10 12 14 16
Days after fertilization
1 22 36 40 45 48 50
Stage no.

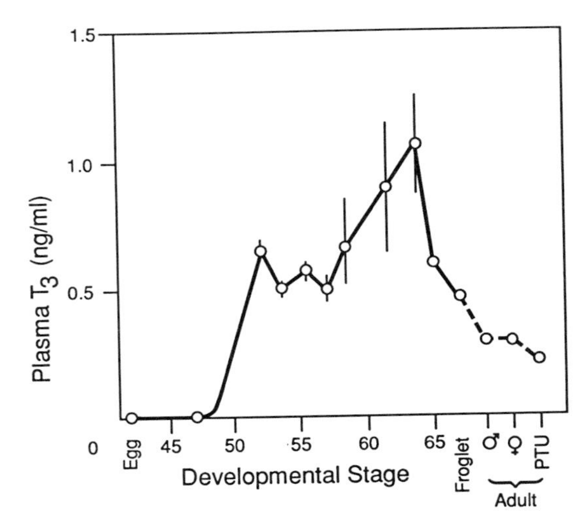

Fig. 3. Plasma T_3 concentration in *Xenopus* tadpoles at different developmental stages and in adult males and females measured from two to six separate assays on batches of 50–230 tadpoles and 2–6 adults. No T_3 was detected in homogenates of eggs or tadpoles up to stage 47. PTU indicates plasma from adult male *Xenopus* treated for 12 days with 90 mg propylthiouracil/liter of water to block their thyroid glands. [Reprinted from *J. Ster. Biochem. Molec. Biol.* **46,** Tata *et al.* Autoinduction of Nuclear Receptor Genes and its Significance, 105–119, Copyright (1993), with kind permission from Elsevier Science Ltd., The Boulevard, Langford Lane, Kidlington OX5 1GB, UK.]

D. Programmed Cell Death: An Important Feature of Amphibian Metamorphosis

Upon the onset of natural or induced metamorphosis the amphibian tadpole shows rapid and substantial loss of cell numbers in many tissues until metamorphosis is completed. The substantial loss of mass and cell number is not restricted to tissues that undergo total regression, such as the tail and gills, but there is also extensive cell death in tissues that are restructured, such as the intestine, pancreas, and the central nervous system. These tissues undergo further development and cell proliferation at the terminal stages of metamorphosis (see Shi, Chapter 14, this volume).

Earlier studies to explain tissue regression during amphibian metamorphosis focused on such processes as infiltration by macrophages, lysosomal expansion, or the increase in activity of such lytic enzymes as proteases, nucleases, and phosphatases (Weber, 1969; Frieden and Just,

1970; Tata, 1971, 1994; Beckingham Smith and Tata, 1976b; Atkinson, 1981; Yoshizato, 1989; see Yoshizato, Chapter 19, this volume). However, T_4 or T_3 cannot directly activate these enzymes in their latent forms, nor do agents known to activate lysosomal enzymes induce cell death or tissue regression characteristic of metamorphosis. It thus raises the possibility that the elevation of lytic enzyme activity by TH was caused by an enhanced, selective synthesis of some proteins in tissues programmed for regression. With the establishment of the technique of TH-induced regression of isolated tadpole tails, it was possible to show that T_3 simultaneously augmented the amount of several lytic enzymes and the protein and RNA synthesizing activity of *Xenopus* tadpole tails in organ culture (Fig. 4A) (Tata, 1966). The availability of inhibitors of RNA and protein synthesis also made it possible to test the idea that tissue regression was dependent on synthesis of new proteins. Thus, several laboratories observed that actinomycin D, puromycin, and cycloheximide prevent T_3 from inducing cultured *Xenopus* tails to regress (Tata, 1966; Weber, 1965; Eeckhout, 1966; Beckingham Smith and Tata, 1976b). As shown in Fig. 4B, actinomycin D produced a paradoxical situation in that an agent which would normally provoke cell death protected the tadpole tail programmed for total regression against the signal initiating cell loss. This phenomenon of the requirement of an active process of new RNA and protein synthesis to induce programmed cell death was also demonstrated in other postembryonic developmental systems (Saunders, 1966; Bowen and Lockshin, 1981; Tomei and Cope, 1991).

As regarding the nature of some "early" proteins that are TH induced and would possibly account for the ensuing regression of the tail, differential protein labeling studies suggested that this indeed may be the case. However, no well-defined proteins have been identified by this approach (Beckingham Smith and Tata, 1976b; Ray and Dent, 1986a). More recently, by exploiting the technique of subtractive hybridization, Brown's laboratory has demonstrated that T_3 rapidly and selectively activates or represses certain genes in the *Xenopus* tadpole tail and intestine, although the nature of their products remains to be determined (Wang and Brown, 1993; Shi and Brown, 1993; see Shi, Chapter 14, this volume). However, the cloning of several cell death and anti-cell death genes, the identification of their products, and their high degree of evolutionary conservation (Tomei and Cope, 1991; Ellis *et al.,* 1991; Dexter *et al.,* 1994) now offer possible candidates for "early" proteins involved in programmed cell death during metamorphosis.

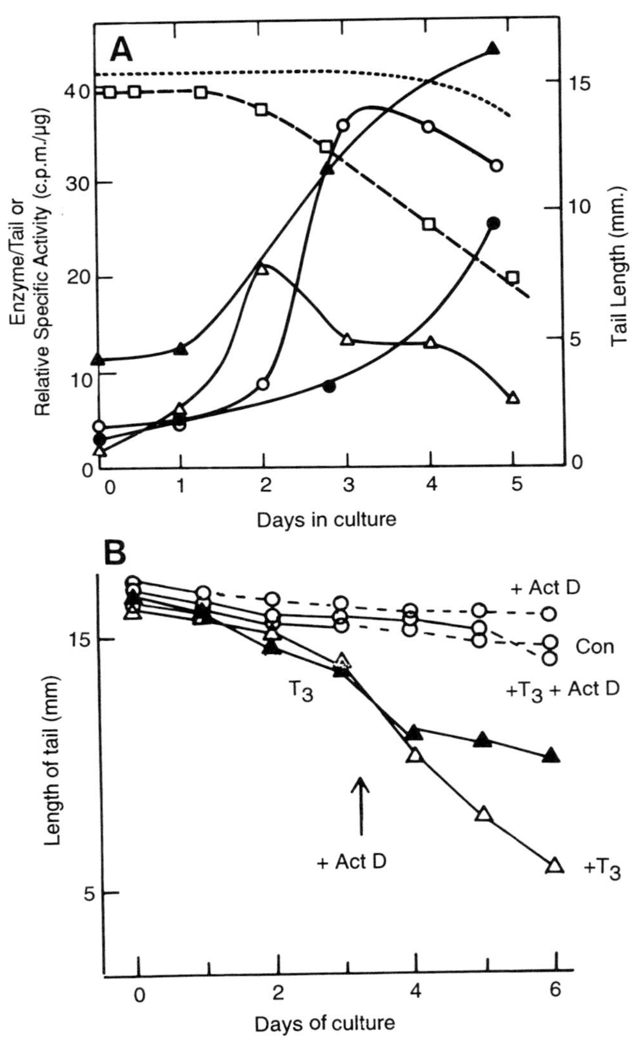

Fig. 4. (A) Accumulation of cathepsin (●) and deoxyribonuclease (▲) and burst of additional RNA and protein synthesis during regression of tadpole tails induced in organ culture with triiodothyronine added to the medium. (○) Incorporation of [^{3}H]uridine into RNA; (△) incorporation of ^{14}C-labeled amino acids into protein. Incorporation on day 0 is the average value for controls; all other points refer to samples to which T_3 was added. (---) Tail length of controls over the duration of the experiment; (□) length of triiodothyronine-treated tails showing a marked onset of regression between the second and third day after culture. [Curves compiled from data of Tata (1966).] (B) Inhibition by actinomycin D of T_3-induced regression of the isolated *Xenopus* tadpole tail. One microgram T_3 and 5 μg actinomycin were present in the medium on day 0 of the culture; the tails were washed free of the medium after different periods of culture and the incubation was continued only with 1 μg T_3 per ml as shown by dashed lines. [Reprinted from Tata (1966) with permission from Academic Press.]

E. Other Hormones Modifying Action of Thyroid Hormones

Although thyroid hormone is the only obligatory signal for the initiation and progression of amphibian metamorphosis, several factors can accelerate or retard the onset and progression of metamorphosis. These include photoperiodicity, neurotransmitters, overcrowding, nutrition, temperature, and other hormones. Best characterized among the latter are glucocorticoid hormone and prolactin. Exogenous CRF (corticotropin releasing factor), ACTH (adrenocorticotropin), and glucocorticoids, are known to accelerate natural metamorphosis in anuran tadpoles, as well as in organ culture (Kikuyama *et al.*, 1993). Both processes of morphogenesis and cell death induced by TH are thus potentiated (see Denver, Chapter 12, this volume).

Many investigators have reported that exogenous mammalian and amphibian PRLs will prevent both natural and thyroid hormone-induced metamorphosis in many different species of amphibian tadpoles (Derby and Etkin, 1968; Nicoll, 1974; White and Nicoll, 1981; Kikuyama *et al.*, 1993; Tata *et al.*, 1991; Baker and Tata, 1992; see Denver, Chapter 12, this volume). More recently, it has been suggested that the action of PRL may be indirect in that a secreted product induced by it in the liver, termed "synlactin" may be the active agent antagonizing the action of TH in tadpoles (see Denver, Chapter 12, this volume). The possibility of the antimetamorphic action of PRL via synlactin or another systemic agent is less likely if one considers that PRL can block TH action *in vitro* when added to organ cultures of tadpole tissues (Derby and Etkin, 1968; Ray and Dent, 1986b; Tata *et al.*, 1991; Baker and Tata, 1992; Kikuyama *et al.*, 1993). As illustrated in Fig. 5, PRL inhibited equally the T_3-induced growth and differentiation of the limb bud and regression of the tail in organ culture (compare with Fig. 1). As regarding the latter tissue, cell death specifically induced by T_3 accounts for its regression and PRL prevents the apoptotic process, as judged by the loss of DNA in cultured tail explants (Fig. 5B).

Unlike the mutual antagonism between juvenile hormone and ecdysteroids in invertebrates (Riddiford and Truman, 1978; Gilbert and Frieden, 1981; see Gilbert *et al.*, Chapter 2, this volume), no clear evidence exists that a similar interplay occurs between endogenous PRL and TH. It has been argued that PRL cannot be considered to be analogous to juvenile hormone in the intact amphibian larva. The argument is largely based on the finding that the PRL protein and mRNA content of the tadpole

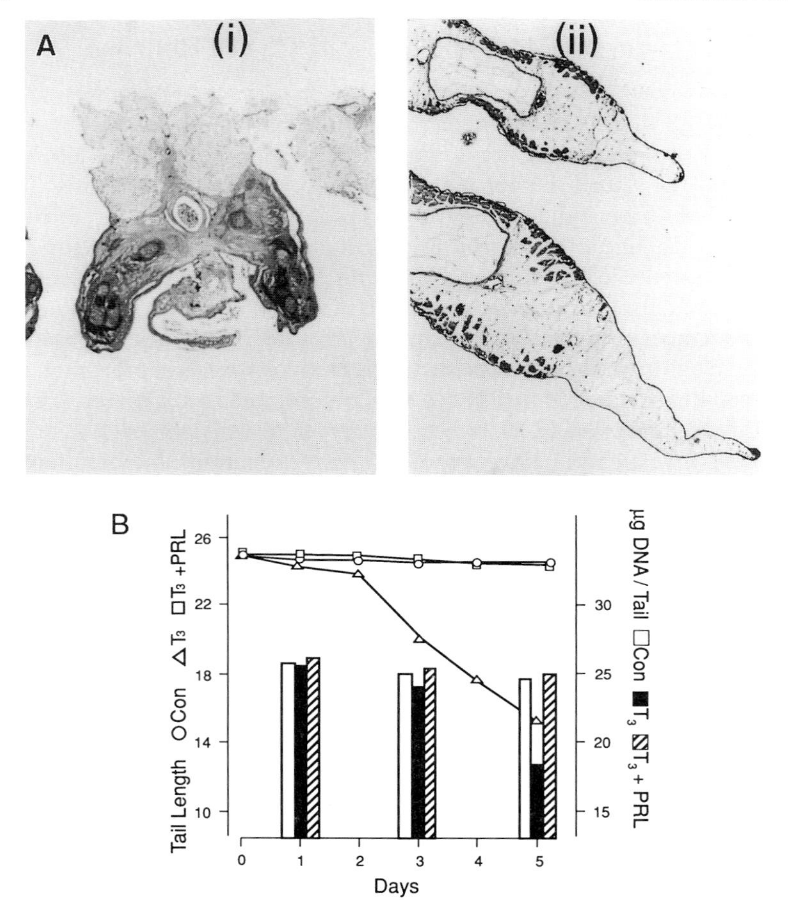

Fig. 5. (A) Prolactin inhibits morphogenesis and regression of *Xenopus* hind limbs (i) and tails (ii) induced by T_3 in organ culture. Data are part of the same experiment and at the same magnification as in Fig. 1. Both limb buds and tails were treated with 2×10^{-9} *M* T_3 in the presence of 0.2 IU of ovine prolactin for 8 days in culture. Other details as in Fig. 1 and Tata *et al.* (1991). (B) Effect of prolactin (PRL) on kinetics of stage 54 *Xenopus* tadpole tail regression (tail length) and cell death (DNA/tail) induced by T_3 in organ culture after 5 days. Geometric symbols refer to tail length whereas vertical bars represent DNA/cell at the end of 5 days of culture. [For other details see Baker and Tata (1992).]

pituitary is low at premetamorphic stages, but elevated during metamorphosis and continues to rise through to adult stages (Buckbinder and Brown, 1993; Kikuyama *et al.,* 1993; Niinuma *et al.,* 1991; see Denver, Chapter 12, this volume). If the pituitary content reflects circulating levels of PRL, it is further argued that, for it to function as a juvenile hormone, PRL levels should be high before metamorphosis and decrease during and after this process. This argument is only valid if PRL has no physiological function in adult amphibia for which a much higher amount of the hormone would be required. Indeed, PRL has multiple physiological actions in a large variety of adult vertebrates—from water drive and osmotic control in amphibia and fish to crop sac growth in birds and lactation in mammals (Bern *et al.,* 1967; Nicoll, 1974; White and Nicoll, 1981; Kikuyama *et al.,* 1993; Baulieu and Kelly, 1990). Furthermore, PRL mRNA and protein are present in the premetamorphic *Xenopus* pituitary at low levels, which may be sufficient to modulate the action of low levels of TH at the onset of metamorphosis. Until more is known about the mechanism of action of PRL, including the nature and function of its receptor, the possibility that prolactin can exert a juvenilizing effect in amphibian development cannot be discounted.

III. THYROID HORMONE RECEPTORS

Although many properties of thyroid hormone receptors were known earlier (Oppenheimer and Samuels, 1983), it was not until the cloning of cellular homolg (c-*erbA*) of the viral oncogene v-erbA from chicken and human tissues (Sap *et al.,* 1986; Weinberger *et al.,* 1986) that the true identity of thyroid hormone receptors was established. The protein products of the cDNAs exhibited similar characteristics as the native TR *in vivo,* and the finding that c-*erbA* was able to mediate the activation of the rat growth hormone gene promoter in the presence of T_3 confirmed its identity. The predicted amino acid sequence of c-*erbA* showed that its structural organization is similar to that of other steroid hormone receptors so that TRs are members of a supergene family of nuclear receptors that act as ligand-inducible transcription factors (Evans, 1988; Green and Chambon, 1988; Parker, 1991, 1993; Chin, 1991; Yen and Chin, 1994).

A. Thyroid Hormone Receptors as Members of Nuclear Steroid/ Thyroid Hormone/Retinoid Multigene Family

In all vertebrates TR exists as two structurally dissimilar receptor isoforms encoded by two genes, termed α and β (Chin, 1991; Chatterjee and Tata, 1992; Yen and Chin, 1994). The transcripts of each of these genes undergo alternate splicing to generate distinct proteins with differing properties and tissue distribution (see Fig. 6). The TRα and β isoforms in all species differ mostly in their N-terminal A/B region whereas the ligand-binding domain D/E/F is almost identical. The middle DNA-binding domain C of TR resembles the same domain in other members of the nuclear steroid/retinoid receptor family, with their highly conserved nine cysteine residues and two "zinc fingers." By taking into account several structural and functional characteristics, the members of this multigene family have been divided into two groups, termed types

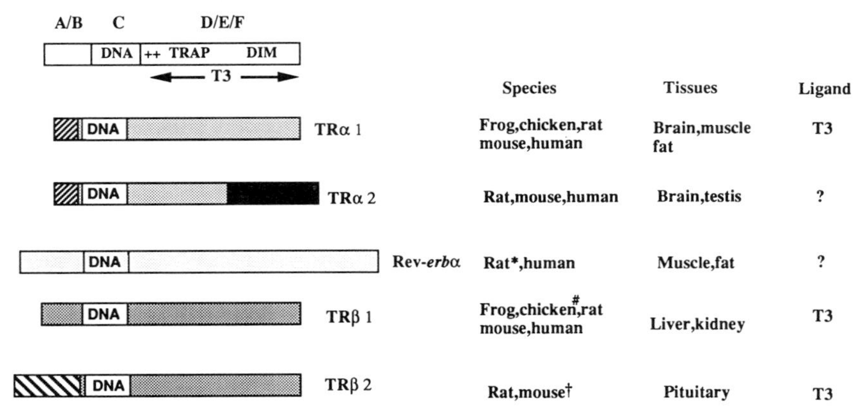

Fig. 6. The structural organization of thyroid hormone receptors and related proteins consists of conserved domains (A to F), which mediate distinct functions such as DNA binding, ligand binding, receptor dimerization (DIM), nuclear localization (++), and interaction with thyroid hormone receptor accessory factor (TRAP). The proteins are aligned via the DNA-binding domains that are most homologous, and regions with divergent sequences are indicated. For each receptor isoform, the putative ligand, tissues in which they are highly expressed, and species from which they have been cloned are indicated. Symbols (asterisk, dagger, number symbol) indicate species in which amino-terminally truncated forms of the proteins shown are expressed. [Reprinted from Chatterjee and Tata (1991), with permission from Cold Spring Harbor Laboratory Press.]

I and II (Stunnenberg, 1993). TR belongs to group II that includes receptors of retinoic acid (RAR), 9-*cis*-retinoic acid (RXR), vitamin D_3 (VDR), and peroxisome proliferators (PPAR).

B. Functional Determinants of Thyroid Hormone Receptors

Unliganded TRs are constitutively located in the nucleus as components of chromatin (Oppenheimer and Samuels, 1983; Chin, 1991; Chatterjee and Tata, 1992). Therefore, unlike the other group of steroid hormone (glucocorticoid, progesterone, and androgen) receptors, the cytoplasmic-nuclear transfer of the liganded receptor is not a major functional determinant. The response elements (TREs) with which the DBD (DNA-binding domain) of TRs (domain C) interact to activate transcriptionally the thyroid hormone target genes have been well characterized; they are present in a wide variety of configuration, depending on the gene, tissue, and species (Chin, 1991; Chatterjee and Tata, 1992; Yen and Chin, 1994). However, a direct repeat separated by four nucleotides (AGGTCAnnnnAGGTCA), termed DR + 4, is considered to be a consensus TRE for most genes that are positvely regulated by thyroid hormones (Umesono *et al.,* 1991; Perlmann *et al.,* 1993). Interestingly, the same direct repeats, but separated by one, two, three, or five nucleotides, act as responsive elements in target genes for receptors for RXR, PPAR, RAR, and VDR, respectively.

The hormone-binding and transactivation functions of TR are present in the D/E/F domains (Fig. 6). The mechanism by which TR regulates gene transcription is not fully understood, but two aspects of this function are worth noting. First, unlike other nuclear receptors, the unliganded TR acts as a transcriptional repressor (Baniahmad *et al.,* 1992; Fondell *et al.,* 1993; Yen *et al.,* 1993). Furthermore, TRα2 and certain forms of truncated receptors or mutations in the ligand-binding domain, as found in human thyroid hormone resistance syndrome (Chatterjee *et al.,* 1991; Lazar, 1993; Yen and Chin, 1994), exert a dominant negative action, just as the viral homolog of the receptor v-*erbA* (Zenke *et al.,* 1990). Such mutant or truncated receptors have substantially or totally lost their ability to bind TH but will prevent the normal liganded TRs from exerting their function. The second feature concerns the suggestion that, although TR monomers and homodimers can bind to TREs, the physiologically active form of the receptor is a heterodimer with retinoid-X receptor

(RXR) whose ligand is 9-*cis*-retinoic acid (Kliewer *et al.,* 1992; Zhang and Pfahl, 1993; Yen and Chin, 1994). Both these features of TRs are relevant to the important role played by thyroid hormones in the regulation of amphibian metamorphosis.

C. Amphibian Thyroid Hormone Receptors

The major forms of TR receptors in *Xenopus* are TRα1 (or TRαA1) or TRβ1 (TRβA). Brooks *et al.* (1989) were the first to isolate from *Xenopus* ovarian cDNA library the amphibian homolog of the mammalian TRα gene. Detailed studies from the laboratory of D. D. Brown have characterized both TRα and β genes of *Xenopus* (Yaoita *et al.,* 1990; Shi *et al.,* 1992; Shi, 1994). They exhibit both similarities and differences between mammalian and amphibian TR genes and their products. *Xenopus* TRα and β each exist as two genes. The two TRα genes are very similar to their mammalian counterparts but no alternatively spliced TRα mRNAs were detected. In contrast, *Xenopus* TRβ mRNA exhibits a complex pattern of alternative splicing within the 5′-untranslated region, as well as multiple transcriptional start sites. Yaoita *et al.* (1990) also described eight exons that are alternatively spliced, giving rise to at least two amino termini for each of the two TRβ proteins. Both TRα and β contain multiple AUG codons with short open reading frames, but the functional significance of this multiplicity is not known. The non-TH-binding isoform TRα2 has not yet been detected in amphibia. It should, however, be emphasized that the major TR transcripts in *Xenopus* tissues occur as a single major species for the α and β isoforms, termed xTRαA and xTRβA (Yaoita *et al.,* 1990; Shi *et al.,* 1992). Although not analyzed in the same detail as *Xenopus* TR genes and their transcripts, TR mRNAs have also been described in tissues of the tadpole of the bullfrog *Rana catesbeiana* (Helbing *et al.,* 1992).

The most impressive diversity of metamorphic responses of different tissues to a single hormonal signal, illustrated in Table I, raises the important question of tissue specificity of thyroid hormone receptor action. It is unlikely that such varied responses at the gene level could be solely attributed to receptor isoform diversity at the tissue level, as has been described (Chin, 1991; Chatterjee and Tata, 1992; Yen and Chin, 1994). It is more likely that different combinations of multiple transcription factors, both ubiquitous and tissue specific, acting in concert with TRs, are responsible for generating the differential expression of a genetic program (Diamond *et al.,* 1990; McKnight and Yamamoto, 1992; Parker, 1991, 1993).

IV. EXPRESSION OF THYROID HORMONE RECEPTOR GENES DURING METAMORPHOSIS

The expression of *Xenopus* TR genes has been studied in detail in the laboratories of D. D. Brown and J. R. Tata and shown to be both developmentally and hormonally regulated, which has led to the concept that the autoregulation of TR genes by thyroid hormones is crucial to the hormonal regulation of metamorphosis itself.

A. Developmental Regulation of *Xenopus* TR Genes

The onset of normal metamorphosis at stage 54 *Xenopus* tadpoles and its accelerated progression both correlate well with the first detection and rapid buildup of circulating thyroid hormones (Fig. 3) (Leloup and Buscaglia, 1977; Tata *et al.,* 1993). However, the *Xenopus* tadpole is known to acquire response to exogenous T_3 by stage 45 (1 week after fertilization), which means that functional TR is present well before the secretion of endogenous hormone. It is therefore not surprising that TR mRNA is detected at or before tadpoles reach stage 44 (Baker and Tata, 1990; Yaoita and Brown, 1990; Kawahara *et al.,* 1991; Banker *et al.,* 1991).

Figure 7 shows that the relative amount of TRα and β mRNAs in three regions of the *Xenopus* tadpole is developmentally regulated. There is a very low, but significant, level of TR transcripts in the fertilized egg, embryo, and tadpole until stage 44, when there is an abrupt increase in TR mRNA accumulation. The fact that endogenous TH is not secreted until much later in development means that both TR genes are constitutively expressed and are not hormonally controlled during early development. It should also be noted in Fig. 7 that (a) TRα transcripts are present at levels that are about 10 times those of TRβ, and (b) the two messages do not accumulate coordinately in the different regions or tissues. That the developmental accumulation of TRα and β mRNAs is differentially regulated, both as regarding temporal pattern and tissue

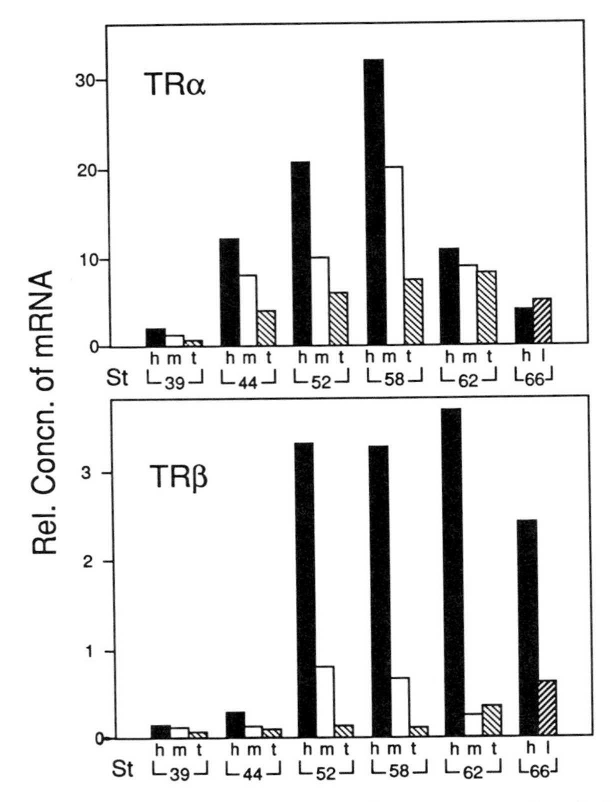

Fig. 7. Developmental regulation of TRα and β gene expression during *Xenopus* metamorphosis. RNA was extracted from the head (h), middle (m), tail (t), and liver (l) of *Xenopus* tadpoles and froglets at different stages (St) before, during, and after metamorphosis. Note the 10-fold difference in scale for the relative values in ordinates for TRα and β mRNAs. [For other experimental details see Kawahara *et al.* (1991).]

distribution, is also observed in the developing chick embryo (Forrest *et al.*, 1990).

In situ hybridization (Kawahara *et al.*, 1991) confirmed the presence of TR mRNA in individual tissues of the stage 44 *Xenopus* tadpole. The transcripts could be visualized in most tissues but were predominantly located in the CNS, spinal cord, intestinal epithelium, tail, liver, and limb bud, i.e., in tissues programmed for remodeling, morphogenesis, and cell death during metamorphosis. On completion of metamorphosis, there is a rapid drop in both TR mRNAs and circulating TH, although low levels of these persist in adult *Xenopus* tissues. The latter may explain why responsiveness to T_3 persists in adult *Xenopus* hepatocytes (Tata *et al.,* 1993; Rabelo and Tata, 1993; Ulisse and Tata, 1994).

B. TR Genes Autoregulated by Thyroid Hormone

After stage 54 and until the completion of metamorphosis there is good correlation between the accumulation of TR transcripts and the circulating level of thyroid hormones in *Xenopus* tadpoles (compare Figs. 3 and 7). It raises the question as to whether or not exogenous TH would precociously up-regulate TR gene expression. A biochemical analysis of RNA and *in situ* hybridization clearly confirmed that this indeed is the case (Yaoita and Brown, 1990; Kawahara *et al.,* 1991; Rabelo *et al.,* 1994). Figure 8A is an example of the enhanced TR mRNA signals observed by *in situ* hybridization in brain, intestine, and liver of stage 52 *Xenopus* tadpoles exposed to low concentrations of T_3 for 4 days. These tissues undergo substantial structural and biochemical remodeling in response to T_3. Autoinduction of TR is also visualized in tissues programmed for growth and morphogenesis, as shown for hind limb buds of stage 53 *Xenopus* tadpoles in Fig. 8B. The same phenomenon is observed for the tadpole tail which is programmed for T_3-induced total regression (see Section IV,D).

A few features of autoinduction of TR that are relevant to the significance of this process are worth emphasizing. The extent of autoinduction of TR is dependent on the developmental stage of the tadpole. It does not occur before stage 44 and its magnitude increases until just past the mid-metamorphic and climax stages of 58–61 for *Xenopus* tadpoles. This correlates well with the well-known increase in sensitivity to T_3 during metamorphosis. Autoinduction is also more marked for TRβ mRNA than for TRα, as depicted for stage 52 *Xenopus* tadpoles in Fig. 9A. The relative amount of TRα transcripts was elevated 2- to 4-fold 48 hr after the addition of 10^{-9} M T_3, whereas TRβ mRNA increased 20- to 30-fold during the same period. As in all organisms, the absolute amount of *Xenopus* TRα mRNA is 10–20 times higher than that of the β isoform. An up-regulation of TRβ mRNA can be seen as early as 4 hr after the exposure of tadpoles to exogenous T_3 (Fig. 9A). This is among the most

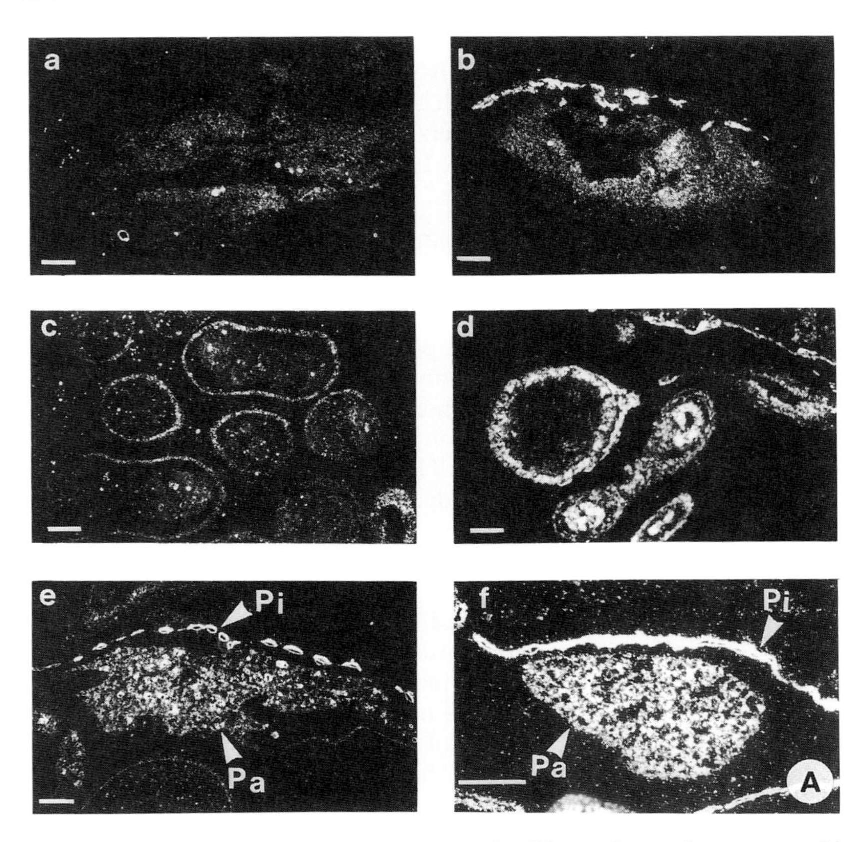

Fig. 8. (A) Up-regulation by T_3 of TR mRNA in different tissues of prematamorphic *Xenopus* tadpoles. Dark-field imaging of localization by *in situ* hybridization of *Xenopus* TRα mRNA in saggital sections of brain (a,b), intestine (c,d), and liver (e,f) of stage 52 tadpoles. (a,c,e) Control tadpoles. (b,d,f) Tadpoles treated for 4 days with 10^{-9} M T_3. Arrows in liver sections (e,f) indicate an artifact produced in black and white photography by pigmentation (Pi) surrounding the bulk of the parenchymal cells (Pa) of the liver. (B) Autoinduction by T_3 of TR mRNA in newly emerging hind limb of stage 53 *Xenopus* tadpole. (a,c) Bright- and (b,d) dark-field imaging of a saggital section of a region comprising the limb bud or growing hind limb (indicated by arrowheads) of a stage 53 tadpole after 5 days of immersion in 10^{-9} M T_3. (a,b) Untreated control; (c,d) T_3-treated. The section of untreated control shows that the limb bud is small and poorly developed in tadpoles at this stage. Note that 5 days of T_3 treatment at stage 53 has accelerated hind limb bud development to reach a stage equivalent to stage 54/55 of untreated tadpoles. The sections were exposed for 1 week for autoradiography. Bars: 100 μm. [Reprinted from Rabelo *et al.* (1994), with permission from Elsevier Science Ireland Ltd.]

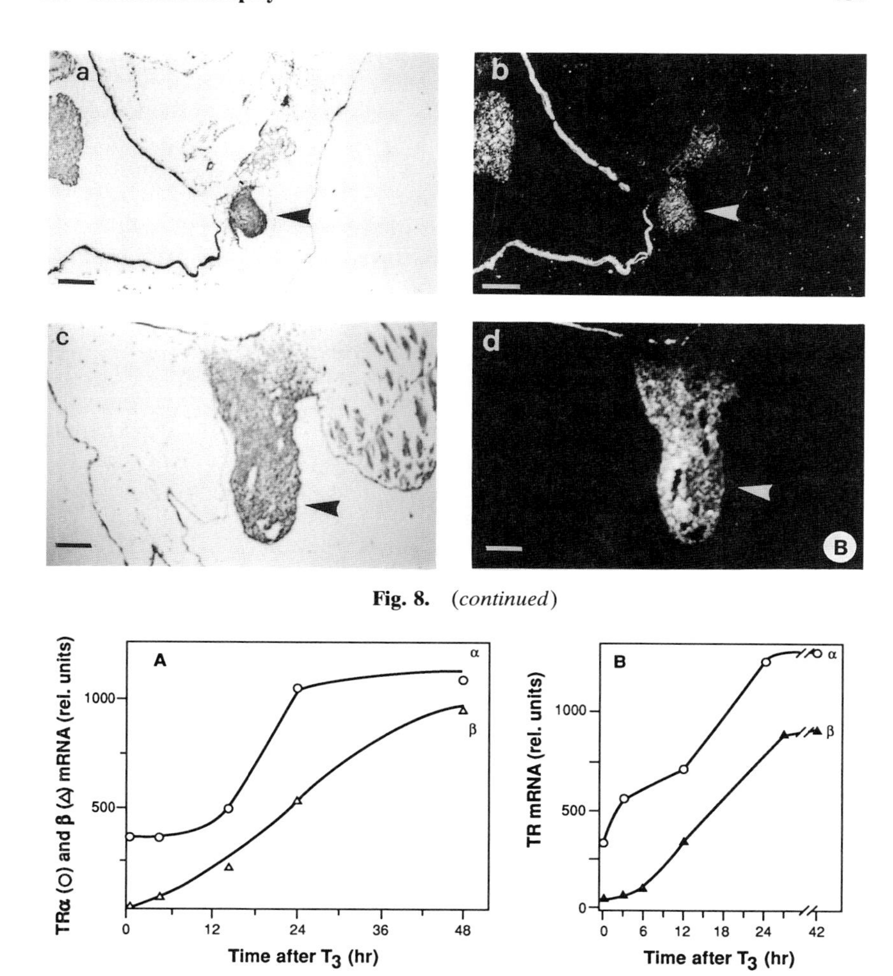

Fig. 8. *(continued)*

Fig. 9. Kinetics of autoinduction by T_3 of thyroid hormone receptor α and β (TRα, TRβ) in (A) whole *Xenopus* tadpoles or (B) XTC-2 cells in culture. The accumulation of TRα and β mRNAs was measured by RNase protection assays, and relative amounts of mRNA were determined by densitometric scanning of autoradiograms. (A) Batches of 24 stage 50/51 tadpoles were immersed in 2×10^{-9} *M* T_3 for different periods of time and total RNA was prepared from whole body homogenates. (B) XTC-2 cells were treated with 10^{-9} *M* T_3 for different periods of time before extraction of total RNA. [Reprinted from *J. Ster. Biochem. Molec. Biol.* **46,** Tata *et al.* Autoinduction of Nuclear Receptor Genes and its Significance, 105–119, Copyright (1993), with kind permission from Elsevier Science Ltd., The Boulevard, Langford Lane, Kidlington OX5 1GB, UK.]

rapid biochemical responses of *Xenopus* tadpoles to T_3 (Table II) (Shi, 1994, and Chapter 14, this volume) and raises the possibility that the up-regulation of TR is a requirement for TH induction of metamorphosis.

Autoinduction of TRα and β mRNAs has also been reproduced in two *Xenopus* cultured cell lines: XL-177 and XTC-2 (Kanamori and Brown, 1992, 1993; Machuca and Tata, 1992). Remarkably, the kinetics of up-regulation of the two transcripts in the cell lines were very similar to those in whole tadpoles. This is illustrated for XTC-2 cells in Fig. 9B. Investigating cultured cells also made it possible to demonstrate that T_3 not only induces TR mRNA but also the receptor protein by transfecting XTC-2 cells with a construct of chloramphenicol acetyltransferase (CAT) reporter fused to the promoter of albumin gene (Tata *et al.,* 1993). The latter gene is a well-known target for T_3 during amphibian metamorphosis (Weber, 1967; Cohen, 1970; Frieden and Just, 1970; Beckingham Smith and Tata, 1976a). When the transfected cells were exposed to T_3, under conditions in which the hormone induces TRα and β mRNAs, there was a significant increase in CAT activity from the albumin promoter, thus indirectly confirming that autoinduction of TR mRNA leads to the production of more functional receptor protein. It should again be emphasized that all studies on TR autoinduction during metamorphosis, or in any other system, are restricted to the transcripts and that, as yet, TRα or β proteins have not been directly measured.

TABLE II

Time Required for Activation of Different Genes in *Xenopus* Tadpoles Induced to Metamorphose Precociously with T_3[a]

Gene	Latent period (hr)
TRβ	3
TRα	6
Albumin	40
L-Arginase	70
63-kDa keratin	100

[a] Stage 52 tadpoles were exposed to 10^{-9} *M* T_3 for different periods of time and the concentration of different mRNAs was measured by RNase protection assay. The results are expressed as time required before the *de novo* appearance of RNA or ~10% increase in amount of RNA already expressed (Tata *et al.,* 1993).

Comparative studies on the response to exogenous thyroid hormones of facultative (axolotl) and obligatory (*Necturus*) neotenic amphibia, i.e., species that do not normally undergo metamorphosis, indirectly point to the association of TR autoinduction with metamorphosis. As indicated in Table III, administration of TH to larval axolotls, whose thyroid glands are nonfunctional, can be induced to undergo metamorphosis, as judged by morphological changes such as keratinization of the skin and regression of tail fin and external gills. In contrast, *Necturus,* which have a functioning thyroid gland, will not respond to even high doses of TH. An explanation for this contrasting behavior of neotenic amphibia was provided when TR mRNAs and their autoinduction by exogenous T_3 were measured (Yaoita and Brown, 1990; Tata *et al.,* 1993). Low levels of TR mRNAs were detected in axolotl tissues but not in those of the obligatory neotenic *Necturus.* Treatment of axolotl (*Ambystoma mexicanum* and *A. tigrinum*) larvae with T_3 upregulated TR in their tissues but not in *Necturus macalosus* tissues. This autoinduction of axolotl TR genes was accompanied by such external morphological indices as regression of gills and tail fins.

C. Mechanism of Autoinduction of TR Genes

Of the different possible mechanisms underlying the autoinduction phenomenon, the one most amenable to explanation would be that the

TABLE III

Metamorphic Response to Thyroid Hormones Linked to Autoinduction of Thyroid Hormone Receptor Genes[a]

Species	Metamorphosis	Endogenous T_4, T_3	TR genes	
			Expressed[b]	Autoinduced[c]
Xenopus	Spontaneous	Yes	Yes	Yes
Axolotl (*Ambystoma*)	Facultatively neotenous	No	Yes	Yes
Necturus	Obligatorily neotenous	Yes	No	No

[a] Normally metamorphosing (*Xenopus*) and facultatively (*Ambystoma*) or obligatorily (*Necturus*) neotenic amphibia.

[b] TRα and β mRNAs detectable in normal larval forms.

[c] TRα and β mRNAs induced by exogenous thyroid hormone.

liganded TR interacts with TREs in the promoters of its own genes. Although the promoter of the human TRα gene has been cloned (Laudet *et al.,* 1992), that of *Xenopus* has not. However, the promoter of the *Xenopus* TRβA gene has been cloned and partially characterized (Shi *et al.,* 1992; Ranjan *et al.,* 1994; Machuca *et al.,* 1995). Since the expression of the xTRβ gene is modulated to a greater extent by T_3 than that of the α isoform, any interaction between TRα or β protein with xTRβ promoter elements would be easier to interpret.

The author's laboratory has identified, structurally and functionally, several regulatory elements within 1.6 kb of the upstream sequence of the xTRβ gene (Machuca *et al.,* 1995). As schematically depicted in Fig. 10, two DR + 4 TREs (an imperfect distal TRE at −793/−778 and a perfect proximal one at −5/+11), a possible DR + 1 RXRE at −1056/ −1044, as well as five SP1 sites, but no TATA or CAAT boxes, were detected. The functional significance of the TREs was assessed by cell transfection and DNA binding to recombinant xTRα and β proteins. Transfection of *Xenopus* XTC-2 cells (which express both TRα and β) and XL-2 cells (which predominantly express TRβ) (Machuca and Tata, 1992) with various CAT constructs of TR promoter fragments showed that both the distal and proximal DR + 4 TREs were functional, i.e., they responded to T_3. Overexpression of unliganded TRα or TRβ in XTC-2 and XL-2 cells substantially reduced the basal transcriptional activity, which is in line with similar findings with mammalian TRs (Lazar, 1993; Barettino *et al.,* 1993; Fondell *et al.,* 1993). Under these conditions, T_3 was able to enhance CAT activity by 4- to 16-fold in *Xenopus* cells cotransfected with the full-length TRβ promoter.

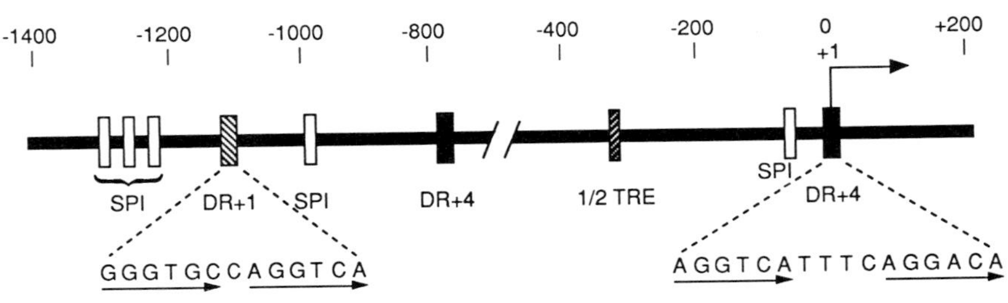

Fig. 10. The *Xenopus* TRβ gene promoter. The positions of the proximal and distal DR + 4 TR-responsive sites, the ½ TRE element, the DR + 1 element (putative RXR-binding site), and five SP1-binding sites are indicated. [Based on data from Machuca *et al.* (1995).]

The just-discussed transfection assays were studied in parallel with electrophoretic mobility shift assays with the same promoter DNA fragments. Using recombinant *Xenopus* TRα and β, as well as RXRα and γ proteins, these assays clearly demonstrated that TR–RXR heterodimers, in any isoform combination, but not TR monomers or homodimers strongly and specifically interact with both the distal and the proximal TREs of the *Xenopus* TRβ gene in a ligand-independent manner; the interaction with the latter site is the strongest (Machuca *et al.,* 1995). The major action of TR is now generally accepted to be through its heterodimerization with RXR (Lazar, 1993; Mader *et al.,* 1993; Leng *et al.,* 1993; Zhang and Pfahl, 1993). Figure 10 also shows that the xTRβ promoter has a DR + 1 RXRE sequence. Although these studies do not rule out the possible intervention of other factors, regulated or not by T_3, it is concluded that both xTRα and β can regulate the expression of the xTRβ gene, thus offering one possible explanation for its autoinduction during T_3-induced metamorphosis (Yaoita and Brown, 1990; Kawahara *et al.,* 1991; Baker and Tata, 1992).

D. Inhibition by Prolactin: Autoinduction of TR Genes Essential for Amphibian Metamorphosis

Does the autoregulation of TR genes in *Xenopus* tadpole have any significance in the thyroid hormonal regulation of amphibian metamorphosis? This question has been addressed in our laboratory by exploiting the antimetamorphic action of prolactin (see Section II,E); see also Denver, Chapter 12, this volume). How PRL would affect the autoinduction of xTRs while inhibiting the action of T_3 in whole *Xenopus* tadpoles and in organ culture of tails and limb buds was determined (Tata *et al.,* 1991; Baker and Tata, 1992; Tata, 1993; Rabelo *et al.,* 1994). A clear correlation existed between the prevention of T_3-induced metamorphosis by PRL with the inhibition of autoinduction of TRα and β mRNAs, as judged by biochemical analysis and *in situ* hybridization. Figure 11 illustrates an example of how PRL blocked the up-regulation of both TRα and β mRNAs in premetamorphic *Xenopus* tadpoles treated with small amounts of T_3. The same phenomenon can be visualized by *in situ* hybridization in whole tadpoles (Fig. 12) and also in organ cultures of tails (Baker and Tata, 1992). It should be noted that PRL alone did not significantly affect the basal level of TR transcripts but that it prevented

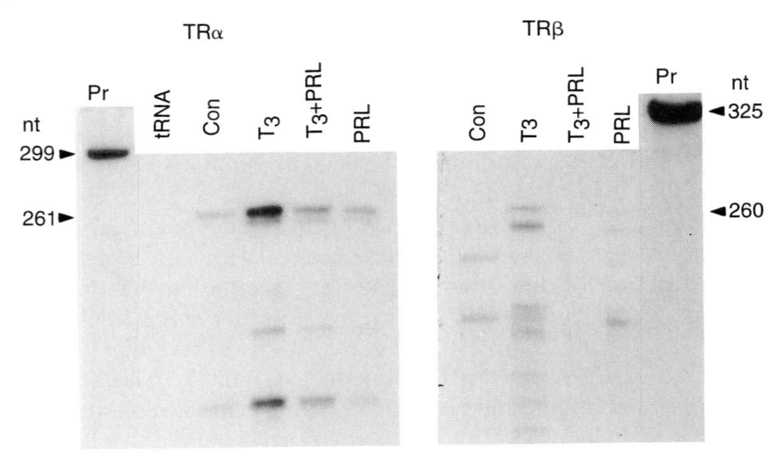

Fig. 11. Inhibition by prolactin of autoinduction by T_3 of TRα and β mRNA. Where indicated, batches of 18 stage 50 *Xenopus* tadpoles were immersed for 2 days in 100 ml of medium containing 0.2 IU/ml of ovine PRL before exposure to 10^{-9} *M* T_3 for a further 2 days with or without PRL. Total RNA was assayed for TRα and β mRNAs by RNase protection assays. Pr, probe; Con, control (no hormone). Arrows indicate the protected bands for TRα and β mRNAs (seen as a doublet for β). [For other experimental details, see Baker and Tata (1992).]

their up-regulation by T_3. This inhibition of TR autoinduction, coupled to the loss of metamorphic response to TH, was also manifested in early *Xenopus* stages, as shown in Table IV, which suggests that PRL receptors in tadpole tissues would be expressed well before natural metamorphosis would commence. This finding also strongly suggests that the onset of metamorphosis is dependent on the autoinduction of TR. If one accepts that up-regulation of TR mRNA leads to that of TR proteins, the conclusion that TR autoinduction is necessary for the activation of T_3 target genes is strongly supported by the inhibition by PRL of the induction of albumin and 63-kDa keratin genes (Baker and Tata, 1992). Thus, although the evidence based on the use of PRL is still indirect, it is compelling enough to suggest that TR autoinduction is an essential requirement for the hormonal initiation and sustenance of amphibian metamorphosis.

How prolactin blocks the metamorphic action of thyroid hormone is not known. Work with mammalian cells suggests that PRL receptor is linked to the tyrosine phosphorylation pathway via JAK phosphokinase

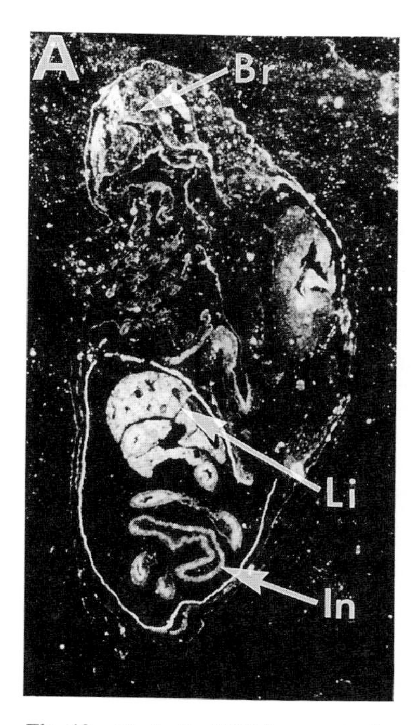

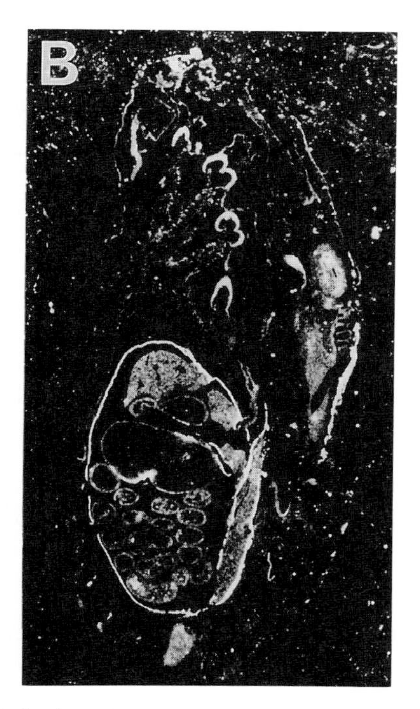

Fig. 12. Prolactin (PRL) prevents the autoinduction of TRα mRNA as visualized by *in situ* hybridization. Stage 54 *Xenopus* tadpoles were treated with (A) T_3 (10^{-9} *M* for 3 days) without PRL or (B) with PRl (0.1 unit/ml for 4 days). Saggital sections were prepared from tadpoles and hybridized to a ^{35}S-labeled antisense TRα cRNA probe, and the distribution of transcripts was visualized by autoradiography. Br, brain; In, intestinal epithelium; Li, liver. [Reprinted from *J. Ster. Biochem. Molec. Biol.* **46,** Tata *et al.* Autoinduction of Nuclear Receptor Genes and its Significance, 105–119, Copyright (1993), with kind permission from Elsevier Science Ltd., The Boulevard, Langford Lane, Kidlington OX5 1GB, UK.]

(protein tyrosine kinase) (Rillema, 1987; Clevenger and Medaglia, 1994; Welte *et al.,* 1994). This pathway is now considered to be a major mechanism for the transduction of extracellular signals to the nucleus by modifying transcription factors and other regulatory elements, including nuclear receptors (Cohen and Foulkes, 1991). Figure 13 is therefore an attempt to bring together an external regulatory signal such as prolactin and thyroid hormones acting via TRs in the nucleus. Obviously, this scheme is an oversimplification of the complex interplay between intracellular responses and the integration of multiple regulatory signals, but it emphasizes the requirement of up-regulation of TR to activate its target

TABLE IV

Relative Accumulation of TRα and β in Early Stages of *Xenopus* Tadpoles Following Treatment with T_3 and Prolactin[a]

Treatment	Relative units	
	TRα	TRβ
None	505	24
T_3	1290	368
T_3 + PRL	799	<10
PRL	405	43

[a] Batches of 20 stage 50 tadpoles were exposed to 2×10^{-9} *M* T_3 and/or 0.1 IU PRL/ml, as indicated, for 4 days before total RNA was extracted from whole larvae and the relative amounts of TRα and β were estimated by quantitative RNase protection assays.

genes. This, in turn, suggests the existence of a differential threshold of receptor concentration to promote autoinduction and target gene activation by thyroid hormones. In other words, levels of TR above a threshold higher than that necessary for TR autoinduction would activate different sets of genes in different tissues depending on different developmental programs, by mechanisms as yet not fully understood. Clearly, much work still needs to be done to elucidate these important issues of differential gene expression.

V. WIDER IMPLICATIONS OF THYROID HORMONAL REGULATION OF AMPHIBIAN METAMORPHOSIS

The interplay between thyroid hormone and prolactin is relevant to the question of "cross-talk" between signals acting via receptors located in the plasma membrane with those in the nucleus, as well as between the various members of the nuclear receptor supergene family. There is now increasing evidence that such signal transduction processes as protein phosphorylation, Ca^{2+} and IP_3 (inositol trisphosphate) signaling, G protein-linked functions, and *jun/fos* oncogene complexes modulate

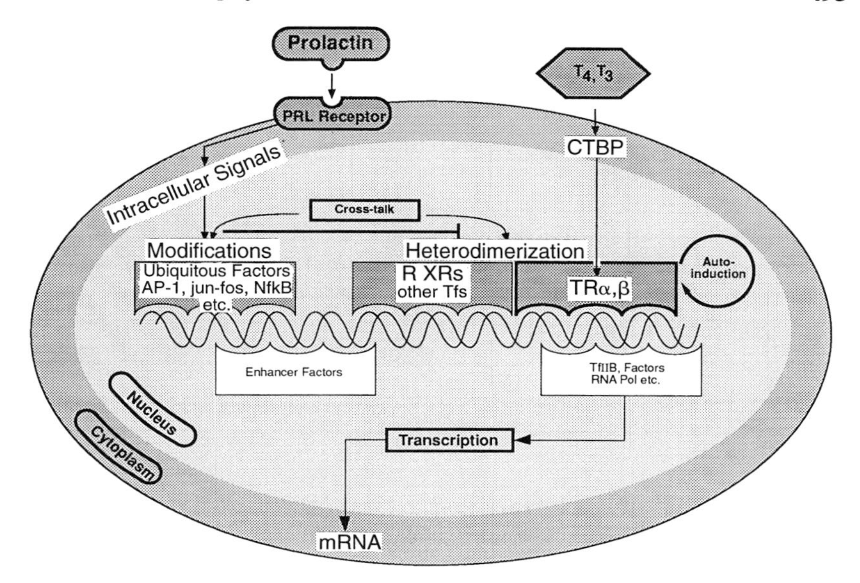

Fig. 13. Possible mechanisms underlying the interplay between prolactin (PRL) and thyroid hormones (T_4, T_3). In this oversimplified diagram, thyroid hormones increase the amount of TRα and β to a higher threshold than the basal, constitutive level by a mechanism of receptor autoinduction. This latter process would be regulated by PRL, via cross-talk with some intracellular or intranuclear signals (e.g., AP-1, *jun-fos,* etc.) generated by transduction from the ligand interacting with the PRL receptor.

nuclear receptor function (Cohen and Foulkes, 1991; Parker, 1991, 1993; McKnight and Yamamoto, 1992). Many recent investigations particularly highlight the modulation by peptide hormones and growth factors of phosphorylation of nuclear receptors, a process considered important for their nuclear transport, dimerization, and ligand binding (Cohen and Foulkes, 1991; Weigel *et al.,* 1993; Parker, 1993). It is interesting to note that the *Xenopus* TRβ promoter has several SP1 sites (see Section IV,C) which are involved in regulation via the *fos/jun* pathway (Curran, 1992). There is also increasing evidence showing cross-regulation of nuclear receptors by their respective ligands (Diamond *et al.,* 1990; Rabelo and Tata, 1993; Rabelo *et al.,* 1994; Ulisse and Tata, 1994). For example, the glucocorticoid hormone is known to potentiate the action of TH in inducing metamorphosis (Galton, 1990; Kikuyama *et al.,* 1993). It is also significant that glucocorticoid hormones can modify TR gene expression, while T_3 does the same for ER gene expression (Rabelo and Tata, 1993; Ulisse and Tata, 1994; Iwamuro and Tata, 1995). An important consequence of interactions among nuclear receptors, and between them and extranuclear factors whose activities are controlled by extracellular signals, would be to provide a combinatorial mechanism for hormone-specific positive and negative modulation of different genes which would not otherwise be operating for individual receptors (McKnight and Yamamoto, 1992; Yamamoto *et al.,* 1992; Cohen and Foulkes, 1991). As cross-talk systems are better defined at the molecular level, it will become increasingly clear that complex intracellular networks of hormonal and nonhormonal signals facilitate a well-coordinated and homeostatically controlled regulation of growth and development.

In a different context, there is the question of how widely does the phenomenon of autoinduction of receptor occur. A survey of recent literature indicates its widespread manifestation during development regulated by ligands of various nuclear receptors (see Tata *et al.,* 1993; Tata, 1994). Previously, cross-regulation of nuclear steroid receptors was well documented for estrogen, progesterone, and glucocorticoids in mammalian tissues. The latter hormone is also known to potentiate keratinization of tadpole skin and tail regression during T_3-induced or natural metamorphosis (see Miller, Chapter 17, this volume). The potentiation by T_3 and dexamethasone of expression of estrogen receptor in larval and adult *Xenopus* hepatocytes has also been reported (Rabelo and Tata, 1993; Rabelo *et al.,* 1994; Ulisse and Tata, 1994). In addition to the example of autoinduction of TR during metamorphosis detailed in this chapter, the up-regulation of estrogen receptor by its own ligand during amphibian vitellogenesis has also been well documented (Perlman *et al.,* 1984; Shapiro *et al.,* 1989). In embryonic mouse tissues, retinoic acid upregulates all three isoforms of the retinoic acid receptor (deThé *et al.,* 1990; Lehmann *et al.,* 1992). More relevant to metamorphosis is the finding that ecdysteroid upregulates the expression of its own receptor transcripts in *Drosophila* cells (Andres and Thummel, 1992; Karim and Thummel, 1992; see Cherbas and Cherbas, Chapter 5, this volume).

The phenomenon of autoinduction of nuclear hormone receptors therefore now appears to be a general feature of hormone-dependent postembryonic growth and development. Although receptor up-regulation is closely associated with the biological activity of its ligand, more direct proof of a cause–effect relationship has still to be obtained. For example, much of the current work on developmental aspects of nuclear receptors, or early response genes during metamorphosis, is restricted to the analysis of their transcripts. In view of the importance of not only receptor heterodimerization, but also protein–DNA interactions and protein–protein interactions in the context of chromatin organization, future work will have to tackle the technically more difficult problem of working with functionally active protein molecules. Nevertheless, it is clear that thyroid hormone receptors play a central role in the initiation and maintenance of amphibian metamorphosis. How precisely their expression and function are integrated into the network of other regulatory elements of extra- and intracellular signaling will be crucial to our future understanding of the process of metamorphosis.

ACKNOWLEDGMENTS

Much of the recent data presented in this chapter is derived from the work of several members of my laboratory, particularly Betty Baker, Graeme Esslemont, Akira Kawahara, Irma Machuca, and Elida Rabelo, to whom I am most grateful. I also thank Ena Heather for preparation of the manuscript.

REFERENCES

Andres, A. J., and Thummel, C. S. (1992). Hormones, puffs and flies: The molecular control of metamorphosis by ecdysone. *Trends Genet.* **8,** 132–138.

Atkinson, B. G. (1981). Biological basis of tissue regression and synthesis. *In* "Metamorphosis: A Problem in Developmental Biology" (L. I. Gilbert and E. Frieden, eds., 2nd ed.), pp. 397–444. Plenum, New York.

Atkinson, B. G. (1994). Metamorphosis: Model systems for studying gene expression in postembryonic development. *Dev. Genet.* **15,** 313–319.

Baker, B. S., and Tata, J. R. (1990). Accumulation of proto-oncogene c-erb-A related transcripts during *Xenopus* development: Association with early acquisition of response to thyroid hormone and estrogen. *EMBO J.* **9,** 879–885.

Baker, B. S., and Tata, J. R. (1992). Prolactin prevents the autoinduction of thyroid hormone receptor mRNAs during amphibian metamorphosis. *Dev. Biol.* **149,** 463–467.

Baniahmad, A., Tsai, S.-Y., O'Malley, B. W., and Tsai, M. J. (1992). Kindred thyroid hormone receptor is an active and constitutive silencer and a repressor for thyroid hormone and retinoic acid responses. *Proc. Natl. Acad. Sci. U.S.A.* **89,** 10633–10637.

Banker, D. E., Bigler, J., and Eisenman, R. N. (1991). The thyroid hormone receptor gene (c-*erb*Aα) is expressed in advance of thyroid gland maturation during the early embryonic development in *Xenopus laevis. Mol. Cell. Biol.* **11,** 5079–5089.

Barettino, D., Bugge, T. H., Bartunek, P., Vivanco-Ruiz, M., Sonntag-Buck, V., Beug, H., Zenke, M., and Stunnenberg, H. G. (1993). Unliganded T3R, but not its oncogenic variant, v-erbA, suppresses RAR-dependent transactivation by titrating out RXR. *EMBO J.* **12,** 1343–1354.

Baulieu, E.-E., and Kelly, P. A., eds. (1990). "Hormones: From Molecules to Disease." Hermann, Paris.

Beckingham Smith, K., and Tata, J. R. (1976a). The hormonal control of amphibian metamorphosis. *In* "Developmental Biology of Plants and Animals" (C. Graham and P. F. Wareing, eds.), pp. 232–245. Blackwell, Oxford.

Beckingham Smith, K., and Tata, J. R. (1976b). Cell death. Are new proteins synthesized during hormone-induced tadpole tail regression? *Exp. Cell Res.* **100,** 129–146.

Bern, H. A., Nicoll, C. S., and Strohman, R. C. (1967). Prolactin and tadpole growth. *Proc. Soc. Exp. Biol. Med.* **126,** 518–520.

Bowen, I. D., and Lockshin, R. A., eds. (1981). "Cell Death in Biology and Pathology." Chapman & Hall, London.

Brooks, A. R., Sweeney, G., and Old, R. W. (1989). Structure and functional expression of a cloned *Xenopus* hormone receptor. *Nucleic Acids Res.* **17,** 9395–9405.

Buckbinder, L., and Brown, D. D. (1993). Expression of the *Xenopus laevis* prolactin and thyrotropin genes during metamorphosis. *Proc. Natl. Acad. Sci. U.S.A.* **90,** 3820–3824.

Burd, G. D. (1990). Role of thyroxine in neural development of the olfactory system. *Proc. Int. Symp. Olfaction Taste, 10th* pp. 196–205.

Chatterjee, V. K. K., and Tata, J. R. (1992). Thyroid hormone receptors and their role in development. *Cancer Surv.* **14,** 147–167.

Chatterjee, V. K. K., Nagaya, T., Madison, L. D., Datta, S., Rentoumis, A., and Jameson, J. L. (1991). Thyroid hormone resistance syndrome: Inhibition of normal receptor function by mutant thyroid hormone receptors. *J. Clin. Invest.* **87,** 1977–1984.

Chin, W. W. (1991). Nuclear thyroid hormone receptors. *In* "Nuclear Hormone Receptors" (M. Parker, ed.), pp. 79–102. Academic Press, San Diego.

Clevenger, C. V., and Medaglia, M. V. (1994). The protein tyrosine kinase $p59^{fyn}$ is associated with prolactin (PRL) receptor and is activated by PRL stimulation of T-lymphocytes. *Mol. Endocrinol.* **8,** 674–681.

Cohen, P. P. (1970). Biochemical differentiation during amphibian metamorphosis. *Science* **168,** 533–544.

Cohen, P., and Foulkes, J. G., eds. (1991). "The Hormonal Control of Gene Transcription." Elsevier, Amsterdam.

Curran, T. (1992). Fos and Jun: Intermediary transcription factors. *In* "The Hormonal Control of Gene Transcription" (P. Cohen and J. G. Foulkes, eds.), pp. 295–308. Elsevier, Amsterdam.

Derby, A., and Etkin, W. (1968). Thyroxine induced tail resorption in vitro as affected by anterior pituitary hormones. *J. Exp. Zool.* **169,** 1–8.

de Thé, H., Vivanco-Ruiz, M., Tiollais, P., Stunnenberg, H., and Dejean, A. (1990). Identification of a retinoic acid responsive element in the retinoic acid receptor beta gene. *Nature (London)* **343,** 177–180.

Dexter, T. M., Raff, M. C., and Wyllie, A. H. (1994). Death from inside out: The role of apoptosis in development, tissue homeostasis and malignancy. *Philos. Trans. R. Soc. B* **345,** 231–333.

Diamond, M. I., Miner, J. N., Yoshinaga, S. K., and Yamamoto, K. R. (1990). Transcription factor interactions: Selectors of positive or negative regulation from a single DNA element. *Science* **249,** 1266–1272.

Eeckhout, Y. (1966). Aspects biochimiques de la métamorphose. *Rev. Quest. Sci.* **3,** 377–393.

Ellis, R. E., Yuan, J., and Horvitz, H. R. (1991). Mechanisms and functions of cell death. *Annu. Rev. Cell Biol.* **7,** 663–698.

Etkin, W. (1964). Metamorphosis. *In* "Physiology of Amphibia" (J. A. Moore, ed.), pp. 427–468. Academic Press, New York.

Evans, R. M. (1988). The steroid and thyroid hormone receptor superfamily. *Science* **240,** 889–895.

Fondell, J. D., Roy, A. L., and Roeder, R. G. (1993). Unliganded thyroid hormone receptor inhibits formation of a functional preinitiation complex: Implications for active repression. *Genes Dev.* **7,** 1400–1410.

Forrest, D., Sjöberg, M., and Vennstrom, B. (1990). Contrasting developmental and tissue specific expression of α and β thyroid hormone receptor genes. *EMBO J.* **9,** 1519–1528.

Frieden, E., and Just, J. J. (1970). Hormonal responses in amphibian metamorphosis. *In* "Biochemical Actions of Hormones" (G. Litwack, ed.), pp. 2–52. Academic Press, New York.

Galton, V. A. (1983). Thyroid hormone action in amphibian metamorphosis. *In* "Molecular Basis of Thyroid Hormone Action" (J. H. Oppenheimer and H. H. Samuels, eds.), pp. 445–483. Academic Press, New York.

Galton, V. A. (1990). Mechanisms underlying the acceleration of thyroid hormone-induced tadpole metamorphosis by corticosterone. *Endocrinology* (*Baltimore*) **127,** 2997–3002.

Gilbert, L. I., and Frieden, E., eds. (1981). "Metamorphosis: A Problem in Developmental Biology," 2nd ed. Plenum, New York.

Green, C. D., and Chambon, P. (1988). Nuclear receptors enhance our understanding of transcription regulation. *Trends Genet.* **4,** 309–314.

Gudernatsch, J. F. (1912). Feeding experiments on tadpoles. *Wilhelm Roux Arch. Entwicklungsmech. Org.* **35,** 457.

Helbing, C., and Atkinson, B. G. (1994). 3,5,3′-Triiodothyronine–induced carbamyl phosphate synthetase gene expression is stabilized in the liver of *Rana catesbeiana* tadpoles during heat shock. *J. Biol. Chem.* **269,** 11743–11750.

Helbing, C., Gergely, G., and Atkinson, B. G. (1992). Sequential up-regulation of thyroid hormone β receptor, ornithine transcarbamylase, and carbamyl phosphate synthetase mRNAs in the liver of *Rana catesbeiana* tadpoles during spontaneous and thyroid hormone-induced metamorphosis. *Dev. Genet.* **13,** 289–301.

Hoskins, S. G. (1990). Metamorphosis of the amphibian eye. *J. Neurobiol.* **21,** 970–989.

Iwamuro, S., and Tata, J. R. (1995). Contrasting patterns of expression of thyroid hormone and retinoid $\times$ receptor genes during hormonal manipulation of *Xenopus* tadpole tail regression in culture. *Mol. Cell. Endocrinol.* **113,** 235–243.

Ishizuya-Oka, A., and Shimozawa, A. (1991). Induction of metamorphosis by thyroid hormone in anuran small intestine cultured organotypically in vitro. *In Vitro Cell. Dev. Biol.* **27A,** 853–857.

Kaltenbach, J. C. (1953). Local action of thyroxin on amphibian metamorphosis. I. Local metamorphosis in *Rana pipiens* larvae effected by thyroxin-cholesterol implants. *J. Exp. Zool.* **122,** 21–39.

Kanamori, A., and Brown, D. D. (1992). The regulation of thyroid hormone receptor β genes by thyroid hormone in *Xenopus laevis. J. Biol. Chem.* **267,** 739–745.

Kanamori, A., and Brown, D. D. (1993). Cultured cells as a model for amphibian metamorphosis. *Proc. Natl. Acad. Sci. U.S.A.* **90,** 6013–6017.

Karim, F. D., and Thummel, C. S. (1992). Temporal coordination of regulatory gene expression by the steroid hormone ecdysone. *EMBO J.* **11,** 4083–4093.

Kawahara, A., Baker, B., and Tata, J. R. (1991). Developmental and regional expression of thyroid hormone receptor genes during *Xenopus* metamorphosis. *Development* **112,** 933–943.

Kikuyama, S., Kawamura, K., Tanaka, S., and Yamamoto, K. (1993). Aspects of amphibian metamorphosis. Hormonal control. *Int. Rev. Cytol.* **145,** 105–148.

Kliewer, S. A., Umesono, K., Heyman, R. A., Mangelsdorf, D. J., Dyck, J. A., and Evans, R. M. (1992). Retinoid X receptor interacts with nuclear receptors in retinoic acid, thyroid hormone and vitamin D_3 signalling. *Nature (London)* **355,** 446–449.

Kollros, J. J. (1984). Growth and death of cells of the mesencephalic fifth nucleus in *Rana pipiens* larvae. *J. Comp. Neurol.* **224,** 386–394.

Laudet, V., Vanacker, J. M., Adelmant, G., Begue, A., and Stehelin, D. (1992). Characterization of a functional promoter for the human thyroid hormone receptor alpha c-erbA-1 gene. *Oncogene* **8,** 975–982.

Lazar, M. A. (1993). Thyroid hormone receptors: Multiple forms, multiple possibilities. Endocr. Rev. **14,** 184–193.

Lehmann, J. M., Zhang, X.-K., and Pfahl, M. (1992). RARγ2 expression is regulated through a retinoic acid response element embedded in Spl sites. *Mol. Cell. Biol.* **12,** 2976–2985.

Leloup, J., and Buscaglia, M. (1977). La triiodothyronine, hormone de la métamorphose des Amphibiens. *C. R. Hebd. Seances Acad. Sci., Ser. D.* **284,** 2261–2263.

Leng, X., Tsai, S. Y., O'Malley, B. W., and Tsai, M.-J. (1993). Ligand-dependent conformational changes in thyroid hormone and retinoic acid receptors are potentially enhanced by heterodimerization with retinoic X receptor. *J. Steroid Biochem. Mol. Biol.* **46,** 643–661.

Machuca, I., and Tata, J. R. (1992). Autoinduction of thyroid hormone receptor during metamorphosis is reproduced in *Xenopus* XTC-2 cells. *Mol. Cell. Endocrinol.* **87,** 105–113.

Machuca, I., Esslemont, G., Fairclough, L., and Tata, J. R. (1995). Analysis of structure and expression of the *Xenopus* thyroid hormone receptor β (xTRβ) gene to explain its autoinduction. *Mol. Endocrinol.* **9,** 96–107.

Mader, S., Chen, J.-Y., Chen, Z., White, J., Chambon, P., and Gronemeyer, H. (1993). The patterns of binding of RAR, RXR and TR homo- and heterodimers to direct repeats are dictated by the binding specificities of the DNA binding domains. *EMBO J.* **12,** 5029–5041.

McKnight, S. L., and Yamamoto, K. R., eds. (1992). "Transcriptional Regulation." Cold Spring Harbor Laboratory, Cold Spring Harbor, New York.

Nicoll, C. S. (1974). Physiological actions of prolactin. *In* "Handbook of Physiology" (E. Knobil and W. H. Sawyer, eds.), Sect. 7, Vol. 4, Part 2, pp. 253–292. American Physiological Society, Washington, D.C.

Niinuma, K., Yamamoto, K., and Kikuyama, S. (1991). Changes in plasma and pituitary prolactin levels in toad (*Bufa japonicus*) larvae during metamorphosis. *Zool. Sci.* **8,** 97–101.

Oppenheimer, J. H., and Samuels, H. H., eds. (1983). "Molecular Basis of Thyroid Hormone Action." Academic Press, New York.

Parker, M., ed. (1991). "Nuclear Hormone Receptors." Academic Press, San Diego.

Parker, M. G., ed. (1993). "Steroid Hormone Action." IRL Press, Oxford.

Perlman, A. J., Wolffe, A. P., Champion, J., and Tata, J. R. (1984). Regulation by estrogen receptor of vitellogenin gene transcription in *Xenopus* hepatocyte cultures. *Mol. Cell. Endocrinol.* **38,** 151–161.

Perlmann, T., Rangarajan, P. N., Umesono, K., and Evans, R. M. (1993). Determinants for selective RAR and TR recognition of direct repeat HREs. *Genes Dev.* **7,** 1411–1422.

Rabelo, E. M., and Tata, J. R. (1993). Thyroid hormone potentiates estrogen activation of vitellogenin genes and autoinduction of estrogen receptor in adult *Xenopus* hepatocytes. *Mol. Cell. Endocrinol.* **96,** 37–44.

Rabelo, E. M. L., Baker, B., and Tata, J. R. (1994). Interplay between thyroid hormone and estrogen in modulating expression of their receptor and vitellogenin genes during *Xenopus* metamorphosis. *Mech. Dev.* **45,** 49–57.

Ranjan, M., Wong, J., and Shi, Y.-B. (1994). Transcriptional repression of *Xenopus* TRβ gene is mediated by a thyroid hormone response element located near the start site. *J. Biol. Chem.* **269,** 24699–24705.

Ray, L. B., and Dent, J. N. (1986a). An analysis of the influence of thyroid hormone on the synthesis of proteins in the tail fin of bullfrog tadpoles. *J. Exp. Zool.* **240,** 191–201.

Ray, L. B., and Dent, J. N. (1986b). Observations on the interaction of prolactin and thyroxine in the tail of the bullfrog tadpole. *Gen. Comp. Endocrinol.* **64,** 36–43.

Riddiford, L. M., and Truman, J. W. (1978). Biochemistry of insect hormones and insect growth regulators. *In* "Biochemistry of Insects" (M. Rockstein, ed.), pp. 307–356. Academic Press, New York.

Rillema, J. A., ed. (1987). "Actions of Prolactin on Molecular Processes." CRC Press, Boca Raton, Florida.

Sap, J., Munoz, A., Damm, K., Goldberg, Y., Ghysdael, J., Leutz, A., Beug, H., and Vennstrom, B. (1986). The c-erb-A protein is a high-affinity receptor for thyroid hormone. *Nature (London)* **324,** 635–640.

Saunders, J. W., Jr. (1966). Death in embryonic systems. *Science* **154,** 604–612.

Shapiro, D. J., Barton, M. C., McKearin, D. M., Chang, T.-C., Lew, D., Blume, J., Nielsen, D. A., and Gould, L. (1989). Estrogen regulation of gene transcription and mRNA stability. *Recent Prog. Horm. Res.* **45,** 29–58.

Shi, Y.-B. (1994). Molecular biology of amphibian metamorphosis. A new approach to an old problem. *Trends Endocrinol. Metab.* **5,** 14–20.

Shi, Y.-B., and Brown, D. D. (1993). The earliest changes in gene expression in tadpole intestine induced by thyroid hormone. *J. Biol. Chem.* **268,** 20312–20317.

Shi, Y.-B., Yaoita, Y., and Brown, D. D. (1992). Genomic organization and alternative promoter usage of the two thyroid hormone receptor beta genes in *Xenopus laevis. J. Biol. Chem.* **267,** 733–738.

Stunnenberg, H. G. (1993). Mechanisms of transactivation by retinoic acid receptors. *BioEssays* **15,** 309–315.

Tata, J. R. (1966). Requirement for RNA protein synthesis for induced regression of the tadpole tail in organ culture. *Dev. Biol.* **13,** 77–94.

Tata, J. R. (1968). Early metamorphic competence of *Xenopus* larvae. *Dev. Biol.* **18,** 415–440.

Tata, J. R. (1971). Hormonal regulation of metamorphosis. *Symp. Soc. Exp. Biol.* **35,** 163–181.

Tata, J. R. (1993). Gene expression during metamorphosis: An ideal model for post-embryonic development. *BioEssays* **15,** 239–248.

Tata, J. R. (1994). Auto- and cross-regulation of nuclear receptor genes. *Trends Endocrinol. Metab.* **5,** 283–290.

Tata, J. R. (1995). Hormonal regulation of programmed cell death during amphibian metamorphosis. *Biochem. Cell Biol.* **72,** 581–586.

Tata, J. R., Kawahara, A., and Baker, B. S. (1991). Prolactin inhibits both thyroid hormone-induced morphogenesis and cell death in cultured amphibian larval tissues. *Dev. Biol.* **146,** 72–80.

Tata, J. R., Baker, B. S., Machuca, I., Rabelo, E. M. L., and Yamauchi, K. (1993). Autoinduction of nuclear receptor genes and its significance. *J. Steroid Biochem. Mol. Biol.* **46,** 105–119.

Tomei, L. D., and Cope, F. O., eds. (1991). "Apoptosis: The Molecular Basis of Cell Death." Cold Spring Harbor Laboratory, Cold Spring Harbor, New York.

Ulisse, S., and Tata, J. R. (1994). Thyroid hormone and glucocorticoid independently regulate the expression of estrogen receptor in male *Xenopus* liver cells. *Mol. Cell. Endocrinol.* **105,** 45–53.

Umesono, K., Murakami, K. K., Thompson, C. C., and Evans, R. M. (1991). Direct repeats as selective response elements for the thyroid hormone retinoic acid and vitamin D3 receptors. *Cell (Cambridge, Mass.)* **65,** 1255–1266.

Wang, Z., and Brown, D. D. (1993). Thyroid hormone-induced gene expression program for amphibian tail resorption. *J. Biol. Chem.* **268,** 16270–16278.

Weber, R. (1965). Inhibitory effect of actinomycin on tail atrophy in *Xenopus* larvae at metamorphosis. *Experientia* **21,** 665–666.

Weber, R. (1967). Biochemistry of amphibian metamorphosis. *In* "The Biochemistry of Animal Development" (R. Weber, ed.), pp. 227–301. Academic Press, New York.

Weber, R. (1969). Tissue involution and lysosomal enzymes during anuran metamorphosis. *In* "Lysosomes in Biology and Pathology" (J. T. Dingle and H. B. Fell, eds.), Vol. I, pp. 437–461. North-Holland, Amsterdam.

Weigel, N. L., Poletti, A., Beck, C. A., Edwards, D. P., Carter, T. H., and Denner, L. A. (1993). Phosphorylation and progesterone receptor function. *In* "Steroid Hormone Receptors: Basic and Clinical Aspects" (V. K. Moudgil, ed.), pp. 309–332. Birkhäuser Boston, Boston.

Weinberger, C., Thompson, C. C., Ong, E. S., Lebo, R., Gruol, D. J., and Evans, R. M. (1986). The c-erb-A gene encodes a thyroid hormone receptor. *Nature (London)* **324,** 641–646.

Welte, T., Garimorth, K., Philipp, S., and Doppler, W. (1994). Prolactin-dependent activation of a tyrosine phosphorylated DNA binding factor in mouse mammary epithelial cells. *Mol. Endocrinol.* **8,** 1091–1102.

White, B. A., and Nicoll, C. S. (1981). Hormonal control of amphibian metamorphosis. *In* "Metamorphosis: A Problem in Developmental Biology" (L. I. Gilbert and E. Frieden, eds.), 2nd ed., pp. 363–396. Plenum, New York.

Yamamoto, K. R., Pearce, D., Thomas, J., and Miner, J. N. (1992). Combinatorial regulation at a mammalian composite response element. *In* "Transcriptional Regulation" (S. L. McKnight and K. R. Yamamoto, eds.), pp. 1169–1192. Cold Spring Harbor Laboratory, Cold Spring Harbor, New York.

Yaoita, Y., and Brown, D. D. (1990). A correlation of thyroid hormone receptor gene expression with amphibian metamorphosis. *Genes Dev.* **4,** 1917–1924.

Yaoita, Y., Shi, Y., and Brown, D. D. (1990). *Xenopus laevis* α and β thyroid hormone receptors. *Proc. Natl. Acad. Sci. U.S.A.* **87,** 7090–7094.

Yen, P. M., and Chin, W. W. (1994). New advances in understanding the molecular mechanisms of thyroid hormone action. *Trends Endocrinol. Metab.* **5,** 65–72.

Yen, P. M., Spanjaard, R. A., Sugawara, A., Darling, D. S., Nguyen, V. P., and Chin, W. W. (1993). Orientation and spacing of half-sites differentially affect T_3-receptor (TR) monomer, homodimer, and heterodimer binding to thyroid hormone response elements (TREs). *Endocrin. J.* **1,** 461–466.

Yoshizato, K. (1989). Biochemistry and cell biology of amphibian metamorphosis with a special emphasis on the mechanism of removal of larval organs. *Int. Rev. Cytol.* **119,** 97–149.

Zenke, M., Munoz, A., Sap, J., Vennstrom, B., and Beug, H. (1990). V-erbA oncogene activation entails the loss of hormone-dependent regulator activity of c-erbA. *Cell (Cambridge, Mass.)* **61,** 1035–1049.

Zhang, X.-K., and Pfahl, M. (1993). Regulation of retinoid and thyroid hormone action through homodimeric and heterodimeric receptors. *Trends Endocrinol. Metab.* **4,** 156–162.

14

Thyroid Hormone-Regulated Early and Late Genes during Amphibian Metamorphosis

YUN-BO SHI

Laboratory of Molecular Embryology
National Institute of Child Health and Human Development
Bethesda, Maryland

I. INTRODUCTION

Many amphibians undergo a biphasic developmental process. Following embryonic development, the animal exists as a tadpole for a relatively short period during which it undergoes mostly growth with little further differentiation. Subsequently, the tadpole metamorphoses into its adult form in a process initiated by the synthesis of endogenous thyroid hormone (TH) (Gudernatsch, 1912; Beckingham Smith and Tata, 1976; Dodd and Dodd, 1976; Gilbert and Frieden, 1981). This metamorphic process systematically transforms all tissues or organs of a tadpole. Among them, limb development and tail resorption represent two extreme cases. The first case involves *de novo* development of a new structure during the early stages of metamorphosis, which involves rapid proliferation of many cell types, including connective tissue, muscle, and epidermal cells, and their subsequent differentiation. In contrast, the tail, which is a tadpole-specific organ comprising essentially the same types of cells as the developing limb, completely degenerates toward the end of metamorphosis (see Yoshizato, Chapter 19, this volume). However, most other tissues or organs, which are present both in tadpoles and frogs, undergo remodeling, which involves both specific cell death of tadpole tissues and selective cell proliferation and differentiation of adult cells. For example, as an aquatic tadpole is transformed into a terrestrial frog, the liver is remodeled so that the activity of urea-cycle enzymes is elevated as the animal changes from ammonotelism to ureotelism (Cohen, 1970; Atkinson *et al.,* Chapter 15, this volume). Similarly, extensive remodeling in the digestive system accompanies the change from herbivorous to carnivorous feeding habits (Smith-Gill and Carver, 1981; Dauca and Hourdry, 1985; Yoshizato, 1989).

Although the timing and specific changes during metamorphosis are highly tissue dependent, all are controlled by TH (White and Nicoll, 1981; Kikuyama *et al.,* 1993; Kaltenbach, Chapter 11, this volume). Thus a simple addition of exogenous TH to the rearing water of premetamorphic tadpoles can induce precocious metamorphosis, whereas inhibiting the synthesis of endogenous TH blocks this transition. Furthermore, the response of different organs to TH are autonomous since organs such as limb, tail, and intestine can be induced to undergo metamorphosis with TH even when cultured *in vitro* (Dodd and Dodd, 1976; Ishizuya-Oka and Shimozawa, 1991; Tata *et al.,* 1991).

While it is unknown how TH regulates these tissue-specific transformations, TH has been shown to affect gene expression in metamorphosing tissues. Thus, it has been assumed that following embryonic development and the subsequent tadpole growth period, the rising concentration of endogenous TH (Leloup and Buscaglia, 1977; White and Nicoll, 1981) induces a new period of gene expression in different tissues/organs that leads to the removal or remodeling of larval organs and differentiation of adult tissues. Many of these TH response genes have been isolated and characterized. This chapter summarizes some of these studies. The expression of the TH-regulated genes is correlated with the metamorphic transitions in different tissues and potential functions of these genes are discussed.

II. THYROID HORMONE RECEPTORS

TH is believed to exert its effect on metamorphosis through its nuclear receptors, i.e., thyroid hormone receptors (TRs). TRs bind TH with much higher affinity than cellular or plasma TH-binding proteins (Galton, 1983). More importantly, these receptors are TH-dependent transcription factors that regulate the transcription of target genes which contain thyroid hormone response elements (TREs) (Evans, 1988; Green and Chambon, 1988; Tsai and O'Malley, 1994).

TR belongs to the steroid hormone receptor superfamily. Members of this family also include receptors for glucocorticoids, androgen, estrogen, and retinoic acid and share similar structural domains. Each of them has a DNA-binding domain located in the amino-terminal half of the protein and a hormone-binding domain in the carboxyl half (Fig. 1). The DNA-binding domains of different receptors share considerable homology and are involved in their specific binding to hormone response elements (e.g., TRE for TR) present in the genes regulated by the corresponding hormones. The hormone-binding domain is, on the other hand, unique to each receptor. Likewise, the other regions of the proteins are more divergent than the DNA-binding domain among different receptors (see Tata, Chapter 13, this volume).

Like other receptors encoded by this gene family, TRs bind TREs as homo- or heterodimers and regulate gene transcription in response to TH. TR can heterodimerize with receptors for retinoic acids and vitamin D (Forman and Samuels, 1990; Yu *et al.,* 1991). Among them, the best dimerization partners are retinoid X receptors, which bind 9-*cis*-retinoic acid (Heyman *et al.,* 1992). More importantly, retinoid X receptors are present in many tissues, and their heterodimers with TR seem to confer the specificity of gene regulation by TH, suggesting that they are the *in vivo* partners of TR (Blumberg *et al.,* 1992; Perlmann *et al.,* 1993; Kuro-

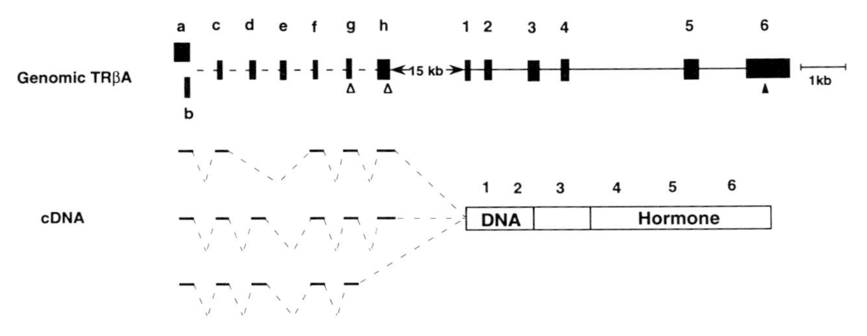

Fig. 1. Organization of TRβ genes in *Xenopus laevis.* The two TRβ genes have similar structures and only TRβA is shown here. (Top) Genomic structure. The exons a–h encode mostly the 5′-untranslated region and exons 1–6 encode the protein-coding region from zinc finger 1 of the DNA-binding domain (exon 1) to the carboxyl terminus. The translation initiation and termination sites are indicated by open and solid triangles, respectively. The relative location of exons a and b are unknown and thus are shown at the same position. The exons are shown as solid bars and introns as lines. The dashed lines indicate introns of unknown sizes. (Bottom) Three examples of TRβA mRNAs derived from alternative splicing of the 5′ exons. The 5′ exons present in the mRNA are indicated by short bars underneath the corresponding exons in the genomic DNA. No evidence has been reported for the existence of alternative splicing involving exons 1–6 (Shi *et al.,* 1992).

kawa *et al.,* 1993; Zhang and Pfahl, 1993; Yen and Chin, 1994; Tsai and O'Malley, 1994).

As in the case of birds and mammals, two subfamilies of TRs, TRα and TRβ, are present in amphibia, which have been cloned in both *Xenopus laevis* and *Rana catesbeiana* (Brooks *et al.,* 1998; Yaoita *et al.,* 1990; Helbing *et al.,* 1992; Schneider and Galton, 1991). These receptor genes are expressed during metamorphosis (Yaoita and Brown, 1990; Kawahara *et al.,* 1991; Schneider and Galton, 1991; Helbing *et al.,* 1992; Tata, Chapter 13, this volume; Atkinson *et al.,* Chapter 15, this volume), suggesting that they are likely to mediate the effect of TH during amphibian metamorphosis.

TRα and TRβ genes are differentially regulated during development. This is best demonstrated in *X. laevis.* While the two TRα genes are activated shortly after tadpole hatching and the effect of TH on their expression is small and kinetically slow, the two TRβ genes are expressed at very low levels until stage 54 when they are activated by the rising concentration of endogenous TH (Yaoita and Brown, 1990; Kawahara *et al.,* 1991). Furthermore, the response of TRβ genes to TH treatment is fast, within several hours, and is independent of new protein synthesis, suggesting that they respond to TH directly at the transcriptional level (Kanamori and Brown, 1992; Wang and Brown, 1993; Shi and Brown, 1993). On the gene structural level, each of the two TRβ genes spans at least 70 kb of genomic DNA and consists of at least 12 exons, about half of which encode the 5′-untranslated region (Fig. 1) (Shi *et al.*, 1992). Alternative splicing of these 5′ exons yields multiple isoforms of TRβ mRNAs that encode proteins that differ only in a small region of the amino terminus (Fig. 1) (Yaoita *et al.*, 1990). In contrast, no evidence of alternatively spliced mRNA has been reported for TRα genes and the genomic structure of the TRα genes is still unknown. Another interesting feature of TRβ genes is the existence of two different promoters for each gene (Fig. 1) (Shi *et al.*, 1992; Kanamori and Brown, 1992). One of the promoters is located upstream of exon a and is constitutively active, although at very low levels. The other, located upstream of exon b, requires the presence of TH for its activity; at least one TRE is present in its regulatory region (Ranjan *et al.*, 1994). While the functional significance of the alternative mRNA splicing is unclear, the alternative promoter usage allows the TRβ genes to be expressed constitutively at low levels in the absence of TH and at much higher levels in the presence of TH during metamorphosis.

III. GENE REGULATION BY THYROID HORMONE DURING METAMORPHOSIS

The transcription regulatory properties of TH through its receptors suggest that TH induces a gene regulation cascade in each tissue that undergoes metamorphosis. The presumption is that TH binds to TR homo- and/or heterodimers and that the resulting complexes activate or repress a set of genes at the transcriptional level, without a requirement for new protein synthesis. If the products of these direct response genes are transcription factors, they are expected to regulate yet another set of genes. If the products of the direct response gene are not transcription factors, they could still be indirectly involved in regulating genes later in the process through other signal transduction mechanisms. An unknown number of these regulatory periods will be required to effect the morphological changes. According to this model, tissue-specific changes during metamorphosis are determined by the activation or repression of tissue-specific genes at various steps of the cascade.

A. Early Thyroid Hormone Response Genes

The gene regulation cascade model just outlined predicts the existence of genes that respond directly to TH by changing their expression at the transcriptional level. Clearly, the characterization of these early TH response genes is crucial in order to understand the molecular mechanisms underlying the hormonal regulation of metamorphosis. These genes, however, appear to be expressed at very low levels or regulated by TH in a tissue-specific manner since a traditional differential hybridization method failed to identify any such genes using RNAs isolated from total premetamorphic tadpoles treated with TH (Shi and Brown, 1990). The isolation of the early response genes has been made possible by the development of a polymerase chain reaction (PCR)-based subtractive differential screening method (Wang and Brown, 1991). Systematic analysis of genes that are regulated by TH within 24 hr in the hind limb (Buckbinder and Brown, 1992), tail (Wang and Brown, 1991, 1993), and intestine (Shi and Brown, 1993) of *X. laevis* tadpoles has revealed that these genes are indeed expressed at low levels and/or regulated in a tissue-specific manner by TH. The following provides a brief summary of these differential screens. (For the discussion here, the early response genes refer to those genes that are regulated by TH within 24 hr, which include the direct response genes, whereas the late response genes require TH treatment of premetamorphic tadpoles for 1 day or longer to alter their mRNA levels).

1. Intestine

Intestinal remodeling represents an important organ transformation that occurs during metamorphosis. It involves specific degeneration of larval (tadpole) tissues and concurrent development of adult ones. In *X. laevis,* the most dramatic period is around stages 60–64, when extensive cell death occurs in the primary epithelium while secondary epithelial cells proliferate and differentiate (Fig. 2) (McAvoy and Dixon, 1977). Anatomically, the length of the small intestine reduces by as much as 90% from stage 58 to stage 64, and the epithelium changes from a simple

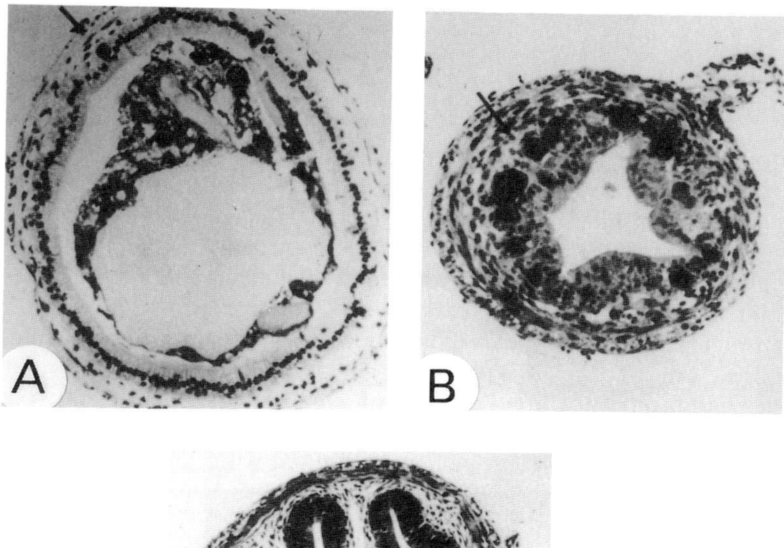

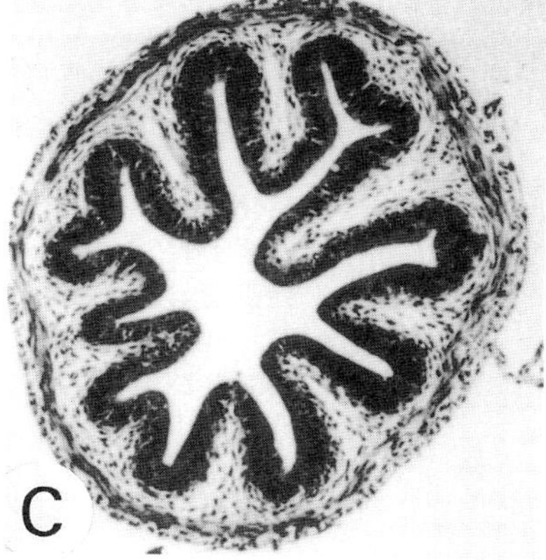

Fig. 2. Histological changes in the small intestine of *X. laevis* during metamorphosis. At stage 60 (A), the epithelial cell islets (arrow) start to develop, which will eventually form the secondary epithelium. By stage 61 (B), the islets are more prominent and exist between the degenerating primary epithelium and developing connective tissue. Within the next 10 days and by stage 66 (C), these islet cells will differentiate into a monolayer, multiply folded structure, completely replacing the primary epithelial cells. The connective tissue and muscles are also well developed by stage 66 (C). [Modified from Cell proliferation and renewal in the small intestinal epithelium of metamorphosing and adult *Xenopus laevis,* McAvoy and Dixon Copyright © (1977). Reprinted by permission from John Wiley & Sons, Inc.] Not shown here is the typhlosole which is the single epithelial fold present in the anterior one-third of the small intestine in tadpoles of stage 60 or younger, where the tadpole connective tissue is abundant (Marshall and Dixon, 1978; Ishizuya-Oka and Shimozawa, 1987) (see also Fig. 8).

tubular structure to a multiply folded form (Fig. 2) (Marshall and Dixon, 1978; Ishizuya-Oka and Shimozawa, 1987). These processes are controlled by TH. In fact, the endogenous TH concentration has been shown to peak around this period (Leloup and Buscaglia, 1977). To identify genes that may play a role in the remodeling process, a subtractive differential screen was conducted using intestinal RNAs isolated from stage 54 premetamorphic tadpoles that had been treated for 18 hr with or without 5 nM T_3, close to the endogenous plasma T_3 concentration at the metamorphic climax (stage 62) (Leloup and Buscaglia, 1977). A total of 22 T_3 upregulated and a single downregulated genes were isolated (Table I). Most of the genes respond to T_3 treatment very quickly and their regulation by T_3 appears to be independent of new protein synthesis (Tables I and II, Fig. 3), suggesting that they are direct TH response genes and represent the first period of gene regulation induced by TH.

The identities of many of the upregulated genes have been revealed by sequence analysis (Table II). In agreement with the gene regulation cascade model noted earlier, several of the direct response genes encode transcription factors and thus will likely be involved in directly activating or repressing the transcription of intermediate and/or late TH response genes. In addition, several genes encoding proteins varying from a transmembrane protein to extracellular enzymes are also found to be among the early response genes in the intestine. These results suggest that TH simultaneously induces many intra- and extracellular events, which in turn cooperate to effect intestinal remodeling.

The first indication that these genes are likely to play important roles in tissue remodeling comes from their dramatic developmental regulation in the intestine during metamorphosis (e.g., Fig. 4). Furthermore, the same expression patterns can be reproduced for most of these genes when premetamorphic tadpoles are treated with 5 nM T_3 (Fig. 4), a treatment that induces remodeling, e.g., intestinal length reduction and epithelial folding (Shi and Hayes, 1994). Developmentally, these early response genes fall into three general classes. The genes in the first class, e.g., TRβ, a basic leucine zipper motif-containing transcription factor (Fig. 4), the extracellular matrix-degrading metalloproteinase stromelysin-3, and a nonhepatic arginase gene (Table II), are expressed strongly only at the climax of intestinal remodeling (around stages 60–62). Much lower levels of expression of these genes are present in pre- or postmetamorphic intestine (Fig. 4). The second class of genes, e.g., NF-1 type of transcription factors and a transmembrane protein (Table II), is activated during metamorphosis and their expression re-

TABLE I

Early Thyroid Hormone Response Genes in Limb, Tail, and Intestine of *Xenopus laevis*

Parameter	Hind limb[a]	Intestine[b]	Tail[c]
Morphological changes	De novo development	Remodeling	Resorption
Major events	Stages 51–56: cell proliferation, morphogenesis (digit formation)	Stages 58–62: cell death in primary epithelium, length reduction, development of the connective tissue and muscles	Stages 62–66: cell death, removal of degenerated tissues by phagocytosis, etc.
	Stages 56–66: limb growth and differentiation	Stages 60–66: proliferation and differentiation of cells of the secondary epithelium, connective tissue and muscles	
Number of upregulated genes isolated	14	22	15
Number of upregulated direct response genes[d]	3	15	11
Regulation during metamorphosis	Moderate	Very dramatic	Very dramatic
Magnitude of T_3 regulation[e] (average)	3- to 12-fold (5)	3- to >20-fold (≥9)	6- to >20-fold (≥11)
Number of downregulated genes isolated	5	1	4

[a] TH treatment: 24 hr at 5 n*M* T_3 on stage 54 tadpoles (Buckbinder and Brown, 1992).
[b] TH treatment: 18 hr at 5 n*M* T_3 on stage 54 tadpoles (Shi and Brown, 1993).
[c] TH treatment: 24 hr at 100 n*M* T_3 on stage 54 tadpoles (Wang and Brown, 1991, 1993).
[d] Based on the resistance of the regulation of these genes to protein synthesis inhibition.
[e] The numbers refer to the factors of activation after respective treatment (a–c). The slightly higher magnitude of activation for the tail genes could be due to the higher concentration of T_3 used.

TABLE II

Early TH Upregulated Genes of Known Identity

Gene name	Homologous genes	Isolated from	TH regulation in[d]			Response to TH[e]	Possible function
			Limb	Tail	Intestine		
Tail 1	Zinc finger region of Sp1	Tail[b]	+	+	+	Direct	
IU16/33	NF-1	Intestine[c]	+	+	+	Direct	Transcription factors: regulating gene transcription[b,c,g]
TRβ		Tail,[b] intestine[c]	+	+	+	Direct	
Tail 8/9	bZip region of E4BP4	Tail,[b] intestine[c]	+	+	+	Direct	
Tail 11	Collagenase	Tail[b]	+	+	+	ND	Metalloproteinase: extracullar matrix remodeling[b,c]
Tail 14	Stromelysin-3	Tail,[b] intestine[c]	+	+	+	Direct	
Tail 15	Type III 5-deiodinase	Tail[b]	+	+		Direct	Conversion of T_4 and T_3 to rT_3 and T_2[b,f]
IU22	Nonhepatic arginase	Intestine[c]			+	ND	Proline biosynthesis from arginine, etc.[h]
IU24	Na^+/PO^-_4 cotransporter	Intestine[c]			+	Direct	PO^-_4 transport[g]
IU12	Transmembrane protein	Intestine[c]	+		+	ND	Signal transduction[g]
B	Heat-shock proteins	Limb[a]	+	−	+	Direct	
I	Heat-shock proteins	Limb[a]	+			ND	
M	Heat-shock proteins	Limb[a]	+			ND	
N	Heat-shock proteins	Limb[a]	+		+	ND	Rapid cell growth[a]
H	Yeast MCM3 Mouse P1	Limb[c]	+		+	ND	
J	Mouse eIF-4A	Limb[a]	+		+	ND	
P	Rat clathrin B chain	Limb[a]	+			ND	Vesicular intracellular transport[a]

[a] Buckbinder and Brown (1992). [b] Wang and Brown (1993). [c] Shi and Brown (1993). [d] +, upregulated; space, no regulation; −, down-regulation. [e] A "direct response indicates that the regulation is resistant to protein synthesis inhibition; ND, not determinable. [f] St. Germain *et al.* (1994). [g] M. Puzianowska-Kuznicki, M. Stolow, and Y.-B. Shi (unpublished observations). [h] Patterton and Shi (1994).

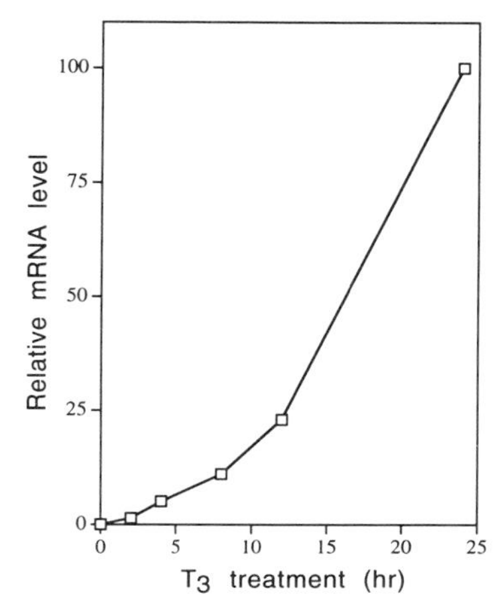

Fig. 3. Kinetics of the up-regulation by T_3 of an early response gene, Na^+/PO^-_4 cotransporter (IU24), in the intestine. [Based on data of Shi and Brown (1993).]

mains high in postmetamorphic frog intestine. Finally, two genes, including the Na^+/PO_4^- cotransporter (Table II), are in the third class. They are expressed at high levels immediately before or after the climax of metamorphosis but minimally at the actual climax.

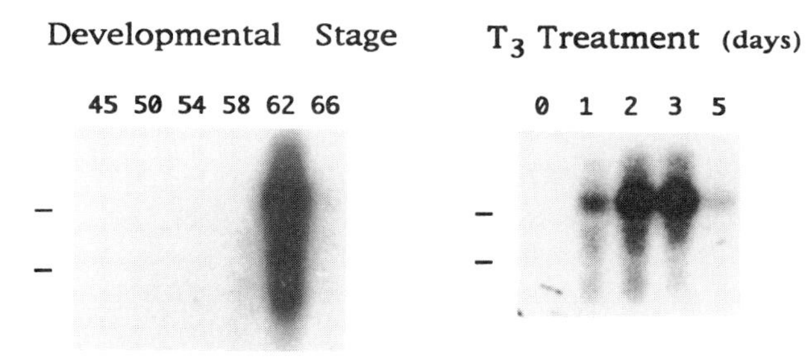

Fig. 4. The expression of an early response gene, a bZIP-containing transcription factor (tail 9, Table II), during normal and T_3-induced metamorphosis in the intestine (stage 54 tadpoles were used for TH treatment. The high levels of expression correlate with the period of dramatic intestinal remodeling. [Modified with permission after Shi and Brown (1993), with permission from the American Society for Biochemistry and Molecular Biology, Inc.]

2. *Tail*

The resorption of the tadpole tail is one of the last changes to be completed during metamorphosis. While some events such as the resorption of the fail fin occur earlier, the reduction in tail length is minimal until about stage 62 in *X. laevis* (Nieuwkoop and Faber, 1956; Dodd and Dodd, 1976), when the endogenous TH is at its peak concentration (Leloup and Buscaglia, 1977). However, tail resorption can be induced by premetamorphic tadpoles by TH treatment. Of the genes that are regulated by TH within the first day of treatment, 15 upregulated and 4 downregulated ones have been isolated (Wang and Brown, 1991, 1993). The majority of these early upregulated genes appear to be direct response genes (Tables I and II). Among the genes identified to date by their sequences are TRβ genes, two other putative transcription factors (one containing a basic leucine zipper motif and the other a zinc finger protein), and genes encoding metalloproteinases and a type III deiodinase. Some of these genes were independently isolated from the intestinal screen discussed earlier (Table II). Developmentally, all but one of the upregulated genes are expressed strongly only during tail resorption. The exception is the 5-deiodinase, whose expression peaks around stages 60–61, immediately before massive tail resorption takes place (for more discussion on this gene, see Section IV,A). All of the downregulated genes show a dramatic reduction in their expression as the tail resorbs. These developmental profiles suggest that the genes participate in the resorption process. For example, as tail resorption takes place, the activities of many macromolecule-degrading enzymes increase dramatically (Weber, 1967; Dodd and Dodd, 1976; see Tata, Chapter 13, this volume; Yoshizato, Chapter 19, this volume). The activation of the collagenase gene (Table II) is likely to be responsible for the well-documented increase in collagen-degrading activity in the resorbing tail fin (Gross, 1966).

3. *Hind Limb*

Hind limb development is one of the earliest events to occur during metamorphosis (Nieuwkoop and Faber, 1956). In *X. laevis,* the hind limb buds are detected by stage 48, when there is little endogenous TH. However, further development requires TH. This is initiated when the endogenous plasma TH concentration starts to rise around stage 54 (Leloup and Buscaglia, 1977) and is essentially completed by the time the TH concentration reaches a peak at climax (stage 62). A subtractive differential screen isolated 19 genes, 14 of which were upregulated and 5 were downregulated in the hind limb within 1 day by TH (Buckbinder and Brown, 1992). With the exception of a few genes, the activation of the upregulated genes occurs only after at least 12 hr of TH treatment, suggesting that the response to TH is delayed compared to the direct response genes. In general, the magnitude of their up-regulation by TH is considerably less than that of the genes isolated from the tail and intestine (Table I). Similarly, their expression levels during hind limb development (stage 54–66) vary only moderately compared to those for genes in the tail and intestine. However, sequence analysis revealed that many of these genes encode proteins that are expected to be involved in cell proliferation and tissue growth (Table II). This is in agreement with their potential functions during limb development, as hind limb morphogenesis takes place mostly before stage 56 and after that the hind limb undergoes rapid growth and maturation.

Most of the early response genes are not tissue specific. Many of the genes isolated from tail and intestine could be activated by TH treatment in other tissues, including the hind limb, even though their mRNA levels do not change appreciably during natural limb development (Wang and Brown, 1993; Shi and Brown, 1993) (Table II). Similarly, some of the limb genes were regulated in the intestine (Buckbinder and Brown, 1992), although their activation by TH in the intestine is to a lesser extent than that of the genes isolated from the intestinal screen. In addition, many of the early response genes also respond to T_3 with similar kinetics in tissue culture cells, and conversely many TH response genes isolated from the tissue culture cells are also TH response genes in tadpole tissues (Kanamori and Brown, 1993).

For an unknown reason, all three organs described earlier were found to have much fewer early response genes that were downregulated by TH within 24 hr (Table I). It is possible that the mRNAs of the downregulated genes are very stable and that their levels change slowly even though the corresponding genes have been repressed by TH. Consequently, the differential screen would not identify them. However, no new downregulated genes were found in the tail when a differential screen was performed using stage 54 tadpoles treated with TH for 2 days, a treatment known to be sufficient to induce tail resorption (Wang and Brown, 1993). Thus, the treatment produced proper regulation of all genes that are crucial for tail resorption and, at least in the tail, there are likely only a small number of early downregulated genes. At present, much less is known about the downregulated genes. It is difficult to assess their importance during metamorphosis.

B. Late Thyroid Hormone Response Genes

Although TH apparently controls the metamorphic transition in all tissues/organs, different tissues undergo drastically different changes. These changes take place over a period of about 1 month in *X. laevis* and undoubtedly involve alterations in the expression of many tissue-specific genes (see Tata, 1993, and Chapter 13, this volume; Kaltenbach, Chapter 11, this volume). Similar changes in the expression of these genes are expected if premetamorphic tadpoles are treated with TH. Some of these genes will respond to TH quickly, i.e., the early response genes described earlier. Others will change their expression only after at least 1 day of TH treatment. These latter genes are, thus, responding to TH indirectly and require the synthesis of missing proteins for their regulation. These genes are considered to be the late TH response gene according to the gene regulation cascade model described earlier. While no systematic studies has been carried out to isolate and characterize these late response genes, many such genes have been identified over the years (Table III). For example, as the herbivorous tadpole metamorphoses into a carnivorous frog, the digestive system is remodeled. This is accompanied by drastic changes in the expression of the intestinal fatty acid-binding protein (IFABP) gene and the exocrine-specific genes encoding pancreatic enzymes such as trypsin (Table III). Both the trypsin and IFABP genes are expressed in the respective larval and adult organs (see Fig. 5 for IFABP). However, their mRNA levels are repressed to a minimum at the climax of metamorphosis, when the exocrine pancreatic or primary intestinal epithelial cells undergo degeneration. This suppression can be induced in premetamorphic tadpoles by prolonged (>1 day) treatment with TH (Fig. 5). At least for the IFABP gene, this downregulation appears to be an instance of gene repression or selective degradation of IFABP mRNA since it occurs before cell death in the primary epithelium in which the gene is expressed (Shi and Hayes, 1994). Both the trypsin and IFABP gene are reactivated at the adult exocrine pancreatic or intestinal epithelial cells differentiate (Fig. 5) (see also Ishizuya-Oka *et al.,* 1994).

TABLE III

Some Known Late TH Response Genes

Gene	Tissue	Frog species	Regulation by TH	Length of TH Treatment	Reference
Carbamoylphosphate synthase I	Liver	*Rana catesbeiana*	Up	≥1 day	Morris (1987), Helbing *et al.* (1992), Galton *et al.* (1991)
Argininosuccinate synthase	Liver	*R. catesbeiana*	Up	≥2 days	Morris (1987)
Arginase	Liver	*Xenopus laevis* *R. catesbeian*	Up	≥2 days	Xu *et al.* (1993), Helbing and Atkinson (1994), Atkinson *et al.* (1994)
Ornithine carbamoyltransferase	Liver	*R. catesbeiana*	Up	≥1 day	Helbing *et al.* (1992)
N-CAM	Liver	*X. laevis*	Up	≥2 days	Levi *et al.* (1990)
Albumin	Liver	*R. catesbeiana* *X. laevis*	Up	≥2 days	Schultz *et al.* (1988), Moskaitis *et al.* (1989)
Myosin heavy chain	Limb	*X. laevis*	Up	≥3 days	Buckbinder and Brown (1992)
Keratin	Epidermis	*X. laevis*	Up	≥2 days	Mathisen and Miller (1989)
Magainin	Skin	*X. laevis*	Up	7 days	Reilly *et al.* (1994)
Trypsin	Pancreas	*X. laevis*	Down	≥2 days	Shi and Brown (1990)
Intestinal fatty acid-binding protein	Intestine	*X. laevis*	Down	≥1 day	Shi and Hayes (1994)
Gelatinase A	Intestine	*X. laevis*	Up	≥3 days	Patterton *et al.* (1995)

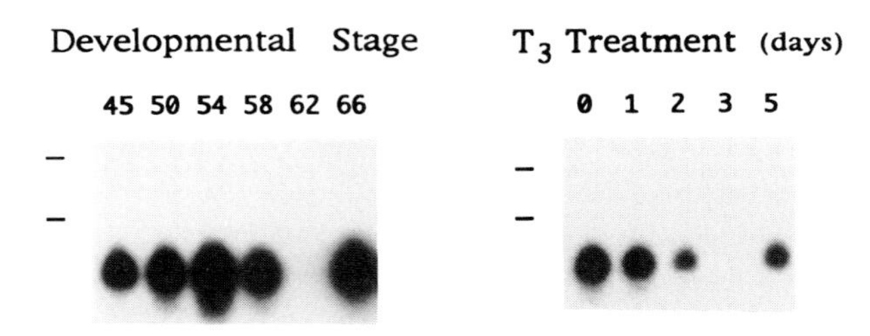

Fig. 5. Regulation of a late TH response gene, the intestinal fatty acid-binding protein gene, during normal and T_3-induced metamorphosis in the intestine (stage 54 tadpoles were used for TH treatment). The gene is not expressed when the primary epithelium is degenerating and adult epithelial cells are not yet differentiated at stage 62 or after 3 days of hormone treatment. [Modified after Shi and Hayes (1994).]

One of the best-studied systems is the activation of genes encoding the urea cycle enzymes in the liver on the transition from ammonotelism to urotelism in many anuran amphibians. The enzymatic activities of these enzymes have long been known to be coordinately upregulated during metamorphosis or by TH treatment (Cohen, 1970; Dodd and Dodd, 1976). It has been shown more recently that at least four of the enzyme-encoding genes, i.e., carbamoyl-phosphate synthase I, argininosuccinate synthase, ornithine carbamoyltransferase, and arginase genes, are activated by prolonged TH treatment of premetamorphic tadpoles (Table III). Although information on the fifth enzyme, arginine synthase, is not available, it is very likely that it too is a late TH response gene and that the up-regulation in its activity is due to an increased mRNA level caused by TH during metamorphosis. In addition to the urea cycle enzymes, a few other genes have also been shown to be late TH response genes in the liver (Table III). [For more information on gene regulation in the liver, see Chapter 15 by Atkinson *et al.* (this volume).]

In the developing hind limb, the genes of the connective tissue and muscles, such as the collagen and myosin heavy chain genes, are highly expressed as the organ differentiates (A. Kanamori, Y.-B. Shi, and D. D. Brown, unpublished observations; also see Buckbinder and Brown, 1992). These genes can also be activated in the hind limb when premetamorphic tadpoles are treated with TH. However, such activation occurs only after prolonged treatment, suggesting that these genes are late TH response genes.

Like the hind limb, the maturation of frog skin is also associated with the activation of adult-specific genes. Of these genes, the epidermal keratin gene can be activated by TH in premetamorphic tadpoles (Table III) (Mathisen and Miller, 1989). Furthermore, a short treatment of larval skin can result in the activation of the adult keratin gene several days later even if the skin explant is subsequently maintained in TH-free medium, demonstrating that TH induces an irreversible gene regulation cascade that leads to the change in keratin expression (for more details see Miller, Chapter 17, this volume). Another family of genes that is highly expressed in the adult skin are the magainin antimicrobial peptide genes (Bevins and Zasloff, 1990). The granular glands of adult frog skin are the predominant sites for synthesis and storage of these polypeptides. At least two of the genes in this peptide family have been shown to be activated at the climax of metamorphosis in *X. laevis* (Reilly *et al.,* 1994). As one would expect from the causative role of TH on metamorphosis, prolonged treatment of tadpoles (e.g., 7 days) with TH leads to premature activation of these genes (Table III). Although it is unclear what is the shortest treatment period for the activation of these genes, they are probably late TH response genes as they are expressed in the differentiated cells in the granular glands.

C. Tissue-Dependent Variation in Gene Expression Program Induced by Thyroid Hormone

The tissue-specific transformations during metamorphosis would argue that TH likely controls a different gene regulation cascade in each different tissue. The systematic analysis of the early response genes in three contrasting tissues and available information on some late TH response gene have indeed revealed drastically different gene regulation profiles for the intestine, tail, and hind limb of *X. laevis.*

The total regression of the tail appears to be the simplest response to TH to explain among the changes in all organs. Although it consists of a variety of cell types, they are all destinated to undergo cell death and resorption in the presence of TH. To get a sense of the complexity of the gene expression program induced by TH in tail resorption, an estimate of total number of genes regulated by TH within the first 48 hr was made based on how often different small PCR fragments from a single gene were isolated during the differential screening process (Wang and Brown, 1993). Within the first 24 hr after the addition of T_3, approximately 25 genes are predicted to be activated by TH, of which 15 have been isolated. Although the accuracy of the gene number is limited by the difficulty to take into account the size of the mRNA, abundance, and the magnitude of up-regulation of an individual gene by TH, it clearly indicates that only a finite member of upregulated genes exist in the resorbing tail within the first 24 hr. Furthermore, a similar estimate predicts that only about 35 genes are induced within the first 48 hr by TH. This treatment has been shown to be sufficiently long for tail resorption to occur even if the tail is subsequently maintained in a medium containing protein synthesis inhibitors (Wang and Brown, 1993). On the other hand, blocking the RNA and/or protein synthesis at the onset of hormone treatment is known to inhibit tail resorption in organ cultures (Tata, 1966). These observations suggest that the 35 or so upregulated genes together with the downregulated genes are sufficient to effect tail resorption (Wang and Brown, 1993). It is important to realize, however, that there are possibly other genes which are crucial for the resorption process but are not included in the estimated 35 genes due to their low abundance (<10 copies of the mRNA per cell) and/or low magnitude of their up-regulation by TH ($<$ sixfold by 24 hr of 100 n*M* T_3 treatment) (Wang and Brown, 1993).

In sharp contrast to the tail, the hind limb appears to have a much more complex gene regulation program (Buckbinder and Brown, 1992). It has been estimated that over 120 genes are upregulated by TH within 24 hr. While most of the 14 isolated genes are not likely to be direct TH response genes, many direct TH response genes isolated from the tail and intestine are also upregulated in the hind limb by T_3 (Table II) (Buckbinder and Brown, 1992; Wang and Brown, 1993). These results indicate the presence of at least two periods of gene up-regulation within the first 24 hr, which is then followed by a third period that involves genes required for cell proliferation as reflected by a significant increase in cells in the S phase of the cell cycle (Buckbinder and Brown, 1992). Finally, as cells differentiate after 3 days or longer, another period of gene regulation leads to the activation of genes of the differentiated adult cells such as myosin heavy chain gene (Table III).

Intestinal remodeling, on the other hand, involves changes that are similar in part to both tail resorption and limb development. The tadpole intestine comprises mostly primary epithelium. One of the major events during intestinal remodeling is the degeneration of the primary epithelium. Thus, it might be anticipated that at least some early response genes are common in the tail and intestine. In fact, many of the early response genes described earlier were found to be regulated in both organs (Tables I and II). Furthermore, several genes were independently isolated from both tissues. The total number of genes that are upregulated by TH within the first 24 hr was also estimated to be similar to that in the tail (Shi and Brown, 1993), contrasting with the large number of genes involved in the hind limb. Finally, the magnitude of the activation of the intestinal and tail genes by TH or during metamorphosis in these tissues is similar and considerably greater than that of the genes isolated from the hind limb (Table I).

While the early gene regulation program shares similarity between the intestine and the tail, intestinal remodeling resembles limb development in a few aspects. The latter two organs eventually differentiate into adult-type tissues in response to TH. It is, therefore, not surprising that the gene regulation cascade is more complex in the intestine and limb than that in the tail. Like hind limb development, intestinal remodeling involves multiple periods of gene activation and repression. Many early response genes are eventually repressed toward the end of metamorphosis while others remain to be expressed at relatively high levels (Shi and Brown, 1993). On the other hand, the genes that are specific to differentiated epithelial cells like the IFABP gene are repressed as the epithelium undergoes cell death. Eventually, e.g., after several days of hormonal treatment, many of these epithelial genes are expected to be reactivated, like the IFABP gene (Fig. 5), as secondary epithelial cells differentiate to form the adult organ.

Another level of complexity in the gene expression program is present in all tissues. That is, the direct TH response genes identified so far belong to very diverse groups, e.g., genes encoding transcription factors and extracellular matrix-degrading enzymes. This diversity could be in part due to the fact that all of these different organs themselves are very complex, consisting of many different cell types. Different cells could respond to TH differently by altering the expression of different genes, for example, the activation of stromelysin-3 genes in only the fibroblastic cells of the intestine (Patterton *et al.,* 1995; Yoshizato, Chapter 19, this volume). On the other hand, a given gene could respond to TH in a similar manner in a specific cell type in different organs, thus allowing many of the same genes to be regulated by TH in organs that undergo drastically different transformations. These complex gene expression profiles contrast sharply with a much simpler but possible mechanism in which TH triggers a tissue-specific gene regulation cascade through sequential activation of a series of tissue-specific genes. Instead, the current information indicates that multiple intra- and extracellular processes are concurrently activated by TH in a given organ.

IV. CORRELATION OF GENE EXPRESSION WITH TISSUE REMODELING AND FUNCTION OF EARLY RESPONSE GENES

The developmental regulation of early TH response gene during metamorphosis in different tissues strongly suggests that they participate directly or indirectly in the regulation of downstream genes and/or other intra- and extracellular events. Currently there is no direct evidence for such functions for any of the genes so far isolated. However, the identification of many of the early response genes through sequence analysis has provided considerable information on the roles that these genes may play during metamorphosis (Table II). For example, the transcription factors will undoubtedly activate or repress the transcription of downstream genes involves in the gene regulation cascade. The following discussion focuses in two aspects of metamorphosis in which some of these early response genes may participate.

A. Tissue-Dependent Temporal Regulation of Metamorphosis

It has long been known that different tissues not only undergo tissue-specific transformations but also exhibit metamorphic changes at different development stages. Thus, in *Xenopus,* hind limb development is one of the earliest changes while tail resorption is one of the last (Nieuwkoop and Faber, 1956; Atkinson, 1981). The exact mechanism for this temporal order is still unknown. However, several genes have been implicated to be involved based on their tissue-dependent developmental expression profiles. The mRNA levels for these genes in different tissues during development are shown in Fig. 6.

The first genes in this regard are the two TRα genes. The genes are highly expressed in the hind limb around stages 54–56 when limb morphogenesis takes place. They are subsequently repressed to low levels. In the tail, the receptor gene expression increases gradually to high levels at the climax of metamorphosis (around stage 62). The TRβ genes, which are direct TH response genes (Table II), have similar expression profiles (Yaoita and Brown, 1990; Wang and Brown, 1993).

Two genes have been implicated in the control of free TH levels in different tissues, thus affecting the metamorphic process indirectly. The first one is a direct TH response gene and encodes a type III 5-deiodinase

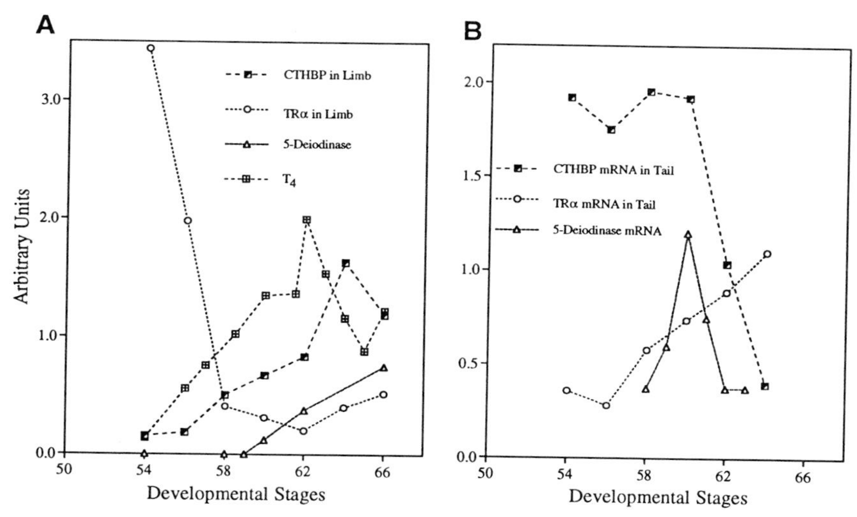

Fig. 6. Expression of TRα, CTHBP, and 5-deiodinase genes in the hind limb (A) and tail (B) during metamorphosis. TRα gene expression correlates with limb morphogenesis (stages 54–56) and tail resorption (stages 62–64) (Nieuwkoop and Faber, 1956) whereas that of CTHBP and 5-deiodinase, the proteins encoded by both of which could reduce the effective TH concentration, show a reverse correlation. The plasma T_4 concentration was from Leloup and Buscaglia (1977) and a similar profile exists for T_3 (see Tata, Chapter 13, this volume). CTHBP and TRα mRNA levels were from Shi *et al.* (1994; see also Wang and Brown, 1993; Yaoita and Brown, 1990). The mRNA levels for 5-deiodinase were based on those of Wang and Brown (1993).

(Table II) which converts T_4 and T_3 to rT_3 and T_2, respectively (St. Germain, 1994; St. Germain *et al.*, 1994). Both T_2 and rT_3 are biologically inactive. Thus, the action of 5-deiodinase reduces the effective TH concentration. During development the 5-deiodinase gene is not expressed in the hind limb until stage 59 or later after limb morphogenesis is essentially completed (Wang and Brown, 1993) (Fig. 6). In the tail, it is expressed at all stages but the mRNA level increases severalfold at the onset of climax (stage 60). The mRNA level drops when massive tail resorption begins (Fig. 6). The second gene encodes a cytosolic thyroid hormone-binding protein (CTHBP) (Shi *et al.*, 1994). The homologous protein in mammals has been shown to bind TH (Kato *et al.*, 1989; Ashizawa *et al.*, 1991). This binding reduces the free cellular TH concentration and consequently inhibits TH-dependent transcriptional activation by TRs (Ashizawa and Cheng, 1992). The expression of the *X. laevis* CTHBP gene is inversely correlated with metamorphosis in both the tail and hind limb (Fig. 6). The CTHBP mRNA levels are low in the hind limb during the period of morphogenesis (around stage 56 or earlier) and increase thereafter, paralleling hind limb growth. In contrast, in the tail, the gene is highly expressed until stage 62 when it is downregulated, coinciding with the rapid reduction in tail length (Nieuwkoop and Faber, 1956).

While these developmental expression profiles are based on the mRNA levels for these genes, analysis of TRα and TRβ proteins in *Xenopus* indicates that higher levels of the receptors do correspond to higher levels of their mRNAs (Eliceiri and Brown, 1994). Thus it is expected, although remains to be proven, that similar correlations exist between the protein levels and tissue remodeling. If such correlations could be established in those proteins, they would suggest several possibilities that could determine the developmental timing of tissue transformation. Thus, in the hind limb at stages 54–56, even though the plasma TH levels are low, the reduced expression of CTHBP and the absence of 5-deiodinase expression would allow the accumulation of a reasonable concentration of free cellular TH. This, together with the high levels of TR, could in turn lead to the activation of an early TH response gene such as stromelysin-3 (Fig. 7) and the subsequent induction of limb morphogenesis. In the tail of tadpoles up to stage 60, the high levels of CTHBP and 5-deiodinase reduce the effective cellular TH concentration, thus preventing tail resorption. However, some early response genes such as the deiodinase gene itself and TRβ genes begin to be activated by the remaining available free TH. It is unclear what causes the reduction in 5-deiodinase gene after stage 60 in the tail. Nonetheless, the result of this reduction together with the repression of CTHBP expression would likely increase the free cellular TH, thereby activating the tail resorption process.

B. Cell–Cell and Cell–Extracellular Matrix Interactions in Tissue Remodeling

The identification of genes encoding enzymes involved in extracellular matrix (ECM) degradation as early TH response genes suggests that ECM plays an important role during metamorphosis. This should not

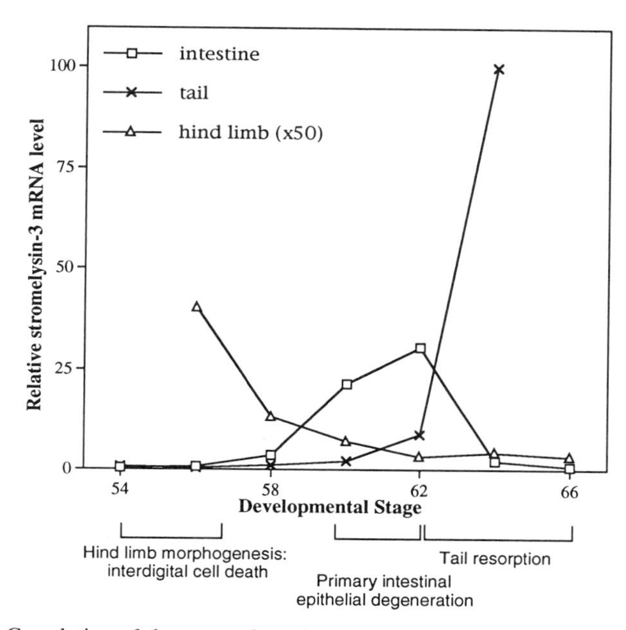

Fig. 7. Correlation of the expression of matrix metalloproteinase gene stromelysin-3 with tissue remodeling involving cell death during metamorphosis of *X. laevis*. The mRNA levels were from Patterton *et al.* (1995).

be surprising as ECM is known to influence development and cell differentiation (Hay, 1991, 1993; Johnson *et al.*, 1992). This is especially true for the intestine. The intestinal epithelium is separated from the mesenchyme by a special ECM, the basal lamina (basement membrane), whose major components include laminin, entactin, type IV collagen, and proteoglycan. While the epithelium is responsible for the major biological function of the organ, its development requires the participation of the connective tissue and the ECM as best shown in birds and mammals (Simon-Assmann and Kedinger, 1993; Louvard *et al.*, 1992; Weiser *et al.*, 1990).

Evidence is also accumulating that cell–cell and cell–ECM interactions are important, at least for intestinal remodeling during metamorphosis. This process can be induced with exogenous TH in organ cultures. By using fragments derived from different regions of the intestine, it has been demonstrated that the development of adult epithelium requires the presence of larval connective tissue, although primary epithelial cell death can be induced by TH in its absence (Ishizuya-Oka and Shimozawa, 1992). Such a requirement is also consistent with the temporal regulation of the developing connective tissue and adult epithelium. The connective tissue is minimal in premetamorphic tadpole intestine and develops extensively during metamorphosis as the primary epithelial cells die and adult epithelial cells proliferate and differentiate (Fig. 2) (McAvoy and Dixon, 1977; Ishizuya-Oka and Shimozawa, 1987). Extensive cell contacts are present between the epithelium and the connective tissue. In addition, precursors of adult epithelial cells proliferate as cell nests or islets between the developing mesenchyme and degenerating larval epithelium.

It is likely that ECM is involved in mediating the mesenchymal–epithelial cell interactions. ECM could also directly regulate cellular events through its physical contacts with cells. Such interactions are likely to be altered by the expression of extracellular matrix-degrading enzymes (metalloproteinases) (Table II and III). In fact, evidence suggests that different metalloproteinases play different roles during metamorphosis. First of all, the type I collagenase gene (Table II) can be induced by TH in limb, tail, and intestine but its expression in the limb and intestine is much weaker than that in the tail (Wang and Brown, 1993). This agrees with the fact that there is little connective tissue degeneration in either the limb or the intestine during metamorphosis. Both the collagenase and the stromelysin-3 genes are early TH response genes, suggesting their participation in the early periods of tissue remodeling. A detailed analysis of the developmental expression of stromelysin-3 mRNA suggests that stromelysin-3 may play a role in tissue remodeling that involves cell death (Fig. 7) (Patterton *et al.,* 1995). The gene is highly expressed in the intestine and tail when massive cell death occurs in these tissues. Even in the hind limb it is expressed, although at much weaker levels than that in the metamorphosing intestine or tail, around stages 54–56 when cell death takes place in the interdigital region of the hind limb as morphogenesis occurs. More importantly, the activation of the stromelysin-3 gene at least in both the tail and intestine occurs before the onset of cell death. In addition, *in situ* and Northern blot hybridizations have revealed that stromelysin-3 is expressed only in the fibroblastic cells of the developing connective tissue (Fig. 8) (Patterton *et al.,* 1995). This reinforces the idea that connective tissue plays an important role in epithelial development.

It is interesting to compare the expression of the stromelysin-3 gene during amphibian metamorphosis to that in mammals. The gene was first cloned from a human breast cancer carcinoma and was found to be

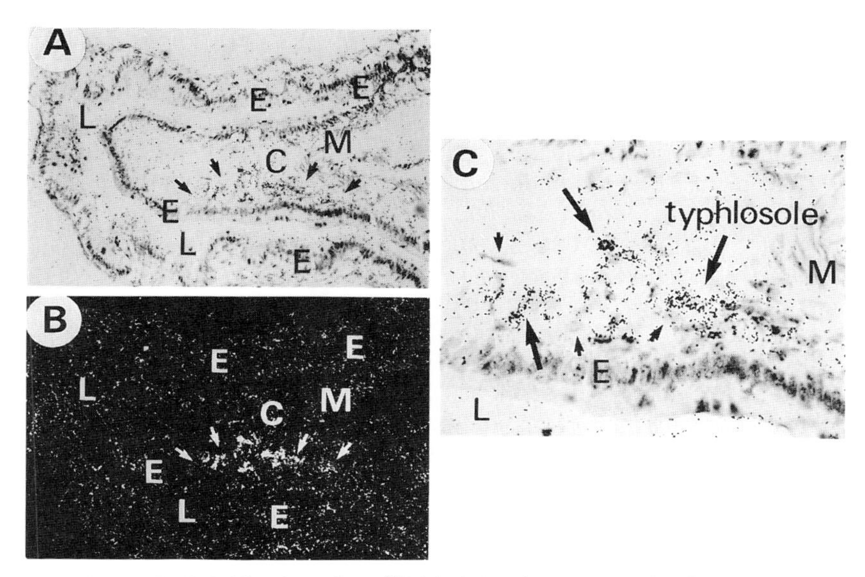

Fig. 8. *In* situ hybridization using a ^{35}S-labeled antisense cRNA localizes stromelysin-3 expression to the connective tissue of the *Xenopus* tadpole small intestine after TH treatment. (A) Bright-field; (B) dark-field. The stromelysin-3 expressing cells (arrows) are loosely clustered in the connective tissue (C) within the typhlosole (the only epithelial fold in tadpoles epithelium). No expression is detected in the epithelium (E), muscle (M), or lumen (L). (C) Higher magnification of the bright-field. Individual stromelysin-3 expressing cells appear to be polarized and fibroblast like (large arrows) in the typhlosole, where they are mixed with other fibroblast-like cells (small arrows) that do not express detectable stromelysin mRNA. [Modified after Patterton *et al.* (1994).]

expressed specifically in all primary carcinomas analyzed, unlike many other metalloproteinase genes (Basset *et al.*, 1990). Its unique expression profiles in different human carcinomas have led to the suggestion that it facilitates tumor cell invasion, although its precise function is still unknown (Basset *et al.*, 1990; Wolf *et al.*, 1992; Muller *et al.*, 1993). However, it is only slightly expressed in most normal mammalian tissues. Those tissues that do express the gene at high levels often involve cell death, for example, the mouse mammary gland during its involution and the interdigital regions of the developing limb (Basset *et al.*, 1990; Lefebvre *et al.*, 1992). Again, the expression is restricted to the fibroblastic cells. These results suggest that at least during normal development, the cell-specific expression and biological function of stromelysin-3 are well conserved from frog to human.

In sharp contrast to the stromelysin-3 gene, two other putative metalloproteinase genes, detected by using the human gelatinase A and stromelysin 1 cDNA probes, have very different expression profiles (Table III) (Patterton *et al.,* 1995). While stromelysin-1 mRNA is present throughout development in *Xenopus* tail and intestine and TH has only a small effect on its expression, the gelatinase A gene is a late TH response gene (Table III) and its expression during development is delayed relative to that of stromelysin-3, i.e., after cell death has begun. These results suggest that stromelysin-1 may be involved in ECM turnover during normal tissue maintenance and metamorphosis. Gelatinase A may participate in degrading ECM after cell death has occurred as suggested by its delayed expression, assuming again that the protein levels correlate with the mRNA concentrations. In contrast, stromelysin-3 is likely involved in the modification of ECM during early stages of tissue remodeling. This change in ECM could in turn influence cell fate in response to TH, i.e., cell death of larval tissues and proliferation and differentiation of adult ones.

V. COMPARISON WITH INSECT METAMORPHOSIS

The hormone ecdysone plays an essentially identical role in insects metamorphosis as TH does in amphibians (Gilbert and Frieden, 1981; Atkinson, 1994; Gilbert *et al.,* Chapter 2, this volume). Based on studies of the puffing patterns of the polytene chromosome in the salivary gland of *Drosophila* in response to pulses of ecdysone during development, Ashburner *et al.* (1974) proposed a gene regulation cascade model in which the hormone first activates genes located at the early puffs through its nuclear receptor (see also Russell and Ashburner, Chapter 3, this volume; Lezzi, Chapter 4, this volume). The products of these early, direct response genes were predicted to activate the expression of late response genes as well as suppress their own expression. This model has been substantiated and refined in recent years through the molecular cloning of the ecdysone receptor (Koelle *et al.,* 1991) and the characterization of genes at some of the early puffs in *Drosophila* (Andres and Thummel, 1992; Cherbas and Cherbas, Chapter 5, this volume). Like TH receptors, the ecdysone receptor is a member of the steroid hormone receptor superfamily and can form a heterodimer with the *Drosophila* protein ultraspiracle, the *Drosophila* homolog of vertebrate receptors for 9-*cis*-retinoic acid (Yao *et al.,* 1992). The heterodimer has strong affinity for ecdysone response elements, which, upon binding ecdysone, can activate and/or repress the transcription of direct ecdysone response genes. The early puffs are expected to encode such upregulated direct response genes that are expressed only transiently in response to an ecdysone pulse. At least some of these encode transcription factors and thus could be involved in the repression of their own expression and the activation of late response genes, as predicted by the Ashburner model. However, as found in *Xenopus,* not all of the direct response genes in *Drosophila* are transcription factors (Andres and Thummel, 1992; Bayer *et al.,* Chapter 9, this volume), suggesting that the actual gene regulation program is more complex than the Ashburner model.

While apparently similar gene regulation cascades are manifested, a few distinct differences exist between amphibian and insect metamorphosis. First of all, amphibian metamorphosis requires the continuous presence of TH, whereas in insects, several ecdysone pulses are required to ensure proper development (Riddiford and Truman, 1978; Richards, 1981; see also Gilbert *et al.,* Chapter 2, this volume; Cherbas and Cherbas, Chapter 5, this volume). The expression of the direct early response genes is only transient in response to each hormone pulse in *Drosophila* (see, e.g., Paine-Saunders *et al.,* 1990; Karim and Thummel, 1992). In *Xenopus,* the early genes are expressed for a much longer period, if not through the entire metamorphic process (references in Table I). Although several early response genes in the tadpole intestine are eventually repressed, it is unclear whether this repression involves the products of these early genes themselves as suggested for insect metamorphosis. The direct ecdysone response genes change their expression very quickly (within 1 hr) upon ecdysone addition to cultures of *Drosophila* imaginal discs (Paine-Saunders *et al.,* 1990) or to organ cultures of *Drosophila* larvae (Karim and Thummel, 1992). In *Xenopus* a lag of several hours exists between the addition of TH and the increase in the mRNA levels of the direct TH response genes in the tail, intestine, or hind limb (see references in Table I). The mechanisms behind these different induction kinetics remain to be investigated.

VI. CONCLUSION

It is becoming increasingly clear that TH controls amphibian metamorphosis by regulating a complex, tissue-dependent gene expression program. As summarized in a simplified scheme in Fig. 9, the hormone first binds to its nuclear receptors (TRs). The homo- or heterodimers of these receptors with other nuclear receptors or transcription factors could then activate or repress a set of direct response genes. These early response genes encode either transcription factors or other protein, which could be either tissue specific or ubiquitous. The transcription factors are presumed to control the expression of downstream genes directly, while other proteins are likely to assert their effect more indirectly. For example, they could regulate and/or participate in signal transduction by growth factors, modify ECM or cell surface, thus influencing cell–cell and cell–matrix interactions. These complex intra- and extracellular processes eventually cooperate to determine the tissue-specific transformations.

Such a simplified scheme immediately raises many questions. First of all, it remains to be demonstrated that the homo- or heterodimers of TRs or both are indeed the mediators of the effects of TH during metamorphosis *in vivo*. If heterodimers are involved, the roles of 9-*cis*-retinoic acid receptors and other possible dimerization partners need to be investigated.

Many of the early response genes are ubiquitous, suggesting that at least some common changes may take place at early stages of the response of different organs to TH. It will be interesting to determine how

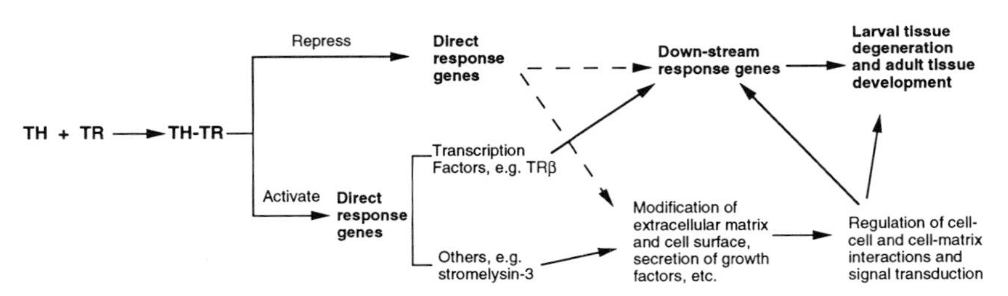

Fig. 9. A simplified gene regulation cascade model for the regulation of tissue remodeling by TH. Currently, there is little information on the downregulated direct response genes, thus the possible effects of their down-regulation on downstream events are indicated by dashed arrows. TRα mRNAs are expressed at higher levels than TRβ mRNAs and, therefore, TRα is likely to play a major role in the regulation of the direct response genes during the early stages of tissue remodeling. TRs are presumed to function as either homo- or heterodimers with other nuclear receptors, especially 9-*cis*-retinoic acid receptors. Not shown is the possibility that the products of some of the TH response genes could in turn influence the expression of the direct TH response genes.

the same genes expressed in different tissues undergoing vastly different changes can exert their biological effect. It is likely that they, together with tissue-specific but TH-independent factors, regulate tissue-specific downstream genes. However, very few such downstream genes are known. Even though some late TH response genes have been identified, it is not clear whether they are regulated by any of the early response genes or whether other intermediate factors are required for their regulation. In this regard, it is interesting to note that many early TH response genes are also regulated in tissue culture cells by TH (Kanamori and Brown, 1993). Thus, tissue culture cells may offer a good model system to study the regulation and function of these genes (see Tata, Chapter 13, this volume).

Some of the early response genes, such as those encoding stromelysin-3 and nonhepatic arginase (Patterton *et al.,* 1995; Patterton and Shi, 1994), are also expressed during embryogenesis. It is unknown whether they play similar roles during embryogenesis and metamorphosis. It will also be interesting to determine how these genes are regulated during these two very distinct developmental periods, one requiring TH and the other not.

It would also be important to know how extracellular processes regulated by TH influence intracellular events. For example, these could lead to changes in the levels of growth factors or other diffusable substances or alter signal transduction through cell surface moleuecles such as integrins (for a review on integrins in signal transduction, see Damsky and Werb, 1992). The ability to induce metamorphic changes by thyroid hormone, especially in organ cultures, and the availability of many genes involved in the process should make it possible to directly address these questions.

ACKNOWLEDGMENTS

The author thanks Drs. D. D. Brown, J. R. Tata, and B. G. Atkinson for very helpful comments and suggestions on the manuscript and Ms. T. Vo for preparing it.

REFERENCES

Andres, A. J., and Thummel, C. S. (1992). Hormones, puffs and flies: The molecular control of metamorphosis by ecdysone. *Trends Genet.* **8,** 132–138.

Ashburner, M., Chihara, C., Meltzer, P., and Richards, G. (1974). Temporal control of puffing activity in polytene chromosome. *Cold Spring Harbor Symp. Quant. Biol.* **38,** 655–662.

Ashizawa, K., and Cheng, S.-Y. (1992). Regulation of thyroid hormone receptor-mediated transcription by a cytosol protein. *Proc. Natl. Acad. Sci. U.S.A.* **89,** 9277–9281.

Ashizawa, K., McPhie, P., Lin, K.-H., and Cheng, S.-Y. (1991). An in vitro novel mechanism of regulating the activity of pyruvate kinase M_2 by thyroid hormone and fructose 1, 6-bisphosphate. *Biochemistry* **30,** 7105–7111.

Atkinson, B. G. (1981). Biological basis of tissue regression and synthesis. *In* "Metamorphosis: A Problem in Developmental Biology" (L. I. Gilbert and E. Frieden, eds.), 2nd ed., pp. 397–444. Plenum, New York.

Atkinson, B. G. (1994). Metamorphosis: Model systems for studying gene expression in postembryonic development. *Dev. Genet.* **15,** 313–319.

Atkinson, B. G., Helbing, C., and Chen, Y. (1994). Reprogramming of gene expression in the liver of *Rana catesbeiana* tadpoles during spontaneous and thyroid hormone induced metamorphosis. *In* "Perspectives in Comparative Endocrinology" (K. S. Davey, S. S. Tobe, and R. G. Peters, eds.), pp. 416–423. National Research Council of Canada, Ottawa.

Basset, P., Bellocq, J. P., Wolf, C., Stoll, I., Hutin, P., Limacher, J. M., Podhajcer, O. L., Chenard, M. P., Rio, M. C., and Chambon, P. (1990). A novel metalloproteinase gene specifically expressed in stromal cells of breast carcinomas. *Nature* (*London*) **348,** 699–704.

Beckingham Smith, K., and Tata, J. R. (1976). The hormonal control of amphibian metamorphosis. *In* "Developmental Biology of Plants and Animals" (C. Graham and P. F. Wareing, eds.), pp. 232–245. Blackwell, Oxford.

Bevins, C. L., and Zasloff, M. (1990). Peptides from frog skin. *Annu. Rev. Biochem.* **59,** 395–414.

Blumberg, B., Mangelsdorf, D. J., Dyck, J. A., Bittner, D. A., Evans, R. M., and De Robertis, E. M. (1992). Multiple retinoid-responsive receptors in a single cell: Families of retinoid "X" receptors and retinoic acid receptors in the *Xenopus* egg. *Proc. Natl. Acad. Sci. U.S.A.* **89,** 2321–2325.

Brooks, A. R., Sweeney, G., and Old, R. W. (1989). Structure and functional expression of a cloned *Xenopus* thyroid hormone receptor. *Nucleic Acids Res.* **17,** 9395–9405.

Buckbinder, L., and Brown, D. D. (1992). Thyroid hormone-induced gene expression changes in the developing frog limb. *J. Biol. Chem.* **267,** 25786–25791.

Cohen, P. P. (1970). Biochemical differentiation during amphibian metamorphosis. *Science* **168,** 533–543.

Damsky, C. H., and Werb, Z. (1992). Signal transduction by integrin receptors for extracellular matrix: Cooperative processing of extracellular information. *Curr. Biol.* **4,** 772–781.

Dauca, M., and Hourdry, J. (1985). Transformations in the intestinal epithelium during anuran metamorphosis. *In* "Metamorphosis" (M. Balls and M. Bownes, eds.), pp. 36–58. Clarendon, Oxford.

Dodd, M. H. I., and Dodd, J. M. (1976). The biology of metamorphosis. *In* "Physiology of the Amphibia" (B. Lofts, ed.), pp. 467–599. Academic Press, New York.

Eliceiri, B. P., and Brown, D. D. (1994). Quantitation of endogenous thyroid hormone receptors α and β during embryogenesis and metamorphosis in *Xenopus laevis*. *J. Biol. Chem.* **269,** 22459–22465.

Evans, R. M. (1988). The steroid and thyroid hormone receptor superfamily. *Science* **240,** 889–895.

Forman, B. M., and Samuels, H. H. (1990). Interaction among a subfamily of nuclear hormone receptors: The regulatory zipper model. *Mol. Endocrinol.* **4,** 1293–1301.

Galton, V. A. (1983). Thyroid hormone action in amphibian metamorphosis. *In* "Molecular Basis of Thyroid Hormone Action" (J. H. Oppenheimer and H. H. Samuels, eds.), pp. 445–483. Academic Press, New York.

Galton, V. A., Morganelli, C. M., Schneider, M. J., and Yee, K. (1991). The role of thyroid hormone in the regulation of hepatic carbamyl phosphate synthetase activity in *Rana catesbeiana. Endocrinology* (*Baltimore*) **129,** 2298–2304.

Gilbert, L. I., and Frieden, E., eds. (1981). "Metamorphosis: A Problem in Developmental Biology," 2nd ed. Plenum, New York.

Green, S., and Chambon, P. (1988). Nuclear receptors enhance our understanding of transcription regulation. *Trends Genet.* **4,** 309–313.

Gross, J. (1966). How tadpoles lose their tails. *J. Invest. Dermatol.* **47,** 274–277.

Gudernatsch, J. F. (1912). Feeding experiments on tadpoles. I. The influence of specific organs given as food on growth and differentiation: A contribution to the knowledge of organs with internal secretion. *Wilhelm Roux Arch. Entwicklungsmech. Org.* **35,** 457–483.

Hay, E. D. (1991). "Cell Biology of Extracellular Matrix," 2nd ed. Plenum, New York.

Hay, E. D. (1993). Extracellular matrix alters epithelial differentiation. *Curr. Biol.* **5,** 1029–1035.

Helbing, C. C., and Atkinson, B. G. (1994). 3,5,3′-Triiodothyronine–induced carbamyl-phosphate synthetase gene expression is stabilized in the liver of *Rana catesbeiana* tadpoles during heat shock. *J. Biol. Chem.* **269,** 11743–11750.

Helbing, C. C., Gergely, G., and Atkinson, B. G. (1992). Sequential up-regulation of thyroid hormone β receptor, ornithine transcarbamylase, and carbamyl phosphate synthetase mRNAs in the liver of *Rana catesbeiana* tadpoles during spontaneous and thyroid hormone-induced metamorphosis. *Dev. Genet.* **13,** 289–301.

Heyman, R. A., Mangelsdorf, D. J., Dyck, J. A., Stein, R. B., Eichele, G., Evans, R. M., and Thaller, C. (1992). 9-*cis* Retinoic acid is a high affinity ligand for the retinoid X receptor. *Cell* (*Cambridge, Mass.*) **68,** 397–406.

Ishizuya-Oka, A., and Shimozawa, A. (1987). Development of the connective tissue in the digestive tract of the larval and metamorphosing *Xenopus laevis. Anat. Anz.* **164,** 81–93.

Ishizuya-Oka, A., and Shimozawa, A. (1991). Induction of metamorphosis by thyroid hormone in anuran small intestine cultured organotypically in vitro. *In Vitro Cell. Dev. Biol.* **27A,** 853–857.

Ishizuya-Oka, A., and Shimozawa, A. (1992). Connective tissue is involved in adult epithelial development of the small intestine during anuran metamorphosis in vitro. *Wilhelm Roux's Arch Dev. Biol.* **201,** 322–329.

Ishizuya-Oka, A., Shimozawa, A., Takeda, H., and Shi, Y.-B. (1994). Cell-specific and spatiotemporal expression of intestinal fatty acid-binding protein gene in the small intestine during amphibian metamorphosis. *Wilhelm Roux's Arch. Dev. Biol.* **204,** 150–155.

Johnson, K. E., Boucaut, J. C., and DeSimone, D. W. (1992). Role of the extracellular matrix in amphibian gastrulation. *Curr. Top. Dev. Biol.* **27,** 91–127.

Kanamori, A., and Brown, D. D. (1992). The regulation of thyroid hormone receptor β genes by thyroid hormone in *Xenopus laevis. J. Biol. Chem.* **267,** 739–745.

Kanamori, A., and Brown, D. D. (1993). Cultured cells as a model for amphibian metamorphosis. *Proc. Natl. Acad. Sci. U.S.A.* **90,** 6013–6017.

Karim, F. D., and Thummel, C. S. (1992). Temporal coordination of regulatory gene expression by the steroid hormone ecdysone. *EMBO J.* **11,** 4083–4093.

Kato, H., Fukuda, T., Parkison, C., McPhie, P., and Cheng, S.-Y. (1989). Cytosolic thyroid hormone-binding protein is a monomer of pyruvate kinase. *Proc. Natl. Acad. Sci. U.S.A.* **86,** 7861–7865.

Kawahara, A., Baker, B. S., and Tata, J. R. (1991). Developmental and regional expression of thyroid hormone receptor genes during *Xenopus laevis* metamorphosis. *Development* **112,** 933–943.

Kikuyama, S., Kawamura, K., Tanaka, S., and Yamamoto, K. (1993). Aspects of amphibian metamorphosis: Hormonal control. *Int. Rev. Cytol.* **145,** 105–148.

Koelle, M. R., Talbot, W. S., Segraves, W. A., Bender, M. T., Cherbas, P., and Hogness, D. S. (1991). The *Drosophila* EcR gene encodes an ecdsyone receptor, a new member of the steroid receptor superfamily. *Cell (Cambridge, Mass)* **67,** 59–77.

Kurokawa, R., Yu, V. C., Naar, A., Kyakumoto, S., Han, Z., Silverman, S., Rosenfeld, M. G., and Glass, C. K. (1993). Differential orientations of the DNA-binding domain and carboxy-terminal dimerization interface regulate binding site selection by nuclear receptor heterodimers. *Genes Dev.* **7,** 1423–1435.

Lefebvre, O., Wolf, C., Limacher, J. M., Hutin, P., Wendling, C., LeMeur, M., Basset, P., and Rio, M. C. (1992). The breast cancer-associated stromelysin-3 gene is expressed during mouse mammary gland apoptosis. *J. Cell Biol.* **119,** 997–1002.

Leloup, J., and Buscaglia, M. (1977). La triiodothyronine, hormone de la métamorphose des Amphibiens. *C. R. Hebd. Seances Acad. Sci., Ser. D* **284,** 2261–2263.

Levi, G., Broders, F., Dunon, D., Edelman, G. M., and Thiery, J. P. (1990). Thyroxine-dependent modulations of the expression of the neural cell adhesion molecule N-CAM during *Xenopus laevis* metamorphosis. *Development* **109,** 681–692.

Louvard, D., Kedinger, M., and Hauri, H. P. (1992). The differentiating intestinal epithelial cell: Establishment and maintenance of functions through interactions between cellular structures. *Annu. Rev. Cell Biol.* **8,** 157–195.

Marshall, J. A., and Dixon, K. E. (1978). Cell specialization in the epithelium of the small intestine of feeding *Xenopus laevis*. *J. Anat.* **126,** 133–144.

Mathisen, P. M., and Miller, L. (1989). Thyroid hormone induces constitutive keratin gene expression during *Xenopus laevis* development. *Mol. Cell. Biol.* **9,** 1823–1831.

McAvoy, J. W., and Dixon, K. E. (1977). Cell proliferation and renewal in the small intestinal epithelium of metamorphosing and adult *Xenopus laevis*. *J. Exp. Zool.* **202,** 129–138.

Morris, S. M., Jr. (1987). Thyroxine elicits divergent changes in mRNA levels of two urea cycle enzymes and one gluconeogenic enzyme in tadpole liver. *Arch. Biochem. Biophys.* **259,** 144–148.

Moskaitis, J. E., Sargent, T. D., Smith, L. H. Jr., Pastori, R. L., and Schoenberg, D. R. (1989). *Xenopus laevis* serum albumin: Sequence of the complementary deoxyribonucleic acids encoding the 68- and 74-kilodalton peptides and the regulation of albumin gene expression by thyroid hormone during development. *Mol. Endocrinol.* **3,** 464–473.

Muller, D., Wolf, C., Abecassis, J., Millon, R., Engelmann, A., Bronner, G., Rouyer, N., Rio, M. C., Eber, M., Methlin, G., Chambon, P., and Basset, P. (1993). Increased stromelysin 3 gene expression is associated with increased local invasiveness in head and neck squamous cell carcinomas. *Cancer Res.* **53,** 165–169.

Nieuwkoop, P. D., and Faber, J. (1956). "Normal Table of *Xenopus laevis*." North-Holland Amsterdam.

Paine-Saunders, S., Fristrom, D., and Fristrom, J. W. (1990). The *Drosophila* IMP-E2 gene encodes an apically secreted protein expressed during imaginal disc morphogenesis. *Dev. Biol.* **140,** 337–351.

Patterton, D., and Shi, Y.-B. (1994). Thyroid hormone-dependent differential regulation of multiple arginase genes during amphibian metamorphosis. *J. Biol. Chem.* **269,** 25328–25334.

Patterton, D., Hayes, W. P., and Shi, Y.-B. (1995). Transcriptional activation of the matrix metalloproteinase gene stromelysin-3 coincides with thyroid hormone induced cell death during frog metamorphosis. *Dev. Biol.* **167,** 252–262.

Perlmann, T., Rangarajan, P. N., Umesono, K., and Evans, R. M. (1993). Determinants for selective RAR and TR recognition of direct repeat HREs. *Genes Dev.* **7,** 1411–1422.

Ranjan, M., Wong, J., and Shi, Y.-B. (1994). Transcriptional repression of *Xenopus* TRβ gene is mediated by a thyroid hormone response element located near the start site. J. Biol. Chem. **269,** 24699–24705.

Reilly, D. S., Tomassini, N., and Zasloff, M. (1994). Expression of magainin antimicrobial peptide genes in the developing granular glands of *Xenopus* skin and induction by thyroid hormone. *Dev. Biol.* **162,** 123–133.

Richards, G. (1981). The radioimmune assay of ecdysteroid titres in *Drosophila melanogaster. Mol. Cell. Endocrinol.* **21,** 181–197.

Riddiford, L. M., and Truman, J. W. (1978). Biochemistry of insect hormones and insect growth regulators. *In* "Biochemistry of insects" (M. Rockstein, ed.), pp. 307–356. Academic Press, New York.

Schneider, M. J., and Galton, V. A. (1991). Regulation of C-erbA–α messenger RNA species in tadpole erythrocytes by thyroid hormone. *Mol. Endocrinol.* **5,** 201–208.

Schultz, J. J., Price, M. P., and Frieden, E. (1988). Triiodothyronine increases translatable albumin messenger RNA in *Rana catesbeiana* tadpole liver. *J. Exp. Zool.* **247,** 69–76.

Shi, Y.-B., and Brown, D. D. (1990). Developmental and thyroid hormone dependent regulation of pancreatic genes in *Xenopus laevis. Genes Dev.* **4,** 1107–1113.

Shi, Y.-B., and Brown, D. D. (1993). The earliest changes in gene expression in tadpole intestine induced by thyroid hormone. *J. Biol. Chem.* **268,** 20312–20317.

Shi, Y.-B., and Hayes, W. P. (1994). Thyroid hormone-dependent regulation of the intestinal fatty acid-binding protein gene during amphibian metamorphosis. *Dev. Biol.* **161,** 48–58.

Shi, Y.-B., Yaoita, Y., and Brown, D. D. (1992). Genomic organization and alternative promoter usage of the two thyroid hormone receptor β genes in *Xenopus laevis. J. Biol. Chem.* **267,** 733–788.

Shi, Y.-B., Liang, V. C.-T., Parkison, C., and Cheng, S.-Y. (1994). Tissue-dependent developmental expression of a cytosolic thyroid hormone binding protein gene in *Xenopus:* Its role in the regulation of amphibian metamorphosis. *FEBS Lett.* **355,** 61–64.

Simon-Assmann, P., and Kedinger, M. (1993). Heterotypic cellular cooperation in gut morphogenesis and differentiation. *Semin. Cell Biol.* **4,** 221–230.

Smith-Gill, S. J., and Carver, V. (1981). Biochemical characterization of organ differentiation and maturation. *In* "Metamorphosis: A Problem in Developmental Biology" (L. I. Gilbert and E. Frieden, eds.), 2nd ed., pp. 491–544. Plenum, New York.

St. Germain, D. L. (1994). Iodothyronine deiodinases. *Trends Endocrinol. Metab.* **5,** 36–42.

St. Germain, D. L., Schwartzman, R. A., Croteau, W., Kanamori, A., Wang, Z., Brown, D. D., and Galton, V. A. (1994). A thyroid hormone regulated gene in *Xenopus laevis* encodes a type III iodothyronine 5-deiodinase. *Proc. Natl. Acad. Sci. U.S.A.* **91,** 7767–7771.

Tata, J. R. (1966). Requirement for RNA and protein synthesis for induced regression of the tadpole tail in organ culture. *Dev. Biol.* **13,** 77–94.

Tata, J. R. (1993). Gene expression during metamorphosis: An ideal model for post-embryonic development. *BioEssays* **15,** 239–248.

Tata, J. R., Kawahara, A., and Baker, B. S. (1991). Prolactin inhibits both thyroid hormone-induced morphogenesis and cell death in cultured amphibian larval tissues. *Dev. Biol.* **146,** 72–80.

Tsai, M.-J., and O'Malley, B. W. (1994). Molecular mechanisms of action of steroid/thyroid receptor superfamily members. *Annu. Rev. Biochem.* **63,** 451–486.

Wang, Z., and Brown, D. D. (1991). A gene expression screen. *Proc. Natl. Acad. Sci. U.S.A.* **88,** 11505–11509.

Wang, Z., and Brown, D. D. (1993). The thyroid hormone-induced gene expression program for amphibian tail resorption. *J. Biol. Chem.* **268,** 16270–16278.

Weber, R. (1967). Biochemistry of amphibian metamorphosis. *In* "The Biochemistry of Animal Development" (R. Weber, ed.), pp. 227–300. Academic Press, New York.

Weiser, M. M., Sykes, D E., and Killen, P. D. (1990). Rat intestinal basement membrane synthesis. *Lab. Invest.* **62,** 325–330.

White, B. A., and Nicoll, C. S. (1981). Hormonal control of amphibian metamorphosis. *In* "Metamorphosis: A Problem in Developmental Biology" (L. I. Gilbert and E. Frieden, eds.), 2nd ed., pp. 363–396. Plenum, New York.

Wolf, C., Chenard, M. P., de Grossouvre, P. D., Bellocq, J. P., Chambon, P., and Basset, P. (1992). Breast-cancer–associated stromelysin-3 gene is expressed in basal cell carcinoma and during cutaneous wound healing. *J. Invest. Dermatol.* **99,** 870–872.

Xu, Q., Baker, B. S., and Tata, J. R. (1993). Developmental and hormonal regulation of the *Xenopus* liver-type arginase gene. *Eur. J. Biochem.* **211,** 891–898.

Yao, T. P., Segraves, W. A., Oro, A. E., McKeown, M., and Evans, R. M. (1992). *Drosophila* ultraspiracle modulates ecdysone receptor function via heterodimer formation. *Cell (Cambridge, Mass.)* **71,** 63–72.

Yaoita, Y., and Brown, D. D. (1990). A correlation of thyroid hormone receptor gene expression with amphibian metamorphosis. *Genes Dev.* **4,** 1917–1924.

Yaoita, Y., Shi, Y.-B., and Brown, D. D. (1990). *Xenopus laevis* α and β thyroid hormone receptors. *Proc. Natl. Acad. Sci. U.S.A.* **87,** 7090–7094.

Yen, P. M., and Chin, W. W. (1994). New advances in understanding of thyroid hormone action. *Trends Endocrinol. Metab.* **5,** 65–72.

Yoshizato, K. (1989). Biochemistry and cell biology of amphibian metamorphosis with a special emphasis on the mechanism of removal of larval organs. *Int. Rev. Cytol.* **119,** 97–149.

Yu, Y. C., Delsert, C., Andersen, B., Holloway, J. M., Devary, O. V., Naar, A. M., Kim, S. Y., Boutin, J. M., Glass, C. K., and Rosenfeld, M. G. (1991). RXRβ: A coregulator that enhances binding of retinoic acid, thyroid hormone, and vitamin D receptors to their cognate response elements. *Cell (Cambridge, Mass.)* **67,** 1251–1266.

Zhang, X.-K., and Pfahl, M. (1993). Regulation of retinoid and thyroid hormone action through homodimeric and heterodimeric receptors. *Trends Endocrinol. Metab.* **4,** 156–162.

15

Reprogramming of Genes Expressed in Amphibian Liver during Metamorphosis

BURR G. ATKINSON, CAREN HELBING, AND YUQING CHEN

Molecular Genetics Unit
Department of Zoology
Western Science Centre
University of Western Ontario
London, Ontario, Canada N6A 5B7

I. INTRODUCTION

Postembryonic development in most amphibians involves biochemical, physiological, and morphological changes which prepare the aquatic tadpole for its transition to a fully or partially terrestrial adult, a process termed metamorphosis (for earlier reviews on this subject, see Etkin and Gilbert, 1968; Gilbert and Frieden, 1981). Many of these changes are striking (e.g., atrophy of the tail and gills, and growth and development of the limbs and lungs) and are manifested through selective cell death and/or stem cell mobilization (Atkinson, 1981). Other changes are morphologically less dramatic and, in some organs/tissues, appear to involve cell-specific reprogramming of gene expression in a fixed population of cells. The changes which occur in the liver of a metamorphosing tadpole are thought to be an example of the latter case. The liver in the metamorphosing tadpoles does not show marked morphological changes, but does undergo extensive cytological, biochemical, and molecular remodeling which is thought to involve both a reprogramming of gene expression and the terminal differentiation of the resident hepatocytes (see Beckingham Smith and Tata, 1976; Cohen *et al.,* 1978; VanDenbos and Frieden, 1976; Chen *et al.,* 1994). While the occurrence of these postembryonic changes in the liver cells of a metamorphosing tadpole is dependent on increasing circulating levels of thyroid hormone and, in fact, can be induced precociously by administration of thyroid hormone to a premetamorphic tadpole, the means by which this hormone initiates these changes are only now beginning to be understood. Thus, the purpose of this chapter is to summarize results characterizing some of the biochemical and molecular changes in gene expression initiated by thyroid hormone in the tadpole liver. In addition, the chapter presents in more detail, results from more recent studies focused on the molecular mechanisms by which this hormone is thought to initiate and orchestrate postembryonic changes in gene expression which eventually lead to an adult, amphibian liver phenotype.

II. EXPRESSION OF PROTEINS CHARACTERISTIC OF THE ADULT LIVER PHENOTYPE

The liver in spontaneously metamorphosing tadpoles produces proteins which characterize the adult liver phenotype.

The liver of adult amphibians is the source of serum proteins, including albumin, and, since most adult amphibians are ureotelic, the source and site of the enzymes required for the biosynthesis of urea (Atkinson, 1994a; Frieden, 1967). The liver of most amphibian tadpoles, however, produces little or no serum albumins and, since the fully aquatic tadpoles are ammonotelic, is not normally involved in producing urea (Atkinson, 1994a; Frieden, 1967). During the postembryonic development/metamorphosis of most amphibians, the tadpole liver begins to assume the adult liver phenotype by synthesizing new and/or enhanced levels of albumin and other serum proteins (Herner and Frieden, 1960; Chen, 1970) and by marked elevations in the activities and levels of the enzymes involved in urea biosynthesis (Cohen, 1966). Indeed, by metamorphic climax [stage XX (Taylor and Kollros, 1946)] the levels (Feldhoff, 1971; Nagano *et al.*, 1973) and synthesis (Ledford and Frieden, 1973) of serum albumin have increased 6-fold in *Rana catesbeiana,* and the activities of the ornithine–urea cycle enzymes (Fig. 1), carbamoyl-phosphate syn-

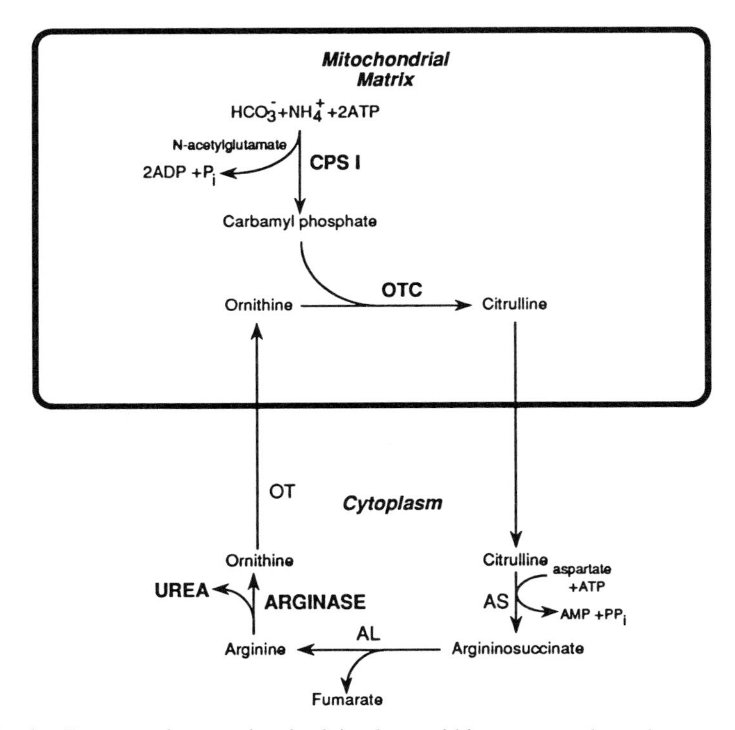

Fig. 1. Enzymes that are involved in the ornithine–urea cycle pathway and are responsible for the production of urea in the liver cells of *Rana catesbeiana* are shown. Two of the urea cycle enzymes, carbamoyl-phosphate synthase-I (CPS-I) and ornithine carbamoyltransferase (OTC), are localized in the mitochondrial matrix whereas the other three enzymes, argininosuccinate synthase (AS), argininosuccinate lyase (AL), and arginase, are cytoplasmic. OT is an ornithine transporter. The enzymes shown in bold represent the products of genes whose expressions are emphasized in this chapter.

thase-1 (CPS-1), ornithine carbamoyltransferase (OTC), argininosuccinate synthase (AS), argininosuccinate lyase (AL), and arginase (Arg), are elevated 30-, 8-, 35-, 20-, and 30-fold, respectively (Brown *et al.,* 1959; Paik and Cohen, 1960; Cohen, 1966, 1970; Cohen *et al.,* 1978). Measurements of the urea cycle enzyme activities in the liver (see Cohen *et al.,* 1978) and of albumin concentrations in the serum of spontaneously metamorphosing *R. catesbeiana* tadpoles (Feldhoff, 1971; Nagano *et al.,* 1973) demonstrate that both the urea cycle enzyme activities and the serum albumin levels increase at a time in development when a measurable increase in plasma thyroid hormone levels is detectable (Galton and St. Germain, 1985) and degradation of the tadpoles' tail is just beginning to be evident. These biochemical changes appear to occur in anticipation of new demands, namely, the subsequent transition from a fully aquatic tadpole to a fully or partially terrestrial adult. As discusssed in subsequent sections, these changes are dependent on and tightly controlled by thyroid hormone.

The liver in thyroid hormone-induced metamorphosing tadpoles produces proteins which characterize the adult liver phenotype.

Although the obligatory control of amphibian metamorphosis by thyroid hormones was established in the early 1900s, it was not until the 1950s that investigators demonstrated that administration of this hormone to premetamorphic tadpoles results in a precocious increase in the levels of serum proteins (Frieden *et al.,* 1957), in urea excretion (Munro, 1953), and in liver arginase activity (Dolphin and Frieden, 1955; Brown *et al.,* 1959). Subsequent studies have confirmed these results and established that administration of thyroid hormone, thyroxine T_4 or 3,5,3′-triiodothyronine (T_3), to premetamorphic tadpoles also induces the precocious synthesis of albumin (Ledford and Frieden, 1973) and a concerted increase in the levels of activity of all of the ornithine–urea cycle enzymes in the tadpole liver (Cohen, 1966, 1970; Wixom *et al.,* 1972; Cohen *et al.,* 1978). Moreover, this last group of studies revealed that each of the ornithine–urea cycle enzymes attained approximately the same relative level of activity (based on a ratio of metamorphic to premetamorphic activity) in the liver of thyroid hormone-induced tadpoles as found in the liver of spontaneously metamorphosing tadpoles (Cohen, 1966; Cohen *et al.,* 1978).

The striking increase in CPS-I activity prompted Cohen and colleagues to investigate the means by which thyroid hormone effects the activity of this urea cycle enzyme. Purification of CPS-I from *R. catesbeiana,* and preparation of an antibody against it, permitted these investigators to demonstrate that the increase in CPS-I activity in the liver of tadpoles exposed to thyroid hormone is the result of *de novo* synthesis (Metzenberg *et al.*, 1961). This observation has been confirmed by others (Kim and Cohen, 1968; Pouchelet and Shore, 1981; Helbing *et al.*, 1992) and, coupled with the studies by Ledford and Frieden (1973) on the thyroid hormone induction of albumin synthesis in this tadpole's liver, clearly implicates a role for thyroid hormone in both albumin and urea biosynthesis and in establishing the adult liver phenotype in this amphibian.

III. EXPRESSION OF THYROID HORMONE RECEPTOR mRNAs AND OF mRNAs CHARACTERISTIC OF THE ADULT LIVER PHENOTYPE

The liver in both spontaneously and thyroid hormone-induced metamorphosing tadpoles accumulates translatable mRNAs which encode proteins characteristic of the adult liver phenotype.

Since the administration of thyroid hormone to premetamorphic *R. catesbeiana* tadpoles results in the precocious synthesis and accumulation of serum albumin and liver CPS-I protein, studies were undertaken to establish if the thyroid hormone-induced synthesis of these proteins is coordinated with a rise in the hepatic mRNAs encoding them. Investigations by Schultz *et al.* (1988), using monospecific antibodies for immunoprecipitating albumin synthesized from *in vitro* translations of *R. catesbeiana* tadpole liver mRNAs, supported the contention that thyroid hormone administration precociously upregulates the level of hepatic mRNAs encoding serum albumin in this organism. Results from Northern and slot-blot hybridization studies, using a cDNA encoding a portion of a *R. catesbeiana* albumin (Averyhart-Fullard and Jaffe, 1990) and a cDNA encoding domains II and III of *Xenopus laevis* albumin (Moskaitis *et al.*, 1989), have confirmed the observations of Schultz and colleagues. Moreover, studies of albumin mRNA levels in the liver of spontaneously metamorphosing tadpoles (Schultz *et al.*, 1988; Moskaitis *et al.*, 1989) reveal that the amount of albumin mRNAs increases as the amount of endogenous thyroid hormone increases in the plasma, is maximal during metamorphic climax, and decreases after metamorphic climax. Unfortunately, none of the reported studies has made a direct correlation between the timing of albumin synthesis and the up-regulation in the level of albumin mRNAs, nor have they addressed the question of why the level of albumin mRNAs appears to decrease after metamorphic climax. Whatever the case, these reports clearly indicate that thyroid hormone, directly or indirectly, influences mRNAs encoding serum albumin to accumulate in the liver of these tadpoles.

Studies by Mori *et al.* (1979) and by Pouchelet and Shore (1981), using antibodies to immunoprecipite CPS-I synthesized by *in vitro* translations of tadpole liver mRNAs, have shown that administration of thyroid hormone to premetamorphic tadpoles results in a premature up-regulation in the levels of hepatic mRNAs encoding CPS-I. Results obtained from hybridization analyses, using nucleotide sequences encoding a portion of a rat CPS-I cDNA (Morris, 1987; Galton *et al.,* 1991; Helbing *et al.,* 1992), have supported data obtained from the *in vitro* translations. Helbing and Atkinson (1994) have isolated and characterized cDNAs encoding a complete *R. catesbeiana* CPS-I and have used specific portions of the nucleotide sequence as probes to reassess the studies reported with the rat CPS-I probes. Results from both hybridization and reverse transcription–polymerase chain reaction (RT–PCR) analyses (Helbing and Atkinson, 1994; Atkinson *et al.,* 1994; Helbing, 1993) confirmed the earlier studies and, as summarized in Fig. 2, demonstrated that by 48 hr after the administration of thyroid hormone the level of CPS-I mRNA is increased six-fold and remains elevated.

Fig. 2. Relationship of TRα, TRβ, CPS-I, OTC, and arginase (Arg) mRNA levels in the liver of thyroid hormone-induced metamorphosing *R. catesbeiana* tadpoles to some of the morphological and physiological changes occurring in these tadpoles. (A) Summary bar graphs depicting the accumulation of TRα, TRβ, CPS-I, OTC, and Arg mRNAs in the liver of stage VI (Taylor and Kollros, 1946) tadpoles that had been injected (time 0) with 3,5,3′-triiodothyronine (T_3) at a dose of 3×10^{-10} moles/g body weight. The amount of urea–nitrogen excreted by similarly treated tadpoles is superimposed on the bar graphs. The average relative amounts of the mRNA transcripts were determined by laser densitometric quantitation of dot-blot hybridizations and RT–PCR products, and the plotted values were corrected based on the abundance of mRNA (either actin or ferritin) which had been shown not to change during this treatment (Helbing *et al.,* 1992; Atkinson *et al.,* 1994; Helbing and Atkinson, 1994). (B) The amount of urea–nitrogen excreted by T_3-induced metamorphosing tadpoles at various times after the hormone treatment is compared with a morphological change in the tadpoles (tail loss; a ratio of tail length to body length or TB) and with decreases in the relative amount of the T_3 originally injected in each tadpole [the last comparison is plotted from data obtained from tadpoles maintained at 25°C as reported by Yamamoto *et al.* (1966)].

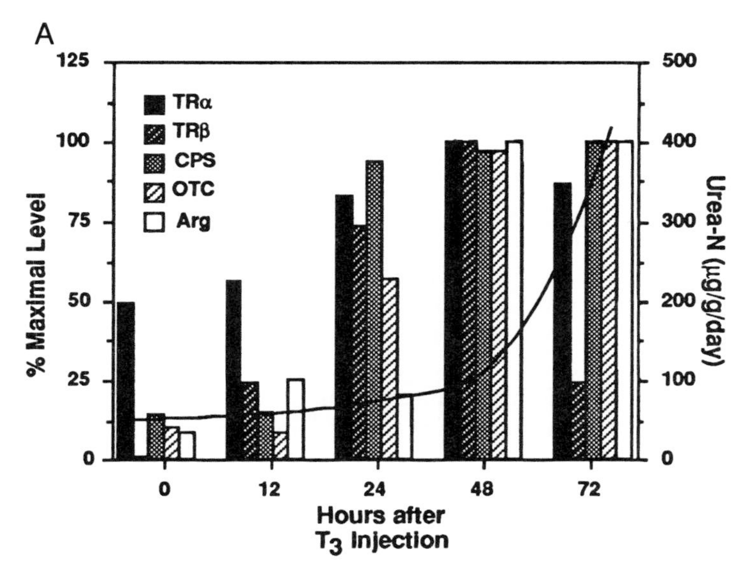
A
125
100
75
50
25
0
500
400
300
200
100
0
TRα
TRβ
CPS
OTC
Arg
% Maximal Level
Urea-N (μg/g/day)
0
12
24
48
72
Hours after
T3 Injection

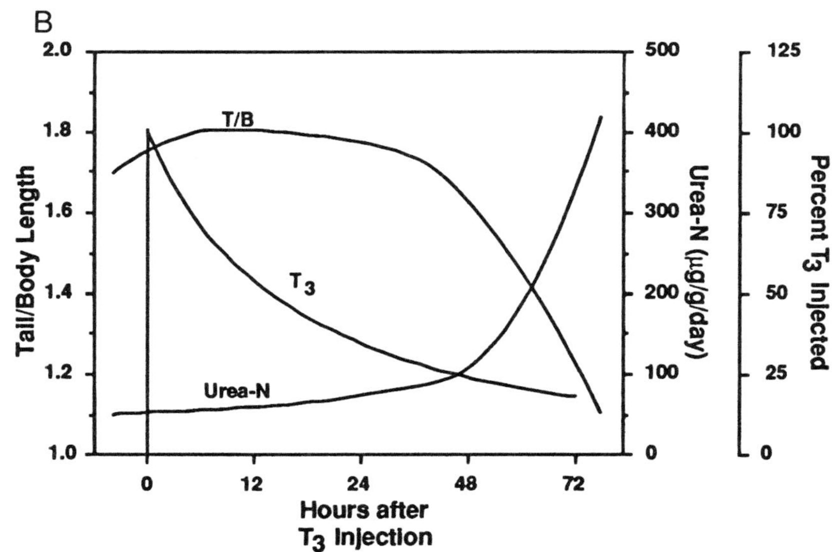
B
2.0
1.8
1.6
1.4
1.2
1.0
500
400
300
200
100
0
125
100
75
50
25
0
T/B
T3
Urea-N
Tail/Body Length
Urea-N (μg/g/day)
Percent T3 Injected
0
12
24
48
72
Hours after
T3 Injection

While these observations support the conjecture that thyroid hormone regulates the expression of the genes encoding the enzymes involved in urea biosynthesis, they were limited to studies with only CPS-I. However, the isolation and characterization of amphibian cDNAs encoding two other enzymes in the ornithine–urea cycle, OTC and arginase in *R. catesbeiana* and arginase in *X. laevis* (Helbing *et al.*, 1992; Atkinson *et al.*, 1994; Xu *et al.*, 1993), have enabled this hypothesis to be more thoroughly investigated. Results from RT–PCR analyses, using PCR primers specific for each of these mRNAs, and from Northern and dot-blot hybridization analyses, using nucleotide sequences specific for OTC and arginase mRNAs as probes (Helbing *et al.*, 1992; Atkinson *et al.*, 1994), substantiate that administration of thyroid hormone to a premetamorphic *R. catesbeiana* tadpole induces a marked up-regulation in the level of the hepatic mRNAs encoding OTC and arginase (Fig. 2). Studies of CPS-I, OTC, and arginase mRNA levels in the liver of spontaneously metamorphosing tadpoles (Helbing *et al.*, 1992; Atkinson *et al.*, 1994; Helbing and Atkinson, 1994) reveal that the amount of each of these mRNAs increases as the level of endogenous thyroid hormone increases in the plasma, is maximal during metamorphic climax, and, thereafter, remains, more or less, at the level attained during metamorphic climax (Fig. 3).

The liver in both thyroid hormone-induced and spontaneously metamorphosing tadpoles accumulates thyroid hormone receptor mRNAs prior to accumulating mRNAs characteristic of the adult liver phenotype.

The major mechanism of thyroid hormone action is thought to be at the transcriptional level and is mediated through the binding of this hormone to a nuclear-localized receptor protein (see Tata, Chapter 13, this volume). Since the amount of thyroid hormone receptor proteins (TRs) and the mRNAs encoding them have been shown to increase in tadpole tissues following thyroid hormone treatment (Kistler *et al.*, 1975; Galton and St. Germain, 1985; Baker and Tata, 1990; Kawahara *et al.*, 1991; Schneider and Galton, 1991; Yaoita and Brown, 1991), investigators hypothesized that an accumulation of the mRNAs encoding the TRs should precede or be coincidental with the up-regulation of CPS-I, OTC, and arginase mRNAs in the liver of thyroid hormone-treated tadpoles. The isolation and characterization of cDNAs encoding the *R. catesbeiana* TRs, TRα and TRβ (Schneider and Galton, 1991; Helbing *et al.*, 1992; Davey *et al.*, 1994), enabled these studies to be conducted. Hybridization and RT–PCR analyses of RNA isolated from the liver of control and thyroid hormone-treated tadpoles, using nucleotide sequences specific

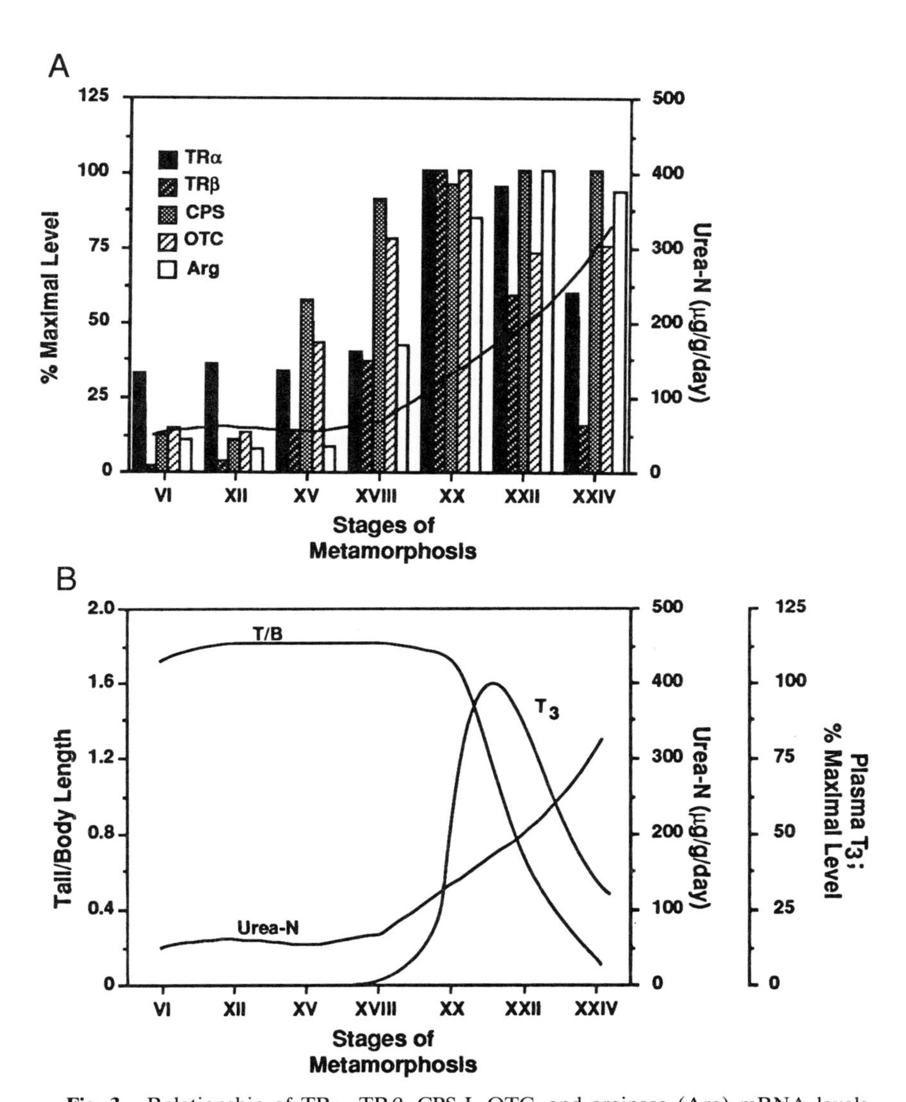

Fig. 3. Relationship of TRα, TRβ, CPS-I, OTC, and arginase (Arg) mRNA levels in the liver of spontaneously metamorphosing *R. catesbeiana* tadpoles to some of the morphological and physiological changes occurring in the tadpoles. (A) Summary bar graphs depicting the accumulation of TRα, TRβ, CPS-I, OTC, and Arg mRNAs in the liver of tadpoles undergoing spontaneous metamorphosis. The amount of urea–nitrogen excreted by these tadpoles at various stages (Taylor and Kollros, 1946) of their postembryonic development is superimposed on the bar graphs. The average relative amounts of the mRNAs were determined as in Fig. 2, and the plotted values were corrected based on the abundance of actin mRNA (Helbing *et al.,* 1992; Atkinson *et al.,* 1994; Helbing and Atkinson, 1994). (B) Comparison of the amount of urea–nitrogen excreted by spontaneously metamorphosing tadpoles with a morphological change in the tadpoles (tail loss; a ratio of tail length to body length or TB) and with the relative amount of endogenous thyroid hormone reported in the plasma of staged tadpoles [the last comparison is plotted from data reported by Galton and St. Germain (1985)].

for each of the *R. catesbeiana* TRs, revealed that the mRNAs encoding these TRs accumulate in the liver of hormone-treated animals (Helbing *et al.,* 1992, 1995). TRβ mRNA accumulation, which is detectable 12 hr after administration of thyroid hormone, shows an 11-fold increase by 48 hr and drops substantially by 72 hr. TRα mRNAs, which are 10 times more abundant than TRβ mRNAs in the liver of control animals, follow an accumulation pattern similar to the TRβ mRNAs except that there is never more than 2-fold increase in the level of TRα mRNAs. As expected, the up-regulation of the mRNAs encoding the TRs, as well as the ornithine–urea cycle enzymes, occurs well before the enhanced synthesis and excretion of urea and before any gross morphological changes (e.g., tail reduction) occur in the tadpoles (Fig. 2). The most interesting observation, however, is that the up-regulation of the TR mRNAs is detectable within 12 hr after administration of thyroid hormone to the tadpole, whereas little or no accumulation of the CPS-I, OTC, or arginase mRNAs is detectable until 24 hr after administration of the hormone. The early accumulation of TR mRNAs in response to thyroid hormone suggests that they are encoded by genes which may be directly regulated by thyroid hormone, and the more extended time period (~24 hr) between the administration of the hormone and the up-regulation of the CPS-I, OTC, and arginase mRNAs—the so-called "lag" phase—suggests that these mRNAs are encoded by genes which are not directly regulated by thyroid hormone (see Shi, Chapter 14, this volume). This latter suggestion has caused the molecular events which occur during the lag phase to become the focus of investigations which will be discussed in subsequent sections of this chapter.

Hybridization and RT–PCR analyses of RNA isolated from the liver of *R. catesbeiana* tadpoles at various stages of spontaneous metamorphosis disclose that no TRβ mRNA is detectable in the liver of tadpoles until stage XV. After stage XV, however, TRβ mRNA levels increase, attain maximal levels (~7-fold increase) at the beginning of metamorphic climax (stage XX), and thereafter, decrease (Helbing *et al.,* 1992). While TRα mRNAs are detectable in tadpole liver in all of the postembryonic developmental stages studies (VI–XXIV), they also appear to accumulate (2- to 3-fold) at metamorphic climax (Helbing, 1993). It is obvious from the data summarized in Fig. 3 that mRNAs encoding both the TRs and ornithine–urea cycle enzymes accumulate before urea excretion is elevated and before any reduction in the tadpole tail is evident. While low but critical increases in the levels of thyroid hormone, undetectable by current technologies (see Galton, 1983; Galton and ST. Germain,

1985; Miller, Chapter 17, this volume), may be responsible for the initial up-regulation of these mRNAs in the liver of stage XVI tadpoles, the correlation between the accumulation of these mRNAs (Fig. 3A) and marked increases in the detectable levels of endogenous thyroid hormone in the plasma of spontaneously metamorphosing *R. catesbeiana* tadpoles is striking. Indeed, the latter observations support the contention that the thyroid hormone-induced precocious accumulation of these mRNAs is a reflection of the natural sequence of events that occur during the postembryonic development/metamorphosis of this organism.

The accumulation of mRNAs encoding the TRs and the ornithine–urea cycle enzymes in the liver of thyroid hormone-induced tadpoles is neither dependent on other extrahepatic factors nor cell replication.

Early studies on the thyroid hormone induction of albumin synthesis and the enhanced activity and synthesis of CPS-I in the liver of *Rana* tadpoles were concerned with the amount of time that lapsed between the administration of the hormone and the detection of a change in the synthesis of albumin and in the synthesis and/or activity of CPS-I (Paik and Cohen, 1960; Metzenberg *et al.,* 1961; Tata, 1967; Shambaugh *et al.,* 1969; Ledford and Frieden, 1973), the so-called "lag" phase. Investigators, attempting to shorten the response time (i.e., reduce the lag phase), used methods which included administration of thyroid hormone by injection (rather than by immersing the tadpole in a solution of the hormone), use of higher doses of the hormone (T_3 and T_4), and/or maintenance of the tadpoles at a higher environmental temperature. All of these changes in protocol led to a reduction of the lag phase; however, a lag phase of ~24–48 hr still appeared to be required. The observation that a lag phase (~ 24 hr) is also evident in the thyroid hormone-induced up-regulation of the mRNAs encoding CPS-I, OTC, and argianse in the liver of tadpoles maintained at 25°C (see Fig. 2) has renewed interest in elucidating the molecular events which occur during this time.

The possibility that this lag phase may represent a period of time required for the synthesis and/or interaction of other, perhaps thyroid hormone-inducible, extrahepatic factors necessary for this response has been addressed in serum-free, tadpole liver culture systems (Shambaugh *et al.,* 1969; Helbing *et al.,* 1992; Atkinson *et al.,* 1994). These culture systems have been used to assess the response of CPS-I synthesis and the up-regulation of the mRNAs encoding CPS-I, OTC, and arginase, as well as the TRα and TRβ mRNAs. Administration of thyroid hormone to cultures of liver from premetamorphic tadpoles results in an early (~12 hr) up-regulation of TRβ and TRα mRNAs; a delayed (~24 hr;

similar to that observed *in vivo*) up-regulation of CPS-I, OTC, and arginase mRNAs; and the enhanced synthesis of CPS-I (Shambaugh *et al.*, 1969). These data support the contention that the responses observed *in vivo* are effected by thyroid hormone and are not dependent on other extrahepatic factors. While these results do not clarify whether the thyroid hormone directly or indirectly affects the expression of the genes encoding the TRs and/or the ornithine–urea cycle enzymes, they clearly establish that the lag phase does not reflect a period of time required for the synthesis and/or interaction of other extrahepatic factors.

Whether the molecular changes accompanying liver metamorphosis occur in a fixed population of hepatocytes or result from a proliferation of stem cells competent to express genes characeristic of the adult liver phenotype has been a subject of controversy for some time (see reviews by Beckingham Smith and Tata, 1976; Cohen *et al.*, 1978; Smith-Gill and Carver, 1981). Indeed, the possibility that the lag phase might represent a period of time required for the replication and production of new, terminally differentiated liver cells capable of expressing genes characteristic of the adult liver phenotype has been raised and evaluated. Chen *et al.* (1994) assessed this possibility by *in situ* hybridization studies using riboprobes encoding sense and antisense nucleotide sequences specific for *R. catesbeiana* CPS-I and OTC mRNAs. They reasoned that if the stem cell proliferation theory was correct then *in situ* hybridization analyses of liver from thyroid hormone-treated tadpoles should detect the expression of CPS-I and OTC transcripts in discrete populations of liver cells, particularly in the 12- to 24-hr period after thyroid hormone treatment of the tadpoles. If, on the other hand, the reprogramming of resident hepatocytes theory was correct, they felt that CPS-I and OTC transcripts should be detectable in all of the liver parenchymal cells. The results from their studies, shown in Fig. 4, support the "reprogramming of resident hepatocytes" hypothesis by demonstrating that the expression of CPS-I and OTC transcripts is *not* confined to discrete populations of hepatocytes. They are, in fact, discernible in all of the parenchymal cells present in the liver of tadpoles treated 15, 24, or 48 hr earlier with thyroid hormone. Indeed, the uniformity of this response among *all* of the hepatocytes in liver from tadpoles which had been treated 15 hr earlier with thyroid hormone (see Fig. 4) clearly supports the contention by Cohen *et al.* (1978) that these liver cells are undergoing molecular and biochemical differentiation in the absence of cell division. Moreover, the fact that thyroid hormone appears to cause terminal differentiation and reprogram gene expression in the resident hepatocytes, rather than in any subpopulation of them, disqualifies the lag phase as a period of time required for cell division.

IV. ACCUMULATION OF mRNAs ENCODING FACTORS IMPLICATED IN TERMINAL DIFFERENTIATION AND LIVER-SPECIFIC GENE EXPRESSION

The accumulation of mRNAs encoding the ornithine–urea cycle enzymes in the liver of thyroid hormone-induced tadpole is preceded by the accumulation of mRNAs encoding factors implicated in terminal differentiation and liver-specific gene expression.

Neither the synthesis and/or interaction of extrahepatic factors, other than thyroid hormone, nor the generation of a new population of liver cells is required for the thyroid hormone-induced expression of the genes encoding the ornithine–urea cycle enzymes. Consequently, these phenomena cannot account for the lag phase. These observations have resulted in the conjecture that the lag phase may represent a period of time in which a cascade of molecular events occur that are required for initiating the terminal differentiation of the resident hepatocytes and, subsequently, orchestrating a reprogramming of gene expression which enables genes characteristic of the adult liver phenotype to be expressed (Chen *et al.,* 1994; Atkinson, 1994b). This hypothesis implies that thyroid hormones does not directly effect the expression of the ornithine–urea cycle enzyme genes but, instead, is responsible for initiating molecular events which lead to the terminal differentiation of the liver cells which, subsequently, results in the expression of the ornithine–urea cycle enzyme genes. Since direct thyroid hormone action on a target gene is thought to be mediated through its interaction with a receptor protein (a TR) bound to a specific nucleotide sequence/element on that gene, Chen and Atkinson (1995) isolated and searched the promoter regions of the *R. catesbeiana* CPS-I and OTC genes for TR-binding elements (TREs). The fact that no sequences corresponding to known TREs were found in the promoter regions (up to 1000 bp upstream from the transcription initiation sites) of these genes supports, albeit circumstantially, the contention that thyroid hormone does not directly regulate the expression of these genes.

The absence of known TREs in the promoter regions of the CPS-I and OTC genes and the notion that other, perhaps thyroid hormone

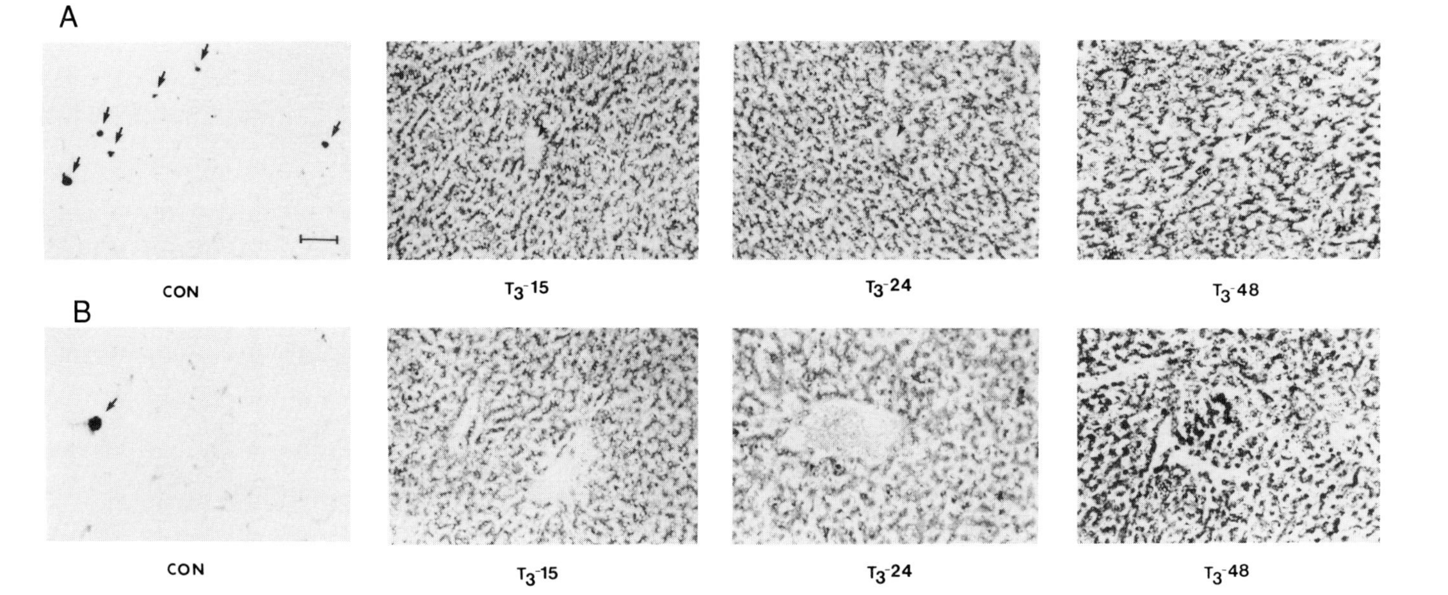
A
CON
T_3-15
T_3-24
T_3-48
B
CON
T_3-15
T_3-24
T_3-48

inducible, transcription factors may be directly responsible for regulating the expression of these genes prompted a search in the promoter regions of these genes for regulatory elements corresponding to known transcription factors and/or factors thought to be involved in the terminal differentiation of cells. The search revealed that the promoters of both genes contained CCAAT/enhancer core-binding protein (C/EBP) elements (Chen and Atkinson, 1995). Since C/EBPs are transcription factors which appear to play a role both in the terminal differentiation of mammalian hepatocytes and in the expression of liver-specific genes (Umek *et al.,* 1991; Friedman *et al.,* 1989; Sladek and Darnell, 1992), Chen *et al.* (1994) explored the possibility that a C/EBP-like transcription factor might be involved in the terminal differentiation of this amphibian's liver cells and/or required for upregulating the expression of the ornithine–urea cycle enzyme genes. With this end in mind, they isolated and characterized cDNAs encoding *R. catesbeiana* homologs of the mammalian C/EBPα (Landschulz *et al.,* 1988; Lincoln *et al.,* 1994) and C/EBPδ (Williams *et al.,* 1991). Northern and dot-blot hybridization analyses revealed that mRNA transcripts encoding one of these transcription factors, the *Rana* homolog of C/EBPα (RcC/EBP-1), accumulate in the liver of spontaneously and thyroid hormone-induced metamorphosing tadpoles. Similar results had been obtained by Xu and Tata (1992) in the liver and other tissues of *Xenopus* tadpoles at the onset of spontaneous metamorphosis. In fact, the thyroid hormone-induced up-regulation of RcC/EBP-1 mRNAs was apparent well before the up-regulation of the mRNAs encoding the ornithine–urea cycle enzymes CPS-I and OTC

Fig. 4. *In situ* hybridization analyses of CPS-I (A) and OTC (B) mRNA accumulation in the liver of *R. catesbeiana* tadpoles [stage VI (Taylor and Kollros, 1946)] that had not been exposed to thyroid hormone (CON) or had been treated with 3,5,3′-triiodothyronine 15 (T_3-15), 24 (T_3-24), or 48 (T_3-48) hr earlier. Serial sections of liver tissue were hybridized with antisense and sense DIG-labeled RNA probes. No hybridization was evident in the sections treated with the sense DIG-labeled RNA probes or in RNase A-treated sections (not shown). (A) Arrows in CON indicate pigment granules, and arrowheads in subsequent panels denote the central vein in a liver lobule. The clear circles are unstained nuclei, and the unstained areas between the cords of liver cells are the hepatic sinusoids. Magnification is the same in all photographs and the bar represents 10 μm. [Modified from Chen *et al.* (1994).]

(Fig. 5). The observation that RcC/EBP-1 mRNAs begin accumulating in the liver of tadpoles, as well as in serum-free cultures of tadpole liver (not shown), prior to or coincidental with the TRβ mRNAs and well before any accumulation of CPS-I or OTC mRNAs suggests that this hormone may be directly effecting the expression of the genes encoding both TRβ and RcC/EBP-1. Whatever the case, the presence of C/EBP-binding elements in the promoter regions of this amphibian's CPS-I and OTC genes, coupled with the demonstration that mRNAs encoded from the RcC/EBP-1 gene accumulate during the period that precedes the accumulation of the mRNAs encoding CPS-I and OTC, supports the contention that the product from the thyroid hormone-induced expression of the RcC/EBP-1 mRNAs may be required for the expression of the CPS-I and OTC genes. Moreover, if RcC/EBP-1 has both antimitotic and transcriptional regulatory properties of the mammalian C/EBPα transcription factor (Vasseur-Cognet and Lane, 1993; Umek *et al.,* 1991), then it also could be envisioned to play an early role, perhaps in concert with a TR or other transcription factors, in orchestrating both the terminal differentiation and the reprogramming of gene expression in this tadpole's liver cells during both spontaneous and thyroid hormone-induced metamorphosis.

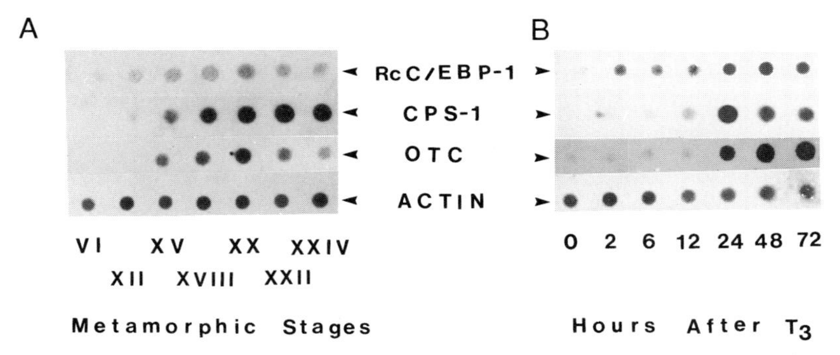

Fig. 5. Dot-blot hybridization analyses comparing the levels of RcC/EBP-1, CPS-I, and OTC mRNA transcripts in the liver of *R. catesbeiana* tadpoles undergoing spontaneous (A) and thyroid hormone-induced metamorphosis (B). Total RNA was isolated from the liver of tadpoles at various stages of metamorphosis (Taylor and Kollros, 1946) and from the liver of stage VI tadpoles at various times after thyroid hormone treatment of the tadpoles. Aliquots (10 μg) of the RNA were dot-blotted to a Zetaprobe membrane and probed with ^{32}P-labeled DNA fragments specific for RcC/EBP-1, CPS-I, OTC, and cytoskeletal actin. [Portions of the figure are from Chen *et al.* (1994).]

The temperature-dependent expression of the ornithine–urea cycle enzyme genes in the liver of thyroid hormone-induced tadpoles provides a means for separating and studying primary and secondary responses to this hormone.

The influence of environmental temperature on the rates of development and, ultimately, the metamorphosis of Amphibia have been known for some time (for a historical review see Frieden *et al.,* 1965). Maintenance of tadpoles at 5°C severely retards the postembryonic/metamorphic changes which occur at higher temperatures in both spontaneously and thyroid hormone-induced metamorphosing tadpoles. Moreover, successively decreasing the environmental temperature from 25°C to 7°C has been shown to progressively decrease the rate of tail regression and urea excretion in thyroid hormone-induced tadpoles (Ashley *et al.,* 1968). The concept, emerging in the late 1960s (for historical reviews see Tata, 1966; Atkinson, 1994b), that thyroid hormone action might be at the nuclear level and be mediated through its interaction with nuclear receptor proteins prompted the suggestion that a hormone-transport process might be involved in the temperature sensitivity. In fact, a preliminary study by Griswold *et al.* (1972) supported the idea that the failure of a temperature-sensitive system for the transport of thyroid hormone into the nucleus was the underlying mechanism responsible for the low temperature inhibition of thyroid hormone-induced metamorphic events. However, more recent reports (Kistler *et al.,* 1975; Galton, 1980, 1983; Toth and Tabachnick, 1980) clearly demonstrate that while transport and nuclear binding of thyroid hormone (T_3 or T_4) is slower in the liver of tadpoles maintained at 4°C than in ones maintained at 25°C, the amount of thyroid hormone bound to saturable, nuclear-binding sites (i.e., the TRs) is comparable 72 hr after hormone administration. Thus, these latter results clearly established that the failure of metamorphic events to occur in tissues of thyroid hormone-treated tadpoles maintained at 4–5°C cannot be attributed to a failure of this hormone to bind to the TRs. Indeed, as suggested by Galton (1980), it seems likely that one or more of the processes involved in the response to the hormone–receptor complex is affected by the low temperature.

The demonstration that an environmental temperature of 5°C can completely inhibit detectable metamorphic events, such as urea excretion and tail degradation, in thyroid hormone-treated tadpoles was augmented by the demonstration that shifting of these tadpoles from 5° to 25°C releases the inhibition and permits these metamorphic transitions to occur (Frieden *et al.,* 1965; Ashley *et al.,* 1968). Thyroid hormone-treated tadpoles can be maintained at 5°C for extensive periods (80–110 days), and when shifted to 25°C, unlike similarly maintained tadpoles not treated with thyroid hormone, undergo metamorphosis. These results raised questions regarding the longevity of the exogenous hormone in tadpoles maintained at 5°C. Determinations (Yamamoto *et al.,* 1966). of the half-life of thyroid hormone injected into tadpoles maintained at this low temperature (20 days for T_4 and 14 days for T_3 at 6°C) led to estimates that less than 1% of the original injected dose would be present in the tadpoles after 60–80 days at 5°C (Frieden, 1968). From these estiamtes, Frieden and colleagues concluded that the level of hormone remaining after 60–80 days at 5°C could not account for the metamorphic transitions which occur after the shift to 25°C. Thus, they proposed that thyroid hormone had left an "imprint" (Frieden *et al.,* 1965; Frieden, 1968) on susceptible cells by initiating certain subtle, metabolic changes at 5°C which enabled the metamorphic transition to be completed at a permissive temperature.

The conjecture that "subtle metabolic changes" are initiated by thyroid hormone in tissues of tadpoles mantained at 5°C and the notion that these changes enable metamorphic transitions to be completed at a permissive temperature (Frieden *et al.,* 1965; Frieden, 1968) have prompted investigations aimed at defining these "metabolic" or preparatory changes at the molecular level. Since urea production and the new and/or enhanced expression of genes encoding the ornithine–urea cycle enzymes are postembryonic/metamorphic changes which are inducible by thyroid hormone and characteristic of the adult phenotype, we explored the possibility that the temperature dependency of urea production might, in fact, reflect a temperature dependency in the expression of the genes encoding the ornithine–urea cycle enzymes. A report, published in 1960 by Paik and Cohen, demonstrated that the synthesis of CPS-I in the liver of thyroid hormone-treated tadpoles was depressed when a tadpole was placed at a lower temperature (15°C versus 25°C). This provided an additional impetus for selecting the genes encoding the ornithine–urea cycle enzymes as a system for assessing the temperature dependence of thyroid hormone-induced changes at the molecular level.

Our initial investigations assessed the relatives synthesis and accumulation of CPS-I and OTC proteins in liver from tapoles maintained at 5° or 25°C and in liver from tadpoles which had been maintained at 5°C for 21 days and, subsequently, shifted to 25°C. The results from these experiments showed that little synthesis and no apparent accumulation of CPS-I or OTC protein occur in the liver of any tadpoles maintained at 5°C. No change in the synthesis or accumulation of CPS-I or OTC was evident in liver from tadpoles not treated with thyroid hormone, regardless of whether they were maintained at 25°C or subjected to a 5° to 25°C shift in temperature. However, marked increases in the synthesis and accumulation of both CPS-I and OTC were evident in liver from tadpoles that had been treated with thyroid hormone and either maintained at 25°C or subjected to the 5° or 25°C shift in temperature (Helbing, 1993; Helbing *et al.,* 1995). These results clearly reflect the temperature dependence of the thyroid hormone-induced changes observed at the morphological (tail reduction) and physiological (urea production) levels. Furthermore, these results validate the use of this system for investigations at the molecular level.

It was necessary to clarify whether this temperature dependence in the thyroid hormone-induced expression of the ornithine–urea cycle enzyme genes is wholly at the translational level or, in fact, involves molecular events occurring prior to or at the level of transcription. Therefore, the relative levels of the mRNAs encoding CPS-I and OTC in the liver of tadpoles subjected to the same temperature regimens as described for assessing the CPS-I and OTC proteins were determined. Quantitative dot-blot hybridization and RT–PCR analyses, using nucleotide sequences from cDNAs encoding *R. catesbeiana* CPS-I and OTC (Helbing *et al.,* 1992; Helbing and Atkinson, 1994), revealed that the level of these mRNAs does not change in the liver of tadpoles maintained for 3 weeks at 5°C, regardless of whether or not they were treated with thyroid hormone. Shifting tadpoles from 5° to 25°C results in a marked, *sustained* increase in the level of these mRNAs in liver from thyroid hormone-treated tadpoles and a slight, albeit transient, increase in the level of these mRNAs in liver from tadpoles not treated with the hormone (Fig. 6). The cause for the accumulation of CPS-I and OTC mRNAs in the liver of tadpoles not treated with thyroid hormone is not known; however, the fact that it is transient suggests that it may be a temporary response to the stress of shifting the tadpoles from 5° to 25°C. Whatever the case, the results from these investigations suggest that the temperature dependency in the thyroid hormone-induced expression of the ornithine–urea cycle enzyme genes occurs at a molecular level which precedes the transcription of these genes. Moreover, these results imply that if thyroid hormones has made an "imprint" to enable the expression of the ornithine–urea cycle enzyme genes at a permissive temperature, then that imprint is also at a molecular level which precedes the transcription of these genes.

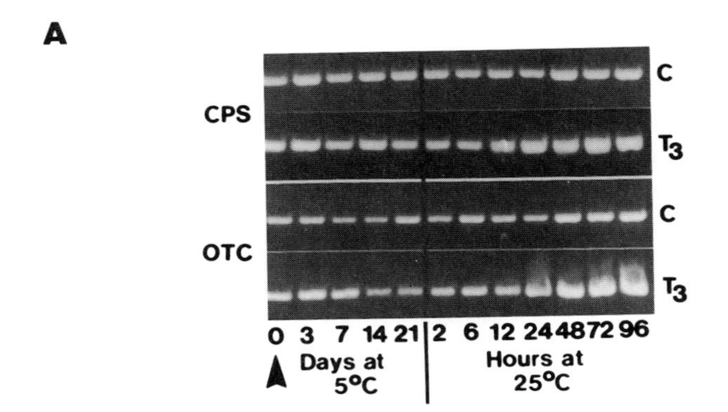

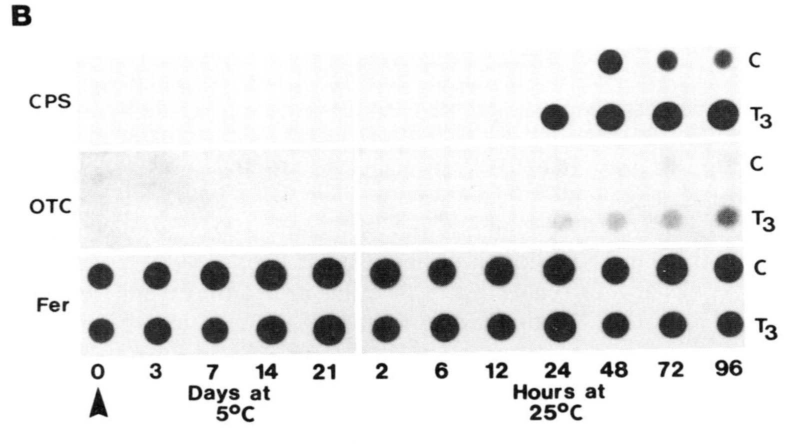

Fig. 6. RT–PCR (A) and dot-blot hybridization (B) analyses of CPS-I and OTC mRNAs in the liver of thyroid hormone-induced stage VI–VIII (Taylor and Kollros, 1946) tadpoles at various times during their maintenance at 5°C and after they had been shifted to 25°C. Tadpoles were acclimated to 5°C for 1 week prior to being injected with a vehicle solution (C) consisting of 200 μM NaOH or with 3×10^{-10} moles/g body weight of T_3 (T_3). After the injection the tadpoles were maintained for 21 days at 5°C before being shifted to 25°C. [Modified from Helbing (1993).]

Since binding of thyroid hormones to its nuclear-localized TRs is crucial for thyroid hormone action (see Tata, Chpater 13, this volume) and has been shown to occur at 4° to 6°C (Galton, 1980, 1983; Toth and Tabachnick, 1980), we questioned whether this low temperature interaction would result in an up-regulation of the TR mRNAs similar to that detected in the liver of tadpoles maintained at 25°C (see Fig. 2). Quantitative dot-blot hybridization and RT–PCR analyses, using nucleotide sequences from cDNAs encoding *R. catesbeiana* TRα (Schneider and Galton, 1991) and TRβ (Helbing *et al.,* 1992), were conducted on the same RNA samples as used in the temperature-shift experiments for assessing the levels of mRNAs encoding CPS-I and OTC. The results of these experiments (Helbing *et al.,* 1995; Helbing, 1993) revealed that the level of TRα mRNAs does not change in the liver of tadpoles maintained for 3 weeks at 5°C, regardless of whether or not they were treated with thyroid hormone. Shifting of the tadpoles from 5° to 25°C resulted in a 2-fold increase in the level of TRα mRNAs only in liver from the tadpoles that had been treated with thyroid hormone. Little or no accumulation of TRβ mRNAs occurred in the liver of tadpoles not treated with thyroid hormone, regardless of whether they were maintained at 5°C or subjected to a temperature shift. However, liver from tadpoles treated with thyroid hormone accumualted TRβ mRNAs at 5°C (4-fold after 21 days at 5°C) and the level continued to increase after they were shifted to 25°C [an additional 2-fold increase occurred by 2 hr after the shift to 25°C and culminated in a 28-fold increase 24 hr after the shift to 25°C (Helbing, 1993; Helbing *et al.,* 1995)]. These results indicate that the thyroid hormone-induced accumulation of TRβ mRNAs is not under the same temperature dependence as demonstrated for the accumulation of TRα and the ornithine–urea cycle enzyme mRNAs. The notion that the TRβ gene is a direct target gene for thyroid hormone (see Tata, Chapter 13, this volume; Shi, Chapter 14, this volume), coupled with the demonstration that the mRNAs encoded from this gene are upregulated by thyroid hormone at a temperature which does not permit the up-regulation of mRNAs from genes encoding the ornithine–urea cycle enzymes, suggests that the accumulated TRβ mRNAs may be one of the so-called "imprints" of Frieden *et al.* (1965). Whatever the case, it is apparent that the differences in the responses of these genes to temperature provide a valuable means for separating the interaction that a hormone–receptor complex has with a target gene from the cascade of molecular events responsible for orchestrating the expression of tissue-specific, late responsive genes, such as those encoding the ornithine–urea cycle enzymes.

V. SUMMARY

The results presented in this chapter highlight the obligatory requirement of thyroid hormone to initiate postembryonic changes in the amphibian tadpole liver which result in the expression of genes characteristic of the adult liver phenotype. Although earlier reviews have summarized many of the protein and enzymatic changes that occur in the liver of spontaneously and thyroid hormone-induced metamorphosing tadpoles (see Cohen *et al.,* 1978), very few of these reported changes have been explored at the molecular level. Since one purpose of this chapter was to provide a critique of studies concerned with the molecular basis by which these postembryonic changes occur, we centered our chapter on changes in gene expression which are well studied at the biochemical level and are currently being exploited at the molecular level. This chapter focuses on the expression of some of the genes (particularly CPS-I and OTC) encoding enzymes responsible for urea biosynthesis in the liver of an anuran, *R. catesbeiana,* which, during its metamorphosis, undergoes a transition from an ammonotelic tadpole to a ureotelic adult. This transition relies on the new and/or enhanced expression of the genes encoding the ornithine–urea cycle enzymes in the liver of this tadpole. Urea production and the synthesis of these enzymes can be induced precociously by administration of thyroid hormone to a premetamorphic tadpole or to serum-free cultures of liver from a premetamorphic tadpole. The fact that the hormone-induced, precocious synthesis of these proteins is preceded, both *in vivo* and in culture, by a marked up-regulation in the level of mRNAs encoding them implies that thyroid hormone is effecting the expression of the genes encoding these proteins.

The precise molecular means by which thyroid hormone exerts this tissue-specific response is unknown. The demonstration that it appears to involve a reprogramming of gene expression in a fixed population of hepatocytes has focused recent investigations on the molecular events that occur in these cells during the period between the administration of the hormone and the up-regulation of the mRNAs encoding these tissue-specific proteins—the so-called "lag" phase. The fact that the lag time in the response of the CPS-I and OTC genes is lengthy (~24 hr), and the demonstration that neither the CPS-I nor the OTC genes appear to have thyroid hormone response elements in their promoter regions, suggests that these genes are not diectly regulated by thyroid hormone. Studies suggest that the expression of the ornithine–urea cycle enzyme

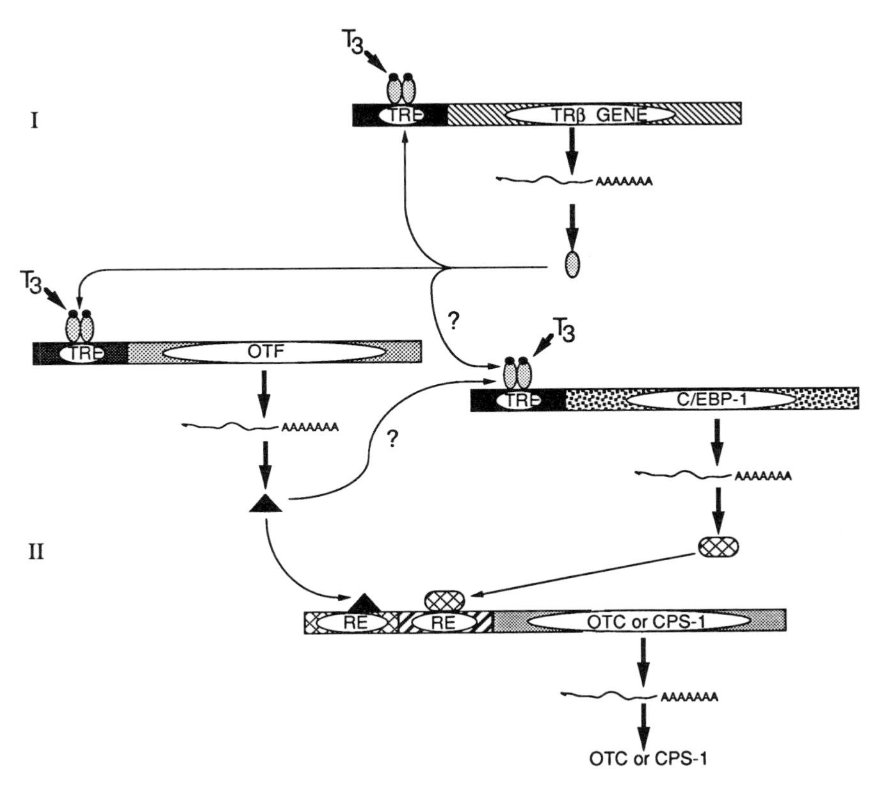

Fig. 7. A putative molecular mechanism(s) to account for the thyroid hormone-induced expression of the genes encoding the ornithine–urea cycle enzymes, CPS-I and OTC, in the liver of *R. catesbeiana* tadpoles. The molecular mechanism(s) proposed involves a cascade of molecular events in which this hormone is thought to initially upregulate the expression of genes encoding transcription factors, including TRβ, which in turn upregulate the expression of the genes encoding the ornithine–urea cycle enzymes CPS-I and OTC. For simplicity, this cascade has been separated into two phases: the first phase (I) depicts molecular events thought to occur during the lag period and is based on the assumption that promoters of genes encoding C/EBP-1 and/or other transcription factors contain functional TREs, and the second phase (II) shows a scheme in which the products, such as the *Rana* homolog of C/EBPα (RcC/EBP-1) and other transcription factors (OTF), bind to the appropriate response elements (REs) in the promoter regions of the CPS-I and OTC genes and upregulate their expression. The molecular events occurring during the lag period (phase I) are poorly understood and may or may not (see question marks) involve thyroid hormone directly regulating the expression of each of these early responding genes.

genes may, in fact, rely on the thyroid hormone induction of the TRs and/or other transcription factors, such as the *Rana* homolog of C/EBPα, which, in turn, initiate the terminal differentiation of the hepatocytes and the expression of these genes. Although studies to establish the *cis* elements and *trans*-acting factors involved in the thyroid hormone induction of these genes are still in their infancy, a hypothetical model depicting a mechanism(s) by which thyroid hormone may influence the expression of this cascade of genes is shown in Fig. 7. The observation that the expression of some late responding genes is temperature dependent provides a valuable means for (1) separating and studying the primary and secondary responses to this hormone, and (2) elucidating the molecular means by which this hormone initiates the postembryonic expression of genes characteristic of the adult, amphibian liver.

ACKNOWLEDGMENTS

The authors express their gratitude to Dr. Rob Dean and Dr. Cheryl Ketola for their helpful suggestions in the preparation of this chapter and to Roger Frappier and Huimin Hu for their excellent technical assistance. Y. Chen was a recipient of a Danish Research Academy Fellowship and an Ontario Graduate Research Fellowship. C. Helbing was a recipient of a 1967 NSERC Scholarship and a Dr. Marion Elder Grant Fellowship from the Canadian Federation of University Women, and is currently a recipient of a Leukemia Research Fellowship for postdoctoral studies at the University of Calgary, Calgary, Alberta, Canada. The work from the author's laboratory (B. G. Atkinson) was supported by funds from the Natural Sciences and Engineering Research Council of Canada.

REFERENCES

Ashley, H., Katti, P., and Frieden, E. (1968). Urea excretion in the bullfrog tadpole: Effect of temperature, metamorphosis, and thyroid hormones. *Dev. Biol.* **17,** 293–307.

Atkinson, B. G. (1981). Biological basis of tissue regression and synthesis. *In* "Metamorphosis: A Problem in Developmental Biology" (L. I. Gilbert and E. Frieden, eds.), 2nd ed., pp. 397–444. Plenum, New York.

Atkinson, B. G. (1994a). Molecular aspects of urogenesis in amphibians. *In* "Nitrogen Metabolism and Excretion" (P. J. Walsh and P. A. Wright, eds.), pp. 133–146. CRC Press, Boca Raton, Florida.

Atkinson, B. G. (1994b). Metamorphosis: Model systems for studying gene expression in postembryonic development. *Dev. Genet.* **15,** 313–319.

Atkinson, B. G., Helbing C., and Chen, Y. (1994). Reprogramming of gene expression in the liver of *Rana catesbeiana* tadpoles during spontaneous and thyroid hormone-induced metamorphosis. *In* "Perspectives in Comparative Endocrinology" (K. S. Davey, S. S. Tobe, and R. G. Peter, eds.), pp. 416–423. National Research Council of Canada, Ottawa.

Averyhart-Fullard, V., and Jaffe, R. C. (1990). Cloning and thyroid hormone regulation of albumin mRNA in *Rana catesbeiana* tadpole liver. *Mol. Endocrinol.* **4,** 1556–1563.

Baker, B. S., and Tata, J. R. (1990). Accumulation of proto-oncogene c-erbA related transcripts during *Xenopus* development: Association with early acquisition of response to thyroid hormone and estrogen. *EMBO J.* **9,** 879–885.

Beckingham Smith, K., and Tata, J. R. (1976). The hormonal control of amphibian metamorphosis. *In* "Developmental Biology of Plants and Animals" (C. Graham and P. F. Wareing, eds.), pp 232–245. Blackwell, Oxford.

Brown, G. W., Jr., Brown, W. R., and Cohen, P. P. (1959). Comparative biochemistry of urea synthesis. *J. Biol. Chem.* **234,** 1775–1780.

Chen, P. S. (1970). Patterns and metamorphic changes of serum proteins in amphibia. *Wilhelm Roux's Arch. Dev. Biol.* **165,** 132–149.

Chen, Y., and Atkinson, B. G. (1995). Characterization of the promoter regions of the carbamyl phosphate synthetase and ornithine transcarbamylase genes from *Rana catesbeiana.* Submitted.

Chen, Y., Hu, H., and Atkinson, B. G. (1994). Characterization and expression of C/EBP-like genes in the liver of *Rana catesbeiana* tadpoles during spontaneous and thyroid hormone-induced metamorphosis. *Dev. Genet.* **15,** 366–377.

Cohen, P. P. (1966). Biochemical aspects of metamorphosis: Transition from ammonotelism to ureotelism. *Harvey Lect.* **60,** 119–154.

Cohen, P. P. (1970). Biochemical differentiation during amphibian metamorphosis. *Science* **168,** 533–543.

Cohen, P. P., Brucker, R. F., and Morris, S. M. (1978). Cellular and molecular aspects of thyroid hormone action during amphibian metamorphosis. *In* "Hormonal Proteins and Peptides" (C. H. Li, ed.), pp. 273–381. Academic Press, New York.

Davey, J. C., Schneider, M. J., and Galton, V. A. (1994). Cloning of a thyroid hormone responsive *Rana catesbeiana* c-erbA-β gene. *Dev. Genet.* **15,** 339–346.

Dolphin, J., and Frieden, E. (1955). Biochemistry and amphibian metamorphosis. II. Arginase activity. *J. Biol. Chem.* **217,** 735–744.

Etkin, W., and Gilbert, L. I., eds. (1968). "Metamorphosis: A Problem in Developmental Biology." Appleton-Century-Crofts, New York.

Feldhoff, R. C. (1971). Quantitative changes in plasma albumin during bullfrog metamorphosis. *Comp. Biochem. Physiol. B* **40B,** 733–739.

Frieden, E. (1967). Thyroid hormones and the biochemistry of amphibian metamorphosis. *Recent Prog. Horm. Res.* **23,** 139–194.

Frieden, E. (1968). Biochemistry of amphibian metamorphosis. *In* "Metamorphosis: A Problem in Developmental Biology" (W. Etkin and L. I. Gilbert, eds.), pp. 349–398. Appleton-Century-Crofts, New York.

Frieden, E., Herner, A., Fish, L., and Lewis, J. C. (1957). Changes in serum proteins in amphibian metamorphosis. *Science* **126,** 559–560.

Frieden, E., Wahlborg, A., and Howard, E. (1965). Temperature control of the response of tadpoles to triiodothyronine. *Nature* (*London*) **205,** 1173–1176.

Friedman, A. D., Landschulz, W. H., and McKnight, S. L. (1989). CCAAT/enhancer binding protein activates the promoter of the serum albumin gene in cultured hepatoma cells. *Genes Dev.* **3,** 1314–1322.

Galton, V. A. (1980). Binding of thyroid hormones *in vivo* by hepatic nuclei of *Rana catesbeiana* tadpoles. *Endocrinology* (*Baltimore*) **106,** 859–866.

Galton, V. A. (1983). Thyroid hormone action in amphibian metamorphosis. *In* "Molecular Basis of Thyroid Hormone Action" (J. H. Oppenheimer and H. H. Samuels, eds.), pp. 445–483. Academic Press, New York.

Galton, V. A., and St. Germain, D. L. (1985). Putative nuclear triiodothyronine receptors in tadpole liver during metamorphic climax. *Endocrinology* (*Baltimore*) **117,** 912–916.

Galton, V. A., Morganelli, C. M., Schneider, M. J., and Yee, K. (1991). The role of thyroid hormone in the regulation of hepatic carbamyl phosphate synthetase activity in *Rana catesbeiana. Endocrinology* (*Baltimore*) **129,** 2298–2304.

Gilbert, L. I., and Frieden, E., eds. (1981). "Metamorphosis: A Problem in Developmental Biology," 2nd ed. Plenum, New York.

Griswold, M. D., Fischer, M. S., and Cohen, P. P. (1972). Temperature-dependent intracellular distribution of thyroxine in amphibian liver. *Proc. Natl. Acad. Sci. U.S.A.* **69,** 1486–1489.

Helbing, C. (1993). "Thyroid Hormone-Induced Changes in Gene Expression in the Bullfrog, *Rana catesbeiana,* Tadpole Liver," Ph.D. thesis. University of Western Ontario, London, Ontario.

Helbing, C., and Atkinson, B. G. (1994). 3,5,3′-Triiodothyronine-induced carbamyl phosphate synthetase gene expression is stabilized in the liver of *Rana catesbeiana* tadpoles during heat shock. *J. Biol. Chem.* **269,** 11743–11750.

Helbing, C., Gergely, G., and Atkinson, B. G. (1992). Sequential up-regulation of thyroid hormone β receptor, ornithine transcarbamylase, and carbamyl phosphate synthetase mRNAs in the liver of *Rana catesbeiana* tadpoles during spontaneous and thyroid hormone-induced metamorphosis. *Dev. Genet.* **13,** 289–301.

Helbing, C., Chen, Y., and Atkinson, B. G. (1995). T_3-induced gene expression in the liver of *Rana catesbeiana* tadpoles is altered by shifts in temperature. Submitted.

Herner, A. E., and Frieden, E. (1960). Biochemistry of anuran metamorphosis. *J. Biol. Chem.* **235,** 2845–2851.

Kawahara, A., Baker, B. S., and Tata, J. R. (1991). Developmental and regional expression of thyroid hormone receptor genes during *Xenopus* metamorphosis. *Development* **112,** 933–943.

Kim, K. H., and Cohen, P. P. (1968). Actinomycin D inhibition of thyroxine-induced synthesis of carbamyl phosphate synthetase. *Biochim. Biophys. Acta* **166,** 574–577.

Kistler, A., Yoshizato, K., and Frieden, E. (1975). Binding of thyroxine and triiodothyronine by nuclei and isolated tadpole liver cells. *Endocrinology* (*Baltimore*) **97,** 1036–1042.

Landschulz, W. H., Johnson, P. F., Adashi, E. Y., Graves, B. J., and McKnight, S. T. (1988). Isolation of a recombinant copy of the gene encoding C/EBP. *Genes Dev.* **2,** 786–800.

Ledford, B. E., and Frieden, E. (1973). Albumin synthesis during induced and spontaneous metamorphosis in the bullfrog *Rana catesbeiana. Dev. Biol.* **30,** 187–197.

Lincoln, A. J., Williams, S. C., and Johnson, P. F. (1994). A revised sequence of rat c/ebp gene. *Genes Dev.* **8,** 1131–1132.

Metzenberg, R. L., Marshall, M., Paik, W. K., and Cohen, P. P. (1961). The synthesis of carbamyl phosphate synthetase in thyroxin-treated tadpoles. *J. Biol. Chem.* **236,** 162–165.

Mori, M., Morris, S. M., and Cohen, P. P. (1979). Cell-free translation and thyroxine induction of carbamyl phosphate synthetase-1 messenger RNA in tadpole liver. *Proc. Natl. Acad. Sci. U.S.A.* **76,** 3179–3183.

Morris, S. M. (1987). Thyroxine elicits divergent changes in mRNA levels for two urea cycle enzymes and one gluconeogenic enzyme in tadpole liver. *Arch. Biochem. Biophys.* **259,** 144–148.

Moskaitis, J. E., Sargent, T. D., Smith, L. H., Jr., Pastori, R. L., and Schoenberg, D. R. (1989). *Xenopus laevis* serum albumin: Sequence of the complementary deoxyribonucleic acids encoding the 68- and 74-kilodalton peptides and the regulation of albumin gene expression by thyroid hormone during development. *Mol. Endocrinol.* **3,** 464–473.

Munro, A. F. (1953). The ammonia and urea excretion of different species of amphibian during their development and metamorphosis. *Biochem. J.* **54,** 29–36.

Nagano, H., Shimada, T., and Shukuya, R. (1973). Increase in serum albumin during bullfrog metamorphosis. *J. Biochem (Tokyo)* **73,** 1307–1309.

Paik, W. K., and Cohen, P. P. (1960). Biochemical studies on amphibian metamorphosis. I. The effect of thyroxine on protein synthesis in the tadpole. *J. Gen. Physiol.* **43,** 683–696.

Pouchelet, M., and Shore, G. (1981). Chloramphenicol inhibits hormone-dependent induction of cytoplasmic mRNA coding for the mitochondrial enzyme, carbamyl phosphate synthetase, in *Rana catesbeiana* tadpoles. *Biochim. Biophys. Acta* **654,** 67–76.

Schneider, M. J., and Galton, V. A. (1991). Regulation of c-erbA-α messenger RNA species in tadpole erythrocytes by thyroid hormone. *Mol. Endocrinol.* **5,** 201–208.

Schultz, J. J., Price, M. P., and Frieden, E. (1988). Triiodothyronine increases translatable albumin messenger RNA in *Rana catesbeiana* tadpole liver. *J. Exp. Zool.* **247,** 69–76.

Shambaugh, G. E., III, Balinsky, J. B., and Cohen, P. P. (1969). Synthesis of carbamyl phosphate synthetase in amphibian liver *in vitro.* The effect of thyroxine. *J. Biol. Chem.* **244,** 5295–5308.

Sladek, F. M., and Darnell, J. E. (1992). Mechanisms of liver-specific gene expression. *Curr. Opin. Genet. Dev.* **2,** 256–259.

Smith-Gill, S. J., and Carver, V. (1981). Biochemical characterization of organ differentiation and maturation. *In* "Metamorphosis: A Problem in Developmental Biology" (L. I. Gilbert and E. Frieden, eds.), 2nd ed. pp 491–544. Plenum, New York.

Tata, J. R. (1966). Hormones and the synthesis and utilization of ribonucleic acids. *Prog. Nucleic Acid Res. Mol. Biol.* **5,** 191–205.

Tata, J. R. (1967). The formation, distribution and function of ribosomes and microsomal membranes during induced amphibian metamorphosis. *Biochem. J.* **105,** 783–801.

Taylor, A. C., and Kollros, J. J. (1946). Stages of the normal development of *Rana pipiens* larvae. *Anat. Rec.* **94,** 7–24.

Toth, E., and Tabachnick, M. (1980). Effect of temperature on the transport of triiodothyronine (T_3) into liver nuclei of *Rana catesbeiana* tadpoles *in vivo. Gen. Comp. Endocrinol.* **42,** 57–62.

Umek, R. M., Friedman, A. D., and McKnight, S. L. (1991). CCAAT-enhancer binding protein: A component of a differentiation switch. *Science* **251,** 288–292.

VanDenbos, G., and Frieden, E. (1976). DNA synthesis and turnover in the bullfrog tadpole during metamorphosis. *J. Biol. Chem.* **251,** 4111–4114.

Vasseur-Cognet, M., and Lane, M. D. (1993). Trans-acting factors involved in adipogenic differentiation. *Curr. Opin. Genet. Dev.* **3,** 238–245.

Williams, S. C., Cantwell, C. A., and Johnson, P. F. (1991). A family of C/EBP-related proteins capable of forming covalently linked leucine zipper dimers *in vitro. Genes Dev.* **5,** 1553–1567.

Wixom, R. L., Reddy, M. K., and Cohen, P. P. (1972). A concerted response of the enzymes of urea biosynthesis during thyroxine-induced metamorphosis of *Rana catesbeiana*. *J. Biol. Chem.* **247,** 3684–3692.

Xu, Q., and Tata, J. R. (1992). Characterization and developmental expression of *Xenopus* C/EBP gene. *Mech. Dev.* **38,** 69–81.

Xu, Q., Baker, B. S., and Tata, J. R. (1993). Developmental and hormonal regulation of the *Xenopus* liver-type arginase gene. *Eur. J. Biochem.* **211,** 891–898.

Yamamoto, K., Kanski, D., and Frieden, E. (1966). The uptake and excretion of thyroxine, triiodothyronine and iodide in bullfrog tadpoles after immersion or injection at 25°C and 6°C. *Gen. Comp. Endocrinol.* **6,** 312–324.

Yaoita, Y., and Brown, D. D. (1991). A correlation of thyroid hormone receptor gene expression with amphibian metamorphosis. *Genes Dev.* **4,** 1917–1924.

16

Switching of Globin Genes during Anuran Metamorphosis

RUDOLF WEBER

Zoological Institute
Division of Cell and Developmental Biology
University of Berne
Switzerland

I. INTRODUCTION

Sequential changes in hemolobin phenotype are a common feature of vertebrate development and provide instructive cases of gene regulation.

In mammals and birds, multiple switches in hemoglobin synthesis occur during embryonic life, involving only partial replacement of globin chains. In contrast, a single transition from larval to adult hemoglobin occurs in anuran amphibians at metamorphosis and is characterized by complete replacement of globin chains. Since metamorphosis is controlled by thyroid hormones, the anuran globin system is amenable to experimental manipulation and is thus most suitable for investigating the mechanism of developmentally controlled gene expression.

This chapter critically evaluates the present knowledge on anuran hemoglobin transition. So far, most information stems from studies on Ranidae and mainly refers to biochemical and cellular aspects. Since this work has already been covered by excellent reviews (Frieden, 1968; Broyles, 1981), it will be only briefly outlined. In contrast, more emphasis will be given to studies on the South African clawed frog (*Xenopus laevis*), that by the advent of recombinant DNA technology had provided the first insight into the molecular aspects of anuran hemoglobin transition.

II. EARLY STUDIES ON HEMOGLOBIN TRANSITION IN ANURAN DEVELOPMENT

Early studies on the properties and the developmental changes in anuran hemoglobins refer almost exclusively to the Ranidae. The biochemical characteristics of anuran hemoglobins have been reviewed by Frieden (1968) and by Sullivan (1974), whereas the cellular basis of hemoglobin transition in *Rana catesbeiana* has been extensively discussed by Broyles (1981).

Some typical properties of *R. catesbeiana* tadpole and frog hemoglobins are compared in Table I. The higher capacity of oxygen binding of the tadpole hemoglobin, first described by McCutcheon (1936), may be interpreted as an adaptation to the hypoxic aquatic habitat of the larval stage. On the other hand, the physiological significance of the Bohr effect, characteristic of frog hemoglobin, is more difficult to explain because there is apparently no significant change in the O_2 dissociation curve throughout the circulatory system (Florey, 1966).

Electrophoretic and chromatographic analyses of hemolysates from tadpoles and frogs have revealed multiple hemoglobins, showing individ-

TABLE I

Properties of Tadpole and Frog Hemoglobins of *Rana catesbeiana*[a]

Characteristic	Hemoglobin	
	Tadpole	Frog
Oxygen-binding capacity	High	Low
Bohr effect	Absent	Typical
Sensitivity to 2,3-diphosphoglyceric acid regulation	No	Yes
Hemoglobin components (electrophoresis, column chromatography)	3–5	3–5
Hemoglobin subunits	$\alpha^{T}{}_{2}\beta^{T}{}_{2}$	$\alpha^{F}{}_{2}\beta^{F}{}_{2}$
Average molecular weight	68,000	68,000

[a] Modified after Frieden (1968).

ual and geographical variation, but it is not yet clear to what extent such variation actually represents protein polymorphism.

Amino acid composition and peptide mapping, as well as immunological data, suggest that tadpole and frog hemoglobins comprise different subunits. So far the primary structure of an α- and a β-globin chain of the major *R. catesbeiana* tadpole hemoglobin has been determined, showing respectively 80 and 55% homology with the corresponding human globin chains (Maruyama *et al.*, 1980; Watt *et al.*, 1980). Moreover, the amino acid sequence of a β-globin from the adult *R. esculenta* has been reported by Chauvet and Acher (1972).

The first evidence of a change in hemoglobin of metamorphosing tadpoles was provided by Trader *et al.* (1963), based on a sedimentation analysis of *R. grylio* hemoglobin, and by data of Baglioni and Sparks (1963) on the electrophoretic separation of *R. catesbeiana* hemolysates. The latter authors also reported that the induction of metamorphosis by thyroxine causes precocious appearance of adult hemoglobin. Despite numerous studies involving different species of Ranidae, there is still disagreement with regard to the timing of hemoglobin transition and the role of thyroid hormones in the control of this process (Sullivan, 1974).

Despite much effort, attempts to relate hemoglobin transition to changes in morphology and/or biosynthetic activities of the erythrocytes are as yet not conclusive. Broyles *et al.* (1981) desribed two morphologically distinct types of erythrocytes of *R. catesbeiana* tadpoles that contain different larval hemoglobin subunits and apparently originate from the liver and the kidney, respectively. From a study on the newly synthesized hemoglobin in erythrocytes, isolated by density-gradient centrifugation, Dorn and Broyles (1982) have concluded that hemoglobin transition in *R. catesbeiana* involves the appearance of new erythrocytes, producing exclusively adult hemoglobin, and the selective removal of erythrocytes, containing larval hemoglobin.

According to Broyles (1982), the developmental changes in erythropoiesis of *R. catesbeiana* comprise the following sequence of erythrocyte population transitions:

Embryo: The ventral blood islands produce a transient population of primitive erythrocytes with as yet unknown hemoglobin type.

↓

Premetamorphosis: Tadpoles contain two types of larval erythrocytes that differ in hemoglobin subunits and originate from the mesonephros (type 1) and the liver (type 2), respectively.

↓

Metamorphosis: During climax, immature erythrocytes containing adult hemoglobin appear in the tadpole liver.

Adult frog: Spleen and bone marrow produce adult erythrocytes with all the typical adult types of hemoglobin.

In this model the liver represents the most important site of anuran erythropoiesis. Moreover, it is evident that the metamorphic transition from larval- to adult-specific erythroid cells—in contrast to current views on hemoglobin switching in human development (Karlsson and Nienhuis, 1985)—apparently does not involve a change in the erythropoietic site. Obviously, erythropoiesis in the Ranidae deserves further investigation. In particular, elucidation of the lineage and the molecular characterization of hemoglobin phenotypes of the different erythrocyte populations would be highly desirable because this information is essential for investigating the mechanism of hemoglobin switching.

It should be stressed that the present knowledge on hemoglobin transition in Ranidae is based exclusively on morphological, biochemical, and immunological studies. These studies demonstrate that larval and adult hemoglobins are immunologically distinct and that the metamorphic switch in hemoglobin phenotype involves replacement of the erythrocyte populations. However, to acquire further insight into the mechanism of hemoglobin switching, it is critically important to extend the analysis to the molecular level in terms of specific gene activities.

III. MOLECULAR AND CELLULAR ANALYSIS OF HEMOGLOBIN TRANSITION IN *XENOPUS LAEVIS*

The South African clawed frog (*X. laevis*) is a most convenient organism for investigating the ontogeny of amphibian hemoglobins because not only does it reproduce under laboratory conditions, but genetically defined strains are also available. Moreover, this organism has become a favorite model in molecular biology for investigating the developmental control of gene expression. This may explain why our present knowledge regarding the genetic basis of the amphibian hemoglobin transition is exclusively derived from studies on *X. laevis*.

A. Characterization of Larval and Adult Hemoglobins

The occurrence of multiple forms of larval and adult hemoglobins of *X. laevis* was first demonstrated by Maclean and Jurd (1971), using chromatographic and electrophoretic separation of hemolysates. In a more detailed analysis, Just *et al.* (1977) have shown that the electrophoretic pattern of hemolysates from single animals comprises five to nine components with marked differences in mobility in the case of tadpoles, but only three components of lower and less variable mobility in frogs. From the observation that the mobility of the major larval and adult hemoglobin components is affected to the same extent by changes in concentrations of the acrylamide gels, these authors have concluded that the electrophoretically resolved fractions most likely represent charge isomers. Just *et al.* (1977) have further shown that the electrophoretic patterns of hemolysates from individual isogenic tadpoles, obtained from the interspecific hybrid *X. laevis* × *X. gilli* (Kobel and Du Pasquier, 1975), were identical, thus suggesting that variations in the hemoglobin pattern of normal *X. laevis* tadpoles apparently represent protein polymorphism.

Immunological studies by different procedures have shown that tadpole and frog hemoglobins do not share common antigenic determinants and, therefore, must represent distinct proteins (Jurd and Maclean, 1969; Just *et al.,* 1980). As will be shown in this chapter, this view has been confirmed by the molecular characterization of the larval and adult globin chains.

Very little is known regarding the physiological properties of the *X laevis* hemoglobins. As in Ranidae, adult hemoglobin of *X. laevis* has a substantial Bohr effect, which seems to be lacking in the tadpole. The oxygen affinity of the adult *X. laevis* hemoglobin is much higher when compared to Ranidae. This latter property may represent an adaptation to the hypoxic aquatic habitat of the adult *X. laevis* (Sullivan, 1974). According to Brunori *et al.* (1985), the absence of a Bohr effect in the tadpole hemoglobin may be due to specific differences in the primary structure of the β chains, but the biological significance of this finding remains to be established.

B. Metamorphic Transition from Larval to Adult Hemoglobin

The developmental switch in hemoglobin type was reported first by Jurd and Maclean (1969), who determined by immunological procedures the levels of larval and adult hemoglobins as well as their distribution in the erythrocytes throughout larval development. Their data suggest that the transition occurs only after completion of metamorphosis. In contrast, the electrophoretic analysis of hemolysates by Just *et al.* (1977) has clearly revealed that the switch from larval to adult hemoglobin sets in during the metamorphic climax (stage 62) and is complete by about 8 weeks after metamorphosis (Fig. 1). This time course has been further confirmed by Just *et al.* (1980) who determined the relative abundance of larval and adult hemoglobins using a radial immunodiffusion assay. It is likely that the late and rather extended transition, observed by Jurd and Maclean (1969), may reflect starvation, which is known to inhibit hemoglobin transition efficiently (J. J. Just, personal communication).

Electrophoretic separation of hemolysates on acid urea acrylamide gels, which resolves individual globin chains, has provided conclusive evidence that the transition from larval to adult hemoglobin involves the complete replacement of globin chains (Hentschel *et al.,* 1979; Sandmeier *et al.,* 1988). As shown in Fig. 2, premetamorphic tadpoles are exclusively characterized by a set of larval globin chains of low mobility, which disappear during metamorphosis and are absent in the adult frog. The adult globin chains, showing greater and more variable mobility, appear during the metamorphic climax (stage 63) and persist throughout adult life. It is further noted that very early tadpole stages contain two

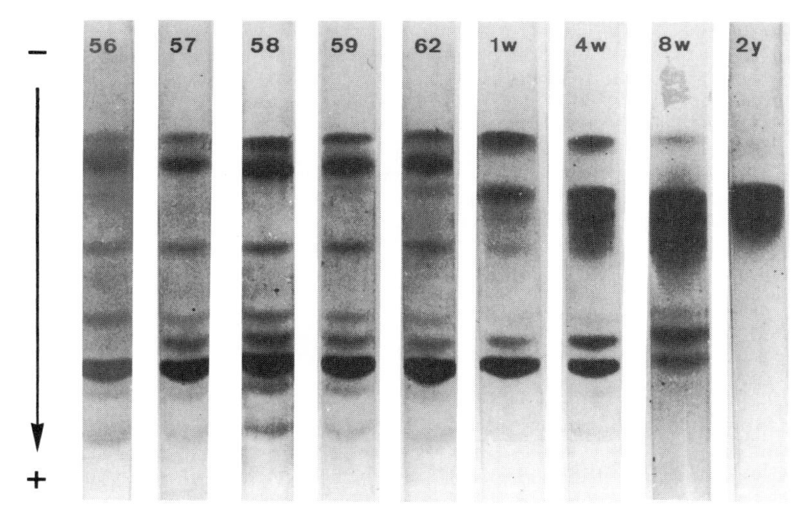

Fig. 1. Developmental changes in the hemoglobin pattern of *X. laevis*. Electropherograms of pooled hemolysates on acrylamide gels after staining with amidoblack. Stages 56, 57, 58, prometamorphic tadpoles; stages 59, 62, metamorphosing tadpoles; postmetamorphic stages are designated by weeks (w) or years (y) after metamorphosis. [Data from Just *et al.* (1977).]

additional minor globin components of unkown identity, already described by Kobel and Wolff (1983).

As shown by Maclean and Turner (1976), inhibition of metamorphosis by propylthiouracil results in giant tadpoles, which are morphologically blocked at stage 54, but nevertheless undergo a delayed hemoglobin transition. This surprising observation has been confirmed by Just *et al.* (1977), who found that thiourea causes a developmental arrest at stage 53 and a delay of 21–24 weeks in hemoglobin transition. Therefore, one is led to conclude that hemoglobin transition is not strictly correlated with the morphological transformation of tadpoles. Because the levels of endogenous thyroid hormone have not been determined in these experiments, it would be premature, however, to conclude that the transition from larval to adult hemoglobin is not controlled by this hormone. In fact, preliminary observations (J. J. Just, personal communication) indicate that thyroxine causes precocious appearence of adult hemoglobin after a latency period of about 3 weeks, thus suggesting a rather indirect mechanism of hormone action.

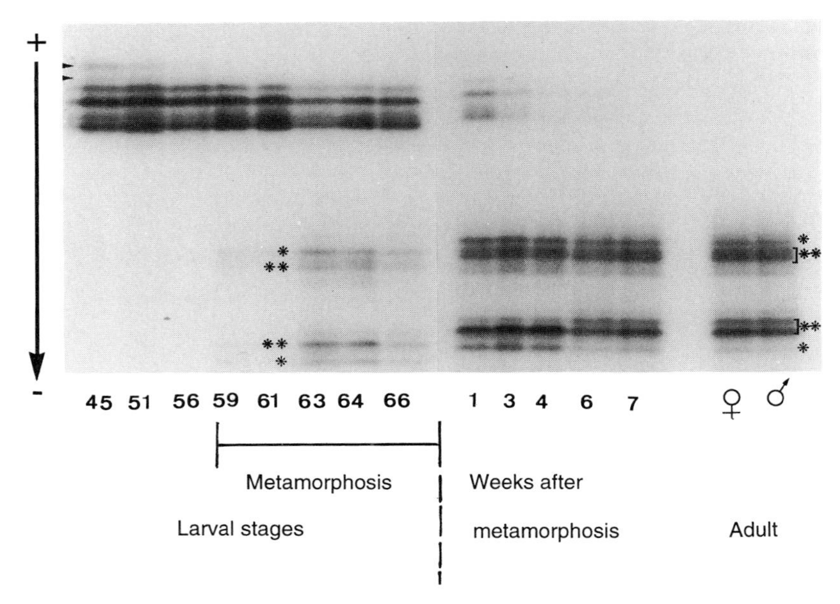

Fig. 2. Developmental changes in globin chains of *X. laevis*. Globin pattern obtained by electrophoresis of hemolysates on acid urea acrylamide gels and stained with Coomassie blue. Larval stages according to Nieuwkoop and Faber (1967); postmetamorphic stages in weeks after metamorphosis. ♂, ♀, adult frogs from which the larval stages were obtained. *, minor; **, major globin chains; arrowheads, transient early globin chains. [Reprinted with permission from *Roux's Arch Dev. Biol.* E. Sandmeier *et al.*, **197,** 406–412, (1988) © Springer-Verlag.]

In summary, it is noteworthy that *X. laevis,* which retains the aquatic habitat after metamorphosis, also exhibits the same type of hemoglobin transition as those anuran species that change from aquatic to terrestrial life at metamorphosis.

C. Characterization of Stage-Specific Globin mRNA Sequences by cDNA Cloning

To elucidate the genetic basis of the complex globin patterns, Widmer *et al.* (1981) analyzed cDNA libraries, which were derived from 9S poly(A)-containing RNA of erythrocytes from stage 56 tadpoles and adult frogs. From these libraries eight abundant cDNA sequences have been isolated, four of which were determined to be larval specific and the remaining four adult specific. Larval and adult sequences were found to comprise two pairs of each closely related α and β sequences, respectively. As shown in Table II, melting experiments with heteroduplexes, obtained by cross-hybridization of related cDNA clones, have further revealed a divergence of 13–14% for the related larval α and β sequences, but only 6–8% divergence for the corresponding adult globin sequences.

The pairwise occurrence of closely related globin sequences in both the larval and the adult stage is consistent with the hypothesis of Bisbee *et al.* (1977), according to which *X. laevis* may have evolved by genome duplication or tetraploidization. This contention is also supported by the finding that the closely related *Xenopus* vitellogenin (Wahli *et al.,* 1979) and albumin sequences (Westley *et al.,* 1981) show a very similar divergence to that of the adult globin sequences. The greater divergence between the related larval globin sequences may be explained by more complex evolutionary changes in the larval globin genes or less selection pressure on the larval globin polypeptides (Hosbach *et al.,* 1982).

Apart from this comparative study several investigators have contributed to the identification of globin mRNA sequences. Table III presents the globin mRNAs for which the most complete data are available. Although a high degree of homology is noted between the coding regions of sequences within the same subgroup, it is not yet known as to whether the closely related mRNAS are derived from the same transcription unit. Furthermore, the larval α-globin mRNA sequences are unique in showing a marked difference in length at the 3′ end, thus allowing easy discrimination.

TABLE II

Relatedness of Abundant *Xenopus laevis* Globin mRNA Sequences

Stage	Type[a]	Divergence	Type[a]	Divergence
Larval	$\alpha_I{}^L$	~14%	$\beta_I{}^L$	~13%
	$\alpha_{II}{}^L$		$\beta_{II}{}^L$	
Adult	$\alpha_I{}^A$	~6%	$\beta_I{}^A$	~8%
	$\alpha_{II}{}^A$		$\beta_{II}{}^A$	

[a] Designation based on Hosbach *et al.* (1983) and Widmer *et al.* (1983).

TABLE III

***Xenopus laevis* Globin mRNA Sequences Derived from cDNA Clones**

Stage	mRNA type	Codons	Homology	3′ noncoding region (nucleotides)	Reference[a]
Larval	α_I^L	141	98%	282	(*1*,*2*)
	αT3	141	97%	263	(*2*)
	αT4	141		261	(*2*)
	α_{II}^L	23–141	99.7%	Incomplete	(*3*)
	αT5	141		93	(*2*)
	β_I^L	146	99.3%	106	(*4*)
	βT1	146		106	(*5*)
	β_{II}^L	146	—	117	(*4*)
Adult	α_I^A	141	99.8%	112	(*1*)
	α1	141		110	(*6*)
	α_{II}^A	141	—	110	(*7*)
	β1	145	—	130	(*8*)
	β_{II}^A	23–146	—	110	(*7*)

[a] Key to references:
(*1*) Andres *et al.* (1984); (*2*) Banville and Williams (1985a); (*3*) G. Micheli *et al.* (unpublished observations); (*4*) Knöchel *et al.* (1985); (*5*) Banville *et al.* (1983); (*6*) Kay *et al.* (1983); (*7*) Knöchel *et al.* (1983); (*8*) Williams *et al.* (1980).

To relate the different globin mRNAs to specific globin chains, Sandmeier *et al.* (1988) analyzed the *in vitro* translation products of hybrid-selected mRNAs from erythrocytes of tadpoles (stage 56) and adult frogs. This procedure has allowed the assignment of each of the prominent translation products to a specific larval or adult globin mRNA. It is evident from Fig. 3A that the translation products of the hybrid-selected adult globin mRNAs encode the most prominent α- and β-globin chains of the adult pattern. As shown by Fig. 3B, the pattern obtained from the larval mRNAs is somewhat more complex. It comprises a slowly migrating minor component, which must be derived from an α-like mRNA, cross-hybridizing with the α_I^L, but not with the α_{II}^L cDNA. Moreover, the translation products of the larval β_I^L and β_{II}^L mRNAs

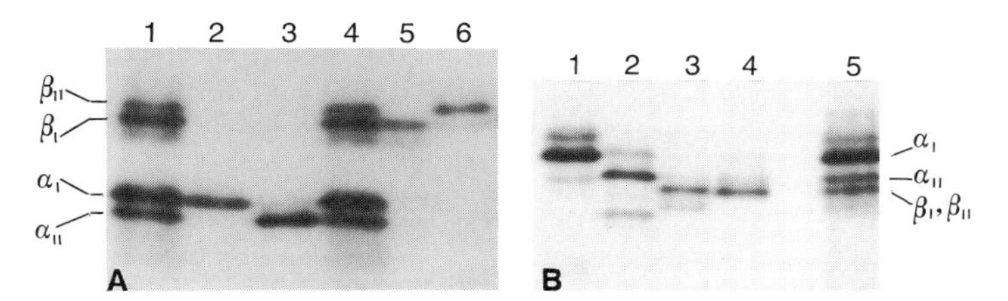

Fig. 3. Identification of individual *X. laevis* globin chains. Autoradiograms of electrophoretically separated *in vitro* translation products. (A) Adult globin chains from total adult erythrocyte RNA (lanes 1 and 4), mRNAs selected with $\alpha_I{}^A$ (lane 2), $\alpha_{II}{}^A$ (lane 3), $\beta_I{}^A$ (lane 5), and $\beta_{II}{}^A$ cDNA (lane 6), respectively. (B) Larval globin chains from total larval erythrocyte RNA (lane 5), mRNAs selected with $\alpha_I{}^L$ (lane 1), $\alpha_{II}{}^L$ (lane 2), $\beta_I{}^L$ (lane 3), and $\beta_{II}{}^L$ cDNA (lane 4), respectively. [Reprinted with permission from *Roux's Arch Dev. Biol.* E. Sandmeier *et al.,* **197,** 406–412, (1988) © Springer-Verlag.]

show the same electrophoretic mobility and, therefore, appear as a single component in the larval pattern. This study also demonstrates that for either stage the type I mRNAs encode major globin chains and the type II mRNAs minor ones.

It is clear that we have now at our disposal well-defined probes for investigating the developmental pattern of globin gene expression as well as the organization of these genes.

D. Developmental Pattern of Globin mRNA Accumulation

The only information so far on developmental changes in globin mRNA accumulation stems from the work of Banville and Williams (1985a,b) and is based on Northern blot and primer extension analysis of RNA from whole animals. The results are summarized in Fig. 4, showing quite different patterns of accumulation for α-like (Banville and Williams, 1985a) and β-like sequences (Banville and Williams, 1985b). αT3 and αT4 transcripts, which represent α_I-like sequences and αT5 transcripts, which are related to α_{II} sequences, appear in prefeeding tadpoles (stages 40–41). The αT3 and αT4 transcripts are equally abundant and persist until metamorphosis, whereas the αT5 transcripts begin to decrease already in premetamorphic tadpoles (stage 50). On the other hand, βT1 transcripts, which represent a larval β_I sequence, are already detected in prehatching embryos (stages 33–36) and by stage 41 are the

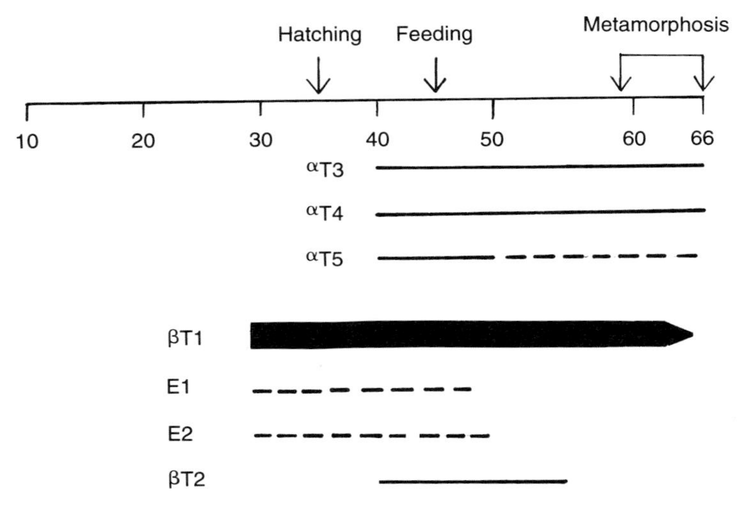

Fig. 4. Developmental pattern of larval globin mRNA accumulation in *X. laevis*. Graphical presentation of data from Banville and Williams (1985a,b) for α- and β-like globin mRNAs, respectively. Thickness of bars indicates the relative abundance of individual mRNAs in each group. The relationship of these mRNAs to those identified by Widmer *et al.* (1983) is indicated in Table III.

most abundant larval β-globin mRNA, but decline at metamorphosis. Moreover, two minor transcripts βE1 and βE2, which are related to the larval β_I sequence, accumulate transiently in early tadpoles (stages 33–49). It is not excluded that these minor transcripts may correspond to the minor globin components of early tadpoles (cf. Fig 2). βT2 transcripts, which represent the larval B_{II} sequence, appear in early tadpoles (stages 39–40) and persist at a relatively low level throughout larval life. Only trace amounts of adult β-globin mRNA are detectable in premetamorphic tadpoles (stages 49–50), but high levels of accumulation occur after metamorphosis.

In evaluating these results, which obviously merit further investigation, it should be stressed that they are based on the analysis of pooled mRNA populations, which may contain transcripts from different allelic variants of globin genes. Therefore it is not possible to determine exactly how many globin sequences are expressed in a given tadpole. Nevertheless, present data suggest that the larval α- and β-globin genes are differentially expressed with respect to time and extent, thus suggesting apparently independent regulatory mechanisms.

E. Characterization of *Xenopus laevis* Globin Gene Family

1. Arrangement of Globin Genes

The elucidation of the organization of the *X. laevis* globin genes was initiated by Jeffreys *et al.* (1980). Based on restriction mapping of genomic DNA, these authors have shown that the adult α- and β-globin genes of *X. laevis* are linked and are present as two closely related pairs of α_1 and β_1 and α_2–β_2 genes. Moreover, they found that *X. tropicalis,* a recent representative of an ancestral *Xenopus,* has only one pair of α- and β-globin genes, thus supporting the hypothesis of Bisbee *et al.* (1977), according to which *X. laevis* may have evolved by genome duplication or tetraploidization.

Based on restriction mapping of a large number of genomic clones and genetic analysis of restriction site polymorphism, Hosbach *et al.* (1983) have proposed a tentative model for the chromosomal arrangement of the globin gene family. As shown by Fig. 5A, the *X. laevis* genome is assumed to encompass eight larval and four adult globin genes that are symmetrically arranged in two clusters in the order $5'$-$\alpha^L{}_a$-$\alpha^L{}_b$-α^A-β^A-$\beta^L{}_a$-$\beta^L{}_b$-$3'$, all showing the same transcriptional polarity. From the identification of the globin chains, given in Fig. 3, it is clear that the genes of cluster I code for the major, and those of cluster II for the minor globin chains.

The linkage of α and β genes is a unique characteristic of *X. laevis* and differs from the arrangement of the globin genes of higher vertebrates, whose α- and β-globin genes occur in separate clusters that are located on different chromosomes (Collins and Weissman, 1984). Although the model, presented earlier, needs further corroboration, it suggests that the pairwise occurrence of coordinately expressed genes in *X. laevis* provides a unique opportunity for investigating the regulatory mechanisms of cell- and stage-specific gene expression.

2. Internal Organization of Globin Genes

Although most members of the *X. laevis* gene family have been partially characterized by various techniques, complete sequence data are

A

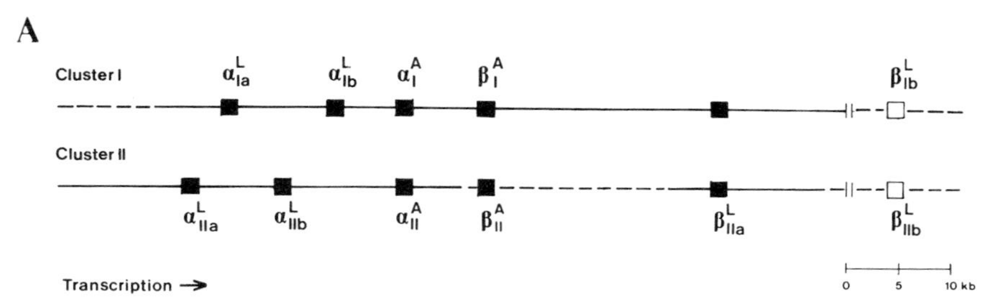

Fig. 5. Arrangement and internal structure of the *X. laevis* globin genes. (A) Putative organization of the *X. laevis* globin gene family. Physically identified genes are represented by black boxes and hypothetical ones by open boxes. Lines correspond to DNA segments, reconstructed from restriction mapping of genomic clones, and dashed lines represent unknown DNA segments. [Modified with permission after Hosbach *et al.* (1983).] (B) Internal organization of some adult and larval globin genes of *X. laevis.* Graphical presentation of data from Stalder *et al.* (1988) for adult α_I and α_{II} genes, Patient *et al.* (1983) for the adult $\beta_I$gene, and Meyerhof *et al.* (1984, 1986) for larval β_I and β_{II} genes. Stippled boxes indicate the 5′- and the 3′-noncoding regions, open boxes represent the first and second introns of each gene, and solid boxes represent the coding regions. The codon numbers at the boundaries of each exon are shown below the genes.

still scanty and comprise the $\alpha_I{}^A$ and $\alpha_{II}{}^A$ genes (Stalder *et al.,* 1986, 1988), the $\beta_I{}^A$ gene (Patient *et al.,* 1983), and the $\beta_I{}^L$ and $\beta_{II}{}^L$ genes (Meyerhof *et al.,* 1984, 1986). From Fig. 5B it is clear that these genes are characterized by the same exon–intron pattern and similar differences in length between and α and β genes as are known from the corresponding genes of higher vertebrates (Collins and Weissman, 1984).

The closely related $\alpha_I{}^A$ and $\alpha_{II}{}^A$ genes comprise 1078 and 1092 bp, respectively, each coding for 141 amino acids. Both genes contain a CCAAT motive and a TATA box in the promoter and a polyadenylation signal in the 3′-noncoding region. The $\beta_I{}^A$ gene, which is longer than the corresponding $\alpha_I{}^A$ gene, encompasses 1649 bp and encodes 145 amino acids. The promoter region is characterized by a CCAT motive and an ATA element, whereas the 3′-noncoding region contains the polyadenylation signal at a typical position. The closely related $\beta_I{}^L$ and $\beta_{II}{}^L$ genes comprise 2183 and 1822 bp, respectively, but code for 146 amino acids each. The promoter region of the $\beta_I{}^L$ gene contains a TATA box and a CAAT motive, but only a TATA box in the $\beta_{II}{}^L$ gene. The 3′-noncoding region of both genes comprises a polyadenylation signal. As compared to the α-globin genes, the β genes are characterized by introns

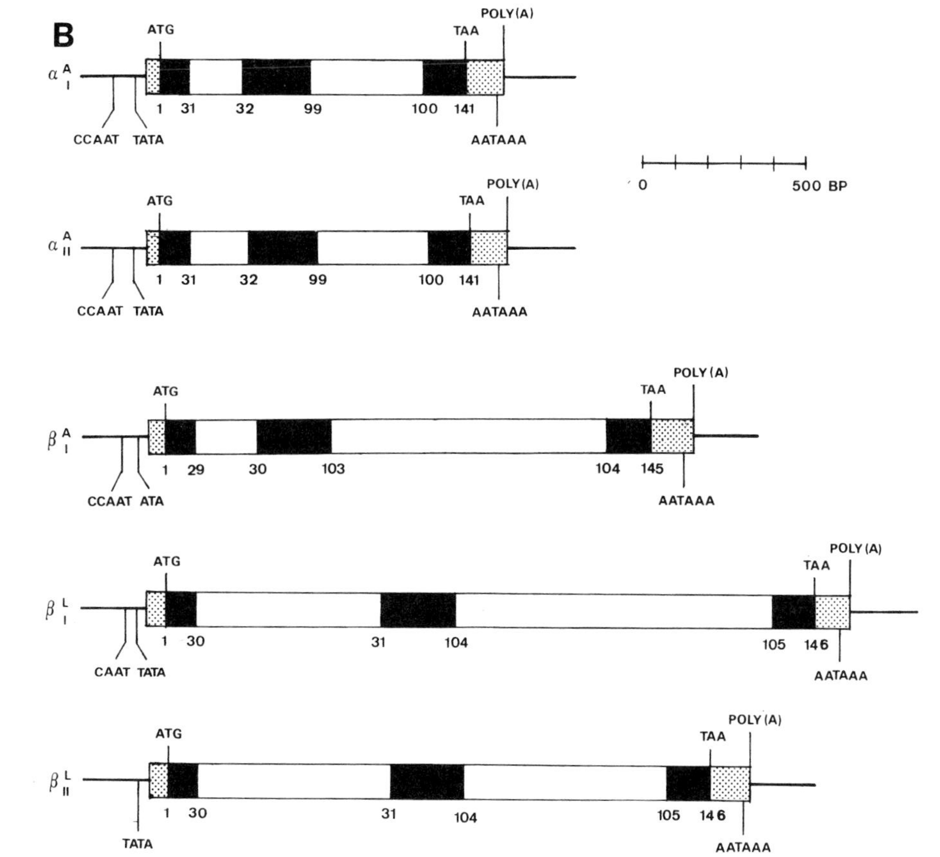

Fig. 5. *(continued)*

of greater and more variable length, which accounts for the differences in size of these genes.

In summary, these findings indicate that the *X. laevis* globin genes are characterized by the same internal organization, but a different chromosomal arrangement, as compared to the globin genes of higher vertebrates. This suggests that the evolutionary changes in structure and chromosomal location of these genes have occurred independently.

3. *Sequence Conservation in Upstream Region of Adult α-Globin Genes*

Since the *X. laevis* globin system comprises pairs of closely related genes that are expressed in a coordinate fashion, the search for structural homologies in putative regulatory regions of these genes might disclose potential regulatory elements. With this objective in mind, Stalder *et al.* (1986) have analyzed 1.5 kb of the 5′ flanking region of the coordinately expressed $\alpha_I{}^A$ and $\alpha_{II}{}^A$ genes. The results, summarized in Fig. 6, demonstrate that the upstream region of these genes contains, beyond the highly conserved promoter (83% homology), seven additional conserved elements (76–95% homology) that are separated by either more divergent sequences or insertions/deletions. The occurrence of similar conserved elements in the α^A-globin gene of *X. tropicalis* (Stalder *et al.*, 1988), a recent representative of an ancestral form of *X. laevis,* indicates

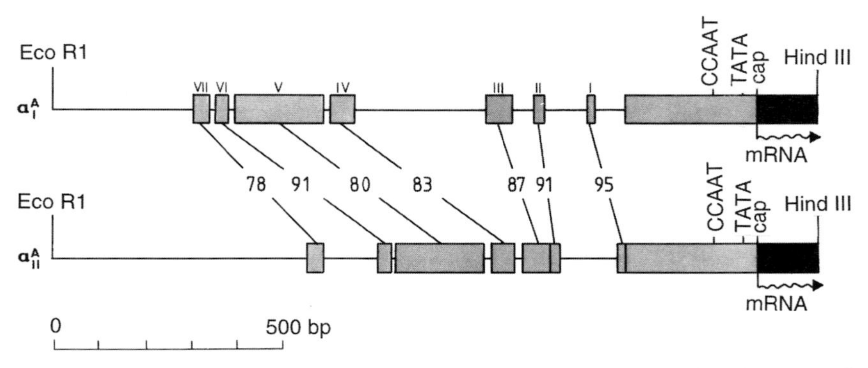

Fig. 6. Sequence conservation in the 5′ flanking regions of the coordinately expressed adult α_I- and α_{II}-globin genes of *X. laevis.* The conserved regions I–VII are represented by shaded boxes, and homologies, expressed as percentages, are indicated by arabic numerals. [Reprinted from Stalder *et al.* (1986) with permission from Academic Press.]

that these upstream elements most likely are under functional constraint and, therefore, may include regulatory elements.

F. Functional Studies on Globin Genes

1. *Cell- and Stage-Specific DNase I Hypersensitivity in Upstream Region of Coordinately Expressed $\alpha_I{}^A$ and $\alpha_{II}{}^A$ Genes*

For eukaryotic genes it has been shown that the initiation of transcription coincides with the appearance of DNase I hypersensitivity in the proximal upstream region, suggesting specific changes in chromatinconfiguration (Elgin, 1984). To determine whether such changes also occur in the adult α-globin genes, Stalder *et al.* (1986) have investigated DNase I hypersensitivity in the chromatin of nuclei from tadpole erythrocytes and liver, where the adult α-globin genes are silent, and from nuclei of adult erythrocytes, where these genes are expressed. By Southern analysis of the DNA, extracted from DNase I-treated nuclei, these authors have provided conclusive evidence that cell- and stage-specific DNase I hypersensitivity occurs in the proximal upstream region of the adult α_I- and α_{II}-globin genes, suggesting specific changes in the chromatin configuration to coincide with transcriptional activation. The DNase I hypersensitive region in both genes was localized within about 200 bp preceding the cap site, showing two preferential cutting sites to the 5′ side of the TATA box and the CCAAT motive, respectively.

2. *Experiments on Specific Transcriptional Activation of Xenopus laevis Globin Genes*

As a first step in the functional analysis of the 5′ flanking region of the adult α_I-globin gene, Stalder (1988) has constructed a hybrid gene by fusing 1.5 kb of the upstream region of the adult α_I gene to the CAT structural gene as a marker. By transfection of *X. laevis* kidney cells with this hybrid gene and subsequent fusion of the transfected cells with adult erythroblasts, Stalder (1988) demonstrated an apparently specific activation of the adult α_I-globin gene promoter. This result suggests that trans-acting factors of the adult erythrocytes may intervene in cell-specific transcriptional activation of the adult α_I-globin gene.

Another approach has been attempted by Dillon *et al.* (1991) by producing transgenic mice which carried the adult $\alpha 1$ and $\beta 1$ genes as well as the larval $\beta 1$ gene of *X. laevis.* The adult genes were transcribed in erythroid tissue of adult mice, whereas the tadpole-specific $\beta 1$ gene was only expressed in the blood from 12.5-day embryos. The differential expression of adult and larval globin genes according to the developmental stage of the host suggests that factors, which might be involved in regulating stage-specific gene expression, have apparently been conserved in vertebrate evolution.

To assess the role of trans-acting regulatory factors, Ramseyer *et al.* (1989) have studied protein and RNA synthesis in transient heterokaryons obtained by fusion of different types of erythroid cells, expressing either adult or larval globin genes. Fusion of erythrocytes from adult *X. laevis* and *R. catesbeiana* tadpoles resulted in cells that expressed adult *R. catesbeiana* globin genes. Similarly, cells obtained by fusion of *R. catesbeiana* tadpole erythrocytes and dimethyl sulfoxide-activated murine erythroleukemia cells also expressed adult *R. catesbeiana* globin genes. In both cases the transient heterokaryons accumulated adult α-globin mRNA. From these findings, it was concluded that stage-specific trans-acting regulatory factors for globin gene expression can function across vertebrate classes.

In conclusion, it should be stressed that the molecular characterization of the *X. laevis* globin genes and their developmental pattern of expression is still at a rather preliminary state. In fact, the organization and the chromosomal location of the globin gene family, including the nature of the larval globin-like sequences, deserve further elucidation. Moreover, it would be of interest to have more extensive data on the primary structure, including putative regulatory elements, of the globin genes, which are a prerequisite for functional assays. To achieve this, there is also an urgent need for establishing stable culture systems and selection procedures for transfected *X. laevis* cells.

IV. CELLULAR BASIS OF HEMOGLOBIN TRANSITION IN *XENOPUS LAEVIS*

Functional assays with isolated genes and presumptive regulatory DNA elements certainly provide important cues on the respective role of cis-acting sequence elements and trans-acting cellular factors in controlling cell-specific gene expression. Yet, in order to understand the mechanism of the developmental switch in globin gene expression, it is equally important to also consider the underlying cellular events.

A. Distribution of Larval and Adult Hemoglobins in Erythrocytes during Spontaneous Metamorphosis

Because the mature erythrocytes of amphibians are nucleated, the transition from larval to adult hemoglobin synthesis could occur either by a switch in gene expression within the same cell or by differential expansion of erythroid cell populations, resulting in the replacement of the larval by adult-specific erythrocytes.

As mentioned in Section I, the cellular aspects of hemoglobin transition have been studied most extensively in *R. pipiens* and *R. catesbeiana*. These investigations have led to the conclusion that the switch in hemoglobin phenotype is mediated by replacement of the larval erythrocytes by a new population of adult erythroid cells. From similar studies by immunofluorescence on *X. laevis* erythrocytes, Jurd and Maclean (1969, 1970) have reported that as many as 25% of the circulating erythrocytes in metamorphosing tadpoles contain both larval and adult hemoglobin, suggesting gene switching in individual erythrocytes. Figure 7 summarizes data from an immunological analysis by Weber *et al.* (1989) on the distribution of hemoglobin types in the circulating erythrocytes during larval development. It shows that the percentages of erythrocytes containing either larval- or adult-specific hemoglobin change inversely during metamorphosis, with their sum never exceeding 100%. This result strongly suggests that larval and adult hemoglobin are present in different cells. This result is in contradiction with previous data of Widmer *et al.* (1983), who from *in situ* hybridization of erythrocytes of anemic tadpoles have concluded that about 15% of the circulating erythrocytes may contain both larval and adult globin mRNA. This conclusion was based on the observation that at an early stage of anemia-induced globin transition about 25% of the erythrocytes were positive for larval and 90% for adult globin mRNA. However, the larval positive cells were only weakly labeled, whereas the adult positive cells were heavily labeled. On closer

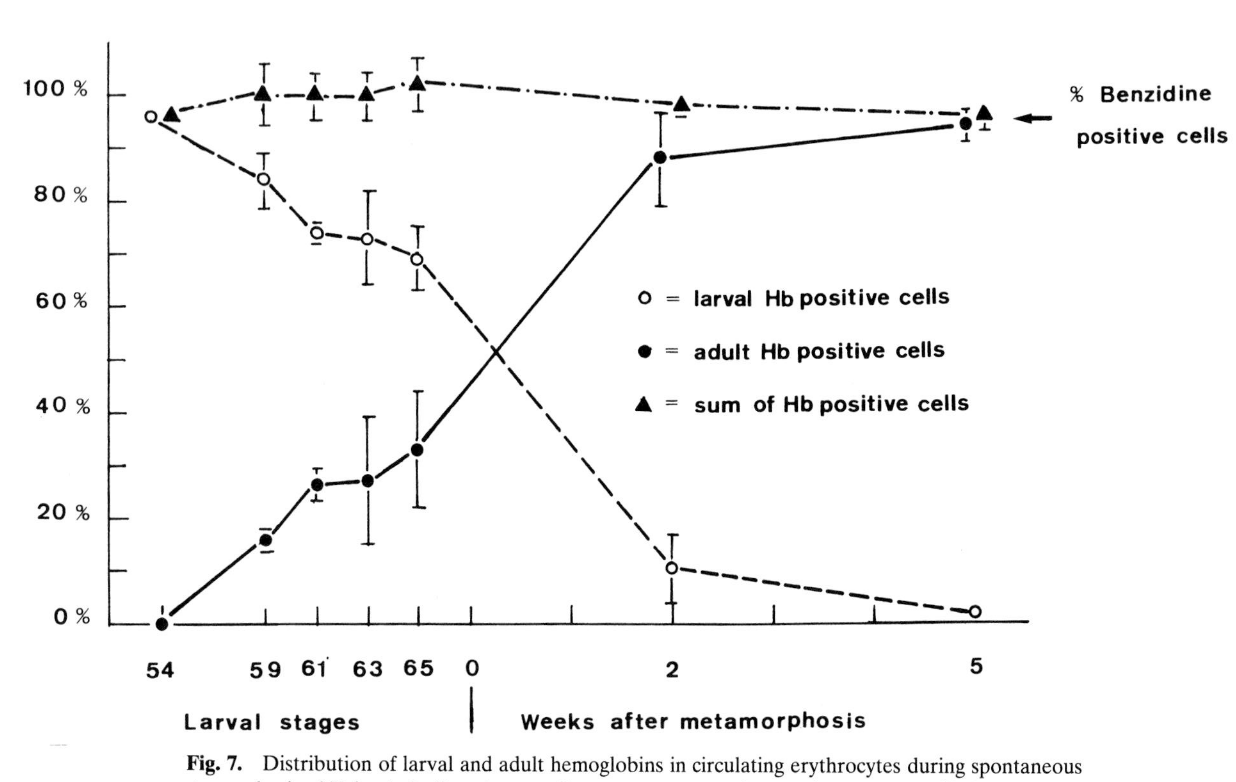

Fig. 7. Distribution of larval and adult hemoglobins in circulating erythrocytes during spontaneous metamorphosis of *X. laevis*. Indirect immunofluorescence was used to monitor stage-specific hemoglobins (Hb), using polyclonal antibodies. The abundance of positive cells is expressed as a percentage of the scored blood cells. Each point represents the mean of three independent experiments ± SD. The arrow indicates the percentage of benzidine or hemoglobin positive cells, which remains unchanged during transition. [Reprinted from *Roux's Arch. Dev. Biol.* Weber *et al.*, **198,** 57–64 (1989) © Springer-Verlag.]

scrutiny this result is rather puzzling because at early transition one would actually expect stronger signals for larval than for adult globin mRNA. Obviously, this result deserves further clarification by discriminatory double-labeling experiments.

On the whole, these findings indicate that the metamorphic switch in the hemoglobin phenotype of *X. laevis* follows the typical anuran pattern, involving replacement of the larval by adult erythrocytes, and, therefore, must be regulated by differential growth characteristics of the precursor cells.

B. Effect of Anemia on Erythroid Cell Populations

Induction of anemia by phenylhydrazine (Flores and Frieden, 1968, 1972), resulting in a rapid regeneration of the circulating erythrocytes, has been frequently used to explore the origin and the developmental properties of the larval erythrocytes. In a study on the effects of anemia on globin gene expression in *X. laevis,* Widmer *et al.* (1983) have not found any qualitative changes in the transcribed globin mRNA sequences. In frogs, however, the abundance of the minor globin mRNA sequences relative to the major ones was increased, as already noted by Battaglia and Melli (1977). In contrast to an earlier observation by Maclean and Jurd (1971), Widmer *et al.* (1983) have found no evidence for reactivation of larval globin gene expression in anemic frogs. On the other hand, these authors have presented data suggesting a precocious appearance of adult hemoglobin in anemic tadpoles, as reported earlier by Flores and Frieden (1972) for *R. catesbeiana.* As shown in Fig. 8, this finding has been confirmed by a more detailed investigation from the author's laboratory on the effect of anemia on hemoglobin transition in *X. laevis.* This work (R. Weber *et al.,* unpublished observations) has further shown that accumulation of larval globin mRNA, synthesis of larval globin polypeptides, and DNA replication are greatly reduced in the erythrocytes of anemic tadpoles. These findings, therefore, suggest that the larval erythropoietic system has a limited capacity for self-renewal, and by its decline may trigger the outgrowth of the adult erythroid cell population.

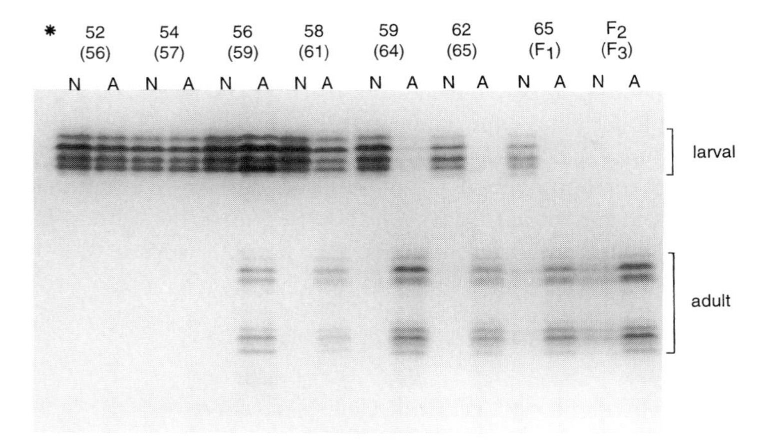

Fig. 8. Precocious appearance of adult globin chains after induction of anemia by phenylhydrazine during larval development of *X. laevis.* Electrophoresis on acid urea acrylamide gels of hemolysates from normal (N) and anemic (A) animals. Larval stages 52–65 after Nieuwkoop and Faber (1967). F2, froglet at the beginning of treatment; (), stages reached 6 days after induction of anemia (E. Sandmeier, personal communication).

C. Sites of Larval and Adult Erythropoiesis

1. Origin and Molecular Characterization of Larval Erythrocytes

Because differential growth of erythroid cell populations seems to be a major factor in determining the switch in globin gene expression, a study of the developmental fate of the larval and the adult erythroid cells should provide insight into the regulatory mechanism.

By grafting experiments on *X. laevis* embryos, Maéno *et al.* (1985) have shown that erythropoietic precursor cells predominantly originate from the ventral mesoderm (= ventral blood islands), but in part also from the dorsolateral mesoderm, which includes the prospective mesonephric region. Combining the *in vitro* culture of fragments from prehatching embryos and subsequently monitoring larval-specific globin mRNAs in nucleic acid extracts, Weber *et al.* (1991) have confirmed the grafting experiments by demonstrating that the ventral mesoderm, and to some degree also the dorsolateral mesoderm, are the primary source of erythropoiesis. As shown by Fig. 9, these experiments have also revealed that the first generations of erythrocytes undergoing functional maturation in the emerging circulatory system already express the most

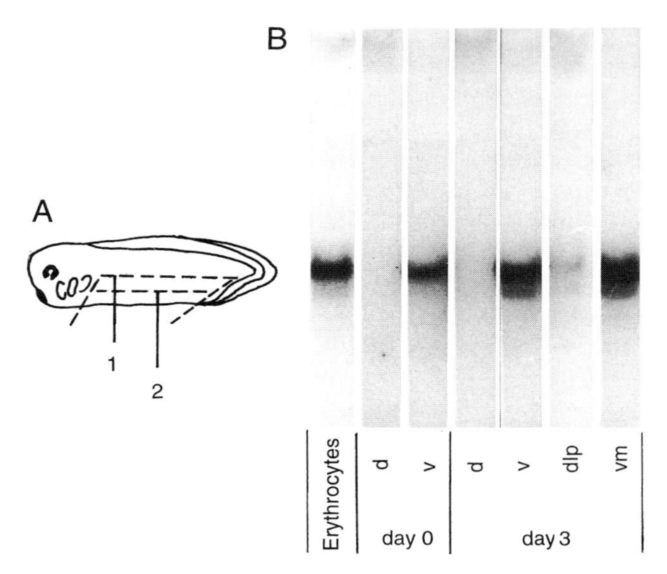

Fig. 9. Distribution of larval α-globin mRNAs in dorsal and ventral fragments of stage 32 *X. laevis* embryos and the corresponding *in vitro* differentiation products. (A) Dissection at level 1 resulted in dorsal (d) and ventral (v) fragments containing the dorsolateral and ventral mesoderm, whereas dissection at level 2 yielded dorsal fragments including the dorsolateral mesoderm (dlp) and fragments containing only the ventral mesoderm (vm). (B) Autoradiogram of Northern blot of total nucleic acid extracts from erythrocytes (stage 54 tadpoles), dorsal and ventral fragments of stage 32 embryos (day 0), and the *in vitro* differentiation products of fragments, as indicated in A. For identification of mRNAs, a mixture of ^{32}P-labeled $\alpha_I{}^L$ and $\alpha_{II}{}^L$ cDNAs was used. [Reprinted from Weber *et al.* (1991) with permission from Company of Biologists Ltd.]

abundant larval α-globin mRNAs. Although the present results suggest an even earlier appearance of the larval α-globin mRNAs, as has been reported by Banville and Williams (1985a), they are consistent with their demonstration that larval α_I-like sequences represent the predominating larval α-globin mRNAs.

The early appearance of tadpole-specific mRNA in the erythrocytes of prehatching embryos is in contradiction to the current concepts of anuran erythropoiesis. As mentioned in Section II, it is assumed that the ventral blood islands give rise to a transient population of primitive erythrocytes, which are replaced by the definitive larval erythrocytes released by the tadpole liver. This concept is based on morphological and biochemical characteristics as well as on grafting experiments with erythropoietic primordia, but does not take into account the type of globin genes that are expressed in these cells.

In this context it is of interest that grafting experiments with genetically marked ventral blood islands, reported by Rollins-Smith and Blair (1990), have shown that erythrocytes, originating from the graft, persist even in postmetamorphic froglets. Although it is not known which type of globin mRNA is expressed in these erythrocytes, the result is nevertheless of great interest because it suggests that the grafted blood islands most likely contain self-perpetuating hematopoietic stem cells, which at metamorphosis may give rise to adult erythroid precursor cells.

2. *Origin of Adult Erythrocytes*

To detect the emergence of the adult erythrocytes during larval development of *X. laevis,* Weber *et al.* (1991) analyzed the distribution of larval and adult α-globin mRNAs in nucleic acid extracts from putative erythropoietic primordia at metamorphosis. The highest levels of larval and adult α-globin mRNAs were found in the circulating blood cells and the liver, but only trace amounts were detected in the mesonephros. A more detailed analysis of erythrocytes and liver, shown in Fig. 10, has revealed that larval α-globin mRNAs are most abundant in the blood cells of premetamorphic tadpoles and decline during the metamorphic

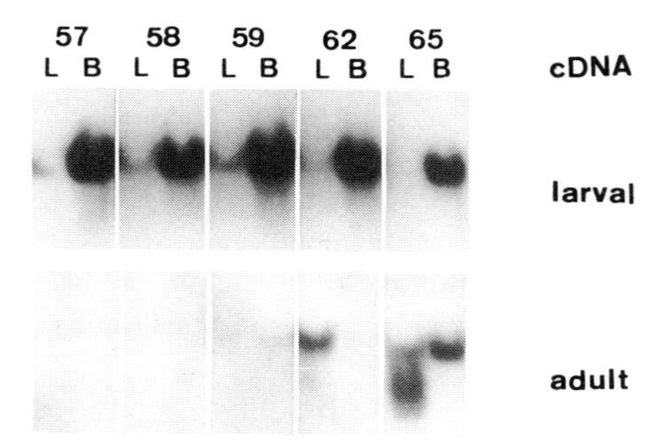

Fig. 10. Developmental change in the distribution of larval and adult α-globin mRNAs in liver (L) and erythrocytes (B) during spontaneous metamorphosis of *X. laevis.* Autoradiogram of Northern blot of nucleic acid extracts from equivalent tissue samples. For detection of mRNAs, ^{32}P-labeled cDNA probes for the major larval and adult α-globin mRNAs were used. [Reprinted from Weber *et al.* (1991) with permission from Company of Biologists Ltd.]

climax. This analysis has further shown that the adult α-globin mRNAs first appear in the liver and, with the progress of metamorphosis, also become detectable in the circulating blood cells. These findings, therefore, suggest that the adult erythrocytes emerge and undergo functional maturation in the liver.

D. Tentative Model for Developmental Regulation of Globin Gene Switching during Anuran Metamorphosis

Based on data on the origin of erythroid cells and the molecular characterization of the emerging erythrocyte populations in *X. laevis,* a tentative model for the developmental control of globin gene expression is proposed in Fig. 11.

It is assumed that the ventral blood islands and eventually also the dorsolateral mesoderm give rise to pluripotent, self-renewing hematopoietic stem cells. It is further proposed that the larval and the adult erythroid precursor cells represent successive progenies of the hematopoietic stem cell line and develop in distinct compartments. The larval erythroid precursor cells and their progeny occur in the developing circulatory system of prehatching embryos and in the blood of early larval stages, whereas the adult erythroid precursor cells are confined to the liver capillaries, where they normally divide and undergo terminal differentiation.

At metamorphosis a metabolic decline occurs in the larval erythrocytes leading to their active elimination (Kistler and Weber, 1975). A similar explanation may be advanced for the effect of anemia which, by causing depletion of the larval erythropoietic system, may initiate precocious outgrowth of the adult erythroid cells in the liver. Since the growth potential of the larval erythropoietic system is limited, the regeneration of erythrocytes is now provided by the adult erythroid precursor cells of the tadpole liver. This view implies that the switch in globin gene expression is mediated by differential growth of the erythroid cell populations. Commitment of these cells to express either larval or adult globin genes is assumed to occur in the precursor cells under the influence of the compartment in which they are located.

Although thyroid hormones are known to influence globin transition, their mode of action is as yet not clear. Depending on the target cells, the hormone could elicit either the aging process in the circulating larval

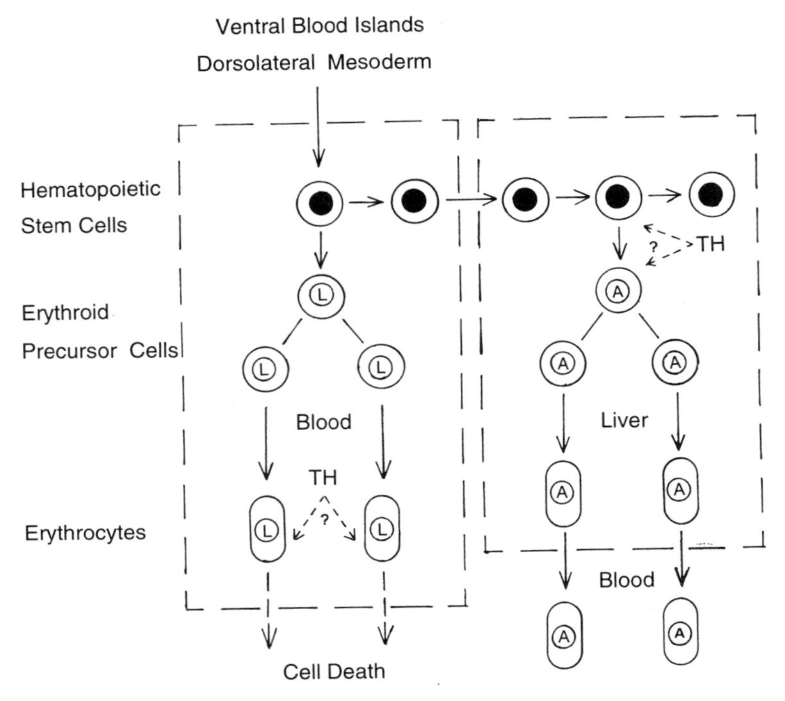

Fig. 11. Tentative model for the developmental regulation of globin gene expression in *X. laevis.* L, larval, B, adult erythroid precursor cells and erythrocytes; TH, thyroid hormone. [Modified after Weber *et al.* (1991).]

erythrocytes and/or initiate the outgrowth of the adult erythroid cell population.

In this model the developmental switch in globin gene expression is attributed to changing erythrocyte populations. Therefore, our understanding of the mechanism of developmental regulation will greatly depend on the elucidation of the factors that are involved in the control of larval and adult erythropoiesis.

V. CONCLUDING COMMENTS

The transition from larval to adult hemoglobin during anuran metamorphosis has been studied most extensively in Ranidae. Data on morphological, biochemical, and immunological characteristics of erythrocytes have led to the conclusion that larval and adult hemoglobins are distinct and that the metamorphic transition of hemoglobin involves replacement of larval-specific erythrocytes by adult-specific ones.

Taking advantage of recombinant DNA technology, more recent work on *X. laevis* has provided information on the structure and the arrangement of the globin genes. Yet, the structural analysis of larval and adult globin genes should be extended and completed by functional assays in order to establish the role of cis-acting sequence elements and trans-acting cellular factors in regulating transcriptional activation of the larval and the adult globin genes.

The molecular approach has also allowed us to follow the developmental changes in globin gene expression and to identify erythrocytes unequivocally by their gene products. By combining molecular and cellular studies it has been shown that the metamorphic switch in globin gene expression is mediated by differential growth characteristics of the larval and the adult erythropoietic systems.

Although evidence exists that thyroid hormones may be involved in this process, the mode of action is not known. Moreover, the lineage of the erythroid cells and the sites of larval and adult erythropoiesis should be experimentally confirmed, using *in vitro* systems and molecular tools for cell identification. Finally, another intriguing problem is to determine the factors that are involved in regulating larval and adult erythropoiesis because such information may provide the key to our understanding of the developmental control of globin gene expression.

ACKNOWLEDGMENTS

The work from the author's laboratory has been supported by the "Schweiz. Nationalfonds." The author is most grateful to research associates and graduate students for their valuable contributions to the *Xenopus* globin project.

REFERENCES

Andres, A.-C., Hosbach, H. A., and Weber, R. (1984). Comparative analysis of the cDNA sequence derived from the larval and the adult α_I-globin mRNAs of *Xenopus laevis*. *Biochim. Biophys Acta* **781,** 294–301.

Baglioni, C., and Sparks, C. E. (1963). A study of hemoglobin differentiation in *Rana catesbeiana. Dev. Biol.* **8,** 272–285.

Banville, D., and Williams, J. G. (1985a). The pattern of expression of the *Xenopus laevis* tadpole α-globin genes and the amino acid sequence of the three major tadpole α-globin polypeptides. *Nucleic Acids Res.* **13,** 5407–5421.

Banville, D., and Williams, J. G. (1985b). Developmental changes in the pattern of larval β-globin gene expression in *Xenopus laevis.* Identification of two early larval β-globin mRNA sequences. *J. Mol. Biol.* **184,** 611–620.

Banville, D., Kay, R. M., Harris, R., and Williams, J. G. (1983). The nucleotide sequence of the mRNA encoding a tadpole β-globin polypeptide of *Xenopus laevis. J. Biol. Chem.* **258,** 7924–7927.

Battaglia, P., and Melli, M. (1977). Isolation of globin messenger RNA of *Xenopus laevis. Dev. Biol.* **60,** 337–350

Bisbee, C. A., Baker, M. A., Wilson, A. C., Hadji-Azimi, I., and Fischberg, M. (1977). Albumin phylogeny for clawed frogs (*Xenopus*). *Science* **195,** 785–787.

Broyles, R. H. (1981). Changes in the blood during amphibian metamorphosis. *In* "Metamorphosis: A Problem in Developmental Biology" (L. I. Gilbert and E. Frieden, eds.), 2nd ed., pp. 461–490. Plenum, New York.

Broyles, R. H., Johnson, G. M., Maples, P. B., and Kindell, G. R. (1981). Two erythropoietic microenvironments and two larval red cell lines in bullfrog tadpoles. *Dev. Biol.* **81,** 299–314.

Brunori, M., Condò, S. G., Bellelli, A., Giardina, B., and Micheli, G. (1985). Tadpole *Xenopus laevis* hemoglobin. Correlation between structure and functional properties. *J. Mol. Biol.* **181,** 327–329.

Chauvet, J.-P., and Acher, R. (1972). Phylogeny of hemoglobins. β chain of frog (*Rana esculenta*) hemoglobin. *Biochemistry* **11,** 916–927.

Collins, F. S., and Weissman, S. M. (1984). The molecular genetics of human hemoglobin. *Prog. Nucleic Acid Res. Mol. Biol.* **31,** 315–437.

Dillon, N., Kollias, G., Grosveld, F., and Williams, J. G. (1991). Expression of adult and tadpole specific globin genes from *Xenopus laevis* in transgenic mice. *Nucleic Acids Res.* **19,** 6227–6230.

Dorn, A. R., and Broyles, R. H. (1982). Erythrocyte differentiation during the metamorphic hemoglobin switch of *Rana catesbeiana. Proc. Natl. Acad. Sci. U.S.A.* **79,** 5592–5596.

Elgin, S. C. R. (1984). Anatomy of hypersensitive sites. *Nature* (*London*) **309,** 213–214.

Flores, G., and Frieden, E. (1968). Induction and survival of hemoglobinless tadpoles and young bullfrogs. *Science* **159,** 103–105.

Flores, G., and Frieden, E. (1972). Hemolytic effect of phenylhydrazine during amphibian metamorphosis. *Dev. Biol.* **27,** 406–418.

Florey, E. (1966). "An Introduction to General and Comparative Physiology." Saunders, Philadelphia.

Frieden, E. (1968). Biochemistry of amphibian metamorphosis. *In* "Metamorphosis: A Problem in Developmental Biology" (W. Etkin and L. I. Gilbert, eds.), pp. 349–398. North-Holland, Amsterdam/Appleton-Century-Crofts, New York.

Hentschel, C. C., Kay, R. M., and Williams, J. G. (1979). Analysis of *Xenopus laevis* globins during development and erythroid cell maturation and the construction of recombinant plasmids containing sequences derived from adult globin mRNA. *Dev. Biol.* **72,** 350–363.

Hosbach, H. A., Widmer, H. J., Andres, A.-C., and Weber, R. (1982). Expression and organization of the globin genes in *Xenopus laevis. In* "Embryonic Development" (M. M. Burger and R. Weber, eds.), Part A, pp. 115–125. Liss, New York.

Hosbach, H. A., Wyler, T., and Weber, R. (1983). The *Xenopus laevis* globin gene family: Chromosomal arrangement and gene structure. *Cell (Cambridge, Mass.)* **32,** 45–53.

Jeffreys, A. J., Wilson, V., Wood, D., Simons, J. P., Kay, M. R., and Williams, J. G. (1980). Linkage of adult α- and β-globin genes in *X. laevis* and gene duplication by tetraploidization. *Cell (Cambridge, Mass.)* **21,** 555–564.

Jurd, R. D., and Maclean, N. (1969). The investigation of *Xenopus laevis* hemoglobins during development by a fluorescent antibody. *Experientia* **25,** 626–628.

Jurd, R. D., and Maclean, N. (1970). An immunofluorescent study of the hemoglobins in metamorphosing *Xenopus laevis. J. Embryol. Exp. Morphol.* **23,** 299–309.

Just, J. J., Schwager, J., and Weber, R. (1977). Hemoglobin transition in relation to metamorphosis in normal and isogenic *Xenopus. Wilhelm Roux's Arch. Dev. Biol.* **183,** 307–323.

Just, J. J., Schwager, J., Weber, R., Fey, H., and Pfister, H. (1980). Immunological analysis of hemoglobin transition during metamorphosis of normal and isogenic *Xenopus. Wilhelm Roux's Arch. Dev. Biol.* **198,** 57–64.

Karlsson, S., and Nienhuis, A. W. (1985). Developmental regulation of human globin genes. *Annu. Rev. Biochem.* **54,** 1071–1108.

Kay, R. M., Harris, R., Patient, R. K., and Williams, J. G. (1983). Complete nucleotide sequence of a cloned cDNA derived from the major adult α-globin mRNA of *X. laevis. Nucleic Acids Res.* **11,** 1537–1542.

Kistler, A., and Weber, R. (1975). A combined biochemical and morphological study on tissue changes in *Xenopus* larvae during induced metamorphosis. *Mol. Cell. Endocrinol.* **2,** 261–288.

Knöchel, W., Meyerhof, W., Hummel, S., and Grundman, U. (1983). Molecular cloning and sequencing of mRNAs coding for minor adult globin polypeptides of *Xenopus laevis. Nucleic Acids Res.* **11,** 1543–1553.

Knöchel, W., Meyerhof, W., Stalder, J., and Weber, R. (1985). Comparative nucleotide sequence analysis of two types of larval β-globin mRNAs of *Xenopus laevis. Nucleic Acids Res.* **13,** 7899–7908.

Kobel, H. R., and Du Pasquier, L. (1975). Production of large clones of histocompatible fully identical clawed toads. *Immunogenetics* **2,** 87–91.

Kobel, H. R., and Wolff, J. (1983). Two transitions of hemoglobin expression in *Xenopus:* From embryonic to larval and from larval to adult. *Differentiation (Berlin)* **24,** 24–26.

Maclean, N., and Jurd, R. D. (1971). The haemoglobins of healthy and anemic *Xenopus laevis. J. Cell Sci.* **9,** 509–528.

Maclean, N., and Turner, S. (1976). Adult hemoglobin in developmentally retarded tadpoles of *Xenopus laevis. J. Embryol. Exp. Morphol.* **35,** 261–266.

Maéno, M., Todate, A., and Katagiri, C. (1985). The localization of precursor cells for larval and adult hemopoietic cells of *Xenopus laevis* in two regions of embryos. *Dev., Growth Differ.* **27,** 137–148.

Maruyama, T., Watt, K. W. K., and Riggs, A. (1980). Hemoglobins of the tadpole of the bullfrog, *Rana catesbeiana* (amino acid sequence of the alpha chain of a major component). *J. Biol. Chem.* **255,** 3285–3293.

McCutcheon, F. H. (1936). Hemoglobin function during the life history of the bullfrog. *J. Cell. Comp. Physiol.* **8,** 63–81.

Meyerhof, W., Klinger-Mitropoulos, S., Stalder, J., Weber, R., and Knöchel, W. (1984). The primary structure of the larval β_I-globin gene of *Xenopus laevis* and its flanking regions. *Nucleic Acids Res.* **12,** 7705–7719.

Meyerhof, W., Köster, M., Stalder, J., Weber, R., and Knöchel, W. (1986). Sequence analysis of the larval β_{II}-globin gene of *Xenopus laevis. Mol. Biol. Rep.* **11,** 155–161.

Nieuwkoop, P. D., and Faber, J. (1967). "Normal Table of *Xenopus laevis* (Daudin)," 2nd ed. North-Holland, Amsterdam.

Patient, R. K., Harris, R., Walmsley, M. E., and Williams, J. G. (1983). The complete nucleotide sequence of the major adult β-globin gene of *Xenopus laevis. J. Biol. Chem.* **258,** 8521–8523.

Ramseyer, L. T., Barker-Harrel, J., Smith, D. J., McBride, K. A., Jarman, R. N., and Broyles, R. H. (1989). Intracellular signals for developmental hemoglobin switching. *Dev. Biol.* **133,** 262–271.

Rollins-Smith, L. A., and Blair, P. (1990). Contribution of ventral blood island mesoderm to hematopoiesis in postmetamorphic and metamorphosis-inhibited *Xenopus laevis. Dev. Biol.* **142,** 178–183.

Sandmeier, E., Gygi, D., Wyler, T., Nyffenegger, U., and Weber, R. (1988). Developmental pattern and molecular identification of globin chains in *Xenopus laevis. Roux's Arch. Dev. Biol.* **197,** 406–412.

Stalder, J. (1988). Erythroid specific activation of the *Xenopus laevis* adult α-globin promoter in transient heterokaryons. *Nucleic Acids Res.* **16,** 1027–1045.

Stalder, J., Meyerhof, W., Wirthmüller, U., Gruber, A., Wyler, T., Knöchel, W., and Weber, R. (1986). Conserved sequences and cell-specific DNase I hypersensitive sites upstream from the co-ordinately expressed α_I and α_{II}-globin genes of *Xenopus laevis. J. Mol. Biol.* **188,** 119–128.

Stalder, J., Wirthmüller, U., Beck, J., Gruber, A., Meyerhof, W., Knöchel, W., and Weber, R. (1988). Primary structure and evolutionary relationship between the adult α-globin genes and their 5′ flanking regions of *Xenopus laevis* and *Xenopus tropicalis. J. Mol. Evol.* **28,** 64–71.

Sullivan, B. (1974). Amphibian hemoglobins. *In* "Chemical Zoology" (M. Florkin and B. T. Scheer, eds.), Vol. 9, pp. 77–122. Academic Press, New York.

Trader, C. D., Wortham, J. S., and Frieden, E. (1963). Hemoglobin: Molecular changes during anuran metamorphosis. *Science* **139,** 918–919.

Wahli, W., Dawid, I. B., Weber, R., and Ryffel, G. U. (1979). Vitellogenin in *Xenopus laevis* is encoded in a small family of genes. *Cell (Cambridge, Mass.)* **16,** 535–549.

Watt, K. W. K., Maruyama, T., and Riggs, A. (1980). Hemoglobins of the tadpole of the bullfrog, *Rana catesbeiana* (amino acid sequence of the beta chain of a major component). *J. Biol. Chem.* **255,** 3294–3301.

Weber, R., Geiser, M., Müller, P., Sandmeier, E., and Wyler, T. (1989). The metamorphic switch in hemoglobin phenotype of *Xenopus laevis* involves erythroid cell replacement. *Wilhelm Roux's Arch. Dev. Biol.* **198,** 57–64.

Weber, R., Blum, B., and Müller, P. R. (1991). The switch from larval to adult globin gene expression in *Xenopus laevis* is mediated by erythroid cells from distinct compartments. *Development* **112,** 1021–1029.

Westley, B., Wyler, T., Ryffel, G., and Weber, R. (1981). *Xenopus laevis* serum albumins are encoded in two closely related genes. *Nucleic Acids Res.* **9,** 3557–3574.

Widmer, H. J., Andres, A.-C., Niessing, J., Hosbach, H. A., and Weber, R. (1981). Comparative analysis of cloned larval and adult globin cDNA sequences of *Xenopus laevis. Dev. Biol.* **88,** 325–332.

Widmer, H. J., Hosbach, H. A., and Weber, R. (1983). Globin gene expression in *Xenopus laevis:* Anemia induces precocious globin transition and appearance of adult erythroblasts during metamorphosis. *Dev. Biol.* **99,** 50–60.

Williams, J. G., Kay, R. M., and Patient, R. K. (1980). The nucleotide sequence of the major β-globin mRNA from *Xenopus laevis. Nucleic Acids Res.* **8,** 4247–4258.

17

Hormone-Induced Changes in Keratin Gene Expression during Amphibian Skin Metamorphosis

LEO MILLER

Department of Biological Sciences
University of Illinois at Chicago
Chicago, Illinois

I. INTRODUCTION

The larval and adult forms of anuran amphibians, such as *Xenopus laevis,* bear little resemblance to each other. They are subject to different selective pressures and express very different genetic programs. Metamorphosis is the process of transition between these two programs. All of the developmental processes which occur during early development—determination, differentiation, growth, induction, and morphogenesis—also take place during metamorphosis. The major differences are that maternal factors are no longer present and most of the progressive steps in determination leading to the adult program of differentiation have already occurred. Cells in the larval skin, brain, liver, etc. have become committed to a stable program of differentiation which maintains the tadpole for months or years depending on the species, until metamorphosis. During metamorphosis the larval program is converted to the adult program of differentiation. This switch is controlled by the thyroid hormones (TH) 3,5,3′-triiodo-L-thyronine (T_3) and L-thyroxine (T_4).

Early studies of metamorphosis demonstrated not only the dependence on TH but also the tissue, temporal, and spatial specificity of the TH response (Kollros, 1961; Gilbert and Frieden, 1981; Yoshizato, 1989; Tata, 1993; Atkinson, 1994; Kaltenbach, Chapter 11, this volume). While a great deal has been learned about metamorphosis, the major question, now being studied at the molecular level, centers on how T_3 initiates multiple tissue-specific responses which occur in a defined spatial pattern and temporal sequence. This chapter concentrates on the transformation that occurs in the *Xenopus* epidermis. In particular, the T_3-induced changes which regulate the activation of the 63-kDa keratin gene during the transition of the larval epidermis to the adult epidermis. The rationale for studying a late, tissue-specific gene at the molecular level is that such an analysis will provide important insights into the biological processes occurring during late organogenesis, with the assumption that the molecular mechanisms controlling this gene will identify a general scheme by which many tissue-specific genes will be regulated during metamorphosis.

II. MORPHOGENESIS AND DIFFERENTIATION OF SKIN

The skin of *Xenopus* is composed of a heterogeneous array of cells. Fox (1985) reviewed the origin and differentiation of more than 20 cell types in the skin of larval and adult amphibians and concluded that while the origin of most larval cell types is not dependent on TH, the development of the adult skin, including the epidermis, the multicellular epidermal-derived glands, and the dermis, is dependent on TH. This conclusion is based on the observations that adult skin normally appears during metamorphic climax when TH levels are high and that adult skin can be precociously induced by adding TH to premetamorphic tadpoles (Kollros and Kaltenbach, 1952; Kemp, 1963; Vanable, 1964). In *Xenopus,* the formation of the adult skin in the head region begins at stage 57 (Nieuwkoop and Faber, 1967) as areas of larval epidermis begin to stratify and differentiate. These areas expand in size as strips of larval skin between them degenerate. The temporal and spatial patterns of larval skin differentiation and degeneration are quite complex and require the entire period of metamorphosis, about 1 month, to be completed (Nieuwkoop and Faber, 1967; Nishikawa *et al.,* 1992). Understanding how these global changes in morphogenesis and differentiation of the skin and other organs are controlled and coordinated is the major goal of the current molecular studies.

Based on studies of the mammalian epidermis, the basal layer of the adult epidermis includes stem cells, transit amplifying cells, and postmitotic cells (Jones and Watt, 1993). Stem cells are capable of renewing themselves and giving rise to transit amplifying cells which are capable of a limited number of cell divisions, but are committed to differentiate. Postmitotic cells leave the basal layer and begin the process of terminal differentiation (cornification), including the production of keratin filaments and the development of an impermeable cell membrane. The adult *Xenopus* epidermis has about five cell layers (Nishikawa *et al.,* 1992). The outermost layer (stratum corneum) is composed of dead cornified cells which are periodically sloughed off the body. The larval epidermis is a bilayered epithelium which apparently does not contain stem cells. That is, cell divisions occur in both layers to accommodate growth (Wright, 1973, 1977), but stratification, terminal differentiation, or shedding of epidermal cells does not occur during larval life. Only the basal layer of epidermal cells is involved in the formation of the adult epidermis, whereas the upper periderm-like layer is sloughed off near the end of metamorphosis (Neiuwkoop and Faber, 1967; Yoshizato, Chapter 19, this volume). TH induces the basal cells to form both granular and mucous glands as well as the surface epidermis (Fig. 1). A very similar transformation occurs at the embryo–fetal transition in humans when glands, hair follicles, and a stratified epidermis are formed from the basal cells of a bilayered epidermis (Dale *et al.,* 1985).

Previous studies demonstrated that one of the earliest effects of TH on basal epidermal cells is the induction of cell proliferation (Neiuwkoop and Faber, 1967; Nishikawa *et al.,* 1989; Wright, 1973, 1977). In the hind

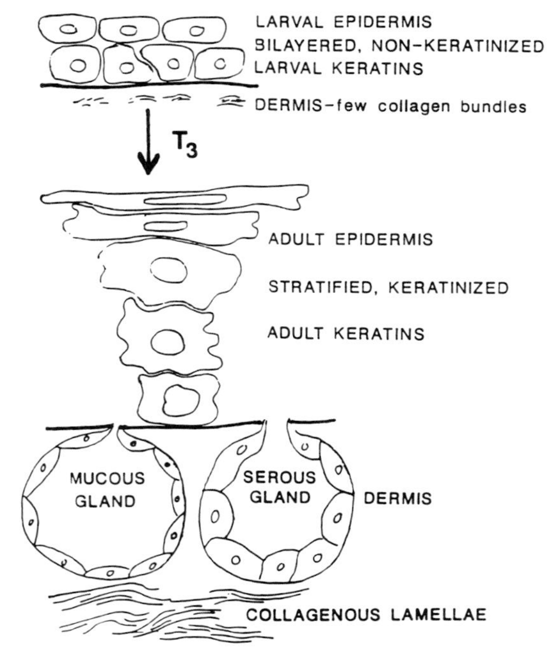

Fig. 1. The change from the larval skin to the adult skin induced by T_3 during metamorphosis. Basal cells of the larval epidermis form the mucous and serous (granular) glands, as well as the surface epidermis. Terminal differentiation (keratinization) only occurs in the adult epidermis.

limb epidermis of *Rana pipiens* (Wright, 1977) the proliferative pool of cells increases from 22% to almost 100% by 48 hr after TH treatment, suggesting that most basal cells take part in the proliferative response. In *Xenopus* (Buckbinder and Brown, 1992) the fraction of thymidine-labeled nuclei of all hind limb bud cells increases by 50% between 24 and 36 hr after T_3 injection. The proliferative response elicited by TH is quite different in the inner and outer cell layers of the larval epidermis. Wright (1973, 1977) has shown that during metamorphosis there is a gradual restriction in proliferation from both cell layers to the basal layer of cells. Likewise, TH also has opposite effects on the proliferation of tail and body epidermal cells (Nishikawa *et al.*, 1989). Clearly, the response of the epidermis to TH during metamorphosis not only shows temporal and spatial specificity but also cell-type specificity. The outer layer of larval epidermal cells stops dividing and differentiates into a periderm layer, which is sloughed off near the end of metamorphosis, while basal cells proliferate and form the adult glands and epidermis. Buckbinder and Brown (1992) show that the increase in replication is preceded by a T_3-induced up-regulation of growth response genes, similar to those induced in the serum response of quiescent mammalian fibroblasts. Thus, the induction of cell proliferation complicates the search for T_3-induced genes since it is difficult to distinguish the genes required for cell proliferation and those which mediate the conversion to the adult program of differentiation.

An interesting but unanswered question is how the larval basal cells give rise to adult cells with three very distinct programs of differentiation. The larval cells may be pluripotent and in response to TH they may enter different pathways of differentiation, including the formation of epidermal stem cells. Alternatively, distinct adult precursor cells for the glands and epidermis may already be present, and TH induces their proliferation and differentiation. A tractable question is whether TH is involved in programming (i.e., determination), readout of the program (i.e., differentiation), or both. This chapter argues that TH is required for the last step in the progressive determinative events leading to the adult program of differentiation and does not appear to be involved directly in the readout of that program.

III. CHANGES IN EPIDERMAL KERATINS DURING METAMORPHOSIS

The structural changes, described earlier, that occur during the conversion of the larval to the adult epidermis are correlated with changes in the pattern of keratins expressed. Keratins form the intermediate filaments of the cytoskeleton in all epithelial cells. Larval keratins gradually disappear and are replaced by a set of keratins typical of the adult epidermis (Ellison *et al.,* 1985; Shimizu-Nishikawa and Miller, 1992). This switch in keratin gene expression is dependent on T_3; it occurs during the metamorphic climax when natural TH levels are very high and can be prevented by the antithyroid drug propylthiouracil (Mathisen and Miller, 1987). The most abundant keratin in the adult epidermis is the 63-kDa keratin, a basic type II keratin (Hoffmann *et al.,* 1985; Ellison *et al.,* 1985; French *et al.,* 1994). Keratin filaments are formed from two keratins, one acidic and one basic, which are often expressed together (Sun *et al.,* 1984). Based on the response of the 63- and 51-kDa keratins to calcium levels and the similarity of these two proteins to keratins 1 (K1) and 10 (K10) of mammals, it seems likely that the 63- and 51-kDa keratins of *Xenopus* are filament-forming partners (Shimizu-Nishikawa and Miller, 1991). The stratification and differentiation of the epidermis occurring during metamorphosis are correlated with a dramatic increase in the expression of the 63-kDa keratin gene (Mathisen and Miller, 1987, 1989). In the periderm-like layer of the larval epidermis, which does not take part in the morphogenesis of the adult skin, the 63-kDa keratin gene is not upregulated (Nishikawa *et al.,* 1992). Similarly, when embryonic basal cells of the human epidermis begin to stratify and differentiate, the keratin gene K1 is activated and, like the 63-kDa keratin, is not expressed in the embryonic periderm (Dale *et al.,* 1985). Thus, 63-kDa keratin gene expression during metamorphosis serves as a molecular marker of epidermal stratification and differentiation. To link the morphogenesis of the epidermis with TH action, we have determined how TH controls the expression of the 63-kDa keratin gene.

IV. TWO REGULATORY STEPS NECESSARY FOR 63-kDa KERATIN GENE EXPRESSION

Thyroid hormones are first identified in the bloodstream near the beginning of prometamorphosis (Buscaglia *et al.,* 1985; Tata *et al.,* 1993). The concentrations of T_3 and T_4 increase throughout metamorphic climax and reach their highest levels at stages 61–62. Since skin differentiation is dependent on thyroid hormones, it was somewhat surprising to find that the 63-kDa keratin could be detected in head skin at stage 48, more than 2 weeks before T_4 appears in the tadpole and more than 1 month before stratification and terminal differentiation occur in the skin (Mathisen and Miller, 1987, 1989). Although some variability exists between tadpoles, the 63-kDa keratin or its mRNA is never observed prior to stage 48 and they are always present by stage 52. The 63-kDa keratin is easier to detect than its mRNA, probably because it integrates into the existing cytoskeleton of the larval basal cells [similar to human K1 (Kartasova *et al.,* 1993)]. These results strongly suggest that the initial activation of this gene is independent of T_3, but such *in vivo* studies do not eliminate the possibility that very low levels of TH are required. Indeed, previous work suggested that small amounts of TH are produced in premetamorphic bullfrog tadpoles (for a review see Galton, 1983). This is an important issue because as it turns out (see Section XI), most of the genes induced by T_3 are already expressed at low levels. Whether T_3 upregulates gene expression from low to high levels or activates genes previously repressed bears directly on the molecular mechanism of T_3 action during metamorphosis.

To verify that the initial activation of the keratin gene was independent of T_3, cultures of larval skin were used. Skin explants and purified epidermal cells of both larval and adult frogs have been cultured. These cultures are extremely valuable because they allow an analysis of the relative importance of various hormones and growth factors involved in the differentiation of the epidermis. The cultures also provided a way to demonstrate that the action of T_3 and other hormones is directly on the epidermis and not mediated via the dermis or other tadpole tissues. Cultures of skin from stage 40–43 larvae, which have not yet developed a thyroid gland (Jayatilaka, 1978), initiate low level 63-kDa keratin synthesis in culture media lacking serum, TH, or any other hormones. Interestingly, the basal level expression cannot be detected until siblings reach stage 48. Addition of T_3 to these cultures or to premetamorphic tadpoles demonstrates that T_3 precociously induces (10- to 20-fold) high level keratin gene expression. However, T_3 has no effect prior to stage 48, indicating again that epidermal cells of the tadpole first become capable of expressing the keratin gene at this stage. The simultaneous acquisition of the competence to respond to T_3 and the initiation of basal level expression at stage 48 suggest that T_3 responsiveness may be dependent on basal level activation (see Section XI). It is not known what changes occur in the epidermis at stage 48 to bring about low level expression and responsiveness to T_3, but the T_3 receptors (TRα and TRβ) do not appear to be limiting since they are present and can be upregulated by stage 40 (Yaoita and Brown, 1990; see Tata, Chapter 13, this volume; Shi, Chapter 14, this volume).

During normal development the 63-kDa keratin gene is upregulated in head epidermis at stage 57. As skin differentiation progresses throughout metamorphosis, the relative level of keratin gene mRNA increases more than 100-fold (Mathisen and Miller, 1987). Using a monospecific antibody for the 63-kDa keratin gene it was shown that the up-regulation of the keratin gene in head skin at stage 57 occurs in cells which move out of the basal layer and begin to differentiate (Nishikawa *et al.,* 1992). Thus up-regulation occurs concomitantly with stratification and differentiation. The up-regulation of the 63-kDa keratin in *Xenopus* and the activation of K1 in suprabasal cells of the mammalian epidermis (Roop *et al.*, 1987) suggest that these high molecular weight keratins are required for terminal differentiation and that their function has been conserved throughout vertebrate evolution. Although it is known that the development of the skin in birds is also dependent on TH (Wessells, 1961), it is not known if TH has a role in the development of the skin of mammals.

The expression analysis of the keratin gene indicates that its full activaton requires at least two distinct steps. At stage 48, basal expression of the keratin gene is initiated and the gene becomes responsive to exogenous T_3. Basal expression continues until stage 57, even though low levels of TH can be detected near the beginning of prometamorphosis. At stage 57, after 2 weeks during which TH levels gradually increase, the basal cells in certain areas of the larval head skin have acquired the adult program of differentiation and they begin the process of terminal differentiation, including high level keratin gene expression, stratification, cornification, and sloughing off of the cornified cells.

V. THYROID HORMONE UP-REGULATION OF KERATIN GENE EXPRESSION

Since the T_3–receptor complex modulates tissue-specific gene expression in mammals by directly interacting with regulatory sequences of the gene, a similar direct action on the keratin gene was expected. That is, the induction of the keratin gene would be rapid, require the continuous presence of the hormone to maintain gene expression, and the gene would remain competent to respond, reversibly, over a long period. The results indicate, however, that none of these expectations are upheld (Mathisen and Miller, 1989). Epidermal cells require at least 2 days of T_3 treatment before an increase in keratin mRNA can be detected. Once the adult program of differentiation has been induced by T_3, the up-regulation of the keratin gene, as well as stratification and differentiation of the epidermis, become unresponsive to T_3 treatment and they remain unresponsive throughout life. Cultures of epidermal cells from individuals in metamorphic climax or adults continue to express the 63-kDa keratin at high levels and differentiate in the absence of T_3; if T_3 is added to these cultures it has no effect on either process.

The 2-day latent period after T_3 treatment, before up-regulation of the keratin gene and differentiation of the epidermis occur, followed by the T_3-independent phase in the program of differentiation, strongly suggests that the T_3–TR complex does not interact directly with the keratin gene. Furthermore, it was shown that a transient T_3 treatment will induce the keratin gene. In one such experiment, stage 40 skin cultures were treated with T_3 for 8 hr at the beginning of a 6-day culture period. Although the keratin gene was not induced by day 4 (siblings at stage 47), even if T_3 was present continuously, it was induced to a high level by day 6 when siblings were at stage 48 (Mathisen and Miller, 1989). Thus even though the keratin gene cannot be induced until stage 48, the T_3-induced event required for keratin gene expression can occur at stage 40 and its memory stably transmitted through cell division to daughter cells which activate high level keratin expression long after the original inducer, T_3, is washed out of the culture. Similarly, when skin cultures of stage 54 larvae are treated with T_3 for 1 day followed by 4 days without hormone, keratin gene expression is upregulated about 15-fold (Fig. 2A). Although the interpretation of experiments using a transient T_3 treatment is not straightforward, i.e., it might not be possible to completely wash T_3 out of the cells of a monolayer culture (for a discussion see Mathisen and Miller, 1989), the results overall clearly indicate that T_3 does not have a direct action on the keratin gene.

In contrast to the *in vitro* results described earlier, a transient treatment of premetamorphic larvae with T_3 for short periods (up to 24 hr) followed by development to stage 48 in water lacking T_3 had no effect on keratin gene expression, while after 48 hr, the keratin gene was upregulated (Mathisen and Miller, 1987, 1989). Wang and Brown (1993) have reported similar results for tail degeneration, and Etkin (1964) has shown that morphological transformations are started, but not completed, if the thyroid gland of the tadpole is removed during metamorphic climax. The difference between the *in vitro* and *in vivo* results may be explained by factors, either systemic or cellular, which inhibit the T_3 response *in vivo* but are not retained in culture. The *in vivo* results clearly indicate that T_3 must be present for a relatively long time (up to 48 hr) to induce the new program of differentiation. Once the new program is fully developed, however, T_3 is unnecessary.

Thus, when assayed by keratin gene expression, the epidermis is responsive to T_3 only during a short window of time during development. Data indicate that T_3 has no apparent direct function in the morphogenesis, terminal differentiation, and tissue-specific gene expression of the adult epidermis. T_3 is essential to acquire the adult program, but is not necessary for readout of the adult program. The readout of the adult epidermal program of differentiation is regulated in part by calcium, glucocorticoids, and other factors in *Xenopus* as it is in mammals (Shimizu-Nishikawa and Miller, 1991, 1992). These results suggest that the developmental function of T_3 in amphibian metamorphosis is to switch cells from the larval program of differentiation to the adult program, possibly by activating developmentally important regulatory genes. In general, tissue-specific programs of differentiation are thought to be initiated and maintained by specific networks of regulatory genes (Gurdon, 1992; Weintraub, 1993). The stability of these programs can be explained by autoregulation and/or cross-regulation of these factors. Because of these properties, stable programs can be induced by transient external signals.

VI. SIMILARITIES BETWEEN THYROID HORMONE ACTION AND EMBRYONIC INDUCTION

The molecular mechanism by which T_3 regulates gene expression is likely to be very similar, if not identical, in all vertebrates. However, the functions of the genes activated by T_3 during development and in the adult are undoubtedly very different. In adult mammalian tissues and cell lines, T_3 modulates the expression of specialized genes such as the growth hormone and apolipoprotein-A1 genes (Lavin *et al.,* 1988; Romney *et al.,* 1992). When the concentration of T_3 is high, these genes are upregulated and if T_3 is withdrawn they are downregulated. In contrast to this reversible gene activity, T_3 induces stable gene expression during metamorphosis, probably by upregulating control genes such as transcription factors. Thus, the role of T_3 during metamorphosis is similar in many aspects to the role of embryonic inducers during early development. Gurdon (1987) has extensively reviewed embryonic induction and compared it to other cell signaling processes. Embryonic induction is unique in that the competence of tissues to respond to an inducer disappears soon after the signal is received and the timing of the response is dependent on the properties of the responding cells. In these aspects, T_3 action during metamorphosis mimics embryonic induction rather than hormone action, including T_3 action, in adult mammals (Oppenheimer and Samuels, 1983). The epidermis is only competent to respond to T_3

for a short time during its life span. Competence begins at stage 48 and is lost soon after adult differentiation begins during metamorphosis. Like embryonic induction the time between reception of the T_3 stimulus and the response is determined by the responding cells. The skin response, for example, occurs first in the hind limb epidermis, then in the body epidermis, and finally in the tail epidermis about 3 weeks after TH appears in the bloodstream (Nishikawa *et al.,* 1992). In contrast, in adult mammalian tissues the ability to respond to T_3 is maintained for years and the timing of the response is controlled by hormone release.

Embryonic induction can be classified as instructive or permissive; the distinction being the extent of preprogramming of the responding tissue. Gurdon (1987) has noted that instructive inductions generally occur early in development when the responding tissue is, relatively speaking, uncommitted and has numerous options open. As development proceeds and options become more restricted, inductions become progressively more permissive. T_3 action during metamorphosis is clearly similar to a global, permissive embryonic induction; most of the progressive steps in determination have occurred and T_3 induces the last step necessary to become adult cells. As with permissive inductions, T_3 induction is almost entirely determined by the attributes of the responding cells rather than the properties of the inducer, and the T_3 response leads to irreversible changes in the program of differentiation. In adult mammals, hormones, including T_3, modulate the properties of responding cells which have already acquired the adult program of differentiation.

VII. SPATIAL AND TEMPORAL SPECIFICITY IN UP-REGULATION OF KERATIN GENE EXPRESSION

Using a monospecific antibody to a unique peptide in the carboxy terminus of the 63-kDa keratin, the appearance of the 63-kDa keratin was found to vary dramatically in different regions of larval *Xenopus* skin (Nishikawa *et al.,* 1992). The onset of high level expression started at stage 54 in the hind limb bud epidermis, at stage 57 in the head epidermis, and at stage 63 in the tail. The low level expression of the keratin gene detected biochemically in the head epidermis prior to stage 57 (Mathisen and Miller, 1987) could not be detected with the antikeratin antibody, suggesting that it is due to low level expression in most cells and not to high level expression in a few cells. The presence of cells with high levels of 63-kDa keratin beginning at stage 57 in head epidermis correlates precisely with the previous estimate based on mRNA content for the beginning of high level expression of the 63-kDa keratin gene. However, the appearance of keratin-positive cells at stage 54 in hind limb and stage 63 in tail was unexpected. T_4 is present in the bloodstream but at extremely low concentrations at stage 54 (Buscaglia *et al.,* 1985; Tata *et al.,* 1993). On the other hand, TRα mRNA is present at extremely high concentrations in hind limb buds at stage 54 and then drops about 20-fold by the time the hind limbs finish their development (Wang and Brown, 1993). Thus, the acute sensitivity of hind limbs to T_3, noted by early workers (e.g., Kollros, 1961), is correlated with a high concentration of TRα and may explain the precocious development of the hind limb skin.

The appearance of keratin-positive cells in the tail epidermis was unexpected because the tail skin degenerates during metamorphosis. Even though T_3 induces proliferation in body epidermis and growth inhibition in the tail epidermis (Nishikawa *et al.,* 1989; Yoshizato, Chapter 19, this volume), some tail epidermal cells undergo the same program of terminal differentiation (i.e., stratification and cornification) as body cells. It was previously shown that tail epidermal cells precociously respond to exogeneous T_3 by upregulating keratin gene expression and that a combination of T_3 and corticosteroid synergistically induces keratin gene expression and differentiation of tail cells *in vitro* (Mathisen and Miller, 1987, 1989; Nishikawa *et al.,* 1990, 1992). The major difference between head and tail epidermal cells is that about four times as many head cells upregulate keratin gene expression, so a much stronger keratin gene response to T_3 is detected in the head epidermis.

One of the most important determinants regulating the response of epidermal cells, including the timing and regional specificity of differentiation, is the dermis. It is likely that reciprocal signaling between dermal and epidermal cells, both of which are sensitive to TH, will be important in the conversion of the larval skin to the adult skin during metamorphosis. For example, skin grafting experiments have shown that larval tail epidermis combined with body skin dermis develops into adult-type stratified epidermis with glands, whereas body epidermis combined with dermis of the tail and grafted to the body degenerates during metamorphosis (Kinoshita *et al.,* 1986). These results suggest that tail epidermal cells have a different propensity to differentiate than body epidermal cells. All tail epidermal cells may have the ability to differentiate in response to T_3, but the only cells that do actually differentiate during normal tail degeneration are those closer to the body (Nishikawa *et al.*, 1992). One possibility is that a decreasing posterior to anterior gradient of sensitivity to T_3 exists in the dermis and the threshold of the dermal response to induce death of the epidermis is set below the T_3 sensitivity of the epidermis to differentiate, except in the area near the body. Thus, distal tail epidermal cells die before they have a chance to differentiate while proximal epidermal cells differentiate before dying.

It is interesting to note that the up-regulation of 63-kDa keratin expression which begins at very different times in the hind limb, head, and tail was always correlated with differentiation of the epidermis. Thus, the up-regulation of keratin gene expression is tightly associated with stratification and cornification of the epidermis. Such coupling among differentiation, tissue-specific gene expression, and morphogenesis normally occurs during organogenesis. However, in many situations it is clear that they can be uncoupled. Low level keratin gene expression occurs about 1 month prior to stratification and differentiation. When cultures of epidermal cells of premetamorphic larvae are treated with T_3, the keratin gene is always upregulated, but differentiation and sloughing off of cornified cells, which occurs in high calcium media, does not occur in low calcium conditions (Shimizu-Nishikawa and Miller, 1991). On the other hand, differentiation of larval epidermal cells in cultures treated with T_3 and corticosteroid occurs before up-regulation of keratin gene expression (Shimizu-Nishikawa and Miller, 1992). Thus, the regulation of keratin gene expression can be uncoupled from morphogenesis and terminal differentiation, especially when precocious T_3 treatment is utilized. Extensive evidence indicates that the competence to respond to T_3 changes as development proceeds (Kollros, 1961). The ability to carry out different parts of the adult program of epidermal differentiation is gradually built into the response of larval basal cells. Thus the uncoupling of keratin gene expression with morphogenesis and differentiation by precocious T_3 treatment is probably due to the immaturity of the responding epidermal cells. These results emphasize the overriding importance of the preprogram to dictate the response to T_3 and to coordinate the global events occurring during metamorphosis. In a similar vein, it has been shown that muscle actin and desmin, but not myogenesis, can be induced in *Xenopus* ectoderm cells by etopic expression of myoD (Hopwood and Gurdon, 1990). Gittes and Rutter (1992) have also shown that low level expession of insulin and glucagon genes occurs in mammalian gut cells before morphogenesis and differentiation of the prancreas. They suggest that such low level expression may reflect a "premorphogenesis" phase during the process of commitment to an endocrine cell lineage.

VIII. EFFECTS OF STEROIDS AND OTHER FACTORS ON KERATIN GENE EXPRESSION

In 1955, Frieden and Nail showed that glucocorticoids accelerate T_3-induced metamorphosis. Because cell cultures were not previously used to analyze the interactions of T_3 and glucocorticoids, it was not known whether glucocorticoids act directly on target cells or indirectly via other tissues and organs. Hydrocortisone (HC) by itself does not induce metamorphic changes in larval skin, but HC in combination with T_3 has a synergistic effect on keratin gene expression and terminal differentiation of purified larval epidermal cells *in vitro* (Shimizu-Nishikawa and Miller, 1992). HC does not reduce the lag time necessary for T_3 to induce high level expression of the keratin gene. In fact, T_3 treatment was required for 3 days before the synergism with T_3 and HC was observed. However, it appears that HC acts more directly than T_3 on keratin gene expression because it acts more rapidly (<1 day) than T_3 (3 days). Furthermore, T_3 and HC together inhibited larval keratin synthesis in premetamorphic tadpoles, whereas T_3 alone did not. These results demonstrate that, in combination with T_3, glucocorticoids have direct effects on epidermal cells during metamorphosis but they probably do not effect the early genes directly induced by T_3. Thus, glucocorticoids and other members of the steroid hormone superfamily may be important for controlling the timing of differentiation.

In contrast to glucocorticoids, other hormones inhibit keratin gene expression. Both prolactin and retinoic acid inhibit the T_3 induction of keratin gene expression in cultures of stage 54 and stage 57 skin (Mathisen, 1987). In these experiments, cultures were treated with T_3 for only 1 day; after T_3 was washed out, the cultured epidermal cells were treated with prolactin or retinoic acid for 4 more days. Both retinoic acid and prolactin reduced by 90% or more the level of keratin mRNA found in controls treated 1 day with T_3 followed by 4 days without any hormones (Fig. 2). Since the T_3-responsive early genes are expected to be upregulated during the first 24 hr of T_3 treatment (see Shi, 1994, and Chapter 14, this volume), it seems probable that prolactin and retinoic acid are acting downstream of the direct response early genes induced by T_3.

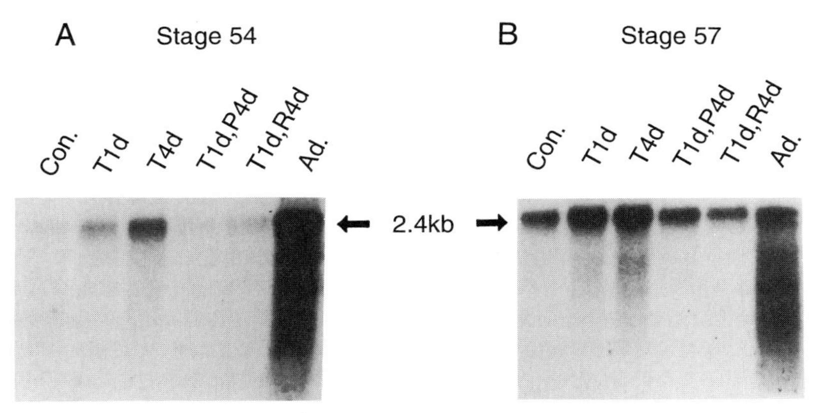

Fig. 2. Prolactin and retinoic acid suppress the T_3-induced expression of the 63-kDa keratin gene. Skin explants of stage 54 (A) or stage 57 (B) larvae were cultured for 6 days in serum-free media. Hormones were not added to control cultures (Con.). T_3 (10^{-8} *M*) was added on the second day for a period of 1 day or the last 4 days (T4d) of culture. After 1 day in T_3 the cultures were washed three times and incubated in hormone-free media (T1d), in media containing 10 μg/ml ovine prolactin (T1d, P4d), or in 10^{-7} *M* retinoic acid (T1d, R4d) for the last 4 days of culture. RNA was extracted from the cultures or from the adult epidermis (Ad.), resolved by formaldehyde gel electrophoresis, and hybridized with a probe (pM7) which specifically identifies the 63-kDa keratin mRNA at 2.4 kb (see Mathisen and Miller, 1987).

While these results are suggestive (see also Baker and Tata, 1992), the effects of hormones on premetamorphic tissues, *in vitro,* may not have real significance unless they mirror *in vivo* results. Buckbinder and Brown (1993) have shown that prolactin mRNA in the *Xenopus* pituitary gland increases late in metamorphosis in response to high T_3 concentrations, suggesting that prolactin may be induced by T_3 and has some role late in metamorphosis or in adult tissues (see Tata, Chapter 13, this volume). Because the skin begins to differentiate at stage 57, when prolactin mRNA levels are very low, it is possible that the effect of prolactin, like the glucocorticoids, is part of the cascade of gene action required to coordinate the temporal and spatial specificity of the readout of the adult program of skin differentiation. As suggested for prolactin mRNA (Buckbinder and Brown, 1993), the production of various hormones and their receptors may be relatively late events induced by T_3, although the rapid up-regulation of TRβ mRNA has been extensively documented (see Tata, Chapter 13, this volume; Shi, Chapter 14, this volume).

IX. THYROID HORMONE-INDUCED FORMATION OF STEM CELLS

It is important to identify the cells of a specific tissue or organ that are the relevant target of T_3 and to perform studies on that cell population. In renewing tissue such as the epidermis, a few precursor cells may be induced by T_3 to form adult stem cells which then proliferate to form the adult tissue. If T_3 only affects a small percentage of cells, it will be very difficult to identify the relevant cells and genes involved in the T_3-induced transformation. For example, the process of hemoglobin (Hb) switching during metamorphosis is accomplished by a switch at the precursor cell level; terminally differentiated red blood cells (RBCs) with both larval and adult Hb are not found (Broyles *et al.,* 1994; Weber, Chapter 16, this volume). In this case it is likely that T_3 induces a change from larval-type to adult-type precursor cells. Consequently, the transcription factors present in the downstream RBCs, which may be several cell generations removed, may be very different than those induced by T_3 in the precursors. In contrast, in the bullfrog a large percentage of hepatocytes are converted from the larval program of differentiation to the adult program in the presence of T_3 and inhibitors of DNA replication (Kawahara *et al.,* 1989). Furthermore, the mRNA for the transcription factor C/EBP-1 is induced by T_3 in the same hepatocytes as specialized late genes such as carbamyl-phosphate synthetase (Chen *et al.,* 1994). Thus, even though hepatocytes arise from a stem cell population (Sell, 1994), the evidence strongly suggests that the reprogramming of genes for both transcription factors and tissue-specific proteins occurs in the same cells. Since responsive hepatocytes comprise a substantial percentage of the total population, the T_3-induced changes can be readily followed (see Atkinson *et al.,* Chapter 15, this volume). The increase in the growth fraction of epidermal cells and the histological studies discussed earlier also indicate that the T_3-induced reprogramming of larval epidermal cells occurs in a large population of basal cells. If reprogramming occurs in the same cells that express tissue-specific proteins, it may be expected that both larval-specific and adult-specific genes would be active in the same cells. This is exactly what the experimental observations on the epidermis suggest (Nishikawa *et al.,* 1992). In premetamorphic tadpoles, both larval-specific keratins and the adult 63-kDa keratin are expressed in basal cells, albeit at very different levels. T_3 then converts these cells to adult basal cells in which the larval keratins are replaced by adult keratins. Thus, although it has not been formally demonstrated, the keratin switch seems to occur in the same cells that are the target of T_3, making it easier to trace the T_3 response from the putative T_3-induced regulatory factors to the high level expression of the 63-kDa keratin gene. In fact, the low level expression of the keratin gene in cells expressing larval functions may be a necessary step for the T_3-induced reprogramming.

Even though large numbers of cells respond to T_3 in hepatocytes and the epidermis, it must be considered that they may be responding differently. For example, in the larval epidermis, basal cells may be a diverse population because of their preprogramming and microenvironment. In response to T_3, some basal cells may be induced to form adult epidermal stem cells while others acquire a program that leads to degeneration or terminal differentiation which bypasses the adult stem cells. The diversity of the responding cells, and consequently the variety of genes induced by T_3, may make it difficult to identify the cascade of gene expression of interest. The use of purified populations of larval cells (Nishikawa *et al.,* 1990) and larval cell lines (Kanamori and Brown, 1993) should help alleviate these difficulties. Further studies linking the molecular aspects of T_3 action with the cellular changes occurring during morphogenesis are essential in understanding the role of T_3 in metamorphosis.

X. EARLY GENES UPREGULATED BY THYROID HORMONE

Since T_3 alters the program of differentiation rather than the readout of the program it is expected that the early genes induced by T_3 will include transcription factors. Two approaches have been used to identify the genes induced by T_3. Wang and Brown (1991) have developed a cDNA subtraction technique based on the polymerase chain reaction (PCR) to isolate genes induced during the lag period between the addition of T_3 and the readout of the program of differentiation. Using this technique they have identified two classes of *Xenopus* genes activated by T_3 during the lag period: direct response (early) genes and delayed response genes (see Shi, 1994, and Chapter 14, this volume). Direct response genes were upregulated within 8 hr of adding T_3 whereas delayed response genes were upregulated by 24 hr after T_3 treatment. The genes identified so far include both putative transcription factors such as TRβ and structural genes.

Another approach to identify the genes induced by T_3 is to work backward from the structural genes activated at the end of the lag period rather than forward from the T_3 stimulus. The rationale for this approach is that some of the transcription factors which regulate structural genes are likely to be activated during the lag period and thus comprise part of the T_3-induced cascade of gene expression. Once the structural gene is cloned, the regulatory sites and transcription factors controlling its expression can be identified. Analysis of the region upstream from the start sites of the 63-kDa keratin gene has identified putative regulatory elements for known transcription factors (French *et al.,* 1994). These include the binding sites for transcription factors, such as AP-2 and SP1, which are known to regulate keratin gene expression in mammals (Leask *et al.,* 1991; Blumenberg, 1993), as well as in early *Xenopus* embryos (Snape *et al.,* 1990, 1991). When stage 54 larvae are treated with T_3, AP-2 mRNA levels in skin are upregulated during the early part of the lag period (French *et al.,* 1994). Mobility shift assays (Fig. 3) have shown that an AP-2 site at −300 to −290 bp and an SP1 site centered at −97 bp upstream of the 63-kDa keratin gene transcription start site specifically bind purified human AP-2 and SP1, respectively. Furthermore, *in vivo* footprinting using ligation mediated PCR has shown that the AP-2 and SP1 sites are occupied throughout metamorphosis (Warshawsky and Miller, 1995).

The upregulation of a C/EBP-like transcription factor during metamorphosis has also been demonstrated in *Xenopus* (Xu and Tata, 1992) and *Rana* (Chen *et al.,* 1994). The identification of genes induced by T_3 in the early portion of the lag period and those which code for transcription factors regulating late gene expression will eventually lead to a resolution of the pathways which characterize the T_3 response of various tissues.

XI. LOW-LEVEL EXPRESSION OF GENES PRIOR TO APPEARANCE OF THYROID HORMONE

One of the interesting aspects of T_3 action is that the response is tissue specific; one hormone induces multiple tissue-specific programs of differentiation. This specificity is most likely the result of preprogramming of the T_3-responsive larval cells. In the larval epidermis the preprogram is so complete that the most abundant protein of the adult epidermis, the 63-kDa keratin, is already expressed almost 2 weeks before T_3

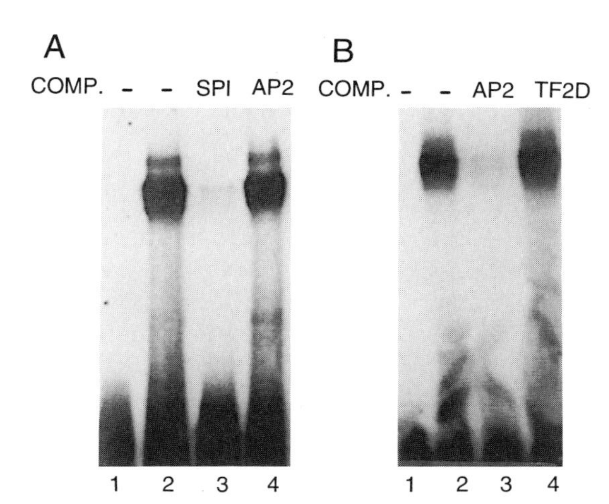

Fig. 3. Mobility shift assays with purified SP1 protein and with an *E. coli* extract containng AP-2. (A) A 40-bp fragment (−83 to −122 bp) which contains two potential SP1-binding sites was end-labeled and incubated in binding buffer in the absence of protein (lane 1), with purified human SP1 (lane 2), with purified SP1 in the presence of a 100-fold molar excess of an oligonucleotide which binds human SP1 (lane 3), and with purified SP1 in the presence of a 100-fold molar excess of a noncompeting oligonucleotide which binds AP-2 (lane 4). (B) A 25-bp fragment (−283 to −307 bp) which contains a potential AP-2-binding site was end-labeled and incubated in binding buffer in the absence of protein (lane 1), with an *E. coli* extract containing human AP-2 (lane 2), with an *E. coli* extract containing AP-2 in the presence of a 100-fold molar excess of an oligonucleotide which binds human AP-2 (lane 3), and with an *E. coli* extract containing human AP-2 in the presence of a 100-fold molar excess of a noncompeting oligonucleotide which binds TFIID (lane 4).

can be detected in the bloodstream! This indicates that many, if not all, of the transcription factors required for 63-kDa keratin gene expression are present in premetamorphic larvae. In fact, most of the direct response and delayed response genes identified in cDNA subtraction libraries (see Shi, 1994, and Chapter 14, this volume) and transcription factors such as TR (see Tata, Chapter 13, this volume; Shi, Chapter 14, this volume), AP-2 (French *et al.,* 1994), and C/EBP (Xu and Tata, 1992; Chen *et al.,* 1994) are also expressed at low levels in premetamorphic larval tissues. Therefore, many of the transcription factors that control the adult program of differentiation are already expressed in larval cells. This immature adult regulatory program is sufficient to activate and maintain the expression of some tissue-specific structural genes, such as the 63-kDa keratin gene, at low levels. The relative completeness of the regulatory program may also explain why it is possible to activate some specialized structural genes such as stromelysin-3 (Wang and Brown, 1993) within 8 hr of T_3 treatment.

It is possible that the extensive preprogramming which occurs independently of T_3 provides a clue for the mechanism of T_3 action during metamorphosis. One possibility is that low level expression of adult-specific genes in cells expressing the larval program of differentiation may be a necessary step for T_3-induced reprogramming. A change from one relatively stable program of differentiation to another, which in this case involves the inactivation of larval-specific genes and the activation of adult-specific genes, does not normally occur in the same cell generation (for a discussion of the stability of the differentiated state, see Gurdon, 1992). The activation of regulatory genes in cells performing differentiated functions of larvae may only be possible to achieve if they were already expressed at low levels. In this view, TH may be required to change the concentration of regulatory factors, the balance of which controls whether larval-specific or adult-specific genes will be expressed. Such changes in the expression pattern of regulatory genes and tissue-specific genes, such as the 63-kDa keratin gene, can occur at multiple levels, including the activation of enhancer elements or at the level of coupling between enhancers and promoters. The early genes directly upregulated by T_3 include both transcription factors such as TRβ and structural genes (Shi, Chapter 14, this volume). A major emphasis for future work, which will require the development of a functional assay, will be to determine the role of thyroid hormone receptors and other direct response genes in the activation of secondary response and late response genes.

XII. SUMMARY

Amphibian metamorphosis is a unique phenomenon because profound changes in body form and function occur abruptly. Since late events in the genesis of all organs are concentrated into a short time period, under the control of TH, metamorphosis is an excellent paradigm for studying the biochemistry and molecular biology of the late stages of vertebrate organogenesis. Furthermore, it is likely that these late developmental events, which have received very little attention compared to early developmental processes, are maintained throughout vertebrate evolution. For example, the processes of development and terminal differentiation of the epidermis are highly conserved from amphibians to mammals. The changes occurring in the epidermis during metamorphosis, from a simple and relatively undifferentiated bilayered epithelium to a complex stratified epithelium, with terminally differentiated keratinocytes and glands, are also observed in human development at the embryo–fetal transition. It is not known whether T_3 has any role in the developmental changes occurring in mammalian skin, but the similarities in development and keratin gene expression during terminal differentiation are striking. Based on this conservation it is reasonable to expect that the isolation of T_3-induced regulatory genes, which control the development of the amphibian epidermis, will allow the identification of the homologous genes of mammals.

A number of lessons have been learned from the analysis of keratin gene expression during metamorphosis. Most important is the discovery that two independent regulatory events are required for full activation of the keratin gene and that only the second step is dependent on TH. The first step, low level expression of the keratin gene, begins during the preprogramming events prior to metamorphosis and is independent of T_3. Subsequent work has shown that other genes which are upregulated during metamorphosis are also expressed at low levels prior to the detection of T_3 in the bloodstream. This includes transcription factors such as AP-2 and C/EBP and putative regulatory factors identified by subtraction hybridization screens. As yet, it has not been ruled out that the basal expression of these transcription factors is due to an extremely low level of TH in premetamorphic larvae. However, the *in vitro* analysis of the keratin gene and the normal development of thyroidectomized larvae, up to prometamorphosis, strongly suggest that the basal expression of many adult genes in premetamorphic larvae is independent of T_3 and not required for larval functions. These results focus the attention for the molecular mechanism of T_3 action during metamorphosis on the processes required for the up-regulation of basal level gene expression.

The second step required for full activation of the keratin gene is its up-regulation from low to high levels. Although this step is strictly dependent on T_3, it is very unlikely that the T_3–TR complex directly activates this or other late genes. The long latent period between T_3 addition and keratin gene up-regulation, the positive correlation of up-regulation and terminal differentiation, and the subsequent stability and

T_3 independence of the adult program of differentiation all indicate that T_3 is essential to acquire the adult program, but is not necessary for readout of that program throughout adult life. Thus T_3 action during metamorphosis is similar to the developmental role of embryonic inducers—it induces a change in commitment from the larval program to the adult program. Like embryonic induction, a transient exposure to T_3 induces stable changes in the program of differentiation which are heavily dependent on the preprogram of the responsive cells. Therefore, the developmental function of T_3 is to induce the final step in the series of determinative events which lead to the establishment of the adult program. In contrast to this developmental function, T_3 modulates the properties of fully differentiated cells in adult mammals. It is interesting to note that embryonic inducers belonging to the transforming growth factor β and fibroblast growth factor families also function as modulators of gene expression in adult tissues. This suggests a conservation of developmental signals, such as hormones and growth factors, which act as embryonic inducers early in the life cycle and modulators of gene expression in the adult. Since T_3 has a critical developmental role in mammals, the analysis of T_3 action during metamorphosis should provide valuable information for understanding the developmental role of T_3 in humans and other mammals.

The remarkable differences in the onset of keratin gene up-regulation and terminal differentiation in different regions of the skin demonstrate the well-known tissue, temporal, and spatial specificity of T_3 action. This differential response to T_3 can be controlled at various levels between the reception of the T_3 signal and readout of the adult program. While T_3 and its receptors are clearly at the heart of the conversion to the adult program of epidermal differentiation, other hormones (corticosteroids, prolactin, and retinoic acid) and their receptors modulate the T_3 response. Work on keratin gene expression has shown that these hormones do not function alone, their action is dependent on the initial cellular response to T_3. Together with T_3, these hormones are involved in coordinating the global changes in gene expression, morphogenesis, and differentiation characteristic of amphibian metamorphosis.

ACKNOWLEDGMENT

The research described in this chapter was supported by a grant from the National Institutes of Health (HD-24438).

REFERENCES

Atkinson, B. G. (1994). Metamorphosis: Model systems for studying gene expression in postembryonic development. *Dev. Genet.* **15,** 313–319.

Baker, B. S., and Tata, J. R. (1992). Prolactin prevents the autoinduction of thyroid hormone receptor mRNAs during amphibian metamorphosis. *Dev. Biol.* **149,** 463–467.

Blumenberg, M. (1993). Molecular biology of human keratin genes. *In* "Molecular Biology of Skin" (M. Darmon and M. Blumenberg, eds.), pp. 1–32. Academic Press, San Diego.

Broyles, R. H., Ramseyer, L. T. H., Do, T. H., McBride, K. A., and Barker, J. C. (1994). Hemoglobin switching in *Rana/Xenopus* erythroid heterokaryons: Factors mediating the metamorphic hemoglobin switch are conserved. *Dev. Genet.* **15,** 347–355.

Buckbinder, L., and Brown, D. (1992). Thyroid hormone-induced gene expression changes in the developing frog limb. *J. Biol. Chem.* **267,** 25786–25791.

Buckbinder, L., and Brown, D. D. (1993). Expression of *Xenopus laevis* prolactin and thyrotropin genes during metamorphosis. *Proc. Natl. Acad. Sci. U.S.A.* **90,** 3820–3824.

Buscaglia, J., Leloup, J., and deLuze, A. (1985). The role and regulation of monodeiodination of thyroxine to 3,5,3′-triiodothyronine during amphibian metamorphosis. *In* "Metamorphosis" (M. Balls and M. Bownes, eds.), pp. 273–293. Oxford University Press, New York.

Chen, Y., Hu, H., and Atkinson, B. B. (1994). Characterization and expression of C/EBP-like genes in the liver of *Rana catesbeiana* tadpoles during spontaneous and thyroid hormone-induced metamorphosis. *Dev. Genet.* **15,** 366–377.

Dale, B. A., Holbrook, K. A., Kimball, J. R., Hoff, M., and Sun, T.-T. (1985). Expression of epidermal keratins and filaggrin during human fetal skin development. *J. Cell Biol.* **101,** 1257–1269.

Ellison, T. R., Mathisen, P. M., and Miller, L. (1985). Developmental changes in keratin patterns during epidermal maturation. *Dev. Biol.* **112,** 329–337.

Etkin, W. (1964). Metamorphosis. *In* "Physiology of the Amphibia" (J. A. Moore, ed.), pp. 427–468. Academic Press, New York.

Fox, H. (1985). Changes in amphibian skin during larval development and metamorphosis. *In* "Metamorphosis" (M. Balls and M. Bownes, eds.), pp. 59–87. Oxford University Press, New York.

French, R. P., Warshawsky, D., Tybor, L., Mylniczenko, N., and Miller, L. (1994). Upregulation of AP-2 in the skin of *Xenopus laevis* during thyroid hormone induced metamorphosis. *Dev. Genet.* **15,** 356–365.

Frieden, E., and Nail, B. (1955). Biochemistry of amphibian metamorphosis. I. Enhancement of induced metamorphosis by glucocorticoids. *Science* **121,** 37–39.

Galton, V. A. (1983). Thyroid hormone action in amphibian metamorphosis. *In* "Molecular Basis of Thyroid Hormone Action" (J. H. Oppenheimer and H. H. Samuels, eds.), pp. 445–483. Academic Press, New York.

Gilbert, L. I., and Frieden, E., eds. (1981). "Metamorphosis: A Problem in Developmental Biology," 2nd ed. Plenum, New York.

Gittes, G. K., and Rutter, W. J. (1992). Onset of cell-specific gene expression in the developing mouse pancreas. *Proc. Natl. Acad. Sci. U.S.A.* **89,** 1128–1132.

Gurdon, J. B. (1987). Embryonic induction: Molecular prospects. *Development* **99,** 285–306.

Gurdon, J. B. (1992). The generation of diversity and pattern in animal development. *Cell (Cambridge, Mass.)* **68,** 185–199.

Hoffmann, W., Franz, J. K., and Franke, W. (1985). Amino acid sequence microheterogeneities of basic (type II) cytokeratins of *Xenopus laevis* epidermis and evolutionary conservativity of helical and non-helical domains. *J. Mol. Biol.* **184,** 713–724.

Hopwood, N. D., and Gurdon, J. B. (1990). Activation of muscle genes without myogenesis by ectopic expression of Myo D in frog embryo cells. *Nature* (*London*) **347,** 197–200.

Jayatilaka, A. D. P. (1978). An ultrastructural study of the thyroid gland in premetamorphic *Xenopus laevis* (Daudin) tadpoles. *J. Anat.* **125,** 579–591.

Jones, P. H., and Watt, F. M. (1993). Separation of human epidermal stem cells from transit amplifying cells on the basis of differences in integrin function and expression. *Cell* (*Cambridge, Mass.*) **73,** 713–724.

Kanamori, A., and Brown, D. (1993). Cultured cells as a model for amphibian metamorphosis. *Proc. Natl. Acad. Sci. U.S.A.* **90,** 6013–6017.

Kartasova, T., Roop, D. R., Holbrook, K. A., and Yuspa, S. H. (1993). Mouse differentiation-specific keratin-1 and keratin-10 require a preexisting keratin scaffold to form a filament network. *J. Cell Biol.* **120,** 1251–1261.

Kawahara, A., Kohara, S., and Amano, M. (1989). Thyroid hormone directly induces hepatocyte competence for estrogen-dependent vitellogenin synthesis during the metamorphosis of *Xenopus laevis. Dev. Biol.* **132,** 73–80.

Kemp, N. E. (1963). Metamorphic changes of dermis in skin of frog larvae exposed to thyroxine. *Dev. Biol.* **7,** 244–254.

Kinoshita, T., Sasaki, F., and Watanabe, K. (1986). Regional specificity of anuran larval skin during metamorphosis: Transdifferentiation of tadpole tail-epidermis. *J. Exp. Zool.* **238,** 201–210.

Kollros, J. J. (1961). Mechanisms of amphibian metamorphosis: Hormones. *Am. Zool.* **1,** 107–114.

Kollros, J. J., and Kaltenbach, J. C. (1952). Local metamorphosis of larval skin in *Rana pipiens. Physiol. Zool.* **25,** 163–170.

Lavin, T. N., Baxter, J. O., and Horita, S. (1988). The thyroid hormone receptor binds to multiple domains of the rat growth hormone 5′-flanking sequence. *J. Biol. Chem.* **263,** 9418–9426.

Leask, A., Byrne, C., and Fuchs, E. (1991). Transcription factor AP-2 and its role in epidermal-specific gene expression. *Proc. Natl. Acad. Sci. U.S.A.* **88,** 7948–7952.

Mathisen, P. M. (1987). "The Induction of Adult Keratin Gene Expression during Xenopus Development," Ph.D. thesis. University of Illinois at Chicago.

Mathisen, P. M., and Miller, L. (1987). Thyroid hormone induction of keratin genes: A two-step activation of gene expression during development. *Genes Dev.* **1,** 1107–1117.

Mathisen, P. M., and Miller, L. (1989). Thyroid hormone induces constitutive keratin gene expression during *Xenopus* development. *Mol. Cell. Biol.* **9,** 1823–1831.

Nieuwkoop, P. D., and Faber, J. (1967). "Normal Table of *Xenopus laevis* (Daudin)." Elsevier/North-Holland, Amsterdam.

Nishikawa, A., Kaiho, M., and Yoshizato, K. (1989). Cell death in the anuran tadpole tail: Thyroid hormone induces keratinization and tail-specific growth inhibition of epidermal cells. *Dev. Biol.* **131,** 337–344.

Nishikawa, A., Shimizu-Nishikawa, K., and Miller, L. (1990). Isolation, characterization and *in vitro* culture of larval and adult epidermal cells of the frog *Xenopus laevis. In Vitro Cell. Dev. Biol.* **26,** 1128–1134.

Nishikawa, A., Shimizu-Nishikawa, K., and Miller, L. (1992). Spatial, temporal, and hormonal regulation of epidermal keratin expression during development of the frog *Xenopus laevis. Dev. Biol.* **151,** 145–153.

Oppenheimer, J. H., and Samuels, H. H., eds. (1983). "Molecular Basis of Thyroid Hormone Action." Academic Press, New York.

Romney, J. S., Chan, J., Carr, F. E., Mooradian, A. D., and Wong, N. C. W. (1992). Identification of the thyroid hormone-responsive messenger RNA spot 11 as apolipoprotein-A1 messenger RNA and effects of the hormone on the promoter. *Mol. Endocrinol.* **6,** 943–950.

Roop, D. R., Huitfeld, H., Kilkenny, A., and Yupsa, S. H. (1987). Regulated expression of differentiation-associated keratins in cultured epidermal cells detected by monospecific antibodies to unique peptides of mouse epidermal keratins. *Differentiation (Berlin)* **35,** 143–150.

Sell, S. (1994). Liver stem cells. *Mod. Pathol.* **7,** 105–112.

Shi, Y.-B. (1994). Molecular biology of amphibian metamorphosis. A new approach to an old problem. *Trends Endocrinol. Metab.* **5,** 14–20.

Shimizu-Nishikawa, K., and Miller, L. (1991). Calcium regulation of epidermal cell differentiation in the frog, *Xenopus laevis. J. Exp. Zool.* **260,** 165–169.

Shimizu-Nishikawa, K., and Miller, L. (1992). Hormonal regulation of adult type keratin gene expression in larval epidermal cells of the frog, *Xenopus laevis. Differentiation (Berlin)* **49,** 77–83.

Snape, A., Jonas, E., and Sargent, T. (1990). KTF-1, a transcription activator of *Xenopus* embryonic keratin expression. *Development* **109,** 157–165.

Snape, A. M., Winning, R. S., and Sargent, T. D. (1991). Transcription factor AP-2 is tissue-specific in *Xenopus* and is closely related or identical to keratin transcription factor 1 (KTF-1). *Development* **113,** 283–293.

Sun, T.-T., Eichner, R., Schermer, A., Cooper, D., Nelson, W. G., and Weiss, R. A. (1984). Classification, expression, and possible mechanisms of evolution of mammalian epithelial keratins: A unifying model. *In* "Cancer Cells: The Transformed Phenotype" (A. Levine, W. Topp, G. Vande Woude, and J. D. Watson, eds.), Vol. 1, pp. 169–176. Cold Spring Harbor Laboratory, Cold Spring Harbor, New York.

Tata, J. R. (1993). Gene expression during metamorphosis: An ideal model for post-embryonic development. *BioEssays* **15,** 239–248.

Tata, J. R., Baker, B. S., Machuca, I., Rabelo, E. M. L., and Yamauchi, K. (1993). Autoinduction of nuclear receptor genes and its significance. *J. Steroid Biochem. Mol. Biol.* **46,** 105–119.

Vanable, J. W. (1964). Granular gland development during *Xenopus laevis* metamorphosis. *Dev. Biol.* **10,** 331–357.

Wang, Z., and Brown, D. D. (1991). A gene expression screen. *Proc. Natl. Acad. Sci. U.S.A.* **88,** 11505–11509.

Wang, Z., and Brown, D. (1993). Thyroid hormone-induced gene expression program in amphibian tail resorption. *Mol. Cell. Biol.* **9,** 1044–1051.

Warshawsky, D., and Miller, L. (1995). Tissue-specific *in vivo* protein-DNA interactions at the promoter region of the 63-kDa keratin gene during metamorphosis. Nucleic Acids Research Vol. 23, No. 21, (in press).

Weintraub, H. (1993). The Myo D family and myogenesis: Redundancy, networks, and thresholds. *Cell (Cambridge, Mass.)* **75,** 1241–1244.

Wessells, N. E. (1961). Thyroxine initiation of epidermal differentiation as studied *in vitro* in chemically defined medium. *Exp. Cell Res.* **24,** 131–142.

Wright, M. L. (1973). DNA synthesis during differentiation of tadpole shank epidermis. *J. Exp. Zool.* **186,** 237–256.

Wright, M. L. (1977). Regulation of cell proliferation in tadpole limb epidermis by thyroxine. *J. Exp. Zool.* **202,** 223–234.

Xu, Q., and Tata, J. R. (1992). Characterization and developmental expression of *Xenopus* C/EBP gene. *Mech. Dev.* **38,** 69–81.

Yaoita, Y., and Brown, D. D. (1990). A correlation of thyroid hormone receptor gene expression with amphibian metamorphosis. *Genes Dev.* **4,** 1107–1113.

Yoshizato, K. (1989). Biochemistry and cell biology of amphibian metamorphosis with a special emphasis on the mechanism of removal of larval organs. *Int. Rev. Cytol.* **119,** 97–149.

18

Metamorphosis: An Immunologically Unique Period in the Life of the Frog

LOUISE A. ROLLINS-SMITH* AND NICHOLAS COHEN†

* Departments of Microbiology and Immunology and of Pediatrics
Vanderbilt University School of Medicine
Nashville, Tennessee

† Department of Microbiology and Immunology
University of Rochester Medical Center
Rochester, New York

I. INTRODUCTION

To the extent that the South African clawed frog, *Xenopus laevis,* can be considered representative of all Anura, the immune system of an adult frog is remarkably similar to that of mammals (reviewed in Du Pasquier *et al.,* 1989). For example, *Xenopus* exhibits T- and B-lymphocyte heterogeneity (Bleicher and Cohen, 1981); class I and class II major histocompatibility complex (MHC)-encoded molecules (Flajnik and Du Pasquier, 1990) that function as restricting elements in T- and B-cell cooperation (Blomberg *et al.,* 1980) and in antigen-specific T-cell proliferation (Harding *et al.,* 1993); immunoglobulin isotype heterogeneity; and leukocyte-derived cytokines.* Although *Xenopus* lacks the mammalian equivalent of lymph nodes and a lymphopoietic bone marrow, it does have a thymus and spleen, important central and peripheral lymphoid organs, respectively.

Free-living larval amphibians, like their postmetamorphic counterparts, have an immune system capable of self-/nonself-discrimination (reviewed by Du Pasquier *et al.,* 1989). Although such defenses are no doubt needed to cope with potential pathogens in their aquatic environment, the immunocompetence of the tadpoles has the potential to be a double-edged sword for the following reason. Metamorphosis is characterized by the development of molecules and structures unique to the adult stage of life, including adult hemoglobin (Just *et al.,* 1977, 1980; Widmer *et al.,* 1981; see Weber, Chapter 16, this volume), adult-type keratin (Ellison *et al.,* 1985; Mathisen and Miller, 1987), the urea cycle enzyme L-arginase (Xu *et al.,* 1993; see Atkinson *et al.,* Chapter 15, this volume), and vitellogenin (Wahli *et al.,* 1981; Knowland, 1985). Since immune responses are designed to recognize and eliminate such nonself molecular entities, one might expect, at least theoretically, that amphibians should "self-destruct" during metamorphosis when their larval immune system encounters these adult-specific novel antigens. Since amphibian metamorphosis is not characterized by autoimmune tissue destruction, larvae must have acquired a strategy that "allows" their developing immune system to "accommodate" new adult-specific molecules. Given that metamorphosis is driven by complex neuroendocrine

* Three immunoglobulin isotypes (homologous with mammalian IgM, IgG, and IgA) and at least three leukocyte-produced cytokines, interleukin (IL)-1-like (Watkins *et al.,* 1987), IL-2-like (Watkins and Cohen, 1987; Haynes and Cohen, 1993a), and transforming growth factor-β (TGF-β) (Haynes and Cohen, 1993b), have been identified thus far.

changes, we and others have sought to discover possible neuroendocrine–immune system interactions occurring during metamorphosis that could explain these accommodating changes in the developing immune system. These studies are the focus of this chapter.†

II. DIFFERENCES IN IMMUNE RESPONSE CAPABILITY OF LARVAL AND ADULT ANURANS

Research from a number of laboratories in the United States, Europe, and Japan has provided a fairly clear picture of the immune response capabilities of larval anurans as compared with adults of the same species. Most of these immunological studies have relied on *X. laevis* because of the availability of inbred strains and isogeneic clones (Kobel and Du Pasquier, 1977; Katagiri, 1978) and because of the relative ease with which this frog can be successfully reared through larval and adult life. Therefore, unless another species is specifically identified, the following studies have been conducted using *Xenopus*.

A. Allograft Recognition

Tadpoles of all anuran species studied thus far can reject skin allografts differing by known or presumed‡ MHC-encoded antigens (Hildemann and Haas, 1959; Bovbjerg, 1966; Horton, 1969; Chardonnens and Du Pasquier, 1973; DiMarzo and Cohen, 1982a,b; Obara *et al.,* 1983). However, allografts from adult donors that differ from MHC-compatible larval recipients by minor histocompatibility (H) locus antigens enjoy either prolonged or indefinite survival relative to the survival of such minor H-locus disparate grafts on adult hosts (DiMarzo and Cohen, 1982a,b; Obara *et al.,* 1983). This prolonged or indefinite survival reflects an antigenically specific tolerance (Cohen *et al.,* 1985; Tochinai, 1993).

† For an earlier review that also addresses changes in the immune system of *Xenopus* during metamorphosis, see Flajnik *et al.* (1987).

‡ The MHC of *Xenopus* has been defined in terms of its genetic organization, biochemistry, nucleotide and amino acid sequence, cell surface expression, and tissue distribution (Flajnik and Du Pasquier, 1990). To our knowledge, no one has attempted to identify the MHC of other anurans in a similar fashion.

When late stage larvae were thymectomized, the frequency of perimetamorphic tolerance was decreased (Barlow and Cohen, 1983), suggesting that larval allotolerance is thymus dependent. That T cells are involved in tolerance (e.g., active suppression as opposed to clonal deletion or anergy) was also suggested by the adoptive transfer of larval thymocytes to syngeneic adults (Du Pasquier and Bernard, 1980). These experiments notwithstanding, little is known about the mechanism(s) responsible for the acquisition and maintenance of allotolerance or whether this mechanism(s) is also responsible for tolerance to self-antigens that must occur during metamorphosis. Available data, however, suggest that the immune system, in general, might be downregulated during metamorphosis. For example, as discussed in Section II,D, a well-documented loss of splenic and thymic lymphocytes occurs during the later stages of metamorphosis (Du Pasquier and Weiss, 1973; Rollins-Smith *et al.,* 1984). Moreover, an apparent decrease in the ability of T cells from metamorphosing larvae to respond *in vitro* to alloantigens (Du Pasquier and Chardonnens, 1975; Du Pasquier *et al.,* 1979) and to the T-cell mitogens, phytohemagglutinin (PHA) and concanavalin A (Con A), has also been reported (Rollins-Smith *et al.,* 1984).

The cytokine TGF-β is involved in a variety of mammalian physiologic processes, including the suppression of immune responses (Roberts and Sporn, 1990). Haynes and Cohen (1993a) reported that mitogen-stimulated lymphocytes from adult *Xenopus* produce TGF-β5 and that adult *Xenopus* splenic lymphoblasts cultured with TGF-β5 exhibit diminished IL-2-stimulated growth. To test the hypothesis that overproduction of TGF-β5 by minor H antigen-stimulated larval lymphocytes plays a role in the establishment and maintenance of perimetamorphic allotolerance, we (N. Cohen and M. Crosby, unpublished observations) attempted to abrogate the development of skin graft tolerance by injecting LG-15 stage 65 tadpole recipients of minor H locus-incompatible LG-17 skin with antibodies against TGF-β5.* Conversely, we attempted to promote tolerance in a donor–host MHC-incompatible combination, where rejection was the normative larval alloimmune response, by treating larval recipients with recombinant TGF-β5. Finally, the effect of recombinant TGF-β5 on the survival of minor H locus-incompatible allografts in postmetamorphic frogs was examined. Although these studies are clearly

* Perimetamorphic tadpoles were injected intraperitoneally on the day of surgery and every other day afterward (total of 12 injections) with 1 or 5 μg goat anti-TGF-β5 antibody; controls were injected with normal goat IgG.

preliminary in that the amount, frequency, and route of cytokine and antibody injected, and the immunogenetic disparities between donors and hosts were not varied parametrically, neither the anticytokine antibody nor the recombinant cytokine had any discernible effect on either graft tolerance or rejection. Indeed, tolerance was firmly established as judged not only by the survival of the initial grafts, but by the survival of second-set grafts transplanted weeks or months after the recipients had metamorphosed.

B. Antibody Responses

Antibody responses have been studied in larval and adult *X. laevis* and *Rana catesbeiana.* Larval antibodies to the dinitrophenyl hapten are less diverse and generally of a lower affinity than those of adults (Du Pasquier and Haimovich, 1976; Du Pasquier *et al.,* 1979; Hsu and Du Pasquier, 1984b, 1992). Unlike adults, young tadpoles do not produce high titers of low molecular weight (IgY) antibodies to thymus-dependent antigens (Du Pasquier *et al.,* 1979; Hsu and Du Pasquier, 1984a). It was shown in *Xenopus* that terminal deoxynucleotidyl transferase (TdT), an enzyme responsible, in mammals, for the addition of template-independent nucleotides (called N regions) to antibody and T-cell receptor DNA sequences, emerges only at metamorphic climax (Lee and Hsu, 1994). Such N region additions are thought to contribute to the greater diversity of the adult antibody and T-cell recognition repertoires (Lee *et al.,* 1993; Desravines and Hsu, 1994).

C. Expression of Major Histocompatibility Complex Antigens

In murine systems, the MHC is intimately involved in the negative and positive selection of developing T lymphocytes for self and nonself antigen recognition. Studies involving fusion of chromosomally tagged anterior portions of tail bud stage amphibian embryos with posterior portions of other embryos have revealed that all T-cell precursors are derived from the posterior portion of the embryo (Deparis and Jaylet, 1976; Tochinai, 1978; Turpen *et al.,* 1979, 1982; Volpe *et al.,* 1979; Flajnik *et al.,* 1984, 1985). In *Xenopus,* these precursors have been shown to become "educated" to recognize self- and foreign antigens in the context of the MHC antigens expressed by cells of the thymus through which they have trafficked (Flajnik *et al.,* 1985). Because of this essential role of the MHC in the establishment of immunological tolerance, it is important to understand when in ontogeny MHC antigens are expressed and what regulates their expression. The MHC of *Xenopus* has been well characterized (Flajnik and Du Pasquier, 1990). Class I and class II molecules homologous to those described in mammalian species have been identified, characterized, and cloned. Monoclonal and polyclonal antibodies that bind specifically to each class of MHC antigens have been used to examine the ontogeny of expression of MHC antigens; several distinct differences between the expression pattern for larval and adult frogs have been described.

1. MHC Class I Ontogeny

Cell surface expression of MHC class I antigens, as assessed by immunoprecipitation, fluorescence microscopy, and microfluorimetry, is a relatively late event in the ontogeny of *Xenopus* (Flajnik *et al.,* 1986; Flajnik and Du Pasquier, 1988; Rollins-Smith *et al.,* 1994). Class I antigens are first detected on erythrocytes and on splenic leukocyte populations at prometamorphic and climax stages of metamorphosis (Flajnik and Du Pasquier, 1988; Rollins-Smith *et al.,* 1994). Microfluorimetric analysis of class I antigen expression suggests that although tadpole cells can express class I, the relative intensity of expression is low in comparison with adult cells (Flajnik and Du Pasquier, 1988; L. A. Rollins-Smith, unpublished observations). Although class I expression in the periphery begins when the tadpole reaches the prometamorphic stages, onset of class I expression is, in fact, independent of those changes in thyroid hormone that drive metamorphosis (Rollins-Smith *et al.,* 1994).

Since MHC antigens displayed on epithelial and stromal cells of the thymus play an essential role in the positive and negative selection of T lymphocytes, it has been of particular importance to define the ontogenetic pattern of MHC class I expression in this central lymphoid organ. Using fluorescence microscopy, Du Pasquier and colleagues could not detect class I expression in the thymus until after metamorphosis (Flajnik *et al.,* 1986). Flow cytometric studies of class I expression on thymocytes in *Xenopus* (predominantly developing T cells rather than epithelial/stromal elements) have confirmed that expression of MHC class I antigens is not detectable on thymocytes until metamorphic climax (Du Pasquier and Flajnik, 1990; Rollins-Smith *et al.,* 1994).

2. MHC Class II Ontogeny

During larval life, MHC class II antigen expression in the periphery is restricted to B lymphocytes and accessory cells (Du Pasquier and Flajnik, 1990; Rollins-Smith and Blair, 1990a). This is the cell type distribution seen in adult mice and rats. However, after metamorphosis, class II antigens are expressed constitutively on virtually all mature peripheral T cells as well as B cells (Du Pasquier and Flajnik, 1990; Rollins-Smith and Blair, 1990a). When metamorphosis is inhibited by treatment of tadpoles with 0.1% sodium perchlorate (see Section IV,A), the adult-type MHC class II-positive peripheral T-cell population does not develop. Thus, maturation of this population appears to depend on normal metamorphosis (Rollins-Smith and Blair, 1990a).

D. Evidence for Distinct Larval and Adult Lymphocyte Populations

The differences between the immune response capabilities of tadpoles and adults suggest that developing tadpoles might "accommodate" new adult-specific antigens by largely discarding their larval immune system and developing a new one during and after metamorphosis. This would permit another opportunity for thymic positive and negative selection in the face of newly emerging adult-specific antigens, including MHC class I determinants. If this idea is valid, then there should be a major loss of larval-type lymphocytes during metamorphosis that is followed by lymphopoiesis and an expansion of the adult-type populations. As highlighted below, several lines of evidence reveal that there is indeed a major turnover of lymphocytes at metamorphosis. (1) Lymphocyte numbers in the spleen, thymus, and liver of normally developing *Xenopus* decline by at least 40% during metamorphosis (Du Pasquier and Weiss, 1973; Rollins-Smith *et al.,* 1984; Cohen *et al.,* 1985). (2) A Triploid (3N) thymus implanted into a diploid (2N) host before metamorphosis is colonized by host-derived precursor cells that completely replace the larval cells around the time of metamorphosis (Turpen and Smith, 1989; Rollins-Smith *et al.,* 1992). (3) Pre-B cells in the liver and spleen of developing *Xenopus* increase and decrease in two waves. The second increase is associated with metamorphosis (Hadji-Azimi *et al.,* 1982, 1990). (4) The spectrotype of anti-DNP antibodies detected in the LG lines of isogeneic *X. laevis* × *gilli* is unique for the larval period. A different spectrotype emerges in the adult members of the same clone (Du Pasquier *et al.,* 1979; Hsu and Du Pasquier, 1984b), suggesting that the adult and larval B-cell repertoires differ. (5) Larval and adult splenic T cells have different phenotypes by virtue of their expression of MHC class II antigens only after metamorphosis (Du Pasquier and Flajnik, 1990; Rollins-Smith and Blair, 1990a).

It should be noted that *Xenopus* larvae primed to an antigen during larval life exhibit a classic memory response to that antigen when they are challenged as adults (Du Pasquier and Haimovich, 1976; DiMarzo and Cohen, 1982b; Barlow and Cohen, 1983; Cohen *et al.,* 1985; Manning and Al Johari, 1985). Such studies demonstrating that immunological memory persists through metamorphosis make it clear that not all components of the larval immune system are discarded during metamorphosis.

III. NEUROENDOCRINE CHANGES ASSOCIATED WITH METAMORPHOSIS

Neuroendocrine changes associated with metamorphosis have been studied far more intensively than the metamorphic changes associated with the immune system and are detailed elsewhere in this volume (see Kaltenbach, chapter 11, this volume; Denver, chapter 12, this volume). This chapter describes briefly what is known about catecholamines and three "sets" of hormones insofar as they affect the immune system during metamorphosis.

A. Thyroid Hormones

The seminal studies of Gudernatsch (1912) clearly established that the thyroid gland plays a critical role in metamorphosis. The thyroid hormones, thyroxine (T_4) and triiodothyronine (T_3), have been studied during the ontogeny of a number of amphibian species (reviewed by White and Nicoll, 1981; Kikuyama *et al.,* 1993). Circulating levels of hormones are low at premetamorphic stages, increase during prometamorphosis, peak during climax, and decline at the end of metamorphosis. [For studies with *X. laevis,* see Leloup and Buscaglia (1977).]

B. Corticosteroid Hormones

Circulating levels of the corticosteroid hormones, corticosterone and aldosterone, are significantly increased during climax of metamorphosis in bullfrogs (Jaffe, 1981; Krug *et al.,* 1983; Kikuyama, 1986) and *Xenopus* (Jaudet and Hatey, 1984). In *Xenopus,* a high affinity serum corticosterone-binding factor is more abundant in adults than in tadpoles. Thus, elevated corticosterone, in excess of serum-binding factors, results in high concentrations of freely available corticosterone only during metamorphosis (Jolivet-Jaudet and Leloup-Hatey, 1986).

C. Prolactin and Growth Hormone

The involvement of growth hormone (GH) and prolactin (PRL) in development and metamorphosis in amphibian species is an active area of study. Both hormones have growth-promoting activity (reviewed by White and Nicoll, 1981; Kikuyama *et al.,* 1993; see Denver, Chapter 12, this volume) and both are elevated at metamorphosis in bullfrogs (Clemons, 1976; Clemons and Nicoll, 1977). In *Xenopus,* it is not clear which is the predominant growth-promoting hormone during larval development. An analysis of the expression of GH and PRL mRNA during ontogeny suggests that GH is present at premetamorphic stages, decreases at Nieuwkoop and Faber (1967) stage 61, increases again at the end of climax, and remains high in young postmetamorphic frogs. In contrast, PRL expression is very low at premetamorphic and prometamorphic stages, and increases dramatically as climax proceeds (Buckbinder and Brown, 1993).

Early studies of immunodeficient pituitary dwarf mice established a connection between the immune system and the pituitary (Baroni, 1967; Baroni *et al.,* 1967). Immunodeficiencies in these mice could be reversed by the administration of PRL or GH and thyroid hormones (Baroni *et al.,* 1969, 1971; Fabris *et al.,* 1971a,b). Cell surface PRL receptors have been demonstrated for virtually all lymphoid cell populations in mice and humans (Russell *et al.,* 1985; O'Neal *et al.,* 1991; Dardenne *et al.,* 1991; Gagnerault *et al.,* 1993) as well as in amphibians. Anti-PRL antibodies can inhibit the proliferation of mitogen- or alloantigen-stimulated T cells in mammalian (Hartman *et al.,* 1989; Clevenger *et al.,* 1990) as well as in amphibian species (L. A. Rollins-Smith, unpublished observations; N. Cohen, unpublished observations).

D. Catecholamines

In mammals, primary and secondary lymphoid organs are innervated with noradrenergic postganglionic sympathetic nerve fibers and peptidergic nerve fibers that form close neuroeffector junctions with lymphocytes and macrophages (Madden and Felten, 1995). Neurotransmitters released from these nerves diffuse to act at distant sites, further extending the potential for neural–immune interactions. Lymphocytes, monocytes/macrophages, and granulocytes express receptors for these neurotransmitters (e.g., norepinephrine, substance P, neuropeptide Y). Ablation of sympathetic nerve fibers *in vivo* results in significant immunomodulatory effects (reviewed by Madden and Felten, 1995).

The spleen of the adult *Xenopus,* like that of mammals, is separated into clearly defined compartments of red pulp and white pulp (Manning, 1991). Also, as in mammals, innervation of the *Xenopus* spleen is noradrenergic and peptidergic, and confined to the white pulp (Clothier *et al.,* 1991; Kinney *et al.,* 1994).

A developmental analysis has revealed that the *Xenopus* spleen becomes innervated during the metamorphic period. Indeed, the pattern and intensity of staining for splenic catecholamines of stage 62 animals are comparable to those seen in postmetamorphic animals; staining at stages 58 and 54, however, is barely detectable (K. S. Kinney and N. Cohen, unpublished observations).

Early larval thymectomy, which eliminates peripheral T lymphocytes (Horton and Manning, 1972), does not affect innervation of the adult spleen, and chemical ablation of the developing splenic noradrenergic innervation with the selective neurotoxin 6-hydroxydopamine (6-OHDA) does not affect the development of splenic compartmentation (Kinney *et al.,* 1993).

IV. NEUROENDOCRINE INFLUENCES ON THE IMMUNE SYSTEM AT METAMORPHOSIS

A. Thyroid Hormones

Because metamorphosis is strictly regulated by the availability of TH, manipulation of TH levels makes it possible to accelerate or inhibit metamorphosis and thereby determine the extent to which development of the adult pattern of immunity in frogs requires a normal metamorphic transition. By raising *Xenopus* larvae from early larval stages (e.g., stages 50–53) in a 0.1% solution of sodium perchlorate [CLO_4-inhibits iodine uptake by thyroid follicle cells (Capen, 1994)] to arrest phenotypic development, we and others have shown that the development of some structural and functional components of the adult *Xenopus* immune system is dependent on TH and a normal metamorphic transition, whereas others are metamorphosis independent. Studies with perchlorate-blocked larvae have revealed that the adult pattern of antibody responses to a specific hapten does not emerge (Hsu and Du Pasquier, 1984b); adult-type MHC class II^+ T lymphocytes do not appear in the periphery (Rollins-Smith and Blair, 1990a); thymus morphology does not become adult type (Clothier and Balls, 1985); lymphocytes in the thymus and spleen do not achieve the expanded cell numbers characteristic of postmetamorphic adults (Rollinis-Smith and Blair, 1990a,b); adult-type skin allograft rejection does not replace tolerogenic responses (N. Cohen and M. Crosby, unpublished observations); and splenic innervation and compartmentation of the spleen into a clearly defined red and white pulp is dramatically delayed* (K. S. Kinney and N. Cohen, unpublished observations). These observations have led to the conclusion that the normal ontogeny of adult-type immunity in frogs requires a postmetamorphic lymphocyte expansion of both T and B lymphocytes that results in adult-type allograft rejection capacity and a different antibody repertoire.

* Innervation is also delayed in nonperchlorate-treated tadpoles whose development lags behind the standard stage progressions delineated by Nieuwkoop and Faber (1967) due to crowding, low temperature, insufficient food, insufficient iodine, genetic variability, etc. If one studies the ontogeny of splenic innervation in such developmentally retarded larvae that do not go through metamorphosis "on schedule," one could arrive at the erroneous conclusion that innervation and compartmentation are associated only with metamorphic climax or early postmetamorphic life.

Immune system changes that are independent of TH levels (i.e., they occur in perchlorate-blocked tadpoles) are the onset of expression of MHC class I antigens (Rollins-Smith *et al.,* 1994); the immigration and expansion of T-cell precursors in the thymus (Rollins-Smith *et al.,* 1992); and the development of high-titer IgY antibody production (Hsu and Du Pasquier, 1984b).

TH appears to drive the terminal maturation of larval erythrocytes (Galton, 1984; Galton and St. Germain, 1985) and the expansion of a separate adult erythrocyte population (Dorn and Broyles, 1982; Just *et al.,* 1977, 1980; Flajnik and Du Pasquier, 1988; see Weber, Chapter 16, this volume). However, since TH have no deleterious effects on lymphocyte viability and inconsistent effects on proliferation (Rollins-Smith and Blair, 1993), it seems doubtful that TH are directly responsible for the loss of larval lymphocytes at metamorphosis.

B. Corticosteroid Hormones

In short-term viability studies, both larval and adult lymphocytes are very sensitive to being killed *in vitro* by physiological concentrations (i.e., those seen at metamorphosis) of corticosterone. Corticosterone and aldosterone significantly inhibit the PHA-induced proliferation of larval and adult spleen cells (Marx *et al.,* 1987; Rollins-Smith and Blair, 1993; L. A. Rollins-Smith, unpublished observations), and corticosterone induces apoptotic death of larval splenocytes at concentrations as low as 1–10 n*M* (K. S. Barker and L. A. Rollins-Smith, unpublished observations). Adult lymphocytes may be protected *in vivo* from the destructive effects of corticosterone since there is less total corticosterone in the circulation and most is in a bound state (Jolivet-Jaudet and Leloup-Hatey, 1986).

Since aldosterone levels do not appear to be affected by specific serum-binding factors, the observed elevated levels of aldosterone in plasma of metamorphosing and adult frogs reflect the level of freely available aldosterone. Because larval lymphocyte function is affected by aldosterone at these physiologically relevant concentrations (L. A. Rollins-Smith, unpublished observations), it could play a role in the loss of larval lymphocytes.

Based on these *in vitro* sensitivities to corticosterone and aldosterone, we hypothesize that increasing levels of these hormones cause the death of a significant number of larval lymphocytes. This would be especially important in the peripheral lymphoid compartments (i.e., peripheral blood and spleen) of metamorphosing frogs because were these immunocompetent cells not eliminated, they could potentially respond to adult-specific molecules. The evidence that larval splenocytes are affected by corticosteroid hormones (CH) can be summarized as follows. The *in vitro* inhibition of PHA-stimulated proliferation by CH can be blocked by the specific CH-receptor antagonist, RU486 (K. S. Barker and L. A. Rollins-Smith, unpublished observations). Preliminary studies involving *in vivo* treatment of stage 57–58 larvae with RU486 demonstrate a dose-dependent inhibition of normal splenocyte cell loss (L. A. Rollins-Smith, unpublished observations). Thus, corticosterone and/or aldosterone appear to be important regulators of peripheral splenocyte populations at metamorphosis in *Xenopus.*

The role of increasing levels of CH on the viability and development of thymocytes is more difficult to evaluate. We have found that viability of larval thymocytes cultured in 1–10 μg/ml corticosterone is significantly reduced (Rollins-Smith and Blair 1993); this reduction can be inhibited by RU486 *in vitro* (K. S. Barker and L. A. Rollins-Smith, unpublished observations). However, other studies have reported significant apoptosis of cultured thymocytes from all stages of larval development and metamorphosis that does not appear to be increased by culture with dexamethasone (Ruben *et al.,* 1994). Regardless of outcome, these *in vitro* studies are difficult to interpret because thymocyte development is critically dependent on stromal/epithelial microenvironmental influences. Thus, in the intact animal, CH could play a more significant role in the induction of apoptosis at metamorphosis than it does *in vitro.*

To complicate this issue further, there is now evidence from mammalian systems that thymic epithelial components can synthesize corticosteroids (Vacchio *et al.,* 1994). Thus, thymocytes may constantly receive local corticosteroid signals that influence their viability. We are currently performing *in vivo* studies that involve treatment of metamorphosing animals with RU486; preliminary data suggest that survival of thymocytes in this critical period can be affected by CH.

C. Prolactin and Growth Hormone

Although a role for PRL and GH in mammalian immunity is now well established (Kelly *et al.,* 1992), very little is known about the involvement of either hormone in the development and function of cells of the amphibian immune system. Hypophysectomized tadpoles continue to grow and can usually reject skin or organ allografts differing by presumed or defined MHC antigens (Rollins and Cohen, 1980; Rollins-Smith and Cohen, 1982; Maéno and Katagiri, 1984). However, like control tadpoles, they fail to reject grafts differing by minor H antigens or by organ-specific antigens (Rollins-Smith and Cohen, 1982; Maéno and Katagiri, 1984).

To examine the possible involvement of PRL in lymphocyte function, we have independently observed, in preliminary studies, that anti-PRL antibodies suppress the proliferation of mitogen-stimulated (lipopolysaccharide, PHA, Con A) larval and adult splenic lymphocytes (L. A. Rollins-Smith, unpublished observations; N. Cohen, unpublished observations). Thus, PRL may be an essential lymphocyte growth factor that developed early in vertebrate evolution.

Although PRL can be provided by serum sources, cultures in serum-free conditions (or with a serum supplement that lacks PRL) demonstrate significant proliferation that can be inhibited with anti-PRL antibodies (L. A. Rollins-Smith, unpublished observations). This suggests that like mammalian leukocytes (Pellegrini *et al.,* 1992), anuran lymphocytes and/or accessory cells may produce their own PRL in response to mitogenic stimulation.

D. Catecholamines

Chemical sympathectomy of rodents with 6-OHDA has been reported to have a variety of effects on *in vitro* and *in vivo* parameters of immunity (Madden and Felten, 1995). To determine whether the postmetamorphic appearance of rich sympathetic innervation of the spleen is causally related to the acquisition of an adult-type alloimmune recognition/rejection response, *Xenopus* larvae were injected intraperitoneally with 6-OHDA (100 μg/g body weight) and grafted 2 days later with skin from adult donors that were incompatible either at MHC plus multiple minor H loci or just minor H loci. Frogs received additional weekly injections of 6-OHDA to prevent the regeneration of sympathetic nerve fibers; control frogs received weekly injections of 0.1% ascorbate vehicle or desipramine 30 min prior to 6-OHDA to prevent the destruction of noradrenergic nerve terminals by 6-OHDA. Despite the fact that immunocytochemistry of the spleen of 6-OHDA-treated animals revealed a complete absence of sympathetic nerve fibers, that HPLC revealed a striking loss of splenic norepinephrine, and regardless of the donor–host immunogenetic disparity and the developmental stage of the larval recipient, no difference in survival times of grafts on sympathectomized and control animals was noted (Kinney *et al.,* 1994; N. Cohen and K. Kinney, unpublished observations).

V. AN ATTEMPT AT SYNTHESIS: EFFECTS OF NEUROENDOCRINE CHANGES ON THE DEVELOPING IMMUNE SYSTEM

The information reviewed in the preceding sections of this chapter leads to the following conclusions and suggests the following testable hypotheses. Larval *Xenopus* have a smaller number of lymphocytes that are phenotypically and functionally distinct from the population of lymphocytes that develop postmetamorphically. It seems reasonable to assume that this numerical limitation is responsible for the restricted diversity of the antibody repertoire that characterizes larval humoral immunity (Du Pasquier *et al.,* 1979; Hsu and Du Pasquier, 1984a,b). In addition, larvae and young postmetamorphic frogs show little or no N region diversification in the third complementarity-determining region (CDR3) of the immunoglobulin heavy chain (Lee *et al.,* 1993; Desravines and Hsu, 1994), perhaps due to the absence of TdT before metamorphosis (Lee and Hsu, 1994). The same reasoning leads to the prediction that when the diversity of larval and adult T-cell receptors is determined, the larval T-cell repertoire will be more limited. Based on the fact that MHC class I antigens are only expressed at low levels late in development, we also predict that class I-restricted T-cell cytotoxicity is deficient relative to adults. Whether class II serves as a restricting element during larval life is unknown.

At metamorphosis, levels of TH are elevated, resulting in increased pituitary function as the hypothalamo-pituitary connections mature. Adrenocorticotropin-releasing hormone may, in turn, rise at metamorphosis due to TH-driven events. As a consequence, corticosteroid hormones are elevated. This elevation may be necessary for gluconeogenesis to provide enough energy for metamorphic changes at a time of fasting (Jaudet and Hatey, 1984). As a consequence, some larval lymphocytes may be killed apoptotically in the periphery and in the thymus. Because immunological memory persists through metamorphosis, some components of the immune system, perhaps memory cells, must be protected from the effects of glucocorticoids.

The adult pattern of immunity emerges because the lymphocyte population expands dramatically after metamorphosis. After metamorphosis, there is an opportunity to fine-tune the T-cell and B-cell repertoires by selection (Du Pasquier, 1994). Class I MHC expression is upregulated, perhaps allowing "better" interactions with MHC and better selection for recognition of self- and foreign antigens. Whether expansion and continual function of the adult lymphocyte population are dependent on hormones such as PRL and GH remains to be determined.

A significant body of data has been published that deals with the complex neuroendocrine–immune system interactions that are proving so important for the maintenance of homeostasis in mammals (Ader *et al.,* 1991; Madden and Felten, 1995). In our opinion, amphibian metamorphosis offers a unique opportunity to study some of the most dramatic examples of these interactions in a developmental and phylogenetic context. We have just begun to understand that prolactin, thyroxine, and corticosteroid hormones influence the immune system of *Xenopus;* we anticipate that future studies will reveal much more about the roles that these and other hormones, neurotransmitters, and neuropeptides play in shaping the amphibian immune system during metamorphosis.

ACKNOWLEDGMENTS

Research cited in this paper from the authors' laboratories has been supported by grants from the National Science Foundation (DCB-8710235, DCB-9004666, and MCB-9421349 to LRS) and the National Institutes of Health (HD-07901 to NC). RU486 was a gift of Roussel Uclaf.

REFERENCES

Ader, R., Felten, D. L., and Cohen, N., eds. (1991). "Psychoneuroimmunology," 2nd ed. Academic Press, San Diego.

Barlow, E. H., and Cohen, N. (1983). The thymus dependency of transplantation allotolerance in the metamorphosing frog, *Xenopus laevis. Transplantation* **35,** 612–619.

Baroni, C. (1967). Thymus, peripheral lymphoid tissues and immunological responsiveness of the pituitary dwarf mouse. *Experientia* **23,** 282–283.

Baroni, C., Fabris, N., and Bertoli, G. (1967). Age dependency of the primary immune response in the hereditary pituitary dwarf and normal Snell–Bagg mouse. *Experientia* **23,** 1059–1060.

Baroni, C. D., Fabris, N., and Bertoli, G. (1969). Effects of hormones on development and function of lymphoid tissues. Synergistic action of thyroxine and somatotropic hormone in pituitary dwarf mice. *Immunology* **17,** 303–314.

Baroni, C. D., Pesando, P. C., and Bertoli, G. (1971). Effects of hormones on development and function of lymphoid tissues. III. Delayed development of immunological capacity in pituitary dwarf mice. *Immunology* **21,** 455–461.

Bleicher, P. A., and Cohen, N. (1981). Monoclonal anti-IgM can separate T-cell from B-cell proliferative responses in the frog, *Xenopus laevis. J. Immunol.* **127,** 1549–1555.

Blomberg, B., Bernard, C. C. A., and Du Pasquier, L. (1980). *In vitro* evidence for T–B lymphocyte collaboration in the clawed toad, *Xenopus. Eur. J. Immunol.* **10,** 869–876.

Bovbjerg, A. M. (1966). Rejections of skin homografts in larvae of *Rana pipiens. J. Exp. Zool.* **162,** 69–80.

Buckbinder, L., and Brown, D. D. (1993). Expression of *Xenopus laevis* prolactin and thyrotropin genes during metamorphosis. *Proc. Natl. Acad. Sci. U.S.A.* **90,** 3820–3824.

Capen, C. C. (1994). Mechanisms of chemical injury of thyroid gland. *Prog. Clin. Biol. Res.* **387,** 173–191.

Chardonnens, X., and Du Pasquier, L. (1973). Induction of skin allograft tolerance during metamorphosis of the toad *Xenopus laevis:* A possible model for studying generation of self tolerance to histocompatibility antigens. *Eur. J. Immunol.* **3,** 569–573.

Clemons, G. K. (1976). Development and preliminary application of a homologous radioimmunoassay for bullfrog growth hormone. *Gen. Comp. Endocrinol.* **30,** 357–363.

Clemons, G., and Nicoll, C. S. (1977). Development and preliminary application of a homologous radioimmunoassay for bullfrog prolactin. *Gen. Comp. Endocrinol.* **32,** 531–535.

Clevenger, C. V., Russell, D. H., Appasamy, P. M., and Prystowsky, H. B. (1990). Regulation of interleukin-2-driven T-lymphocyte proliferation by prolactin. *Proc. Natl. Acad. Sci. U.S.A.* **87,** 6460–6464.

Clothier, R. H., and Balls, M. (1985). Structural changes in the thymus glands of *Xenopus laevis* during development. *In* "Metamorphosis" (M. Balls and M. Bownes, eds.), pp. 332–359. Oxford University Press, Oxford.

Clothier, R. H., Ruben, L. N., Balls, M., and Greenhalgh, L. (1991). Morphological and immunological changes in the spleen of *Xenopus laevis* during metamorphosis. *Res. Immunol.* **142,** 360–362.

Cohen, N., DiMarzo, S., Rollins-Smith, L., Barlow, E., and Parsons, S. (1985) The ontogeny of allo-tolerance and self-tolerance in larval *Xenopus laevis. In* "Metamorphosis" (M. Balls and M. Bownes, eds.), pp. 388–419. Oxford University Press, Oxford.

Dardenne, M., Kelly, P. A., Bach, J.-F., and Savino, W. (1991). Identification and functional activity of prolactin receptors in murine lymphoid cells in normal and autoimmune situations. *Proc. Natl. Acad. Sci. U.S.A.* **88,** 9700–9704.

Deparis, P., and Jaylet, A. (1976). Thymic lymphocyte origin in the newt *Pleurodeles waltlii* studied by embryonic grafts between diploid and tetraploid embryos. *Ann. Immunol. (Paris)* **127c,** 827–831.

Desravines, S., and Hsu, E. (1994). Measuring CDR3 length variability in individuals during ontogeny. *J. Immunol. Methods* **168,** 219–225.

DiMarzo, S. J., and Cohen, N. (1982a). An *in vivo* study of the ontogeny of alloreactivity in the frog, *Xenopus laevis. Immunology* **45,** 39–48.

DiMarzo, S. J., and Cohen, N. (1982b). Immunogenetic aspects of *in vivo* allotolerance induction during the ontogeny of *Xenopus laevis. Immunogenetics* **16,** 103–116.

Dorn, A. R., and Broyles, R. H. (1982). Erythrocyte differentiation during the metamorphic hemoglobin switch of *Rana catesbeiana. Proc. Natl. Acad. Sci. U.S.A.* **79,** 5592–5596.

Du Pasquier, L. (1994). The evolution of the immune response. *Dev. Comp. Immunol.* **18,** S31–S32.

Du Pasquier, L., and Bernard, C. C. A. (1980). Active suppression of allogeneic histocompatibility reactions during metamorphosis of the clawed toad, *Xenopus. Differentiation (Berlin)* **16,** 1–7.

Du Pasquier, L., and Chardonnens, X. (1975). Genetic aspects of tolerance to allografts induced at metamorphosis in the toad *Xenopus laevis. Immunogenetics* **2,** 431–440.

Du Pasquier, L., and Flajnik, M. F. (1990). Expression of MHC class II antigens during *Xenopus* development. *Dev. Immunol.* **1,** 85–95.

Du Pasquier, L., and Haimovich, J. (1976). The antibody response during amphibian ontogeny. *Immunogenetics* **3,** 381–391.

Du Pasquier, L., and Weiss, N. (1973). The thymus during the ontogeny of the toad, *Xenopus laevis:* Growth, membrane-bound immunoglobulins and mixed lymphocyte reaction. *Eur. J. Immunol.* **3,** 773–777.

Du Pasquier, L., Blomberg, B., and Bernard, C. C. A. (1979). Ontogeny of immunity in amphibians: Changes in antibody repertoires and appearance of adult major histocompatibility antigens in *Xenopus. Eur. J. Immunol.* **9,** 900–906.

Du Pasquier, L., Schwager, J., and Flajnik, M. F. (1989). The immune system of *Xenopus. Annu. Rev. Immunol.* **7,** 251–275.

Ellison, T. R., Mathisen, P. M., and Miller, L. (1985). Development changes in keratin patterns during epidermal maturation. *Dev. Biol.* **112,** 329–337.

Fabris, N., Pierpaoli, W., and Sorkin, E. (1971a). Hormones and the immunological capacity. III. The immunodeficiency disease of the hypopituitary Snell–Bagg dwarf mouse. *Clin. Exp. Immunol.* **9,** 209–225.

Fabris, N., Pierpaoli, W., and Sorkin, E. (1971b). Hormones and the immunological capacity. IV. Restorative effects of developmental hormones or of lymphocytes on the immunodeficiency syndrome of the dwarf mouse. *Clin. Exp. Immunol.* **9,** 227–240.

Flajnik, M. F., and Du Pasquier, L. (1988). MHC class I antigens as surface markers of adult erythrocytes during the metamorphosis of *Xenopus. Dev. Biol.* **128,** 198–206.

Flajnik, M. F., and Du Pasquier, L. (1990). The major histocompatibility complex of frogs. *Immunol. Rev.* **113,** 47–63.

Flajnik, M. F., Horan, P. K., and Cohen, N. (1984). A flow cytometric analysis of the embryonic origin of lymphocytes in diploid/triploid chimeric *Xenopus laevis. Dev. Biol.* **104,** 247–254.

Flajnik, M. F., Du Pasquier, L., and Cohen, N. (1985). Immune responses of thymus/lymphocyte embryonic chimeras: Studies on tolerance and major histocompatibility complex restrictions in *Xenopus. Eur. J. Immunol.* **15,** 540–547.

Flajnik, M. F., Kaufman, J. F., Hsu, E., Manes, M., Parisot, R., and Du Pasquier, L. (1986). Major histocompatibility complex-encoded class I molecules are absent in immunologically competent *Xenopus* before metamorphosis. *J. Immunol.* **137,** 3891–3899.

Flajnik, M. F., Hsu, E., Kaufman, J. F., and Du Pasquier, L. (1987). Changes in the immune system during metamorphosis of *Xenopus. Immunol. Today* **8,** 58–64.

Gagnerault, M. C., Touraine, P., Savino, W., Kelly, P. A., and Dardenne, M. (1993). Expression of prolactin receptors in murine lymphoid cells in normal and autoimmune situations. *J. Immunol.* **150,** 5673–5681.

Galton, V. A. (1984). Putative nuclear triodothyronine receptors in tadpole erythrocytes: Regulation of receptor number by thyroid hormone. *Endocrinology* (*Baltimore*) **114,** 735–742.

Galton, V. A., and St. Germain, D. (1985). Putative nuclear triodothyronine receptors in tadpole erythrocytes during metamorphic climax. *Endocrinology* (*Baltimore*) **116,** 99–104.

Gudernatsch, J. F. (1912). Feeding experiments on tadpoles. I. The influence of specific organs given as food on growth and differentiation. A contribution to the knowledge of organs with internal secretion. *Wilhelm Roux Arch. Entwicklungsmech. Org.* **35,** 457–483.

Hadji-Azimi, I., Schwager, J., and Thiebaud, C. (1982). B-lymphocyte differentiation in *Xenopus laevis* larvae. *Dev. Biol.* **90,** 253–258.

Hadji-Azimi, I., Coosemans, V., and Canicetti, C. (1990). B-lymphocyte populations in *Xenopus laevis. Dev. Comp. Immunol.* **14,** 69–84.

Harding, F. A., Flajnik, M. F., and Cohen, N. (1993). MHC restriction of T-cell proliferative responses in *Xenopus. Dev. Comp. Immunol.* **17,** 425–437.

Hartman, D. P., Holaday, J. W., and Bernton, E. W. (1989). Inhibition of lymphocyte proliferation by antibodies to prolactin. *FASEB J.* **3,** 2194–2202.

Haynes, L., and Cohen, N. (1993a). Transforming growth factor beta (TGF-β) is produced by and influences the proliferative response of *Xenopus laevis* lymphocytes. *Dev. Immunol.* **3,** 223–230.

Haynes, L., and Cohen, N. (1993b). Further characterization of an interleukin-2–like cytokine produced by *Xenopus laevis* T lymphocytes. *Dev. Immunol.* **3,** 231–238.

Hildemann, W. H., and Haas, R. (1959). Homotransplantation immunity and tolerance in the bullfrog. *J. Immunol.* **83,** 478–485.

Horton, J. D. (1969). Ontogeny of the immune response to skin allografts in relation to lymphoid organ development in the amphibian *Xenopus laevis. J. Exp. Zool.* **170,** 449–466.

Horton, J., and Manning, M. J. (1972). Response to skin allografts in *Xenopus laevis* following thymectomy at early stages of lymphoid organ maturation. *Transplantation* **14,** 141–154.

Hsu, E., and Du Pasquier, L. (1984a). Ontogeny of the immune system in *Xenopus.* I. Larval immune response. *Differentiation* (*Berlin*) **28,** 109–115.

Hsu, E., and Du Pasquier, L. (1984b). Ontogeny of the immune system in *Xenopus.* II. Antibody repertoire differences between larvae and adults. *Differentiation* (*Berlin*) **28,** 116–122.

Hsu, E., and Du Pasquier, L. (1992). Changes in the amphibian antibody repertoire are correlated with metamorphosis and not with age or size. *Dev. Immunol.* **2,** 1–6.

Jaffe, R. C. (1981). Plasma concentration of corticosterone during *Rana catesbeiana* tadpole metamorphosis. *Gen. Comp. Endocrinol.* **44,** 314–318.

Jaudet, G. J., and Hatey, J. H. (1984). Variations in aldosterone and corticosterone plasma levels during metamorphosis in *Xenopus laevis* tadpoles. *Gen. Comp. Endocrinol.* **56,** 59–65.

Jolivet-Jaudet, G., and Leloup-Hatey, J. (1986). Corticosteroid binding in plasma of *Xenopus laevis.* Modifications during metamorphosis and growth. *J. Steroid Biochem.* **25,** 343–350.

Just, J. J., Schwager, J., and Weber, R. (1977). Hemoglobin transition in relation to metamorphosis in normal and isogenic *Xenopus. Wilhelm Roux's Arch. Dev. Biol.* **183,** 307–332.

Just, J. J., Schwager, J., Weber, R., Fey, H., and Pfister, H. (1980). Immunological analysis of hemoglobin transition during metamorphosis of normal and isogeneic *Xenopus laevis. Wilhelm Roux's Arch. Dev. Biol.* **188,** 75–80.

Katagiri, C. (1978). *Xenopus laevis* as a model for the study of immunology. *Dev. Comp. Immunol.* **2,** 5–14.

Kelly, K. W., Arkins, S., and Li, Y. M. (1992). Growth hormone, prolactin and insulin-like growth factor; new jobs for old players. *Brain. Behav. Immun.* **6,** 317–326.

Kikuyama, S., Suzuki, M. R., and Iwamuro, S. (1986). Elevation of plasma aldosterone levels in tadpoles at metamorphic climax. *Gen. Comp. Endocrinol.* **63,** 186–190.

Kikuyama, S., Kawamura, K., Tanaka, S., and Yamamoto, K. (1993). Aspects of amphibian metamorphosis: Hormonal control. *Int. Rev. Cytol.* **145,** 105–148.

Kinney, K. S., Felten, S. Y., Horton, J. D., and Cohen, N. (1993). Chemical sympathectomy and thymectomy effects on splenic anatomy in the frog *Xenopus laevis. Soc. Neurosci. Abstr.* **19,** (Part 2), 945.

Kinney, K. S., Cohen, N., and Felten, S. Y. (1994). Chemical sympathectomy with 6-OHDA does not alter skin graft rejection in the frog, *Xenopus laevis. Soc. Neurosci. Abstr.* **20,** 105.

Kinney, K. S., Cohen, N., and Felten, S. Y. (1994). Noradrenergic and peptidergic innervation of the amphibian spleen: Comparative studies. *Dev. Comp. Immunol.* **18,** 511–521.

Knowland, J. (1985). Vitellogenesis and levels of oestrogen receptor in *Xenopus* liver during metamorphosis. *In* "Metamorphosis" (M. Balls and M. Bownes, eds.), pp. 108–116. Oxford University Press, Oxford.

Kobel, H. R., and Du Pasquier, L. (1977). Strains and species for immunological research. *In* "Developmental Immunobiology" (J. B. Solomon and J. D. Horton, eds.), pp. 299–306. Elsevier/North-Holland, Amsterdam.

Krug, E. C., Honn, K. V., Battista, J., and Nicoll, C. S. (1983). Corticosteroids in serum of *Rana catesbeiana* during development and metamorphosis. *Gen. Comp. Endocrinol.* **52,** 232–241.

Lee, A., and Hsu, E. (1994). Isolation and characterization of the *Xenopus* terminal deoxynucleotidyl transferase. *J. Immunol.* **152,** 4500–4507.

Lee, A., Desravines, S., and Hsu, E. (1993). IgH diversity in an individual with only one million B lymphocytes. *Dev. Immunol.* **3,** 211–222.

Leloup, J., and Buscaglia, M. (1977). La triiodothyronine, hormone de la métamorphose des Amphibiens. *C. R. Hebd. Seances Acad. Sci., Ser. D* **284,** 2261–2263.

Madden, K. S., and Felten, D. L. (1995). Experimental basis for neural–immune interactions. *Physiol. Rev.* **75,** 77–106.

Maéno, M., and Katagiri, C. (1984). Elicitation of weak immune response in larval and adult *Xenopus laevis* by allografted pituitary. *Transplantation* **38,** 251–255.

Manning, M. J. (1991). Histological organization of the spleen: Implications for immune functions in amphibians. *Res. Immunol.* **142,** 355–359.

Manning, M. J., and Al Johari, G. M. (1985). Immunological memory and metamorphosis. *In* "Metamorphosis" (M. Balls and M. Bownes, eds.), pp. 420–433. Oxford University Press, Oxford.

Marx, M., Ruben, L. N., Nobis, C., and Duffy, D. (1987). Compromised T cell regulatory functions during anuran metamorphosis: The role of corticosteroids. *In* "Developmental and Comparative Immunology" (E. L. Cooper, C. Langlet, and J. Bierne, eds.), pp. 129–140. Liss, New York.

Mathisen, P. M., and Miller, L. (1987). Thyroid hormone induction of keratin genes: A two step activation of gene expression during development. *Genes Dev.* **1,** 1107–1117.

Nieuwkoop, P. D., and Faber, J. (1967). "Normal Table of *Xenopus laevis* (Daudin)." North-Holland, Amsterdam.

Obara, N., Kawahara, H., and Katagiri, C. (1983). Response to skin grafts exchanged among siblings of larval and adult gynogenetic diploids in *Xenopus laevis. Transplantation* **36,** 91–95.

O'Neal, K. D., Schwarz, L. A., and Yu-Lee, L. Y. (1991). Prolactin receptor gene expression in lymphoid cells. *Mol. Cell. Endocrinol.* **82,** 127–135.

Pellegrini, I., Lebrun, J. J., Ali, S., and Kelly, P. A. (1992). Expression of prolactin and its receptor in human lymphoid cells. *Mol. Endocrinol.* **6,** 1023–1031.

Roberts, A. B., and Sporn, M. B. (1990). The transforming growth factor βs. *In* "Peptide Growth Factors and Their Receptors I" (M. B. Sporn and A. B. Roberts, eds.), pp. 417–472. Springer-Verlag, New York.

Rollins, L. A., and Cohen, N. (1980). On tadpoles, transplantation, and tolerance. *In* "Development and Differentiation of Vertebrate Lymphocytes" (J. D. Horton, ed.), pp. 203–214. Elsevier/North-Holland, Amsterdam.

Rollins-Smith, L. A., and Blair, P. (1990a). Expression of class II major histocompatibility complex antigens on adult T cells in *Xenopus* is metamorphosis-dependent. *Dev. Immunol.* **1,** 97–104.

Rollins-Smith, L. A., and Blair, P. (1990b). Contribution of ventral blood island mesoderm to hematopoiesis in postmetamorphic and metamorphosis-inhibited *Xenopus laevis. Dev. Biol.* **142,** 178–183.

Rollins-Smith, L. A., and Blair, P. J. (1993). The effects of corticosteroid hormones and thyroid hormones on lymphocyte viability and proliferation during development and metamorphosis of *Xenopus laevis. Differentiation (Berlin)* **54,** 155–160.

Rollins-Smith, L. A., and Cohen, N. (1982). Self-pituitary grafts are not rejected by frogs deprived of their pituitary anlagen as embryos. *Nature (London)* **299,** 820–821.

Rollins-Smith, L. A., Parsons, S. C. V., and Cohen, N. (1984). During frog ontogeny, PHA and Con A responsiveness of splenocytes preceeds that of thymocytes. *Immunology* **52,** 491–500.

Rollins-Smith, L. A., Blair, P. J., and Davis, A. T. (1992). Thymus ontogeny in frogs T-cell renewal at metamorphosis. *Dev. Immunol.* **2,** 207–213.

Rollins-Smith, L. A., Flajnik, M. F., Blair, P., Davis, A. T., and Green, W. F. (1994). Expression of MHC class I antigens during ontogeny in *Xenopus* is metamorphosis-independent. *Dev. Comp. Immunol.* **18,** S95.

Ruben, L. N., Proochista, A., Johnson, R. O., Buchholz, D. R., Clothier, R. H., and Shiigi, S. (1994). Apoptosis in the thymus of developing *Xenopus laevis. Dev. Comp. Immunol.* **18,** 343–352.

Russell, D. H., Kibler, R., Matrisian, L., Larson, D. F., Poulos, B., and Magun, B. E. (1985). Prolactin receptors on human T and B lymphocytes: Antagonism of prolactin binding by cyclosporine. *J. Immunol.* **134,** 3027–3031.

Tochinai, S. (1978). Thymocyte stem cell inflow in *Xenopus laevis* after grafting diploid thymic rudiments into triploid tadpoles. *Dev. Comp. Immunol.* **2,** 627–635.

Tochinai, S. (1993). Strictly thymus-dependent tolerance induction in immunologically competent *Xenopus* larvae. *Zool. Sci.* **10,** 855–858.

Turpen, J. B., and Smith, P. B. (1989). Precursor immigration and thymocyte succession during larval development and metamorphosis in *Xenopus. J. Immunol.* **142,** 41–47.

Turpen, J. B., Turpen, C. J., and Flajnik, M. (1979). Experimental analysis of hematopoietic cell development in the liver of larval *Rana pipiens. Dev. Biol.* **69,** 466–479.

Turpen, J. B., Cohen, N., Deparis, P., Jaylet, A., Tompkins, R., and Volpe, E. P. (1982). Ontogeny of amphibian hemopoietic cells. *In* "The Reticuloendothelial System" (N. Cohen and M. M. Sigel, eds.), Vol. 3, pp. 569–588. Plenum, New York.

Vacchio, M. S., Papadopoulos, V., and Ashwell, J. D. (1994). Steroid production in the thymus: Implications for thymocyte selection. *J. Exp. Med.* **179,** 1835–1846.

Volpe, E. P., Tompkins, R., and Reinschmidt, D. (1979). Clarification of studies on the origin of thymic lymphocytes. *J. Exp. Zool.* **208,** 57–66.

Wahli, W., Dawid, I. B., Ryffel, G. U., and Weber, R. (1981). Vitellogenesis and the vitellogenin gene family. *Science* **212,** 298–304.

Watkins, D., and Cohen, N. (1987). Mitogen-activated *Xenopus laevis* lymphocytes produce a T-cell growth factor. *Immunology* **62,** 119–125.

Watkins, D., Parsons, S. C., and Cohen, N. (1987). A factor with interleukin-1–like activity is produced by peritoneal cells from the frog, *Xenopus laevis. Immunology* **62,** 669–673.

White, B. A., and Nicoll, C. S. (1981). Hormonal control of amphibian metamorphosis. *In* "Metamorphosis: A Problem in Developmental Biology" (L. I. Gilbert and E. Frieden, eds.), 2nd ed., pp. 363–396. Plenum, New York.

Widmer, H. J., Andres, A. C., Niessing, J., Hosbach, H. A., and Weber, R. (1981). Comparative analysis of cloned larval and adult globin cDNA-sequences of *Xenopus laevis. Dev. Biol.* **88,** 325–332.

Xu, Q., Baker, B. S., and Tata, J. R. (1993). Developmental and hormonal regulation of *Xenopus* liver-type arginase gene. *Eur. J. Biochem.* **211,** 891–898.

19

Cell Death and Histolysis in Amphibian Tail during Metamorphosis

KATSUTOSHI YOSHIZATO

Yoshizato MorphoMatrix Project of ERATO
JRDC
Saijo Misonou 242-37, Higashihiroshima 739, Japan
and
Department of Biological Science
Hiroshima University
Kagamiyama 1-3-1, Higashihiroshima 739, Japan

I. INTRODUCTION

Anura occupy a unique position as experimental animals for studying the mechanism of expression of the body plan which governs the formation of species-specific body shapes (Gilbert, 1991). The synthesis of new tissues along with the deconstruction of preformed tissues are equally important in the process of body shaping for an animal (Yoshizato, 1989). The life cycle of Anura contains a larval period between the embryo and adult, during which the larvae undergo drastic remodeling to change their body to an adult form (metamorphic remodeling) (Just *et al.*, 1981).

The larval tissues stop functioning and fall into the histolysis. Likewise, the larval cells undergo apoptosis (programmed cell death) (Yoshizato, 1992). Cell death and histolysis occur at a specific time in a specific location, suggesting that some specific program (death program) controls the whole process of tissue destruction (Weber, 1965; Tata, 1966; Beckingham Smith and Tata, 1976; Atkinson, 1981; Baker and Tata, 1992; Wang and Brown, 1993).

This chapter describes the progress that has been made on the metamorphic program of cell death and histolysis of the amphibian larval tail at cellular and molecular levels. More general reviews have been previously presented on the process of tail regression (Atkinson, 1981; Yoshizato, 1986, 1989, 1992).

II. TADPOLE TAIL AS UNIQUE TARGET FOR STUDYING PROGRAMMED CELL DEATH AND HISTOLYSIS

The tail is a larval-specific organ and has a complex structure composed of skin, connective tissue, muscle, nerve tissue, blood, and notochord (Yoshizato, 1986). The tail is large in both size and volume, occupying more than half of the whole body and, during metamorphosis, is resorbed in a short period in a strictly regulated and ordered manner (Yoshizato, 1989).

The metamorphic remodeling of larval tissues progresses in a region- and thyroid hormone (TH)-dependent manner (see Kaltenbach, Chapter 11, this volume; Shi, Chapter 14, this volume). In a series of studies on metamorphic remodeling, the focus has been on the larval skin because of its unique character. The skin in the early premetamorphic stage does not show the region-dependent difference in its structure and morphology (Izutsu *et al.,* 1993). During the progress of metamorphosis it gains regional specificity and shows quite different responses to the stimuli of TH. The head and trunk skin (collectively called body skin) is subject to remodeling to the adult form and the tail skin to cell death and histolysis. This chapter focuses on the process in which larval skin acquires the region-dependent characters at premetamorphosis and expresses its phenotype at the climax stage of metamorphosis.

III. METAMORPHIC FATE OF TAIL SKIN DETERMINED BY MESENCHYME AND THYROID HORMONE

The larval skin of the bullfrog, *Rana catesbeiana,* is suitable for morphological study because it gives clear histological views to the identity of cells and extracellular matrix structures when it is subjected to ordinary histological sectioning and staining. The skin of *Xenopus laevis* tadpoles, which are commonly and often utilized in laboratories as experimental animals for the study of metamorphosis, does not give such a clear histology.

The larval skin of bullfrogs has a uniform structure over the entire body in both the epidermis and the underlying mesenchyme in the early premetamorphic stages of larval development which corresponds to TK stage I to around TK stage IV defined by Taylor and Kollros (1946). No body- or tail-specific structures exist in the skin. The epidermis is composed of two types of cells: apical and skein (Robinson and Heintzelman, 1987). Skein cells are larva-specific cells and contain a unique structure of huge bundles of intermediate filaments called the figure of Eberth (Izutsu *et al.,* 1993). The skein cell can be easily identified through a fluorescent microscope because the figure of Eberth has strong and clear fluorescence when the tissues are fixed in Bouin's solution and stained with both hematoxylin and eosin (see Fig. 3) (Kawai *et al.,* 1994). Bundles of collagen are conspicuous in histological sections and are inserted between the basement membrane and the mesenchyme.

Outstanding changes in morphology occur in the body skin around TK stage IV (Fig. 1). The connective tissues are newly formed between the basement membrane and the collagen layer. Basal cells appear for the first time in the body skin. These changes do not occur in the tail skin whose basic structures remain unchanged through the premetamorphic stages. Thus, at this particular time of larval development tadpole skin becomes heterogeneous along the head–tail axis (Fig. 2).

It seems that the appearance of basal cells in the body skin is closely related to that of new connective tissues. Figure 1 shows a TK stage III tadpole that contains basal cells in one region of the body skin but not in another of the same skin. The former region contains the new connective tissue whereas the latter does not. At present, it seems possible that two possibilities exist for the mechanism of appearance of basal cells: (1) the new connective tissue plays a pivotal role in this mechanism and exerts an influence to induce the basal cells, and (2) the collagen layer has an important function and inhibits the appearance of basal cells. The transformation of the body skin at this stage is referred to as early premetamorphic remodeling or the first remodeling.

The change in the body skin appears to be TH dependent because tadpoles kept in tap water containing thiourea fail to show early premetamorphic remodeling (Fig. 3). As has been well established, TH is detect-

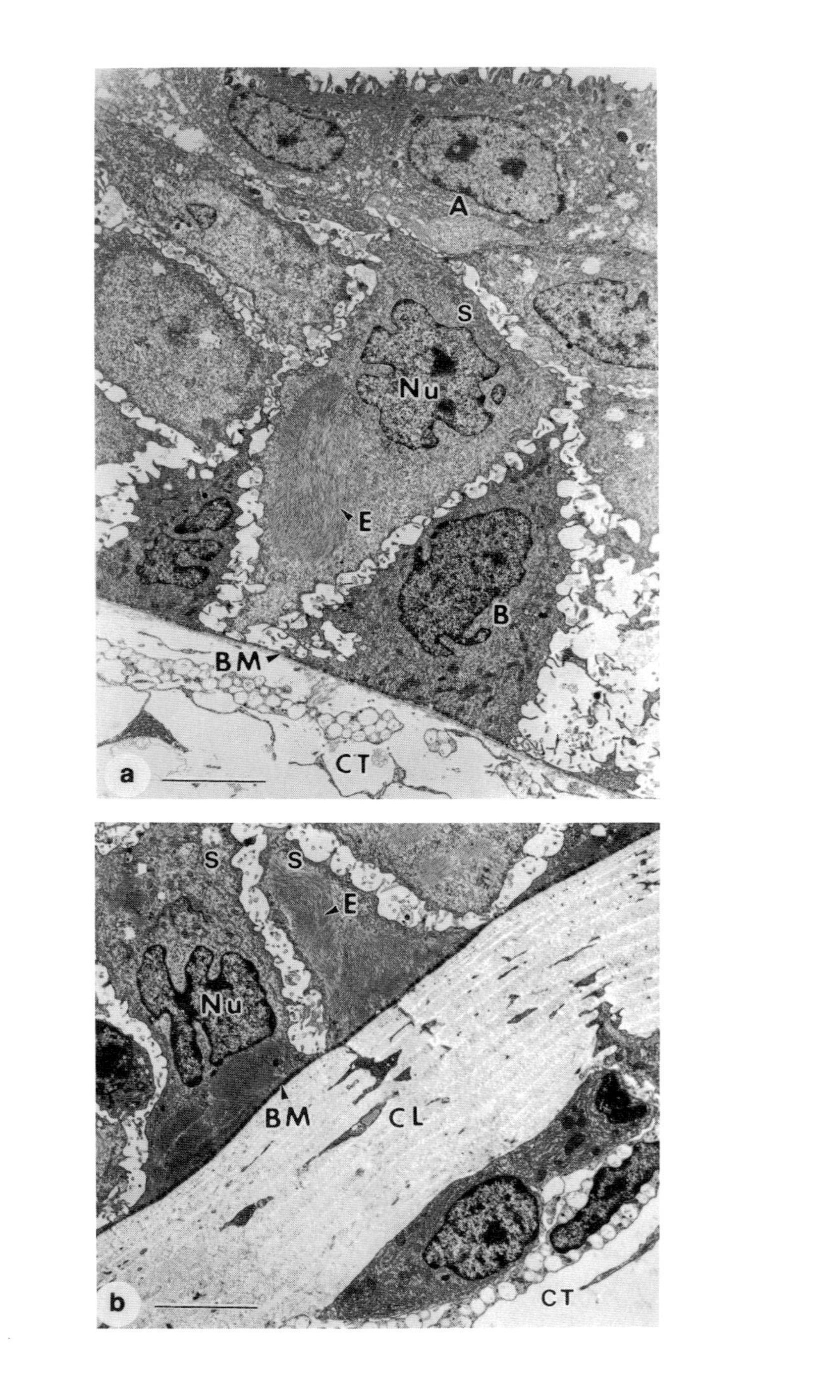
A
S
Nu
E
B
BM
CT
a
S
S
E
Nu
BM
CL
CT
b

able by the standard radioimmunoassay only around the late premetamorphic stages through climax stages (Regard *et al.*, 1978; Suzuki and Suzuki, 1981), although the earliest stage of detection of small amounts of TH at the onset of metamorphosis would depend on the sensitivity of the assay. Therefore, it is assumed that a trace of TH is required to induce the early premetamorphic skin remodeling. Actually, Hanaoka *et al.* (1973) demonstrated that thyroid glands function at the early stage of larval development and incorporate iodine. Furthermore, this speculation is supported by Kemp's (1963) study. He demonstrated that connective tissues (stratum spongiosum) were formed between collagen lamella and epidermis when larvae of *Rana pipiens* that had not formed the connective tissue were immersed in solution containing TH.

As stated earlier, the early premetamorphic skin remodeling takes place in the body but not in the tail. This regional cutaneous transformation contains the changes in both epidermis and mesenchyme. The mesenchyme of skin seems to play a leading role in determining the character of the epidermis, as suggested by a series of studies utilizing the heterotypically recombined skin described next.

Small flaps of skin were obtained from the body or the tail of bullfrog tadpoles and were enzymatically separated into the epidermis and the mesenchyme. The tissues thus prepared were recombined heterotypically (experimental combinations) or homotypically (control and normal combinations): body epidermis combined with tail mesenchyme (B/T), tail epidermis with body mesenchyme (T/B), and two normal recombinants (B/B, T/T). The recombinant skins were autografted to host tadpoles which were then allowed to metamorphose (Fig. 4). As was shown with tadpoles of *Rana japonica* (Kinoshita *et al.*, 1986a,b, 1989), the metamorphic changes of the epidermis are influenced by the mesenchyme combined. The body epidermis is transformed to the tail epidermis when it is associated with the tail mesenchyme and vice versa. The

---

Fig. 1. Body skin at late TK stage III. (a) A region where basal cells were identified. Note that new connective tissue is being formed under the basement membrane. (b) A region where basal cells were not observed. No new connective tissue exists between the basement membrane and the collagen layer. A, S, and B indicate apical, skein, and basal cells, respectively. E and Nu represent the figure of Eberth and the nucleus, respectively. BM, CL, and CT are the basement membrane, collagen layers, and connective tissues, respectively. Bar: 4 μm.

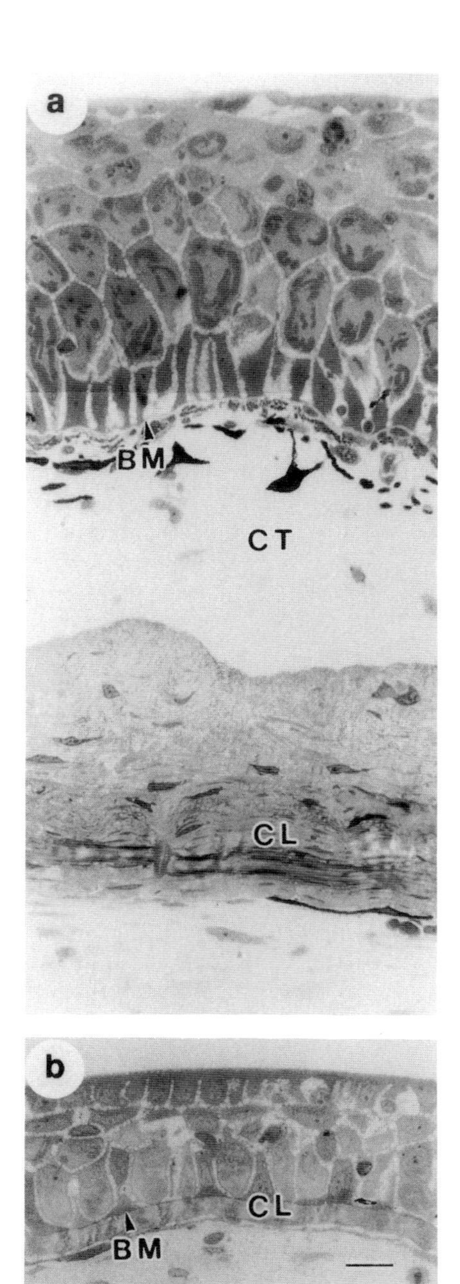
a
BM
CT
CL
b
CL
BM

underlying mesenchyme exerts an inducing activity on the epidermis and determines its region-specific character, i.e., body or tail.

Utilizing tadpoles of the bullfrog, this inducing activity of mesenchyme has been shown to be dependent on the stage of larval development (Kawai *et al.*, 1994). The mesenchyme of tadpoles at early premetamorphic stages shows the inducing activity (Fig. 5). The tail epidermis that normally does not produce basal cells transforms into the back epidermis when it is recombined with the back mesenchyme and produces the basal cells. Likewise, the back epidermis transforms into the tail epidermis when it is recombined with the tail mesenchyme and fails to produce the basal cells. The tail epidermis was also shown to express adult-specific antigens when it had been recombined with the back mesenchyme. Likewise, the back epidermis loses the ability to produce the adult-specific antigen when it is recombined with the tail mesenchyme. In contrast, the mesenchyme of larvae at middle premetamorphic stages (stage TK X) loses the inducing activity (Fig. 6). The recombined skin transplants of B/T and T/B do not change their originally determined characters as the body and the tail epidermis, respectively, i.e., the body epidermis of B/T does not lose basal cells and the tail epidermis of T/B does not produce basal cells. It has been hypothesized that the mesenchyme exerts the inducing activity at stages before the early premetamorphic skin remodeling. After this critical stage, the mesenchymal activity appears to be abolished or the reactivity of epidermis toward the mesenchymal stimulation is lost. It can be said that the important event occurs in the skin at the early premetamorphic stage that is responsible for the determination of region-dependent characters of the skin. TH does not seem to be involved in this determination because it has been shown that the inducing activity is traced to the dermatome at the embryonic stage preceding tail bud formation (Kinoshita *et al.*, 1989). As stated earlier, the traces of TH induce skin remodeling. Therefore, it is assumed

---

Fig. 2. Region-specific morphology of tadpole skin. The histology of skin of the bullfrog tadpole at TK stage X is shown for the body (a) and the tail(b). The body skin is histologically quite different from the tail skin. Connective tissue (CT) can be observed in the body skin between the basement membrane (BM) and the dense collagenous lamellae (CL). Basal cells are aligned along the BM. In the tail skin, no CT is formed and the BM is in direct contact with the CL. No basal cells are seen on the BM and instead skein cells are on it. Bar: 20 μm.

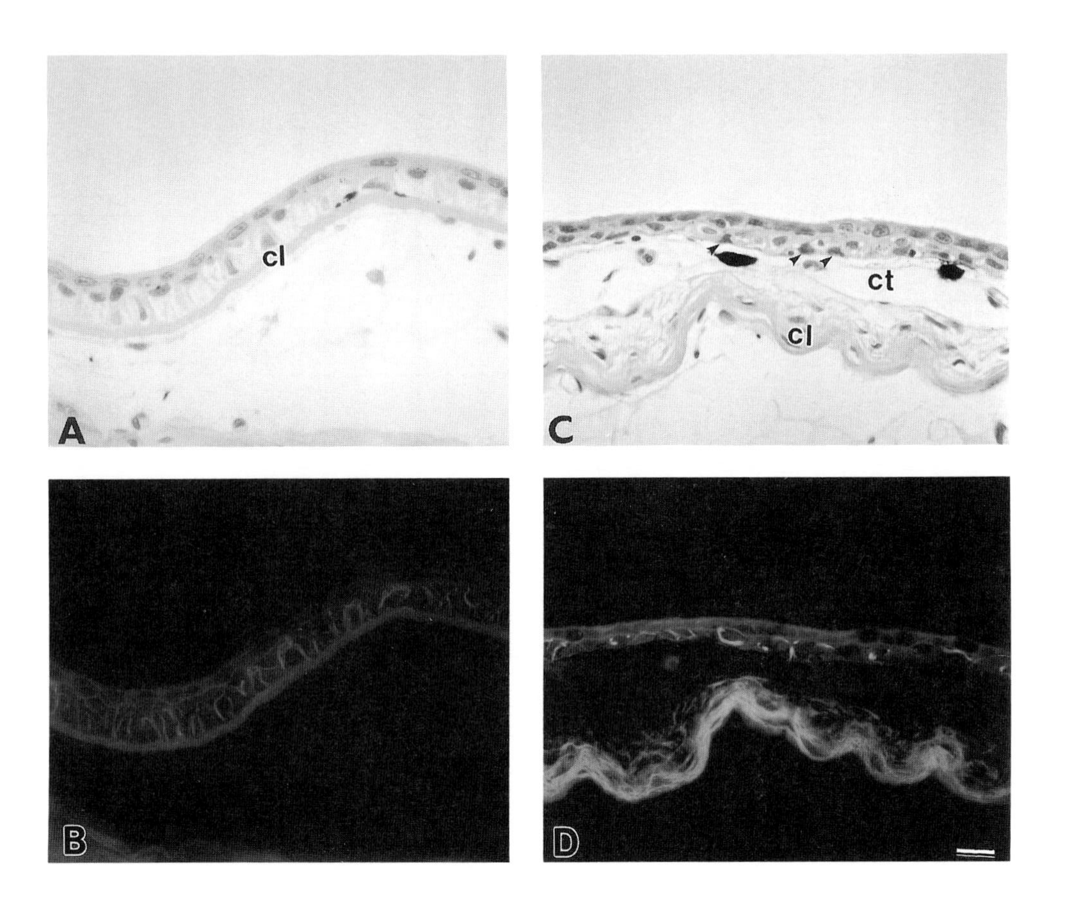
cl
A
ct
cl
C
B
D

that the structural change in skin has some connection to the loss of the mesenchymal inducing activity or of the epidermal reactivity.

All the data presently available suggest the presence of at least three identifiable steps in which the skin of tadpole undergoes remodeling to the adult skin, as summarized in Fig. 7.

Step I is characterized as "early premetamorphic induction" and is from the tail bud stage to around TK stage IV. Step I proceeds without the involvement of TH. The mesenchyme exerts its inducing activity toward the epidermis and the regional specificity of epidermis (body or tail) is determined. The determined regionality is expressed not in Step I, but in Step II and Step III. The structure of both epidermis and mesenchyme is identical in both the body and the tail. The epidermis contains skein and apical cells, but not basal cells. The determined regionality of epidermis is potentially changeable to the other type of regionality (from the tail to the body or vice versa), depending on the origin of mesenchyme which contacts the epidermis.

Step II is featured by changes in the body skin and is called "premetamorphic remodeling" or the first remodeling. Changes do not occur in the tail skin whose basic characteristics are identical to those of the preceding step. However, it should be pointed out that the tail mesenchyme loses its inducing activity. A trace of TH (less than 10^{-10} M) induces the first remodeling of both the epidermis and the mesenchyme. Basal cells appear for the first time in the body epidermis. Notable changes also occur in the body mesenchyme. New connective tissue (stratum spongiosum) is formed between the basement membrane and the thick collagen layer (the stratum compactum). The appearance of basal cells is important for the transition of larval skin to adult type because these cells are believed to transform into the germinative epidermal cell during the climax stages of metamorphosis (Robinson and Hein-

---

Fig. 3. Thiourea inhibits the early premetamorphic remodeling of tadpole skin. Bullfrog tadpoles that had not undergone the early premetamorphic remodeling were treated with 0.005% thiourea. Tissues were sectioned, stained with hematoxylin–eosin, and observed through a light microscope (A, C) or a fluorescent microscope (B, D). Skein cells are easily identified in B and D as cells that contain fluorescent bodies in the cytoplasm. (A and B) Thiourea-treated animals. (C and D) Control animals. cl, collagen laminae; ct, connective tissue. Bar: 20 μm. The original is in color and can be found in Kawai *et al.* (1994). Reproduced with permission from Academic Press.

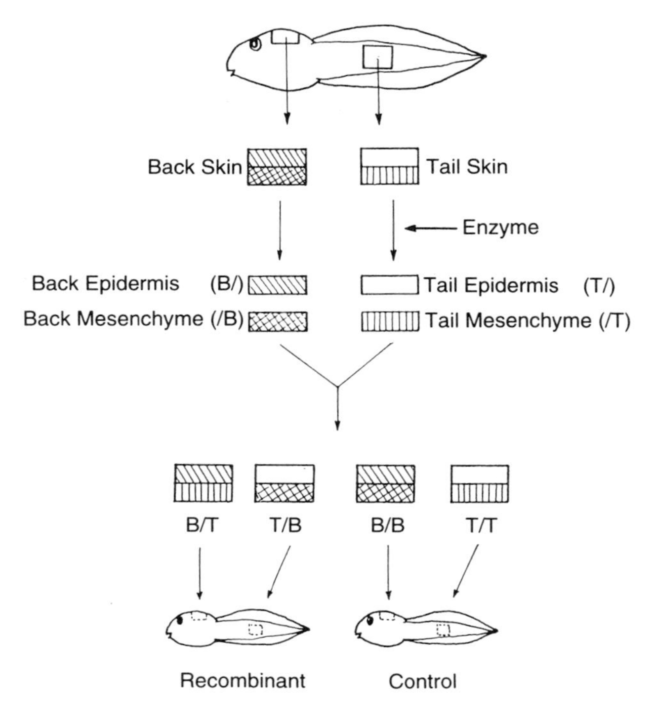

Fig. 4. Reconstruction and transplantation of the skin. Bullfrog tadpoles at stages IV and X were anesthetized in 0.05% MS222 (ethyl *m*-aminobenzoate methanesulfonate), and flaps of skin (ca. 5×5 mm^2 for stage IV and ca. 7×7 mm^2 for stage X) were dissected from back and tail regions. Flaps at stage IV were digested with 0.3% dispase and those at stage X with 0.3% collagenase and 0.1% hyaluronidase. The epidermis and the mesenchyme were separated from the digested tissues. Skins were reconstructed homotypically or heterotypically by these pieces of isolated epidermis and mesenchyme. Reconstructs were autografted to the region of epidermal origin of the transplant. The animals were reared in tap water, and the changes in the histology of the transplants were examined.

tzelman, 1987; Yoshizato, 1989; Izutsu *et al.,* 1993; Kawai *et al.,* 1994). Therefore, in order to know the origin of basal cells, it is important to understand the mechanism of metamorphic transformation of larval skin, which, however, still remains unclear.

In step III the body skin completely transforms into the adult skin and the tail skin falls into histolysis and is called "metamorphic remodel-

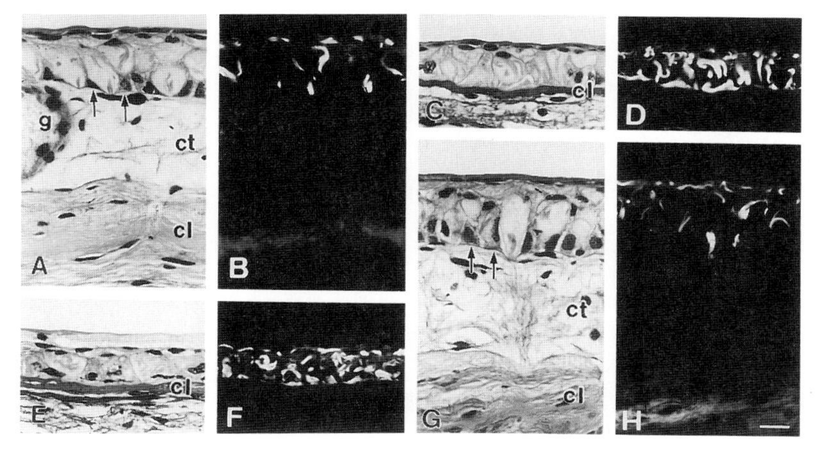

Fig. 5. Inductive activity of the subepidermal mesenchyme of tadpoles at a very early metamorphic stage. The inductive activity was shown by mesenchyme-dependent histological changes of the epidermis. Skins were reconstructed with epidermis and mesenchyme obtained from tadpoles at stage IV which had not undergone the first remodeling of skin and were autografted. Grafts were examined after 4 months when the host with heterotypical transplants developed to stage XVII and those with homotypical ones to stage XIV. Grafts were fixed and stained as in Fig. 3. (A, B) The tail epidermis with the back mesenchyme (T/B). (C, D) T/T. (E, F) B/T. (G, H) B/B. (A,C,E,G) Light microscopy; (B,D,F,H) fluorescent microscopy. Arrows indicate basal cells. cl, collagen laminae, ct, connective tissue; g, secretory gland. Bar: 20 μm.

ing'' or the second remodeling. A higher level of TH (around 10^{-9} *M* as triiodo-L-thyronine) is required for the initiation of this step and triggers the expression of predetermined region-specific characteristics of the skin.

IV. CELL DEATH AND HISTOLYSIS

Two major events that occur in the skin during step III are cell death and cell proliferation, which are apparently contradictory to each other. As described earlier, the larval skin contains two types of cells: larva-specific cells and larva-to-adult cells. Skein and apical cells are the former type and basal cells belong to the latter. Cell death occurs specifically and only in larva-specific cells. Larva-specific epidermal cells are present in both the body and the tail, as described in the previous sections, and

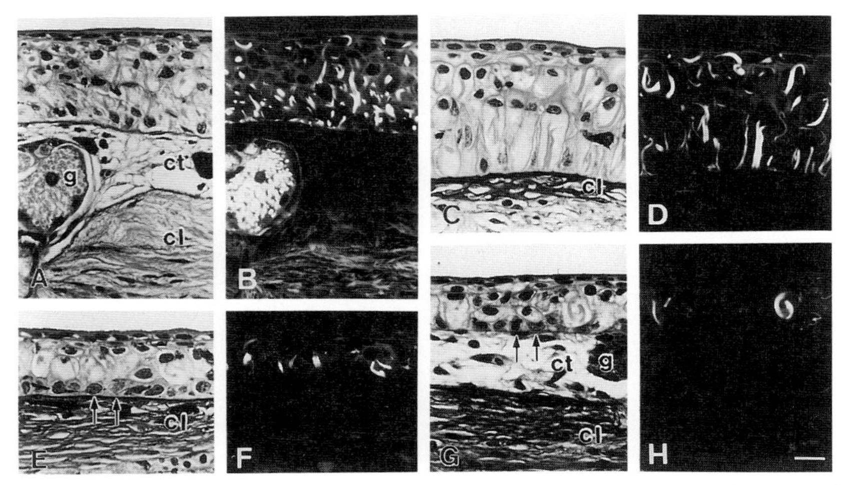

Fig. 6. Mesenchyme-independent changes of epidermis of recombinant skin prepared from tadpoles at stage TK X. The epidermis and the mesenchyme were separated by treating the skin with colagenase and hyaluronidase. Grafts were examined as in Fig. 5 after 3 weeks of transplantation. (A, B) T/B. (C, D) T/T. (E, F) B/T. (G, H) B/B. (A,C,E,G) Light microscopy, (B,D,F,H) fluorescent microscopy. Arrows in E and G indicate basal cells. cl, ct, g: the same as in Fig. 5. Bar: 20 μm.

are equally subject to cell death (Fig. 8). There seems to be a difference in the time of death of larval cells between the body and the tail. As shown in Fig. 8, there is a delay in cell death for the tail larval epidermal cells, some of which remain alive near the end of metamorphosis. TH is directly responsible for inducing the death of larva-specific cells because the tail epidermal cells, whose major population is the skein cells, begin to die when they are cultured in the presence of TH (Nishikawa and Yoshizato, 1985, 1986; Nishikawa *et al.,* 1989). However, the mechanism is not known by which this delay in death of larval epidermal cells occurs in the tail.

As described earlier, TH exerts direct effects on cultured skein cells and seems to induce cell death. Cells in the culture dish decrease in number as the culture period is increased in a TH concentration-dependent manner. It is not clear how the hormone exerts these actions on larval cells at the molecular level (see Tata, Chapter 13, this volume). DNA synthesis of the cells is greatly decreased by TH. It is assumed

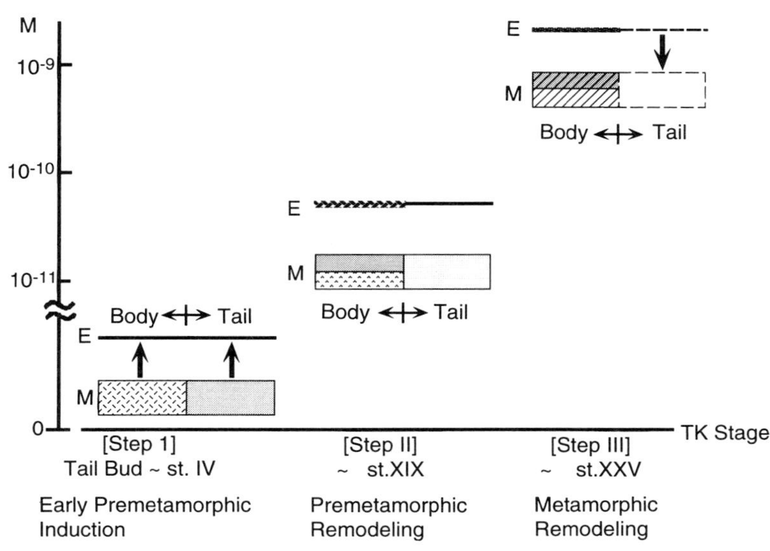

Fig. 7. A three-step mechanism of action of TH and mesenchyme in the transition of larval skin to adult skin. At least three steps are identifiable in which the skin of tadpoles undergoes remodeling to the adult skin. These steps are schematically drawn as steps I, II, III. Each step is characterized by two parameters: *y* axis, the TH level that is required for the transition from step I to step II and from step II to step III; *x* axis, [Step I] Early premetamorphic induction of the epidermis with the mesenchyme. This step proceeds without the involvement of TH. The mesenchyme exerts its inducing activity toward the epidermis and the regional specificity of epidermis (body or tail) is determined. [Step II] A trace of TH less than 10^{-10} M induces the premetamorphic remodeling (the first remodeling) of both the epidermis and the mesenchyme which results in disappearance of the inducing activity of mesenchyme. [Step III] A high level of TH (around $10^{-9}M$) triggers the expression of predetermined characters of epidermis, leading to the metamorphic remodeling (the second remodeling) of the skin. E, epidermis; M, mesenchyme. No morphological difference exists between the mesenchymes of body and tail at the "early premetamorphic induction stage." However, they are functionally different in that they show the region-specific inducing activity. The boundary between body (←) and tail (→) is shown by vertical lines. The vertical arrows in step I represent the region-dependent inducing activity of the mesenchyme. ▬ : The new connective tissue formed in the back skin at the premetamorphic remodeling. ▦ and ▨ : Body skin at step III representing the epidermis and the subepidermal mesenchyme (dermis) that are expressing adult characters of the corresponding tissues, respectively. The broken line of tail epidermis and the dotted rectangle of tail mesenchyme in the "metamorphic remodeling stage" indicate that they are undergoing deconstruction due to the high level of TH.

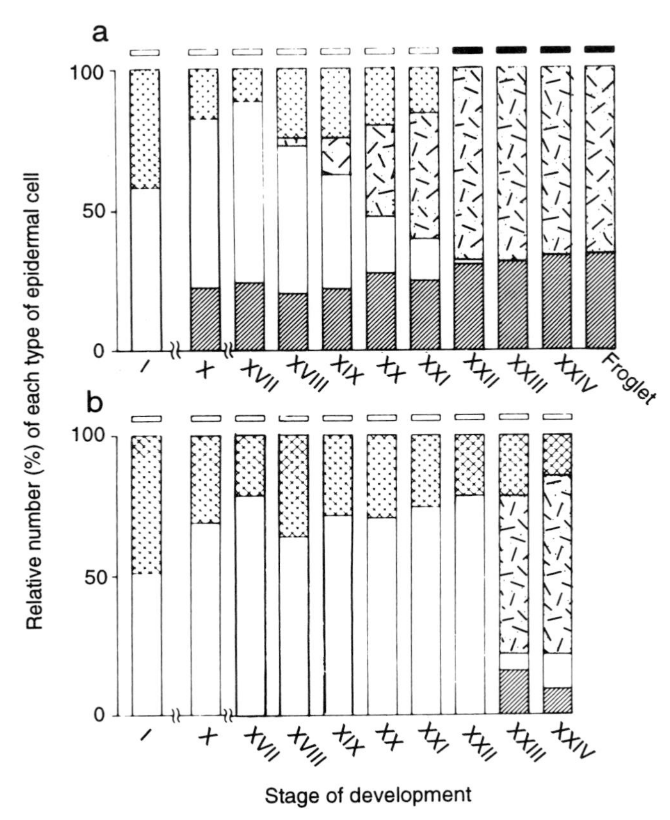

Fig. 8. Replacement of larval cells with adult cells in the skin during metamorphosis. A total of 200 to 300 cells at each metamorphic stage were surveyed on histological sections prepared from tadpoles at stages indicated, and the cells were identified as skein, apical, and granular. The relative number of each type of cell in the epidermis is expressed as a percentage. (a) Back skin. Cornified cells were not counted and their presence (▬) or absence (▭) is indicated at the top of the histograms. (b) Tail skin. ▒ , apical cells; ▭, skein cells; ▨ , basal cells; ⊠, granular cells.

that general cellular synthetic activities, including RNA and protein synthesis, are shut off by a direct action of TH. However, early studies on organ culture of *X. laevis* tadpole tails indicated that their TH-induced regression requires *de novo* protein synthesis (Weber, 1965; Tata, 1966; Beckingham Smith and Tata, 1976). With the advent of recombinant

DNA technology, it has been shown that TH activates or upregulates the expression of several genes in the tadpole tail (Baker and Tata, 1992; Wang and Brown, 1993).

Larval basal cells are not subject to cell death at step III, but survive this step and are thought to transdifferentiate into the epidermal germinative cells. The transdifferentiation requires the prior proliferation of basal cells. It has been shown that bromodeoxyuridine-positive cells increase in number substantially when premetamorphic tadpoles are treated with TH (M. Chono and K. Yoshizato, unpublished observations, 1994) or when tadpoles spontaneously go into the climax stage of metamorphosis (Kinoshita and Sasaki, 1994). Basal cells also respond to TH *in vitro* and increase their rate of DNA synthesis (Nishikawa *et al.,* 1989; Kinoshita and Sasaki, 1994).

These arguments suggest dual roles for TH in controlling cell death and cell proliferation. TH induces the death of larval-specific cells and the proliferation of larva-to-adult cells. It will be most interesting to unravel the mechanism underlying this contradictory dual action of TH on tadpole cells.

V. COLLAGENASE AND ITS GENE

The tadpole tail contains several tissues such as muscle, notochord, interstitial connective tissue, nerve, blood, and epidermis (Yoshizato, 1986). These tissues are eventually degraded at step III of metamorphosis. There is no doubt that many enzymes are responsible for degrading proteins, proteoglycans, RNA, and DNA at the site of histolysis. Actually, such enzymes were studied by biochemists who wanted to understand the complex phenomenon of tail regression. Several types of DNase and RNase have been reported in the regressing tail (Weber, 1977; Coleman, 1963; Hickey, 1971). Others measured the activity of hyaluronidase (Silbert *et al.,* 1965) and β-glucuronidase (Kubler and Frieden, 1964) or *N*-acetyl-β-hexosaminidase (Stuart *et al.,* 1978) for the degradation of sugar polymers. Cathepsins were also of major interest in considering the breakdown of proteins of the tail (Weber, 1977). Indeed, novel new proteinases were discovered in the course of studying the histolysis of the larval tissues such as collagenase (Gross and Lapiere, 1962; Nagai *et al.,* 1966) and a unique protease (protease T_1) that preferentially degrades actin (Yoshizato and Nakajima, 1982; Motobayashi and Yoshizato, 1986; Motobayashi *et al.*, 1986).

Among all the proteinases, collagenase merits description in this chapter because the enzyme has been well-characterized biochemically and its gene has been cloned. The gene-encoding collagenase can be considered a useful model for understanding the mechanism of TH-dependent regulation of metamorphosis-associated genes.

Specific polyclonal antibodies against bullfrog tadpole collagenase were prepared (Oofusa and Yoshizato, 1991). Western blot analysis showed that the amount of collagenase increases substantially in the tail skin of T_3-treated tadpoles. The enzyme was also detectable in the control tail skin. No positive signals were obtained from the control body skin but an intense reactivity was observed in body samples from T_3-treated tadpoles.

Immunohistochemistry with anticollagenase antibodies identified the cells that produce collagenase in response to TH (Oofusa and Yoshizato, 1991). Although epidermal cells and mesenchymal cells were both positively stained with antibodies, the former cells reacted much more intensely. This result is interesting for the following reasons: (1) very few reports have been available which describe the epidermal collagenase in mammals and (2) the epidermis is poor in the substrate proteins of collagenase (collagens). It is not known whether collagenase secreted from epidermal cells participates in the degradation of interstitial collagens in the mesenchyme.

Complementary DNAs and genes for collagenase were isolated from some mammalian species, such as human (Goldberg *et al.*, 1986; Collier *et al.*, 1988), porcine (Krebs *et al.*, 1990), and rabbit (Fini *et al.*, 1987), but not from nonmammalian vertebrates. Both the cDNA and the gene for amphibian collagenase have been cloned and their unique properties as compared to the mammalian counterparts have been revealed.

A cDNA clone (Tcl), comprising the full-length mRNA of bullfrog tadpole collagenase, was isolated from an expression library using the antisera described earlier (Oofusa *et al.*, 1994). The nucleotide and deduced amino acid sequences of Tcl reveal that collagenase is highly conserved from anurans through humans. Collagenase is a member of the matrix metalloproteinases (MMPs); the sequence of the active center is VAAHELGHSLGLSHS and is completely conserved in bullfrog. Some unique features are also revealed in tadpole collagenase (Fig. 9). Four cysteine residues are present in both Tcl and porcine collagenase, which contrasts with human collagenase which contains three cysteine

Fig. 9. Comparison of amino acid sequences of bullfrog tadpole collagenase with mammalian collagenases. Amino acid sequences of bullfrog collagenase are compared with those of porcine (Krebs *et al.*, 1990) and human enzymes (Goldberg *et al.*, 1986). The amino acids of mammalian enzymes identical to those of tadpole are shown by dots. The sequences of proenzymes are compared among the three because sequences of prepeptide of the porcine enzyme are not available. Gap sequences in the tadpole enzyme are shown by dashes. The first amino acid of active enzyme is indicated by an arrow. Cysteine residues and RGD sequences are marked by arrowheads and are boxed, respectively. The active center is underlined.

residues. Conversely, several sequences of mammalian collagenases are missing in the tadpole enzyme. Collagenases of these three species contain the RGD sequence in the same position. Interestingly, the tadpole enzyme has an additional RGD sequence. A polyadenylation site is present in the 3′ flanking region. However, Tcl lacks a poly(A) sequence in its 3′ end region.

As expected, the expression of the collagenase gene was increased as tadpoles metamorphose spontaneously (Fig. 10). In the experiment shown in Fig. 10, expression of the collagen gene was also determined using radiolabeled cDNA for the human type I collagen α 1 chain. The body skin that transforms into the adult skin showed increased collagen gene expression during the late climax stages of metamorphosis, whereas the expression is markedly depressed in the tail skin. Identical changes

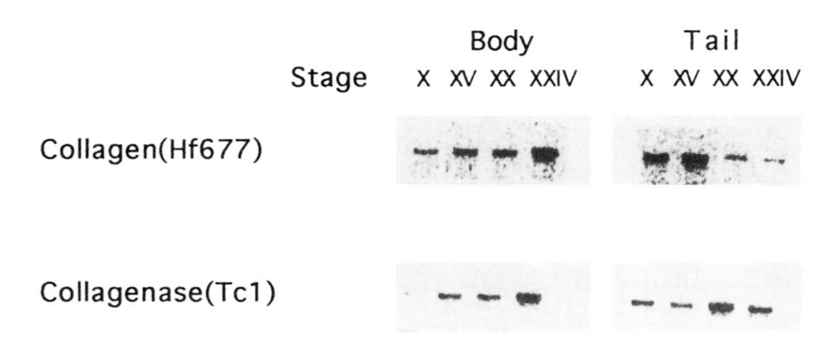

Fig. 10. Expression of genes of collagen and collagenase in spontaneously metamorphosing tadpoles. Five micrograms of total RNA was electrophoresed for RNA blot analysis with ^{32}P-labeled Hf677 in the upper line or ^{32}P-labeled Tcl in the lower one. The left column shows RNA extracted from the body skin whereas the right column shows that of the tail skin. Roman numerals at the top of each lane represent the developmental stage (TK stage) of the tadpoles. Hf677 is a cDNA clone of human type I collagen α 1 chain (Chu *et al.*, 1982).

were observed during the T_3-induced metamorphosis. At present the mechanism is not known by which TH exerts dual actions that are opposite to each other: up-regulation of collagen genes in the body skin versus down-regulation of the same gene in the tail skin and up-regulation of collagenase genes in the tail skin versus down-regulation of collagen genes in the same tissue. It is assumed that region-specific transcription factors are involved in the mechanism of the former case and that TH-responsive and gene-specific transcription factors play important roles in the mechanism of the latter case.

The collagenase gene (*TcgM718*) was cloned from bullfrog genomic DNA using Tc1 as a probe and sequenced from −1 to +4 kbp. Coding regions were found to be composed of four exons and three introns. Sequences corresponding to the putative active center of the enzyme were located in exon 2.

Comparison of DNA sequences of the exon–intron junctions of the bullfrog collagenase gene with that of the human fibroblast collagenase gene is given in Table I, which reveals the structural uniqueness of the bullfrog enzyme gene. Genes of mammalian collagenase reported hitherto (Collier *et al.,* 1988; Fini *et al.,* 1987) contain 10 exons with the sequence of the active center in exon 5. It is inferred that the exons have been split which generates diversities in the structure during the course of vertebrate evolution.

TABLE I

Comparison of Genes of Collagenase of Bullfrog and Human DNA Sequence of Exon (Coding Region)/Intron (Noncoding Region) Boundaries of Collagenase Genes[a]

Bullfrog skin collagenase				Human fibroblast collagenase			
Exon	Position in cDNA (being–end)	Size (bp)	5′ border/coding region/3′ border	Exon	Position in cDNA (begin–end)	Size (bp)	5′ border/coding region/3′ border
1	1–456	456	TCAATCA/ATGTT. . .GGGAG/GTAGAA	1–3	1–567	567	GGCCAGT/ATGCA. . .GGGAG/GTAAGT
2	457–717	261	TTTTCAG/ATCAT. . .TGATG/GTAAGT	4, 5	568–849	282	CTTTCAG/ATCAT. . .ATATG/GTGAGT
3	718–921	204	CCTTTAG/GACCT. . .CAAAG/GTGGAC	6, 7	850–1101	252	TCCTTAG/GACGT. . .CAAAG/GTAATC
4	922–1520	599	TGGGCAG/GTAAC. . .ATTAA/ATGGGA	8–10	1102–1970	869	TTGGCAG/GGAAT. . .GTTTT/ATACAC

[a] The whole sequence of bullfrog collagenase cDNA (Tc1) is shown in the work of Oofusa *et al.* (1994). Sequences of the human gene are from Collier *et al.* (1988) in which, however, the whole sequence of introns is not available. The numbering of nucleotides is based on these sequences. See Oofusa and Yoshizato (1995) for the whole sequence of the bullfrog collagenase gene.

To look for the cis-acting element that regulates the expression of the bullfrog collagenase gene, its 5′ upstream region was characterized (Fig. 11). A potential binding site for SpI (CCGCCC) is present at −32 to −27 and two sites for AP-1 (TGAGTCAG) at −22 to −15 and at −81 to −74. These sequences are identical to those of corresponding binding sites found in mammalian genes.

As described earlier, several lines of evidence strongly suggest that the expression of the collagenase gene is regulated by TH. Brown and colleagues tried to identify genes whose expression is under control of TH by the polymerase chain reaction subtractive method (gene screen method) and cloned the gene of matrix metalloproteinase as a middle late TH-responsive gene (Wang and Brown, 1991). Therefore, it is expected that a thyroid-responsive element (TRE) should be present in the transcriptional regulatory regions of the collagenase gene.

Umesono *et al.* (1991) proposed that one of several possible TREs consists of a tandem repeat of AGGTCA-like half-sites separated by four nucleotides (Chatterjee and Tata, 1992). Up until now, three such TRE direct repeat sequences were found: GGGTTA xxxx AGGACA in the 5′ upstream transcriptional regulatory region of the malic enzyme gene (Petty *et al.*, 1990); AGGTGA xxxx AGGACA in the 5′ upstream region of the myosin heavy chain gene (Flink and Morkin, 1990); and AGGTAA xxxx AGGTAA in the third intron of growth hormone gene (Sap *et al.*, 1990). A search was made for the possible TRE sequence in the 5′ upstream region of human MMP genes according to this criterion and no such candidate sequence was found. The gene for amphibian collagenase, however, contains a unique sequence of AGGTAA xxxx AGGATA which conforms to the proposal by Umesono and colleagues described earlier (Fig. 11). Therefore, the intriguing possibility is raised that the amphibian collagenase gene is a thyroid hormone-responsive gene but that mammalian collagenase genes are not.

VI. CONCLUDING REMARKS

The resorption of anuran tadpole tail is one of the most conspicuous events in the process of metamorphosis and requires a series of very complex biochemical reactions to proceed normally. At the cell level, the tail is mostly composed of larva-specific cells and contains few larva-to-adult cells, which should be the major reason why the tail is degraded.

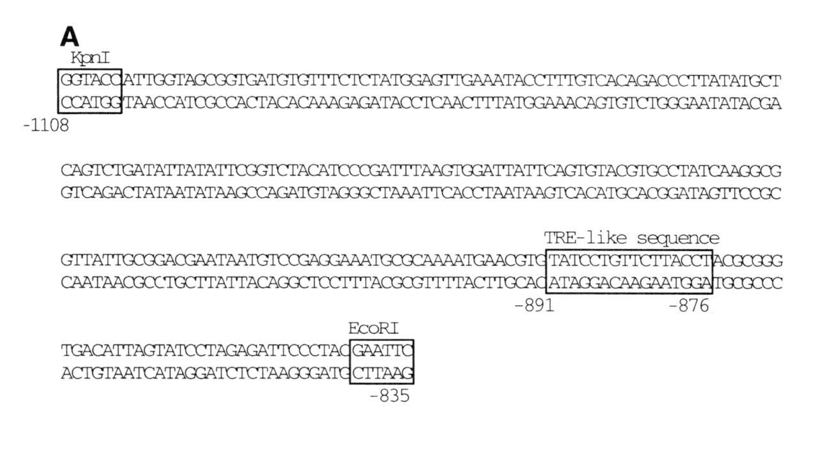

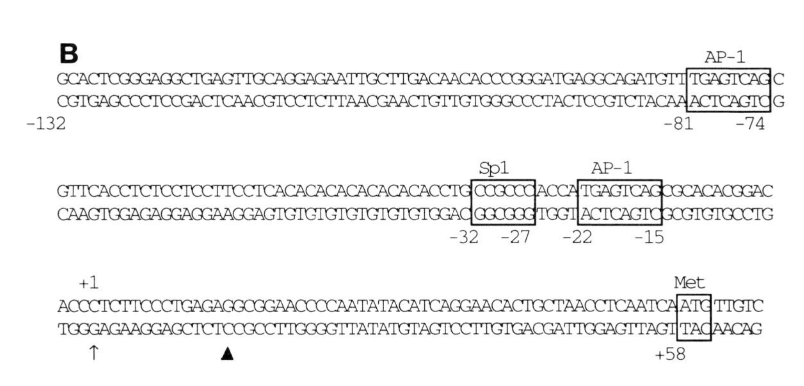

Fig. 11. Nucleotide sequence of the 5′ upstream region of the bullfrog collagenease gene. (A) Nucleotide sequence of the *Kpn*I–*Eco*RI fragment of the bullfrog collagenase gene that contains a TRE-like sequence. Restriction sites of *Kpn*I and *Eco*RI and the TRE-like sequence are boxed. (B) Nucleotide sequence flanking the transcription initiation site of the bullfrog gene. The arrow with +1 indicates the initiation start site identified by the primer extension, and the arrowhead indicates the 5′ end of the tadpole collagenase cDNA (Tcl). The putative binding sites for transcription factors AP-1, Sp1, and a TRE-like sequence, the translation initiation codon for methionine are boxed.

Thus, the most obvious question is "What is the mechanism by which a cell becomes larva specific?" As described in this chapter, the regionality of epidermal cells (body or tail) is determined by the subepidermal mesenchyme. Normally the tail epidermis does not contain the basal cell, a larva-to-adult cell. However, its regional character is totally changed when it is associated with the body mesenchymal tissue at an appropriate developmental stage (step I) to give rise to the basal cell. Therefore, the above question is closely related to a still unknown mesenchymal factor(s) which determines the epidermal regionality. Elucidation of the chemical entity of the factor is one of the most important future tasks for understanding the molecular mechanism of programmed cell death and histolysis.

There is no doubt about the key role of TH in triggering step III. Molecular biologists have been engaged in cloning TH-responsive genes from tadpoles and have succeeded in obtaining biologically significant genes. However, characterization of transcription regulatory regions of such TH-responsive genes has just begun. It is expected that more will be known in the near future about region-specific transcription factors which regulate the expression of TH-responsive genes. This, in turn, will contribute to our understanding of the molecular basis of dual roles of TH described in this chapter.

REFERENCES

Atkinson, B. G. (1981). Biological basis of tissue regression and synthesis. *In* "Metamorphosis: A Problem in Developmental Biology" (L. I. Gilbert and E. Frieden, eds.), 2nd ed., pp. 397–444. Plenum, New York.

Baker, B. S., and Tata, J. R. (1992). Prolactin prevents the autoinduction of thyroid hormone receptor mRNAs during amphibian metamorphosis. *Dev. Biol.* **149,** 463–467.

Beckingham Smith, K., and Tata, J. R. (1976). Cell death. Are new proteins synthesized during hormone-induced tadpole tail regression? *Exp. Cell Res.* **100,** 129–146.

Chatterjee, V. K. K., and Tata, J. R. (1992). Thyroid hormone receptors and their role in development. *Cancer Surv.* **14,** 147–167.

Chu, M.-L., Myers, J. C., Barnard, M. P., Ding, J.-F., and Ramirez, F. (1982). Cloning and characterization of five overlapping cDNAs specific for the human pro 1(1) collagen chain. *Nucleic Acids Res.* **10,** 5925–5934.

Coleman, J. R. (1963). Acid deoxyribonuclease activity in amphibian metamorphosis. *Biochim. Biophys. Acta* **68,** 141–143.

Collier, I. E., Smith, J., Kromberger, A., Bauer, E. A., Wilhelm, S. M., Eisen, A. Z., and Goldberg, G. I. (1988). The structure of the human skin fibroblast collagenase gene. *J. Biol. Chem.* **263,** 10711–10713.

Fini, M. E., Plucinska, I. M., Mayer, A. S., Gross, R. H., and Brinckerhoff, C. E. (1987). A gene for rabbit synovial cell collagenase: Member of a family of metalloproteinases that degrade the connective tissue matrix. *Biochemistry* **26,** 6156–6165.

Flink, I. L., and Morkin, E. (1990). Interaction of thyroid hormone receptors with strong and weak cis-acting elements in the human α-myosin heavy chain gene promoter. *J. Biol. Chem.* **265,** 11233–11237.

Gilbert, F. (1991). Cell interactions at a distance: Hormones as mediators of development. *In* "Developmental Biology," 3rd ed., pp. 685–717. Sinauer, Sunderland, Massachusetts.

Goldberg, G. I., Wilhelm, S. M., Kronberger, A., Bauer, E. A., Grant, G. A., and Eisen, A. Z. (1986). Human fibroblast collagenase. Complete primary structure and homology to an oncogene transformation-induced rat protein. *J. Biol. Chem.* **261,** 6600–6605.

Gross, J., and Lapiere, C. M. (1962). Collagenolytic activity in amphibian tissues: A tissue culture assay. *Proc. Natl. Acad. Sci. U.S.A.* **48,** 1014–1022.

Hanaoka, Y., Miyashita-Koya, S., Kondo, Y., Kobayashi, Y., and Yamamoto, K. (1973). Morphological and functional maturation of the thyroid during early development of anuran larvae. *Gen. Comp. Endocrinol.* **21,** 410–423.

Hickey, E. D. (1971). Behavior of DNA, protein and acid hydrolases in response to thyroxine in isolated tail tips of *Xenopus* larvae. *Wilhelm Roux's Arch. Dev. Biol.* **166,** 303–330.

Izutsu, Y., Kaiho, M., and Yoshizato, K. (1993). Differential distribution of epidermal basal cells in the anuran larval skin correlates with the skin's region-specific fate at metamorphosis. *J. Exp. Zool.* **267,** 605–615.

Just, J. J., Kraus-Just, J., and Check, D. A. (1981). Survey of chordate metamorphosis. *In* "Metamorphosis: A Problem in Developmental Biology" (L. I. Gilbert and E. Frieden, eds.), 2nd ed., pp. 265–326. Plenum, New York.

Kawai, A., Ikeya, J., Kinoshita, T., and Yoshizato, K. (1994). A three-step mechanism of action of thyroid hormone and mesenchyme in metamorphic changes of anuran larval skin. *Dev. Biol.* **166,** 477–488.

Kemp, N. E. (1963). Metamorphic changes of dermis in skin of frog larvae exposed to thyroxine. *Dev. Biol.* **7,** 244–254.

Kinoshita, T., and Sasaki, F. (1994). Body specific proliferation of adult precursor cells in *Xenopus* larval epidermis. *Histochemistry* **101,** 397–404.

Kinoshita, T., Sasaki, F., and Watanabe, K. (1986a). Regional specificity of anuran larval skin during metamorphosis: Transdifferentiation of tadpole tail-epidermis. *J. Exp. Zool.* **238,** 201–210.

Kinoshita, T., Sasaki, F., and Watanabe, K. (1986b). Regional specificity of anuran larval skin during metamorphosis: Dermal specificity in development and histolysis of recombined skin grafts. *Cell Tissue Res.* **245,** 297–304.

Kinoshita, T., Takayama, H., Sasaki, F., and Watanabe, K. (1989). Determination of cell death in the developmental process of anuran larval skin. *J. Exp. Zool.* **251,** 37–46.

Krebs, M. P., Rajbhandary, U. L., and Khorana, H. G. (1990). Nucleotide sequence of ISH 11, a new *Halobacterium halobium* insertion element isolated from the plasmid pGRB1. *Nucleic Acids Res.* **18,** 6699–6703.

Kubler, H., and Frieden, E. (1964). The increase in β-glucuronidase of the tadpole tail during metamorphosis and its relation to lysosomes. *Biochim. Biophys. Acta* **93,** 635–643.

Motobayashi, Y. N., and Yoshizato, K. (1986). Partial purification and characterization of an actin-degrading proteinase from the metamorphosing tadpole tail. *Zool. Sci.* **3,** 83–89.

Motobayashi, Y. N., Horiguchi, T., and Yoshizato, K. (1986). Mode of degradation of myofibrils and muscle tissues by the actin-degrading enzyme. *Zool. Sci.* **3,** 91–96.

Nagai, Y., Lapiere, C. M., and Gross, J. (1966). Tadpole collagenase: Preparation and purification. *Biochemistry* **5,** 3123–3130.

Nishikawa, A., and Yoshizato, K. (1985). Epidermal cells of the anuran tadpole tail: Its isolation and characterization *in vitro*. *Zool. Sci.* **2,** 201–211.

Nishikawa, A., and Yoshizato, K. (1986). Hormonal regulation of growth and life span of bullfrog tail epidermal cells cultured *in vitro*. *J. Exp. Zool.* **237,** 221–230.

Nishikawa, A., Kaiho, M., and Yoshizato, K. (1989). Cell death in the anuran tadpole tail: Thyroid hormone induces keratinization and tail-specific growth inhibition of epidermal cells. *Dev. Biol.* **131,** 337–344.

Oofusa, K., and Yoshizato, K. (1991). Biochemical and immunological characterization of collagenase in tissues of metamorphosing bullfrog tadpoles. *Dev., Growth Differ.* **33,** 329–339.

Oofusa, K., Yomori, S., and Yoshizato, K. (1994). Regionally and hormonally regulated expression of genes of collagen and collagenase in the anuran larval skin. *Int. J. Dev. Biol.* **38,** 345–350.

Oofusa, K., and Yoshizato, K. (1995). Thyroid hormone-dependent expression of bullfrog tadpole collagenase gene. *Roux's Arch. Dev. Biol.* (in press).

Petty, K. J., Desvergne, B., Mitsuhashi, T., and Nikodem, V. M. (1990). Identification of a thyroid hormone responsive element in the malic enzyme gene. *J. Biol. Chem.* **265,** 7395–7400.

Regard, E., Taurog, A., and Nakashima, T. (1978). Plasma thyroxine and triiodothyronine level in spontaneously metamorphosing *Rana catesbeiana* tadpoles and in adult anuran amphibia. *Endocrinology* (*Baltimore*) **102,** 674–684.

Robinson, D. H., and Heintzelman, M. B. (1987). Morphology of ventral epidermis of *Rana catesbeiana* during metamorphosis. *Anat. Rec.* **217,** 305–317.

Sap, J., Magistris, L., Stunnenberg, H., and Vennstrom, B. (1990). A major thyroid hormone responsive element in the third intron of the rat growth hormone gene. *EMBO J.* **9,** 887–896.

Silbert, J. E., Nagai, Y., and Gross, J. (1965). Hyaluronidase from tadpole tissue. *J. Biol. Chem.* **240,** 1509–1511.

Stuart, E., Everett, G. B., and Fisher, M. S. (1978). Multiple forms of N-acetyl-β-hexosaminidase in developing tadpole fins. *J. Exp. Zool.* **205,** 255–260.

Suzuki, S., and Suzuki, M. (1981). Changes in thyroidal and plasma iodine compounds during and after metamorphosis of bullfrog, *Rana catesbeiana*. *Gen. Comp. Endocrinol.* **45,** 74–81.

Tata, J. R. (1966). Requirement for RNA and protein synthesis for induced regression of the tadpole tail in organ culture. *Dev. Biol.* **13,** 77–94.

Taylor, A. C., and Kollros, J. J. (1946). Stages in the normal development of *Rana pipiens* larvae. *Anat. Rec.* **94,** 7–23.

Umesono, K., Murakami, K. K., Thompson, C. C., and Evans, R. M. (1991). Direct repeats as selective response elements for the thyroid hormone, retinoic acid and vitamin D3 receptors. *Cell* (*Cambridge, Mass.*) **85,** 1255–1266.

Wang, Z., and Brown, D. D. (1991). A gene expression screen. *Proc. Natl. Acad. Sci. U.S.A.* **88,** 11505–11509.

Wang, Z., and Brown, D. D. (1993). Thyroid hormone-induced gene expression program for amphibian tail resorption. *J. Biol. Chem.* **268,** 16270–16278.

Weber, R. (1965). Inhibitory effect of actinomycin D on tail atrophy in *Xenopus* larvae at metamorphosis. *Experientia* **21,** 665–666.

Weber, R. (1977). Biochemical characteristics of tail atrophy during anuran metamorphosis. *Colloq. Int. C.N.R.S.* **266,** 137–145.

Yoshizato, K. (1986). How do tadpoles lose their tails during metamorphosis? *Zool. Sci.* **3,** 219–226.

Yoshizato, K. (1989). Biochemistry and cell biology of amphibian metamorphosis with a special emphasis on the mechanism of removal of larval organs. *Int. Rev. Cytol.* **119,** 97–149.

Yoshizato, K. (1992). Death and transformation of larval cells during metamorphosis of anura. *Dev., Growth Differ.* **34,** 607–612.

Yoshizato, K., and Nakajima, Y. (1982). Actin degradation in metamorphosing tadpole tail. *Dev., Growth Differ.* **24,** 553–562.

Index

F

G

H

I

J

N

O

U

W

X

Z